Energy units

Energy	*Physical Interpretation*
eV	Energy scale of the outer electrons in atoms
keV $\equiv 10^3$ eV	Energy scale of the inner electrons in heavy atoms
MeV $\equiv 10^6$ eV	Energy scale of neutrons and protons inside nuclei
GeV $\equiv 10^9$ eV	Energy scale of quarks inside protons
TeV $\equiv 10^{12}$ eV	Energy scale to be studied by the next generation of particle physics experiments

Conversions

$1 \text{ eV} = 1.60 \times 10^{-19}$ J

$1 \text{ u} = 1.66 \times 10^{-27}$ kg $= 931.5 \text{ MeV}/c^2$

Mass Energies

electron	0.511 MeV	alpha	3730 MeV
proton	938 MeV	W	80 GeV
neutron	940 MeV	Z^0	91.2 GeV

Greek Alphabet

alpha	A	α	iota	I	ι	rho	P	ρ
beta	B	β	kappa	K	κ	sigma	Σ	σ
gamma	Γ	γ	lamda	Λ	λ	tau	T	τ
delta	Δ	δ	mu	M	μ	upsilon	Υ	υ
epsilon	E	ε	nu	N	ν	phi	Φ	ϕ
zeta	Z	ζ	xi	Ξ	ξ	chi	X	χ
eta	H	η	omicron	O	o	psi	Ψ	ψ
theta	Θ	θ	pi	Π	π	omega	Ω	ω

MODERN

PHYSICS

from α to Z^0

James William Rohlf

Professor of Physics
Boston University

John Wiley & Sons, Inc.

New York   Chichester
Brisbane   Toronto   Singapore

Chapter Opener Photo Credits

Chapters 1, 2, 3, 4, 5, 6, 7, 9, 11, and 14 Courtesy American Institute of Physics.
Chapter 8 Courtesy Ursula Lamb.
Chapter 10 Courtesy University of Wisconsin–Madison Archives.
Chapter 12 Courtesy Cavendish Laboratory.
Chapters 13 and 15 Courtesy AT&T Bell Laboratories.
Chapter 16 Courtesy Cornell University Archives, Photo by Sol Goldberg.
Chapter 17 Courtesy Murray Gell–Mann.
Chapter 18 Courtesy Steven Weinberg.
Chapter 19 Courtesy California Institute of Technology.

ACQUISITIONS EDITOR   Cliff Mills
MARKETING MANAGER   Catherine Faduska
SENIOR PRODUCTION EDITOR   Katharine Rubin
DESIGNER   Kevin Murphy
MANUFACTURING MANAGER   Andrea Price
PHOTO RESEARCHER   Hilary Newman
ILLUSTRATION COORDINATOR   Jaime Perea
DIGITAL PRODUCTION   Jennifer Dowling

This book was set in Times Roman by Digital Production and printed
and bound by Hamilton Printing Company. The cover was printed by Hamilton Printing Company.

Recognizing the importance of preserving what has been written, it is a
policy of John Wiley & Sons, Inc. to have books of enduring value published
in the United States on acid-free paper, and we exert our best
efforts to that end.

Library of Congress Cataloging in Publication Data:

Rohlf, James William.
    Modern physics from [alpha] to Z⁰ / James William Rohlf. -- 1st
ed.
        p.   cm.
    Includes index.
    ISBN 0-471-57270-5 (cloth)
    1. Physics.   I. Title.
QC21.2.R62   1994
539--dc20   93-48737
                                    CIP

Printed in the United States of America
10 9 8 7 6 5 4 3 2 1

To Tanya

chi non risica, non rosica . . .

Modern Physics from α to Z⁰ is written for an introductory course in modern physics taken by physics majors and engineering students, usually during the second year. The primary goal of the book is to explain the observed basic properties of atoms. The prerequisites are calculus-based introductory mechanics and electromagnetism.

The intention is to bring the student to the exciting frontiers of physics in a simple, comprehensible manner, while at the same time providing enough detail to satisfy the intellectual curiosity of a hungry student. This approach has an advantage for the student who will have a ready reference for an introduction to many advanced concepts. It is an advantage for the professor who has the flexibility to choose the pace and content of the course. In this sense I believe that "more is better."

The text begins with an introduction to particles and forces, in order to make a connection with basic mechanics and electromagnetism, as well as to give a broad overview of physics. The most important part of special relativity needed for the rest of the course, mass and binding energy, is introduced in Chapter 1. Distribution functions are introduced in Chapter 2. The time spent here will pay dividends when particle wave functions are discussed in the context of the Schrödinger equation. It is also impossible to grasp the significance of energy quantization, discovered by Planck (Chapter 3), without first understanding the Maxwell-Boltzmann distribution. Special relativity (Chapter 4) is included after a discussion of the photoelectric effect, when it is needed to explain the results of scattering experiments (*e.g.*, Compton scattering).

The text is divided into three parts. Chapters 1–9 comprise the core. Chapter 10 is a short discussion of molecules and Chapter 11 covers the basics of nuclear physics. Chapters 12–15 are an introduction to condensed matter physics and Chapters 16–19 are an introduction to particle physics and cosmology. My experience in teaching a one semester course is that the core material in Chapters 1–9 can be covered in 10–12 weeks. The remaining time can be used to cover parts of Chapters 10 and 11 and then concentrate on either topics in condensed matter physics or topics in particle physics. The instructor that wishes to go more slowly may choose to spend the entire semester on the core material and perhaps assign other chapters as optional reading. The more ambitious instructor may well choose to cover the core more rapidly, depending on the background of the students.

Material appearing between asterisks is marked "challenging." These sections contain pertinent material that is not ordinarily covered in the first modern physics course. Much of this material may be omitted on the first reading, if desired, without loss of continuity. The list of references and suggestions for further reading at the end of each chapter are intended to serve as a starting point for those wishing to delve deeper into a subject. The questions and problems are an important part of the book. These vary in degree of difficulty with the most challenging denoted with an asterisk.

I have received much sound advice from the following persons who patiently read early drafts of this book:

Professor Gordon J. Aubrecht (Ohio State Univ.),
Professor Bernard Chasan (Boston Univ.),
Professor Harris Kagan (Ohio State Univ.),

and

Professor John W. Northrip (Southwest Missouri State Univ.).

In addition, the following persons have made valuable suggestions on one or more chapters: Professors Steve Ahlen (Boston Univ.), Ed Booth (Boston Univ.), Sekhar Chivukula (Boston Univ.), Marcus Price (Univ. of New Mexico), Sidney Rudolf (Univ. of Utah), William Skocpol (Boston Univ.), and T. A. Wiggins (Pennsylvania State Univ.). I am also indebted to several students who have read and critiqued the text, especially Ian Goepfert, Eric Hawk, and John Ross.

It is a pleasure to acknowledge the expert contribution made by the staff of John Wiley & Sons, especially Clifford Mills (physics acquisition editor), Cathy Donovan (editorial assistant), Julia Salsbury (editorial assistant), Katharine Rubin (senior production editor), Ishaya Monokoff (illustration), Jaime Perea (illustration coordinator), Stella Kupferberg (photo research), Hilary Newman (photo research), Ann Berlin (production), Paul Constantine (digital production), Jennifer Dowling (digital production), Kevin Murphy (designer), and Cathy Faduska (marketing manager).

James William Rohlf
Brookline, Massachusetts

CONTENTS

**SURVEY OF
PARTICLES AND
FORCES**

If in some cataclysm, all of scientific knowledge were to be destroyed, and only one sentence passed on to the next generations of creatures, what statement would contain the most information in the fewest words? I believe it is the *atomic hypothesis* (or the atomic *fact*, or whatever you wish to call it) that *all things are made of atoms—little particles that move around in perpetual motion, attracting each other when they are a little distance apart, but repelling upon being squeezed into one another*. In that one sentence, you will see, there is an enormous amount of information about the world, if just a little imagination and thinking are applied.

Richard P. Feynman

Matter is made of atoms. The properties of atoms are quite remarkable. Consider an ordinary rock. Try pulling the atoms in a rock apart or squeezing them together. It is not easy to do so! The atoms in the rock are remarkably stable. The discovery of atoms and the measurement of their properties have paved the way for our present understanding of the universe. The idea of matter being composed of atoms is the single most important concept in all of science. The atomic composition of matter explains such apparently diverse phenomena as why the sky looks blue, why a rock feels hard, why a rose smells fragrantly, why a violin sounds mellow, and why a lime tastes sour. Our story of modern physics begins by tracing the important ideas and experiments leading to the discovery of atoms.

1-1 DISCOVERY OF ATOMS

About 2400 years ago, the Greek philosopher Anaxagoras invented the idea that matter was composed of tiny invisible seeds, or spermata. This concept was expanded a few years later by Democritus, who called the indivisible particles of matter *atoms*. The atomic hypothesis had its renaissance in the nineteenth century as scientists made the famous classification of the elements in the form of the periodic table. The idea of explaining the properties of a complex object with elementary building blocks has survived from the ancient Greeks into modern science. We *know* that matter is composed of atoms because we have developed the experimental techniques needed to *test* the atomic hypothesis. The ancient Greeks did not have the necessary experimental tools; this is why there was no advance in the understanding of atoms for more than 2000 years!

Atomic Mass Numbers

The experimental foundation of the atomic theory is the *law of definite proportions*: Whenever a given compound is formed from two elements, the ratio of the combining masses of the elements is observed to be a constant. This result holds for every compound although the mass ratio is different for each compound. If a compound is made up of more than two elements, then the ratio of masses of any two elements is constant.

In 1807, John Dalton postulated that atoms of each element had a unique mass. Dalton's atomic theory contained a simple prediction for the case where the same two elements combine to form two different compounds: For a given mass of one of the combining elements, the masses

of the other element needed to make the two compounds must be in the ratio of two small integers. The Dalton atomic theory was quickly proven to be correct by experiment. (For example, 16 g of oxygen combines with 12 g of carbon to form carbon monoxide and 32 g of oxygen combines with 12 g of carbon to form carbon dioxide. The ratio of oxygen masses needed to make the two compounds is 2/1.) This result is known as the *law of multiple proportions*. According to the theory of Dalton, each element was assigned an integer *atomic mass number* (A). Scientists of the early nineteenth century faced the formidable problem of determining both the atomic masses of the elements and the chemical formulas of compounds.

A great leap forward in the understanding of the structure of matter was made in 1811 by Amedeo Avogadro. Avogadro correctly hypothesized that the particles of a gas were small in size compared to the distance between the particles. Avogadro determined that the particles of the gas were often made up of more than one atom bound together into *molecules* and that at a fixed temperature and pressure, equal volumes of a gas contained equal numbers of molecules. This important result, which will be discussed in much more detail in Chapter 2, is the basis of the *ideal gas law*.

The *molecular mass number* is defined to be the sum of the atomic mass numbers of the atoms that make up the molecule. Relative molecular mass numbers of compounds were determined by measuring the masses of equal volumes of gases at fixed temperature and pressure. Together with the assumption that the simplest molecules contained only one atom of certain elements, the discovery of Avogadro provided a systematic method for measurement of the atomic mass numbers.

The Periodic Table

In 1869, Dmitri Mendeleev made the first classification of the elements according to their chemical properties and their atomic mass numbers. The elements were ordered with increasing atomic mass number and placed in several columns according to their chemical properties. Starting with hydrogen, an integer serial number was assigned sequentially to each element. This serial number is called the *atomic number* (Z). For hydrogen Z = 1, for helium Z = 2, and so on. In his periodic table, Mendeleev discovered some gaps that allowed him to correctly predict the existence of undiscovered elements, the ultimate goal of a theoretician! The missing elements were soon discovered. All was fine with the periodic table until William Ramsay and Lord Rayleigh discovered the element argon in 1894.

Argon had no place in the theoretical classification of the elements; such a discovery is the ultimate goal of an experimentalist! The periodic table was modified by adding a whole extra column to accommodate argon and other inert gases that were soon discovered. All the great advancements in science have been made through such interplay between theory and experiment. The modern periodic table of the elements is shown in Figure 1-1.

Avogadro's Number

Once the atomic mass numbers of the elements were known, scientists had a very powerful atomic relationship: There are equal numbers of atoms in A grams of any element, where A is the atomic mass number of the element. For example, 1 g of hydrogen, 12 g of carbon, and 238 g of uranium all contain the same number of atoms (see Figure 1-1). The number of atoms in A grams of any element is called *Avogadro's number* (N_A). The quantity of matter comprising Avogadro's number of atoms is called one *mole*. The next great experimental challenge was to determine the value of Avogadro's number. *Just how many atoms are there in one gram of hydrogen?*

Measuring the Size of an Atom

Consider the measurement of the size of an object using light as a probe. Suppose that the object to be measured is the width of a narrow slit, as illustrated in Figure 1-2. Rays of light are allowed to pass through the slit, and the intensity of the light is measured at a large distance from the slit. The image of the narrow slit is not infinitely sharp because the rays of light bend or *diffract* on passing through the slit. Diffraction is a fundamental property of waves. The location of the maxima and minima of the diffraction pattern may be deduced by tracing rays of light through the slit. Destructive interference occurs when rays have path lengths that differ by an amount (ΔL) equal to one-half of the wavelength of the light rays (λ_{light}):

$$\Delta L = \frac{\lambda_{light}}{2}. \tag{1.1}$$

If the width of the slit is d, then the path length difference is related to the angle at which the intensity is a minimum (θ_{min}) by

$$\Delta L = \frac{d}{2} \sin \theta_{min}. \tag{1.2}$$

Combining these results gives

$$\lambda_{light} = d \sin \theta_{min}. \tag{1.3}$$

Measurement of θ_{min} determines the size d of the slit. The sharpness of the intensity pattern, which is governed by diffraction, is directly proportional to the wavelength of the light.

For $\lambda_{light} = d$, $\theta_{min} = \pi/2$ and destructive interference is not measurable. We cannot measure the size of the slit using light that has a wavelength larger than the size of the slit. For this case, all we can experimentally determine is an upper limit on the slit size, $d < \lambda_{light}$.

As a result of diffraction, measurement of the size of an object is limited by the wavelength of the light used in the measurement. Two points separated by a distance d can be resolved only if the wavelength of the light does not exceed d. A consequence of this is that a single atom cannot be resolved with an ordinary microscope. This has nothing to do with the quality of the microscope, but rather with the fundamental limit imposed by diffraction. The wavelength of light, defined by the sensitivity of the eye, is in the range

$$400 \text{ nm} < \lambda_{light} < 700 \text{ nm}. \tag{1.4}$$

One nanometer (nm) is equal to 10^{-9} meters. The diameter of an atom (d_{atom}) is much smaller than the wavelength of light:

$$d_{atom} \ll \lambda_{light}. \tag{1.5}$$

The microscope *was* used, however, to make the first determination of the size of an atom! This grew out of the discovery in 1828 by Robert Brown that small particles suspended in a liquid have a small but measurable random motion. This *Brownian* motion is caused by molecules of the liquid colliding randomly with the suspended particles. The average displacement as a function of time depends on the rate at which molecules strike the suspended particle. The rate at which the molecules strike the suspended particle depends on the number of molecules in the liquid.

In 1905, Albert Einstein published a famous paper on the molecular theory of heat. From his molecular theory, Einstein deduced a formula for the time (t) dependence of the average displacement (R) of a sphere of known radius (r_0),

$$R = C \sqrt{\frac{t}{N_A r_0}}, \tag{1.6}$$

where C is a constant for a given liquid at a fixed temperature. (The meaning of temperature is an important concept that is the subject of Chapter 2.)

Periodic Table of the Elements

Legend:

atomic number (Z)	atomic mass (A)
	name
	symbol
	density (10³ kg/m³)

1 1.01 hydrogen **H** 0.0708																	2 4.00 helium **He** 0.125
3 6.94 lithium **Li** 0.542	4 9.01 beryllium **Be** 1.82											5 10.8 boron **B** 2.47	6 12.0 carbon **C** 3.52	7 14.0 nitrogen **N** 0.808	8 16.0 oxygen **O** 1.14	9 19.0 fluorine **F** 1.11	10 20.2 neon **Ne** 1.21
11 23.0 sodium **Na** 1.01	12 24.3 magnesium **Mg** 1.74											13 27.0 aluminum **Al** 2.70	14 28.1 silicon **Si** 2.33	15 31.0 phosphorous **P** 1.82	16 32.1 sulfur **S** 2.07	17 35.5 chlorine **Cl** 1.56	18 40.0 argon **Ar** 1.40
19 39.1 potassium **K** 0.910	20 40.1 calcium **Ca** 1.53	21 45.0 scandium **Sc** 2.99	22 47.9 titanium **Ti** 4.51	23 50.9 vanadium **V** 6.09	24 52.0 chromium **Cr** 7.19	25 54.9 manganese **Mn** 7.47	26 55.9 iron **Fe** 7.87	27 58.9 cobalt **Co** 8.9	28 58.7 nickel **Ni** 8.91	29 63.6 copper **Cu** 8.93	30 65.4 zinc **Zn** 7.13	31 69.7 gallium **Ga** 5.91	32 72.6 germanium **Ge** 5.32	33 74.9 arsenic **As** 5.77	34 79.0 selenium **Se** 4.81	35 79.9 bromine **Br** 3.12	36 83.8 krypton **Kr** 3.09
37 85.5 rubidium **Rb** 1.63	38 87.6 strontium **Sr** 2.58	39 88.9 yttrium **Y** 4.48	40 91.2 zirconium **Zr** 6.51	41 92.9 niobium **Nb** 8.58	42 95.9 molybdenum **Mo** 10.2	43 98 technetium **Tc** 11.5	44 101 ruthenium **Ru** 12.4	45 103 rhodium **Rh** 12.4	46 106 palladium **Pd** 12.0	47 108 silver **Ag** 10.5	48 112 cadmium **Cd** 8.65	49 115 indium **In** 7.29	50 119 tin **Sn** 5.76	51 122 antimony **Sb** 6.69	52 128 tellurium **Te** 6.25	53 127 iodine **I** 4.95	54 131 xenon **Xe** 3.52
55 133 cesium **Cs** 2.00	56 137 barium **Ba** * 3.59	71 175 lutetium **Lu** 9.84	72 178 hafnium **Hf** 13.2	73 181 tantalum **Ta** 16.7	74 184 tungsten **W** 19.3	75 186 rhenium **Re** 21.0	76 190 osmium **Os** 22.6	77 192 iridium **Ir** 22.6	78 195 platinum **Pt** 21.5	79 197 gold **Au** 19.3	80 201 mercury **Hg** 14.3	81 204 thallium **Tl** 11.9	82 207 lead **Pb** 11.3	83 209 bismuth **Bi** 9.80	84 209 polonium **Po** 9.31	85 210 astatine **At**	86 222 radon **Rn**
87 223 francium **Fr**	88 226 radium **Ra** †	103 260 lawrencium **Lr**	104 261 unnilquadium **Unq**	105 262 unnilpentium **Unp**	106 263 unnilhexium **Unh**	107 262 unnilseptium **Uns**	108 265 unniloctium **Uno**	109 266 unnilennium **Une**									

* Lanthanide series

57 139 lanthanum **La** 6.17	58 140 cerium **Ce** 6.77	59 141 praseodymium **Pr** 6.78	60 144 neodymium **Nd** 7.00	61 145 promethium **Pm**	62 150 samarium **Sm** 7.54	63 152 europium **Eu** 5.24	64 157 gadolinium **Gd** 7.89	65 159 terbium **Tb** 8.27	66 163 dysprosium **Dy** 8.53	67 165 holmium **Ho** 8.80	68 167 erbium **Er** 9.04	69 169 thulium **Tm** 9.32	70 173 ytterbium **Yb** 6.97

† Actinide series

89 227 actinium **Ac** 10.1	90 232 thorium **Th** 11.7	91 231 protactinium **Pa** 15.4	92 238 uranium **U** 19.1	93 237 neptunium **Np** 20.5	94 244 plutonium **Pu** 19.8	95 243 americium **Am** 11.9	96 247 curium **Cm**	97 247 berkelium **Bk**	98 251 californium **Cf**	99 252 einsteinium **Es**	100 257 fermium **Fm**	101 258 mendelevium **Md**	102 259 nobelium **No**

FIGURE 1-1 Periodic table of the elements.
Elements in the same column have similar properties. The lanthanide and actinide series elements have properties similar to La and Ac. Each element has a unique atomic number (Z). For a given element, the atomic mass number (A) is not unique. The value of A in the table corresponds to the weighted average of the abundance on earth. If the element is not stable, the value of A corresponds to the isotope with the longest lifetime.

In 1908, Jean-Baptiste Perrin developed a technique for manufacturing small resin spheres of uniform size ($r_0 \approx 10^{-6}$ m $= 1$ μm). Perrin accurately measured the Brownian motion of these spheres and used Einstein's result to determine N_A. He checked the validity of Einstein's formula (1.6) by varying the size of the sphere and the composition of the liquid in which the spheres were suspended. Measurements made by Perrin on a single sphere are shown in Figure 1-3. Figure 1-3a shows the (x,y) position of the sphere at 30-second intervals. Figure 1-3b shows the net displacement versus the square root of time. From the measurement the displacement (1.6) of many spheres, Perrin found Avogadro's number to be about 6×10^{23}.

Perrin (with help from Einstein!) had determined the number of atoms in A grams of any element.

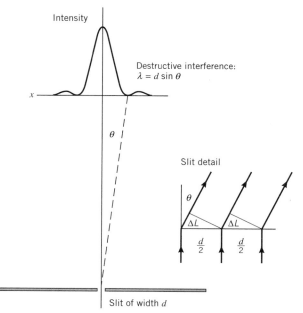

FIGURE 1-2 Diffraction of light.
An image of a narrow slit is made by measuring the intensity distribution of light that reaches a detector placed a large distance from the slit. The image is not perfectly sharp because of the diffraction of the light waves.

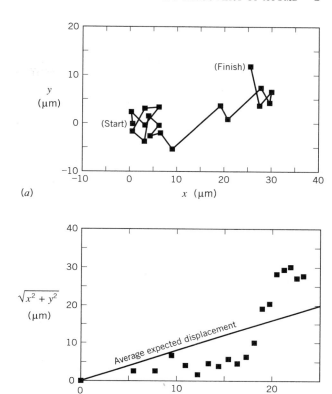

FIGURE 1-3 Data of Perrin on Brownian motion.
A tiny sphere of resin ($r_0 = 0.5$ μm) is suspended in a liquid. The sphere is observed to move when viewed under a microscope because of collisions with molecules of the liquid. (a) The measured position (x,y) of a single sphere is plotted at 30-second intervals. (b) The net displacement for the sphere is plotted versus the square root of time. Einstein predicted that the average net displacement is proportional to the square root of time. The constant of proportionality contains $N_A^{-1/2}$. The deviations from the average for any one sphere may be quite large, as indicated. The results of many measurements by Perrin proved that Einstein's prediction was correct and provided an accurate determination of Avogadro's number. Data are taken from J.-B. Perrin, *Atoms*, D. Van Nostrand (1923).

To be precise, Avogadro's number has been defined to be the number of atoms in 12 g of carbon ($A = 12$). Avogadro's number may be accurately determined by measuring the mass of a single carbon atom and then dividing 12 g by the atomic mass. The result is

$$N_A = 6.02 \times 10^{23}. \tag{1.7}$$

(Precise values of physical constants may be found in Appendix A.) The atomic mass numbers of all the other elements are defined so that A grams of every element contains exactly N_A atoms. If we assume that the mass of a molecule is equal to the sum of the masses of the atoms that make up the molecule, then for any compound there are N_A molecules in M grams, where M is the sum of the atomic mass numbers of the atoms making up one molecule.

Avogadro's number is

$$N_A = 6.02 \times 10^{23}.$$

For any element there are N_A atoms in A grams, where A is the atomic mass number.

EXAMPLE 1-1

Use Avogadro's number to calculate the mass of the hydrogen atom.

SOLUTION:

Since $A = 1$, 1 g of hydrogen contains N_A atoms. The mass of a single atom (m_H) is

$$m_H = \frac{A(10^{-3} \text{ kg})}{N_A} = \frac{(1)(10^{-3} \text{ kg})}{6.02 \times 10^{23}} \approx 1.7 \times 10^{-27} \text{ kg.} \ \blacksquare$$

Matter in a condensed state (solid or liquid) is incompressable. Combining Avogadro's number with the density of a solid or liquid gives us the approximate size of an atom.

EXAMPLE 1-2

The density of liquid hydrogen is about 71 kg/m³. Use Avogadro's number to estimate the size of the hydrogen atom.

SOLUTION:

There are N_A atoms in 1 g of hydrogen. In the liquid state, the atoms are packed closely together. The volume (V) occupied by 1 g of hydrogen is

$$V = \frac{1.0 \times 10^{-3} \text{ kg}}{\rho}.$$

If we divide this quantity by Avogadro's number, we get the volume occupied by one atom (V_{atom}):

$$V_{atom} = \frac{V}{N_A} \approx \frac{1.0 \times 10^{-3} \text{ kg}}{\rho N_A}$$

$$\approx \frac{1.0 \times 10^{-3} \text{ kg}}{(71 \text{ kg/m}^3)(6.0 \times 10^{23})} \approx 2.3 \times 10^{-29} \text{ m}^3.$$

This is the volume taken up by a single atom in liquid hydrogen. If we approximate the liquid as a collection of closely packed spheres, then the diameter of the atom (d_{atom}) is approximately

$$d_{atom} \approx V_{atom}^{1/3} \approx (2.3 \times 10^{-29} \text{ m}^3)^{1/3}$$

$$\approx 3 \times 10^{-10} \text{ m} = 0.3 \text{ nm} \qquad \blacksquare$$

EXAMPLE 1-3

Estimate the size of a uranium atom. The density of uranium is 1.9×10^4 kg/m³.

SOLUTION:

The atomic mass number of uranium is 238. Following the previous example, we have

$$V_{atom} \approx \frac{(238)(10^{-3} \text{ kg})}{(1.9 \times 10^4 \text{ kg/m}^3)(6.0 \times 10^{23})}$$

$$\approx 2.1 \times 10^{-29} \text{ m}^3,$$

and

$$d_{atom} \approx V_{atom}^{1/3} \approx (2.1 \times 10^{-29} \text{ m}^3)^{1/3}$$

$$\approx 3 \times 10^{-10} \text{ m} = 0.3 \text{ nm.} \qquad \blacksquare$$

The sizes of the atoms in solids and liquids calculated in the fashion of Examples 1-2 and 1-3 are shown in Figure 1-4. These data show that atoms of the different elements are all of the same order of magnitude in size, even though they vary in mass by more than two orders of magnitude. The diameter of an atom (d_{atom}) is approximately

$$d_{atom} \approx 3 \times 10^{-10} \text{ m} = 0.3 \text{ nm.} \qquad (1.8)$$

Feynman has given us a clever way to visualize the size of an atom:

if an apple is magnified to the size of the earth, then the atoms in the apple are approximately the size of the original apple.

The diameter of an atom (d_{atom}) is approximately equal to

$$d_{atom} \approx 3 \times 10^{-10} \text{ m} = 0.3 \text{ nm.}$$

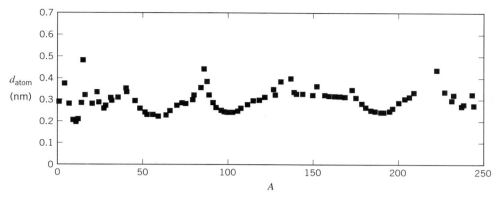

FIGURE 1-4 Approximate diameters of atoms in condensed matter as a function of the atomic mass number.

1-2 CLASSICAL ELECTROMAGNETISM

The cornerstone for the description of the electric force is Coulomb's law, which gives us an expression for the force (**F**) between two charges (q_1 and q_2) *at rest*:

$$\mathbf{F} = \frac{kq_1q_2}{r^2}\mathbf{i_r}. \qquad (1.9)$$

The force is directed along the axis ($\mathbf{i_r}$) of the two charges. The electric force constant (k) together with the value of the electric charges specifies the strength of the interaction. Coulomb's law holds only for charges at rest. If the charges are moving, the expression for the force is much more complicated! Fortunately, a very ingenious method was invented that greatly simplifies the description of the force through the concept of electric and magnetic fields. Once the fields are known in some region of space, they may be used to calculate the force acting on a charge q moving with a velocity **v**:

$$\mathbf{F} = q(\mathbf{E} + \mathbf{v} \times \mathbf{B}). \qquad (1.10)$$

This expression is called the *Lorentz force law*. An electric field of 1 volt per meter produces the same magnitude of force on a charge as a magnetic field of 1 tesla when the charge has a speed of 1 meter per second: 1 V/m = 1 T·m/s.

The Superposition Principle

Coulomb's law gives us the electromagnetic force between two charges at rest. *How do we find the forces between three charges at rest?* The electromagnetic force has the very remarkable property that the force between

two charges does not depend on the presence of the third charge! The resultant force on one charge is the vector sum of the forces due to the other two charges. This important experimental result is known as the *superposition principle*. The superposition principle can also be stated in terms of electric and magnetic fields. The net electric and magnetic fields due to many charges ($\mathbf{E}_{net}$ and $\mathbf{B}_{net}$) are given by the vector sum of the fields ($\mathbf{E}_1, \mathbf{E}_2, \mathbf{E}_3 \ldots$ and $\mathbf{B}_1, \mathbf{B}_2, \mathbf{B}_3 \ldots$) due to the individual charges:

$$\mathbf{E}_{net} = \mathbf{E}_1 + \mathbf{E}_2 + \mathbf{E}_3 + ..., \qquad (1.11)$$

and

$$\mathbf{B}_{net} = \mathbf{B}_1 + \mathbf{B}_2 + \mathbf{B}_3 + ... \qquad (1.12)$$

We may use the superposition principle and the Lorentz force law (1.10) to calculate the total force on a charge q moving with velocity **v**:

$$\mathbf{F} = q(\mathbf{E}_{net} + \mathbf{v} \times \mathbf{B}_{net}). \qquad (1.13)$$

Maxwell's Equations

The determination of the expressions for the electric and magnetic fields of moving charges, by Faraday, Ampère, and others, was a great scientific accomplishment. The crowning achievement was due to James Clerk Maxwell, who provided a unified set of four equations relating the charges and fields. Maxwell's equations together with the Lorentz force law summarize all that was known about the electromagnetic force at the end of the nineteenth century. Table 1-1 gives a brief summary of Maxwell's equations in integral form. These equations are also often written as differential equations (see Appendix B).

TABLE 1-1
SUMMARY OF MAXWELL'S EQUATIONS.

Gauss's Law	Gauss's Law for Magnetic Fields	Faraday's Law	Ampère's Law
$\oiint da \cdot \mathbf{E} = 4\pi k q_{\text{tot}}$	$\oiint da \cdot \mathbf{B} = 0$	$\oint d\mathbf{l} \cdot \mathbf{E} = -\dfrac{\partial}{\partial t} \oiint da \cdot \mathbf{B}$	$\oint d\mathbf{l} \cdot \mathbf{B} = \dfrac{4\pi k I}{c^2} + \dfrac{1}{c^2}\dfrac{\partial}{\partial t}\oiint da \cdot \mathbf{E}$
Flux of electric field through any closed surface is proportional to the electric charge contained inside the volume enclosed by that surface	There are no magnetic charges (*monopoles*)	Line-integral of the electric field around a closed loop is equal to the negative of the time-rate of change of the magnetic flux through the surface enclosed by the loop	Line-integral of the magnetic field around a closed loop is the sum of two terms, one proportional to the current through that loop and the second proportional to the time-rate of change of the electric flux through the loop

The Wave Equation

All of classical electrodynamics is contained in the four Maxwell equations plus the Lorentz force law. Unlike Newton's Second Law ($\mathbf{F} = m\mathbf{a}$), Maxwell's equations are relativistically correct. They have the same form even when the speeds of the charges are not small compared to the speed of light. It is not hard to write down Maxwell's equations or to find them on a tee-shirt. To appreciate the implication of Maxwell's equations, that an accelerated charge radiates electromagnetic waves, requires a much more sophisticated level of understanding.

An electromagnetic wave consists of oscillating electric and magnetic fields, which are perpendicular to each other and to the direction of travel of the wave (see Figure 1-5). The wave is able to propagate in vacuum because the changing electric field creates a magnetic field (Ampère's law) and the changing magnetic field creates an electric field (Faraday's law). The wave equation relates the second-order spatial and time derivatives of each component of the electric and magnetic fields. The wave equation may be derived from the Maxwell equations. The result (see Appendix B) is

$$\frac{\partial^2 F}{\partial x^2} + \frac{\partial^2 F}{\partial y^2} + \frac{\partial^2 F}{\partial z^2} = \frac{1}{c^2}\frac{\partial^2 F}{\partial t^2}. \quad (1.14)$$

An equation of the same form is satisfied for every component of both $\mathbf{E}$ and $\mathbf{B}$, so that the $F(x,y,z,t)$ in the wave equation (1.14) can represent $E_x, E_y, E_z, B_x, B_y,$ or B_z. The speed of propagation of the wave is c, the speed of light in vacuum. It was the appearance of c in Ampère's law that led Maxwell to the brilliant deduction that light is an electromagnetic wave.

Consider an electromagnetic wave propagating in the z direction. The wave equation (1.14) becomes

$$\frac{\partial^2 F}{\partial t^2} = c^2 \frac{\partial^2 F}{\partial z^2}. \quad (1.15)$$

The solution is a function of the form $F(z-ct)$ or $F(z+ct)$ as can be seen by direct differentiation. Taking the electric field to be in the x direction, we may write the solution as

$$\mathbf{E} = F(z-ct)\mathbf{i}_x. \quad (1.16)$$

From Faraday's law the magnetic field must be in the y direction and have a strength

$$\mathbf{B} = \frac{\mathbf{E}}{c} = \frac{1}{c}F(z-ct)\mathbf{i}_y. \quad (1.17)$$

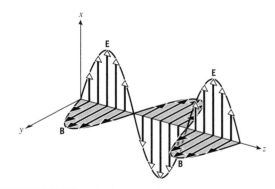

FIGURE 1-5 The electromagnetic wave.
The wave consists of oscillating electric and magnetic fields that are perpendicular to each other and to the direction of propagation (z direction).

Consider the case where the electric field has the simple form

$$\mathbf{E} = E_0 \cos(kz - \omega t)\mathbf{i}_x. \qquad (1.18)$$

A wave of this form is called a *plane wave* because the magnitudes of the electric and magnetic fields do not depend on the x or y coordinates. When we substitute this expression for the electric field into the wave equation, we get a relationship between the constants k, ω and the wave speed c:

$$c = \frac{\omega}{k}. \qquad (1.19)$$

The constant ω is called the *angular frequency* and the constant k is called the *wave number*. When we make the transformation,

$$z \rightarrow z + \frac{2\pi}{k}, \qquad (1.20)$$

the electric field does not change because

$$\cos(kz - \omega t) = \cos(kz - \omega t + 2\pi). \qquad (1.21)$$

The amplitude of the electric field is periodic in the coordinate z. The *wavelength* (λ) of the wave is defined to be

$$\lambda \equiv \frac{2\pi}{k}, \qquad (1.22)$$

so that the transformation

$$z \rightarrow z + \lambda, \qquad (1.23)$$

leaves the electric field unchanged. Similarly if we make the transformation

$$t \rightarrow t + \frac{2\pi}{\omega}, \qquad (1.24)$$

the electric field is also unchanged. The *period* (T) of the wave is defined to be

$$T \equiv \frac{2\pi}{\omega}, \qquad (1.25)$$

and the inverse of the period is defined to be the *frequency* (*f*)

$$f \equiv \frac{1}{T} = \frac{\omega}{2\pi}. \qquad (1.26)$$

From the relationship (1.19) between k and ω, we have

$$c = \lambda f. \qquad (1.27)$$

The speed of the wave propagation is equal to the wavelength multiplied by the frequency. Electromagnetic waves may have a wavelength of any size. The electromagnetic spectrum is summarized in Table 1-2.

> The speed of a plane wave is equal to the wavelength times the frequency,
> $$c = \lambda f.$$

Using the superposition principle, an arbitrary wave may always be constructed from plane waves of different amplitudes (A_n) and wave numbers (k_n):

$$F(z - ct) = \sum_{n=0}^{\infty} A_n \cos(k_n z - c k_n t). \qquad (1.28)$$

This expression is also a solution of the wave equation.

1-3 LOOKING INSIDE THE ATOM: ELECTRONS AND A NUCLEUS

X Rays, Alpha and Beta Particles

At the end of 1895, Wilhelm Röntgen discovered *x rays*, a mysterious radiation that penetrated matter. The discovery of Röntgen started a revolution in physics. (Röntgen's discovery will be taken up in more detail in Chapter 3.) In 1896, Antoine Henri Becquerel set out to investigate if the phenomena of fluorescence and phosphorescence (light

TABLE 1-2
THE ELECTROMAGNETIC SPECTRUM.

There is no convention for the exact wavelengths of the boundaries.

Wavelength	Name
less than 10^{-12} m	γ ray
10^{-12} to 10^{-9} m	x ray
10^{-9} to 4×10^{-7} m	ultraviolet
4×10^{-7} m to 7×10^{-7} m	light
7×10^{-7} m to 10^{-3} m	infrared
10^{-3} m to 0.1 m	microwave
0.1 m to 10^3 m	radio
greater than 10^3 m	ultra low frequency

emission by certain substances when exposed to radiation) produced x rays. In this investigation Becquerel discovered a new type of radiation. Marie and Pierre Curie investigated the properties of the Becquerel radiation and called the phenomenon *radioactivity*. Natural radioactivity is the spontaneous emission of radiation from certain heavy elements such as uranium. (Radioactivity is discussed in detail in Chapter 11.) This natural radiation was discovered to be *quantized* in the form of particles. Ernest Rutherford observed that the particles emitted from atoms were of two types: *alpha* particles (α), which did not penetrate matter, and *beta* particles (β), which easily penetrated matter (see Figure 1-6). The α and β particles were destined to play a crucial role in the understanding of the atom. These particles were used to probe the structure of the atom. Furthermore, the quest for the understanding of their existence would lead to the discovery of two new forces!

Discovery of the Electron

A type of radiation, called *cathode rays*, was observed to be emitted from metallic surfaces when voltage was applied. At the end of the nineteenth century there was much theoretical speculation about the fundamental properties of the cathode rays. One school of thought held the belief that cathode rays were particles. The main evidence for the particle hypothesis was the observation that cathode rays were deflected by magnetic fields. The main obstacle to this interpretation was the lack of observation of the deflection of cathode rays by electric fields. The other school of thought held the belief that cathode rays were a wave phenomenon dependent on the medium of space (*ether* or *aether*). The wave interpretation was supported by the observation that cathode rays could pass through metal foils without deflection. It is interesting to note that physicists were troubled by the apparent contradiction of both a particle and a wave interpretation. This important issue would arise again in 30 years.

In 1897, Joseph John Thomson performed a definitive set of experiments that proved that cathode rays had a particle behavior. The situation was summed up by Thomson in the introduction of the paper reporting his results:

The experiments discussed in this paper were undertaken in the hope of gaining some information as to the nature of Cathode Rays. The most diverse opinions are held as to these rays; according to the almost unanimous opinion of German physicists they are due to some process in the aether to which—inasmuch as in a uniform magnetic field their course is circular and not rectilinear—no phenomenon hitherto observed is analogous; another view of these rays is that, so far from being wholly aetherial, they are in fact wholly material, with negative electricity. It would seem at first sight that it ought not to be difficult to discriminate between views so different, yet experience shows that this is not the case, as amongst the physicists who have deeply studied the subject can be found supporters of either theory.

The Experiment of J. J. Thomson

The key to the success of the Thomson experiment was the development of the technique necessary to observe the deflection of cathode rays in an electric field. This led to the interpretation of cathode rays as charged particles, commonly known as *electrons*. In his experiment, Thomson accelerated electrons in an electric field and measured the curvature of their trajectories in a magnetic field, clearly demonstrating that they were particles with a negative electric charge. The apparatus developed by Thomson is called a *mass spectrometer*. Thomson used his spectrometer to measure the charge-to-mass ratio of the electron.

The Thomson spectrometer is illustrated in Figure 1-7. A stream of electrons emitted from a cathode passed through collimators into a region of two parallel plates of length L, separated by a distance d. A voltage (V) was applied to the plates, creating an electric field (E):

$$E = \frac{V}{d}. \tag{1.29}$$

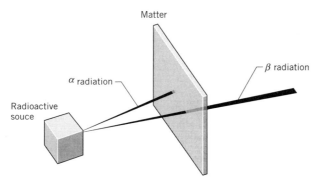

FIGURE 1-6 Rutherford's classification of the radiation discovered by Becquerel.
The α particle does not penetrate matter, whereas the β particle readily penetrates matter.

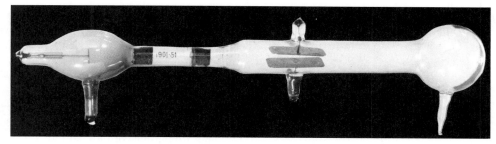

(a)

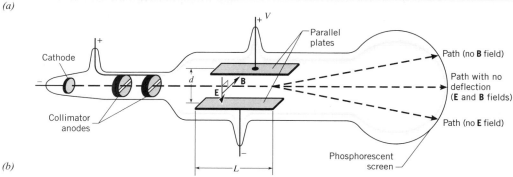

(b)

FIGURE 1-7 Spectrometer used by J. J. Thomson to measure the charge-to-mass ratio of the electron. (a) Photograph courtesy of Science Museum Library, London. (b) Schematic fashioned after J. J. Thomson, "Cathode Rays," *Phil. Mag.* **44**, 293 (1897).

The apparatus was enclosed in an evacuated glass tube to reduce collisions of the electrons with molecules in the air. The vacuum was important because collisions of the electrons with molecules produced ions (charged particles). The ions would collect on the plates and cancel the effect of the applied electric field! This is why the deflection of cathode rays by electric fields had escaped detection.

The acceleration (a) of the electron is in the y direction with a magnitude equal to the electric force divided by the electron mass (m). Taking the electron charge to be q, we have

$$a = \frac{F}{m} = \frac{qE}{m} = \frac{qV}{md}. \tag{1.30}$$

The time (t_p) that the electron spends between the plates is inversely proportional to the x component of the electron velocity (v_x):

$$t_p = \frac{L}{v_x}. \tag{1.31}$$

The y component of velocity (v_y) is the product of the acceleration (1.30) and the time (1.31),

$$v_y = at_p = \frac{qVL}{mdv_x}. \tag{1.32}$$

This gives the following expression for the charge-to-mass ratio (q/m) of the electron:

$$\frac{q}{m} = \frac{dv_y v_x}{VL}. \tag{1.33}$$

Since d, L and V, were known quantities, measurement of v_x and v_y would determine q/m. The position of the stream of electrons was measured on a phosphorescent screen. The ratio of v_x and v_y is a relatively easy quantity to measure because it is the tangent of the deflection angle (θ):

$$\frac{v_y}{v_x} = \tan\theta. \tag{1.34}$$

The expression for q/m, however, involves the product $v_y v_x$, and the electron velocity is not an easy quantity to measure directly. Thomson thought of a clever method to determine the electron velocity. Thomson put a magnetic field perpendicular to the electric field and adjusted it such that there was no deflection. In the case for no deflection, the electric and magnetic forces are balanced:

$$qE = qv_x B. \qquad (1.35)$$

Thus, v_x is determined by measurement of E and B:

$$v_x = \frac{E}{B} = \frac{V}{dB}. \qquad (1.36)$$

The expression for q/m (1.33) may be written as

$$\frac{q}{m} = \frac{d\left(\dfrac{v_y}{v_x}\right)v_x^2}{VL} = \frac{V\tan\theta}{dLB^2}, \qquad (1.37)$$

where θ is the deflection angle with no magnetic field, and B is the magnetic field which produces no deflection. Thomson measured the quantities on the right-hand side of the expression for q/m (1.37). Thomson's data are plotted in Figure 1-8 in the form of $\tan\theta$ versus B^2L/E. Thomson's early result for the electron charge-to-mass ratio was about 10^{11} C/kg. This ratio was observed to be constant, that is, all cathode rays gave the same value. The value of q/m for the electron determined by Thomson was substantially smaller (more than three orders of magnitude) than the values of q/m determined by electrolysis,

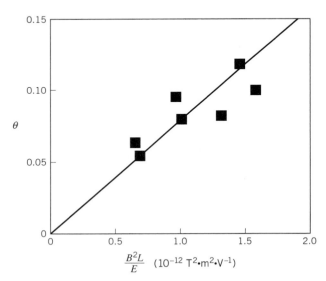

that is, q/m for the electron is much smaller than for ionized atoms. There were two extreme possibilities: (1) The electron charge is much smaller than the charge of an ionized atom, or (2) the electron mass is much smaller than the mass of an ionized atom (or both!). The ability of the electron to penetrate matter led Thomson to believe that the mass of the electron was much smaller than the mass of an atom.

Thomson's pioneering result was systematically low by about a factor of two due to the neglect of magnetic fields outside the deflecting plates in his spectrometer in measuring the deflection angle. Thomson knew of this effect and addressed it in his paper. An accurate measurement of the charge-to-mass ratio of the electron with the Thomson technique gives

$$\frac{q}{m} = 1.76 \times 10^{11} \text{ C/kg}. \qquad (1.38)$$

EXAMPLE 1-4

The electric field in the Thomson spectrometer is set at 10^4 V/m and the deflection angle is observed to be 0.10 radians after passing through a distance of $L = 0.050$ m when there is no magnetic field. Calculate the speed of the electron.

SOLUTION:

First we calculate the magnetic field strength needed to produce no deflection,

$$B = \sqrt{\frac{E\tan\theta}{L\left(\dfrac{q}{m}\right)}}$$

$$= \sqrt{\frac{\left(10^4 \text{ V/m}\right)\left(0.1\right)}{\left(0.05\,\text{m}\right)\left(1.76\times10^{11}\text{ C/kg}\right)}}$$

$$= 3.4 \times 10^{-4} \text{ T}.$$

The electron speed is the electric field divided by the magnetic field,

$$v = \frac{E}{B} = \frac{10^4 \text{ V/m}}{3.4\times10^{-4}\text{ T}} \approx 2.9\times10^7 \text{ m/s}. \qquad \blacksquare$$

The Millikan Oil-Droplet Experiment

In 1909, Robert Millikan made the first accurate measurement of the electron charge. The experiment of Millikan is

FIGURE 1-8 Thomson's data on the charge-to-mass ratio of the electron.
The data are expected to fall on a straight line passing through the origin. The slope gives q/m. Data are from J. J. Thomson, "Cathode Rays," *Phil. Mag.* **44**, 293 (1897).

sketched in Figure 1-9. Tiny droplets of oil are sprayed between two conducting plates and viewed under a microscope. The oil droplets fall toward the earth due to gravity, and they experience a frictional *drag* force that is proportional to the speed of the droplet. The drag force is due to the collisions of the oil droplet with the air molecules. The droplet quickly reaches the equilibrium condition where the net force on the droplet is zero. The acceleration of the droplet is then equal to zero and the droplet falls with a constant speed called the *terminal* speed (v_T). The gravitational force (mg) is balanced by the drag force (bv_T)

$$mg = bv_T. \qquad (1.39)$$

The coefficient b in the drag force term is directly proportional to the radius of the droplet (R). The coefficient b also depends on how "sticky" or *viscous* the air is. We write the drag force constant as

$$b = 6\pi\eta R, \qquad (1.40)$$

where η is the coefficient of viscosity of air. This result is known as Stokes's law. The mass of the droplet (m) and density of the oil (ρ) are related by

$$\rho = \frac{3m}{4\pi R^3}. \qquad (1.41)$$

Eliminating b and m in the force equation (1.39), we may solve for the radius of the droplet:

$$R = \sqrt{\frac{9v_T\eta}{2g\rho}}. \qquad (1.42)$$

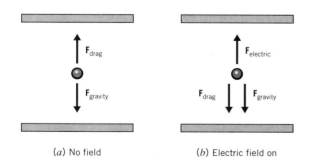

(a) No field   (b) Electric field on

FIGURE 1-9 The Millikan oil-droplet experiment.
(a) An oil droplet in free fall reaches a terminal speed, with the frictional drag force opposing the force of gravity. (b) An electric field is present, giving an upward force on a charged oil droplet that overcomes the force of gravity. The oil droplet reaches a terminal speed with the electric force opposing both the drag force and gravity.

Measurement of the terminal speed of the droplet together with knowledge of the density of the oil and coefficient of viscosity of air gives the size of the droplet. The typical size of an oil droplet in the Millikan experiment was 3×10^{-6} m (3 μm).

In the process of spraying the oil, individual droplets may acquire one or more excess electrons. When an electric field is switched on, there is an electrical force on a charged droplet. If the electric field is directed downward, then the electrical force on the droplet is upward, against the force of gravity. If the electric force is stronger than the gravitational force, then the droplet moves upward and the drag force is now directed downward. The droplet reaches a terminal velocity (v_E) given by

$$qE = mg + bv_E. \qquad (1.43)$$

Eliminating b from the force equations (1.39) and (1.43), we have

$$q = \frac{mg\left(1 + \dfrac{v_E}{v_T}\right)}{E}. \qquad (1.44)$$

Since the size of the droplet (1.42) can be calculated and the density of the oil is readily measured, the mass of the droplet is also known. Thus, measurement of the two terminal speeds, v_T with no field and v_E with an electric field E applied, gives the electric charge on the droplet.

Millikan measured the trajectory of a *single* droplet over a long period (several minutes), turning on the electric field when the droplet was near the bottom plate and turning off the electric field when the droplet was near the top plate. By repeated measurements of the droplet speeds v_E and v_T, Millikan made a determination of the charge (1.44) on the oil droplet. The result of numerous measurements by Millikan was that the charge on the droplets was always an integer multiple of 1.6×10^{-19} C. Furthermore, Millikan observed that occasionally a droplet would gain or lose a charge of an integer multiple of 1.6×10^{-19} C as it drifted through the air. Millikan deduced that the change in charge (Δq) on the droplet was caused by collisions of the droplet with air molecules, resulting in the gain or loss of one or more electrons. Thus, Millikan determined that the magnitude of the charge of the electron was 1.6×10^{-19} C. The data of Millikan from a single droplet are shown in Figure 1-10, where we plot the fractional deviation of Δq from the nearest integer multiple of the electron charge (e).

On the numerical value of the electron charge, Millikan wrote:

Perhaps these numbers have little significance to the general reader who is familiar with no electrical units save those in which his monthly light bills are rendered. If these latter seem excessive, it may be cheering to reflect that the number of electrons contained in the quantity of electricity which courses every second through a common sixteen-candle-power electric-lamp filament, and for which we pay 1/ 100,000 of 1 cent, is so large that if all the two and one-half million inhabitants of Chicago were to begin to count out these electrons and were to keep on counting them out each at the rate of two a second, and if no one of them were ever to stop to eat, sleep, or die, it would take them just twenty thousand years to finish the task.

The result of Millikan marked the discovery of *charge quantization*. Charge is an intrinsic property of the electron. It is not possible to remove the charge from an electron; there is no such thing as an electron without its charge. The fundamental unit of charge in modern physics is called *e*. The electron is defined to have an electric charge of *minus e*:

$$\text{electron charge} = -e . \qquad (1.45)$$

Millikan's measurement of *e* was systematically low by about 0.4% due to inaccurate knowledge of the coefficient of viscosity. Accurate measurement of the value *e* gives

$$e = 1.602 \times 10^{-19} \text{ C} . \qquad (1.46)$$

Particles that are *electrically* attracted to an electron are assigned a positive charge, and particles that are electrically repelled by an electron are assigned a negative charge. Particles that are neither attracted nor repelled electrically are assigned a charge of zero. All free particles are observed to have values of electric charge (*q*) equal to an integer times the fundamental charge *e*:

$$q = ne , \qquad (1.47)$$

where $n = \ldots -3, -2, -1, 0, 1, 2, 3 \ldots$ The integer *n* is called the electric charge *quantum number*. A free particle has never been observed with a charge unequal to an integer times the electron charge.

The experimental results of Thomson and Millikan may be combined to yield the mass of the electron:

$$m = q\left(\frac{m}{q}\right) = 9.11 \times 10^{-31} \text{ kg} . \qquad (1.48)$$

The electron mass is much smaller than the mass of the hydrogen atom (see Example 1-1).

The Wilson Cloud Chamber

In 1906, Charles T. R. Wilson, a student of Thomson's, made a brilliant discovery on the detection of charged particles. When a charged particle such as an electron passes through any material, there is an electromagnetic interaction between the charged particle and the electrons in the atoms of the material. This interaction is strong enough to remove electrons or *ionize* atoms in the material. The ionization occurs along the trajectory of the moving charged particle.

Imagine that the material being ionized by an energetic incoming electron is damp air. If the air is suddenly expanded, droplets of condensation form. Wilson's discovery was this: The droplets of condensation form around the ions! The droplets may be photographed, revealing a picture of the ionization trail created by the electron. This device is called a *cloud chamber*.

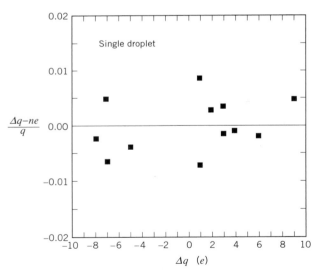

FIGURE 1-10 Data of Millikan.
The terminal speeds (v_T and v_E) of a single oil droplet were measured over a time interval of 45 minutes and the electric charge was computed. Occasionally the charge of the droplet changed by collision with air molecules as evidenced by a change in v_E. The data show the change in charge of the droplet (Δq) is equal to an integer (*n*) times a fundamental charge (*e*). Data from R.A. Millikan, *Electrons (+ and −), Protons, Photons, Mesotrons, and Cosmic Rays*, University of Chicago Press (1947).

The cloud chamber was an extremely important tool in discovering particles and measuring their properties. The amount of ionization observed in the cloud chamber depends on the speed of the charged particle. A particle with a lower speed spends more time in the vicinity of individual atoms and produces more ionization. Thus, the density of droplets in the cloud chamber gives information on the particle speed. If the cloud chamber is placed in a magnetic field, the curvature of the particle trajectory may be measured to determine the particle momentum. If the speed and momentum are known, the mass of the particle can be calculated.

The charge-to-mass ratio of the β particle discovered by Becquerel was determined to be the same as that of the electron. The β particle *is* the electron! Wilson's discovery allows us to see the trajectory of an electron. An early picture of an electron track in a cloud chamber made by C. T. R. Wilson is shown in Figure 1-11. The source of the electron is the radioactive decay of radium, the β particles discovered by Becquerel.

Discovery of the Nucleus

One early hint that atoms might have structure (i.e., contain other particles) was the fact that there are so many elements. If each atom was a fundamental particle, then there would be more than 100 types of fundamental particles. This could have been the case, but physicists were suspicious of having so many fundamental particles in nature. Furthermore, groups of atoms (see Figure 1-1) have similar properties. If atoms contain electrons and the net charge of atoms is zero, the atoms must also contain positive charge.

In 1912, Ernest Rutherford and his associates discovered that the positive charge of the atom is concentrated in a *nucleus*. (The Rutherford experiment is discussed in detail in Chapter 6.) Rutherford discovered the nucleus by experimenting with the α particles discovered by Becquerel. The charge of the α particle was determined to be $2e$ and the mass of the α particle was determined to be about four times the mass of the hydrogen atom.

Rutherford directed α particles through thin foils of material and observed that they are occasionally scattered at very large angles ($\theta > \pi/2$). Since the α particle has a large mass compared to the electron mass, collisions with electrons cannot produce large scattering angles. Rutherford ingeniously deduced that the large scattering angles are the result of a large electric force between the α particle and the nucleus and that the large force is

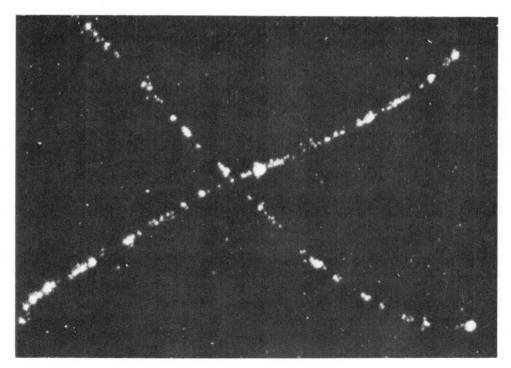

FIGURE 1-11 Electron tracks in a cloud chamber made by C. T. R. Wilson.
Courtesy Science Museum Library, London.

possible only at short distances, which in turn is possible only if the nucleus is concentrated at a "point."

The experiments of Rutherford also showed that the nucleus contains almost all the mass of the atom and that each element has a nucleus with a unique electric charge. The charge-to-mass ratio of the helium nucleus was measured to be the same as that of the α particle. The α particle *is* the helium nucleus.

The nucleus and the electrons are attracted to each other by the electromagnetic force. The strength of this force determines the size of atoms. The size of an atom is given by the typical distance between an outer electron and the nucleus. The stronger the force, the closer the electrons are to the nucleus, on the average. *What keeps electrons from being attracted all the way to the nucleus?* This question baffled physicists for a number of years. The answer will be explained in Chapter 5. The atom does not collapse because the electron behaves as a wave. In an atom, an electron cannot be localized to infinite precision. There are two competing processes in the electron–nucleus interaction: The electromagnetic force is pulling the electron to the nucleus and the electron is waving itself away.

Experiments like those of Thomson show us that electrons are *identical* particles in each atom. The measurement of the charge of the nucleus shows that there are different numbers of electrons in each type of atom. In discovering this important fact, we have made a great step in understanding the structure of matter, because the chemical properties of the elements may be explained by the *number* of electrons in each atom. The properties of carbon differ greatly from the properties of oxygen because the carbon atom has 6 electrons and oxygen has 8, but the electrons in the two atoms are identical particles. We have also raised a fundamental question because the nucleus is different for each element. *Why are there so many different nuclei?*

1-4 LOOKING INSIDE THE NUCLEUS: PROTONS AND NEUTRONS

One early hint that the nucleus is not a fundamental constituent of matter is that there are so many of them! The nucleus of the lightest atom (hydrogen) is called the *proton*. The charge of the proton is measured to be e. A new particle with zero electric charge was discovered by bombarding beryllium atoms with α particles. James Chadwick showed that the new particle, called the *neutron*, had mass nearly equal to that of the proton. (The

discovery of the neutron and the measurement of its properties are described in Chapter 11.)

Through detailed study of many nuclear interactions, neutrons and protons were found to be components of every nucleus. The nucleus of each atom is made up of identical protons and identical neutrons. Nuclei of the elements differ only by their *numbers* of protons and neutrons. The elements are classified by the number of protons (Z) and the total number of protons plus neutrons (A). The mass of an atom is dominated by the neutrons and protons. This is why the atomic mass is very nearly proportional to an integer (A).

We should make a distinction between the atomic mass number (an integer) and the slightly different atomic mass for which we use the same symbol (A). The atomic mass number is defined to be the number of neutrons plus protons in the nucleus. The atomic mass is defined to be the number of grams that correspond to N_A atoms. The atomic mass is not exactly equal to the number of neutrons plus protons for several reasons: (1) The electrons have some mass, (2) the proton and neutron have a small mass difference, and (3) the neutrons and protons interact with each other. These effects have a relatively small influence on the atomic mass. For example, the atomic mass number of the isotope of oxygen that makes up 99.8% of its natural abundance is equal to 16 (eight neutrons and eight protons). The atomic mass of this atom is equal to 15.994915, very nearly equal to 16.

It is possible for the nuclei of two atoms to have the same number of protons and different numbers of neutrons. Two atoms with the same atomic number (Z) but different atomic mass number (A) are called isotopes. Obviously, the atomic masses of isotopes differ. The value of A listed in a periodic table is usually the weighted average of the isotopes as they occur in nature. For example, carbon may be seen listed as $A = 12.01$, reflecting the fact that the most common isotope of carbon is $A = 12$, but there is a tiny percentage of the isotopes with $A = 13$ and $A = 14$.

The composition of atoms is explained by the electrical attraction of a positively charged nucleus and negatively charged electrons. A force other than the electric force is needed to explain the attraction of the protons and neutrons to one another in the formation of nuclei. This force is called the *strong force*. The strong force between any combination of two neutrons or protons is attractive and of equal strength. The strong force between an electron and a proton or neutron is zero. In the same manner that the strength of the electromagnetic force determines the size of atoms, the strength of the strong force determines the size of the nucleus. All nuclei are the same order of magnitude in size. The approximate diameter of a proton (d_{proton}) is 2 femtometer (fm):

$$d_{proton} \approx 2 \times 10^{-15} \text{ m} = 2 \text{ fm}. \qquad (1.49)$$

1-5 MASS AND BINDING ENERGY

Energy Units

The energy units of modern physics are derived from the unit of potential difference in electricity, the volt (V). The definition of the volt is

$$1 \text{ V} \equiv 1 \text{ J/C}. \qquad (1.50)$$

The unit of energy is called the electronvolt (eV). One electronvolt is defined to be the amount of kinetic energy that an electron acquires when it is accelerated through a potential difference of 1 volt. Since we have defined the magnitude of the electron charge to be e, the joule and the electronvolt are related by

$$1 \text{ J} = (1 \text{ V})(1 \text{ C})\left(\frac{e}{1.602 \times 10^{-19} \text{ C}} \right)$$
$$= \frac{e(1 \text{ V})}{1.602 \times 10^{-19}} = \frac{1 \text{ eV}}{1.602 \times 10^{-19}}, \qquad (1.51)$$

or

$$1 \text{ eV} = 1.602 \times 10^{-19} \text{ J}. \qquad (1.52)$$

The electric force on a particle depends on the charge of the particle but not on its mass. Thus, a proton accelerated through 1 V also acquires a kinetic energy of 1 eV.

By definition of the electron charge (1.46) and the electronvolt (1.52),

$$\frac{1 \, e}{1 \text{ C}} = \frac{1 \text{ eV}}{1 \text{ J}}. \qquad (1.53)$$

EXAMPLE 1-5

The electric force constant k in Coulomb's law is measured to be 8.99×10^9 J·m/C^2. Calculate the quantity ke^2 in units of J·m and eV·nm.

SOLUTION:

We have

$$ke^2 = \left(8.99 \times 10^9 \text{ J·m/C}^2 \right)\left(1.60 \times 10^{-19} \text{ C} \right)^2$$
$$= 2.30 \times 10^{-28} \text{ J·m},$$

and

$$ke^2 = \left(\frac{2.30 \times 10^{-28} \text{ J·m}}{1.60 \times 10^{-19} \text{ J/eV}} \right)\left(\frac{10^9 \text{ nm}}{\text{m}} \right) = 1.44 \text{ eV·nm}. \blacksquare$$

EXAMPLE 1-6

Calculate the strength of the electric field at a distance of 0.1 nm from a proton.

SOLUTION:

The electric field is

$$E = \frac{ke}{r^2} = \frac{ke^2}{er^2} = \frac{1.44 \text{ eV·nm}}{e(0.1 \text{ nm})^2}$$
$$= 144 \text{ V/nm} = 1.44 \times 10^{11} \text{ V/m}.$$

The electric fields are gigantic on atomic scales compared to fields accessible in the laboratory! $\blacksquare$

EXAMPLE 1-7

The magnetic field at the center of a current-loop is $(2\pi kI)/(c^2 R)$, where I is the current in the loop and R is the radius of the loop. If the hydrogen atom is modeled as an electron moving in a circular orbit with $R = 0.1$ nm at a speed of 3×10^6 m/s, calculate the strength of the magnetic field inside the atom.

SOLUTION:

A moving charge constitutes a current ($I = dq/dt$). A charge moving in a circle of radius R with speed v makes one revolution in the time

$$T = \frac{2\pi R}{v}.$$

The corresponding current is

$$I = \frac{e}{T} = \frac{ve}{2\pi R}.$$

The magnetic field is

$$B = \frac{2\pi kI}{c^2 R} = \frac{ke^2 v}{ec^2 R^2}$$
$$= \frac{\left(1.44 \times 10^{-9} \text{ eV·m} \right)\left(3 \times 10^6 \text{ m/s} \right)}{(e)\left(3 \times 10^8 \text{ m/s} \right)^2 \left(10^{-10} \text{ m} \right)^2} \approx 5 \text{ T}.$$

This is a large magnetic field by laboratory standards.

■

The electromagnetic force constant (see Example 1-5),

$$ke^2 = 1.44 \text{ eV} \cdot \text{nm}, \qquad (1.54)$$

is worth remembering because it specifies the strength of the electric force and is often used in calculations. The strength of the electric force on the atomic scale is given directly by the numerical value of ke^2. An electron and proton separated by a fraction of a nanometer will require a few electronvolts of energy in order to separate them. The electronvolt is a convenient unit because it is the characteristic energy scale of electrons in atoms (see Table 1-3).

> The strength of the electric force (F) between an electron and a proton that are separated by a distance r is
>
> $$F = \frac{ke^2}{r^2} = \frac{1.44 \text{ eV} \cdot \text{nm}}{r^2}.$$

Mass Energy

Consider the β particles (electrons) from the spontaneous decay of a heavy nuclei first observed by Becquerel. These electrons did not exist before the nuclei decayed! The electrons are *created* in the decay process! The electrons are produced with a typical kinetic energy of an MeV. *Where does this energy come from?*

TABLE 1-3
ENERGY UNITS IN MODERN PHYSICS.

Energy	Physical Interpretation
eV	Energy scale of the outer electrons in atoms
keV $\equiv 10^3$ eV	Energy scale of the inner electrons in heavy atoms
MeV $\equiv 10^6$ eV	Energy scale of neutrons and protons inside nuclei
GeV $\equiv 10^9$ eV	Energy scale of quarks inside protons
TeV $\equiv 10^{12}$ eV	Energy scale to be studied by the next generation of particle physics experiments

The energy that the electron acquires was stored in the form of *mass energy* of the decaying nucleus. The *total energy* (E) of any particle is defined to be the sum of two parts: energy due to motion, called the *kinetic energy* (E_k), and energy stored as mass, called the *mass energy* (E_0). Energy is defined in this manner,

$$E \equiv E_k + E_0, \qquad (1.55)$$

because this quantity is observed to be conserved in all particle interactions. Einstein was the first to deduce that the mass energy of a particle is equal to the mass (m) times the speed of light squared,

$$E_0 = mc^2. \qquad (1.56)$$

The cornerstone of Einstein's theory is that the speed of light is an absolute constant that does not depend on the motion of the source. This is verified by experiment. (The theory of special relativity is the subject of Chapter 4.) The speed of light (c) is measured to be

$$c \approx 3.00 \times 10^8 \text{ m/s}. \qquad (1.57)$$

The electron mass energy is

$$
\begin{aligned}
E_0 &= mc^2 \\
&\approx \left(9.11 \times 10^{-31} \text{ kg}\right)\left(3.00 \times 10^8 \text{ m/s}\right)^2 \\
&\quad \times \left(\frac{1 \text{ eV}}{1.60 \times 10^{-19} \text{ J}}\right) \\
&\approx 5.11 \times 10^5 \text{ eV} = 0.511 \text{ MeV}.
\end{aligned}
\qquad (1.58)
$$

Similarly, the proton mass energy is

$$E_0 \approx 938 \text{ MeV}. \qquad (1.59)$$

A scale of energies found in nature together with their mass equivalents is shown in Figure 1-12.

> The mass energy of a particle (E_0) is the mass times the speed of light squared,
>
> $$E_0 = mc^2.$$

EXAMPLE 1-8
Calculate the amount of energy stored in 1 kg of matter.

SOLUTION:
The amount of energy is

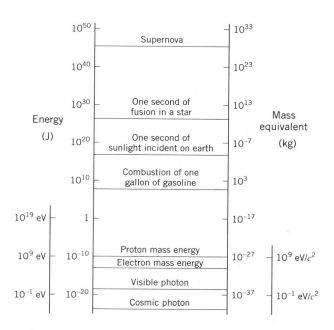

FIGURE 1-12 Energies in our universe and their equivalent masses.

$$E_0 = mc^2 = (1 \text{ kg})(3.00 \times 10^8 \text{ m/s})^2 = 9 \times 10^{16} \text{ J}.$$

This is an enormous amount of energy! If this energy could be released and harnessed, it could provide a kilowatt of power for about 3 million years! *Just how much energy is this?* At a cost of 10 cents per kilowatt-hour, the energy stored in one kilogram of dirt is worth about 2.5 billion dollars! ∎

Most of the energy in the universe is stored in the form of mass energy. This was not always the case. About 10 billion years ago, most of the energy of the universe was in the form of kinetic energy. The physics of the early universe is discussed in Chapter 19.

The fundamental relationship between mass energy and mass of a particle (1.56) invites the use of a particularly useful mass unit, the MeV/c^2. By defining this natural mass unit, it saves us the trouble of dividing by the speed of light squared. It can be a tremendous convenience not to have to do the division. For more massive particles, we define the analogous mass unit, the GeV/c^2.

EXAMPLE 1-9

Calculate the electron mass in units of MeV/c^2.

SOLUTION:

The mass energy of the electron is

$$E_0 = 0.511 \text{ MeV}.$$

Using the relationship between mass and mass energy, we have

$$m = \frac{E_0}{c^2} = 0.511 \text{ MeV}/c^2.$$

That was easy! ∎

EXAMPLE 1-10

Show explicitly that MeV/c^2 has units of mass by expressing it in kilograms.

SOLUTION:

We have

$$1 \text{ MeV} = (10^6 \text{ eV})(1.60 \times 10^{-19} \text{ J/eV})$$
$$= 1.60 \times 10^{-13} \text{ kg} \cdot \text{m}^2/\text{s}^2.$$

Dividing by c^2, we get

$$1 \text{ MeV}/c^2 = \frac{1.60 \times 10^{-13} \text{ kg} \cdot \text{m}^2/\text{s}^2}{(3.00 \times 10^8 \text{ m/s})^2}$$
$$= 1.78 \times 10^{-30} \text{ kg}. \quad ∎$$

EXAMPLE 1-11

The atomic number of hydrogen is $A = 1.0078$. Calculate the mass energy of the carbon atom ($A = 12$) in electronvolts.

SOLUTION:

The mass energy of the hydrogen atom is the sum of the masses of the electron and proton (neglecting the atomic binding energy):

$$m_H c^2 = 938.3 \text{ MeV} + 0.5 \text{ MeV} = 938.8 \text{ MeV}.$$

The relationship between the *atomic mass unit* (u) and MeV/c^2 is

$$1 \text{ u} = \frac{938.8 \text{ MeV}/c^2}{1.0078} = 931.5 \text{ MeV}/c^2.$$

The mass energy of the carbon atom ($A = 12$) is

$$m_C c^2 = (12)(931.5 \text{ MeV})$$
$$= 1.118 \times 10^4 \text{ MeV} = 11.18 \text{ GeV}. \quad ∎$$

Binding Energy

Consider an electron and proton bound together to form an atom of hydrogen. The energy (ΔE) required to separate the electron and proton to a large distance, work done against the Coulomb force, is measured to be

$$\Delta E = 13.6 \text{ eV}. \qquad (1.60)$$

What happens to this energy?

The energy needed to ionize the hydrogen atom has been converted into mass energy. The mass energy of the hydrogen atom ($m_H c^2$) is smaller than the sum of the mass energies of the electron ($m_e c^2$) plus proton ($m_p c^2$) by an amount equal to 13.6 eV:

$$m_H c^2 + 13.6 \text{ eV} = m_e c^2 + m_p c^2. \qquad (1.61)$$

The difference between the mass energy of the components (the electron and the proton) and the mass energy of the composite object (the hydrogen atom) is called the *binding energy* (E_b):

$$E_b = m_e c^2 + m_p c^2 - m_H c^2 = 13.6 \text{ eV}. \qquad (1.62)$$

The binding energy is the amount of energy that must be provided in order to break the atom into its components. When this happens, the fractional change in mass ($\Delta m_H / m_H$) of the atom is

$$\frac{\Delta m_H}{m_H} = \frac{E_b}{m_H c^2} \approx \frac{13.6 \text{ eV}}{9.39 \times 10^8 \text{ eV}} \approx 10^{-8}. \qquad (1.63)$$

For most processes, we may neglect this change in the mass of the atom.

1-6 ATOMS OF THE TWENTIETH CENTURY: QUARKS AND LEPTONS

The proton and neutron have similar masses, and both have a strong interaction of identical strength. This is a hint that they might be built of the same constituents. In 1964, Murray Gell-Mann and George Zweig independently made a classification of all known strongly interacting particles, called *hadrons*. All known hadrons could be constructed from objects that Gell-Mann called quarks after a passage in James Joyce's *Finnegans Wake,* "Three quarks for muster Mark." Three elementary building blocks were needed to make a model of the hadrons.

In 1967, Jerome Friedman, Henry Kendall, Richard Taylor, and collaborators experimentally detected the quark structure of the proton and were able to measure the momentum of the quarks inside the proton. This and other related experiments with quarks are discussed in Chapters 6, 17, and 18.

Looking Inside the Quarks

If quarks are the constituents of protons, what are the constituents of quarks? In our attempt to answer this question, we have reached the current limit of experimental resolution. The reason we are experimentally limited is the same reason that an atom cannot be resolved in an optical microscope; the wavelength of the probe is too large. We have probed quarks with a wavelength of about 10^{-18} m. This is our current experimental limit.

This is not a fundamental limit but rather a technical limit. We can summarize this by saying that the quarks are *pointlike* particles down to a distance of at least 10^{-18} m. The same is found to be true of the electron. The electron belongs to a class of particles, called *leptons*, that do not participate in the strong interaction. The quarks and the leptons have no detected structure. Today's atoms by the Greek definition are the quarks and leptons. Quarks and leptons are considered the building blocks of all matter because they are indivisible with present technology. Figure 1-13 shows the structure of matter in the universe as a function of distance of observation.

The Periodic Table of Fundamental Particles

The Quarks

Two additional quarks have been discovered since Gell-Mann and Zweig made the classification of the hadrons in 1964. The discovery of these quarks and the measurement of their properties have been crucial to our present understanding of particles and forces. The two additional quarks are much more massive than the original three, and they are often referred to as the *heavy quarks*. There are five known quarks. The original three quarks are named *up* (*u*), *down* (*d*), and *strange* (*s*). The heavy quarks are called *charm* (*c*) and *bottom* (*b*). The quarks are grouped into pairs by their physical properties. A periodic table of the quarks is shown in Figure 1-14.

Ordinary matter (i.e., protons and neutrons) is made of up and down quarks. A decade before its discovery, James Bjorken and Sheldon Glashow predicted that a fourth quark must exist in order to explain the properties of the other three quarks. They gave the name *charm* to the quark they predicted must exist, and they were correct! Notice that there is an empty entry in the position above the bottom quark. The properties of the bottom quark indicate

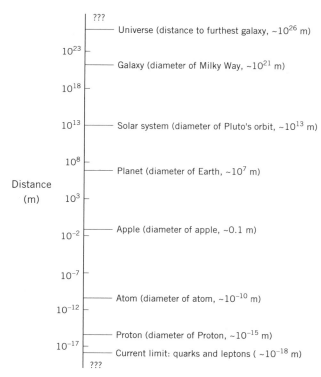

FIGURE 1-13 Sizes of various objects in our universe.
The size of our universe is defined to be the distance to the furthest known object. Within our universe we find objects of varying size: galaxies, solar systems, planets, apples, atoms, protons, and finally, today's ultimate constituents: quarks and leptons.

FIGURE 1-14 Periodic table of the quarks.
The quarks are permanently bound into hadrons. The mass energies shown are deduced from the mass energies of the particles that contain the quarks. The quarks have fractional electric charge, either +2/3 or −1/3. Each quark also has one of three types of strong charge called color: red (R), blue (B), or green (G). The blank box indicates the position expected to be occupied by a sixth quark (top).

that a sixth quark should exist. This quark has not yet been observed, because it has a mass energy beyond the sensitivity of present experiments. The sixth quark is called *top*. (Physicists sometimes refer to the bottom and top quarks by the more imaginative names of *beauty* and *truth*.)

The Leptons

In addition to the quarks, there is another group of fundamental particles called the *leptons*. Leptons are particles that do not interact strongly. The electron is the charter member of this group. The partner of the electron was postulated to exist by Wolfgang Pauli in 1931 in order to explain an apparent nonconservation of energy in decays that produce beta particles (electrons). In the 1930s, Enrico Fermi developed the theory of these decays and gave the name *neutrino* ("little neutral one") to the hypothetical particle. Finally, in 1956, the first direct observation of a neutrino particle interacting with matter was made by Frederick Reines and Clyde Cowan.

A second charged lepton, analogous to the electron, was discovered in cosmic ray experiments in 1937. The discovery of the *muon* caused a tremendous confusion for physicists, because it had no known role in nature. The discovery prompted the physicist Isidor I. Rabi to make the famous remark, "Who ordered that?" In 1961, the partner of the muon, the *muon neutrino*, was demonstrated to exist and have different properties than the partner of the electron, the *electron neutrino*. Finally, in 1975, a third *heavy* lepton was discovered. The complete family of leptons is shown in Figure 1-15.

A summary of the important experiments revealing our present understanding of the structure of matter is given in Table 1-4.

* Challenging

1-7 PROPERTIES OF THE FOUR FORCES

Quantum Nature of the Electromagnetic Force

The discovery of the Lorentz force law and the four Maxwell equations relating electric and magnetic fields was undoubtedly the greatest intellectual achievement

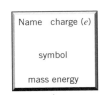

Name	charge (*e*)
symbol	
mass energy	

Group I	Group II	Group III
Electron −1	Muon −1	Tau −1
e	μ	τ
0.5110 MeV	105.7 MeV	1777 MeV
Electron 0 Neutrino	Muon 0 Neutrino	Tau 0 Neutrino
ν_e	ν_μ	ν_τ
< 7 eV	< 0.27 MeV	< 33 MeV

FIGURE 1-15 Periodic table of the leptons.

TABLE 1-4
FAMOUS EXPERIMENTAL DISCOVERIES OF MODERN PHYSICS ON THE STRUCTURE OF MATTER.

Year	Experimenters	Discovery
1895	Röntgen	x rays
1896	Becquerel	α and β particles
1897	Thomson	electron
1909	Rutherford	nucleus
1932	Anderson	antimatter (positron)
1932	Chadwick	neutron
1937	Street, Stevenson, Anderson, Neddermeyer	second lepton (muon)
1956	Reines, Cowan	neutrino
1967	Friedman, Kendall, Taylor et al.	quark structure of proton
1974	Richter, Ting et al.	heavy quark (charm)
1975	Perl et al.	heavy lepton (tau)
1977	Lederman et al.	heavy quark (bottom)
1983	Rubbia et al.	W and Z^0 particles

of the nineteenth century. The Maxwell equations predict that charges will radiate electromagnetic energy when accelerated. This is confirmed by experiment.

Amazingly enough, a fundamental characteristic of electromagnetism remained to be discovered in the twentieth century. This additional physics is not contained in the Maxwell equations. The new phenomenon is the quantization of the electromagnetic wave! This appears in much the same manner as the quantization of electric charge. The quantization of electromagnetic radiation was first deduced by Einstein in 1905 (the same year he published the theories of Brownian motion and special relativity).

A single electromagnetic quantum is called a *photon*. Photons have nonzero energy and momentum but zero mass energy, and always travel at the speed c (in vacuum). An electromagnetic wave of a given frequency contains individual photons, each with the same energy proportional to the frequency. The experimental proof of this is discussed in Chapter 3.

Feynman Diagrams
When two electrons collide elastically,

$$e + e \rightarrow e + e, \qquad (1.64)$$

the momentum of each electron changes in a manner such that the total momentum of the two electrons is unchanged. We may consider the scattering as a two-step process in which an intermediate particle ($\gamma*$) is emitted by one electron and absorbed by the other:

$$e \rightarrow e + \gamma* \quad \text{and} \quad \gamma* + e \rightarrow e. \qquad (1.65)$$

The intermediate particle causes the momentum transfer between the electrons. Since the intermediate particle is not free, it is called a *virtual* particle.

The virtual particle can cause a positive or negative momentum transfer corresponding to whether the two charges are the same or opposite in sign. The virtual particle that transmits the electromagnetic force is the photon! This is the physical mechanism by which one charge is attracted or repelled by a second charge. The two charges exchange photons, causing the transfer of energy and momentum. The theory that describes the interaction of charged particles by photon exchange is called *quantum electrodynamics* (QED). Quantum electrodynamics is the most accurate and successful theory ever constructed. The major advances in this theory were made in the late 1940s by Richard Feynman, Julian Schwinger, Freeman Dyson, and Sin-itiro Tomonaga. The theory is tested in certain cases to 10 decimal places! Nobody has ever observed a violation of any prediction of QED.

A handy pictorial representation of particle interactions was invented by Feynman, called the *Feynman diagram*. A Feynman diagram is a space–time picture for one possible path by which an interaction may occur. In the Feynman diagram, time moves upward, charged particles are represented by solid lines, and photons are represented by wavy lines. Figure 1-16*a* shows a Feynman diagram representing the scattering of two electrons. The picture presented in Figure 1-16*a* represents only one possible path, or *amplitude* for the process. The amplitude for the coupling of a photon to an electron is proportional to the charge of the electron. Photons couple to electric charge. To get the total interaction probability, we must add up all the amplitudes and then square the result. There can be interference between the different paths, analogous to the interference phenomena observed in waves. (And why not, since the photon is a wave!) Often, however, the amplitude is dominated by one or perhaps a small number of paths.

The Feynman diagrams provide a useful representation of the physical process. Figure 1-16*b* shows radiation from an accelerated electron:

$$e + e \rightarrow e + e + \gamma. \qquad (1.66)$$

In this process, called *bremsstrahlung*, a virtual photon causes an electron to be accelerated and a real photon is radiated.

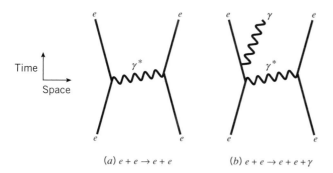

(a) $e + e \rightarrow e + e$ (b) $e + e \rightarrow e + e + \gamma$

FIGURE 1-16 Feynman diagrams for electron scattering. These diagrams give a pictorial representation of the interaction of two electrons. The pictures are read from bottom to top. The solid lines represent electrons and the wavy lines represent photons. (*a*) Two electrons interact by the electromagnetic force through the exchange of a virtual photon (γ^*). The photon is not a free particle but is emitted by one electron and absorbed by the other. The coupling strength of the photon to the electron is proportional to the electric charge of the electron. (*b*) In the interaction of two electrons, a real photon (γ) is radiated.

The Range of the Electromagnetic Force

The electromagnetic force has an infinite range. This is a direct consequence of the role that photons play in mediating the force and the fact that the photon has zero mass. No matter how far apart two charges are, there is a nonvanishing force between them. The charged particles in a distant star exert a force on the electrons in your eyes, causing them to move, and this is how you can see the star even though it is a long distance away!

Introduction to Alpha

We have observed that the strength of the electromagnetic interaction (ke^2) has units of distance times energy. If we choose benchmark distance (R_0) and energy (E_0) scales, then we can represent the strength of the electromagnetic force as a dimensionless number:

$$\text{dimensionless strength} = \frac{ke^2}{R_0 E_0}. \qquad (1.67)$$

The choice of the distance and energy scales is a matter of definition but there is one and only one scale for the product $R_0 E_0$ that occurs *naturally* in the interaction of radiation and matter. The natural scale is determined by the energy and wavelength of radiation quanta that are exchanged in the electromagnetic interaction. We may imagine that shorter wavelength radiation can get closer to the pointlike electron and in the process impart a greater momentum transfer to the electron. The energy of an electromagnetic quantum (E_{photon}) is inversely proportional to its wavelength (λ_{photon}), so that the product is a universal constant:

$$E_{\text{photon}} \lambda_{\text{photon}} = \text{constant}. \qquad (1.68)$$

This relationship gives us a natural energy and distance to evaluate the strength of the force. We shall spend all of Chapter 3 discussing the discovery and the physics of this fundamental constant. We define alpha (α) as

$$\alpha \equiv \frac{2\pi ke^2}{E_{\text{photon}} \lambda_{\text{photon}}}. \qquad (1.69)$$

The factor of 2π is arbitrarily included in the definition of α because it occurs so often in calculations. The energy of a photon that has a wavelength of 1 nm is determined from experiment to be 1240 eV. The numerical value of α is

$$\alpha \equiv \frac{2\pi(1.44 \text{ eV} \cdot \text{nm})}{(1240 \text{ eV})(1 \text{ nm})} = \frac{1}{137}. \qquad (1.70)$$

Alpha is called the *dimensionless electromagnetic coupling strength.*

We said that the natural choice of R_0E_0 was unique. One also might try to construct an energy and distance scale from the electron itself. The choice of energy scale is easy; it is the mass energy of the electron, mc^2. You might think that the distance scale could be the size of the electron. In the same manner that a photon has no definite size because its wavelength depends on its energy, the electron also has no definite size. A higher energy electron appears smaller than a lower energy electron. Furthermore, an electron always has some kinetic energy because there are always photons bumping into it. At the shortest measurable distances, the electron appears like a pointlike particle. It turns out that the electron does have a characteristic size that is determined by its energy, but the product R_0E_0 leads us back to the same constant! This wonderful and fundamental physics is the subject of Chapter 5.

Source of the Forces

Four distinct forces are observed in the interaction between various types of particles: electromagnetic, strong, weak, and gravitational. Each force is caused by some intrinsic property of the particles, analogous to electric charge.

The Strong Force

The existence of the strong force was realized with the discovery that the nuclei of atoms are bound states of neutrons and protons. The mechanism of the strong force was understood only after it was discovered that the neutrons and protons are not fundamental constituents, but are made up of quarks. The quarks have an intrinsic property called strong charge or *color*. There are *three* types of strong charges: red, green, and blue. The proton and neutron have no net strong charge, analogous to an atom having no net electric charge. The proton and neutron attract each other by the strong force in much the same manner as two atoms attract each other by the electromagnetic force to form a molecule. (The physics of how two neutral atoms can attract each other is itself an interesting question that is taken up in Chapter 10.) The quantum theory of the strong interaction is called *quantum chromodynamics* (QCD). The force between two quarks is transmitted by massless particles called *gluons.* The interaction between two quarks is qualitatively described by the Feynman diagram shown in Figure 1-17. Gluons couple to color.

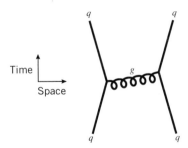

FIGURE 1-17 Feynman diagram for the strong interaction of two quarks.
The quarks (q) interact by the exchange of gluons (g).

The Weak Force

The existence of the weak force was realized with understanding of how the beta particles (electrons) were produced in the decays of nuclei. The weak force is capable of transforming neutrons into protons and vice versa. The transformation occurs because the weak interaction can cause changes in the quark flavors. The weak interaction provides the first step of the reaction chain in which protons are combined into alpha particles inside the sun and other stars. The mass energy that is released makes the sun shine!

The weak force may be described by assigning a property called *weak charge* to all quarks and leptons. The weak force between any combination of quarks and leptons is transmitted by massive particles called W^+, W^- and Z^0. The interaction between a quark and a lepton is qualitatively described by the Feynman diagram shown in Figure 1-18. The W and Z^0 particles couple to weak charge.

Gravity

The first theory of gravity was formulated by Isaac Newton and published in 1683. In spite of the fantastic

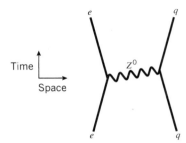

FIGURE 1-18 Feynman diagram for the weak interaction of a quark and a lepton.
The quark and lepton interact by the exchange of Z^0 particles.

success of the gravitational theory, Newton was not able to address the question of what causes the force. We now know that gravity is caused by energy. Energy is our "gravitational charge"! Recall that mass is one form of energy. It is perhaps ironic that gravity is the least understood of the four forces. There is no accepted quantum theory of gravity. The hypothetical particle that would be responsible for transmitting the gravitational force is called the *graviton*. The graviton is expected to be massless and to couple to energy, but unlike the photon, the W and Z particles, and the gluon, the graviton has not been observed!

Relative Strengths of the Forces

The four forces are equally fundamental. Our universe could not exist if any one of them were absent. Amazingly, however, they have strengths that vary by many orders of magnitude. The force between two particles depends on the particle type and also on the distance between the particles. The strength of the force of gravity between two particles depends on the mass of the particles. The gravitational force of attraction between two masses (m_1 and m_2) separated by a distance r is given by

$$\mathbf{F}_g = \frac{Gm_1 m_2}{r^2} \mathbf{i_r}. \qquad (1.71)$$

Strictly speaking, the expression for the gravitational force (1.71) is valid only if the particles are not moving at large speeds (compared to the speed of light). The gravitational coupling ($Gm_1 m_2$) plays the analogous role of the electromagnetic coupling ($kq_1 q_2$). The value of $Gm_1 m_2$, for example in the attraction of the earth to the sun, must be determined from experiment. We find that

$$G = 6.67 \times 10^{-11} \text{ m}^3 \cdot \text{kg}^{-1} \cdot \text{s}^{-2}. \qquad (1.72)$$

EXAMPLE 1-12
Calculate the gravitational force constant $Gm_1 m_2$ between an electron and a proton.

SOLUTION:
The electron mass is

$$m_e = 0.511 \text{ MeV}/c^2 = 9.11 \times 10^{-31} \text{ kg},$$

and the proton mass is

$$m_p = 938 \text{ MeV}/c^2 = 1.67 \times 10^{-27} \text{ kg}.$$

We shall leave one of the masses in kilograms and one in MeV/c^2 to arrive at units of eV·nm:

$$Gm_e m_p = \left(6.67 \times 10^{-11} \text{ m}^3 \cdot \text{kg}^{-1} \cdot \text{s}^{-2}\right)$$
$$\times \left(9.11 \times 10^{-31} \text{ kg}\right)\left(938 \text{ MeV}/c^2\right)$$
$$\times \left(\frac{c}{3.00 \times 10^8 \text{ m/s}}\right)$$
$$= 6.33 \times 10^{-55} \text{ MeV} \cdot \text{m}$$
$$= 6.33 \times 10^{-40} \text{ eV} \cdot \text{nm}. \qquad \blacksquare$$

The ratio of strengths (gravitational to electromagnetic) between an electron and proton is

$$\frac{Gm_e m_p}{ke^2} = \frac{6.33 \times 10^{-40} \text{ eV} \cdot \text{nm}}{1.44 \text{ eV} \cdot \text{nm}}$$
$$\approx 4.4 \times 10^{-40}. \qquad (1.73)$$

The gravitational attraction between an electron and a proton is about 40 orders of magnitude weaker than the electromagnetic force. Gravity is so weak that if it was the force that governed the atomic size, then the hydrogen atom would be larger than the distance to the furthest galaxy (see Figure 1-13)!

The gravitational constant $Gm_1 m_2$ depends on the masses of the particles. For two protons, the gravitational force constant is

$$Gm_p m_p = \left(Gm_e m_p\right)\left(\frac{m_p}{m_e}\right)$$
$$\approx \left(6.33 \times 10^{-40} \text{ eV} \cdot \text{nm}\right)\left(\frac{938 \text{ MeV}}{0.511 \text{ MeV}}\right)$$
$$\approx 1.2 \times 10^{-36} \text{ eV} \cdot \text{nm}. \qquad (1.74)$$

Recall how we specified the strength of the electromagnetic force as a dimensionless constant α by dividing by the fundamental constant (1240 eV·nm). If we do the same thing for the gravitational force between two protons, we can define the coupling strength "alpha-g" (α_g):

$$\alpha_g \equiv \frac{Gm_p m_p}{1240 \text{ eV} \cdot \text{nm}}$$
$$\approx \frac{1.2 \times 10^{-36} \text{ eV} \cdot \text{nm}}{1240 \text{ eV} \cdot \text{nm}} \approx 10^{-39}. \qquad (1.75)$$

The numerical value of α_g depends on the choice of mass or equivalent energy scale ($m_p c^2$). The above value of α_g corresponds to an energy scale of about one GeV.

Besides their widely varying strengths, there is another significant difference between the forces, which is their *range*. The gravitational and electromagnetic forces have infinite ranges, but the strong force and the weak force are observed to have extremely short ranges. The strong force binds protons and neutrons together in a nucleus, but the force between the nuclei of two neighboring atoms is zero. The reason for this is that the proton and neutron have no net strong charge, analogous to an atom having no net electric charge.

The weak force also has an extremely short range, but for a completely different reason than the strong force. The reason that the weak force has a short range is that the weak force is mediated by the exchange of very massive particles called the W and Z^0 bosons. The large masses of the W and Z^0 particles results in a short range for the weak force. The range of the weak force is given by the fundamental energy–distance constant $(1240 \text{ eV} \cdot \text{nm})$ divided by the mass energy of the W and Z^0 particles. The mass energy of these particles is approximately one hundred times the proton mass energy, or roughly 100 GeV/c^2. The range of the weak force is

$$R \approx \frac{1240 \text{ eV} \cdot \text{nm}}{10^{11} \text{ eV}} \approx 10^{-2} \text{ fm}. \qquad (1.76)$$

The range of the weak force is so small that the weak force is referred to as a contact interaction.

The strong force and the weak force have a much different distance dependence than the electromagnetic and gravitational forces. We can still measure the strengths of these forces and then divide by our benchmark energy–distance scale of $(1240 \text{ eV} \cdot \text{nm})$ to define the coupling strengths "alpha-s" (α_s) for the strong force and "alpha-w" (α_w) for the weak force. We have seen that the numerical value of α_g depends on the choice of energy scale. The numerical values of α_s and α_w also depend on the energy scale, *but for different reasons than α_g does.* (The strong and weak forces are the subject of Chapters 17 and 18.) The strong force between two quarks is clearly much greater than the electromagnetic force between a quark and an electron; the experimental evidence for this is that the proton size is about 10^{-15} m and the atomic size is 10^{-10} m. At an energy scale of about one GeV, the size of the dimensionless strong coupling is

$$\alpha_s \approx 1, \qquad (1.77)$$

compared to $\alpha = 1/137$. At the same energy scale of one GeV, the size of the weak coupling is

$$\alpha_w \approx 10^{-6}. \qquad (1.78)$$

The approximate relative strengths of the four forces is summarized in Table 1-5.

Gravity is far, far weaker than the other three forces. Gravity is detectable to us only because there are a large number of protons and neutrons in the earth all pulling in the same direction on us. At distances larger than a nuclear size, the strong and weak forces have zero strength and the electromagnetic force rules! The strengths and ranges of the forces are shown in Figure 1-19.

Time Constants of the Forces

Associated with each force is a time constant for the interaction. Each force is always at work to make every particle decay into a lower energy state. Of course, some particles like the electron and proton never decay, and appropriate conservation laws are invoked to explain this. It may be that the proton or the electron does decay and we have not observed it because the decay lifetime is very long!

Most particles, however, are observed to decay. A stronger force makes the decay happen faster. Often a conservation law can prevent the action of one or more forces. The decay rate is proportional to the square of the relative strength of the force (one power of alpha for each of the two particles that are coupled). The average lifetime of a particle is inversely proportional to the decay rate and therefore inversely proportional to alpha squared. The typical lifetime of a particle that decays by the strong interaction (τ_{strong}) is characterized by the time that it takes a particle traveling at the speed of light to

**TABLE 1-5
COMPARISON OF THE FOUR FUNDAMENTAL FORCES.**

Force	Source	Mediator	Rel. Strength (1 GeV)
Strong	strong charge	gluon	$\alpha_s \approx 1$
EM	electric charge	photon	$\alpha = 1/137$
Weak	weak charge	W and Z^0	$\alpha_w \approx 10^{-6}$
Gravity	energy	graviton ?	$\alpha_g \approx 10^{-39}$

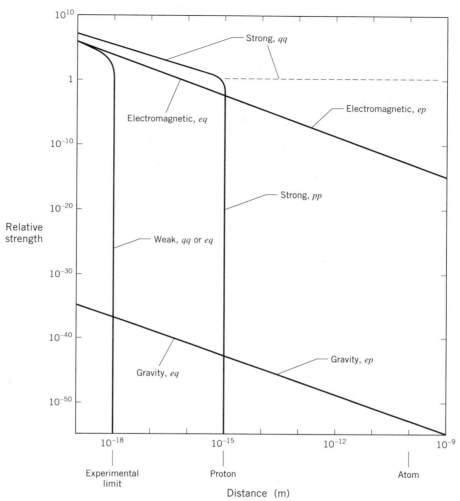

FIGURE 1-19 Strength of the four forces as a function of distance.
At atomic distances there are only two forces, electromagnetism and gravity, and the electromagnetic force between a proton and electron is about 40 orders of magnitude greater than gravity. The force between two quarks, if they could be separated, would be enormous, as shown by the dashed line. At a distance equal to the proton size, the strong force turns on abruptly to a strength about 100 times the electromagnetic force. (The strong force does not affect the electron at all.) At a distance equal to about 1/1000 of the proton size, the weak force turns on abruptly. This is the limit of current experimentation.

travel a distance equal to the size of a proton,

$$\tau_{strong} \sim \frac{\left(10^{-15}\ \text{m}\right)\alpha_s^2}{3 \times 10^8\ \text{m/s}} \sim 10^{-23}\ \text{s}. \quad (1.79)$$

The typical lifetime of a particle that decays by the electromagnetic interaction (τ_{em}) is characterized by a time scale of

$$\tau_{em} \sim \tau_{strong}\ \frac{\alpha_s^2}{\alpha^2} \sim 10^{-19}\ \text{s}. \quad (1.80)$$

The typical lifetime of a particle that decays by the weak interaction (τ_{weak}) is characterized by a time scale of

$$\tau_{weak} \sim \tau_{strong}\ \frac{\alpha_s^2}{\alpha_w^2} \sim 10^{-11}\ \text{s}. \quad (1.81)$$

The above order of magnitude estimates of τ_{strong}, τ_{em}, and τ_{weak} are only very crude values. The actual particle lifetimes vary widely from decay to decay because there are other factors that affect the decay rates. The main factor that sets the scale in each case, however, is "alpha" squared. The time scale of various processes in the universe are indicated in Figure 1-20.

In this introductory chapter, we have given a broad survey of particles and forces. The great majority of this text will be concerned with the physics of the electromagnetic force at the atomic distance scale (see Figure 1-19).   *

CHAPTER 1: PHYSICS SUMMARY

- Matter is composed of atoms: negatively charged electrons attracted to a positively charged nucleus by the electromagnetic force. The characteristic diameter of an atom is about 0.3 nm.

- The mass of an atom is concentrated in a positively charged nucleus made up of protons and neutrons

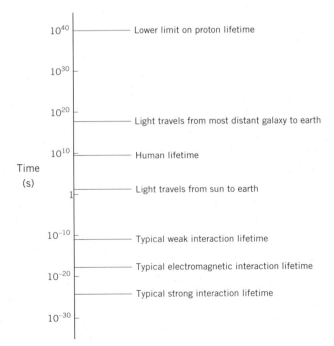

FIGURE 1-20 Time scales of various processes in the universe.

held together by the strong force. The number of neutrons plus the number of protons is called the atomic mass number (A). The mass of an atom is proportional to A. The characteristic size of a nucleus is about 1 fm.

- The number of electrons or protons in an atom is called the atomic number (Z) of the element. The chemical and physical properties of the element are determined by the number of electrons in the atom. A given element may have slightly different numbers of neutrons in the nucleus. Two versions of the same element with different numbers of neutrons are called isotopes.

- Avogadro's number is defined to be the number of atoms in 12 grams of carbon with $A = 12$:

$$N_A = 6.02 \times 10^{23}.$$

For any element there are N_A atoms in A grams, where A is the atomic mass number. For any compound there are N_A molecules in M grams, where M is the sum of the atomic mass numbers of the atoms making up one molecule.

- Electric charge is quantized. The charge observed on any free particle is an integer multiple of the fundamental charge,

$$e = 1.60 \times 10^{-19} \text{ C}.$$

The electric charge of the proton is e, the charge of the electron is $-e$, and the charge of a neutron is zero. The electric charges of the proton, electron, and neutron are intrinsic particle properties that cannot be altered.

- An electromagnetic wave with a frequency f is made up of radiation quanta called photons. These quanta each have an energy proportional to the frequency of the wave and have zero mass. The radiation quanta travel at a speed,

$$c = 3.00 \times 10^8 \text{ m/s}.$$

- The electromagnetic force is mediated by continuous photon exchange between electric charges. The strength of the electromagnetic force is specified by the constant,

$$ke^2 = 1.44 \text{ eV} \cdot \text{nm}.$$

In dimensionless units, obtained by dividing by the energy times the wavelength of a photon, the strength of the force is

$$\alpha = \frac{2\pi k e^2}{E_{photon}\,\lambda_{photon}} \approx \frac{1}{137}.$$

- There are four fundamental forces in nature: strong, electromagnetic, weak, and gravity. The approximate relative strength of the four forces (α_s, α, α_w, α_g) at an energy scale of one GeV are

$$1,\ \frac{1}{137},\ 10^{-6},\ 10^{-39}.$$

The ranges of the electromagnetic and gravitational forces are infinite, while the ranges of the strong and weak forces are extremely short. From distances of 10^{-15} m (the nuclear size) to 10^{26} m (the distance to the furthest visible galaxy) the physics of the universe is dominated by the electromagnetic interaction. The effects of gravity are noticeable when a very large number of particles are involved.

- The mass energy of a particle is potential energy stored in the form of mass. The relationship between mass energy (E_0) and mass (m) is

$$E_0 = mc^2.$$

where c is the speed of light. The mass energy of an electron is 0.511 MeV, and the mass energy of the proton is 938 MeV.

- At 10^{-18} m, the resolution of current experiments, matter is observed to be made up of structureless particles, quarks and leptons.
- The following units are used in modern physics:

Unit	SI	Modern	Conversion
Length	m	nm	1 nm = 10^{-9} m
		fm	1 fm = 10^{-15} m
Energy	J	eV	1 eV = 1.60×10^{-19} J
Charge	C	e	1 e = 1.60×10^{-19} C
Mass	kg	MeV/c^2	1 MeV/c^2 = 1.78×10^{-30} kg

REFERENCES AND SUGGESTIONS FOR FURTHER READING

R.H. Dicke, "The Eötvös Experiment," *Sci. Am.* **205**, No. 6, 84 (1961).

A. Einstein, *Investigations on the Theory of the Brownian Movement,* edited by R. Fürth, trans. A. D. Cowper, E. P. Dutton.

R. P. Feynman, *QED: The Strange Theory of Light and Matter,* Princeton University Press, (1985).

R. P. Feynman, R. B. Leighton, and M. Sands, *The Feynman Lectures on Physics,* Addison-Wesley (1963).

G. E. M. Jauncey, "The Early Years of Radioactivity," *Am. J. Phys.* **14**, 226 (1946).

B. H. Lavenda, "Brownian Motion," *Sci. Am.* **252**, No. 2, 70 (1985).

R. A. Millikan, *The Electron,* The University of Chicago Press (1917); *Electrons (+ and −), Protons, Photons, Neutrons, Mesotrons, and Cosmic Rays,* The University of Chicago Press (1947).

Particle Data Group, "Review of Particle Properties," *Phys. Rev.* **D45**, 1 (1992).

E. M. Purcell, *Electricity and Magnetism,* McGraw-Hill (1985).

F. Richtmyer, E. Kennard, and J. Cooper, *Introduction to Modern Physics,* McGraw-Hill (1969).

E. Segrè, *From X-Rays to Quarks,* Freeman (1980).

J. J. Thomson, "Cathode Rays," *Phil. Mag.* **44**, 293 (1897);

E. C. Watson, "Jubilee of the Electron," *Am. J. Phys.* **15**, 458 (1947).

The World of Physics, ed. A. Beiser, McGraw-Hill (1960).

QUESTIONS AND PROBLEMS

Discovery of atoms

*1. Think about Feynman's challenge. Can you put more scientific information in fewer words?

2. A mass m_1 of carbon combines with a mass m_2 of nitrogen to form a mass M of a compound. Show that m_1/M and m_2/M are both constants.

3. One kg of hydrogen combines with 4.5 kg of carbon. What is the ratio of number of carbon to hydrogen atoms in the hydrocarbon molecule that is formed?

4. One kg of carbon combines with oxygen to make carbon monoxide (CO), carbon dioxide (CO_2), or carbon suboxide (C_3O_2). For each of the three reactions, calculate the mass of the oxide produced.

5. Determine the mass (in kg) of the following atoms: (a) ^{4}He, (b) ^{12}C, and (c) ^{208}Pb.

6. Use the density of graphite, $\rho \approx 2 \times 10^3$ kg/m^3, to make an estimate of the size of a carbon atom.

7. Use the atomic number of copper, $A = 64$, to make an estimate of the density of copper.

8. The density of air at room temperature and atmospheric pressure is about 1.2 kg/m^3. Estimate the number of molecules in one cubic meter of air.

9. Estimate the number of atoms in a pencil. (*Hint:* All atoms have approximately the same size.)

10. Make an estimate of the number of atoms in a person who has a mass of 100 kg.

11. How many molecules are in (a) 1 kg of sodium-chloride (NaCl) and (b) 1 kg of sugar ($C_{12}H_{22}O_{11}$)?

Classical electromagnetism

12. Electrons in a synchrotron are made to travel in a circular orbit at a speed very close to the speed of light (c). If the radius of the orbit is 100 m, how many electrons are needed to make a current of 1 A?

13. An electric field applied to a metal causes a certain number of mobile electrons to move with an average drift speed (v_d). (a) Show that the current per area (J) is given by

$$J = nev_d,$$

where n is the density of mobile electrons. (b) A current of 1 A flows in a copper wire, which has a radius of 1 mm. Assuming that copper has one mobile electron per atom, make an estimate the average drift speed.

14. Consider a simple model of a hydrogen atom in which an electron makes a circular orbit about a stationary proton. (a) Show that the kinetic energy of the electron ($mv^2/2$) is equal to minus 1/2 times the potential energy ($-ke^2/r$). (b) An energy of 13.6 eV is required to separate the electron and proton from an orbit of radius r to some large distance ($\gg r$). What is the size of the orbit? (c) What is the speed of the electron in the orbit?

*15. How would Maxwell's equations be modified if a magnetic monopole was discovered? (*Hint:* What is the equation for conservation of monopoles?) How would the Lorentz force law be modified?

16. If $\mathbf{E}_1$ and $\mathbf{B}_1$ satisfy Maxwell's equations and $\mathbf{E}_2$ and $\mathbf{B}_2$ satisfy Maxwell's equations, show that ($\mathbf{E}_1 + \mathbf{E}_2$) and ($\mathbf{B}_1 + \mathbf{B}_2$) are also solutions of Maxwell's equations.

17. Consider an electromagnetic wave where the electric field has the form

$$\mathbf{E} = E_0 \cos(kz - \omega t)\mathbf{i}_x.$$

Show that the average value of the electric field amplitude squared is $E_0^2/2$.

Looking inside the atom: electrons and a nucleus

18. (a) Why can we neglect the force of gravity in the Thomson experiment? (b) Calculate the gravitational deflection of an electron in the Thomson spectrometer (see Example 1-4).

19. (a) Thomson made measurements with three different cathode materials (aluminum, iron, and platinum). They all gave the same value of q/m. What do we learn from this? (b) Thomson made measurements with three different gases initially present in his tube, which were then evacuated (air, carbon dioxide, and hydrogen). They all gave the same value of q/m. What do we learn from this?

20. In the Thomson experiment a voltage V is applied to parallel plates 0.05 m long separated by a distance of 0.01 m. Electrons are observed to be deflected by an angle of 120 milliradians. When a magnetic field of 10^{-4} T is applied, there is no deflection. (a) Calculate the speed of the electrons. (b) Calculate the increase in kinetic energy that the electrons gain from their acceleration in the electric field.

21. An electron with a speed of 10^6 m/s moves in a uniform magnetic field of 10^{-4} T. Calculate the radius of curvature of the electron's trajectory.

22. Electrons are accelerated from rest through a potential difference of 10^4 V. The electrons are then directed into a magnet that has a uniform field of 10^{-3} T. The magnetic field direction is orthogonal to the electron velocity. Calculate the radius of curvature of the electron's trajectory inside the magnet.

23. An electron and a proton each have the same radius of curvature in a uniform magnetic field. If the electron speed is 10^6 m/s, what is the speed of the proton?

24. In the Millikan oil-droplet experiment, a relatively small metal plate can pull a droplet upward with a stronger force than the whole earth pulling it downward! Do you consider this to be convincing evidence that the electric force is many, many orders of magnitude greater in strength than gravity? Can you think of other common examples that illustrate the relative strengths of electromagnetism and gravity?

25. Suppose that we designed a detection method for electrons. Could the Millikan experiment be performed directly with electrons?

26. Estimate the number of electrons in a Millikan oil droplet.

27. In the Millikan oil-droplet experiment, the mass of a droplet is 10^{-14} kg. (a) Calculate the electric field needed to overcome the force of gravity if the droplet has the charge of one electron. (b) If the area of each parallel plate is 3×10^{-3} m^2, calculate the number of excess charges present on the surface of each plate in order to produce the electric field of (a). Make a rough comparison of the number of excess charges with the number of atoms in the plates.

28. In the Millikan oil-droplet experiment, the free-fall speed of a 10^{-14} kg droplet is observed to be 10^{-3} m/s. An electric field of 3×10^5 V/m is then switched on and the droplet is observed to rise. What are the possible values of terminal rise speeds of the droplet?

29. In the Millikan oil-droplet experiment, calculate the electric field needed to make a droplet rise at the same speed as it free-falls with the field off. Take the mass of the droplet to be 10^{-14} kg and take the charge on the droplet to be the electron charge.

Looking inside the nucleus: protons and neutrons

30. Make an estimate of the mass energy density of nuclear matter.

Mass and binding energy

31. Which does your physical intuition tell you is greater, the mass energy of a mosquito or the kinetic energy of a 747 jumbo jet at cruising speed? Estimate the order of magnitude of each.

32. The energy released in the explosion of one ton of TNT is about 4×10^9 J. (a) Where does this energy come from? (b) The ^{235}U nucleus may be broken apart by bombarding it with neutrons, a process called *nuclear fission*. In the fission process, an energy of about 200 MeV is released. Where does this energy come from? (c) What mass of ^{235}U is needed to produce the equivalent of one megaton of TNT by the fission process?

33. A beam of low-energy antiprotons is directed into a hydrogen target, causing protons and antiprotons to annihilate. Assume that in the annihilation process, most of the mass of the proton and the antiproton is converted into kinetic energy. Calculate the annihilation rate that would produce one watt.

34. The deuterium nucleus (d) is a bound state of one proton and one neutron. (a) Use the masses of the n, p, and d (see Appendix K) to calculate the nuclear binding energy of deuterium. (b) The ^{238}U nucleus is a bound state of 92 protons and 146 neutrons. Calculate the nuclear binding energy of ^{238}U.

35. When an atom of carbon combines with a molecule of oxygen to produce CO_2, an energy of 11.4 eV is released. (a) How much energy is released in the burning of 1 kg of carbon? (b) How much matter is converted to energy?

36. Four protons are combined into an alpha particle by a series of nuclear *fusion* reactions that occur in the sun. The energy released in this process ($4p \rightarrow \alpha$) is about 25 MeV. If the solar luminosity (4×10^{26} W) is dominated by proton fusion, at what rate are protons "burned" in the sun?

Atoms of the twentieth century: quarks and leptons

*37. According to our current understanding of the universe, matter is composed of six quarks with similar properties and six leptons with similar properties. Do you believe that this is an indication that the quarks and leptons might have structure? Explain!

Properties of the four forces

38. When two billiard balls collide, they appear not to exert a force on each other until they actually touch. What force is at work in the scattering of billiard balls?

39. (a) Estimate the size of an atom if the attraction of the electron and proton were due to gravity. (b) What is the typical kinetic energy of an electron in this "gravitational atom"?

40. If a tiny fraction of the molecules in two apples had an excess charge (e), what would this fraction need to be in order for the electric force between the apples to be equal in magnitude to the gravitational force between the apples?

Additional problems

*41. Do you think that there could be a fifth fundamental force that has not yet been observed? Explain.

*42. Consider a 1-mm thick plate of brass. (a) Make an estimate of the number of atomic layers in the plate. Why does a beam of light not penetrate the plate? (b) A beam of energetic neutrons is directed into the plate. Make an estimate of the probability that the neutron will collide with a nucleus. What do you think could happen in such a collision?

43. *Searching for proton decay.* A huge tank containing 10,000 tons (10^7 kg) of water (H_2O) is instrumented to search for proton decay. If the proton lifetime is predicted to be 10^{32} years and the detection efficiency is 50%, how many proton decay events are expected to be observed in one year?

44. Use the strength of the electromagnetic force and the size of an atom to estimate the acceleration of an electron in the hydrogen atom. How many "*g*" is this? ($g = 9.8$ m/s^2)

*45. The strong force between two protons in a nucleus is about 100 times the electromagnetic force, that is, $\alpha_s \approx 100\, \alpha$. (a) What is the approximate strength of the strong force between two protons separated by a distance of 2 fm? (b) Determine the acceleration of a proton in the nucleus. How many "*g*" is this? Compare your answer to the acceleration of an electron in an atom (see problem 44.)

46. Show that the gravitational binding energy (E_b) of a uniform sphere of mass M and radius R is

$$E_b = \frac{3\, GM^2}{5R}.$$

(*Hint*: Start with an infinitesimal mass dM and calculate the energy released as mass is brought in from infinity.)

... we may begin by supposing that the molecules are equal hard elastic spheres, which exert no force upon one another except at the instant of collision. By calling the molecules hard, it is implied that the collisions are instantaneous, and it follows that at any moment the potential energy of the system is negligible in comparison with the kinetic energy. If the volume of the molecules be very small in comparison with the space they occupy, the virial of the impulsive forces may be neglected, and the equation may be written $PV = \frac{1}{3} \sum mv^2$ where P is the pressure exerted upon the walls of the enclosure, V the volume, m the mass, and v the velocity of the molecule. ... In the theory of gases $\sum mv^2$ is proportional to the absolute temperature ...

Lord Rayleigh

Every macroscopic quantity of matter contains a huge number of atoms or molecules. One liter (10^{-3} m³) of water contains about 3×10^{25} molecules, as does one cubic meter of air. The description of matter necessarily involves the physics of a large number of particles. The molecules of matter, whether in the form of gas, liquid, or solid, are in endless motion, continuously colliding with each other. The molecular collisions result in accelerated charges that cause radiation. The radiation collides with the molecules, causing the molecules to be accelerated. A state of equilibrium is reached, provided that the total energy of all the particles, molecules, and photons, is not changing. The 3×10^{25} molecules in one cubic meter of gas (or one liter of water) have a continuous distribution of energies as a result of the collisions with each other and with the photons. It is not possible to measure the energies of 3×10^{25} molecules! We may characterize the system, however, by specifying the *typical* kinetic energy of a molecule or photon. The typical kinetic energy of a molecule or photon in the atmosphere of the earth is measured to be 1/40 eV. This typical kinetic energy is much smaller than the binding energy of electrons in atoms, which is a few electronvolts. Indeed, the typical photon on earth does not have enough energy, by about two orders of magnitude, to knock an electron out of an atom. If it did, we would not have atoms!

Where do the molecules and photons on the earth get their kinetic energy? The energy comes from the sun. In the interior of the sun, mass energy is converted into kinetic energy by nuclear fusion. The typical energy released in a nuclear fusion reaction is about 10^6 eV (1 MeV). Through collisions, kinetic energy from fusion gets spread among a large number of particles in the sun. The collision process also produces radiation because charged particles are accelerated. Some of this radiation escapes the sun. (Indeed, we see the sun shining!)

The typical energy of the photons escaping the sun is 1/2 eV. Some of this radiation strikes the atmosphere of the earth, and kinetic energy is transferred from the photons to the molecules of the atmosphere. These molecules collide with each other, and energy from the sun is spread out among the large number of particles making up the earth. The energy of the earth is not changing; as much energy is radiated as is absorbed. Thus, the earth is an energy transformer: For every photon with an energy of 1/2 eV that arrives from the sun, 20 photons with energies of 1/40 eV are reradiated into space, on the average. Particles on the surface of the sun are about 20 times more energetic than particles on the surface of the earth.

In this chapter we shall derive an expression for the energy distribution of molecules in the atmosphere of the earth. In Chapter 3, we shall derive the energy distribution of the radiation. The energy distributions for the particles and the radiation are not identical, but they have the same characteristic scale (i.e., 1/2 eV for both particles and radiation on the surface of the sun and 1/40 eV for both particles and radiation on the surface of the earth).

Before taking on the description of the physics of large numbers of particles, we shall find it worthwhile to develop some important tools.

2-1 DISTRIBUTION FUNCTIONS

Consider some physically measurable quantity, or *observable*, represented by the variable x. In some cases, the observable may take on only specific values. For example, x may represent the number of charged particles produced in the collision of two energetic protons, or the scores on a college entrance exam. In other instances, the observable is continuous so that within some allowed range, x can have any value. For example, x may represent the lifetime of a cosmic ray muon, or the speed of an oxygen molecule in a room.

The allowed values of x are seldom equally likely to occur. A mathematical function that represents the relative probability of occurrence for different values of x is called a *distribution function* (see Appendix D). The distribution function gives an instant snapshot of how often we expect to find x near any specified value.

The Probability Distribution

The result of the measurement of an observable represented by x is called an *event*. If x is not a continuous variable, then we may represent the allowed values of x by x_i, where i is an integer. The function that gives the probability for an event to occur with x in the interval

$$x_i \leq x < x_i + \Delta x_i, \qquad (2.1)$$

where

$$\Delta x_i = x_{i+1} - x_i, \qquad (2.2)$$

is called the *probability distribution* ($\Delta P_i / \Delta x_i$). The probability distribution is normalized such that

$$\sum_{i=-\infty}^{+\infty} \frac{\Delta P_i}{\Delta x_i} \Delta x_i = 1. \qquad (2.3)$$

If x is a continuous variable, the distribution function dP/dx gives the probability for an event to occur in an infinitesimal interval about some specified value of x. The probability distribution for a continuous variable is the limit where Δx_i is zero:

$$\frac{dP}{dx} = \lim_{\Delta x_i \to 0} \left(\frac{\Delta P_i}{\Delta x_i} \right). \qquad (2.4)$$

The normalization condition is

$$\int_{-\infty}^{+\infty} dx \frac{dP}{dx} = 1. \qquad (2.5)$$

Average Value

For a discontinuous variable, the *average* or *mean* value of x is

$$\langle x \rangle \equiv \sum_{i=-\infty}^{+\infty} x_i \frac{\Delta P_i}{\Delta x_i}. \qquad (2.6)$$

Often we want to know the average value of some other quantity (for example, x^2). The average value of any quantity $h(x)$ is given by the expression

$$\langle h \rangle = \sum_{i=-\infty}^{+\infty} h(x_i) \frac{\Delta P_i}{\Delta x_i}. \qquad (2.7)$$

When x is continuous, the average is computed by the integral

$$\langle x \rangle \equiv \int_{-\infty}^{+\infty} dx \, x \frac{dP}{dx}. \qquad (2.8)$$

The average value of any quantity $h(x)$ is given by the expression

$$\langle h \rangle = \int_{-\infty}^{+\infty} dx \, h(x) \frac{dP}{dx}. \qquad (2.9)$$

EXAMPLE 2-1

A free neutron is unstable because the weak interaction can transform the neutron into a proton (plus an electron and an antineutrino). The distribution of neutron lifetimes (τ) is measured to be $dP/d\tau = Ce^{-\tau/\lambda}$, where λ is the decay constant and C is a normalization constant. The range of τ is from zero to infinity. What is the average lifetime?

SOLUTION:

The constant C is given by the normalization condition (2.5),

$$\int_0^\infty d\tau \frac{dP}{d\tau} = C \int_0^\infty d\tau \, e^{-\tau/\lambda} = 1,$$

or

$$C = \frac{1}{\displaystyle\int_0^\infty d\tau \, e^{-\tau/\lambda}}.$$

The average lifetime is

$$\langle \tau \rangle = \int_0^\infty d\tau \, \tau \frac{dP}{d\tau}$$

$$= C \int_0^\infty d\tau \, \tau \, e^{-\tau/\lambda} = \frac{\displaystyle\int_0^\infty d\tau \, \tau \, e^{-\tau/\lambda}}{\displaystyle\int_0^\infty d\tau \, e^{-\tau/\lambda}}.$$

The integral in the numerator may be evaluated by the technique of integration by parts. We differentiate τ and integrate $e^{-\tau/\lambda}$. The integrated part, $-\lambda \tau e^{-\tau/\lambda}$, evaluated at the limits $\tau = 0$ and $\tau = \infty$, is zero. We are left with

$$\int_0^\infty d\tau \, \tau \, e^{-\tau/\lambda} = \lambda \int_0^\infty d\tau \, e^{-\tau/\lambda}.$$

Therefore,

$$\langle \tau \rangle = \frac{\lambda \displaystyle\int_0^\infty d\tau \, \tau \, e^{-\tau/\lambda}}{\displaystyle\int_0^\infty d\tau \, \tau \, e^{-\tau/\lambda}} = \lambda.$$

The average value of a negative exponential function over the range zero to infinity is the reciprocal of the factor in the exponent. For the neutron, λ is measured to be about 890 s. The average lifetime of an unstable particle is often referred to as simply the particle lifetime. ∎

Standard Deviation

The root-mean-square deviation of x from its average is called the *standard deviation*. The standard deviation is a measure of the how much the distribution deviates from the average. For the discrete case the standard deviation (σ) is given by

$$\sigma \equiv \sqrt{\left\langle \left(x_i - \langle x \rangle \right)^2 \right\rangle}$$

$$= \sqrt{\sum_{i=-\infty}^{+\infty} \left(x_i - \langle x \rangle \right)^2 \frac{\Delta P_i}{\Delta x_i}}. \qquad (2.10)$$

For the continuous case

$$\sigma \equiv \sqrt{\left\langle \left(x - \langle x \rangle \right)^2 \right\rangle}$$

$$= \sqrt{\int_{-\infty}^{+\infty} dx \, (x - <x>)^2 \frac{dP}{dx}}. \qquad (2.11)$$

EXAMPLE 2-2

For the particle lifetime distribution, $dP/d\tau = Ce^{-\tau/\lambda}$, what is the standard deviation from the average lifetime?

SOLUTION:

The average value of $(\tau - \lambda)^2$ is

$$\left\langle (\tau - \lambda)^2 \right\rangle = \int_0^\infty d\tau (\tau - \lambda)^2 \frac{dP}{d\tau}$$

$$= \frac{\int_0^\infty d\tau (\tau - \lambda)^2 e^{-\tau/\lambda}}{\int_0^\infty d\tau e^{-\tau/\lambda}}.$$

The numerator is

$$\int_0^\infty d\tau (\tau - \lambda)^2 e^{-\tau/\lambda} = \int_0^\infty d\tau (\tau^2 - 2\lambda\tau + \lambda^2) e^{-\tau/\lambda}.$$

The first term of this integral may be solved by integration by parts (differentiate τ^2 and integrate $e^{-\tau/\lambda}$) with the result

$$\int_0^\infty d\tau \, \tau^2 e^{-\tau/\lambda} = -\int_0^\infty d\tau (2\tau)(-\lambda) e^{-\tau/\lambda}$$

$$= \int_0^\infty d\tau (2\lambda\tau) e^{-\tau/\lambda}.$$

Therefore,

$$\int_0^\infty d\tau (\tau - \lambda)^2 e^{-\tau/\lambda} = \lambda^2 \int_0^\infty d\tau e^{-\tau/\lambda},$$

and

$$\left\langle (\tau - \lambda)^2 \right\rangle = \lambda^2.$$

The standard deviation is

$$\sigma = \sqrt{\left\langle (\tau - \lambda)^2 \right\rangle} = \lambda.$$

The standard deviation of the lifetime is equal to the average lifetime (see Example 2-1). This is a special property of the exponential distribution. ∎

EXAMPLE 2-3

The distribution function for the x coordinate of a particle is a constant in the range $0 < x < L$. Calculate the standard deviation.

SOLUTION:

The distribution of x is a constant (C),

$$\frac{dP}{dx} = C.$$

The normalization condition is

$$1 = \int_0^L dx \frac{dP}{dx} = C \int_0^L dx = CL,$$

or

$$C = \frac{1}{L}.$$

The average value of x is

$$\langle x \rangle = \int_0^L dx \, x \frac{dP}{dx} = \left(\frac{1}{L} \right)\left(\frac{L^2}{2} \right) = \frac{L}{2}.$$

The standard deviation is

$$\sigma = \sqrt{\left\langle (x - \langle x \rangle)^2 \right\rangle} = \sqrt{\left\langle x^2 - Lx + \frac{L^2}{4} \right\rangle}.$$

The average of $(x^2 - Lx + L^2/4)$ is

$$\left\langle x^2 - Lx + \frac{L^2}{4} \right\rangle = \int_0^L dx \left(x^2 - Lx + \frac{L^2}{4} \right) \frac{dP}{dx}$$

$$= \frac{1}{L} \int_0^L dx \left(x^2 - Lx + \frac{L^2}{4} \right)$$

$$= \frac{1}{L} \left(\frac{L^3}{3} - \frac{L^3}{2} + \frac{L^3}{4} \right) = \frac{L^2}{12}.$$

The standard deviation is

$$\sigma = \sqrt{\frac{L^2}{12}} = \frac{L}{\sqrt{12}}.$$

This is a useful result. The standard deviation of a constant distribution in an interval of length L is equal to L divided by the square root of 12. ∎

* *Challenging*

Examples of Distribution Functions

The *binomial distribution* function, $f_b(x)$, specifies the number of times (x) that an event occurs in n independent trials, where p is the probability of the event occurring in a single trial. The possible values of x are integers from zero to n. When n is very large, the binomial distribution may be treated as a continuous function of x, resulting in the *Gaussian distribution* function, $f_g(x)$. When p is so small that $f_b(x)$ is appreciably different from zero only for very small x, then the binomial distribution reduces to the *Poisson distribution* function, $f_p(x)$. These distribution functions (see Appendix D) provide the foundation for the interpretation of measurements. Table 2-1 summarizes the results.

The Binomial Distribution

The form of the binomial distribution (see Appendix D) $f_b(x)$ is

$$f_b(x) = \frac{n! \, p^x \, (1-p)^{n-x}}{x!(n-x)!}. \qquad (2.12)$$

The normalization condition on the binomial distribution is

$$\sum_{x=0}^{n} f_b(x) = 1. \qquad (2.13)$$

Figure 2-1 shows the binomial distribution for $n = 8$ for four different values of p (0.5, 0.25, 0.1, and 0.01).

EXAMPLE 2-4

Calculate the probability of getting heads exactly 5 times in 10 flips of a coin, if each flip has a 40% chance of producing a head.

SOLUTION:

The probability of getting heads exactly 5 times is given by the binomial distribution:

$$
\begin{aligned}
f_b(5) &= \frac{10!}{(5!)(5!)}(0.4)^5(0.6)^5 \\
&= \frac{(10)(9)(8)(7)(6)}{(5)(4)(3)(2)(1)}(0.4)^5(0.6)^5 \\
&= 252(0.4)^5(0.6)^5 = 0.201.
\end{aligned}
$$

Note that the probability of getting heads one-half of the time would be the same if the chances of getting a head were 60% instead of 40%. ∎

In Appendix D it is shown that the average value of a binomial distribution is

$$\langle x \rangle = np, \qquad (2.14)$$

and the standard deviation is

$$\sigma_x = \sqrt{\left\langle \left(x - \langle x \rangle\right)^2 \right\rangle} = \sqrt{np(1-p)}. \qquad (2.15)$$

The average and standard deviation of a binomial distribution are of fundamental importance in the interpretation of measurements. In an experiment where N events are observed, we may consider the number N to be made up of a very large number of trials (n) times a very small probability (p):

$$N \approx np. \qquad (2.16)$$

TABLE 2-1
SUMMARY OF THE BINOMIAL, GAUSSIAN, AND POISSON DISTRIBUTION FUNCTIONS.

Distribution	Functional Form	Mean	Standard Dev.
Binomial, n trials of probability p	$f_b(x) = \dfrac{n! \, p^x \, (1-p)^{n-x}}{x!(n-x)!}$	np	$\sqrt{np(1-p)}$
Gaussian, large-n limit	$f_g(x) = \dfrac{1}{\sqrt{2\pi\sigma^2}} e^{-(x-a)^2/2\sigma^2}$	a	σ
Poisson, large-n, small-p limit	$f_p(x) = \dfrac{e^{-a} a^x}{x!}$	a	$\sqrt{a}$

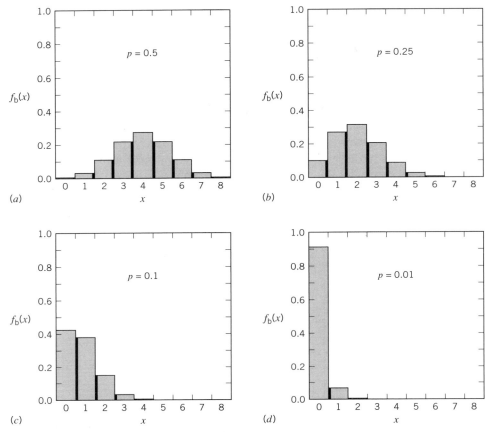

FIGURE 2-1 Binomial probability distribution for $n = 8$.
(a) The probability (p) for a certain event to occur on a single trial is equal to one-half. For eight trials, the number of occurrences can be anywhere between 0 and 8, inclusive, and is distributed as shown. Parts (b), (c), and (d) show the probability distribution for $p = 0.25, 0.1$, and 0.01, respectively.

Since $1 - p$ is nearly unity, the standard deviation is

$$\sigma_x = \sqrt{np(1-p)} \approx \sqrt{N}. \qquad (2.17)$$

The statistical error is given by the standard deviation which is proportional to the square root of the number of events.

The Gaussian Distribution

If n is large, the variable x in the binomial distribution (2.12) may be treated as continuous, provided that the product np is not too small. (The special case where np is very small leads to the Poisson distribution.) Figure 2-2 shows the binomial distribution for $n = 100$ and $p = 1/2$. The large-n limit of the binomial distribution function is the Gaussian distribution $f_g(x)$:

$$f_g(x) = \frac{1}{\sqrt{2\pi\sigma^2}} e^{-(x-a)^2/2\sigma^2}. \qquad (2.18)$$

The Gaussian distribution function is a bell-shaped curve that is symmetric about the mean, $x = a$. The Gaussian function decreases very rapidly for values of x greater than $a + \sigma$, or less than $a - \sigma$. The Gaussian distribution is also referred to as the *normal* distribution. The Gaussian distribution is shown in Figure 2-3. The mean is taken to be zero, and the abscissa is plotted in units of the standard deviation (σ). The plot is normalized to unit area.

EXAMPLE 2-5

The momenta of muons from the decays of Z^0 particles ($Z^0 \rightarrow \mu^+ + \mu^-$) at rest are measured in a spectrometer that

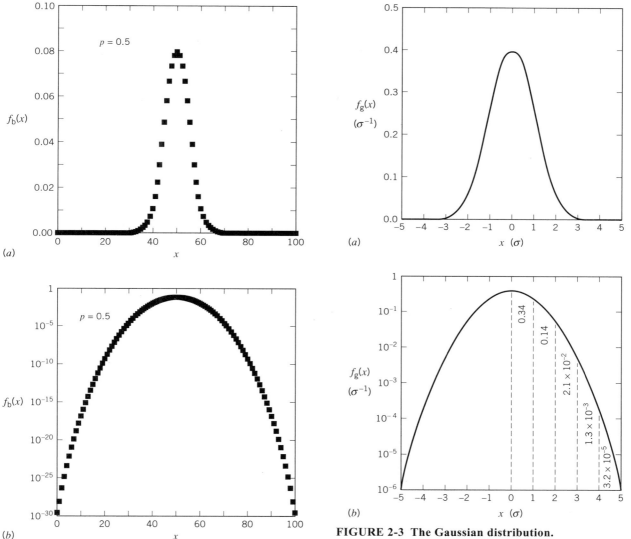

(a)

(b)

FIGURE 2-2 Binomial probability distribution for $n = 100$ and $p = 1/2$.
(a) Linear scale. (b) Logarithmic scale. The binomial distribution is approximately Gaussian when n is large.

FIGURE 2-3 The Gaussian distribution.
The mean is taken to be zero, and the horizontal scale is plotted in units of the standard deviation (σ). The plot is normalized to unit area. (a) Linear scale. (b) Logarithmic scale. The area under different parts of the Gaussian distribution in 1σ intervals is indicated.

has a resolution (σ) of 1%. In the measurements of 10^6 muons, how many muons are expected to have momentum deviating from the mean by more than 3%?

SOLUTION:

Three percent deviation from the mean corresponds to three standard deviations. From Figure 2-3(b), we see that the fraction of the events that are expected to deviate from

the mean by more than $\pm 3\sigma$ is 2.6×10^{-3}. The number of times (n) we expect this in 10^6 measurements is

$$n = \left(10^6 \right)\left(2.6 \times 10^{-3} \right) = 2.6 \times 10^3.$$

This expectation is valid provided that the measurement error is due to random sources. In practice, every measurement will also contain nonrandom errors, which are called *systematic* errors. A non-Gaussian "tail" is a signature of the presence of a systematic error.   ■

The Poisson Distribution

Very often, the event probability p is so small that even for very large values of n, the product np is small. In this case the binomial probability distribution (2.12) reduces to the Poisson distribution $f_p(x)$:

$$f_b(x) \approx \frac{(np)^x e^{-np}}{x!}. \qquad (2.19)$$

The quantity np is the average value of x. If we let a represent the average value of x, then the Poisson distribution may be written

$$f_p(x) = \frac{e^{-a} a^x}{x!}, \qquad (2.20)$$

where x is a nonnegative integer $(0, 1, 2, \ldots)$. The Poisson distribution is normalized to unity:

$$\sum_{x=0}^{\infty} f_p(x) = \sum_{x=0}^{\infty} \frac{e^{-a} a^x}{x!} = 1. \qquad (2.21)$$

Note that the possible values of x are nonnegative integers, while the mean value of x can be any nonnegative number and is not necessarily an integer. When the mean is larger than 7, we may approximate the Poisson distribution function by a continuous distribution function. The standard deviation is

$$\sigma = \sqrt{np(1-p)} \approx \sqrt{np} = \sqrt{a}. \qquad (2.22)$$

The Poisson distribution is useful in the statistical analysis of situations where small numbers of events are observed. The Poisson distribution for a mean of 2.5 is shown in Figure 2-4a. Figure 2-4b shows the Poisson distribution for a mean of 10.

Often when a small number of events are observed, we wish to know an upper limit on a (the average value of x). For example, in a certain experiment in which we are trying to determine the value a, we observe no events. What can we say statistically about a? The meaningful question to ask is: *If we did the same experiment a large number of times, for what value of a would we observe x greater than zero 90% of the time?* This value of a is called the 90% *confidence level.* The answer is found with Poisson statistics. From the definition of 90% confidence level, we have

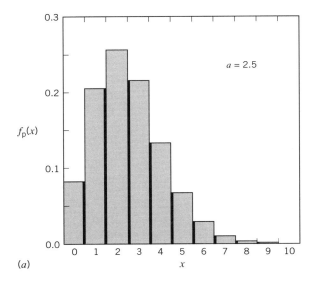

(a)

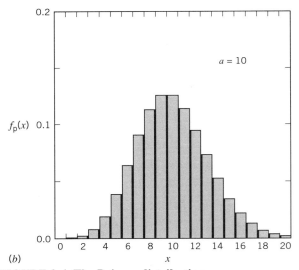

(b)

FIGURE 2-4 The Poisson distribution.
The Poisson distribution is a convenient form of the binomial distribution, valid when the event probability p is small. The probability distribution for observing x events, $(0, 1, 2, \ldots)$ is $f_p(x) = e^{-a} a^x / x!$, where the mean value of the variable x is a. The sum of all the probabilities is unity: $f_p(0) + f_p(1) + f_p(2) + \ldots = 1$. (a) The average value of x is 2.5. (b) The average value of x is 10. When the average value of x is greater than about 7, we may treat the Poisson distribution as a continuous function of x.

$$f_p(1) + f_p(2) + f_p(3) + \ldots = 0.9. \qquad (2.23)$$

Since the Poisson distribution is normalized to unity, we have

$$f_p(0) = 1 - \left[f_p(1) + f_p(2) + f_p(3) + \ldots \right]$$
$$= 0.1. \qquad (2.24)$$

We want to determine what value of a satisfies this condition. The Poisson distribution (2.20) with $x = 0$ is

$$f_p(0) = \frac{e^{-a} a^0}{0!} = e^{-a} = 0.1. \qquad (2.25)$$

Solving for a,

$$a = -\ln(0.1) = \ln(10) = 2.3. \qquad (2.26)$$

Thus, at the 90% confidence level, we can say that a is less than 2.3.

Similarly we could ask: If we did the same experiment a large number of times, for what value of a would $x > 0$ be observed 95% of the time? This is called the 95% confidence level. The Poisson distribution with $x = 0$ is

$$f_p(0) = 0.05 = e^{-a}, \qquad (2.27)$$

corresponding to

$$a = -\ln(0.05) = \ln(20) = 3.0. \qquad (2.28)$$

Thus, at the 95% confidence level, we can say that a is less than 3.0.

Suppose now that our measurement produces one event ($x = 1$). What is the upper limit of a at the 90% confidence level? The sum of the first two terms of the Poisson distribution gives

$$f_p(0) + f_p(1) = e^{-a} + a e^{-a} = 0.1, \qquad (2.29)$$

which corresponds to $a = 3.89$. Thus, at the 90% confidence level, we can say that a is less than 3.89. The 90% and 95% confidence levels given by Poisson statistics are tabulated in Table 2-2.

TABLE 2-2
UPPER LIMITS CALCULATED FROM POISSON STATISTICS WHEN X EVENTS ARE OBSERVED.

x	90% Limit	95% Limit
0	2.30	3.00
1	3.89	4.74
2	5.32	6.30
3	6.68	7.75
4	7.99	9.15
5	9.27	10.51

EXAMPLE 2-6

A huge tank containing 10,000 tons (10^7 kg) of water (H_2O) is instrumented to search for proton decay. If the detection efficiency is 50% and only one "candidate" proton decay event is observed during one year, what can be said about the lower limit on the proton lifetime at 95% confidence level?

SOLUTION:

Since there are 10 protons in each water molecule, the number of protons in the detector is

$$N_p = \frac{N_A (10)(10^7 \text{ kg})}{0.018 \text{ kg}}.$$

The decay rate is

$$R = \frac{1}{(1\,\text{y})(0.5) N_p}$$

$$= \frac{(1)(0.018 \text{ kg})}{(1\,\text{y})(0.5)(6 \times 10^{23})(10)(10^7 \text{ kg})} = 6 \times 10^{-34}\,/\text{y}.$$

If this event was "real" then, our best estimate of the proton lifetime would be $1/R$. From Poisson statistics (see Table 2-2) there is a 5% chance that the true rate could be 4.74 times greater. Therefore, at 95% confidence level, the experiment indicates the proton lifetime is

$$\tau > \frac{1}{(4.74)R} = \frac{1\,\text{y}}{(4.74)(6 \times 10^{-34})} = 3.5 \times 10^{32}\,\text{y}.$$

When the number of observed events is very small and the event signature is not 100% positive, then the results of an experiment are usually quoted as limit. ■*

2-2 TEMPERATURE AND THE IDEAL GAS

A macroscopic quantity of matter contains a huge number of atoms. To physically describe such a system of particles we need global variables, single quantities that describe the complex behavior of the system. One of these global quantities is temperature.

The Zeroth Law of Thermodynamics

Consider an evacuated chamber into which we shoot a large number of energetic molecules. These molecules will bounce off the walls of the chamber and reach a state of equilibrium. The state of equilibrium is brought about by a large number of collisions of the molecules with the walls of the container. In the equilibrium state, the molecules have a distribution of velocities that does not depend on the initial velocity distribution of the molecules. The distribution of velocities has a universal form, called the *Maxwell-Boltzmann* distribution that we shall derive shortly.

In the equilibrium state, the distribution of velocities does not change, and the average kinetic energy of the molecules remains fixed. There is no net energy transfer between the molecules and the chamber. This equilibrium is called *thermal equilibrium.*

The *zeroth law of thermodynamics* states that there is a useful quantity called *temperature,* which is a measure of the equilibrium state. When two systems of particles (for example, the gas molecules and the container walls) are in equilibrium and there is no net energy transfer between them, then the two systems have the same temperature.

Temperature and Kinetic Energy

If the total kinetic energy of the molecules of a gas is not changing, the gas is defined to be at a constant temperature. The gas molecules are in thermal equilibrium with their surroundings. For a gas, temperature (T) may be defined to be a quantity proportional to the average kinetic energy of the molecules:

$$\left\langle E_k \right\rangle \equiv \frac{3}{2} kT. \tag{2.30}$$

The temperature in the above expression for average kinetic energy is called the *absolute temperature* and its units are the kelvin (K). The constant that relates absolute temperature with energy is called the Boltzmann constant (k). The Boltzmann constant,

$$k = 8.617 \times 10^{-5} \text{ eV/K}, \tag{2.31}$$

is obviously determined by the definition of the kelvin temperature scale.

> The average kinetic energy of molecules in an ideal gas is $3kT/2$, where the temperature is expressed on the absolute (kelvin) scale.

The Kelvin Temperature Scale

To define a temperature scale, we need two well-defined states of a system. These states usually are chosen by the equilibrium condition between distinct phases of matter. The *triple point* of water is defined to be the state where the liquid, solid, and vapor phases of water are in equilibrium under one atmosphere of pressure. The *boiling point* of water is defined to be the state where the vapor pressure of water is equal to one atmosphere.

Consider a gas of molecules at constant pressure in thermal equilibrium with a water. The volume of the gas in thermal equilibrium with boiling water (V_{boil}) is larger than the volume of the gas in thermal equilibrium with water at the triple point (V_{triple}). The ratio of the volumes is easily measurable. Experimentally we find (if the pressure is not too large) that the ratio of the volumes is the same for all types of gas. The observed ratio of volumes is

$$\frac{V_{boil}}{V_{triple}} = 1.3661. \tag{2.32}$$

The difference in temperature (ΔT_C) on the *Celsius* scale (units of °C) is defined to be a quantity proportional to the fractional change in the gas volume:

$$\Delta T_C \equiv \frac{a\left(V_{boil} - V_{triple}\right)}{V_{triple}}, \tag{2.33}$$

where a is a constant. Using the definition of the Celsius scale (2.33), we may write the volume ratio (2.32) as

$$\frac{V_{boil}}{V_{triple}} = \left(1 + \frac{\Delta T_C}{a}\right). \tag{2.34}$$

The constant a is fixed by defining the temperature of the *triple point* of water ($T_{C\text{-triple}}$) to be

$$T_{C-triple} \equiv 0\,°C, \tag{2.35}$$

and the temperature of the *boiling point* of water to be

$$T_{C-boil} \equiv 100\,°C. \tag{2.36}$$

Using the definitions of the Celsius scale (2.33), the triple point (2.35), and the boiling point (2.36), the volume ratio (2.32) becomes

$$\frac{V_{boil}}{V_{ice}} = 1.3661 = \left(1 + \frac{100\,°C}{a}\right), \tag{2.37}$$

or

$$a = 273.15\,°C. \tag{2.38}$$

The constant a gives the relationship between the absolute temperature scale and the Celsius temperature scale. The limit where the kinetic energy of the gas molecules is zero (and can't become any smaller!) may be taken as the limit where the gas volume would become zero (and can't become any smaller!). Before the volume becomes zero, the gas condenses into a liquid because attractive forces between the molecules become important. If we ignore this (idealizing to nonattractive molecules) and extrapolate to very low temperatures, we can find the temperature corresponding to zero volume. The temperature difference between the triple point and the zero-volume temperature is given by replacing V_{boil} by zero in the definition of the Celsius scale (2.33), yielding

$$\Delta T_C = -a = -273.15\,°C. \qquad (2.39)$$

Therefore, the definition of the *absolute* or kelvin temperature scale is

$$T \equiv \left(T_C\, /\,°C + 273.15 \right) K. \qquad (2.40)$$

This expression relates the Celsius temperature scale (T_C in °C) with the absolute kelvin scale (T in K).

EXAMPLE 2-7

Calculate the average kinetic energy per gas molecule in a room at $T_C = 20\,°C$ and outside on a very cold day in Minnesota at $T_C = -40\,°C$.

SOLUTION:
At room temperature,

$$T = 20.\,K + 273.\,K = 293\,K.$$

From the definition of temperature (2.30),

$$\langle E_k \rangle = \frac{3}{2} kT$$
$$= \left(\frac{3}{2} \right) \left(8.62 \times 10^{-5}\ eV/K \right) \left(293.\,K \right)$$
$$= 0.038\,eV.$$

For the cold day,

$$T = -40.\,K + 273.\,K = 233\,K,$$

and

$$\langle E_k \rangle = \frac{3}{2} kT$$
$$= \left(\frac{3}{2} \right) \left(8.62 \times 10^{-5}\ eV/K \right) \left(233.\,K \right)$$
$$= 0.030\,eV. \qquad \blacksquare$$

At a temperature of 300 K, the value of kT is

$$kT = 0.02585\,eV. \qquad (2.41)$$

You will find it very handy to remember that the value of kT is about 1/40 of an electronvolt at room temperature. The kinetic energy scale of molecules at room temperature is about two orders of magnitude smaller than the kinetic energy scale of outer electrons in atoms.

> At room temperature,
> $$kT \approx \frac{1}{40}\,eV.$$

EXAMPLE 2-8
The root-mean-square speed is defined to be the square root of the average speed squared. Calculate the root-mean-square speed of a nitrogen molecule at room temperature.

SOLUTION:
Writing the kinetic energy as

$$E_k = \frac{1}{2} mv^2,$$

the definition of temperature (2.30) gives

$$\left\langle \frac{mv^2}{2} \right\rangle = \frac{3}{2} kT,$$

or

$$\sqrt{\langle v^2 \rangle} = \sqrt{\frac{3kT}{m}}.$$

Dividing both sides by c, we have

$$\frac{\sqrt{\langle v^2 \rangle}}{c} = \sqrt{\frac{3kT}{mc^2}}.$$

The nitrogen molecule (N_2) has $A = 28$. The mass energy of the nitrogen molecule (see Example 1-11) is

$$mc^2 = (28)(0.932\ GeV) = 2.61 \times 10^{10}\ eV,$$

so

$$\frac{\sqrt{\langle v^2 \rangle}}{c} = \sqrt{\frac{(3)\left(\frac{1}{40}\,eV \right)}{2.61 \times 10^{10}\ eV}} \approx 1.7 \times 10^{-6},$$

or

$$\sqrt{\langle v^2 \rangle} \approx 510 \text{ m/s}.$$

The root-mean-square speed of molecules in a gas is the same order of magnitude as the speed of sound in the gas. ∎

The relationship between kinetic energy of a system of particles and temperature has universal application. For example, we may speak of the energy or temperature of electrons inside atoms or quarks inside protons, as illustrated in Figure 2-5.

The Ideal Gas Law

Consider a container of gas that has a volume V in thermal equilibrium at a temperature T. The pressure on the walls of the container is caused by collisions of the gas molecules with the walls of the container. When a molecule of mass m traveling in the x direction with speed v_x strikes the wall of the container, the molecule scatters elastically and the momentum transfer (Δp) between the wall and molecule is

$$\Delta p = 2mv_x. \tag{2.42}$$

If the momentum transfer occurs in a time Δt, the force per area (F/A) exerted by the gas molecule on the container wall is

$$\frac{F}{A} = \frac{\left(\dfrac{\Delta p}{\Delta t}\right)}{A} = \frac{2mv_x}{A\Delta t}. \tag{2.43}$$

To get the contribution to the pressure from a single molecule, we need to evaluate the probability that the molecule hits the wall in a time Δt. The molecule will hit the wall in a time Δt only if it is within a distance $v_x\Delta t$ of the wall and headed toward the wall. Since the molecules are distributed at random in the volume, the fraction within $v_x\Delta t$ of the wall is equal to $v_x A\Delta t/V$. Only one-half of the molecules are headed toward the wall at any instant. Therefore, the hit probability (H) is

$$H = \frac{v_x A\Delta t}{2V}. \tag{2.44}$$

The contribution to the pressure from the single molecule P_s is the probability of hitting the wall (2.44) multiplied by the force per area exerted when it hits (2.43):

$$P_s = H\left(\frac{F}{A}\right) = \left(\frac{v_x A\Delta t}{2V}\right)\left(\frac{2mv_x}{A\Delta t}\right) = \frac{mv_x^2}{V}. \tag{2.45}$$

The total pressure is just the total number of molecules (N) times the average x component of velocity squared:

$$P = \frac{Nm\langle v_x^2 \rangle}{V}. \tag{2.46}$$

Since there is no preferred direction inside the container, all components of velocity have the same distribution. Therefore,

$$\langle v^2 \rangle = \langle v_x^2 \rangle + \langle v_y^2 \rangle + \langle v_z^2 \rangle = 3\langle v_x^2 \rangle, \tag{2.47}$$

and the pressure (2.46) becomes

$$P = \frac{1}{3}\frac{Nm\langle v^2 \rangle}{V}, \tag{2.48}$$

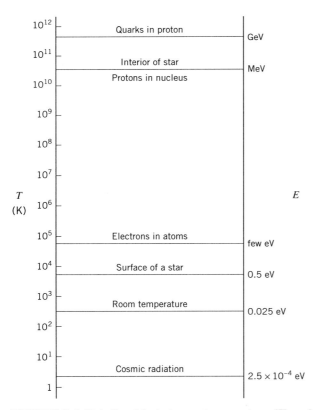

FIGURE 2-5 Relationship between temperature (T) and typical kinetic energy of a particle (kT).

or

$$PV = \frac{2}{3} N \left\langle \frac{mv^2}{2} \right\rangle. \qquad (2.49)$$

From our definition of temperature (2.30), we have

$$PV = NkT . \qquad (2.50)$$

This is the *ideal gas law*. The ideal gas law was established by much experimentation in the seventeenth and eighteenth centuries. Great experimenters included Robert Boyle, who discovered that a gas at a fixed temperature has a pressure that is inversely proportional to the volume, and Joseph Gay-Lussac, who discovered that a gas at fixed pressure has a volume that is proportional to the temperature. If we take the ideal gas law as an empirical fact, we may derive the expression for the average kinetic energy of the gas molecules. The ideal gas law (2.50) and our definition of temperature (2.30) are equivalent.

We now summarize the physics of the ideal gas law: The pressure of a gas enclosed in a container of fixed volume is caused by collisions of the gas molecules with the walls of the container. This pressure is proportional to the square of the speed of the molecules. One power of the speed comes from the momentum transfer of a molecule hitting the wall; another power of the speed comes from the rate at which molecules strike the wall. The average speed squared of the molecules is proportional to the average kinetic energy, and the average kinetic energy is proportional to the temperature.

EXAMPLE 2-9

Calculate the volume occupied by one mole of molecules at a pressure of one atmosphere (1.01×10^5 N/m²) and a temperature of 273 K. This condition is called *standard temperature and pressure* (STP).

SOLUTION:

From the ideal gas law (2.50) the volume is

$$V = \frac{N_A kT}{P} .$$

At T = 273 K, the value of kT is

$$kT = (0.02585 \text{ eV}) \left(\frac{273 \text{ K}}{300 \text{ K}} \right) = 0.0235 \text{ eV} .$$

The volume of the gas is

$$V = \left[\frac{(6.02 \times 10^{23})(0.0235 \text{ eV})}{1.01 \times 10^5 \text{ N/m}^2} \right] (1.60 \times 10^{-19} \text{ J/eV})$$

$$= 0.0224 \text{ m}^3 ,$$

or 22.4 liters.   ■

EXAMPLE 2-10

Estimate the average distance between molecules at standard temperature and pressure.

SOLUTION:

From the previous example, we know that one mole of molecules occupies a volume

$$V = 0.0224 \text{ m}^3 .$$

The average distance between molecules (ℓ) is

$$\ell \approx \left(\frac{V}{N_A} \right)^{1/3} = \left(\frac{0.0224 \text{ m}^3}{6.02 \times 10^{23}} \right)^{1/3}$$

$$\approx 3 \times 10^{-9} \text{ m} = 3 \text{ nm}.$$

The average distance between molecules is an order of magnitude larger than the size of the molecules.   ■

Mean Free Path

From Example 2-10, we have seen that the typical distance (ℓ) between molecules in an ideal gas is a few nanometers. The distance between molecules is an order of magnitude larger than the diameter of a molecule (d_{mol}). Another characteristic distance useful in the description of the motion of the gas molecules is the average distance that a molecule travels before colliding with another molecule. This average distance between molecular collisions is called the *mean free path* (d). There is a simple approximate relationship between the mean free path, the distance between molecules, and the molecular size. Consider a molecule in motion that moves a distance d before colliding with another molecule, as indicated in Figure 2-6. The collision occurs when two molecules pass within a distance d_{mol} of each other. The volume swept out by a cylinder of diameter d_{mol} and length d must be equal to the volume that contains one molecule on the average:

$$\pi d_{mol}^2 d \approx \ell^3 . \qquad (2.51)$$

Solving for d, we get

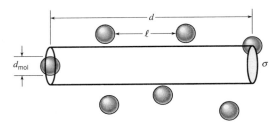

FIGURE 2-6 Qualitative comparison of the molecular diameter (d_{mol}), the average distance between molecules (ℓ), and the mean free path (d).

$$d \approx \frac{\ell^3}{\pi d_{mol}^2}. \qquad (2.52)$$

In terms of the number of molecules per volume (n),

$$n \approx \frac{1}{\ell^3}, \qquad (2.53)$$

we may write the mean free path (2.52) as

$$d \approx \frac{1}{n\sigma}, \qquad (2.54)$$

where

$$\sigma \equiv \pi d_{mol}^2. \qquad (2.55)$$

The quantity σ is called the collision *cross section*, the effective area of a target molecule for producing a collision.

EXAMPLE 2-11
Estimate the mean free path of a molecule at STP.

SOLUTION:
According to Example 2-9, the volume that contains one molecule on the average is 0.0224 m³ divided by Avogadro's number. Taking the molecular diameter to be roughly 0.3 nm, we estimate the mean free path (2.52) to be

$$d \approx \frac{\ell^3}{\pi d_{mol}^2}$$

$$\approx \frac{0.0224 \, \text{m}^3}{\left(6 \times 10^{23}\right)\pi\left(3 \times 10^{-10} \, \text{m}^2\right)^2}$$

$$\approx 10^{-7} \, \text{m} = 100 \, \text{nm}.$$

The mean free path is about two orders of magnitude larger than the average distance between the molecules. ■

Comparison of the mean free path between collisions (d), the average distance between molecules (ℓ), and the molecular size (d_{mol}) shows that

$$d_{mol} \ll \ell \ll d. \qquad (2.56)$$

2-3 THE MAXWELL-BOLTZMANN DISTRIBUTION

Law of Atmospheres

Consider the gas molecules in the vicinity of the earth, our atmosphere. These molecules are gravitationally attracted to the earth and have a continuous random motion. Let us assume a condition of thermal equilibrium. (This does not apply exactly to the real atmosphere, which is colder with increasing height.) Let us also assume that our atmosphere is without turbulence; there are no winds pushing the molecules around. In such an ideal atmosphere, there is a simple relationship for the variation of the density of molecules with height (z). If the molecules each have a mass m and the number per volume is N/V, then the force per area (ΔP) caused by the weight of the molecules in a slice of thickness (Δz) is

$$\Delta P = \frac{mgN\Delta z}{V}. \qquad (2.57)$$

In the limit where $\Delta z \rightarrow 0$, we have

$$\frac{dP}{dz} = -\frac{mgN}{V} = -mgn, \qquad (2.58)$$

where we have defined n to be the number of particles per volume (N/V). The particle density n depends on z. We may determine the z dependence using the ideal gas law (2.50). Writing the ideal gas law as

$$n = \frac{P}{kT}, \qquad (2.59)$$

and differentiating with respect to P, we get

$$\frac{dn}{dP} = \frac{1}{kT}. \qquad (2.60)$$

Using the chain-rule for derivatives and our result for the pressure gradient (2.58), we have

$$\frac{dn}{dz} = \frac{dn}{dP}\frac{dP}{dz} = -\frac{mg}{kT}n. \qquad (2.61)$$

The solution of this differential equation is

$$n = n_0 e^{-mgz/kT}, \tag{2.62}$$

where n_0 is the particle density at $z=0$. The particle density decreases exponentially with increasing altitude. This result is called the *law of atmospheres*. It is a direct consequence of the force of gravity acting on an ideal gas. Note the appearance of the particle mass in the exponent. If the atmosphere is made up of different types of molecules H_2, N_2, O_2, and so on, then the density of the heavier molecules decreases faster with increasing altitude. For this reason, there is almost no hydrogen and very little helium in the earth's atmosphere, even though these two elements are the most abundant in the universe.

EXAMPLE 2-12
Estimate the thickness of the earth's atmosphere.

SOLUTION:
We may characterize the thickness of the atmosphere by the height where the density of gas molecules drops by a factor of $1/e$. This occurs at a height above the earth's surface of

$$z = \frac{kT}{mg},$$

where m is the mass of a gas molecule. The atmosphere is made up mostly of nitrogen; the mass of the nitrogen molecule is

$$m = (28)(1.66 \times 10^{-27} \text{ kg}),$$

so the height is

$$z = \frac{kT}{mg} = \frac{\left(\frac{1}{40} \text{ eV}\right)\left(1.6 \times 10^{-19} \text{ J/eV}\right)}{(28)(1.66 \times 10^{-27} \text{ kg})(9.8 \text{ m/s}^2)}$$

$$\approx 9 \times 10^3 \text{ m}.$$

The atmosphere protects us from high-energy (ultraviolet) solar radiation. The atmosphere is a necessary component of the environment needed for the formation of life on our planet. ∎

Distribution of Velocities

The velocity distribution of particles in a gas may be deduced from the exponential atmosphere. Consider par-

ticles at two altitudes, z_1 and z_2 as indicated in Figure 2-7. We define a variable ζ, whose origin coincides with $z=z_1$. The number of particles per volume as a function of ζ is

$$n(\zeta) = n_0 e^{-mg\zeta/kT}. \tag{2.63}$$

In the following analysis we shall ignore collisions between the particles. (We may consider ζ to be small compared to the mean free path.) By conservation of energy, a particle at $\zeta=0$ that has a z component of velocity (v_z) given by

$$\frac{1}{2}mv_z^2 = mg\zeta \tag{2.64}$$

can travel from z_1 to z_2. Now consider particles at an altitude infinitesimally larger than z_2 $(z_2 + dz = \zeta + d\zeta)$. The number of particles per volume with altitudes between ζ and $\zeta + d\zeta$ is obtained by differentiating n (2.63),

$$-dn = n_0 \left(\frac{mg}{kT}\right) e^{-mg\zeta/kT} d\zeta. \tag{2.65}$$

Particles from $\zeta=0$ reach the interval $\zeta + d\zeta$ only if they have a z component of velocity between v_z and $v_z + dv_z$. The flux (number per time per area) of particles traveling from $\zeta=0$ to $\zeta + d\zeta$ is equal to v_z times the particle density times the fraction of particles (df) with a z component of velocity between v_z and $v_z + dv_z$. Therefore,

$$v_z n_0 df \propto -dn$$

$$= n_0 \left(\frac{mg}{kT}\right) e^{-mg\zeta/kT} d\zeta$$

$$= n_0 \left(\frac{m}{kT}\right) e^{-mv_z^2/2kT} v_z dv_z, \tag{2.66}$$

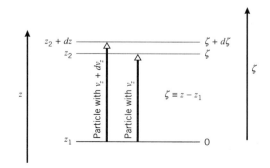

FIGURE 2-7 Particles traveling between two altitudes.
Particles at an altitude z_1 travel to an altitude z_2 if they have z component of velocity equal to v_z. Particles with infinitesimally larger velocity $(v_z + dv_z)$ travel to an infinitesimally larger altitude $(z_2 + dz_2)$.

where we have used the energy relationship (2.64) between v_z and ζ. Therefore, distribution of the z component of velocity (df/dv_z) is

$$\frac{df}{dv_z} = \sqrt{\frac{m}{2\pi kT}}\, e^{-mv_z^2/2kT}. \qquad (2.67)$$

The constant $(m/2\pi kT)^{1/2}$ comes from the normalization condition,

$$\int_{-\infty}^{+\infty} dv_z\, \frac{df}{dv_z} = 1. \qquad (2.68)$$

The distribution df/dv_z is called the *Maxwell velocity* distribution. The distribution of velocities is a Gaussian, and we may appreciate directly why this must be so. A gas molecule gets its velocity from repeated random collisions, and when we add up random numbers we get a Gaussian (see Appendix D). The Maxwell velocity distribution is shown in Figure 2-8 for nitrogen gas at room temperature.

The particle velocity distribution (2.67) does not depend on the acceleration of gravity g. The particles would have the same velocity distribution in the absence of gravity. The effect of gravity results in the exponential behavior of the particle density. The velocity distribution of the particles does not depend on the altitude because we have assumed that the atmosphere is at constant temperature; there is no variation of average particle energy with altitude. Finally we should remark that even though we have ignored collisions of molecules, our result still holds because the *average* number of particles reaching the interval $\zeta + d\zeta$ from $\zeta = 0$ is unchanged by collisions. Collisions do not change the total energy distribution of the particles; indeed our assumption of thermal equilibrium is that the velocity distribution (2.67) is the result of collisions of a fixed number of particles sharing a fixed amount of energy.

The x and y velocity components have identical distribution functions,

$$\frac{df}{dv_x} = \sqrt{\frac{m}{2\pi kT}}\, e^{-mv_x^2/2kT}, \qquad (2.69)$$

and

$$\frac{df}{dv_y} = \sqrt{\frac{m}{2\pi kT}}\, e^{-mv_y^2/2kT}. \qquad (2.70)$$

The combined velocity distribution function is the product of the distribution functions of the three components:

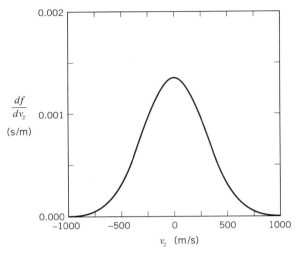

FIGURE 2-8 The Maxwell velocity distribution.
The distribution is shown for nitrogen at room temperature. The average value of any component (v_x, v_y, or v_z) of the velocity is zero. The root-mean-square value is $(kT/m)^{1/2}$.

$$\frac{d^3f}{dv_x\, dv_y\, dv_z}$$

$$= \left(\frac{m}{2\pi kT}\right)^{3/2} e^{-m(v_x^2 + v_y^2 + v_z^2)/2kT}$$

$$= \left(\frac{m}{2\pi kT}\right)^{3/2} e^{-mv^2/2kT}, \qquad (2.71)$$

where

$$v^2 = v_x^2 + v_y^2 + v_z^2. \qquad (2.72)$$

EXAMPLE 2-13
Calculate the average value of v_x.

SOLUTION:
Using the Maxwell velocity distribution, the average value of the x component of velocity is

$$\langle v_x \rangle = \int_{-\infty}^{+\infty} dv_x\, v_x\, \frac{df}{dv_x} = C\int_{-\infty}^{+\infty} dv_x\, v_x e^{-mv_x^2/2kT}$$

$$= C\int_{-\infty}^{0} dv_x\, v_x e^{-mv_x^2/2kT} + C\int_{0}^{+\infty} dv_x\, v_x e^{-mv_x^2/2kT}$$

$$= -C\int_{0}^{+\infty} dv_x\, v_x e^{-mv_x^2/2kT}$$

$$\qquad + C\int_{0}^{+\infty} dv_x\, v_x e^{-mv_x^2/2kT}$$

$$= 0.$$

The integral is zero because df/dv_x is an even function of x, making the integrand $v_x(df/dv_x)$ an odd function of x. The physical interpretation of this result is that there are just as many particles moving in the positive x direction as there are moving in the negative x direction. The average of any component of velocity (v_x, v_y, or v_z) is zero. ∎

EXAMPLE 2-14

Calculate the average value of v_x^2. Give a numerical answer for nitrogen at room temperature.

SOLUTION:

Using the Maxwell velocity distribution, the average value of the square of the x component of velocity is

$$\left\langle v_x^2 \right\rangle = \int_{-\infty}^{+\infty} dv_x \, v_x^2 \, \frac{df}{dv_x} = \sqrt{\frac{m}{2\pi kT}} \int_{-\infty}^{+\infty} dv_x \, v_x^2 \, e^{-mv_x^2/2kT}.$$

The integral we need to evaluate (see Table D-1) is

$$I = \int_{-\infty}^{+\infty} dx \, x^2 e^{-x^2/a} = \frac{\sqrt{\pi}}{2} a^{3/2},$$

where

$$a = \frac{2kT}{m}.$$

Therefore,

$$\left\langle v_x^2 \right\rangle = \sqrt{\frac{m}{2\pi kT}} \frac{\sqrt{\pi}}{2} \left(\frac{2kT}{m} \right)^{3/2} = \frac{kT}{m}.$$

For nitrogen at room temperature, we have

$$mc^2 = 26.1 \, \text{GeV} \,,$$

and

$$kT = \frac{1}{40} \text{eV} \,,$$

so

$$\left\langle v_x^2 \right\rangle = \frac{kTc^2}{mc^2} = \frac{\left(\frac{1}{40} \text{eV} \right) \left(3.00 \times 10^8 \, \text{m/s} \right)^2}{2.61 \times 10^{10} \, \text{eV}}$$
$$= 8.6 \times 10^4 \, \text{m}^2/\text{s}^2 \,. \quad ∎$$

From Example 2-14, we see that the average value of $mv_x^2/2$ is

$$\left\langle \frac{1}{2} mv_x^2 \right\rangle = \frac{kT}{2}. \qquad (2.73)$$

This is the amount of kinetic energy associated with motion in the x direction. This result is known as the *equipartition theorem* and may be more generally stated as follows: *In thermal equilibrium each degree of freedom contributes an amount of average kinetic energy equal to kT/2.* Translational motion in the x, y, and z directions results in a kinetic energy of $3kT/2$, consistent with our definition of temperature (2.30).

Distribution of Speeds

We now derive an expression for the distribution of molecular speeds. We need to make a transformation from the three-dimensional velocity space (v_x, v_y, v_z) to the one-dimensional speed space (v), where the speed is given by

$$v = \sqrt{v_x^2 + v_y^2 + v_z^2} \,. \qquad (2.74)$$

This involves transforming rectangular coordinates into spherical coordinates (see Appendix E). Since the differential volume element is

$$dv_x \, dv_y \, dv_z = (dv)(v \, d\theta)(v \sin\theta \, d\phi), \quad (2.75)$$

the velocity and speed distributions are related by

$$\frac{d^3 f}{dv_x \, dv_y \, dv_z} = \frac{1}{v^2 \sin\theta} \frac{d^3 f}{dv \, d\theta \, d\phi}, \qquad (2.76)$$

or

$$\frac{d^3 f}{dv \, d\theta \, d\phi} = v^2 \sin\theta \frac{d^3 f}{dv_x \, dv_y \, dv_z}. \qquad (2.77)$$

Integrating over the angles gives

$$\frac{df}{dv} = \int_0^\pi d\theta \sin\theta \int_0^{2\pi} d\phi \, v^2 \frac{d^3 f}{dv_x \, dv_y \, dv_z}$$
$$= 4\pi v^2 \frac{d^3 f}{dv_x \, dv_y \, dv_z}. \qquad (2.78)$$

Using the combined velocity distribution (2.71), we arrive at the *Maxwell speed* distribution,

$$\frac{df}{dv} = 4\pi \left(\frac{m}{2\pi kT} \right)^{3/2} v^2 e^{-mv^2/2kT} \,. \qquad (2.79)$$

The number of molecules with speeds between v and $v + dv$, theta between θ and $\theta + d\theta$, and phi between ϕ and $\phi + d\phi$ is the same as the number of molecules with velocity components between v_x and $v_x + dv_x$, v_y and $v_y + dv_y$, and v_z and $v_z + dv_z$. The number with speeds between v and $v + dv$ *at any angle* differs by a factor of $4\pi v^2$. The Maxwell speed distribution is plotted in Figure 2-9 for nitrogen at room temperature.

EXAMPLE 2-15
Calculate the average speed of a gas molecule. Give a numerical answer for nitrogen at room temperature.

SOLUTION:
The average speed is

$$\langle v \rangle = \int_0^{+\infty} dv \, v \frac{df}{dv} = 4\pi \left(\frac{m}{2\pi kT} \right)^{3/2} \int_0^{+\infty} dv \, v^3 e^{-mv^2/2kT} .$$

The integral we need to evaluate (see Table D-1) is

$$I = \int_0^{+\infty} dx \, x^3 e^{-x^2/a} = \frac{a^2}{2} ,$$

where

$$a = \frac{2kT}{m} .$$

We have

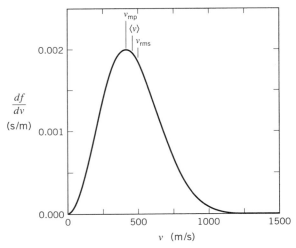

FIGURE 2-9 **The Maxwell speed distribution.**
The distribution is shown for nitrogen at room temperature. The average speed is $(8kT/\pi m)^{1/2} \approx 470$ m/s, the root-mean-square speed is $(3kT/m)^{1/2} \approx 510$ m/s, and the most probable speed is $(2kT/m)^{1/2} \approx 420$ m/s.

$$\langle v \rangle = 4\pi \left(\frac{m}{2\pi kT} \right)^{3/2} \frac{1}{2} \left(\frac{2kT}{m} \right)^2 = \sqrt{\frac{8kT}{\pi m}} ,$$

or

$$\langle v \rangle = c \sqrt{\frac{8kT}{\pi mc^2}}$$

$$\approx \left(3.00 \times 10^8 \text{ m/s} \right) \sqrt{\frac{(8)\left(\frac{1}{40} \text{ eV} \right)}{(\pi)\left(2.61 \times 10^{10} \text{ eV} \right)}}$$

$$\approx 470 \text{ m/s}. \qquad \blacksquare$$

EXAMPLE 2-16
Calculate the most probable speed of a gas molecule. Give a numerical answer for nitrogen at room temperature.

SOLUTION:
The most probable speed (v_{mp}) is obtained by differentiating the speed distribution and setting it equal to zero:

$$\left[\frac{d}{dv} \left(\frac{df}{dv} \right) \right]_{v=v_{mp}} = \left[C(2v)e^{-mv^2/2kT} \right.$$

$$\left. + Cv^2 \left(\frac{-mv}{kT} \right) e^{-mv^2/2kT} \right]_{v=v_{mp}}$$

$$= 0.$$

Solving for v_{mp} gives

$$v_{mp} = \sqrt{\frac{2kT}{m}} .$$

For nitrogen at room temperature, we have

$$mc^2 = 26.1 \text{ GeV} ,$$

and

$$kT = \frac{1}{40} \text{ eV} ,$$

so

$$v_{mp} = c \sqrt{\frac{2kT}{mc^2}}$$

$$\approx \left(3.00 \times 10^8 \text{ m/s} \right) \sqrt{\frac{(2)\left(\frac{1}{40} \text{ eV} \right)}{2.61 \times 10^{10} \text{ eV}}} \approx 420 \text{ m/s}. \qquad \blacksquare$$

EXAMPLE 2-17
Calculate the root-mean-square speed of a gas molecule. Give a numerical answer for nitrogen at room temperature.

SOLUTION:
We have already calculated this in Example 2-8 using the relationship between average kinetic energy and temperature. We now perform the calculation using the Maxwell speed distribution. The average squared speed is

$$\left\langle v^2 \right\rangle = \int_0^{+\infty} dv \, v^2 \, \frac{df}{dv}$$

$$= 4\pi \left(\frac{m}{2\pi kT} \right)^{3/2} \int_0^{+\infty} dv \, v^4 e^{-mv^2/2kT} .$$

The integral we need to evaluate (see Table D-1) is

$$I = \int_0^{+\infty} dx \, x^4 e^{-x^2/a} = \frac{3\sqrt{\pi} \, a^{5/2}}{8} ,$$

where

$$a = \frac{2kT}{m} .$$

We have

$$\left\langle v^2 \right\rangle = 4\pi \left(\frac{m}{2\pi kT} \right)^{3/2} \frac{3\sqrt{\pi}}{8} \left(\frac{2kT}{m} \right)^{5/2} = \frac{3kT}{m} .$$

The root-mean-square speed is

$$v_{\rm rms} = \sqrt{\left\langle v^2 \right\rangle} = \sqrt{\frac{3kT}{m}} .$$

or

$$v_{\rm rms} = c \sqrt{\frac{3kT}{mc^2}}$$

$$\approx \left(3.00 \times 10^8 \text{ m/s} \right) \sqrt{\frac{(3)\left(\frac{1}{40} \text{ eV} \right)}{2.61 \times 10^{10} \text{ eV}}} \approx 510 \text{ m/s.} \quad \blacksquare$$

Example 2-17 shows that the average kinetic energy of the gas molecules is $3kT/2$, in agreement with our definition of temperature (2.30). From Examples 2-15, 2-16, and 2-17, we see that the most probable, average, and root-mean-square speeds are related by

$$v_{\rm mp} < \left\langle v \right\rangle < v_{\rm rms} . \qquad (2.80)$$

We may convert the Maxwell-Boltzmann speed distribution to an energy distribution, as demonstrated in the following example.

EXAMPLE 2-18
Calculate the kinetic energy distribution, df/dE, of particles in an ideal gas from the speed distribution, df/dv.

SOLUTION:
The speed distribution is

$$\frac{df}{dv} = 4\pi \left(\frac{m}{2\pi kT} \right)^{3/2} v^2 e^{-mv^2/2kT} .$$

Make the change of variables

$$E \equiv \frac{1}{2} mv^2 ,$$

$$dE = mv \, dv .$$

The kinetic energy distribution is

$$\frac{df}{dE} = \frac{dv}{dE} \frac{df}{dv} = \frac{4\pi}{mv} \left(\frac{m}{2\pi kT} \right)^{3/2} v^2 e^{-mv^2/2kT} ,$$

or

$$\frac{df}{dE_{\rm K}} = \frac{4\pi}{m} \left(\frac{m}{2\pi kT} \right)^{3/2} \sqrt{\frac{2E}{m}} e^{-E/kT}$$

$$= 2\pi \left(\frac{1}{\pi kT} \right)^{3/2} \sqrt{E} \, e^{-E/kT} . \qquad \blacksquare$$

Example 2-18 shows that the distribution of molecular kinetic energies (E) is

$$\frac{df}{dE} = 2\pi \left(\frac{1}{\pi kT} \right)^{3/2} \sqrt{E} \, e^{-E/kT} . \qquad (2.81)$$

The energy dependence appears in two places, the square-root and the exponential factor. This is a general and important property of energy distribution functions. The energy distribution function is the product of two parts. One part, called the *density of states*, tells us what energy states are available for the particles. For an ideal gas, the density of states is proportional to the square-root of the kinetic energy. The other part, called the *Maxwell-Boltzmann* distribution, tells us the probability that a state with a given energy is occupied. We shall discuss each of these factors in more detail.

Exponential Form of the Maxwell-Boltzmann Distribution

The Maxwell-Boltzmann distribution (f_{MB}) specifies the probability that a state with energy E is occupied:

$$f_{MB} = Ce^{-E/kT}. \qquad (2.82)$$

The second law of thermodynamics states that if the total energy of a system is divided among a large number of particles, all possible divisions of energies are equally likely. This leads to the exponential form of the Maxwell-Boltzmann distribution. Consider the number of ways that a total energy E can be shared by N particles. As a specific example, we choose 3 arbitrary units of energy and 5 particles. We may choose any value of N and any energy unit, but we have purposely chosen small numbers so that the counting of the number of states will be simple. We shall see that even with these small numbers, we already get a good approximation to the Maxwell-Boltzmann distribution. The number of ways that the 5 particles can have 3 units of energy is illustrated in Figure 2-10. The first histogram represents 4 particles with zero energy and one particle with 3 units of energy. If the particles are distinguishable, then this combination occurs 5 times since any one of the 5 particles could be the particle that has 3 units of energy. The second histogram represents the states where 3 particles have zero energy, 1 particle has 1 unit of energy, and 1 particle has 2 units of energy. This combination occurs 20 times, since there are 5×4 unique ways to choose the 2 particles with nonzero energy. The third histogram represents the states where 2 particles have zero energy, and 3 particles have 1 unit of energy. This combination occurs 10 times, since there are $5 \times 4/2$ unique ways to choose the 2 particles with zero energy. The resulting distribution of particle energies (dN/dE) is shown in Figure 2-11, compared with an exponential distribution.

The exponential nature of the Maxwell-Boltzmann distribution has a simple physical interpretation. Consider a part of the total energy that is shared by two particles, one with energy E_1 and another with energy E_2. The probability that particle 1 has energy E_1 we call $P(1)$ and the probability that particle 2 has an energy E_2 we call $P(2)$. The probability that particle 1 has energy E_1 *and* particle 2 has an energy E_2 is the product of the probabilities, $P(1)P(2)$. If all possible energy divisions are equally likely, then the product of the two probabilities is a function only of the total energy ($E_1 + E_2$). For example, 10 units of energy could be divided as (0,10), (1,9), (2,8) … (10,0), each division being equally likely to occur:

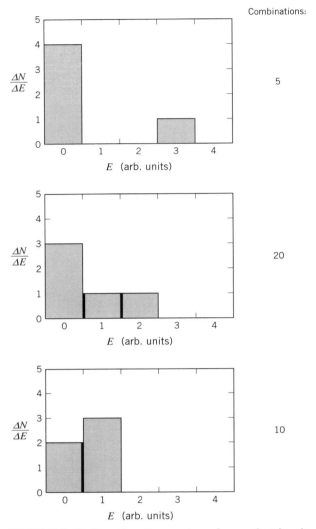

Combinations:

5

20

10

FIGURE 2-10 Counting the number of ways that 3 units of energy can be shared by five particles.
There are three such combinations. The first combination has a degeneracy of 5, the second a degeneracy of 20, and the third a degeneracy of 10 because the particles are assumed to be distinguishable.

$$P(0)P(10) = P(1)P(9) = P(2)P(8) \ldots (2.83)$$

This means that the product of the two probabilities depends only on the sum of the arguments:

$$P(E_1)P(E_2) = f(E_1 + E_2). \qquad (2.84)$$

The exponential function is unique in having this property. The product of two exponentials is obtained by adding the exponents ($e^x e^y = e^{x+y}$). Therefore, we may write the probability distribution function as $Ce^{-E/kT}$, where C is

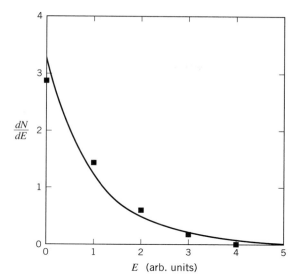

FIGURE 2-11 Approximating the Maxwell-Boltzmann distribution.

Three units of energy are shared by 5 particles. The resulting energy distribution (dN/dE) is plotted, assuming all combinations of energy sharing are equally likely to occur. The sum of the probabilities is normalized to 5. The curve is an exponential, $Ce^{-\beta E}$, where C and β are constants.

a constant. The energy parameter kT is introduced to make the exponent dimensionless, and we have chosen a negative sign corresponding to the fact that higher energy states occur in nature with smaller probabilities. The choice of kT for the energy scale follows from the definition of temperature. By selecting kT as the denominator of the exponent, we get for the average energy of a system of particles,

$$\langle E \rangle = kT, \qquad (2.85)$$

provided that there are an equal number of states available at all energies. The average kinetic energy of a particle in an ideal gas is larger by a factor of 3/2 because more states are available at higher energy. The density of energy states is discussed in the next section.

The probability that a state with energy E is occupied is given by the Maxwell-Boltzmann factor, which depends on the temperature of the system,

$$f_{MB} = Ce^{-E/kT}.$$

EXAMPLE 2-19

In atoms, the electron energy is not a continuous variable.

The energy levels are quantized. In the hydrogen atom, the energy of the first excited state is about 10 eV higher than the ground state energy. (a) At room temperature, estimate the fraction of hydrogen atoms in the excited state. (Neglect the density of states factor, which is of order unity.) (b) At roughly what temperature would 1% of the hydrogen atoms be in the excited state?

SOLUTION:

(a) Let the ground state energy be E_1 and the excited state energy be E_2. The number of hydrogen atoms in the ground state is given by the Maxwell-Boltzmann factor

$$N_1 = Ce^{-E_1/kT}.$$

The number of hydrogen atoms in the excited state is

$$N_2 = Ce^{-E_2/kT}.$$

The ratio is

$$\frac{N_2}{N_1} = \frac{Ce^{-E_2/kT}}{Ce^{-E_1/kT}} = e^{(E_1 - E_2)/kT}.$$

At room temperature,

$$kT \approx \frac{1}{40}\,\mathrm{eV},$$

so that

$$\frac{E_1 - E_2}{kT} \approx \frac{10\,\mathrm{eV}}{\left(\dfrac{1}{40}\,\mathrm{eV}\right)} = 400.$$

The fraction of hydrogen atoms in the excited state is

$$\frac{N_2}{N_1} \approx e^{-400}.$$

There are no atoms in the excited state at room temperature.

(b) We have

$$e^{(E_1 - E_2)/kT} \approx 0.01,$$

or

$$\frac{E_1 - E_2}{kT} = \ln(0.01),$$

and

$$kT = \frac{E_2 - E_1}{\ln(100)} = \frac{10\,\mathrm{eV}}{\ln(100)} \approx 2\,\mathrm{eV}.$$

Since $kT = 1/40$ eV at $T = 300$ K, this corresponds to a temperature of

$$T = \frac{(300\,\text{K})(2\,\text{eV})}{\left(\frac{1}{40}\,\text{eV}\right)} \approx 2 \times 10^{4}\,\text{K}.$$

Such a temperature exists in stars.   ∎

Vacuum Tubes

As an important example of the application of the Maxwell-Boltzmann distribution law, we consider the vacuum diode. The vacuum diode consists of an evacuated glass tube containing two electrodes, the cathode and the plate as shown in Figure 2-12a. The electrodes are accessible to the outside world by means of feed-through electrodes connecting them to pins at the base of the tube. The electrodes inside a vacuum diode come in a variety of geometric shapes.

The purpose of the vacuum is to allow the transport of electrons between the electrodes without absorption by gas molecules. The cathode is made a source of electrons by heating it either directly with an internal current or indirectly with a separate heater element. The electrons are bound in the cathode by electrical attraction, but the most energetic electrons in a heated cathode have enough kinetic energy (kT) to escape. The minimum energy an electron needs to escape is called the *work function* (Φ). The work function of a metal is discussed in more detail in Chapter 3.

The escape of a "hot" electron is called thermionic emission. A hotter cathode means that more electrons escape, exponentially more with increasing temperature. Inside the metal, the electrons behave as a gas and have a distribution of energies. The energy distribution of the most energetic electrons in the metal electrodes obeys the Maxwell-Boltzmann statistical distribution. Since the energy distribution is exponential, most of the electrons that escape barely have enough energy to do so. The expression for the current per area (J) of escaping electrons is known as the Richardson-Dushman equation,

$$J = C(kT)^{2}\,e^{-\Phi/kT}. \tag{2.86}$$

The thermionic emission from a tungsten filament is shown in Figure 2-13. The temperature dependence of the current density J may be understood as follows: the number of electrons with sufficient energy to escape contains the Maxwell-Boltzmann factor ($e^{-\Phi/kT}$), the distribution of speeds of those that escape contain a factor $v^{3} \propto (kT)^{3/2}$, and the current is proportional to the electron speed $v \propto (kT)^{1/2}$. The constant of proportionality is 1.6×10^{14} A $\cdot$ m^{-2} $\cdot$ eV^{-2}.

The rate that electrons escape can be quite large for high temperatures. If the plate voltage is made positive relative to the cathode, then the electrons are attracted to it.

The principle of the vacuum diode was discovered as early as 1883 by Thomas A. Edison. Edison noticed that when he inserted a wire into a light bulb and put a positive

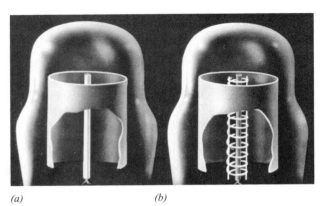

(a)   *(b)*

FIGURE 2-12 Electron vacuum tubes.
(*a*) Diode, consisting of a cathode surrounded by a cylindrical anode called the plate. (*b*) Triode, which contains a third element inserted between the cathode and the plate to control the flow of electrons. Courtesy Westinghouse Electric Corporation.

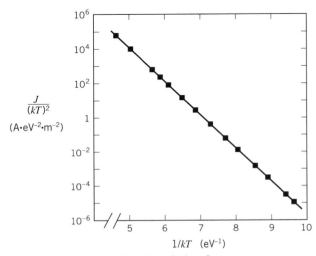

FIGURE 2-13 Thermionic emission from tungsten.
After R. L. Sproull and W. A. Phillips, *Modern Physics: The Quantum Physics of Atoms, Solids, and Nuclei*, Wiley (1980).

voltage on the wire, current flowed in the wire, a result that became known as the *Edison effect*. Edison may have been close to discovering the electron, but alas, he was an inventor, not a physicist. The first practical vacuum tube diode was made by John A. Fleming in 1904. These vacuum diodes became known as the electron valves because of their remarkable ability to conduct current in one direction only.

A powerful extension of the vacuum diode was made in 1907 by Lee DeForest, who discovered that current amplification could be obtained with the addition of a third electrode called the grid (see Figure 2-12b). Such a vacuum tube is called a triode. In the vacuum triode, the grid is inserted between the cathode and the plate and consists of a wire mesh. The purpose of the grid is to provide an additional electric field to control the movement of electrons. One important application of the vacuum triode is an oscillator circuit, which made possible the first radio transmission of speech and music, and all because the motion of the electrons that have enough kT to escape the metal are cleverly controlled with electric fields!

Vacuum tubes have been largely replaced by semiconductor diodes and transistors for most (but not all!) practical applications. The physics of semiconductor devices is discussed in Chapter 14.

2-4 DENSITY OF STATES

The Maxwell-Boltzmann distribution gives the occupation probability for any given energy, but it does not contain any information on which states exist. The number of states that exist for a particle in a given physical system depends on the particle energy. There are more states available at higher energies. There are more ways that a gas molecule can have 10 eV of kinetic energy than there are ways that the molecule can have 1 eV of kinetic energy.

How can this be true? One might be tempted to speculate that there are an infinite number of ways that a gas particle could have either 1 eV or 10 eV of kinetic energy, since

$$E_k = \frac{p_x^2 + p_y^2 + p_z^2}{2m}, \qquad (2.87)$$

and there appear to be an infinite number of choices for the combinations of $p_x, p_y,$ and p_z that satisfy the equation. This is not true! There is a finite, though very large, number of combinations, and the number of combinations increases with increasing energy. The physics of explaining why the number of combinations is finite lies in the fact that our world is quantized. We shall learn later how to calculate the number of states.

We may make an analogy between the number of states of a particle with energy E and the number of ways to make a dollar amount D, using nickels, dimes, and quarters. There are many more ways to compose $D = 10$ dollars, than $D = 1$ dollar. This is because the units of currency are quantized. The number of ways to build one dollar can be determined by counting the number of combinations (C_n) with n nickels, where $n = 0, 1, 2, \ldots 20$. There are three combinations with zero nickels (10 dimes, 5 dimes + 2 quarters, and 4 quarters). There are two combinations with one nickel (7 dimes + 1 quarter, and 2 dimes + 3 quarters). There are also two combinations with 2, 3, 4, 5, and 6 nickels. Continuing in this manner, we find that the total number of combinations (N) to make change for a dollar is

$$N = \sum_{n=0}^{20} C_n = 3 + (6 \times 2) + 1 + 2 + 1 + 2$$
$$+ (6 \times 1) + 0 + 1 + 0 + 1 = 29. \quad (2.88)$$

Now consider making change for 10 dollars. Consider the number of combinations with zero nickels. We could have 100 dimes, 95 dimes + 2 quarters, 90 dimes + 4 quarters, and so on. There are 21 combinations with zero nickels. We need to count all the combinations for 0, 1, 2, ... 200 nickels. It is not too hard to see that the answer is

$$N = 21 + (6 \times 20) + 19 + 20 + 19 + 20$$
$$+ (6 \times 19) + \ldots (6 \times 1) + 0 + 1 + 0 + 1$$
$$= 2081. \qquad (2.89)$$

There are many more combinations (nearly two orders of magnitude) to make 10 dollars than to make one dollar. The reason for this is that the money units are quantized. For example, there is a 10 dollar combination of 1 nickel, 97 dimes, and 1 quarter. The analogous combination for one dollar of 0.1 nickel, 9.7 dimes, and 0.1 quarter does not exist.

The number of states (N_s) available is energy dependent. The density of states, $\rho(E)$, is defined to be the number of states per volume ($n_s = N_s/V$) with energies between E and $E + dE$. The density of states is the derivative of n_s with respect to E:

$$\rho(E) \equiv \frac{dn_s}{dE}. \qquad (2.90)$$

Note that the derivative is a positive number since n_s increases with increasing energy. The density of states is the number of possible states per volume per unit energy. The density of states is energy dependent because the number of states is energy dependent. The density of states function $\rho(E)$ multiplied by the Maxwell-Boltzmann factor f_{MB} gives the number of states per volume (n) with energies between E and $E + dE$ that are *occupied*. This energy distribution of the particles (dn/dE) is equal to the density of states times the occupation probability:

$$\frac{dn}{dE} = \rho(E) f_{MB}. \qquad (2.91)$$

The total number of particles per volume is equal to the integral of dn/dE over all available energies.

In Example 2-18, we calculated the kinetic energy distribution df/dE for an ideal gas. The kinetic energy distributions dn/dE (2.91) and df/dE (2.81) are equivalent except that dn/dE is normalized to the particle density (N/V) and df/dE is normalized to unity. Therefore,

$$\frac{dn}{dE} = \frac{N}{V} \frac{df}{dE}. \qquad (2.92)$$

Therefore, the density of states for an ideal gas is

$$\rho = \frac{dn_s}{dE} = \left(\frac{N}{V}\right)(2\pi)\left(\frac{1}{\pi kT}\right)^{3/2}\sqrt{E}. \qquad (2.93)$$

The important part of the density of states is the square-root dependence on the energy. We have assumed that the density of states is also proportional to the number of particles. This is true as long as the gas is "ideal," meaning that the gas particles are distinguishable, in principle, by their positions. If N/V is large, such that the average distance between particles is no longer large compared to the size of the particles, then the particles can no longer be considered distinguishable. In this case we shall need to modify the expression for the density of states.

We can easily check that this density of states gives the correct average kinetic energy of the particles. The average kinetic energy is

$$\langle E_k \rangle = \frac{\int_0^{+\infty} dE\, E\, \dfrac{dn}{dE}}{\int_0^{+\infty} dE\, \dfrac{dn}{dE}}$$

$$= \frac{\int_0^{+\infty} dE\, E\rho f_{MB}}{\int_0^{+\infty} dE\, \rho f_{MB}} = \frac{\int_0^{+\infty} dE E\sqrt{E}e^{-E/kT}}{\int_0^{+\infty} dE \sqrt{E}e^{-E/kT}}. \qquad (2.94)$$

To do the integral in the numerator, we integrate by parts, by differentiating $E^{3/2}$ and integrating the exponential. The integrated part vanishes when evaluated at the limits. Therefore,

$$\langle E_k \rangle = \frac{\int_0^{+\infty} dE \dfrac{3}{2}\sqrt{E}(kT)Ce^{-E/kT}}{\int_0^{+\infty} dE \sqrt{E}Ce^{-E/kT}}$$

$$= \frac{3}{2}kT. \qquad (2.95)$$

The average kinetic energy of the particles is $3kT/2$ in accordance with the definition of temperature.

CHAPTER 2: PHYSICS SUMMARY

- If an event has probability of occurrence p, and an experiment is performed n times, the probability of observing x events, in any order, is specified by the binomial probability distribution:

$$f_b(x) = \frac{n!\, p^x (1-p)^{n-x}}{x!(n-x)!}.$$

The mean value of x is np and the root-mean-square deviation is

$$\sigma_x = \sqrt{np(1-p)}.$$

- When n is large, x approaches the behavior of a continuous variable and the probability distribution is Gaussian:

$$f_g(x) = \frac{1}{\sqrt{2\pi\sigma^2}}e^{-(x-a)^2/2\sigma^2}.$$

- If p is small, the binomial distribution may be approximated by the Poisson distribution, useful in the interpretation of experiments with small numbers of events:

$$f_p(x) = \frac{e^{-a}a^x}{x!},$$

where $x = 0, 1, 2, \ldots$

- An ideal gas is a collection of particles in which the average distance a particle travels before colliding with another particle is large compared to the size of the particles. The absolute temperature scale is defined in terms of the average kinetic energy of particles of an ideal gas

$$\langle E_k \rangle \equiv \frac{3}{2} kT .$$

The unit of temperature is the kelvin (K). The constant k is called Boltzmann's constant:

$$k = 8.617 \times 10^{-5} \text{ eV/K} .$$

At room temperature,

$$kT \approx \frac{1}{40} \text{ eV} .$$

• The pressure of a gas is proportional to the average speed squared of the particles. The definition of temperature leads to the ideal gas law:

$$PV = NkT .$$

• The Maxwell velocity distribution function gives the distribution of any one component of velocity:

$$\frac{df}{dv_z} = \sqrt{\frac{m}{2\pi kT}} e^{-mv_z^2 / 2 kT} .$$

The form of this distribution is Gaussian as the result of a large number of random collisions. The average velocity is zero and the average kinetic energy per degree of freedom is $kT/2$.

• The distribution of molecular speeds in an ideal gas is

$$\frac{df}{dv} = 4\pi \left(\frac{m}{2\pi kT} \right)^{3/2} v^2 e^{-mv^2 / 2kT} .$$

• The second law of thermodynamics states that when the total energy of a system is divided among a large number of particles, all possible divisions of energies are equally likely. For a system of particles in thermal equilibrium, this leads to the Maxwell-Boltzmann distribution, which specifies the probability that a state with energy E is occupied:

$$f_{MB} = Ce^{-E/kT} .$$

It is valid as long as the particles are distinguishable.

• The density of states for a system of particles whose kinetic energy is given by the expression $E = mv^2/2$ is proportional to the square root of the kinetic energy:

$$\rho \propto \sqrt{E} .$$

• The energy distribution of a system of particles in thermal equilibrium is proportional to the density of states times the Boltzmann factor:

$$\frac{df}{dE} \propto \rho(E) e^{-E/kT} .$$

REFERENCES AND SUGGESTIONS FOR FURTHER READING

R. H. Bacon, "Practical Statistics for Practical Physicists," *Am. J. Phys.* 14, 84 (1946).

S. Chandrasekhar, "Stochastic Problems in Physics and Astronomy," *Rev. Mod. Phys.* 15, 1 (1943).

E. Fermi, *Thermodynamics*, Dover (1957).

S. Goldberg, *Probability, An Introduction*, Prentice Hall (1960).

Lord Rayleigh, "On the Virial of a System of Hard Colliding Bodies," *Nature* 45, 80 (1891).

F. Reif, *Statistical Physics*, Volume 5 of the Berkeley Physics Series, McGraw-Hill (1967).

P. A. Tipler, *Modern Physics*, first edition, Worth (1969).

M. W. Zemansky and R.H. Dittman, *Heat and Thermodynamics,* McGraw-Hill (1981).

QUESTIONS AND PROBLEMS

Distribution functions

1. In plotting the data of a distribution function, what happens if we make the bin size smaller than the experimental resolution? Explain.

2. Can you give an example of a discrete distribution whose possible values are not integer?

3. Define two distribution functions, one discrete and one continuous, that affect your everyday life. Make a qualitative sketch of the distributions, indicating the horizontal and vertical scales and units.

4. Make a histogram of the following 23 test scores:

   58, 61, 63, 68, 69, 71, 71, 72, 72, 74, 75, 76, 77, 78, 79, 79, 79, 80, 82, 82, 86, 91, 98.

   (a) Use your histogram to make a quick estimate of the average score and the standard deviation.

(b) Determine the exact values of the average and standard deviation.

5. A *bunch* of 10^6 particles emerges from an accelerator at $x = 0$ m and is directed at a detector located at $x = 1$ m. The speed of the particles is very nearly equal to the speed of light. If the particle lifetime is 10^{-6} s, how many (N) particles are expected to decay before reaching the detector? Make a sketch of the decay distribution dN/dx versus x.

6. A proportional wire chamber contains an array of parallel wires that are used to detect the x coordinate of a particle that passes through the chamber. (The axis of each wire is along the y direction.) The chamber is used to detect particles that pass through at random locations. An electrical signal is detected on the wire closest to which the particle passes. The measurement of the particle x coordinate is the x coordinate of the wire with the electrical signal. If the wire spacing is 1 mm, what is the position resolution (σ) of the chamber?

7. Show that the full-width at half-maximum (Γ) of a Gaussian distribution is given by

$$\Gamma = 2\sqrt{2\ln 2}\ \sigma \approx 2.35\sigma,$$

where σ is the standard deviation of the Gaussian distribution.

8. A baseball enthusiast goes to his local convenience store to purchase baseball cards, hoping to find players from his favorite team, the Red Sox. He purchases five packages containing 15 cards each. Assuming that the cards are distributed randomly among 26 teams, calculate the probability of getting n Red Sox players cards. Determine n to two significant figures for $n = 0, 1, 2, \ldots 6$. (When the author tried this experiment, he got zero Red Sox players. Was he "unlucky"?)

9. Researchers at the CERN UA1 experiment discovered the Z^0 boson in a sample of a large number of proton–antiproton collisions, where 5 decays of the Z^0 boson into lepton pairs were identified. Calculate the probability of observing at least 2 more Z^0 decays in a second data sample of the same size.

10. Estimate the probability of getting heads exactly 500 times in 1000 unbiased flips of a coin.

11. A hockey goalie lets the puck in the net an average of 2.3 times per game. Make an estimate of the probability per game that the goalie records a shutout (zero goals). What assumptions did you make?

12. In a given number of particle interactions, it is known that there is a 50% probability that a certain type of rare event occurs at least once. What is the average number of the rare events expected in the sample?

13. In the analysis of a sample of 10^6 particle interactions, experimenters find one unexpected event. The experimenters wish to find a second such event. What size sample of interactions do they need to be 90% certain of getting a second event?

14. A particle has a certain rare decay with a probability of 10^{-6}. How many decays must be measured in order to be 95% certain that the rare decay will be observed at least once?

15. Electrons with an energy of 10 GeV are elastically scattered into a detector, where the electron energy is determined with an error (sigma) of 10%. An experimenter makes measurements on 10^3 electrons. (a) Make a sketch of the expected distribution of measured energies, labeling the scale on both the horizontal and vertical axes. (b) How many times does the experimenter expect the measured energy to be greater than 13 GeV?

Temperature and the ideal gas

16. Why is the mean free path of a molecule at STP much larger than the average distance between molecules?

17. Does the mean free path of a molecule in a gas depend on the temperature?

18. Use the ideal gas law to make an estimate of the density of air at STP.

19. Make an estimate of the number of collisions per second that a nitrogen molecule makes with other molecules at STP.

20. Calculate the root-mean-square speed of a nitrogen molecule at a temperature of 1200 K.

*21. The mean free path of a molecule is d. Determine the probability that the molecule travels at least a distance x before it collides.

22. The ideal gas law (2.50) is sometimes written

$$pV = \nu RT,$$

where ν is the number of moles of molecules and R is a constant called the *ideal gas constant*. Determine the numerical value of R.

23. (a) Determine the typical speed of a *thermal* neutron at room temperature, that is, a neutron that has a kinetic energy equal to kT. (b) At what temperature is the typical neutron speed equal to 1% of the speed of light?

24. For protons on the surface of the sun ($T \approx 6000$ K) estimate (a) the average kinetic energy and (b) the most probable speed.

The Maxwell-Boltzmann distribution

25. Why does the earth's atmosphere get colder with increasing altitude?

26. The average speed of the molecules of a certain gas at room temperature is 440 m/s. Identify the gas.

27. In an atmosphere of nitrogen at a constant temperature of 300 K, what change in altitude corresponds to a pressure change of 10%? What is the result at a temperature of 250 K?

28. Consider the energy distribution of particles in an ideal gas. Consider the number of particles in a small interval (ΔE) centered at the average energy compared to the number in the same size interval centered at twice the average energy. What is the ratio?

29. Consider a gas of nitrogen molecules in thermal equilibrium at room temperature. What fraction of the molecules have an x component of velocity within the range

$$(-0.01)\,v_{rms} < v_x < (0.01)\,v_{rms},$$

where v_{rms} is the root-mean-square value of v_x?

30. Consider a gas of helium atoms (He). At what temperature is the average molecular speed equal to 10^3 m/s?

31. Consider a gas of nitrogen molecules in thermal equilibrium at room temperature. (a) Write down an exact expression for the fraction of molecules with speed greater than twice the average speed. (b) Use the plot of Figure 2-9 to make a rough estimate of this fraction. (c) What fraction of the molecules have speeds within the interval ± 50 m/s, centered on twice the average speed?

32. An oxide cathode of a vacuum tube has a work function of 2.0 eV. Calculate the electron flux (number per time per area) from thermal emission when the cathode is heated to 1000 K.

33. Electrons are emitted at the rate of one per second from a cathode of area 5×10^{-4} m² at room temperature. (a) What is the work function of the cathode? (b) At what rate are electrons emitted if the cathode is cooled to liquid nitrogen temperature (77 K)?

Density of states

34. In the hydrogen atom the energy of the electron is usually defined to be the kinetic plus potential energy.

There are two states with an energy of -13.6 eV (the lowest possible energy) and eight states with an energy of -3.4 eV (the first excited state). There are no states with energies between these two values. At what temperature would 1% of the hydrogen atoms be in the first excited state?

Additional problems

35. *The game show problem.* Behind one of three doors is the "grand prize." The contestant chooses one of the doors. The host opens one of the two remaining doors revealing a "booby prize." The host then offers to trade doors. What is the best strategy for the contestant, trade or not trade, or does it not make any difference?

*36. Consider a container of gas molecules in thermal equilibrium at a temperature T. (a) Find an expression for the speed distribution of the molecules that strike the walls of the container in a unit time interval. (b) Suppose that the container has a tiny hole. Calculate the average kinetic energy of those gas molecules that escape through the hole.

37. Protons can escape from the surface of the sun if their kinetic energy is sufficient to overcome the gravitational potential energy. (a) Calculate the minimum speed necessary for escape. (b) The surface temperature of the sun is about 6000 kelvin. Calculate the root-mean-square speed of protons and make a graph of the speed distribution.

*38. A physicist pours a glass of beer and observes that the bubbles formed on the surface of the glass increase in size as they rise to the top of the glass. The reason for this is that the pressure of the dissolved CO_2 in the beer is larger than the pressure of the bubble. Assuming that the pressure difference is approximately constant, the change in the number (N) of CO_2 molecules in a bubble is proportional to the surface area of the bubble,

$$\frac{dN}{dt} = C r^2,$$

where r is the time-dependent radius of the bubble and C is a constant. Use the ideal gas law to show that the radius of the bubble increases linearly with time. (See N. E. Shafer and R. N. Zare, "Through a Beer Glass Darkly," *Phys. Today* **44**, No. 10, 48 (1991).)

*39. Consider N gas molecules in a container of volume V at a temperature T. (a) Show that the flux Φ (number per area per time) of gas molecules hitting the container wall is

$$\Phi = \frac{N\langle v \rangle}{4V},$$

where $\langle v \rangle$ is the average speed of the gas molecules.

(b) A container at room temperature with a volume of one cubic meter has a hole of size 10^{-5} square meters. Assuming that the pressure inside the container is larger than the pressure outside the container, calculate the time for the container pressure to drop by a factor of e.

CHAPTER

3

PLANCK'S CONSTANT

Entropy, according to Boltzmann, is a measure of a physical probability, and the meaning of the second law of thermodynamics is that the more probable a state is, the more frequently it will occur in nature.... To work out these probability considerations the knowledge of two universal constants is required, each of which has an independent meaning, so that the evaluation of these constants from the radiation law could serve as a posteriori test whether the whole process is merely a mathematical artifice or has a true physical meaning. The first constant is of a somewhat formal nature; it is connected with the definition of temperature.... Much less simple than that of the first was the interpretation of the second universal constant of the radiation law, which, as the product of energy and time, I called the elementary quantum of action.... So long as it could be regarded as infinitely small, that is to say for large values of energy or long periods of time, all went well; but in the general case a difficulty arose at some point or other, which became more pronounced the weaker and the more rapid the oscillations. The failure of all attempts to bridge this gap soon placed one before the dilemma: either the quantum of action was only a fictitious magnitude, and, therefore, the entire deduction from the radiation law was illusory and a mere juggling with formulae, or there is at the bottom of this method of deriving the radiation law some true physical concept. If the latter were the case, the quantum would have to play a fundamental role in physics, heralding the advent of a new state of things, destined, perhaps, to transform completely our physical concepts which, since the introduction of the infinitesimal calculus by Leibnitz and Newton, have been founded upon the assumption of the continuity of all causal chains of events. Experience has decided for the second alternative.

Max Planck

Matter and radiation are intimately coupled to each other. Every piece of matter is continually emitting and absorbing radiation. In Chapter 2, we discussed the kinetic energy distribution of atoms or molecules in equilibrium at a given temperature (T). The energy scale of the particles is kT. In this chapter, we shall examine the energy distribution of radiation. The energy scale of radiation is also kT, however, an additional new fundamental constant is needed in order to explain the observed radiation spectra.

3-1 ATOMS AND RADIATION IN EQUILIBRIUM

All matter radiates because the electromagnetic force between electrons in matter causes the electrons to be accelerated. We know from classical electrodynamics that an accelerated charge emits electromagnetic radiation. The inverse process also occurs; electrons absorb radiation and are accelerated. A macroscopic quantity of matter contains a very large number of electrons or "atomic oscillators," which are constantly absorbing and emitting electromagnetic radiation. In the absorption process energy in the form of radiation is converted to kinetic energy of the atomic oscillator, while in the emission process oscillator energy is converted to radiation energy. Since there are so many atomic oscillators present, matter is capable of absorption and emission of radiation of all frequencies or wavelengths.

Blackbody Radiation

The term *blackbody* is used to denote an object that is a perfect absorber of radiation of all wavelengths. A blackbody absorbs all radiation that strikes its surface, and reradiates the energy with a universal wavelength spectrum. We shall derive the formula for the universal radiation spectrum in the next section. The universal (black-

body) spectrum depends only on the temperature of the object, not on its shape, size, or chemical composition!

In practice, an object will not absorb all the radiation that is incident on it; some of this radiation will be reflected without being *thermalized*, that is, without being converted into the universal spectrum. The total wavelength spectrum of the emitted radiation is usually not significantly altered by the reflection because the amount of reflection is small or is confined in a very small wavelength range. To a reasonable approximation, all matter in thermal equilibrium behaves like a blackbody, absorbing and emitting radiation with the universal spectrum that depends only on the temperature. The universal spectrum is referred to as a *thermal* spectrum.

For objects that are not ideal absorbers, we define the *emissivity* (ε) to be the fraction of incident radiation energy that an object absorbs. The emissivity is temperature dependent. Table 3-1 shows emissivities of some common materials.

The Classical Formula of Rayleigh-Jeans

The scientific problem at the beginning of the twentieth century was the lack of understanding of the observed radiation spectrum from an object in thermal equilibrium. Consider a hollow cavity whose walls are held at a fixed temperature. In one of the walls is a tiny hole (see Figure 3-1). The cavity is a blackbody because it absorbs all radiation that enters through the small hole; radiation that enters the hole has no chance to escape before being absorbed by the walls of the cavity. The radiation inside the cavity is in thermal equilibrium with the walls of the cavity and is continually being absorbed and reemitted by the cavity walls. Radiation that happens to be directed toward the hole can escape from the cavity. The spectrum of the escaping radiation is identical to the spectrum inside the cavity. If the temperature of the cavity is constant, then as much energy escapes from the cavity as enters the cavity.

TABLE 3-1
APPROXIMATE EMISSIVITIES OF SOME COMMON MATERIALS.

Material	Temperature (K)	Emissivity
Snow	270	0.95
Fire brick	1000	0.8
Glass	300	0.9
Lamp black	300	0.96
Aluminum (polished)	300	0.05
Aluminum (oxidized)	300	0.1
Graphite	300	0.7
Brass (dull)	200	0.6
Brass (polished)	200	0.03
Soot	300	0.94
Daytime sky	300	0.7

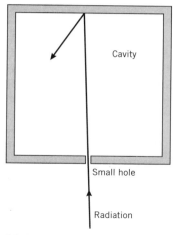

FIGURE 3-1 Model of a blackbody.
The cavity absorbs all radiation that enters a small hole. The radiation inside the cavity is in thermal equilibrium with the walls of the cavity. Radiation that is emitted through the hole has the universal thermal spectrum that depends only on the temperature of the cavity.

* *Challenging*

How much energy per volume in the form of radiation is inside the cavity? We shall calculate the radiation energy inside a cubical cavity. The calculation involves counting the number of electromagnetic waves that the cavity can support by solving the electromagnetic wave equation:

$$\frac{\partial^2 F}{\partial x^2} + \frac{\partial^2 F}{\partial y^2} + \frac{\partial^2 F}{\partial z^2} = \frac{1}{c^2}\frac{\partial^2 F}{\partial t^2}. \tag{3.1}$$

In the wave equation, $F(x,y,z,t)$ represents components of the oscillating electric and magnetic fields. One solution to the wave equation (3.1) may be written as the product of sine functions:

$$F = C\sin(k_1 x)\sin(k_2 y)\sin(k_3 z)\sin(\omega t), \tag{3.2}$$

where C is an arbitrary constant, and the wave number constants (k_1, k_2, k_3) and angular frequency (ω) are constrained by the wave equation. The electric field must vanish at the boundaries of the cube. If we take the cube boundaries to be at $x = 0$, $y = 0$, $z = 0$, $x = L$, $y = L$, and $z = L$, then the constants k_1, k_2, and k_3 may be written

$$k_1 = \frac{n_1 \pi}{L}, \tag{3.3}$$

$$k_2 = \frac{n_2 \pi}{L}, \tag{3.4}$$

and

$$k_3 = \frac{n_3 \pi}{L}, \tag{3.5}$$

where n_1, n_2, and n_3 are positive integers. The frequency may be written in terms of the wavelength (λ):

$$\omega = \frac{2\pi c}{\lambda}. \tag{3.6}$$

Our solution to the wave equation satisfying the boundary conditions of the cube is

$$F = C\sin\left(\frac{n_1 \pi x}{L}\right)\sin\left(\frac{n_2 \pi y}{L}\right)\sin\left(\frac{n_3 \pi z}{L}\right)\sin\left(\frac{2\pi c t}{\lambda}\right).$$

$$\tag{3.7}$$

This type of wave is called a *standing wave* (see Figure 3-2). We may deduce the relationship between the wavelength (λ) and the size of the box (L) by substituting the wave function (3.7) into the wave equation (3.1). The result is

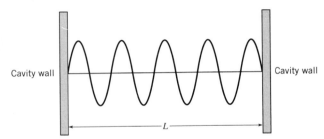

FIGURE 3-2 Standing wave in a box of size L.
An integer number of half-wavelengths must fit into the distance L.

$$\left(\frac{n_1\pi}{L}\right)^2 + \left(\frac{n_2\pi}{L}\right)^2 + \left(\frac{n_3\pi}{L}\right)^2 = \left(\frac{2\pi}{\lambda}\right)^2, \quad (3.8)$$

or

$$n_1^2 + n_2^2 + n_3^2 = \frac{4L^2}{\lambda^2}. \quad (3.9)$$

We now proceed to count the number of standing waves in the cavity. Consider the coordinate system in n space (n_1, n_2, n_3), as shown in Figure 3-3. In these coordinates, each integer value of (n_1, n_2, n_3) corresponds to one standing wave. The volume of a sphere in n space would be the total number of modes if the values of (n_1, n_2, n_3) were allowed to be negative. Since only positive values of (n_1, n_2, n_3) are allowed, we must divide the volume of the sphere by eight. In addition, there is an internal degree of freedom corresponding to the relative orientations of the electric ($\mathbf{E}$) and magnetic ($\mathbf{B}$) fields. The two possible orientations of E and B are referred to as the two *polarizations* of the radiation. Therefore, the number of standing waves (N) in n space is

$$N = \left(\frac{1}{8}\right)(2)\frac{4}{3}\pi(n_1^2 + n_2^2 + n_3^2)^{3/2}$$

$$= \frac{\pi}{3}(n_1^2 + n_2^2 + n_3^2)^{3/2}. \quad (3.10)$$

We may write N in terms of the wavelength by using the expression for $n_1^2 + n_2^2 + n_3^2$ (3.9), which gives

$$N = \frac{8\pi L^3}{3\lambda^3}. \quad (3.11)$$

The number of modes per unit wavelength is obtained by differentiating N with respect to λ:

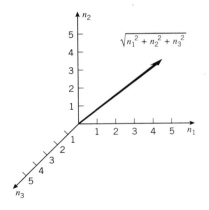

FIGURE 3-3 Coordinate system in (n_1, n_2, n_3) space.
A standing wave exists for each integer value of n_1, n_2, and n_3. The number of standing waves in a sphere of radius $(n_1^2 + n_2^2 + n_3^2)^{1/2}$ is equal to $4\pi(n_1^2 + n_2^2 + n_3^2)^{3/2}/3$.

$$-\frac{dN}{d\lambda} = \frac{8\pi L^3}{\lambda^4}. \quad (3.12)$$

(The minus sign indicates the number of modes decreases with increasing wavelength.) The number modes per unit wavelength per cavity volume is

$$-\left(\frac{1}{L^3}\right)\frac{dN}{d\lambda} = \frac{8\pi}{\lambda^4}. \quad (3.13)$$

If the average energy of a wave is kT, then the energy per volume per unit wavelength ($du/d\lambda$) of radiation in the cavity would be

$$\frac{du}{d\lambda} = \left(\frac{1}{L^3}\right)\frac{dE}{d\lambda} = -(kT)\left(\frac{1}{L^3}\right)\frac{dN}{d\lambda} = \frac{8\pi kT}{\lambda^4}. \quad (3.14)$$

We now relate the energy in the cavity volume to the power per area radiated from the cavity surface. Consider a small area ΔA of cavity surface as indicated in Figure 3-4a. For radiation normal to the surface of the cavity, the time Δt taken for radiation to travel a distance Δx is $\Delta t = \Delta x/c$. The amount of energy per unit wavelength in the volume $\Delta A \Delta x$ is related to the power per area per unit wavelength $(dR/d\lambda)_{\theta=0}$ radiated by the wall by

$$\frac{dE}{d\lambda} = 2\left(\frac{dR}{d\lambda}\right)_{\theta=0} \Delta t\,\Delta A = \left(\frac{dR}{d\lambda}\right)_{\theta=0} \frac{2\Delta x\Delta A}{c}, \quad (3.15)$$

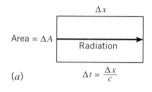

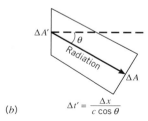

FIGURE 3-4 Energy radiated from the cavity surface.
(*a*) Radiation emitted normal to the surface of the cube. (*b*) Radiation emitted at an angle θ from the normal to the surface of the cube.

where the factor of 2 accounts for the fact that only 1/2 of the radiation is traveling in the x direction. Therefore, if all the radiation was at right angles, we would have

$$\left(\frac{dR}{d\lambda} \right)_{\theta=0} = \left(\frac{dE}{d\lambda} \right)\left(\frac{c}{2\Delta x \Delta A} \right) = \left(\frac{du}{d\lambda} \right)\left(\frac{c}{2} \right). \quad (3.16)$$

We now account for the angle of incidence of the radiation not being normal. The energy per time flowing in each direction across any unit area inside the box is equal to the power per unit area radiated by the walls. Consider radiation from the wall at an angle θ as indicted in Figure 3-4*b*. The x component of velocity of the radiation is $c \cos\theta$, so that the time taken to travel the distance Δx in the x direction is $\Delta t' = \Delta x / (c \cos\theta)$. Radiation that reaches an area ΔA comes from a larger area $\Delta A' = \Delta A / \cos\theta$. For radiation at an angle θ, the power per area per unit wavelength, $(dR/d\lambda)_\theta$, is

$$\left(\frac{dR}{d\lambda} \right)_\theta = \left(\frac{dE}{d\lambda} \right)\left(\frac{1}{2\Delta t' \, \Delta A'} \right)$$

$$= \left(\frac{dE}{d\lambda} \right)\left(\frac{c \cos^2 \theta}{2\Delta x \Delta A} \right)$$

$$= \left(\frac{du}{d\lambda} \right)\left(\frac{c}{2} \right)\cos^2\theta \quad . \quad (3.17)$$

The power per area per unit wavelength is obtained by averaging over all angles:

$$\frac{dR}{d\lambda} = \left(\frac{du}{d\lambda} \right)\left(\frac{c}{2} \right)\langle \cos^2 \theta \, \rangle$$

$$= \left(\frac{du}{d\lambda} \right)\left(\frac{c}{2} \right)\left(\frac{1}{2\pi} \right)\int_{-\pi}^{\pi} d\theta \cos^2 \theta$$

$$= \left(\frac{du}{d\lambda} \right)\left(\frac{c}{2} \right)\left(\frac{1}{2} \right) = \frac{c}{4}\frac{du}{d\lambda}. \quad (3.18)$$

The average over all angles produces a factor of 1/2. The radiated power per area per unit wavelength (3.18) from the cavity walls is equal to the energy per volume per unit wavelength (3.14) of radiation in the cavity times a factor $c/4$:

$$\frac{dR}{d\lambda} = \frac{du}{d\lambda}\left(\frac{c}{4} \right) = \frac{2\pi ckT}{\lambda^4}. \quad (3.19)$$

This is the *Rayleigh-Jeans* formula. Note that $dR/d\lambda$ does not depend on the size of the box.   ✱

Figure 3-5 shows the Rayleigh-Jeans distribution (3.19) for an object at a temperature of 1500 K together with the experimentally observed distribution. The Rayleigh-Jeans distribution predicts that the object is glowing blue whereas our experience tells us that it is glowing red. The Rayleigh-Jeans formula gives a correct description of the measured equilibrium radiation spectra at large wavelengths. At small wavelengths the Rayleigh-Jeans formula blows up as λ^{-4}. This prediction became known as the *ultraviolet catastrophe*. We know that the power radiated does not become large at small wavelengths. Objects in a dark room do not glow blue from their thermal radiation. Something is very wrong. The part of the Rayleigh-Jeans formula that is wrong is the average energy per oscillator. At large wavelengths the average energy per oscillator is indeed equal to kT, but at small wavelengths it is not!

The Empirical Formula of Wien

Near the end of the nineteenth century, Josef Stefan and (independently) Ludwig Boltzmann had deduced from thermodynamics that the total power per area radiated by any object was proportional to the *fourth power of the temperature*. This important result, which we shall derive in this chapter, was in agreement with experiment. Following the work of Stefan and Boltzmann, Wilhelm Wien made the hypothesis that the power per area per unit wavelength ($dR/d\lambda$) radiated from an object at a fixed temperature (T) was of the form,

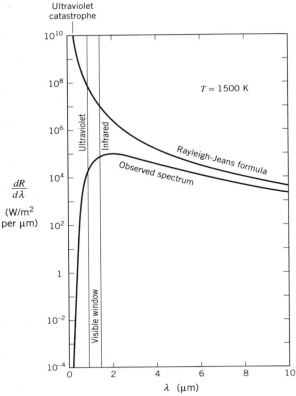

FIGURE 3-5 Observed spectrum of radiation from an object at 1500 K.
The observed spectrum is such that one can just see the low-wavelength edge of the radiation, and the object glows red. The Rayleigh-Jeans classical derivation incorrectly predicts that the power spectrum goes as λ^{-4} at small wavelengths, a phenomenon known as the ultraviolet catastrophe. The observed spectrum does go as λ^{-4} at large wavelengths, but at small wavelengths it vanishes exponentially.

$$\frac{dR}{d\lambda} = \frac{ae^{-b/\lambda T}}{\lambda^5}, \qquad (3.20)$$

where a and b are constants. The radiation spectrum goes to zero exponentially at small wavelengths and goes to zero as λ^{-5} at large wavelengths. The Wien distribution (3.20) predicts two important results: (1) the total power per area radiated is proportional to T^4, and (2) the wavelength where $dR/d\lambda$ is maximum is inversely proportional to the temperature. Although these results are confirmed by experiment, detailed comparison of the Wien formula (3.20) with the measured spectrum shows small deviations at large wavelengths. The Wien distribution is *very nearly* but *not exactly* in agreement with experimental measurements of the spectra.

EXAMPLE 3-1
Show that the Wien distribution (3.20) predicts that the total power varies as the fourth power of the temperature.

SOLUTION:
We make the change of variables

$$x = T\lambda,$$

and

$$dx = Td\lambda.$$

The Wien distribution (3.20) becomes

$$\frac{dR}{d\lambda} = \frac{ae^{-b/\lambda T}}{\lambda^5} = \frac{aT^5 e^{-b/x}}{x^5}.$$

The total power is

$$R = \int_0^\infty d\lambda \frac{dR}{d\lambda} = aT^5 \int_0^\infty \frac{dx}{T} \frac{e^{-b/x}}{x^5}$$

$$= aT^4 \int_0^\infty dx \frac{e^{-b/x}}{x^5}. \qquad ∎$$

EXAMPLE 3-2
Show that the Wien distribution (3.20) predicts that at the wavelength (λ_m) where $dR/d\lambda$ is maximum is inversely proportional to the temperature.

SOLUTION:
To get λ_m, we differentiate the Wien distribution and set it equal to zero:

$$\frac{d}{d\lambda}\left(\frac{dR}{d\lambda}\right) = \left[\frac{(-5)ae^{-b/\lambda T}}{\lambda^6} + \left(\frac{a}{\lambda^5}\right)\left(\frac{be^{-b/\lambda T}}{T\lambda^2}\right)\right]_{\lambda=\lambda_m}$$

$$= 0.$$

The spectrum has a maximum at

$$\lambda_m = \frac{b}{5T}. \qquad ∎$$

3-2 THE THERMAL RADIATION SPECTRUM

Quantization of the Energy Levels

The classical radiation formula of Rayleigh-Jeans (3.19) does not agree with experiment at small wavelengths. The correct solution was found by Max Planck in 1900. Planck made the revolutionary hypothesis that the energy distribution of the atomic oscillators was not continuous. Planck discovered that the measured thermal radiation spectra could be explained if the spacing between energy levels was proportional to the frequency of oscillation. This observation marked the discovery of *energy quantization*. The constant of proportionality, which relates the energy of an oscillator and its frequency of oscillation, is called *Planck's constant* (h). The quantization hypothesis of Planck is

$$E_n = nhf, \qquad (3.21)$$

where n is a nonnegative integer $(0, 1, 2, \ldots)$. The essential ingredient of the quantization condition (3.21) is that for a given frequency, the energy difference between two adjacent oscillator energy levels is

$$E_{n+1} - E_n = (n+1)hf - nhf = hf. \qquad (3.22)$$

If an oscillator of frequency f gives radiation of the same frequency, we may write the energy difference (3.22) in terms of the corresponding radiation wavelength (λ),

$$E_{n+1} - E_n = \frac{hc}{\lambda}. \qquad (3.23)$$

The reason energy quantization can explain the radiation spectrum is conceptually simple. If the distribution of oscillator energies is continuous, the average energy per oscillator is kT. If the energy levels are quantized, then at high frequencies most of the oscillators are in the $n = 0$ state, corresponding to an energy of $E_0 = 0$. The number of oscillators in the $n = 1$ state is proportional to $e^{-hf/kT}$ so that the average energy decreases exponentially, as observed in the data.

EXAMPLE 3-3

Consider a piece of matter at $T = 1500$ K. Suppose that at some relatively large frequency, the energy spacing between atomic oscillator levels is 1 eV. Calculate the average energy per oscillator.

SOLUTION:

At 1500 K, the value of kT is

$$kT = (0.26\,\text{eV})\left(\frac{1500\ \text{K}}{300\ \text{K}}\right) = 0.13\,\text{eV}.$$

The number of atoms in the ground state (N_0) is proportional to $e^{-E_0/kT}$, where E_0 is the ground state energy of the oscillator. By Planck's hypothesis (3.21)

$$E_0 = 0,$$

and

$$N_0 = Ce^{-E_0/kT} = C.$$

where C is some constant related to the size of the object. The number of atoms in the next highest available level, the *first excited state*, is proportional to $e^{-E_1/kT}$, where E_1 is the energy of the oscillator. For

$$E_1 = 1\,\text{eV},$$

the number of oscillators in the first excited state (N_1) is

$$N_1 = Ce^{-E_1/kT} = Ce^{-1\,\text{eV}/0.13\,\text{eV}} = C\left(4.6 \times 10^{-4}\right).$$

The number in the *second excited state* is even smaller. The energy of the second excited state (E_2) is

$$E_2 = 2\,\text{eV},$$

and the number of oscillators in the second excited state (N_2) is

$$N_2 = Ce^{-E_2/kT} = Ce^{-2\,\text{eV}/0.13\,\text{eV}} = C\left(4.6 \times 10^{-4}\right)^2.$$

The average oscillator energy is

$$\begin{aligned}
\langle E \rangle &= \frac{N_0 E_0 + N_1 E_1 + N_2 E_2 + \ldots}{N_0 + N_1 + N_2 + \ldots} \\
&= \frac{C(0) + C\left(4.6 \times 10^{-4}\right)(1\,\text{eV}) + \ldots}{C + C\left(4.6 \times 10^{-4}\right) + \ldots} \\
&\approx 4.6 \times 10^{-4}\,\text{eV}.
\end{aligned}$$

The average energy of a distribution of high-frequency oscillators is much less than kT under the hypothesis of Planck. ∎

The effect of energy quantization of oscillator levels is illustrated in Figure 3-6 for an object at a temperature of 1500 K, or $kT \approx 0.13$ eV. The energy distributions are shown for oscillators of four different frequencies corresponding to hf equal to $kT/10$, $kT/2$, kT, and $2kT$. When the spacing between levels is small compared to kT, the average energy per oscillator is kT. When the spacing between levels is large, the average energy goes to zero.

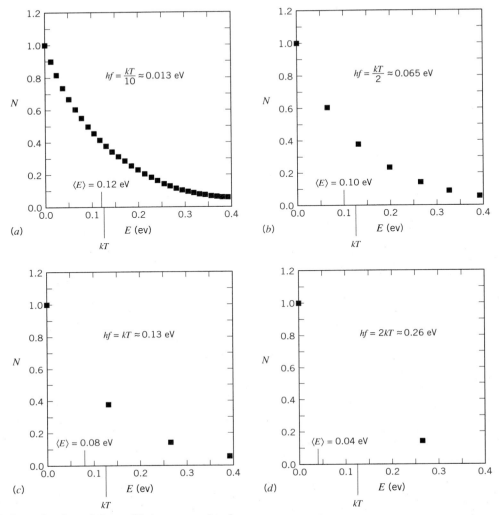

FIGURE 3-6 Quantization of the oscillator energy levels.
For an object at $T = 1500\,K$, we plot the relative number (N) of oscillators having an energy E. The spacing of the oscillator levels depends on the frequency of oscillation and is given by the condition $\Delta E = hf$. The average energy per oscillator depends on the frequency of the oscillator: (a) $hf = kT/10$, (b) $hf = kT/2$, (c) $hf = kT$, (d) $hf = 2kT$.

Planck observed that quantization of the energy levels would provide a cure for the ultraviolet catastrophe and produce a radiation spectrum that agreed with experiment. Planck did not realize, however, *why* the energy levels were quantized. This understanding came many years after Planck first made his brilliant deduction. We now know that the oscillator energy levels are quantized because of the *wave nature* of the electrons in atoms. This is the same reason that the atom does not collapse. These topics are the subject of Chapter 5.

We turn now to the calculation of the average energy per oscillator. The distribution function for the oscillators is discrete:

$$f_n = Ce^{-E_n/kT} = Ce^{-nhc/\lambda kT}, \qquad (3.24)$$

where n is a nonnegative integer. The average energy per oscillator is

$$\langle E \rangle = \frac{\sum_{n=0}^{\infty} E_n f_n}{\sum_{n=0}^{\infty} f_n} = \frac{\sum_{n=0}^{\infty} \left(\frac{nhc}{\lambda} \right) e^{-nhc/\lambda kT}}{\sum_{n=0}^{\infty} e^{-nhc/\lambda kT}}. \qquad (3.25)$$

To calculate the average energy per oscillator, we make the change of variables:

$$x \equiv \frac{hc}{\lambda kT}, \qquad (3.26)$$

and

$$y \equiv e^{-x}. \qquad (3.27)$$

The denominator of the expression for average energy (3.25) is

$$\sum_{n=0}^{\infty} e^{-nhc/\lambda kT} = \sum_{n=0}^{\infty} e^{-nx}$$

$$= 1 + y + y^2 + y^3 + \ldots = \frac{1}{1-y}. \quad (3.28)$$

To see that the last equality of the sum (3.28) holds, multiply each side by $1 - y$. To evaluate the sum in the numerator of the expression for average energy (3.25), we note that when we take the derivative of the exponential e^{-nx}, we produce a factor of $-n$:

$$\frac{d}{dx}\left(e^{-nx} \right) = -ne^{-nx}. \qquad (3.29)$$

Therefore, the numerator may be written

$$\sum_{n=0}^{\infty} \left(\frac{nhc}{\lambda} \right) e^{-nhc/\lambda kT} = -\left(\frac{hc}{\lambda} \right) \frac{d}{dx} \sum_{n=0}^{\infty} e^{-nx}$$

$$= -\left(\frac{hc}{\lambda} \right) \frac{d}{dx} \left(\frac{1}{1-y} \right)$$

$$= -\left(\frac{hc}{\lambda} \right) \left[\frac{-y}{(1-y)^2} \right], \quad (3.30)$$

where we have used the result of the sum in the denominator (3.28). Combining these results gives the average energy per oscillator,

$$\langle E \rangle = \left(\frac{hc}{\lambda} \right) \left(\frac{y}{1-y} \right) = \frac{hc}{\lambda \left(e^{hc/\lambda kT} - 1 \right)}. \quad (3.31)$$

The radiated power per area per unit wavelength may be obtained from the Rayleigh-Jeans formula (3.19), by replacing kT with the average energy per oscillator (3.31):

$$\frac{dR}{d\lambda} = \frac{2\pi hc^2}{\lambda^5 \left(e^{hc/\lambda kT} - 1 \right)}. \qquad (3.32)$$

This is the thermal radiation formula discovered by Planck. Unlike the Rayleigh-Jeans formula (3.19), the *Planck*

formula (3.32) correctly predicts that the radiated power goes to zero for small wavelengths. The Planck formula quantitatively fits the measured radiation spectrum emitted from objects in thermal equilibrium at a temperature T. The thermal radiation formula deduced from the quantization hypothesis agrees with experiment because the atomic oscillator energy levels *are* quantized.

At large wavelengths,

$$\frac{hc}{\lambda} \ll kT, \qquad (3.33)$$

and we may expand the exponential,

$$e^{hc/\lambda kT} \approx 1 + \frac{hc}{\lambda kT}. \qquad (3.34)$$

The Planck formula (3.32) becomes

$$\frac{dR}{d\lambda} = \frac{2\pi ckT}{\lambda^4}. \qquad (3.35)$$

We have recovered the Rayleigh-Jeans formula (3.19). We demand this to be so, because at large wavelengths, we know that the radiated power is inversely proportional to the fourth power of the wavelength. This was a successful prediction of classical electrodynamics, and we should not like to give this up!

The thermal spectrum (3.32), power per unit area per unit wavelength, is plotted in Figure 3-7 for three temperatures, 300 K, 1500 K, and 5900 K. The shape of the spectrum, which has a broad maximum, is identical for each temperature. The area under each curve varies with the fourth power of the temperature. At 300 K, the object is radiating but we cannot see this radiation with our eyes because the intensity in the visible range (400 nm to 700 nm) is so low. If the object is at a temperature of 1500 K, it glows red because there is enough intensity just below 700 nm for the object to be visible. The width of the spectrum is larger than the wavelength range of light, so that for an object of 5900 K, all visible wavelengths are radiated with comparable intensities. The object at a temperature of 5900 K glows "white."

All objects at a temperature T continually emit and absorb radiation. The power per area per unit wavelength spectrum is universal, depending only on the temperature of the object:

$$\frac{dR}{d\lambda} = \frac{2\pi hc^2}{\lambda^5 \left(e^{hc/\lambda kT} - 1 \right)}.$$

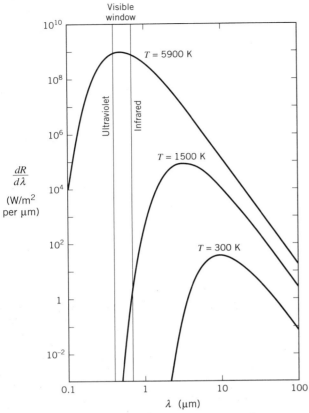

FIGURE 3-7 Radiated power per area per unit wavelength from an object at three different temperatures, 300 K, 1500 K and 5900 K.
Note that the data are presented on a logarithmic scale on both axes. The shape of the spectrum is exactly the same for each temperature. The area under each curve varies with the fourth power of the temperature. The visible window from 400 to 700 nm is indicated. The radiation from the 300 K object is not visible, the 1500 K object is visible just in the red, and the 5900 K object radiates in the entire visible region; it is white hot.

Numerical Value of Planck's Constant

The value of Planck's constant was first determined from the measured thermal radiation spectra. The original calculations of Planck determined h to a precision of 1%. Planck's constant is currently known to about six significant figures and has the approximate value of

$$h = 6.63 \times 10^{-34} \text{ J} \cdot \text{s} = 4.14 \times 10^{-15} \text{ eV} \cdot \text{s}. \quad (3.36)$$

The combination of Planck's constant times the speed of light occurs so often in practical calculations that is well worth the effort to remember its value:

$$hc = 1240 \text{ eV} \cdot \text{nm}. \quad (3.37)$$

The quantity $h/2\pi$ also appears so often that a special symbol called h-bar ($\hbar$) has been defined as

$$\hbar \equiv \frac{h}{2\pi} = 1.05 \times 10^{-34} \text{ J} \cdot \text{s}$$
$$= 6.58 \times 10^{-16} \text{ eV} \cdot \text{s}. \quad (3.38)$$

The value of $\hbar c$ is

$$\hbar c = 197 \text{ eV} \cdot \text{nm}. \quad (3.39)$$

$$hc = 1240 \text{ eV} \cdot \text{m},$$
$$\hbar c = 197 \text{ eV} \cdot \text{nm}.$$

EXAMPLE 3-4
What is the frequency of radiation from an atomic oscillator that has an energy spacing of 1 eV? Can you see the radiation from this oscillator?

SOLUTION:
The Planck quantization condition (3.21) is

$$E_n = nhf.$$

The energy spacing (ΔE) is

$$\Delta E = E_{n+1} - E_n = (n+1)hf - nhf = hf.$$

The frequency of the radiation is

$$f = \frac{\Delta E}{h} = \frac{1 \text{ eV}}{4.14 \times 10^{-15} \text{ eV} \cdot \text{s}} \approx 2.4 \times 10^{14} \text{ s}^{-1}.$$

The wavelength of the radiation is

$$\lambda = \frac{c}{f} = \frac{3 \times 10^8 \text{ m/s}}{2.4 \times 10^{14} \text{ s}^{-1}} = 1.2 \times 10^{-6} \text{ m}.$$

This radiation is in the infrared region (see Table 1-2). ■

The Stefan-Boltzmann Law

The total power per area radiated can be obtained by integrating the thermal radiation formula (3.32) over all wavelengths:

$$R = \int_0^\infty d\lambda \frac{dR}{d\lambda} = 2\pi hc^2 \int_0^\infty d\lambda \frac{1}{\lambda^5 \left(e^{hc/\lambda kT} - 1 \right)}. \quad (3.40)$$

With the change of variables

$$x \equiv \frac{hc}{\lambda kT}, \qquad (3.41)$$

the total power per area (3.40) becomes

$$R = \frac{2\pi(kT)^4}{h^3c^2} \int_0^\infty dx \frac{x^3}{e^x - 1}. \qquad (3.42)$$

The value of the definite integral is $\pi^4/15$. The total power per area radiated is

$$R = \frac{2\pi^5(kT)^4}{15h^3c^2} = \sigma'(kT)^4, \qquad (3.43)$$

where

$$\sigma' \equiv \frac{2\pi^5}{15h^3c^2} = 1.03 \times 10^9 \text{ W} \cdot \text{m}^{-2} \cdot \text{eV}^{-4}. \quad (3.44)$$

The thermal radiation formula contains the important and basic result that the power per area radiated by an object depends on the fourth power of the temperature, and not on any other property of the object. This result is known as the *Stefan-Boltzmann law*. The Stefan-Boltzmann law is commonly written

$$R = \sigma T^4, \qquad (3.45)$$

where

$$\sigma \equiv \frac{2\pi^5 k^4}{15h^3c^2} = 5.67 \times 10^{-8} \text{ W} \cdot \text{m}^{-2} \cdot \text{K}^{-4}. \quad (3.46)$$

> The total power per area radiated by an object is proportional to the fourth power of the temperature.

EXAMPLE 3-5

Estimate the power radiated at room temperature from an object that has a surface area of one meter squared.

SOLUTION:

The power radiated by an object of area A is

$$P = RA = \sigma'(kT)^4 A$$

$$= (1.03 \times 10^9 \text{ W} \cdot \text{m}^{-2} \cdot \text{eV}^{-4})(1\,\text{m}^2)\left(\frac{1}{40}\text{eV}\right)^4$$

$$\approx 400 \text{ W}.$$

A gigantic amount of energy (compared to the typical kinetic energy of a particle) is being exchanged per

second. The object absorbs the same amount of energy per unit time as it radiates provided the temperature is not changing. ∎

EXAMPLE 3-6

A physicist designs a circuit to operate in a room temperature environment in which a 100-Ω resistor carries a current of 1 mA. Estimate the minimum size of the resistor if the temperature is not to exceed 400 K.

SOLUTION:

The power appearing as heat in the resistor is the current squared times the resistance:

$$P = (10^{-3} \text{ A})^2 (100 \ \Omega) = 0.1\,\text{W}.$$

The net power per area radiated by the resistor is the difference between power radiated at the resistor temperature (T_{res}) and power absorbed at room temperature (T_{room}):

$$\frac{P}{A} = \sigma'(kT_{\text{res}})^4 - \sigma'(kT_{\text{room}})^4,$$

where A is the surface area of the resistor. The value of kT_{res} is

$$kT_{\text{res}} = (0.026 \text{ eV})\left(\frac{400\,\text{K}}{300\,\text{K}}\right) \approx 0.035\,\text{eV}.$$

This surface area must be kept at least as large as

$$A = \frac{P}{\sigma'\left[(kT_{\text{res}})^4 - (kT_{\text{room}})^4\right]}$$

$$= \frac{0.1\,\text{W}}{(10^9 \text{ W} \cdot \text{m}^{-2} \cdot \text{eV}^{-4})\left[(0.035\,\text{eV})^4 - (0.026\,\text{eV})^4\right]}$$

$$\approx 10^{-4} \text{ m}^2. \qquad ∎$$

Wien's Law

We now calculate the wavelength (λ_{m}) at which the power per area per unit wavelength is a maximum. To get λ_{m}, we differentiate the thermal radiation formula (3.32) and set it equal to zero:

$$\left[\frac{d}{d\lambda}\frac{dR}{d\lambda}\right]_{\lambda = \lambda_{\text{m}}}$$

$$= 2\pi hc^2 \left[\frac{(-5)}{\lambda^6\left(e^{hc/\lambda kT} - 1\right)} + \frac{e^{hc/\lambda kT}\left(\dfrac{hc}{\lambda^2 kT}\right)}{\lambda^5\left(e^{hc/\lambda kT} - 1\right)^2}\right]_{\lambda = \lambda_{\text{m}}}$$

$$= 0. \qquad (3.47)$$

The spectrum has a maximum at

$$\frac{hc}{\lambda_m kT} = 5\left(1 - e^{-hc/\lambda_m kT}\right). \quad (3.48)$$

The expression for λ_m is a transcendental equation; we cannot solve it in closed form for λ_m. A numerical calculation gives

$$\frac{hc}{\lambda_m kT} = 4.97, \quad (3.49)$$

which we may express as

$$\lambda_m = \frac{hc}{4.97\,kT} = \frac{1240 \text{ eV} \cdot \text{nm}}{4.97\,kT} = \frac{250 \text{ eV} \cdot \text{nm}}{kT}. \quad (3.50)$$

The wavelength where $dR/d\lambda$ is a maximum is inversely proportional to the temperature. This result is known as *Wien's law*. The constant ($250 \text{ eV} \cdot \text{nm}$) that appears in Wien's law relates the wavelength where the radiated power per area per wavelength peaks, with kT of the radiating object. Wien's law is often written

$$\lambda_m = \frac{0.00290 \text{ m} \cdot \text{K}}{T}. \quad (3.51)$$

> The wavelength at which the the power per unit wavelength is maximum is inversely proportional to the temperature.

EXAMPLE 3-7
Near what wavelength is the most energy radiated from an object at room temperature?

SOLUTION:
From Wien's law (3.50) the wavelength near which the most energy is radiated is

$$\lambda_m \approx \frac{250 \text{ eV} \cdot \text{nm}}{kT} = \frac{250 \text{ eV} \cdot \text{nm}}{\left(\frac{1}{40}\text{ eV}\right)} \approx 10^4 \text{ nm} = 10^{-5} \text{ m}.$$

This is the middle of the infrared region. ∎

EXAMPLE 3-8
The eye is most sensitive to a wavelength of 500 nm. At what temperature would the power spectrum radiated from an object be a maximum at this wavelength?

SOLUTION:
The peak of the spectrum $dR/d\lambda$ is at a wavelength of

$$\lambda_m \approx 500 \text{ nm}.$$

Using Wien's law (3.50) this corresponds to

$$kT = \frac{250 \text{ eV} \cdot \text{nm}}{\lambda_m} = \frac{250 \text{ eV} \cdot \text{nm}}{500 \text{ nm}} \approx 0.50 \text{ eV}.$$

The temperature is

$$T = \left(\frac{0.5 \text{ eV}}{0.026 \text{ eV}}\right)(300 \text{ K}) \approx 5800 \text{ K}.$$

This is the approximate temperature on the surface of the sun. ∎

The Blackbody Formula in Terms of Frequency

To get the thermal radiation formula in terms of frequency, we make the change of variables

$$f = \frac{c}{\lambda}. \quad (3.52)$$

The thermal radiation formula (3.32) in terms of frequency is

$$\frac{dR}{df} = \frac{dR}{d\lambda}\frac{d\lambda}{df} = \frac{2\pi hf^3}{c^2\left(e^{hf/kT} - 1\right)}. \quad (3.53)$$

When the wavelength goes from zero to infinity, the frequency goes from infinity to zero, which accounts for the minus sign in the derivative $d\lambda/df$. In Figure 3-8, we make a comparison of $dR/d\lambda$ (3.32) with dR/df (3.53).

EXAMPLE 3-9
Consider an object at room temperature that has a surface area of one meter squared. (a) Use Figure 3-8a to make an estimate of the amount of power radiated in a wavelength interval from 10 to 11 microns. (b) Make the same estimate from Figure 3-8b.

SOLUTION:
(a) The power radiated is

$$P = \frac{dR}{d\lambda}\Delta\lambda A,$$

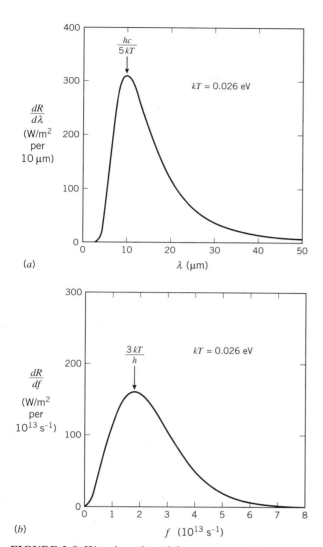

FIGURE 3-8 Wavelength and frequency spectra.
(*a*) Radiated power per area per wavelength ($dR/d\lambda$) for an object at $T = 300\,K$. The distribution $dR/d\lambda$ peaks at a wavelength of about $hc/5kT$. The area under the curve gives the total power per area radiated. (*b*) Radiated power per area per frequency (dR/df) for an object at T = 300 K. The distribution dR/df peaks at a frequency of about $3kT/h$.

where A is the surface area of the object and $\Delta\lambda$ is the wavelength interval. At a wavelength of 10 microns, the value of $dR/d\lambda$ from Figure 3-8*a* is

$$\frac{dR}{d\lambda} \approx \frac{300\ \text{W/m}^2}{10\ \mu\text{m}}.$$

Therefore,

$$P = \frac{dR}{d\lambda}\Delta\lambda A \approx \left(\frac{300\ \text{W/m}^2}{10\ \mu\text{m}}\right)\!\left(1\ \mu\text{m}\right)\!\left(1\ \text{m}^2\right) \approx 30\ \text{W}.$$

(b) The frequency interval (Δf) corresponding to

$$\Delta\lambda = 1\ \mu\text{m},$$

at a wavelength of 10 μm is

$$\Delta f = \frac{c\Delta\lambda}{\lambda^2} = \frac{\left(3\times10^8\ \text{m/s}\right)\!\left(10^{-6}\ \text{m}\right)}{\left(10^{-5}\ \text{m}\right)^2} = 3\times10^{12}\ \text{s}^{-1}.$$

The frequency that corresponds to a wavelength of 10 μm is

$$f = \frac{c}{\lambda} = \frac{3\times10^8\ \text{m/s}}{10^{-5}\ \text{m}\cdot} = 3\times10^{13}\ \text{s}^{-1}.$$

The value of dR/df at this frequency from Figure 3-8*b* is

$$\frac{dR}{df} \approx \frac{100\ \text{W/m}^2}{10^{13}\ \text{s}^{-1}}.$$

The radiated power is

$$P = \frac{dR}{df}\Delta f A$$

$$\approx \left(\frac{100\ \text{W/m}^2}{10^{13}\ \text{s}^{-1}}\right)\!\left(3\times10^{12}\ \text{s}^{-1}\right)\!\left(1\ \text{m}^2\right)$$

$$\approx 30\ \text{W}. \qquad\blacksquare$$

Radiation from the Sun

The rate at which energy from the sun reaches a unit area on the earth is called the *solar constant* (S). The solar constant is measured to be

$$S = 1350\ \text{W/m}^2. \qquad (3.54)$$

(The measurement error on the solar constant is about 2%.) The total power output of the sun, or *solar luminosity* (L_s), is calculated from the solar constant and the distance to the sun (D):

$$L_s = 4\pi D^2 S. \qquad (3.55)$$

The distance to the sun is measured to be 1.50×10^{11} m, which gives

$$L_s = 4\pi D^2 S$$
$$= (4\pi)(1.50 \times 10^{11} \text{ m})^2 (1350 \text{ W/m}^2)$$
$$= 3.83 \times 10^{26} \text{ W}. \tag{3.56}$$

If we make the simple assumption that the sun is an ideal thermal radiator, then we may use the Stefan-Boltzmann law to calculate the surface temperature of the sun. The total power per area (R) radiated by the sun is the luminosity (3.56) divided by the surface area of the sun ($4\pi r_s^2$):

$$R = \frac{L_s}{4\pi r_s^2} = S\left(\frac{D}{r_s}\right)^2. \tag{3.57}$$

Taking the solar radius to be 6.96×10^8 m, we have

$$R = (1350 \text{ W/m}^2)\left(\frac{1.50 \times 10^{11} \text{ m}}{6.96 \times 10^8 \text{ m}}\right)^2$$
$$= 6.27 \times 10^7 \text{ W/m}^2. \tag{3.58}$$

The temperature calculated from the Stefan-Boltzmann law (3.43) is

$$kT = \left(\frac{R}{\sigma'}\right)^{1/4}$$
$$= \left(\frac{6.27 \times 10^7 \text{ W} \cdot \text{m}^{-2}}{1.03 \times 10^9 \text{ W} \cdot \text{m}^{-2} \cdot \text{eV}^{-4}}\right)^{1/4}$$
$$\approx 0.50 \text{ eV}, \tag{3.59}$$

or

$$T = \left(\frac{0.50 \text{ eV}}{0.026 \text{ eV}}\right)(300 \text{ K}) \approx 5800 \text{ K}. \tag{3.60}$$

Since the energy is radiated from the surface of the sun, the temperature that we have just estimated is the *surface temperature* of the sun. The interior temperature of the sun is much greater.

The radiation that reaches the earth comes from within 500 km of the solar surface. (Note that 500 km is much smaller than the radius of the sun.) In this region the temperature varies from about 4300 K to 6600 K. Figure 3-9 shows the solar spectrum measured from a research aircraft. The solar spectrum is approximately that of a blackbody with a temperature near 5800 K. The radiation that reaches the earth's surface does not have a smooth spectrum due to absorption in the earth's atmosphere,

primarily by water molecules. As we shall see in some detail in Chapter 10, water is particularly effective at absorbing radiation *except in the visible region.*

The wavelength at which the solar spectrum has maximum power is given by Wien's law (3.50). Using the calculated value of kT (3.59) we have

$$\lambda_m = \frac{250 \text{ eV} \cdot \text{nm}}{kT}$$
$$\approx \frac{250 \text{ eV} \cdot \text{nm}}{0.50 \text{ eV}} = 500 \text{ nm}. \tag{3.61}$$

The solar spectrum peaks in the visible region. Our eyes have adapted to the sun through evolution!

EXAMPLE 3-10
Estimate the temperature of our planet assuming that the earth is an ideal blackbody absorbing its energy from the sun.

SOLUTION:
Let the radius of the earth be r_e. The power absorbed from the sun by the earth is

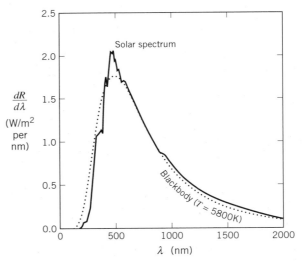

FIGURE 3-9 The solar spectrum.
The intensity of solar radiation ($dR/d\lambda$ in watts per meter squared per nanometer) as measured from a research aircraft is shown (solid line) compared to the blackbody formula at a temperature of about 5800 K (dashed line). Data from "The Solar Constant and Solar Spectum Measured from a Research Aircraft," NASA Technical Report R-351, ed. M.P. Thekaekara, Goddard Space Flight Center (reproduced in the CRC Handbook of Chemistry and Physics.)

$$P = \pi r_e^2 S.$$

This energy is reradiated from the entire surface area of the earth. The power per area (R) radiated is

$$R = \frac{P}{4\pi r_e^2} = \frac{S}{4}.$$

From the Stefan-Boltzmann law, the temperature of the earth is given by

$$kT = \left(\frac{R}{\sigma'}\right)^{1/4} = \left(\frac{S}{4\sigma'}\right)^{1/4}$$

$$= \left[\frac{1350 \text{ W} \cdot \text{m}^{-2}}{(4)(1.03 \times 10^9 \text{ W} \cdot \text{m}^{-2} \cdot \text{eV}^{-4})}\right]^{1/4}$$

$$= 0.024 \text{ eV},$$

or

$$T = \left(\frac{0.024 \text{ eV}}{0.0259 \text{ eV}}\right)(300 \text{ K}) \approx 280 \text{ K}. \quad \blacksquare$$

We have made a good approximation of the temperature of the earth! The measured average temperature of the earth is slightly warmer (about 290 K), indicating that something other than the sun is warming the earth. The additional energy is provided by natural radioactive processes that convert mass energy to kinetic energy.

EXAMPLE 3-11

Suppose that the sun was formed as a hot ball of hydrogen at some very high temperature. Estimate the time for the sun to cool to 6000 K if there were no internal energy source.

SOLUTION:

The power radiated by the sun is given by the Stefan-Boltzmann law:

$$R = -\frac{dE_k}{dt} = 4\pi r_s^2 \sigma' (kT)^4,$$

where r_s is the radius of the sun. In this expression R depends on time because the temperature T depends on time. The total kinetic energy of particles in the sun is proportional to temperature:

$$E_k \approx \frac{3}{2} NkT,$$

where N is the total number of particles in the sun. Using

$$dE_k \approx \frac{3}{2} Nk \, dT,$$

we can write the Stefan-Boltzmann law as

$$dt \approx -\frac{\left(\frac{3}{2} Nk \, dT\right)}{4\pi r_s^2 \sigma' (kT)^4}.$$

The length of time (Δt) for the sun to cool from some very hot initial temperature (T_i) to $T_f = 6000$ K is obtained by integration:

$$\Delta t \approx -\frac{3N}{8\pi r_s^2 \sigma' k^3} \int_{T_i}^{T_f} dT \frac{1}{T^4}$$

$$= -\frac{3N}{8\pi r_s^2 \sigma' k^3} \left[-\frac{1}{3T^3}\right]_{T=T_i}^{T=T_f}$$

$$\approx \frac{N}{8\pi r_s^2 \sigma' (kT_f)^3}.$$

The initial temperature of the sun has a negligible effect on the cooling time provided $T_i \gg T_f$. We estimate the total number of protons in the sun to be the mass of the sun (M_s) divided by the mass of the proton:

$$N \approx \frac{M_s}{m_p},$$

which gives

$$\Delta t \approx \frac{M_s c^2}{8\pi r_s^2 m_p c^2 \sigma' (kT_f)^3}$$

$$\approx \frac{(2 \times 10^{30} \text{ kg})(3 \times 10^8 \text{ m/s})^2}{(8)(\pi)(7 \times 10^8 \text{ m})^2 (10^9 \text{ eV})}$$

$$\times (10^9 \text{ W} \cdot \text{m}^{-2} \cdot \text{eV}^{-4})(0.5 \text{ eV})^3$$

$$\approx 10^{11} \text{ s},$$

or a mere 3000 years. The actual age of the sun is six orders of magnitude greater than our estimated radiation cooling time, which assumes no internal energy source. The sun is heated by proton fusion, which converts matter into kinetic energy. This is discussed further in Chapter 11. $\quad \blacksquare$

Radiation from the Early Universe

The universe is presently expanding and cooling. The observational evidence is overwhelming that the expansion has been occurring for about 10 billion years. In the very hot, early stages of the expansion, the energy and density of radiation were so large they prevented the formation of atoms ($kT \gg 1$ eV). The radiation coupled to the free charged particles in the early universe and remained in thermal equilibrium. At a few times 10^5 years after the start of this expansion, the temperature of the universe had cooled to 3000 K ($kT \approx 1/4$ eV) and hydrogen atoms were formed. With the formation of electrically neutral atoms, the free charges that the photons couple to were greatly diminished. At that time the photons had a thermal distribution with a temperature of 3000 K. Since then, the universe has expanded by a factor of 1000 and cooled further. The radiation that today fills up the entire universe has a thermal spectrum with a temperature near 3 K. From the Wien displacement law (3.50), the energy density of these *cosmic photons* is maximum at a wavelength of

$$\lambda_m = \frac{250 \text{ eV} \cdot \text{nm}}{kT}$$

$$\approx \frac{250 \text{ eV} \cdot \text{nm}}{\left(\frac{1}{40} \text{eV}\right)\left(\frac{3 \text{ K}}{300 \text{ K}}\right)}$$

$$= 10^6 \text{ nm} = 0.001 \text{ m} . \qquad (3.62)$$

This is in the microwave region (see Table 1-2).

The cosmic microwave radiation was discovered in 1965 by Arnio Penzias and Robert W. Wilson while making measurements of extraterrestrial sources of radio waves. Penzias and Wilson measured the radiation energy density at a wavelength of 0.0735 meters. They observed a nonzero value for the microwave radiation in every direction in space. In 1990, the Cosmic Background Explorer (COBE) satellite measured the background radiation spectrum with great precision. The observed cosmic spectrum fits a thermal (blackbody) distribution at a temperature of 2.74 K, nearly perfectly.

The energy density of the cosmic radiation is 2.7×10^5 eV per cubic meter, roughly the same as the average energy density of starlight in our galaxy. This corresponds to an average photon energy of 6.4×10^{-4} eV, and a photon density of 4.2×10^8 per cubic meter. Cosmic photons are everywhere! The cosmic radiation from the early universe is discussed further in Chapter 19.

3-3 QUANTIZATION OF ELECTROMAGNETIC RADIATION

Photoelectric Effect

When electromagnetic radiation is incident on the surface of a metal, it is observed that electrons may be ejected. This phenomenon is called the *photoelectric effect*. The photoelectric effect was first observed in the experiments of Heinrich Hertz in 1887. Subsequently, J. J. Thomson proved that electrons were emitted. In 1902, Philip Lenard observed that the maximum kinetic energy of the electrons does not depend on the intensity of the incident radiation. The kinetic energy of the electrons was measured by determining the voltage required to stop them. Lenard also observed that the maximum kinetic energy increased with frequency, but his data were not accurate enough to measure the functional form.

We have seen that the spectrum of radiation in equilibrium with matter was explained by the quantization of the oscillator levels in the matter. In 1905, Albert Einstein deduced that electromagnetic radiation is also quantized. A quantum of electromagnetic radiation is called a photon (γ). A photon is a particle that carries energy and momentum but has zero mass. Einstein deduced that when a photon of frequency f is absorbed by matter, this occurs together with one of the oscillators changing energy levels. By conservation of energy, the photon energy is given by the difference in energy levels ($E_{n+1} - E_n$) of the oscillator:

$$E_{\text{photon}} = E_{n+1} - E_n = (n+1)hf - nhf$$

$$= hf = \frac{hc}{\lambda}, \qquad (3.63)$$

where h is Planck's constant, the same constant that is deduced from the thermal radiation spectrum (3.32).

EXAMPLE 3-12
Calculate the energy range of visible photons.

SOLUTION:
The range of wavelengths of visible photons is

$$400 \text{ nm} < \lambda < 700 \text{ nm} .$$

The energy of a photon with $\lambda = 400$ nm is

$$E = \frac{hc}{\lambda} = \frac{1240 \text{ eV} \cdot \text{nm}}{400 \text{ nm}} \approx 3.1 \text{ eV} .$$

The energy of a photon with $\lambda = 700$ nm is

$$E = \frac{hc}{\lambda} = \frac{1240\,\text{eV} \cdot \text{nm}}{700\,\text{nm}} \approx 1.8\,\text{eV}.$$

The energy range of visible photons is from 1.8 to 3.1 eV. ∎

> Electromagnetic radiation is quantized. The energy of a single quantum of radiation (E_{photon}) of frequency f is equal to Planck's constant times the frequency,
>
> $$E_{\text{photon}} = hf = \frac{hc}{\lambda}.$$

The quantization of energy in radiation explains the photoelectric effect. The process is

$$\gamma + (e^- \text{ in atom}) \rightarrow e^- + (\text{atom minus } e^-) \quad (3.64)$$

(see Figure 3-10). In a metal the outer electrons are free to move from atom to atom. The electrons behave like a gas with a continuous spectrum of energy levels. The electron absorbs a photon, and if the photon energy is sufficiently large, the electron gains enough energy to be freed from the metal. The photon has disappeared; it has given all its energy to the electron. For an incident photon with a given energy, an electron emerges with a maximum kinetic energy if it happens to reside in the outermost energy level. If an electron in a slightly lower energy level absorbs the photon, then the electron emerges with less kinetic energy. The maximum kinetic energy of the ejected electrons is

$$E_k^{\text{max}} = hf - \phi, \quad (3.65)$$

where ϕ is a property of the metal called the *work function*. The work function is the amount of energy by which an outermost electron is bound in the metal. (Inner electrons have a larger binding energy.) The typical size of ϕ is a few electronvolts. The work functions of some selected materials are given in Table 3-2.

EXAMPLE 3-13

Electromagnetic radiation is incident on a sheet of lead. What is the maximum wavelength of radiation that can cause electrons to be ejected from the lead?

SOLUTION:

Electrons can be ejected by the photoelectric effect if the photon energy (E) is greater than the work function of the material. This condition is

$$\frac{hc}{\lambda} > \phi.$$

TABLE 3-2
WORK FUNCTIONS OF SELECTED ELEMENTS DETERMINED FROM THE PHOTOELECTRIC EFFECT.

Element	Work Function
Aluminum	4.3 eV
Beryllium	5.0 eV
Calcium	2.9 eV
Cesium	2.1 eV
Cobalt	5.0 eV
Copper	4.7 eV
Gold	5.1 eV
Iron	4.5 eV
Lead	4.3 eV
Magnesium	3.7 eV
Mercury	4.5 eV
Nickel	5.2 eV
Niobium	4.3 eV
Potassium	2.3 eV
Platinum	5.7 eV
Silver	4.3 eV
Sodium	2.8 eV
Uranium	3.6 eV
Zinc	4.3 eV

From the *Handbook of Chemistry and Physics*.

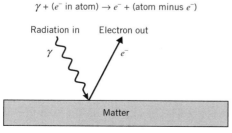

$\gamma + (e^- \text{ in atom}) \rightarrow e^- + (\text{atom minus } e^-)$

Radiation in   Electron out

γ e^-

Matter

FIGURE 3-10 The photoelectric effect.

This means that

$$\lambda < \frac{hc}{\phi}.$$

The work function of lead (see Table 3-2) is

$$\phi = 4.3\,\text{eV}.$$

The maximum photon wavelength is

$$\lambda = \frac{hc}{\phi} = \frac{1240 \ \text{eV} \cdot \text{nm}}{4.3\,\text{eV}} \approx 290\,\text{nm}. \qquad \blacksquare$$

The maximum kinetic energy of electrons from the photoelectric effect can be measured with the simple apparatus depicted in Figure 3-11a. Two parallel plates are connected to a voltage source. One of these plates, called the cathode or *photocathode*, is made of the material to be studied. The other plate is called the anode. Electromagnetic radiation is directed onto the cathode. If the frequency of the electromagnetic radiation is sufficiently high, electrons are ejected from the cathode by the photoelectric effect. If the battery is connected so that the anode is more positive than the cathode, the electrons will be accelerated toward the anode and a current will be observed in the circuit. (The convention for the direction of the current is opposite to the direction the electrons move.) If the anode is made more negative than the cathode, the acceleration of the electrons will be negative. At some negative voltage $(-V_s)$, no electrons will reach the anode because the electric field is sufficiently strong to pull the electrons back to the cathode. The most energetic electrons will be stopped just short of the anode; in the acceleration process these electrons have lost an energy eV_s. The current as a function of the voltage is sketched in Figure 3-11b. A measurement of the voltage that stops all electrons, called the *stopping potential*, determines the maximum kinetic energy of the ejected electrons:

$$eV_s = E_k^{\text{max}}. \qquad (3.66)$$

Using the expression for the maximum kinetic energy of the electron (3.65), we have

$$eV_s = hf - \phi. \qquad (3.67)$$

EXAMPLE 3-14

Ultraviolet radiation with a wavelength of 200 nm is incident on an aluminum cathode. How much voltage is need to stop the electrons?

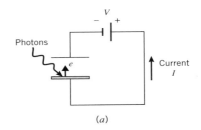

(a)

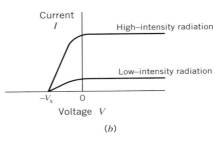

(b)

FIGURE 3-11 Measuring the electron maximum kinetic energy from the photoelectric effect.
Radiation of sufficiently high frequency $(f > \phi/h)$ causes electrons to be ejected from the cathode and a current is observed in the circuit. (a) Circuit used to measure current and voltage. (b) Current versus voltage.

SOLUTION:

The work function of aluminum (see Table 3-2) is about

$$\phi = 4.3\,\text{eV}.$$

The energy of one photon (E) is

$$E = \frac{hc}{\lambda}.$$

The maximum kinetic energy of an ejected electron is

$$E_k^{\text{max}} = E - \phi = \frac{hc}{\lambda} - \phi.$$

The stopping voltage is

$$V_s = \frac{E_k^{\text{max}}}{e} = \left(\frac{1}{e}\right)\left(\frac{hc}{\lambda} - \phi\right)$$

$$= \left(\frac{1}{e}\right)\left(\frac{1240 \ \text{eV} \cdot \text{nm}}{200\,\text{nm}} - 4.3\,\text{eV}\right) \approx 1.9\,\text{V}. \qquad \blacksquare$$

Now consider what happens when the intensity of the radiation is varied. The intensity of the radiation is defined to be the energy per time per area transported by the electromagnetic waves. The units of intensity are watts per meter-squared. The experimental results for two different radiation intensities of the same frequency ($f > \phi/h$) is sketched in Figure 3-11b. When the intensity is larger, more electrons are ejected and the measured current is larger. The same negative voltage ($-V_s$) will stop the current, however, independent of the radiation intensity. This means that the maximum kinetic energy of the electrons does not depend on the radiation intensity provided that the frequency is fixed. A higher intensity radiation field at a fixed frequency is made up of a larger number of photons all of which have an identical energy hf. The photoelectric effect is a quantum phenomenon; a single photon is absorbed by a single electron. A more intense radiation source increases the rate at which electrons are ejected from the metal, provided that the frequency is above threshold, but the intensity does not effect the kinetic energy of the ejected electrons.

The prediction of Einstein (3.67) was first accurately verified by Millikan in 1916 (his second great experiment!). Millikan recognized that it was easiest to work with the alkali metals like sodium because it was easiest to remove their outer electrons. At the time of Millikan's work, the exact values of the work functions were not known. Millikan took great care in the preparation of the electrodes so that the results would not be spoiled by surface effects such as dirt on the metal. Millikan made a plot of stopping voltage versus frequency of incident radiation. Data of Millikan (Figure 3-12) demonstrated that the maximum kinetic energy of the electrons is directly proportional to the frequency. A measurement of the slope gave a determination of h/e. This proved that the constant h relating the frequency and energy of the photon is the same constant that appears in the thermal radiation formula of Planck (3.32).

The actual absorption of the photon occurs instantaneously in the photoelectric process. There is no time lag as expected classically if the intensity of the electromagnetic radiation is low. Consider electromagnetic radiation with intensity I incident on a metal. Since the size of an atom is small, the power absorbed per atom is small when the intensity is low. Classically, the photoelectric effect could not occur until the atom had enough time (Δt) to absorb enough energy to overcome the work function (ϕ).

EXAMPLE 3-15
Radiation with an intensity of 1 mW per square meter is incident on a metal that has a work function of 4 eV. Assuming a uniform and continuous distribution of radia-

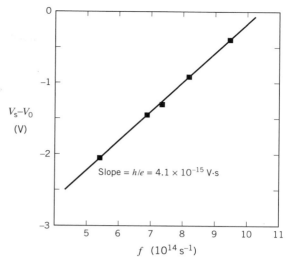

FIGURE 3-12 Photoelectric effect data of Millikan. Photons of varying frequencies are incident on sodium, and the relative voltage needed to stop all electrons from reaching a second electrode is measured. The data prove that the maximum kinetic energy of the electrons is directly proportional to the frequency of the incident photon, as predicted by Einstein. The slope of these data are a direct measurement of h/e. With these data, Millikan determined Planck's constant to an accuracy of 0.3%. Data from R.A. Millikan, *The Electron,* University of Chicago Press (1917).

tion, estimate the time needed for an atom to absorb enough energy to eject an electron.

SOLUTION:
The diameter of an atom is about $d = 0.3$ nm. The energy per time (P) incident on a single atom is

$$P = I d^2$$
$$= \left(10^{-3} \text{ J} \cdot \text{s}^{-1} \cdot \text{m}^{-2}\right)\left(3 \times 10^{-10} \text{ m}^2\right)^2$$
$$\times \left(\frac{1 \text{ eV}}{1.6 \times 10^{-19} \text{ J}}\right)$$
$$\approx 6 \times 10^{-4} \text{ eV/s}.$$

The time taken for an energy of 4 eV to be incident on the atom would be

$$\Delta t = \frac{\phi}{P} \approx \frac{4 \text{ eV}}{6 \times 10^{-4} \text{ eV/s}} \approx 7 \times 10^3 \text{ s},$$

or nearly two hours! Experiments show that electrons are ejected instantaneously, provided that the frequency is above threshold ($f > h/\phi$). There is no time lag. If the radiation frequency is below threshold ($f < h/\phi$), then electrons are never ejected by the photoelectric effect, no matter how long one waits. ∎

Numerical Value of α

We can now appreciate the numerical value of the dimensionless electromagnetic coupling constant α (1.69). Recall that α is defined to be $2\pi ke^2$ divided by the energy (E_{photon}) times the wavelength (λ_{photon}) of a photon. Since

$$E_{\text{photon}} \lambda_{\text{photon}} = \frac{hc}{\lambda_{\text{photon}}} \lambda_{\text{photon}} = hc, \quad (3.68)$$

we have

$$\alpha = \frac{2\pi ke^2}{hc} = \frac{ke^2}{\hbar c} \approx \frac{1.44\,\text{eV} \cdot \text{nm}}{197\,\text{eV} \cdot \text{nm}} \approx \frac{1}{137}. \quad (3.69)$$

Photomultipliers

A photomultiplier (or phototube) is an electronic photon detector whose basic principle of operation is based on the photoelectric effect. A schematic diagram of a phototube is shown in Figure 3-13. The device is contained in an evacuated glass tube. At the front of the tube is a thin quartz window with a photosensitive layer deposited on it, the *photocathode*. The photocathode is usually made of a semiconducting alloy of sodium, potassium, or cesium. Note that these elements have the smallest work functions (see Table 3-2). When a photon of sufficient energy is incident on the photocathode, a single electron may be ejected. This ejected electron is referred to as a *photoelectron*. The probability that a single photon produces a photoelectron that is detected in the phototube is called the *quantum efficiency*. The quantum efficiency is a function of the wavelength of the incident photon. A typical value of the quantum efficiency in a good phototube is about 25% at a wavelength of 400 nm.

The photoelectron is accelerated through a potential difference of about 150 V in the first stage of the photomultiplier. The accelerated electron, which has a kinetic energy of 150 eV, then hits a secondary electrode called a *dynode,* where it frees other electrons by its electromagnetic interaction. These secondary electrons are accelerated and focused onto a second dynode, where the process is repeated, causing a multiplication of electrons. A photomultiplier will typically have 10 or 12 stages, and a total multiplication or *gain* of 10^6 or larger may be achieved.

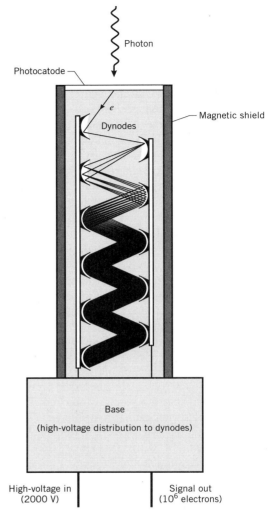

FIGURE 3-13 Schematic diagram of a photomultiplier. Photoelectrons are accelerated and multiplied by a series of dynodes, usually in 10 or 12 stages. Electromagnetic fields focus the electrons from dynode to dynode as they are multiplied. The total voltage for acceleration and multiplication is typically 2000 V and a typical multiplication factor is 10^6.

These 10^6 electrons are easily detected in an external electrical circuit. Photomultiplier tubes are sensitive to the presence of external magnetic fields because as the electrons travel from dynode to dynode, magnetic forces can distort the focusing. For this reason, the photomultiplier tubes are placed inside a *Mumetal* cylindrical shell that shields external magnetic fields. (Mumetal, or Muntz metal, is an alloy of nickel, iron, and copper with a high magnetic permeability.)

A photomultiplier is also a good detector of the time of arrival of the photon, having a resolution of about 1 ns. The time resolution of the phototube is governed by the fluctuations in the time of arrival of the photoelectron at the first dynode. Consider two cases, an electron at rest and an electron with kinetic energy of 1 eV both being accelerated in an electric field of 1.5×10^4 V/m to the first dynode. The difference in the time of arrival of the two electrons is equal to the time it takes to accelerate an electron at rest to a kinetic energy of 1 eV. This time difference is the speed of a 1 eV electron (Δv) divided by the acceleration (a) of the electron:

$$\Delta t = \frac{\Delta v}{a} = \frac{\sqrt{\frac{2E_0}{m}}}{\left(\frac{eE}{m}\right)}$$

$$= \frac{\sqrt{2mE_0}}{eE} = \frac{\sqrt{2mc^2 E_0}}{eEc}$$

$$\approx \frac{\sqrt{10^6 \text{ eV}^2}}{\left(1.5 \times 10^4 \text{ eV/m}\right)\left(3 \times 10^8 \text{ m/s}\right)}$$

$$\approx 0.2 \text{ ns.} \tag{3.70}$$

This time-jitter, the difference in time of arrival of photoelectrons at the first dynode with various starting kinetic energies, is the fundamental limiting factor in the time resolution of a photomultiplier tube. This time information may be used to measure coincidences of events in particle collisions and also to measure particle speeds by time-of-flight measurement.

Photomultipliers are sensitive detectors of single photons and as such are designed to operate in the dark. In our description of the photoelectric effect we have assumed there is a sharp cutoff to the maximum energy of an electron in a metal. This is true to a good approximation; however, a very tiny fraction of the electrons (a number proportional to $e^{-E/kT}$ where E is the electron energy) have a slightly larger energy than the assumed maximum and can be ejected by a photon of slightly smaller energy than our assumed threshold (ϕ). Therefore, a photomultiplier designed to detect photons below a specific wavelength will also be sensitive to photons of slightly longer wavelength but with greatly reduced efficiency. In the following example we make an estimate of the counting rate of a photomultiplier in a dark room.

EXAMPLE 3-16

The quantum efficiency of a certain photomultiplier is 10^{-2} at a wavelength of 1000 nm and rises exponentially with decreasing wavelength. Make a rough estimate of the counting rate in a dark room at 300 K if the area of the photocathode is 10^{-2} m^2.

SOLUTION:

We use the thermal radiation formula (3.32) to calculate the number of photons with wavelengths near 1000 nm incident on the photocathode at room temperature. The scale factor for the exponential term is

$$\frac{hc}{\lambda kT} = \frac{1240 \text{ eV} \cdot \text{nm}}{(1000 \text{ nm})(0.026 \text{ eV})} \approx 48,$$

which gives

$$e^{hc/\lambda kT} \approx e^{48} \approx 5 \times 10^{20}.$$

The exponential factor increases very rapidly with decreasing wavelength. A change in wavelength of 5% causes a change in the spectrum by about a factor of 10. At $\lambda = 1000$ nm, we have

$$\frac{dR}{d\lambda} = \frac{2\pi hc^2}{\lambda^5 \left(e^{hc/\lambda kT} - 1\right)}$$

$$\approx \frac{(2\pi)(1.24 \times 10^{-6} \text{ eV} \cdot \text{m})(3 \times 10^8 \text{ m/s})}{(10^{-6} \text{ m})^5 (5 \times 10^{20})}$$

$$\approx 10^{13} \text{ eV} \cdot \text{s}^{-1} \cdot \text{m}^{-3}.$$

The quantum efficiency (E) also increases rapidly with decreasing wavelength. Taking the wavelength interval to be $\Delta\lambda = 100$ nm, and using the fact that the energy of a photon with $\lambda = 1000$ nm is about 1 eV, the rate r_e that electrons are ejected per second is

$$r_e = \left(\frac{dR}{d\lambda}\right)(\Delta\lambda)(A)\left(\frac{1}{E_{photon}}\right)(\varepsilon)$$

$$\approx \left(10^{13} \text{ eV} \cdot \text{s}^{-1} \cdot \text{m}^{-3}\right)\left(10^{-7} \text{ m}\right)\left(10^{-2} \text{ m}^2\right)$$

$$\times \left(\frac{1}{1 \text{ eV}}\right)\left(10^{-2}\right)$$

$$= 10^2/\text{s}.$$

This is the typical order of magnitude of the background counting rate in a photomultiplier at room temperature.

The actual counting rate will depend on the exact form of the quantum efficiency as a function of wavelength.

■

3-4 ATOMIC SPECTRA AND THE BOHR MODEL

An object at a temperature T, for example a star at 6000 K, emits a continuous spectrum of radiation. The atomic oscillators in the star are so many and so varied, that they are capable of emitting radiation over a wide range of wavelengths.

Decades before Planck deduced the blackbody formula, atoms were known to emit discrete spectra. By discrete spectra we mean, in a specific atom, only very sharply defined wavelengths are emitted in certain portions of the spectrum. These wavelengths were first measured by passing light through a narrow slit and prism. Each wavelength emitted by the atom is called a *line* because it appears in such an experiment as an image of the narrow slit. Each element has its own unique set of lines, a sort of atomic fingerprint (see color plate 1). The experimental apparatus for line measurements has become quite refined.

From the solar spectrum, it was also observed that the intensity was suppressed at certain wavelengths. In addition to the usual bright lines, dark lines were sometimes observed, corresponding to the absence of certain wavelengths. In 1814, Joseph Fraunhofer was able to identify some of these dark lines in the solar spectrum, called *Fraunhofer lines*, with wavelengths emitted by certain atoms. In 1861, a major discovery was made by Gustav R. Kirchhoff and Robert W. Bunsen when they were able to observe Fraunhofer lines in laboratory spectra. Kirchhoff and Bunsen demonstrated that Fraunhofer lines were caused by the absorption of light at specific wavelengths by atoms. Thus, it was known that atoms both emitted and absorbed radiation at specific wavelengths associated with each element.

Tens of thousands of wavelengths emitted by atoms have been measured and cataloged. For example, the photon spectrum emitted by mercury-198 (^{198}Hg) is shown in Figure 3-14. These photon wavelengths are unique to ^{198}Hg. The lines of an atom do not all appear with equal intensity. For ^{198}Hg the strongest (most intense) line is at a wavelength of

$$\lambda = 253.6506 \text{ nm}. \qquad (3.71)$$

This line provides a good signature for ^{198}Hg. If we examine the photon spectrum for iron, we find lines at wavelengths of 253.5607 and 253.6792 nanometers, slightly different from mercury. It is by this technique of finding a specific wavelength of radiation that one can tell how much mercury is in a tuna fish!

The Franck-Hertz Experiment

In 1914, James Franck and Gustav Hertz performed an important experiment on the atomic structure of matter. The goal of the experiment was to determine the interactions of electrons when they passed through a gas of atoms. The apparatus of Franck and Hertz is illustrated in Figure 3-15. Franck and Hertz built a three-electrode (triode) vacuum tube in which they inserted mercury vapor. The source of the electrons was thermal emission from a heated cathode. The electrons were accelerated by

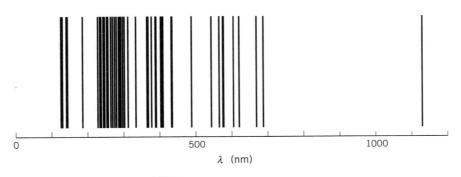

FIGURE 3-14 Photon wavelength spectrum of ^{198}Hg.
These wavelengths are unique to ^{198}Hg. All atoms have their own unique wavelength spectrum providing an atomic "fingerprint."

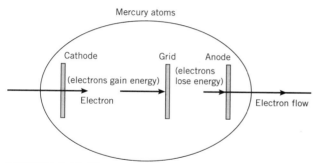

FIGURE 3-15 The experiment of Franck and Hertz.
A tube containing three electrodes ("triode") is filled with mercury vapor. Electrons are accelerated by a voltage V_c between the cathode and the grid. The voltage V_c is varied. A smaller voltage V_a with the opposite polarity is applied between the anode and the grid. The flow of electrons through the tube depends on how they interact with the mercury atoms.

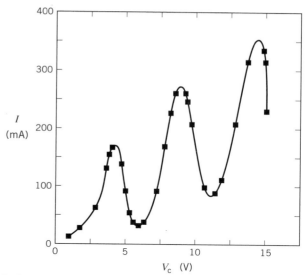

FIGURE 3-16 The data of Franck and Hertz.
The current through the tube is measured as a function of accelerating voltage (V_c). The general trend is for more electrons to reach the anode when the voltage is increased. At $V_c = 4.9$ volts, two times 4.9 volts, and three times 4.9 volts, the current is observed to decrease. This corresponds to an electron losing 4.9 eV per collision with mercury atoms. Data are from J. Franck and G. Hertz, *Verhand. Deut. Phys. Ges.* **16**, 457 (1914).

a voltage difference (V_c) between the cathode and the grid. The acceleration voltage V_c was varied during the experiment. The kinetic energy of the electrons at the grid, if they did not interact with the mercury atoms, is eV_c.

The electrons were accelerated by a small voltage difference (V_a) between the anode and the grid. The voltage V_a was kept constant during the experiment. The kinetic energy of the mercury atoms at the anode, assuming no interaction with the mercury atoms, would be $eV_c - eV_a$.

If the electrons lose some of their energy (ΔE) by interaction with the mercury atoms, then their kinetic energy at the anode may be written

$$E_k^{anode} = eV_c - eV_a - \Delta E. \qquad (3.72)$$

The experimenters made a measurement of anode current as a function of V_c. The result of Franck and Hertz is shown in Figure 3-16. As expected, a larger acceleration voltage gives a larger electron kinetic energy, which means that more electrons reach the anode, resulting in a larger current. Something remarkable happens at a cathode voltage of about 4.9V. The current decreases! This is a signature of an interaction of the electron with the mercury atom at a threshold energy of 4.9 eV. For voltages just above 4.9 V, electrons have this much energy only when they are near the grid. For somewhat larger voltages, they have this energy while further away from the grid. This means that they have a greater chance to interact with a mercury atoms before they reach the grid, resulting in a lower current. The current decreases with electron energy. At higher voltages, the current increases again until the

cathode voltage is 2 times 4.9V. With some probability, the electron has lost an energy of 9.8 eV. The current versus voltage pattern repeats itself with maxima at intervals of 4.9V.

It is possible to observe 10 sequential bumps all with a spacing of 4.9 volts. These data show a quantum threshold effect. The electron either loses no energy or some integer multiple of 4.9 eV. The electron can lose only 4.9 eV to the mercury atom because the energy levels of the mercury atom are quantized (see Figure 3-17). The electron loses an integer multiple of 4.9 eV by colliding with more than one atom.

The mercury atom returns to the ground state by emission of a photon with an energy of 4.9 eV. The wavelength of such a photon is

$$\lambda = \frac{hc}{\Delta E} \approx \frac{1240\,\text{eV} \cdot \text{nm}}{4.9\,\text{eV}} \approx 253\,\text{nm}. \qquad (3.73)$$

In a second experiment, Franck demonstrated that an energy loss of 4.9 eV by the electrons was accompanied by the emission of a photon by mercury of this wavelength (3.73). As pointed out earlier, this is the principle wavelength (3.71) found to be emitted by the mercury atom.

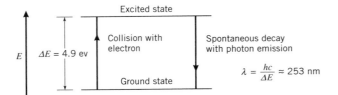

FIGURE 3-17 The two relevant atomic energy levels of the mercury atom.
The lower level is the ground state and the upper level is the excited state corresponding to one of the electrons in the mercury atom having gained an energy of $\Delta E = 4.9$ eV. The atomic electron gains this energy from the collision with an electron in the tube. The atom does not stay in the excited state. When it decays to the ground state, a photon is emitted with an energy of 4.9 eV.

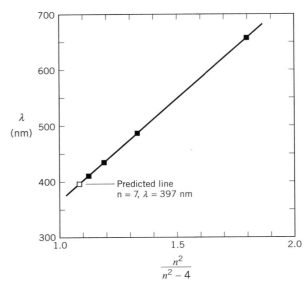

FIGURE 3-18 Prediction of Balmer compared with data of Ångstrom.
Ångstrom accurately measured the wavelengths of four visible lines emitted by the hydrogen atom. Balmer observed that these lines fit the form $\lambda = (364.56\text{ nm})\, n^2/(n^2-4)$. A plot of the measured wavelengths versus $n^2/(n^2-4)$ is linear, and the y intercept gives the constant 364.56 nm. From J.J. Balmer, *Ann. Phys. und Chem.* **25** 80 (1885).

The Balmer Series

In the late nineteenth century, Anders Ångstrom made wavelength measurements of four visible lines emitted by hydrogen accurate to 10^{-11}m. In 1885, Johann Balmer wrote a famous paper giving a simple numerical formula for the visible lines in the hydrogen spectrum as measured by Ångstrom. Balmer's formula was

$$\lambda = \frac{Cn^2}{n^2-4}, \qquad (3.74)$$

where n is an integer greater than or equal to 3, and C is a constant. The Balmer formula is plotted in Figure 3-18 together with Ångstrom's data. The slope of the line gives the constant

$$C = 364.56 \text{ nm}. \qquad (3.75)$$

Furthermore, Balmer predicted that more hydrogen lines would be found corresponding to n equal to 7, 8, 9... The wavelengths of these lines were predicted to be 396.97 nm, 377.02 nm, 374.98 nm, ... The first is barely in the visible region, whereas the others are in the near ultraviolet. Balmer was correct! All the lines do exist and were subsequently discovered.

In addition, Balmer generalized his formula to

$$\lambda = \frac{C_m n^2}{n^2-m^2}. \qquad (3.76)$$

where m and n are two integers ($n > m$ and $m > 0$) and C_m is a parameter that depends only on the value m. Thus,

Balmer predicted that hydrogen would emit an infinite number of series of lines (one series for each value of m), with each series having an infinite number of lines (one line for each value of n) in a regular pattern. Balmer was correct again.

This result of Balmer (3.76) established that the wavelengths of light emitted by the hydrogen atom were given by the differences of squares of integers. This was an important beginning, even though Balmer did not understand the physics of his formula. It represented a great leap forward in the quantitative understanding of the hydrogen atom. The work of Balmer inspired Niels Bohr, who was born the same year that Balmer wrote his famous paper, to propose that angular momentum was quantized in the hydrogen atom.

The observation of Balmer revolutionized spectroscopy. W. Ritz and J. R. Rydberg were able to generalize the Balmer formula to predict *some* lines for all elements with the expression

$$\frac{1}{\lambda} = R\left(\frac{1}{m^2} - \frac{1}{n^2}\right) = \frac{R(n^2-m^2)}{n^2 m^2}, \qquad (3.77)$$

or

$$\lambda = \frac{n^2 m^2}{R\left(n^2 - m^2\right)}, \qquad (3.78)$$

where m is a positive integer and n is an integer greater than or equal to m. The parameter R is different for each element, but varies slowly from element to element. This is a more general formula than found by Balmer because R is the same for all series of a given element. For hydrogen, the value of R is

$$R_{\mathrm{H}} = 1.09677576 \times 10^7 \ \mathrm{m}^{-1}. \qquad (3.79)$$

For heavy elements, the value of R approaches

$$R_{\infty} = 1.09737315 \times 10^7 \ \mathrm{m}^{-1}. \qquad (3.80)$$

The Rydberg-Ritz formula (3.80) works for heavier elements in the same manner that it works for hydrogen. The outermost electron in an excited heavy element feels an electrical force from Z protons plus $Z-1$ electrons, which is approximately the same force as from one proton, as in hydrogen. Thus, a tremendous amount of knowledge about which lines were emitted from various atoms was accumulated before the physics of photon emission from the atom was understood.

The series of hydrogen lines are named after their discoverers, Lyman ($m = 1$), Balmer ($m = 2$), Paschen ($m = 3$), Brackett ($m = 4$) and Pfund ($m = 5$). Figure 3-19 shows the lines emitted by the hydrogen atom, the "hydrogen bar code" (see also color plate 1). One should not have the impression that every line found was understood. In fact, this was not the case. Many lines did not fit into the Rydberg-Ritz formula. Furthermore, there were temperature and pressure variations. What was found was an important regularity of the series of many lines involving the differences of the squares of integers.

The Bohr Model of the Atom

In 1913, Niels Bohr made the first quantitatively successful model of the atom. The physics motivation for the work of Bohr came from three sources:

1. Balmer had demonstrated that the photon energies emitted by the hydrogen atom were given by the differences of the squares of integers.

2. Planck had demonstrated that the quantization of energy levels in atomic oscillators explained thermal radiation spectra.

3. Rutherford and his collaborators had discovered the existence of the nucleus of the atom. (The discovery of the nucleus is the subject of Chapter 6.)

The discovery of the nucleus by Rutherford was singularly important in the motivation of the Bohr model of the atom. The work of Rutherford showed that all the mass and positive charge of the atom was concentrated in a volume much smaller than the size of an atom. This inspired Bohr to make a planetary model of the atom with electrons moving in circular orbits about the nucleus. Bohr's model of the atom seems quite simple in retrospect, but for the time it was a great advancement of science. Bohr's model of the hydrogen atom had two parts. The first part was strictly classical physics. Bohr assumed that electrons move nonrelativistically in circular orbits of radius r under the Coulomb potential,

$$V = -\frac{ke^2}{r}. \qquad (3.81)$$

Since the proton is much more massive than the electron, we may neglect its motion. (We shall show that we can make a small correction to take into account the large but finite proton mass.) Newton's second law for uniform circular motion is

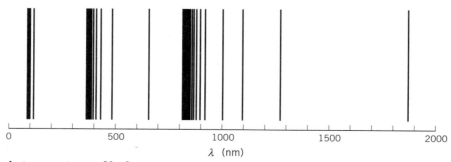

FIGURE 3-19 The photon spectrum of hydrogen.

$$\frac{ke^2}{r^2} = \frac{mv^2}{r}, \qquad (3.82)$$

where m and v are the electron mass and speed. From the potential energy (3.81) and Newton's second law (3.82), we see that the electron kinetic energy (E_k) is one-half of the potential energy but with the opposite sign:

$$E_k = \frac{1}{2}mv^2 = \frac{ke^2}{2r} = -\frac{V}{2}. \qquad (3.83)$$

In the Bohr model, the electron energy is *defined* to be the sum of the kinetic (E_k) and potential (V) energies. With this definition we write the electron energy as

$$E \equiv E_k + V = -\frac{V}{2} + V = -\frac{V}{2} = -\frac{ke^2}{2r}. \qquad (3.84)$$

The energy is negative, indicating that the electron is bound. When discussing atoms, we do not add in the electron or proton mass energies. If we added these mass energies, we would get a positive number for the total energy. Since $E_k + V$ is negative, the mass of the hydrogen atom is smaller than the mass of the proton plus the mass of the electron.

The second part of Bohr's atomic model contains a bold hypothesis of new physics. The new physics recognizes that *angular momentum is quantized* and can take on only certain values. The quantization condition on angular momentum (L) that Bohr assumed was

$$L = mvr = \frac{nh}{2\pi} = n\hbar, \qquad (3.85)$$

where n is a positive integer. Solving for the electron speed, we get

$$v = \frac{n\hbar}{mr}. \qquad (3.86)$$

The electron kinetic energy is

$$E_k = \frac{1}{2}mv^2 = \frac{1}{2}m\left(\frac{n\hbar}{mr}\right)^2 = \frac{ke^2}{2r}. \qquad (3.87)$$

Solving for the electron radius, we get

$$r = \frac{n^2\hbar^2}{mke^2}. \qquad (3.88)$$

Since r can take on only certain radii given by positive integer values of n, we assign r a *serial number* (n), and write

$$r_n = \frac{n^2\hbar^2}{mke^2}. \qquad (3.89)$$

For $n = 1$ the radius has the smallest possible value. This radius is called the *Bohr radius*. The numerical value of the Bohr radius (a_0) is

$$a_0 \equiv r_1 = \frac{\hbar^2}{mke^2} = \frac{\hbar^2 c^2}{mc^2 ke^2} = \frac{\hbar c}{\alpha mc^2}$$

$$= \frac{197\,\text{eV}\cdot\text{nm}}{\left(\dfrac{1}{137}\right)\left(5.11\times10^5\ \text{eV}\right)} \approx 0.053\,\text{nm}. \qquad (3.90)$$

This is the correct order of magnitude for the size of an atom (see Example 1-2)! The energy is quantized and we also give it a serial number (n):

$$E_n = -\frac{ke^2}{2r_n} = -\frac{mc^2\left(ke^2\right)^2}{2n^2\hbar^2 c^2}. \qquad (3.91)$$

Using the definition of α (3.69), we have

$$E_n = -\frac{\alpha^2 mc^2}{2n^2}$$

$$= -\frac{\left(\dfrac{1}{137}\right)^2\left(5.11\times10^5\ \text{eV}\right)}{2n^2}$$

$$= -\frac{13.6\,\text{eV}}{n^2}. \qquad (3.92)$$

The electron speed is also quantized:

$$v_n = \frac{n\hbar}{mr_n} = \frac{ke^2}{n\hbar}. \qquad (3.93)$$

Dividing by c, we get

$$\frac{v_n}{c} = \frac{ke^2}{n\hbar c} = \frac{\alpha}{n}. \qquad (3.94)$$

For $n = 1$, we have

$$v_1 = \alpha c = \frac{c}{137}. \qquad (3.95)$$

The Bohr model predicts that the speed of an electron in an atom is about 1% of the speed of light. Orbits with larger values of the quantum number n have smaller speeds.

By conservation of energy, when an electron moves from a larger orbit (higher energy) to a smaller orbit (lower energy) a photon is emitted. We may write the process as

$$(\text{atom})^* \rightarrow (\text{atom}) + \gamma, \qquad (3.96)$$

where (atom)* denotes an atom in the excited state. By conservation of momentum, the momentum of the atom is equal in magnitude to the momentum of the photon; however, the kinetic energy gained by the atom in the decay is very small because the mass of the atom (M) is large:

$$Mc^2 \gg E_{\text{photon}}, \qquad (3.97)$$

where (E_{photon}) is the energy of the photon. Therefore, we may neglect the change in kinetic energy of the atom and write the photon energy as the change in energy of the electron due to its change in orbit. For a transition from level n_i to n_f, the photon energy is equal to the energy difference of the atomic energy levels (3.92),

$$E_{\text{photon}} = \frac{mc^2 \alpha^2}{2} \left(\frac{1}{n_f^2} - \frac{1}{n_i^2} \right)$$
$$= (13.6 \,\text{eV}) \left(\frac{1}{n_f^2} - \frac{1}{n_i^2} \right), \qquad (3.98)$$

where n_i and n_f are positive integers with ($n_i > n_f$). This expression gives the observed photon energies emitted by hydrogen. In Figure 3-20, we summarize the electron energy levels calculated in the Bohr model.

EXAMPLE 3-17

(a) Calculate the energy of the photon emitted when an electron goes from the first excited state ($n_i = 2$) to the ground state ($n_f = 1$) in hydrogen. (b) What is the largest energy photon emitted or absorbed by the hydrogen atom?

SOLUTION:

(a) The photon energy is

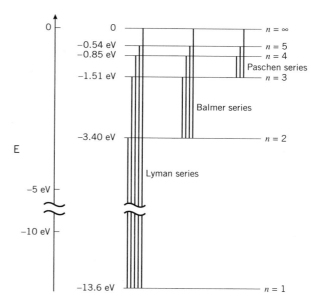

FIGURE 3-20 Atomic transitions in the Bohr model of the hydrogen atom.

$$E_{\text{photon}} = \frac{mc^2 \alpha^2}{2} \left(\frac{1}{1} - \frac{1}{4} \right)$$
$$= \left(\frac{3}{4} \right) (13.6 \,\text{eV}) = 10.2 \,\text{eV}.$$

(b) The emission corresponds to a transition between an excited state corresponding to a very large quantum number ($n_i = \infty$) and the ground state ($n_f = 1$). The photon energy is

$$E_{\text{photon}} = \frac{mc^2 \alpha^2}{2} (1 - 0) = \frac{mc^2 \alpha^2}{2} = 13.6 \,\text{eV}.$$

The absorption of a 13.6 eV photon occurs when the atom goes from $n_i = 1$ to $n_f = \infty$. This is known as the *ionization energy* of hydrogen. ∎

Correspondence Principle

Planck's constant has units of energy-time or angular momentum,

$$(\text{energy}) (\text{time}) = (\text{mass}) (\text{speed})^2 (\text{time})$$
$$= (\text{mass}) (\text{speed}) (\text{length})$$
$$= (\text{angular momentum}). \qquad (3.99)$$

The constant $\hbar$ is the minimum possible change in angular momentum of a particle. This is not noticed in macroscopic systems because the value of $\hbar$ is so small that it is a negligible fraction of the total angular momentum;

angular momentum appears to behave as a continuous variable. In the atom, angular momentum is observed to be quantized. The component of angular momentum in any given direction is an integer multiple of $\hbar$. This is independent of the frequency of the emitted photon. Bohr realized that a quantization condition on angular momentum holds true and is evident when the angular momentum is small.

For large values of the quantum number n, there must be agreement with classical physics, that is, the frequency of radiation from an accelerated charge is equal to the frequency of radiation emitted. (This is supported by a great amount of experimentation!) Consider an electron that goes from an orbit n to an orbit $n-1$ when n is very large. The energy lost by the electron is

$$\Delta E = E_n - E_{n-1} = -\frac{mc^2\alpha^2}{2n^2} + \frac{mc^2\alpha^2}{2(n-1)^2}$$

$$= \frac{mc^2\alpha^2}{2}\left[\frac{1}{(n-1)^2} - \frac{1}{n^2}\right]$$

$$= \frac{mc^2\alpha^2}{2}\left[\frac{n^2-(n-1)^2}{n^2(n-1)^2}\right]$$

$$= \frac{mc^2\alpha^2}{2}\left[\frac{2n-1}{n^2(n-1)^2}\right] \approx \frac{mc^2\alpha^2}{n^3}. \quad (3.100)$$

The frequency of the radiation (f_{rad}) is

$$f_{rad} = \frac{\Delta E}{h} = \frac{mc^2\alpha^2}{hn^3}. \quad (3.101)$$

The orbital frequency (f_{orb}) is

$$f_{orb} = \frac{v_n}{2\pi r_n} = \frac{\left(\dfrac{\alpha c}{n}\right)}{\left(\dfrac{2\pi n^2\hbar c}{mc^2\alpha}\right)} = \frac{mc^2\alpha^2}{hn^3}. \quad (3.102)$$

For large values of n, the orbital frequency is equal to the frequency of the radiation:

$$f_{orb} = f_{rad}. \quad (3.103)$$

For large values of the quantum number n, we get agreement with classical physics. This result is known as the *correspondence principle*.

For small values of n, the situation is different. For $n = 2$ the orbital frequency (3.102) is

$$f_{orb} = \frac{v_2}{2\pi r_2} = \frac{mc^2\alpha^2}{8h}. \quad (3.104)$$

When the electron makes a transition from $n = 2$ to $n = 1$, the energy of the radiation (3.98) is

$$hf_{rad} = E_{photon}$$

$$= \frac{mc^2\alpha^2}{2}\left(\frac{1}{1^2} - \frac{1}{2^2}\right) = \frac{3mc^2\alpha^2}{8}. \quad (3.105)$$

The radiation frequency is

$$f_{rad} = \frac{3mc^2\alpha^2}{8h}. \quad (3.106)$$

Comparison of the radiation (3.106) and orbital (3.104) frequencies shows that they are not equal:

$$f_{rad} = 3f_{orb}. \quad (3.107)$$

Quantization of angular momentum, $L = n\hbar$, leads to energy quantization. For large values of the quantum number n, energy and angular momentum appear continuous and we get agreement with classical physics.

EXAMPLE 3-18
Use the Bohr model to calculate the quantum number n of the earth in its orbit about the sun.

SOLUTION:
The earth is attracted to the sun by the gravitational force. The Bohr model applied to earth has the electron mass replaced by the mass of the earth (M_e) and the electromagnetic force constant ke^2 replaced by the gravitational force constant GM_eM_s, where M_s is the solar mass. The radius of the orbit (3.89) becomes

$$r_n = \frac{n^2\hbar^2}{M_e GM_e M_s} = \frac{n^2\hbar^2}{GM_e^2 M_s}.$$

The quantum number is

$$n = \frac{M_e\sqrt{r_n GM_s}}{\hbar}$$

$$\approx \frac{\left(6\times10^{24}\text{ kg}\right)\sqrt{\left(6\times10^6\text{ m}\right)}}{10^{-34}\text{ J}\cdot\text{s}}$$

$$\times\sqrt{\left(7\times10^{-11}\text{ m}^3\cdot\text{kg}^{-1}\cdot\text{s}^{-2}\right)\left(2\times10^{30}\text{ kg}\right)}$$

$$\approx 2\times10^{72}.$$

This is indeed a very large quantum number! ∎

Collapse of the Classical Atom

We may estimate the time scale for the collapse of a classical atom by using the classical formula for the power (P) radiated by an accelerated charge. The rate at which energy is radiated, according to classical electrodynamics (see Appendix B), is proportional to the acceleration of the charge squared:

$$P = \frac{2\,ke^2\,a^2}{3c^3}. \qquad (3.108)$$

For a classical atom, the acceleration is

$$a = \frac{v^2}{r}, \qquad (3.109)$$

and the radiated power is

$$P = \frac{2\,ke^2\,v^4}{3c^3r^2}. \qquad (3.110)$$

The radiated power is a function of time because the speed and radius are changing with time. Let us examine how the radiated power (3.110) depends on the electron energy. The energy of the electron is proportional to v^2. The radius of the electron orbit is inversely proportional to the energy. Therefore, radiated power varies with the *fourth power* of the energy. The classical atom is expected to collapse very quickly. Of course, the atom does not really collapse! We shall need to learn some more physics to fully appreciate why the atom does not collapse. This is discussed in detail in Chapter 5.

EXAMPLE 3-19

Estimate the time taken for an atom to collapse in the classical model.

SOLUTION:

We first write the power (3.110) in terms of the total energy (E). Recall that E is defined to be negative, and E is getting more negative as the atom collapses. The radius is

$$r = -\frac{ke^2}{2E},$$

and since

$$v^2 = \frac{2E}{m},$$

we have

$$P = -\frac{dE}{dt} = \frac{32\,E^4}{3c^3\,ke^2\,m^2}.$$

Let us take the initial energy to be -14 eV, corresponding to a radius of 0.1 nm. The energy of the collapsed atom is $-\infty$. Integrating the power, we get

$$\int_{-\infty}^{-14\,eV} dE\,\frac{1}{E^4} = \frac{32}{3c^3\,ke^2\,m^2}\int_0^T dt = \frac{32\,T}{3c^3\,ke^2\,m^2},$$

where T is the time for the atom to collapse. Solving for T,

$$T \approx \left[\frac{c^3\,ke^2\,m^2}{32\,E^3}\right]_{E=14\,eV}.$$

Numerically,

$$T = \left[\frac{(1.44\ \text{eV}\cdot\text{nm})(5\times10^5\ \text{eV})^2}{(32)(3\times10^8\ \text{m/s})(14\,\text{eV})^3}\right]\left(\frac{1\,\text{m}}{10^9\ \text{nm}}\right)$$

$$\approx 10^{-11}\ \text{s}.$$

The classical model of the atom predicts a rapid collapse! ∎

Other Atoms with One Electron

For other atoms with one electron, for example singly ionized helium, the Bohr model also applies. Everywhere we had a factor of ke^2 we replace it by Zke^2, where Z is the number of protons in the nucleus. The radii (r_n) are smaller by a factor of Z and the energy levels are multiplied by a factor of Z^2. The Bohr model works perfectly for predicting the energy levels of one-electron atoms. For atoms with more than one electron, the situation is more complicated.

Reduced Mass

We may take into account the finite value of the proton mass and its motion by noticing that the center-of-mass of the electron–proton system does not move. We can transform from the system where the proton motion is very small to the center-of-mass system by making a change of variables from the electron mass to the reduced mass (μ), defined by

$$\mu \equiv \frac{mm_p}{(m+m_p)}, \qquad (3.111)$$

where m is the electron mass and m_p is the proton mass. The value of the reduced mass is

$$\mu \approx 0.9995\,m. \qquad (3.112)$$

The atomic energy levels are proportional to this mass and are slightly shifted when this correction is added to the hydrogen atom.

There is a stable isotope of hydrogen with a nucleus consisting of one proton and one neutron. This atom is called deuterium (the nucleus is called the deuteron) and it has a natural abundance of about 1.5×10^{-4}. Since the deuterium nucleus has roughly twice the mass of the proton, the deuterium electron has a slightly larger reduced mass. The photon energies from deuterium are slightly larger than the photon energies from hydrogen. Deuterium has its own atomic "fingerprint!" This fact was used by Harold C. Urey to establish the existence of deuterium in 1931 by measurement of the Balmer lines.

EXAMPLE 3-20

Calculate the difference of wavelengths between deuterium and hydrogen of the lowest energy Balmer line.

SOLUTION:

This line results from the $n=3$ to $n=2$ transition. For hydrogen, the photon energy (replacing m with the reduced mass) is

$$E_{\text{photon}} = \frac{\mu c^2 \alpha^2}{2}\left(\frac{1}{4} - \frac{1}{9}\right) = \frac{5 m m_p c^2 \alpha^2}{72(m + m_p)}.$$

The photon wavelength for hydrogen is

$$\lambda_{\text{H}} = \frac{hc}{E_{\text{photon}}} = \frac{72 hc(m + m_p)}{5 m m_p c^2 \alpha^2}.$$

For deuterium (D), the wavelength is

$$\lambda_{\text{D}} = \frac{72 hc(m + m_d)}{5 m m_d c^2 \alpha^2}.$$

where m_d is the mass of the deuteron. The difference in wavelengths is

$$\lambda_{\text{H}} - \lambda_{\text{D}} = \frac{72 hc}{5 m c^2 \alpha^2}\left(\frac{m}{m_p} - \frac{m}{m_d}\right)$$

$$= \frac{72 hc(m_d - m_p)}{5 m_p m_d c^2 \alpha^2}.$$

The proton mass is

$$m_p = 938 \text{ MeV/c}^2.$$

The deuteron mass is

$$m_d = 1876 \text{ MeV/c}^2.$$

Thus, we have

$$\lambda_{\text{H}} - \lambda_{\text{D}} = \frac{(72)(1240 \text{ eV} \cdot \text{nm})(938 \text{ MeV})}{(5)(938 \text{ MeV})(1876 \text{ MeV})\left(\frac{1}{137}\right)^2}$$

$$= 0.18 \text{ nm}.$$

The wavelengths of radiation from hydrogen are slightly longer than from deuterium. ∎

X Rays

Discovery of Röntgen

In 1895, Wilhelm Röntgen discovered that when energetic cathode rays interacted with matter, a new type of radiation was observed, which he called *x rays*. Röntgen observed the x rays by fluorescence, the emission of visible light by certain materials when exposed to certain types of radiation. The fluorescent material used by Röntgen to detect the x rays was barium platino-cyanide.

A remarkable property of the "x" radiation was its ability to penetrate matter. In his original paper, Röntgen reported:

> . . . *all bodies are transparent to this agent, though in varying degrees. . . paper is very transparent; behind a book of about one thousand pages I saw the fluorescent screen light up brightly. . . in the same way the fluorescence appeared behind a double pack of cards. . . a single sheet of tin foil is scarcely perceptible. . . pine boards two or three cm thick absorbing only slightly. . . if the hand be held between the discharge-tube and the screen, the darker shadow of the bones is seen within the slightly dark shadow-image of the hand itself*

Röntgen also observed that it was not possible to deflect x rays with very strong magnetic fields and that x rays were not visible to the eye. The observed properties of this radiation, namely that it penetrates matter in inverse proportion to density, led to immediate medical application. An x-ray photograph taken by Röntgen is shown in Figure 3-21.

As the name implies, there was little understanding of the nature of x rays when they were discovered. We now

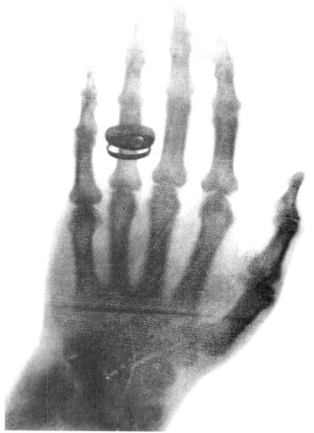

FIGURE 3-21 One of the first x-ray photographs taken by Röntgen. The Bettman Archive.

incoming electron and extending to longer wavelengths. The electron may also remove an inner electron, causing an atomic transition to fill the empty position. In this process radiation of a wavelength *characteristic* of the material is emitted. The exact wavelengths of the characteristic radiation are a property of the element bombarded. The wavelength spectrum of x rays produced by the interaction of energetic electrons in molybdenum is shown in Figure 3-22.

Scattering of X Rays

Since x rays are electromagnetic waves, we expect interference phenomena in scattering experiments. Interference effects will be observable when the scatterer has structure comparable to the wavelength of the radiation. We have seen that the wavelengths of x rays are about the size of an atom. In 1912, William Lawrence Bragg made an analysis of x-ray scattering by regular arrays of atoms. Bragg analyzed x rays scattered from two successive layers of atoms, as shown in Figure 3-23. In the description of x-ray scattering experiments, the scattering angle (θ) is traditionally defined to be the angle that the x rays

know that x rays are electromagnetic radiation in the energy range of a few tens of keV. A typical x ray photon from your dentist has an energy of 40 keV, corresponding to a wavelength of

$$\lambda = \frac{hc}{E} = \frac{1240 \text{ eV} \cdot \text{nm}}{4 \times 10^4 \text{ eV}} \approx 0.03 \text{ nm} . \quad (3.113)$$

The wavelength of a typical x ray is roughly the size of an atom.

X rays are generated by bombarding matter with electrons. When electrons pass though material, they have electromagnetic interactions with the positively charged nuclei. The electrons lose energy by such collisions and in the process radiate x rays. This *bremsstrahlung* or "braking radiation" spectrum is a broad curve that has a cutoff at a wavelength determined by the kinetic energy of the

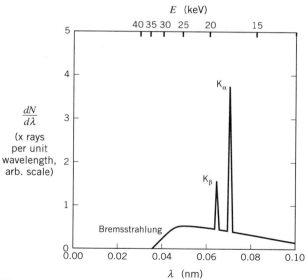

Figure 3-22 X ray wavelength spectrum from the interaction of energetic electrons in molybdenum. The maximum energy x ray is equal to the kinetic energy of the electrons. The smooth portion of the spectrum is due to *bremsstrahlung* radiation from the accelerated electrons and is the same shape for all elements. The sharp spikes are the characteristic radiation caused by the removal of an inner electron. Data from D. Ulrey, *Phys. Rev.* **11**, 405 (1918).

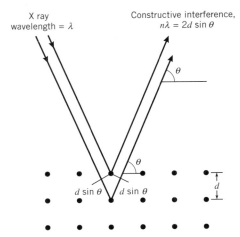

FIGURE 3-23 Bragg scattering of x rays from a crystal of atoms.
The condition for constructive interference is that $n\lambda = 2d\sin\theta$, where λ is the wavelength of the x ray, d is the spacing between atomic layers, θ is the scattering angle, and n is an integer.

make with the plane of the atoms. Thus, $\theta = \pi/2$ is normal incidence. (In optics $\theta = 0$ is usually defined to be normal incidence.) The intensity of the scattered x-ray radiation will be a maximum when the scattered waves arrive in phase. The x rays will be in phase at a scattering angle θ only if

$$n\lambda = 2\,d\sin\theta. \qquad (3.114)$$

William Henry Bragg, the father of W. L. Bragg, built the first spectrometer to test the Bragg condition (3.114). X-ray scattering proved to be an extremely powerful technique. If the x-ray wavelength is known, then the measurement of the scattering angle can be used to determine the atomic spacing. Conversely, if the atomic spacing is known, measurement of the scattering angle serves as a determination of the x-ray wavelength.

Characteristic X Rays

In 1913, Henry Moseley made a systematic study of the x rays produced by all known elements from aluminum to gold. The characteristic x rays were produced by bombardment of the elements with electrons. Moseley measured the wavelength of the atomic x rays through the Bragg technique by scattering the x rays from a potassium ferrocyanide crystal. He determined the spacing between atomic planes, d. Moseley made use of the discovery of Bragg that an x ray of definite wavelength was reflected from the crystal only when incident at a certain angle. Measurement of that scattering angle and knowledge of

the crystal lattice spacing provides an accurate measurement of the x-ray wavelength.

Moseley observed that each element produced two unique x rays of different intensities. The stronger intensity x rays are labeled K_α, and the weaker K_β. The heavier elements produced five additional lines that are labeled L_α, L_β, L_γ, L_δ, and L_ε. Very weak lines due to impurities were also observed, demonstrating that the x-ray technique is a powerful tool for chemical analysis.

Moseley classified the data by assigning an atomic number (Z) to each element and making a plot of Z versus the square root of the x-ray frequency. Some of Moseley's x-ray data are shown in Figure 3-24. These plots show that the square root of the x-ray frequency is proportional to Z. Moseley went further in his analysis and showed that all the x-ray data could be fit by the form

$$f = a(Z - b)^2, \qquad (3.115)$$

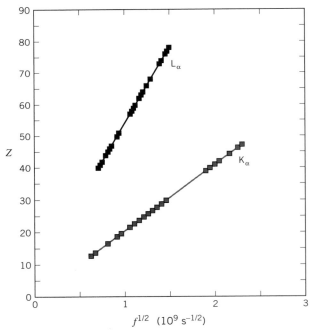

FIGURE 3-24 Moseley's data on high-frequency spectra of the elements.
Moseley bombarded various elements with energetic electrons. The atoms were observed to emit x rays. Moseley measured the energy of the x rays by scattering from a crystal and observed that each element could be assigned an atomic number Z and that the square root of the frequency of the radiation was directly proportional to Z. Moseley's data are shown for the K_α lines and the L_α lines. Data from H.G. Moseley, "High-Frequency Spectra of the Elements," *Phil. Mag.* **27,** 703 (1914).

where a and b are constants. For the K lines, Moseley observed that $b = 1$ for both K_α and K_β, and slightly different values of a were needed for the K_α and K_β lines. Recall that the x rays were generated by bombarding the elements with electrons. For atoms other than hydrogen, there are *two* inner electrons. The bombarding electron can knock out one of the inner electrons from the atom. The x ray is emitted when another electron makes a transition from a larger orbit to the vacant position. The electron making a transition to the vacant position is exposed to a charge of Z protons and one electron. Thus, the energy of the electron is proportional to $(Z - 1)^2$. The K_α data correspond to the transitions $n = 2$ to $n = 1$, whereas the K_β data correspond to the transition $n = 3$ to $n = 1$.

For the L-line data, Moseley deduced that $b = 7.4$. These data correspond to the bombarding electron removing electrons from the next smallest orbits (other than the two innermost electrons). Since the nucleus is screened by the inner electrons, $Z - 7.4$ represents the effective charge felt by an electron making a transition from a larger orbit to the vacant position. The L_α data correspond to the transitions $n = 3$ to $n = 2$, the L_β data correspond to the transition $n = 4$ to $n = 2$, the L_γ data correspond to the transitions $n = 5$ to $n = 2$, and so on.

The data taken by Moseley contained a tremendous amount of physics. Moseley concluded that the atomic theory of a nucleus surrounded by electrons with quantized energy levels was correct. He identified Z as the number of protons in the nucleus, and he concluded that three and only three missing elements existed between aluminum and gold with atomic numbers 43, 61, and 75. Moseley was correct and the missing elements now known as technetium, promethium, and rhenium, were eventually discovered! (Element number 43 has no stable isotope and is not found in nature. It was "discovered" only after it was artificially produced, hence the name technetium.)

EXAMPLE 3-21

Use the Bohr model to calculate the energy of a K_α x ray from the element copper ($Z = 29$), assuming that an electron makes a transition from $n = 2$ to $n = 1$ and that the effective nuclear charge is $Z - 1$. Compare your answer to the measurement of Moseley.

SOLUTION:

The energy of the electron (E_n) is greater than in hydrogen by a factor of the $(Z - 1)^2$, or

$$E_n = \frac{(-13.6\,\text{eV})(Z-1)^2}{n^2}.$$

The photon energy from an $n = 2$ to $n = 1$ transition is

$$E_{\text{photon}} = E_2 - E_1$$
$$= (-13.6\,\text{eV})(28)^2\left(1 - \frac{1}{4}\right) = 8.00\,\text{keV}.$$

The frequency of this photon is

$$f = \frac{E_{\text{photon}}}{h}$$
$$= \frac{8.00 \times 10^3\,\text{eV}}{4.14 \times 10^{-15}\,\text{eV} \cdot \text{s}} = 1.93 \times 10^{18}\,\text{s}^{-1}.$$

The square root of the frequency is

$$\sqrt{f} = 1.39 \times 10^9\,\text{s}^{-1/2}.$$

Examination of Figure 3-24 shows excellent agreement with the data of Moseley. ∎

Successes and Limitations of the Bohr Model

The great success of the Bohr model is its quantitative explanation of the photon spectrum from hydrogen. The size of the hydrogen atom (0.05 nm) obtained in the model as well as the speed of the electron in the ground state ($c/137$) are the correct order of magnitude. The quantization condition on the angular momentum explains why the atom does not collapse!

Despite its great successes, there are two serious deficiencies of the Bohr model of the atom:

1. It cannot predict the relative intensities of the photon radiation from atoms.

2. It does not work for atoms with more than one electron.

Furthermore, we are left with a major question: *Why is angular momentum quantized?*

CHAPTER 3: PHYSICS SUMMARY

- Matter and radiation in thermal equilibrium are continuously exchanging energy in the form of electromagnetic radiation. The wavelength distribution of radiation emitted by an ideal object (blackbody) that absorbs all incident radiation has a universal form that depends only on the temperature. The power per area per unit wavelength is

$$\frac{dR}{d\lambda} = \frac{2\pi hc^2}{\lambda^5 \left(e^{hc/\lambda kT} - 1 \right)}.$$

The radiation energy per volume per unit wavelength is

$$\frac{du}{d\lambda} = \frac{4}{c}\frac{dR}{d\lambda}.$$

• The total power per area radiated is proportional to the fourth power of the temperature:

$$R = \frac{2\pi^5 (kT)^4}{15 h^3 c^2} = \sigma' (kT)^4,$$

where

$$\sigma' \equiv \frac{2\pi^5}{15 h^3 c^2} = 1.03 \times 10^9 \ \text{W/m}^2/\text{eV}^4.$$

• The wavelength at which the power per area per wavelength is a maximum is inversely proportional to the temperature:

$$\lambda_{\text{m}} = \frac{250 \ \text{eV} \cdot \text{nm}}{kT}.$$

• Radiation of frequency f is quantized in the form of individual particles called photons. The energy of a photon is Planck's constant times the radiation frequency,

$$E = hf = \frac{hc}{\lambda}.$$

• The numerical value of Planck's constant times the speed of light is

$$hc = 1240 \ \text{eV} \cdot \text{nm}.$$

The combination $hc/2\pi$ appears often in calculations,

$$\hbar c \equiv \frac{hc}{2\pi} = 197 \ \text{eV} \cdot \text{nm}.$$

The value of the dimensionless electromagnetic coupling strength is

$$\alpha = \frac{ke^2}{\hbar c} = \frac{1}{137}.$$

• The quantization of angular momentum in units of $\hbar$ is the key ingredient of the Bohr model of the hydrogen atom. The Bohr model predicts the allowed energy levels of the atom, including the important fact that there is a minimum energy of the atom and therefore, the atom does not collapse. The ground state radius of the hydrogen atom in the Bohr model is

$$a_0 = \frac{\hbar^2 c^2}{mc^2 ke^2} \approx 0.053 \ \text{nm}.$$

The energy levels in the hydrogen atom are

$$E_n = -\frac{13.6 \ \text{eV}}{n^2}.$$

The speed of an electron in the hydrogen atom at the Bohr radius is

$$v = \alpha c.$$

• The correspondence principle states that for large values of the quantum number n, we enter the classical regime where angular momentum appears as a continuous variable.

REFERENCES AND SUGGESTIONS FOR FURTHER READING

J. Balmer, "Note on the Spectral Lines of Hydrogen," *Ann. Phys. Chem.* **25**, 80 (1885), trans. H. A. Boorse and L. Motz.

M. R. Howells, J. Kirz and D. Sayre, "X-ray Microscopes," *Sci. Am.* **264**, No. 2, 88 (1991).

V. W. Hughes, "The Muonium Atom," *Sci Am.* **214**, No. 4, 93 (1966).

F. Hund, "Paths to Quantum Theory Historically Viewed," *Phys. Today* **19**, No. 8, 23 (1966).

M. J. Klein, "Thermodynamics and Quanta in Planck's Work," *Phys. Today* **19**, No. 11, 23 (1966).

A. C. Melissinos, *Experiments in Modern Physics*, Academic Press (1966).

R. A. Millikan, *The Electron*, The University of Chicago Press (1917).

R. A. Millikan, *Electrons (+ and −), Protons, Photons, Neutrons, Mesotrons, and Cosmic Rays*, The University of Chicago Press (1947).

H. G. Moseley, "The High-Frequency Spectra of the Elements," *Phil. Mag.* **26**, 1024 (1913); *Phil. Mag.* **27**, 703 (1914).

W. Röntgen, "On a New Kind of Rays," translation by G. F. Barker found in *Amer. Jour. of Phys.* **13**, 284 (1945).

A. Rose and P. Weimer, "Physical Limits to the Performance of Imaging Systems," *Phys. Today* **42**, 24 (1989).

V.F. Weisskopf, "How Light Interacts with Matter," *Sci. Am.* **219**, No. 3, 60 (1968).

QUESTIONS AND PROBLEMS

Atoms and radiation in equilibrium

1. Can an object glow *green* from thermal radiation? Why or why not?

2. Can thermal radiation from an object at 300 K be visible to the naked eye if the object is large enough?

3. Can you think of some objects that do not behave like a blackbody? Give an example of radiation that does not have a thermal spectrum.

The thermal radiation spectrum

4. A pottery maker places a small bowl and a large vase in a very hot oven. Which piece glows red first?

5. The peak of the thermal radiation power spectrum $(dR/d\lambda)$ is at a wavelength of about $\lambda_m = hc/5kT$. Why is the peak of the same power spectrum plotted as dR/df not at $f_m = c/\lambda_m = 5kT/h$?

6. Make a rough estimate of the number of photons radiated per second by thermal radiation from the pencil or pen that you are holding in your hand as you write down the solution to this problem.

7. Show that the solar photon spectrum (dN/dE) peaks at an energy of about 0.8 eV.

8. (a) Make a rough estimate of the temperature of a tungsten filament in an ordinary light bulb. (*Hint:* The filament is "white hot.") (b) If the light bulb has a radiated power of 100 W, make a rough estimate of the surface area of the filament. (c) Approximately how many photons are radiated per second?

9. A thermal light source has a temperature of 6000 K and a total radiated power of 100 W. (a) Calculate the power per area per μm $(dR/d\lambda)$ at the peak of the spectrum. (b) Estimate the number of visible photons from the light source that enter a pupil of radius 2 mm at a distance of 1 km.

10. Consider a distant star with the same luminosity and surface temperature as the sun. A person can see the star if 250 visible photons per second pass through the pupil, which has a radius of 2 mm. What is the maximum distance at which the star is visible?

11. A dark room of size $3\text{ m} \times 3\text{ m} \times 3\text{ m}$ is shielded from external light. Estimate the number of visible photons in the room at any given time.

12. A metal foil is heated to a temperature of 1000 K. Determine the surface area of the foil if the radiated power is 1 W.

13. Make an estimate of the fraction of the solar radiation energy that is visible.

14. Suppose that the earth was formed as molten rock at some very high temperature. Estimate the time for the earth to cool to 300 K. (*Note*: This calculation was first performed by Lord Kelvin. The cooling time determined by Kelvin was significantly smaller than the geological evidence for the age of the earth. This generated a great scientific debate. The problem was resolved with the discovery of radioactive elements in the earth. The energy released in radioactive decays increases the cooling time.)

15. Determine an algorithm to solve the expression (3.48) for λ_m. Use a calculator to find the solution.

16. A piece of pottery with a mass of 1 kg is heated in an oven to a temperature of 1500 K. The pottery is taken out of the oven and placed in a cool room. Make a rough estimate of the time for the pottery to reach a temperature of 300 K.

17. A naked person on the South Pole exits a sauna with a body temperature of 311 K. The person encounters an outside temperature very much below the freezing point of water. (a) Estimate the energy lost per time by the person due to radiation cooling. (b) Estimate the radiation loss if the person steps out of the sauna into a cool room $(T_r = 290\text{ K})$ instead of outside. (c) Make an order of magnitude comparison of the energy lost per second calculated in parts (a) and (b) with the total kinetic energy of the atoms inside the person.

18. Use the solar constant (1350 W/m^2) to calculate the energy density of sunlight (a) near the earth and (b) near the surface of the sun.

19. Use the Stefan-Boltzmann law to calculate the energy density of the cosmic background radiation $(T = 2.74\text{ K})$.

*20. (a) For thermal radiation, show that the relationship between the energy per volume per unit wavelength $(du/d\lambda)$ and the number of photons per volume per unit wavelength $(dn/d\lambda)$ is

$$\frac{du}{d\lambda} = \left(\frac{hc}{\lambda}\right)\frac{dn}{d\lambda}.$$

(b) Show that the total number of photons per volume is given by

$$n = 8\pi \left(\frac{kT}{hc} \right)^3 \int_0^\infty dx \, \frac{x^2}{e^x - 1}$$
$$\approx \left(3.17 \times 10^{19} \text{ eV}^{-3} \cdot \text{m}^{-3} \right) (kT)^3.$$

(c) Show that the average photon energy is

$$\langle E \rangle \approx 2.7 \, kT.$$

(d) Use these results to calculate the density of cosmic background photons and their average energy.

21. How many cosmic photons per second per square meter were incident on the antenna of Penzias and Wilson?

Quantization of electromagnetic radiation

22. Estimate the typical energy of photons from (a) an FM radio station, (b) a microwave oven ($\lambda \approx 0.01$ m), (c) the sun, (d) a piece of ceramic heated to $T \approx 1000$ K, and (e) the cosmic photons from the early universe ($T \approx 3$ K).

23. A helium-neon laser produces red light at a wavelength of 633 nm. The laser light is shined on a can of beans in order to read the supermarket bar-code. If the laser output power is 1 milliwatt, what is the rate at which photons strike the can of beans?

24. A photon with an energy equal to the work function is absorbed by an electron and the electron is freed from the atom, but with zero kinetic energy. Where has the energy of the photon gone?

25. Radiation of a certain wavelength is incident on potassium, which has a work function of 2.3 eV. Electrons are observed to be ejected with a maximum kinetic energy of 0.7 eV. What is the wavelength of the radiation?

26. Photons with a wavelength of 200 nm are directed at a silver cathode. What is the maximum kinetic energy of the ejected electrons?

27. Photons with a wavelength of 410 nm are used to eject electrons from a metallic cathode by the photoelectric effect. The electrons are prevented from reaching the anode by applying a stopping potential of 0.88 volt. What is the work function of the cathode material?

28. The maximum wavelength of radiation that can free electrons from a metal is 380 nm. What is the work function of the metal?

29. Photons with energy E are incident on a photocathode and produce electrons with a maximum kinetic energy of 2 eV. When the energy of the photons is doubled, the maximum kinetic energy of the electrons increases to 8 eV. What is the work function of the photocathode?

30. Electromagnetic radiation with a frequency f is incident on a metal, causing the photoelectric effect to occur. When the frequency of the radiation is doubled, the maximum kinetic energy of the ejected electrons is tripled. What is the minimum radiation frequency for initiating the photoelectric effect in the metal?

31. When radiation of wavelength $\lambda = 410$ nm is incident on a cathode with an unknown work function, electrons are observed to be ejected by the photoelectric effect. The minimum voltage necessary to prevent all electrons from reaching the anode, the "stopping voltage," is unknown. The experimenters change the wavelength of the radiation and measure the change in the stopping voltage. What wavelength of radiation would correspond to an increase in the stopping voltage of 1 V?

32. Why was it important for Millikan to have very clean plates for the measurement of h/e by the photoelectric effect?

33. Consider the bombardment of an iron plate ($\phi = 4.5$ eV) with electromagnetic waves of three wavelengths 600 nm, 240 nm, and 200 nm with the relative intensities of 100 to 10 to 1. An external voltage (V) is applied to the plate and a maximum current of 1 nA is observed when $V = 10$ V. Make a careful sketch of the current observed as a function of V, for both positive and negative values of V, indicating the voltage and current scales clearly.

34. Radiation with a wavelength of 300 nm is directed at a metallic surface. Estimate the radiation flux (power per area) necessary to have an average of 1 photon per second strike each atom on the surface of the metal.

35. Light from thermal source ($T = 6000$ K) is filtered so that only photons in the visible region are allowed to strike a photocathode, which has a work function of 2.0 eV. When the intensity of the light that reaches the photocathode is 1 mW, a current of 1 μA is observed in a circuit that detects the photoelectrons. Estimate the quantum efficiency of the photocathode.

Atomic spectra and the Bohr model

36. What is the minimum wavelength of a photon that can be emitted by the hydrogen atom?

37. What is the maximum wavelength of radiation that a hydrogen atom in the first excited state can absorb?

38. In the Bohr model, what is the largest speed that an electron can have in a hydrogen atom? What is the smallest speed?

39. In the Bohr model of the hydrogen atom, compare the orbital frequency of an electron with $n = 3$ to the allowed frequencies that it can radiate.

40. (a) Calculate the wavelength of the emitted radiation in the $n = 3$ to $n = 2$ transition in hydrogen. (b) In singly ionized helium, which transition produces radiation of a wavelength closest to that of the $3 \rightarrow 2$ transition in hydrogen?

*41. (a) Use the Bohr model to calculate the size of the hydrogen atom if gravity were the force responsible for holding the atom together. (b) Calculate the ground state energy of the atom. (c) Calculate the speed of the electron in the ground state.

*42. A cosmic ray muon of negative charge, after stopping in matter may form an exotic atom by being captured by a proton. Use the Bohr model to calculate the energy levels of this muonic atom. Does the muon have a longer or shorter lifetime due to the formation of the muonic atom? (Hint: The proton can "capture" the muon, resulting in the decay $\mu^- p \rightarrow \nu_\mu + n$.)

43. Which has greater energy, a K_α or a K_β x ray from the same element?

44. Estimate the energy of a K_α x ray from the element lead.

45. Which element has an L_α x ray that is closest in wavelength to the K_α x ray from the element manganese ($Z = 25$)?

Additional problems

46. Make an estimate of the fraction of the thermal radiation energy that is visible at room temperature.

47. Water is effective at absorbing radiation over a wide range of wavelengths (except in the narrow wavelength region of the visible). When a cup of soup is placed in a microwave oven, photons with a wavelength of about 0.01 m are absorbed by the soup, which is heated in the process. The cosmic photons from the early universe, which number about 4×10^8 per cubic meter, have a wavelength of about 0.001 m. Why don't these cosmic photons make your blood boil?

*48. Prove that when the universe expands by a factor C, a thermal power spectrum of initial temperature T has the same shape but has a new temperature T/C. (*Hint:* The wavelength of each photon is increased by a factor of C during the expansion.)

49. Estimate the rate at which photons from the sun strike the earth.

50. Show that h/e^2 has units of resistance and calculate the value in ohms.

51. The work function of a metal is equal to the energy threshold for the photoelectric effect at $T = 0$. At room temperature a small fraction of the electrons are energetic enough to be ejected from a metallic surface by thermionic emission. The current per area is given by the Richardson-Dushman equation (2.86) as

$$J = C \left(kT \right)^2 e^{-\phi/kT},$$

where $C = 1.6 \times 10^{14}$ A·m^{-2}·eV^{-2}. Use the Richardson-Dushman equation to estimate the work function of a photocathode if the electrons are ejected at a rate of 10^2 per second from an area of 10^{-2} m^2 at room temperature.

*52. Consider a black box of gas at STP. Make an estimate of the ratio of number of photons to gas particles inside the box. Does the answer depend on the size or shape of the box?

*53. Show that the pressure (P) generated by a gas of photons is equal to 1/3 times the energy density (u).

54. Use the Stefan-Boltzmann law to calculate the heat capacity (dE/dT) of radiation in a cavity.

55. In Example 3-10 the solar constant was used to calculate the temperature of the earth to be 280 K. The actual average temperature is measured to be 290 K. (a) At what rate is energy being produced inside the earth? (b) The energy source is natural radioactive decays. If each decay produces an energy of 6 MeV, at what rate do particles decay? (c) What fraction of the particles in the earth decay each second?

*56. Consider the Bohr model applied to the deuteron, a bound state of a proton and a neutron. Estimate the strength of the force to be 10 times stronger than the electromagnetic force, that is, $\alpha_s \approx 0.1$ compared to $\alpha = 1/137$. (a) Estimate the speed of the proton and neutron. (b) What is the nuclear "Bohr radius"? (c) Estimate the binding energy of the deuteron.

57. Normal body temperature is about 307 K. An infrared detector which measures the power per area (R) radiated from a small section of skin is sensitive to variations in R of 1%. What temperature difference can be detected?

It is not good to introduce the concept of the mass $M = m/(1-v^2/c^2)^{1/2}$ of a moving body for which no clear definition can be given. It is better to introduce no other mass concept than the "rest mass" m. Instead of introducing M it is better to mention the expression for the momentum and energy of a body in motion.

Albert Einstein

Electromagnetic waves are observed to propagate in a vacuum at the speed

$$c \approx 3.00 \times 10^8 \text{ m/s}. \tag{4.1}$$

The speed c is a universal constant, that is, c does not depend on the wavelength. The constant c is called the *speed of light*. We have seen from the photoelectric effect that electromagnetic radiation is quantized. Electromagnetic waves have a particle behavior. The particles of radiation are photons. The energy of an individual photon is directly proportional to the frequency of the wave, but all photons have the same speed (c) when moving in a vacuum, independent of their energies.

The speed (v) of any macroscopic object of mass m found on the earth is much smaller than the speed of light, $v \ll c$. The classical definitions of momentum (**p**),

$$\mathbf{p} \equiv m\mathbf{v}, \tag{4.2}$$

and kinetic energy (E_k),

$$E_k \equiv \frac{1}{2}mv^2 = \frac{p^2}{2m}, \tag{4.3}$$

were established empirically as the quantities that are conserved in all collisions. When a particle has a speed approaching c, the momentum (4.2) and kinetic energy (4.3) are no longer conserved quantities in particle collisions. We shall need a revised description for particles that have very large speeds and large kinetic energies. This description is contained in the theory of special relativity.

4-1 FOUNDATIONS OF SPECIAL RELATIVITY

The Postulates

The theory of special relativity was first formulated by Einstein in 1905, the same year he published his famous papers on Brownian motion and the photoelectric effect. The two postulates of special relativity are as follows:

1. The laws of physics are identical in all inertial frames of reference.

2. The speed of electromagnetic radiation in vacuum is constant, independent of any motion of the source.

The first postulate is identical to that used by Newton in his formulation of the laws of classical physics. The defini-

tion of an inertial frame is a reference frame where a particle has no acceleration unless there is a force acting on the particle. The second postulate may at first sight seem counterintuitive; it is our common experience at low speeds, $v \ll c$, that speeds add linearly.

Consider a particle moving with speed (dx/dt) in the x direction as indicated in Figure 4-1. In a frame that is moving with a speed v in the x direction, the coordinates are

$$t' = t, \tag{4.4}$$

$$x' = x - vt, \tag{4.5}$$

$$y' = y, \tag{4.6}$$

and

$$z' = z. \tag{4.7}$$

This coordinate transformation is called the *Galilean transformation*. Since $dt = dt'$, differentiation of the Galilean transformation of the x coordinate (4.5) with respect to t leads to the velocity addition rule:

$$\frac{dx'}{dt'} = \frac{dx}{dt} - v. \tag{4.8}$$

A headwind of speed v decreases the speed of an airplane with respect to the earth's surface by an amount v. The Galilean transformation works provided that the speeds of the objects are small compared to the speed of light.

The Galilean transformation does not work for very large speeds. You will need to develop a relativistic

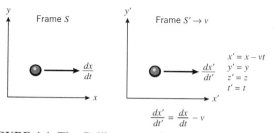

FIGURE 4-1 The Galilean transformation.
Consider the motion of a particle as observed in two frames of reference. In frame S, the particle has a speed dx/dt in the x direction. Frame S' is defined to be a frame that is moving with a speed v in the x direction. The speed of the particle in S' is $dx'/dt' = dx/dt - v$, provided that the coordinates are related by the transformation: $x' = x - vt$, $y' = y$, $z' = z$, and $t' = t$. This transformation is called the Galilean transformation. It holds true for speeds much less than the speed of light.

physics intuition! Consider a light source at rest. Light emitted by the source propagates at the speed c. Now consider a frame that is moving with a speed v in the x direction. In this frame, the light source has a speed v. The Galilean transformation predicts that the speed of light waves moving in the minus x direction in the moving frame is equal to $v + c$. The Galilean transformation is not correct. The speed of the light wave in the moving frame is unchanged (c). This is true even if the speed of the reference frame (v) is very large. The relative speeds of stars in distant galaxies with respect to the earth is an appreciable fraction of the speed of light, but the light from all stars propagates to the earth at a speed c. The speed of electromagnetic radiation is a universal constant.

The Michelson-Morley Experiment

Since sound waves need a medium for propagation, it was natural to suspect that electromagnetic waves might also need a medium for propagation. The hypothetical medium was called the *ether*. Albert Michelson designed an experiment in 1881 that was sensitive to the possible presence of an ether. This experiment was repeated by Michelson and Edward Morley with greater sensitivity in 1887. If the propagation of light waves depended on the ether, then motion with respect to the ether would necessarily have an effect on the speed of light. Search for the existence of an ether is equivalent to testing the second postulate of special relativity. The objective of the Michelson-Morley experiment was to see if the motion of the earth could affect the speed of light. The speed of the earth in its orbit about the sun is 3×10^4 m/s, or

$$\frac{v}{c} \approx \frac{3 \times 10^4 \text{ m/s}}{3 \times 10^8 \text{ m/s}} = 10^{-4}. \qquad (4.9)$$

If the Galilean transformation was valid, then the speed of light from a source that was directed parallel to the motion of the earth would be increased by one ten-thousandth, while speed of light from a source that was directed antiparallel to the motion of the earth would be decreased by one ten-thousandth.

The Michelson-Morley experiment is shown in Figure 4-2. The waves from a light source are directed onto a partially transmitting mirror. The mirror divides the light into two beams that travel at right angles to each other. Each light beam travels a distance of about 1 m and is reflected by a mirror. After multiple reflections, the light beams merge. Suppose for a moment that the total path lengths for the two light beams were *exactly* equal. According to the theory of special relativity, the light beams

arrive at precisely the same time. On the other hand, if the Galilean transformation were to hold true, the speed of light would not be constant along the two paths because of the motion of the earth, and the two light beams would not arrive at the same time. In practice, the two path lengths can never be made *exactly* equal. This means that an interference pattern is observed arising from the unequal path lengths. The light from the two paths arrives out of phase. The experimenters looked for a *change* in the interference pattern when the apparatus was rotated because the rotation causes a change in the direction of travel of the light beams with respect to the direction of motion of the earth. The rotation was accomplished by mounting the entire apparatus on a massive stone that was floated on mercury. The apparatus invented by Michelson is called an *interferometer*. Michelson used interferometers to measure fine structure of spectral lines in 1891 and to compare wavelengths with the standard meter in 1895.

According to the theory of special relativity, the speed of the light beams remains constant and the interference pattern does not change when the apparatus is rotated through 90°.

We now analyze the experiment using the Galilean transformation velocity addition rule (4.8). This analysis will tell us the sensitivity for an experiment to distinguish between the Galilean transformation and the second postulate of special relativity. Consider the case where the light beam is traveling parallel and antiparallel to the motion of the earth. The time (t_1) taken to reach the interferometer is

$$t_1 = \frac{L_1}{c+v} + \frac{L_1}{c-v}, \qquad (4.10)$$

where $2L_1$ is the path length. For the case where the light beam travels perpendicular to the motion of the earth, the time (t_2) taken to reach the interferometer is

$$t_2 = \frac{L_2}{\sqrt{c^2 - v^2}} + \frac{L_2}{\sqrt{c^2 - v^2}}. \qquad (4.11)$$

where $2L_2$ is the path length. The light beams are out of phase (ϕ) at the interferometer by

$$\phi = \frac{|c(t_2 - t_1)|}{\lambda}. \qquad (4.12)$$

FIGURE 4-2 The Michelson-Morley experiment.
(*a*) Sketch of the apparatus. (*b*) Path of the light. (*c*) Support system designed for rotation of the apparatus. After A. A. Michelson and E. M. Morley, *Phil. Mag.* **190**, 449 (1887).

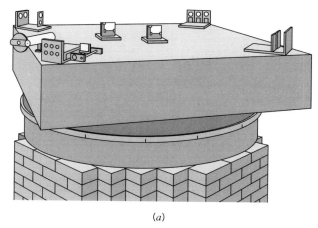

(a)

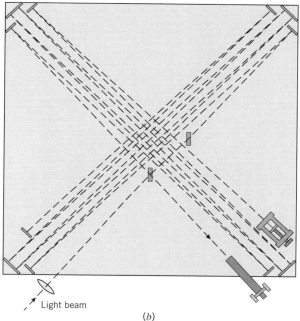

Light beam

(b)

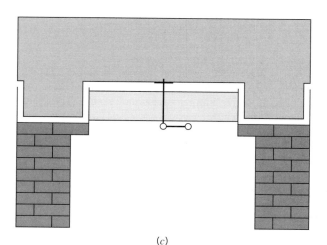

(c)

When the apparatus is rotated by 90°, the change in the phase or *fringe shift* is given by

$$\Delta\phi = \frac{2Lv^2}{\lambda c^2}, \qquad (4.13)$$

where L is the approximate value of the path length, $L = L_1 \approx L_2$ (see problem 4). In the Michelson-Morley experiment $L \approx 10$ m. The wavelength of light is about 500 nm. The expected fringe shift is

$$\Delta\phi = \frac{(2)(10\,\text{m})(10^{-4})^2}{5 \times 10^{-7}\,\text{m}} \approx 0.4 \qquad (4.14)$$

if the light beams have traveled at different speeds because of the earth's motion. No fringe shift was observed. The speed of light does not depend on the motion of the earth. The speed of light is an absolute constant, and there is no ether. Electromagnetic waves can propagate in a vacuum!

The Michelson-Morley experiment has been refined and repeated many times. Several of these results from the period 1881–1930 are summarized in Figure 4-3. On the vertical axis we plot the observed fringe shift and on the horizontal axis we plot the expected fringe shift as calculated from the Galilean transformation. If the speed of light depended on the motion of the earth, then the data would be expected to follow the 45° line. If the speed of light is constant, then zero fringe shift is expected. In practice a small fringe shift (much smaller than that predicted by the Galilean transformation) is observed due to the finite precision of the experimental apparatus. The

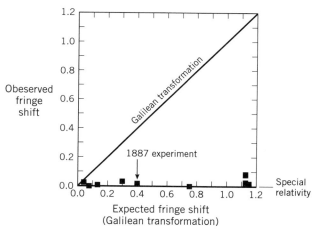

FIGURE 4-3 Results of the Michelson-Morley experiment.
The data are summarized in M. A. Handschy, *Am. J. Phys.* **50**, 987 (1982).

main source of such a tiny fringe shift is a small change in the path length due to temperature variation.

The speed of light in vacuum (c) has been measured to great precision by measurement of the wavelength and frequency of light waves. (This will be discussed further in Chapter 13). In 1983, a new definition of the meter was adopted so as not to limit the accuracy of c. The new definition of the meter is the distance that light travels in vacuum in 1/299792458 second. By definition of the meter, the speed of light is *exactly*

$$c = 2.99792458 \times 10^8 \text{ m/s}. \tag{4.15}$$

Although the speed of light is very large, it is not infinite. It takes a photon a whole nanosecond to travel a distance of 0.3 m, roughly the distance from the page of this book to your eye. It takes 500 seconds for light to travel from the sun to the earth and about 15 billion years for light to travel from the furthest galaxy to the earth!

So far we have been discussing the speed of light in empty space. A medium that is transparent to light, such as air or glass, is characterized by an index of refraction (n). The speed of light is smaller in the medium than in a vacuum by a factor of n. Usually we may neglect this effect in air, where the index of refraction is 1.0003.

> The speed of light (in vacuum) is the same in all frames of reference:
>
> $$c = 2.99792458 \times 10^8 \text{ m/s}.$$

4-2 RELATIONSHIP BETWEEN SPACE AND TIME

Since the definition of speed is the derivative of the space coordinate with respect to time, we see that the second postulate of relativity must have a profound implication on the relationship between space and time. Figure 4-4a shows a plot of the time coordinate of a particle times the speed of light (ct) versus the x coordinate of the particle. This type of plot is called a *space–time diagram*. At $ct = 0$, the present time, the position of the particle is at $x = 0$. The space–time coordinates of the particle may be written as

$$\begin{pmatrix} ct \\ x \end{pmatrix} = \begin{pmatrix} 0 \\ 0 \end{pmatrix}. \tag{4.16}$$

Uniform motion corresponds to a straight line in Figure 4-4a with a slope of (c/v). Since it is impossible for the

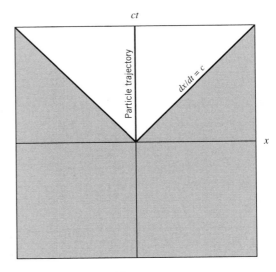

(a) Frame S

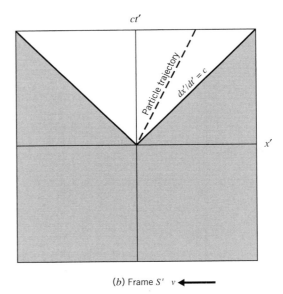

(b) Frame S' v ←

FIGURE 4-4 Space–time diagrams.
The origin corresponds to the present coordinates of a particle. The trajectory of a particle moving at constant speed is a straight line. The 45 degree line represents the speed c. The shaded region is forbidden; the particle can never have those space–time coordinates. (*a*) In frame S the particle is at rest. The particle trajectory is along the vertical axis. (*b*) In a frame S' moving with relative speed v in the negative x direction, the particle has a speed v. The particle trajectory is indicated by the dashed line. Both the time and space coordinates are necessarily changed in this frame; otherwise a large speed would put the particle into the forbidden region.

particle to travel faster than c, the minimum slope is unity. The region below the 45° line corresponds to $v > c$. The particle can never have these space–time coordinates. (You can't get "there" from "here!")

Consider a particle at rest in the frame S. At a later time (t_1), its space–time coordinates are

$$\begin{pmatrix} ct \\ x \end{pmatrix} = \begin{pmatrix} ct_1 \\ 0 \end{pmatrix}. \qquad (4.17)$$

The trajectory of the particle in the space–time diagram is along the vertical axis. The time interval t_1 has passed and the particle has not moved from the position $x = 0$. Now consider the analysis of the particle in the frame S', which is moving with a speed v in the negative x direction. The space–time diagram is shown in Figure 4-4b. The space–time coordinates in this frame are denoted by

$$\begin{pmatrix} ct' \\ x' \end{pmatrix} = \begin{pmatrix} ct_1' \\ x_1' \end{pmatrix}. \qquad (4.18)$$

Both time and the space coordinates of the particle have increased ($t_1' > t_1$ and $x_1' > 0$). In the frame S' the particle has a speed v and follows the trajectory indicated by the dashed line. The particle speed is

$$v = \frac{x_1'}{t_1'}. \qquad (4.19)$$

The time coordinate must increase because if it did not, a large relative speed would put the particle coordinates into the forbidden region. The Galilean transformation is incorrect at large speeds, in part, because it is assumed that the time coordinates in the two frames are identical. Of course, for small speeds, they are very nearly identical.

The Lorentz Transformation

The correct relationship between the coordinates of a stationary frame (t, x, y, and z) and the coordinates in a frame that is moving at a constant speed (t', x', y', and z') was discovered by Hendrik Lorentz in 1890. Lorentz found that a certain transformation of the space and time coordinates left the form of Maxwell's equations unchanged. This transformation is called the *Lorentz transformation*. Maxwell's equations are *invariant* under the Lorentz transformation, that is, Maxwell's equations have the same form in all inertial frames of reference. This is true even for very large speeds, as long as $v < c$. *How could it be that Maxwell's equations in their original form were*

relativistically correct, before the discovery of relativity? The reason is that the magnetic force on a charge moving with $v \ll c$ is so small compared to the electric force, that one needs to include effects of size (v/c) to account for it. This is why the speed of light appears in Ampère's law (see Appendix B).

The Lorentz transformation relating the coordinates t, x, y, and z in a stationary frame to the coordinates t', x', y', and z' in a frame moving with a speed v in the x direction has the following properties:

1. The transformation is a linear function of x and t.
2. The transformation does not change the y and z coordinates.
3. The transformation does not affect the speed of a light wave.
4. Making a second transformation with a speed v in the minus x direction gives back the original space–time coordinates.
5. The transformation reduces to the Galilean transformation at small speeds ($v \ll c$).

The space–time coordinate transformation that has these properties is

$$t' = \gamma \left(t - \frac{xv}{c^2} \right), \qquad (4.20)$$

$$x' = \gamma (x - vt), \qquad (4.21)$$

$$y' = y, \qquad (4.22)$$

$$z' = z, \qquad (4.23)$$

where

$$\gamma \equiv \frac{1}{\sqrt{1 - \dfrac{v^2}{c^2}}}. \qquad (4.24)$$

This is the famous Lorentz transformation. For small speeds ($v \ll c$), we have $\gamma \approx 1$ and the Lorentz transformation reduces to the Galilean transformation. The form of the Lorentz transformation is a direct consequence of the fact that the speed of light is the same in all frames of reference.

We may solve the above equations for t, x, y, and z to get the inverse transformation. We note, however, that the inverse transformation may be obtained by merely exchanging the primed and unprimed variables and also changing the sign of v. The inverse transformation is

$$t = \gamma\left(t' + \frac{x'v}{c^2}\right), \qquad (4.25)$$

$$x = \gamma(x' + vt'), \qquad (4.26)$$

$$y = y', \qquad (4.27)$$

and

$$z = z'. \qquad (4.28)$$

EXAMPLE 4-1

Show that if we transform first in the x direction and then in the minus x direction, with the same speed (v), we end up with the original space–time coordinates.

SOLUTION:

The first transformation gives

$$t' = \gamma\left(t - \frac{xv}{c^2}\right),$$

and

$$x' = \gamma(x - vt)$$

with y and z unchanged. Now let the coordinates ct'', x'', y'', and z'' be the result of the second transformation. The time part is

$$t'' = \gamma\left(t' + \frac{x'v}{c^2}\right)$$

$$= \gamma^2 t - \frac{\gamma^2 xv}{c^2} + \frac{\gamma^2 xv}{c^2} - \frac{\gamma^2 tv^2}{c^2}$$

$$= \gamma^2\left(1 - \frac{v^2}{c^2}\right)t = t.$$

The space part is

$$x'' = \gamma vt' + \gamma x'$$

$$= \gamma^2 vt - \frac{\gamma^2 xv^2}{c^2} + \gamma^2 x - \gamma^2 vt$$

$$= \gamma^2\left(1 - \frac{v^2}{c^2}\right)x = x.$$

The space–time coordinates are unchanged if we transform first in the x direction and then in the minus x direction with the same speed. ∎

The Wave Equation

Lorentz discovered the transformation (4.20–4.23) by analyzing the behavior of the electric and magnetic fields of a charge moving with constant speed. The Lorentz transformation was identified as the transformation of space–time coordinates that did not change the form of Maxwell's equations. Since the electromagnetic wave equation is contained in Maxwell's equations, the Lorentz transformation leaves the form of the wave equation unchanged. The wave equation is

$$\frac{\partial^2 F}{\partial x^2} + \frac{\partial^2 F}{\partial y^2} + \frac{\partial^2 F}{\partial z^2} = \frac{1}{c^2}\frac{\partial^2 F}{\partial t^2}. \qquad (4.29)$$

The Galilean transformation does not preserve the form of the wave equation. Under the Galilean transformation the space derivatives are

$$\frac{\partial F}{\partial x} = \frac{\partial F}{\partial x'}\frac{\partial x'}{\partial x} + \frac{\partial F}{\partial t'}\frac{\partial t'}{\partial x} = \frac{\partial F}{\partial x'}, \qquad (4.30)$$

or

$$\frac{\partial^2 F}{\partial x^2} = \frac{\partial^2 F}{\partial x'^2}, \qquad (4.31)$$

$$\frac{\partial^2 F}{\partial y^2} = \frac{\partial^2 F}{\partial y'^2}, \qquad (4.32)$$

and

$$\frac{\partial^2 F}{\partial z^2} = \frac{\partial^2 F}{\partial z'^2}. \qquad (4.33)$$

The time derivatives are

$$\frac{\partial F}{\partial t} = \frac{\partial F}{\partial t'}\frac{\partial t'}{\partial t} + \frac{\partial F}{\partial x'}\frac{\partial x'}{\partial t} = \frac{\partial F}{\partial t'} - v\frac{\partial F}{\partial x'}, \qquad (4.34)$$

and

$$\frac{\partial^2 F}{\partial t^2} = \left(\frac{\partial}{\partial t'} - v\frac{\partial}{\partial x'}\right)\left(\frac{\partial F}{\partial t'} - v\frac{\partial F}{\partial x'}\right)$$

$$= \frac{\partial^2 F}{\partial t'^2} + v^2\frac{\partial^2 F}{\partial x'^2} - 2v\frac{\partial^2 F}{\partial x'\partial t'}. \qquad (4.35)$$

Using the expressions for the second derivatives (4.31–4.33 and 4.35), the wave equation in the moving frame (primed coordinates) becomes

$$\frac{\partial^2 F}{\partial x'^2} + \frac{\partial^2 F}{\partial y'^2} + \frac{\partial^2 F}{\partial z'^2}$$

$$= \frac{1}{c^2}\left(\frac{\partial^2 F}{\partial t'^2} + v^2\frac{\partial^2 F}{\partial x'^2} - 2v\frac{\partial^2 F}{\partial t'\partial x'}\right). \quad (4.36)$$

The form of the wave equation has changed because of the appearance of the last two terms. Note, however, that we get the wave equation back in the limit $v \to 0$.

The Lorentz transformation, however, preserves the form of the wave equation. Under the Lorentz transformation,

$$\frac{\partial F}{\partial x} = \frac{\partial F}{\partial x'}\frac{\partial x'}{\partial x} + \frac{\partial F}{\partial t'}\frac{\partial t'}{\partial x} = \gamma\frac{\partial F}{\partial x'} - \frac{\gamma v}{c^2}\frac{\partial F}{\partial t'}, \quad (4.37)$$

and

$$\frac{\partial F}{\partial t} = \frac{\partial F}{\partial t'}\frac{\partial t'}{\partial t} + \frac{\partial F}{\partial x'}\frac{\partial x'}{\partial t} = \gamma\frac{\partial F}{\partial t'} - \gamma v\frac{\partial F}{\partial x'}. \quad (4.38)$$

Applying the derivatives twice, the components of the wave equation are

$$\frac{\partial^2 F}{\partial x^2} = \gamma^2\frac{\partial^2 F}{\partial x'^2} + \frac{\gamma^2 v^2}{c^4}\frac{\partial^2 F}{\partial t'^2} - \frac{2\gamma^2 v}{c^2}\frac{\partial^2 F}{\partial x'\partial t'}, \quad (4.39)$$

and

$$\frac{1}{c^2}\frac{\partial^2 F}{\partial t^2} = \frac{\gamma^2}{c^2}\frac{\partial^2 F}{\partial t'^2} + \frac{\gamma^2 v^2}{c^2}\frac{\partial^2 F}{\partial x'^2} - \frac{2\gamma^2 v}{c^2}\frac{\partial^2 F}{\partial x'\partial t'}. \quad (4.40)$$

The y and z derivatives are the same as before. Using our calculated second derivatives (4.32, 4.33, 4.39, and 4.40), we have

$$\left(\gamma^2 - \frac{\gamma^2 v^2}{c^2}\right)\frac{\partial^2 F}{\partial x'^2} + \frac{\partial^2 F}{\partial y'^2} + \frac{\partial^2 F}{\partial z'^2}$$

$$= \left(\gamma^2 - \frac{\gamma^2 v^2}{c^2}\right)\frac{1}{c^2}\frac{\partial^2 F}{\partial t'^2}, \quad (4.41)$$

or

$$\frac{\partial^2 F}{\partial x'^2} + \frac{\partial^2 F}{\partial y'^2} + \frac{\partial^2 F}{\partial z'^2} = \frac{1}{c^2}\frac{\partial^2 F}{\partial t'^2}, \quad (4.42)$$

because

$$\left(\gamma^2 - \frac{\gamma^2 v^2}{c^2}\right) = \gamma^2\left(1 - \frac{v^2}{c^2}\right) = 1 \quad (4.43)$$

from the definition of γ (4.24). The Lorentz transformation preserves the form of the wave equation.

Transformation of Velocities

One important property of the Lorentz transformation is that it contains the experimental fact that the speed of light does not depend on the motion of the source. We now investigate this important property. Consider a particle in the frame S moving with a velocity dx/dt in the x direction. In a frame S', defined to be moving with a velocity v in the x direction relative to the frame S, the same particle has a speed dx'/dt'. The determination of dx'/dt' in terms of dx/dt is obtained from the Lorentz transformation by differentiation of the coordinates. From the Lorentz transformation (4.20 and 4.21), we have

$$\frac{dx'}{dt'} = \frac{dx - v\,dt}{\left(dt - \frac{v\,dx}{c^2}\right)} = \frac{\left(\frac{dx}{dt} - v\right)}{\left(1 - \frac{v}{c^2}\frac{dx}{dt}\right)}. \quad (4.44)$$

The velocity transformation (4.44) is illustrated in Figure 4-5 for the case $v = -dx/dt$.

Let us examine our result in some limiting cases. When the relative speed of the two reference frames is small compared to c, then $v/c \approx 0$ and the particle speed (4.44) is

$$\frac{dx'}{dt'} = \frac{dx}{dt} - v, \quad (4.45)$$

which is the Galilean velocity addition rule (4.8).

If the particle is a photon, then $dx/dt = c$ and the transformed photon speed (4.44) is

$$\frac{dx'}{dt'} = \frac{(c - v)}{\left(1 - \frac{v}{c^2}c\right)} = c. \quad (4.46)$$

The speed of the photon is unchanged. This is the second postulate of special relativity. The Lorentz transformation contains this important physics.

A corollary of the second postulate of relativity is that no particle may move faster than the speed of light (c). No particle has ever been observed to move at a speed faster than c.

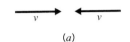

(a)

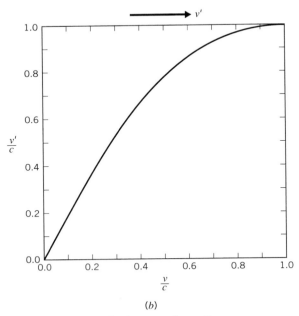

(b)

FIGURE 4-5 The velocity transformation.
(a) Two particles each have a speed v traveling in opposite directions. (b) In the frame where one of the particles is at rest, the other particle has a speed v' given by the velocity transformation (4.44).

> No particle has ever been observed to move at a speed faster than c.

In the limit where v approaches c, we have

$$\frac{dx'}{dt'} = \frac{\left(\dfrac{dx}{dt} - c\right)}{\left(1 - \dfrac{c}{c^2}\dfrac{dx}{dt}\right)} = c. \qquad (4.47)$$

Thus, in a frame that is moving with a speed approaching the speed of light, the particle speed approaches c. In no case is it possible for dx/dt, dx'/dt', or v to be larger than c. This important result is verified directly by many experiments.

We now examine the more general case where the particle has velocity components dx/dt, dy/dt, and dz/dt in the x, y, and z directions in the frame S. In the frame S' the x component of velocity, dx'/dt', is given by the transfor-

mation (4.44) because the coordinates x' and t' do not depend on the coordinates y and z. Although the coordinates y' and z' are unchanged, the velocity components dy'/dt' and dz'/dt' are changed because the time coordinate t' is not equal to the time coordinate t. The Lorentz transformation (4.22) gives $dy' = dy$ and the y component of velocity is

$$\frac{dy'}{dt'} = \frac{dy}{\gamma\left(dt - \dfrac{v\,dx}{c^2}\right)} = \frac{\dfrac{dy}{dt}}{\gamma\left(1 - \dfrac{v}{c^2}\dfrac{dx}{dt}\right)}. \qquad (4.48)$$

The y component of velocity in the moving frame dy'/dt' depends on both the x and y components of velocity in the stationary frame (dx/dt and dy/dt) because the *total* speed of the particle cannot exceed c. Similarly, the z component of velocity is

$$\frac{dz'}{dt'} = \frac{dz}{\gamma\left(dt - \dfrac{v\,dx}{c^2}\right)} = \frac{\dfrac{dz}{dt}}{\gamma\left(1 - \dfrac{v}{c^2}\dfrac{dx}{dt}\right)}. \qquad (4.49)$$

> If a particle has a velocity (dx/dt, dy/dt, dz/dt) in the frame S, then in the frame S', which is moving in the x direction with a speed v, the velocity components of the particle are
>
> $$\frac{dx'}{dt'} = \frac{\left(\dfrac{dx}{dt} - v\right)}{\left(1 - \dfrac{v}{c^2}\dfrac{dx}{dt}\right)},$$
>
> $$\frac{dy'}{dt'} = \frac{\dfrac{dy}{dt}}{\gamma\left(1 - \dfrac{v}{c^2}\dfrac{dx}{dt}\right)},$$
>
> and
>
> $$\frac{dz'}{dt'} = \frac{\dfrac{dz}{dt}}{\gamma\left(1 - \dfrac{v}{c^2}\dfrac{dx}{dt}\right)}.$$

You may have the feeling that the velocity equations (4.44, 4.48 and 4.49) are not physically intuitive. We shall find that in most practical applications of special relativity, the velocity is not a useful variable. The reason for this is that in the extreme relativistic regime, particles move at speeds close to c. It is not easy to see the physical difference between a particle moving at the speed $0.99c$ and $0.999c$, but there is a big difference! The velocity transformation is an important result in that it illustrates the second postulate of special relativity. It is not too useful as a calculational tool. We shall see that the variables of choice for the description of particles are energy and momentum.

EXAMPLE 4-2

In the frame S, two electrons approach each other, each having a speed $v = c/2$. What is the relative speed of the two electrons?

SOLUTION:

The relative speed of the two electrons is the speed of one of the electrons in the frame where the other electron is at rest. Let the frame S' move with a speed $c/2$ in the minus x direction. In the frame S', one of the electrons is at rest and the other electron is moving in the x direction. The speed of the moving electron in the frame S' is

$$\frac{dx'}{dt'} = \frac{\left(\frac{dx}{dt} - v\right)}{\left(1 - \frac{v}{c^2}\frac{dx}{dt}\right)} = \frac{\left(\frac{c}{2} - \frac{-c}{2}\right)}{\left(1 - \frac{-c}{2c^2}\frac{c}{2}\right)} = \frac{4c}{5}. \quad \blacksquare$$

EXAMPLE 4-3

In the frame S, an electron has velocity $c/2$ in the x direction and a photon has a velocity c in the y direction. What is the relative speed of the electron and the photon?

SOLUTION:

We immediately know the answer from the second postulate of special relativity; the relative speed of the electron and photon is c. Let us see that this is so from the velocity transformation. Let the frame S' move with a speed $c/2$ in the x direction. In the frame S', the electron is at rest. The velocity components of the photon in the frame S' (4.44 and 4.48)

$$\frac{dx'}{dt'} = \frac{\left(\frac{dx}{dt} - v\right)}{\left(1 - \frac{v}{c^2}\frac{dx}{dt}\right)} = \frac{\left(0 - \frac{c}{2}\right)}{\left[1 - \frac{c}{2c^2}(0)\right]} = \frac{-c}{2},$$

and

$$\frac{dy'}{dt'} = \frac{\frac{dy}{dt}}{\gamma\left(1 - \frac{v}{c^2}\frac{dx}{dt}\right)}.$$

The gamma factor is

$$\gamma = \frac{1}{\sqrt{1 - \frac{v^2}{c^2}}} = \sqrt{\frac{4}{3}}.$$

The y component of velocity of the photon is

$$\frac{dy'}{dt'} = \frac{c}{\sqrt{\frac{4}{3}}\left[1 - \frac{c}{2c^2}(0)\right]} = \frac{\sqrt{3}}{2}c.$$

The z component of velocity of the photon is zero. The speed of the photon is

$$\sqrt{\left(\frac{dx'}{dt'}\right)^2 + \left(\frac{dy'}{dt'}\right)^2 + \left(\frac{dz'}{dt'}\right)^2}$$
$$= \sqrt{\frac{c^2}{4} + \frac{3c^2}{4} + 0} = c. \quad \blacksquare$$

Time Dilation

Consider a time interval (Δt) measured at a fixed position (x_0) in a stationary frame,

$$\Delta t = t_2 - t_1. \quad (4.50)$$

If we use the Lorentz transformation (4.20) to calculate the time interval $(\Delta t')$ in a frame moving with a speed v, we arrive at the result

$$\Delta t' = t_2' - t_1'$$
$$= \gamma\left(t_2 - \frac{x_0 v}{c^2}\right) - \gamma\left(t_1 - \frac{x_0 v}{c^2}\right)$$
$$= \gamma(t_2 - t_1) = \gamma\Delta t. \quad (4.51)$$

The time interval is longer in the moving frame. This result is called *time dilation*.

Time dilation is a direct consequence of the second postulate of special relativity. The time interval measured in the frame where a clock is at rest is called the *proper time*. Time intervals measured in other frames are longer. The physics of time dilation can be illustrated by the following example. Consider the length of time (Δt) that it takes light to travel the distance d from point A to point B in frame S as shown in Figure 4-6a. The "clock" in this case is apparatus that detects the light. Since the speed of light is c, we have

$$\Delta t = \frac{d}{c}. \tag{4.52}$$

Now consider the time interval for the same event, light traveling from point A to point B, in a moving frame (S') as shown in Figure 4-6b. The second postulate of special relativity states that the speed of light is constant. Examination of the velocity vector diagram shows that since the horizontal component of velocity is v and the net speed is c, that the vertical component of velocity is $(c^2 - v^2)^{1/2}$. Therefore,

$$\Delta t' = \frac{d}{\sqrt{c^2 - v^2}} = \frac{d}{c\sqrt{1 - \dfrac{v^2}{c^2}}} = \frac{\gamma d}{c}. \tag{4.53}$$

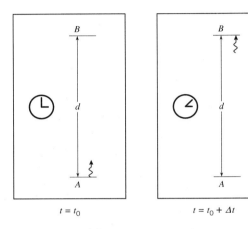

$t = t_0$ $t = t_0 + \Delta t$

(*a*) Clock at rest (frame S)

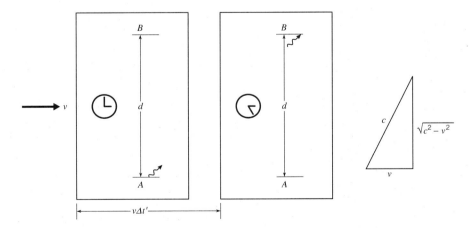

(*b*) Moving clock (frame S')

FIGURE 4-6. Time dilation.
(*a*) The clock is at rest. The time for the light to travel from point A to point B is $\Delta t = d/c$. The time measured in the frame where the clock is at rest is called the proper time. (*b*) The clock is moving with a speed v. Because the speed of light is constant, the time for the light to travel from point A to point B is $\Delta t' = d/(c^2 - v^2)^{1/2}$.

The time interval measured in frame S' is longer than in frame S by a factor of γ,

$$\Delta t' = \gamma \Delta t. \qquad (4.54)$$

The difference between the two frames is that the clock is stationary in frame S and the clock is moving in frame S'. For any time interval measurement, there is one special frame, the frame in which the clock is at rest. In the case of a particle that undergoes spontaneous radioactive decay, the clock is the particle itself and the particle lifetime is shortest in the frame in which the particle is at rest. In all other frames, the particle lifetime is observed to be longer.

EXAMPLE 4-4

A pion is created in a particle collision with a large speed, such that $\gamma = 100$, and it is observed to travel a distance of 300 m before it spontaneously decays. How long does the pion live in its rest frame?

SOLUTION:

Since $\gamma = 100$, the speed of the pion is very nearly equal to the speed of light,

$$\frac{v}{c} = \sqrt{1 - \frac{1}{\gamma^2}} = \sqrt{1 - \frac{1}{10^4}} \approx 1.$$

The lifetime in the frame in which the pion travels 300 m is

$$\Delta t' = \frac{d}{v} = \frac{300\,\text{m}}{3 \times 10^8\,\text{m/s}} = 10^{-6}\,\text{s}.$$

The lifetime of the pion in its rest frame, the proper lifetime, is

$$\Delta t = \frac{\Delta t'}{\gamma} = \frac{10^{-6}\,\text{s}}{100} = 10^{-8}\,\text{s}. \qquad \blacksquare$$

Length Contraction

The second postulate of special relativity also leads to a phenomenon called *length contraction*. The length measured in the frame in which the object is at rest is called the *proper length*. Consider a stick moving with speed v in frame S as shown in Figure 4-7a. The length of the stick (L) may be determined by measuring the time for the stick to pass a stationary clock,

$$L = v \Delta t. \qquad (4.55)$$

We could also make a length measurement (L') in a frame

in which the stick is at rest, as shown in Figure 4-7b. In this frame, the clock is moving with a speed v,

$$L' = v \Delta t'. \qquad (4.56)$$

We know how the two time intervals are related by the time dilation rule: The time interval is longer by a factor of γ in the frame where the clock is moving,

$$\Delta t' = \gamma \Delta t. \qquad (4.57)$$

Therefore,

$$L = L' \frac{\Delta t}{\Delta t'} = \frac{L'}{\gamma}. \qquad (4.58)$$

The length of the stick is longest in the frame where the stick is at rest ($L' > L$). In a frame where the stick is moving in a direction parallel to its length, the stick is measured to be shorter by a factor of γ. Length contraction applies to any two points in space that may be considered connected by an imaginary stick.

EXAMPLE 4-5

A pion is created in a particle collision with $\gamma = 100$, and it is observed to travel a distance of 300 m before it spontaneously decays. What is the decay distance in the pion rest frame?

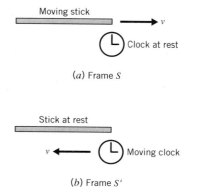

(a) Frame S

(b) Frame S'

FIGURE 4-7 Length contraction.
(a) The length of a stick is measured in a frame S, where it is moving. The measurement is made by measuring the time that it takes to pass from end to end past a stationary clock. (b) The stick is measured in a frame S', where it is stationary. The measurement is made by measuring the time that it takes a moving clock to pass from end to end. The length measured in this frame where the stick is at rest is called the proper length. The length measured in frame S is shorter.

SOLUTION:

Let the pion be created at point A and decay at point B. In the frame where the pion is moving, the distance from A to B is not moving and is 300 m. In the rest frame of the pion, both points A and B are moving, and the distance from A to B (L') is contracted:

$$L' = \frac{L}{\gamma} = \frac{300\,\text{m}}{100} = 3\,\text{m}. \qquad \blacksquare$$

Figure 4-8 illustrates time dilation and length contraction. A particle is created with speed v at position x_1' and spontaneously decays at position x_2'. The particle lives for a time $\Delta t'$ and travels a distance $\Delta x' = (x_1' - x_2')$. This decay-length distance is the proper length in this frame. The speed of the particle is

$$v = \frac{\Delta x'}{\Delta t'}. \qquad (4.59)$$

In the rest frame of the particle, the lifetime is shorter:

$$\Delta t = \frac{\Delta t'}{\gamma}. \qquad (4.60)$$

This decay time is the proper time. The decay length is also shorter:

$$\Delta x = x_1 - x_2 = \frac{x_1' - x_2'}{\gamma} = \frac{\Delta x'}{\gamma}. \qquad (4.61)$$

The speed of the frame is

$$v = \frac{\Delta x}{\Delta t} = \frac{\Delta x'}{\gamma \Delta t} = \frac{\Delta x'}{\Delta t'}. \qquad (4.62)$$

The speed of the frame in Figure 4-8b must be equal to the speed of the particle in Figure 4-8a by definition of the rest frame. We see that the time dilation and length contraction gamma factors cancel each other.

There is an excellent example of time dilation and length contraction occurring constantly in nature all around us. Cosmic ray muons are generated in the upper atmosphere, a few kilometers above the surface of the earth. The muons are created with a large speed ($v \approx c$), corresponding to a $\gamma \approx 20$ on the average. A large flux of these cosmic ray muons are observed at sea level (about $180\,\text{m}^{-2} \cdot \text{s}^{-1}$). The mean proper lifetime of the muon (the average time that a muon lives in its rest frame) is

$$\tau_0 = 2.2\,\mu\text{s}. \qquad (4.63)$$

Without time dilation, the muon could travel a mean distance of only

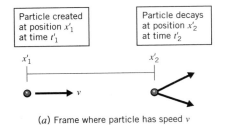

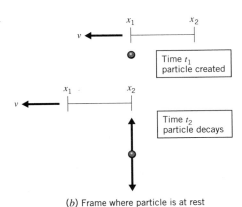

(a) Frame where particle has speed v

(b) Frame where particle is at rest

FIGURE 4-8 Time dilation and length contraction in particle decay.
(a) A particle is created with a speed v at the space–time coordinates (ct_1', x_1'). The particle decays at the coordinates (ct_2', x_2'). The particle speed is given by $v = (x_2' - x_1')/(t_2' - t_1')$. (b) In the frame where the particle is at rest, the proper lifetime is $\Delta t = (t_2 - t_1) = (t_2' - t_1')/\gamma$. The distance $(x_2' - x_1')$ is length contracted to $(x_2 - x_1) = (x_2' - x_1')/\gamma$. The speed of the frame is $v = (x_2 - x_1)/(t_2 - t_1)$, which is the same as the particle speed in the previous frame.

$$L = c\tau_0$$
$$= \left(3.0 \times 10^8\,\text{m/s}\right)\left(2.2 \times 10^{-6}\,\text{s}\right)$$
$$\approx 660\,\text{m}. \qquad (4.64)$$

Including the time dilation factor, the muon travels a mean distance of

$$L_0 = \gamma c\tau_0 \approx (20)(660\,\text{m}) \approx 1.3 \times 10^4\,\text{m}. \qquad (4.65)$$

The cosmic ray muons are observed to travel distances of several thousand meters to reach the surface of the earth, and the Lorentz γ factor is needed to explain this. In the muon rest frame the lifetime is shortest and the earth is moving toward it so the distance to the earth's surface is length contracted by a factor of γ. In the rest frame of the earth, the muon is moving and lives longer than the proper lifetime by a factor of γ. You may find this example of the cosmic ray muon to be a useful tool in sorting out from first principles what times are dilated and what distances are contracted in many problems. It is much more practical than trying to remember an equation!

Simultaneity

The space–time relationship of special relativity forces us to give up the notion of universal simultaneity. Events that occur at the same time in one frame of reference do not necessarily occur at the same time in another frame of reference. At first, this may seem counterintuitive. Consider the following example, as illustrated in Figure 4-9. In the frame S, a pion (π^0) is produced with a speed v in the

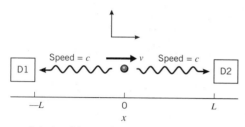

Detectors D1 and D2 are hit at the same time.

(a) Frame S

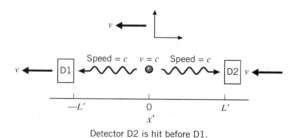

Detector D2 is hit before D1.

(b) Frame S'

FIGURE 4-9 Simultaneity.
(a) In frame S, a particle is produced with a speed v, and when it is at $x = 0$, the particle decays into two photons. Detectors D1 and D2 are located at $x = -L$ and $x = L$. Since the speed of each photon is c, the photons arrive at the two detectors at the same time. (b) The same event is analyzed in the frame where the particle is at rest (frame S'). Detector D2 is moving toward the photon that hits it, and detector D1 is moving away from the photon that hits it. The speed of each photon is equal to c. Therefore, detector D2 gets hit before D1. Events that are simultaneous in frame S are not simultaneous in frame S'.

x direction. When the pion is at the space–time coordinates $(ct, x) = (0,0)$, it spontaneously decays into two photons. One of these photons travels in the negative x direction and strikes a detector (D1) that is placed at the location $x_1 = -L$. The other photon travels in the positive x direction and strikes a detector (D2) that is placed at $x_2 = L$. The speed of each photon is equal to c. The photon traveling in the negative x direction strikes the detector at a time $t_1 = L/c$, and the other photon strikes the other detector at a time $t_2 = L/c$:

$$\Delta t = t_2 - t_1 = \frac{L}{c} - \frac{L}{c} = 0. \qquad (4.66)$$

In the frame S, the two photons strike the detector at the same time.

Now analyze the pion decay in the rest frame of the pion (frame S'). To evaluate the space–time coordinates in the moving frame, we do a Lorentz transformation with a speed v in the negative x direction. We want to evaluate the times t_1' and t_2'. The Lorentz transformation gives

$$ct_1' = \gamma ct_1 + \frac{\gamma v x_1}{c}, \qquad (4.67)$$

and

$$ct_2' = \gamma ct_2 + \frac{\gamma v x_2}{c}. \qquad (4.68)$$

The time difference is

$$\begin{aligned} c\Delta t' &= ct_2' - ct_1' \\ &= \left(\gamma ct_2 + \frac{\gamma v x_2}{c}\right) - \left(\gamma ct_1 + \frac{\gamma v x_1}{c}\right) \\ &= \frac{\gamma v (x_2 - x_1)}{c} = \frac{2Lv\gamma}{c}, \end{aligned} \qquad (4.69)$$

where we have used the fact that $(t_2 - t_1) = 0$. Therefore,

$$\Delta t' = \frac{2Lv\gamma}{c^2}. \qquad (4.70)$$

The photons do not strike the two detectors at the same time when the situation is analyzed in the frame S'. The time t_2' is larger than t_1', so that the detector D2 is struck before the detector D1. The reason for this is that the pion is traveling toward the detector D2 so that in the rest frame of the pion, detector D2 is moving toward the pion and detector D1 is moving away from the pion.

4-3 RELATIONSHIP BETWEEN ENERGY AND MOMENTUM

Classical Approximations

Mass is not a conserved quantity. Particles are created and destroyed in high-energy collisions. In experiments involving the collisions of energetic particles, we observe that kinetic energy may be converted into matter and matter may be converted into kinetic energy. In this sense, matter and energy are interchangeable. Neither mass nor kinetic energy are absolutely conserved.

For analysis of particles with large speeds, we shall need to modify the classical expression for kinetic energy, $mv^2/2$, which is inconsistent with the second postulate of special relativity. A photon has a constant speed (c) and zero mass energy ($mc^2 = 0$). Therefore, $mv^2/2$ for a photon is equal to zero, but a photon does have kinetic energy. We observe that a photon can knock an electron out of an atom.

The classical expression for kinetic energy, $mv^2/2$, does not work for massive particles moving at high speeds, either. The classical expression for kinetic energy predicts a violation of an important experimental observation: No particle has ever been observed to move faster than the speed of light (c).

EXAMPLE 4-6
Using the classical expression for kinetic energy, $E_k = mv^2/2$, calculate the speed of a proton that has a kinetic energy of 300 GeV.

SOLUTION:
The speed of the proton would be given by

$$v = \sqrt{\frac{2E_k}{m}}.$$

Dividing both sides by the speed of light (c), we get

$$\frac{v}{c} = \sqrt{\frac{2E}{mc^2}} \approx \sqrt{\frac{600 \text{ GeV}}{0.94 \text{ GeV}}} \approx 25.$$

The truth is that a proton with a kinetic energy of 300 GeV has a speed of approximately (but not exceeding) c. ∎

The speed of a 300 GeV proton would be 25 times the speed of light if $mv^2/2$ was the correct expression for kinetic energy! The classical expression for kinetic energy, $mv^2/2$, is approximately correct as long as $v/c \ll 1$. It is grossly incorrect for $v/c \approx 1$.

Momentum (**p**) defined as mass times velocity ($m\mathbf{v}$) is a useful quantity in classical mechanics because it is conserved in collisions provided that $v \ll c$. For large speeds, the quantity $m\mathbf{v}$ is not conserved. This can be seen by considering the acceleration of an electron in an electric field. The electrical force causes an increase in momentum given by

$$\mathbf{F} = \frac{d\mathbf{p}}{dt}. \tag{4.71}$$

If the force acts for a long enough time, the momentum would increase without bound. There is no limit to the electron momentum, however, the electron speed can never exceed c. The classical expression ($p = mv$) has a maximum value of mc. The classical expression does not work at large speeds.

Relativistic Momentum

Consider the elastic collision of two particles, A and B, of equal mass (m) as indicated in Figure 4-10. In the frame S' the particles have equal magnitudes of momentum in opposite directions both before and after the collision. Let v represent the speed of each particle in frame S', and let $\beta = v/c$ and $\gamma = (1-\beta^2)^{-1/2}$. Before the collision the velocities of both particles are along the x axis. After the collision the velocities of both particles are along the y axis.

We now analyze the collision in the frame S where particle B is at rest. The speed of particle A before the collision (v_1) is given by the velocity transformation (4.44):

$$v_1 = \frac{v+v}{1+\dfrac{v^2}{c^2}} = \frac{2v}{1+\dfrac{v^2}{c^2}} = \frac{2v}{1+\beta^2}. \tag{4.72}$$

After the collision, the x component of velocity of each particle in the frame S must be equal to the relative speed of the two reference frames (v) according to the velocity transformation (4.44). The quantity mass times velocity is *not conserved* in frame S:

$$mv_1 \neq mv + mv. \tag{4.73}$$

We may see why mass times velocity is not conserved by examining the classical definition of momentum (4.2). The expression $d\mathbf{r}/dt$ is frame dependent; however, the expression $d\mathbf{r}/dt'$, where t' is the proper time, is invariant. The relationship between the differential time intervals is

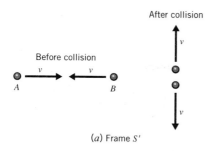

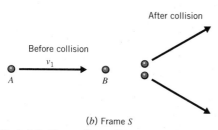

FIGURE 4-10 Particle collision analyzed in two frames.
(a) In frame S' the particles have equal speeds (v) directed in opposite directions. The particles scatter elastically at 90 degrees. (b) In frame S particle B is at rest and the speed of particle A is $v_1 = 2v/(1 + v^2/c^2)$.

$$dt' = \sqrt{1 - \frac{v^2}{c^2}}\, dt = \frac{dt}{\gamma}, \qquad (4.74)$$

We define the *relativistic momentum* to be

$$\mathbf{p} \equiv m\frac{d\mathbf{r}}{dt'} = m\frac{d\mathbf{r}}{dt}\frac{dt}{dt'}. \qquad (4.75)$$

The relativistic momentum (4.75) is observed to be conserved in all processes and is commonly written as

$$\mathbf{p} = \frac{m\mathbf{v}}{\sqrt{1 - \frac{v^2}{c^2}}}, \qquad (4.76)$$

or

$$\mathbf{p} = \gamma m\mathbf{v}, \qquad (4.77)$$

where $\gamma = (1 - v^2/c^2)^{-1/2}$ is defined in terms of the particle speed in the frame where we are evaluating the momentum.

We shall now verify that the relativistic momentum (4.77) is conserved in the collision of Figure 4-10. We may

calculate the gamma factor of particle A before the collision (γ_1) from the speed v_1 (4.72),

$$\gamma_1 = \frac{1}{\sqrt{1 - \frac{v_1^2}{c^2}}} = \frac{1}{\sqrt{1 - \frac{4\beta^2}{(1+\beta^2)^2}}}$$

$$= \frac{1}{\sqrt{\frac{1+\beta^4 + 2\beta^2 - 4\beta^2}{(1+\beta^2)^2}}} = \frac{1+\beta^2}{1-\beta^2}. \qquad (4.78)$$

The total momentum before the collision is

$$p_i^{\text{tot}} = \gamma_1 m v_1$$

$$= \left(\frac{1+\beta^2}{1-\beta^2}\right)(m)\left(\frac{2v}{1+\beta^2}\right) = \frac{2vm}{1-\beta^2}. \qquad (4.79)$$

The magnitude of the y component of velocity of each particle after the collision is $v(1 - v^2/c^2)^{1/2}$. The net speed (v_2) of each particle is

$$v_2 = \sqrt{v^2 + v^2\left(1 - \frac{v^2}{c^2}\right)} = c\beta\sqrt{2 - \beta^2}. \qquad (4.80)$$

We may calculate the gamma factor for each particle after the collision from the speed v_2 (4.80)

$$\gamma_2 = \frac{1}{\sqrt{1 - \frac{v_2^2}{c^2}}} = \frac{1}{\sqrt{1 - \frac{v^2(2-\beta^2)}{c^2}}}$$

$$= \frac{1}{\sqrt{1 - 2\beta^2 + \beta^4}} = \frac{1}{1-\beta^2}. \qquad (4.81)$$

The total momentum after the collision is

$$p_2^{\text{tot}} = \gamma_2 mv + \gamma_2 mv = \frac{2vm}{1-\beta^2}. \qquad (4.82)$$

Therefore,

$$p_1^{\text{tot}} = p_2^{\text{tot}}. \qquad (4.83)$$

The momentum of a particle with nonzero mass (m) is

$$\mathbf{p} = \frac{m\mathbf{v}}{\sqrt{1 - \dfrac{v^2}{c^2}}} = \gamma m\mathbf{v}.$$

Relativistic Energy

Consider a force acting on a particle, for example, an electron in an electric field. The kinetic energy of the particle is given by the expression

$$E_k = \int_0^x dx'\, F' = \int_0^x dx'\, \frac{dp'}{dt'}$$
$$= \int_0^p dp'\, \frac{dx'}{\cdot dt'} = \int_0^p dp'\, v', \qquad (4.84)$$

where the integration variables p' and v' are related by

$$p' = \frac{mv'}{\sqrt{1 - \dfrac{v'^2}{c^2}}}. \qquad (4.85)$$

This expression may be integrated by parts. Using

$$d(p'v') = p'\, dv' + v'\, dp', \qquad (4.86)$$

we have

$$E_k = \left[p'v' \right]_{v'=v} - \int_0^v dv'\, p'$$
$$= \frac{mv^2}{\sqrt{1 - \dfrac{v^2}{c^2}}} - \int_0^v dv'\, \frac{mv'}{\sqrt{1 - \dfrac{v'^2}{c^2}}}. \qquad (4.87)$$

Integrating, we get

$$E_k = \frac{mv^2}{\sqrt{1 - \dfrac{v^2}{c^2}}} - \left[-mc^2 \sqrt{1 - \dfrac{v'^2}{c^2}} \right]_{v'=0}^{v'=v}$$
$$= \frac{mv^2}{\sqrt{1 - \dfrac{v^2}{c^2}}} + mc^2 \sqrt{1 - \dfrac{v^2}{c^2}} - mc^2, \qquad (4.88)$$

or

$$E_k = \frac{mc^2}{\sqrt{1 - \dfrac{v^2}{c^2}}} \left(\frac{v^2}{c^2} + 1 - \frac{v^2}{c^2} \right) - mc^2$$
$$= \frac{mc^2}{\sqrt{1 - \dfrac{v^2}{c^2}}} - mc^2. \qquad (4.89)$$

For small speeds ($v \ll c$) the kinetic energy reduces to the classical form:

$$E_k = \frac{mc^2}{\sqrt{1 - \dfrac{v^2}{c^2}}} - mc^2$$
$$\approx mc^2 \left[1 + \frac{v^2}{2c^2} \right] - mc^2 = \frac{1}{2}mv^2. \qquad (4.90)$$

The momentum and kinetic energy of an electron as a function of its speed are shown in Figure 4-11. At small speeds the momentum is proportional to v and the kinetic energy is proportional to v^2. At large speeds the kinetic energy is equal to pc.

Energy, Mass, and Momentum

The total energy (E) of a particle is defined to be the sum of the mass and kinetic energies,

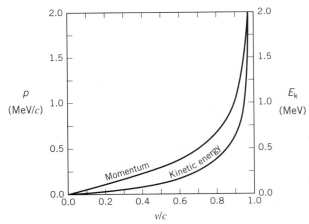

FIGURE 4-11 Momentum and kinetic energy of an electron as a function of speed.

$$E = E_k + mc^2 = \frac{mc^2}{\sqrt{1 - \dfrac{v^2}{c^2}}} = \gamma mc^2. \quad (4.91)$$

The total energy is observed to be conserved in all particle interactions,

$$\sum_{\substack{\text{before} \\ \text{collision}}} E = \sum_{\substack{\text{after} \\ \text{collision}}} E. \quad (4.92)$$

Energy and momentum are closely related because they both contain the factor γm:

$$E = \gamma mc^2 = \frac{pc^2}{v}. \quad (4.93)$$

The particle speed in terms of energy and momentum is

$$\frac{v}{c} = \frac{pc}{E}. \quad (4.94)$$

Thus, the energy may be written

$$E = \gamma mc^2 = \frac{mc^2}{\sqrt{1 - \dfrac{(pc)^2}{E^2}}}. \quad (4.95)$$

Solving for E, we arrive at the expression relating the total energy (E), momentum (p) and mass (m) of a particle:

$$E = \sqrt{(pc)^2 + (mc^2)^2}. \quad (4.96)$$

This is the "master equation" of special relativity. When two out of three of the quantities E, p, and m are known, the energy equation determines the third quantity. The expression for energy (4.96) is universally valid for all particles; there are no exceptions.

The total energy (E), momentum (p) and mass (m) of a particle are related by

$$E = \sqrt{(pc)^2 + (mc^2)^2}.$$

The speed of a particle divided by the speed of light is given by

$$\frac{v}{c} = \frac{pc}{E}.$$

For massless particles ($m = 0$) like the photon, the energy is related to the momentum by

$$E = pc. \quad (4.97)$$

The momentum of a photon is equal to its energy divided by the speed of the photon. This expression for photon momentum is experimentally verified by scattering photons and electrons, a process called *Compton scattering*, and measuring how much momentum is transferred from the photons to the electrons. Compton scattering is discussed later in this chapter.

The energy equation (4.96) invites a convenient unit for momentum of a particle, (eV/c). One eV/c is defined to be one electronvolt divided by the speed of light. By defining this unit, we can save the trouble of dividing by c when calculating momentum from the energy equation. Our use of eV/c as a momentum unit is analogous to our use of eV/c^2 as a mass unit.

EXAMPLE 4-7

A particle has a momentum of 1.0 MeV/c. Express this momentum in SI units.

SOLUTION:

The particle momentum is

$$p = 1.0 \text{ MeV/c}$$

$$= (1.0 \text{ MeV/c}) \left(\frac{1.6 \times 10^{-13} \text{ J}}{1.0 \text{ MeV}} \right) \left(\frac{c}{3 \times 10^8 \text{ m/s}} \right)$$

$$\approx 5.3 \times 10^{-22} \text{ kg} \cdot \text{m/s}. \quad \blacksquare$$

EXAMPLE 4-8

Calculate the momentum of a photon that has an energy of 1 eV.

SOLUTION:

The photon momentum is

$$p = E/c = 1 \text{ eV/c}.$$

Apart from a factor of c, the energy and momentum of a photon are identical. $\blacksquare$

EXAMPLE 4-9

An electron is accelerated through a potential difference of 1.00 megavolts. Calculate the momentum of the electron.

SOLUTION:

The kinetic energy of the electron is

$$E_k = (e)(10^6 \text{ V}) = 1.00 \text{ MeV}.$$

The total energy of the electron is the kinetic energy plus the mass energy:

$$E = E_k + E_0.$$

The momentum of the electron times the speed of light is

$$pc = \sqrt{E^2 - E_0{}^2} = \sqrt{(E_k + E_0)^2 - E_0{}^2}$$
$$= \sqrt{E_0{}^2 + 2E_k E_0},$$

or

$$pc = \sqrt{(1.00 \text{ MeV})^2 + (2)(1.00 \text{ MeV})(0.511 \text{ MeV})}$$
$$= 1.42 \text{ MeV}.$$

The electron momentum is

$$p = 1.42 \text{ MeV}/c.\qquad\blacksquare$$

EXAMPLE 4-10

Calculate the momentum of an electron that has a speed of $c/2$.

SOLUTION:

The gamma factor of the electron is

$$\gamma = \frac{1}{\sqrt{1 - \dfrac{v^2}{c^2}}} = \frac{1}{\sqrt{1 - \dfrac{1}{4}}} = \sqrt{\frac{4}{3}}.$$

The electron momentum is

$$p = \gamma mv = \frac{\gamma mc}{2} = \frac{\gamma mc^2}{2c}$$
$$= \sqrt{\frac{4}{3}}(0.511\,\text{MeV})\left(\frac{1}{2c}\right)$$
$$= 0.295\,\text{MeV}/c.\qquad\blacksquare$$

EXAMPLE 4-11

Calculate the energy of an electron that has a speed of 80% of the speed of light.

SOLUTION:

The gamma factor of the electron is

$$\gamma = \frac{1}{\sqrt{1 - \dfrac{v^2}{c^2}}} = \frac{1}{\sqrt{1 - (0.8)^2}} = \frac{1}{0.6} = \frac{5}{3}.$$

The electron energy is

$$E = \gamma mc^2 = \left(\frac{5}{3}\right)(0.511\,\text{MeV}) = 0.852 \text{ MeV}.\qquad\blacksquare$$

EXAMPLE 4-12

A massless particle has an energy E. Calculate the speed of the particle.

SOLUTION:

The mass energy of the particle is zero

$$mc^2 = 0.$$

The particle energy is

$$E = pc,$$

$$\frac{v}{c} = \frac{pc}{pc} = 1,$$

and

$$v = c.$$

A particle with zero mass always travels at the speed of light. $\blacksquare$

EXAMPLE 4-13

The speed of a particle is c. Calculate the mass of the particle.

SOLUTION:

$$\frac{v}{c} = 1 = \frac{pc}{E}.$$

The particle energy is

$$E = pc.$$

The particle mass energy is

$$mc^2 = \sqrt{E^2 - (pc)^2} = 0.$$

The particle mass is zero. A particle that has a speed equal to c must be massless. Thus, the second postulate of special relativity guarantees that a particle has a speed c if and only if it is massless. ∎

EXAMPLE 4-14

Calculate the total energy, kinetic energy, speed, and gamma factor of an electron that has a momentum of 1.00 MeV/c.

SOLUTION:

The total energy is

$$E = \sqrt{(pc)^2 + (mc^2)^2}$$
$$= \sqrt{(1.00 \text{ MeV})^2 + (0.511 \text{ MeV})^2} = 1.12 \text{ MeV}.$$

The kinetic energy is

$$E_k = E - mc^2 = 1.12 \text{ MeV} - 0.511 \text{ MeV} = 0.61 \text{ MeV}.$$

The speed is given by

$$\frac{v}{c} = \frac{pc}{E} = \frac{1.00 \text{ MeV}}{1.12 \text{ MeV}} = 0.89,$$

or

$$v = 0.89 c.$$

The gamma factor is

$$\gamma = \frac{E}{mc^2} = \frac{1.12 \text{ MeV}}{0.511 \text{ MeV}} = 2.19.$$ ∎

EXAMPLE 4-15

A particle at rest with mass M decays into two particles of equal mass m. Calculate the speed of the two decay particles. Give a numerical answer for the decay of a rho particle ($M = 770$ MeV/c^2) into two charged pions ($m = 140$ MeV/c^2).

SOLUTION:

Let γ be the Lorentz gamma factor for the pions. By conservation of energy,

$$Mc^2 = \gamma mc^2 + \gamma mc^2 = 2\gamma mc^2.$$

The gamma factor is

$$\gamma = \frac{M}{2m}.$$

The speed of the pion is given by

$$\frac{v}{c} = \sqrt{1 - \frac{1}{\gamma^2}} = \sqrt{1 - \frac{4m^2}{M^2}}.$$

For the decay of the rho particle into two pions,

$$\frac{v}{c} = \sqrt{1 - \frac{(4)(0.14 \text{ MeV}/c^2)^2}{(0.77 \text{ MeV}/c^2)^2}} = 0.93.$$ ∎

EXAMPLE 4-16

In the LEP accelerator at CERN, electrons are accelerated to energies of about 50 GeV. By how much do the electron speeds deviate from c?

SOLUTION:

The gamma of the electrons is the electron energy (E) divided by the electron mass energy (mc^2):

$$\gamma = \frac{E}{mc^2} = \frac{50 \text{ GeV}}{0.511 \text{ MeV}} \approx 10^5.$$

The electron speed is

$$\frac{v}{c} = \sqrt{1 - \frac{1}{\gamma^2}} \approx 1 - \frac{1}{2\gamma^2}.$$

The deviation from the speed of light is

$$\frac{c - v}{c} = \frac{1}{2\gamma^2} \approx 5 \times 10^{-11}.$$ ∎

Force and Momentum

The classical relationship between force on a particle and the resulting acceleration ($\mathbf{a} = d\mathbf{v}/dt$), Newton's second law of mechanics, is given by

$$\mathbf{F} = \frac{d\mathbf{p}}{dt} = m\frac{d\mathbf{v}}{dt}. \tag{4.98}$$

The force is parallel to the acceleration. Let us examine the relationship between force and acceleration in light of our new definition of momentum. We have

$$\mathbf{F} = \frac{d\mathbf{p}}{dt} = \frac{d(m\mathbf{v}\gamma)}{dt} = m\frac{d(\mathbf{v}\gamma)}{dt}$$
$$= m\gamma\frac{d\mathbf{v}}{dt} + m\mathbf{v}\frac{d\gamma}{dt}. \tag{4.99}$$

This expression contains the factor

$$\frac{d\gamma}{dt} = \left(1 - \frac{v^2}{c^2}\right)^{-3/2} \frac{v}{c^2}\frac{dv}{dt} = \frac{\gamma^3 v}{c^2}\frac{dv}{dt}. \quad (4.100)$$

Writing the dot product of the force and velocity as

$$\mathbf{F} \cdot \mathbf{v} = m\gamma v\frac{dv}{dt} + \frac{mv^3\gamma^3}{c^2}\frac{dv}{dt}$$

$$= mv\gamma\frac{dv}{dt}\left(1 + \frac{\gamma^2 v^2}{c^2}\right) = mv\gamma^3\frac{dv}{dt}, \quad (4.101)$$

we see that

$$\frac{d\gamma}{dt} = \frac{\mathbf{F} \cdot \mathbf{v}}{mc^2}, \quad (4.102)$$

and the expression for the force (4.99) becomes

$$\mathbf{F} = m\gamma\frac{d\mathbf{v}}{dt} + \frac{\mathbf{v}(\mathbf{F} \cdot \mathbf{v})}{c^2}$$

$$= m\gamma\frac{d\mathbf{v}}{dt} + \boldsymbol{\beta}(\mathbf{F} \cdot \boldsymbol{\beta}). \quad (4.103)$$

Therefore, the acceleration is

$$\mathbf{a} \equiv \frac{d\mathbf{v}}{dt} = \frac{\mathbf{F} - \boldsymbol{\beta}(\mathbf{F} \cdot \boldsymbol{\beta})}{m\gamma}. \quad (4.104)$$

This is the relativistically correct expression that relates force and acceleration (replaces $\mathbf{a} = \mathbf{F}/m$). The acceleration is not parallel to the force at large speeds because the total speed cannot exceed c, and the speed depends on all spatial components, not just on the component in the direction of the force.

Testing the Formula for Momentum

The relativistic expression for momentum is readily verified by experiment. Consider the motion of a charged particle of mass m and charge q in a magnetic field B. The electromagnetic force on the particle is

$$\mathbf{F} = q\mathbf{v} \times \mathbf{B} = \frac{d\mathbf{p}}{dt} = \frac{d(\gamma m\mathbf{v})}{dt}. \quad (4.105)$$

If the force is perpendicular to the velocity, the magnitude of the velocity does not change, only its direction. We have

$$\frac{d(\gamma m\mathbf{v})}{dt} = \gamma m\frac{d\mathbf{v}}{dt} = q\mathbf{v} \times \mathbf{B}. \quad (4.106)$$

The acceleration is

$$\frac{d\mathbf{v}}{dt} = \frac{q\mathbf{v} \times \mathbf{B}}{\gamma m}. \quad (4.107)$$

The orbit is circular and the acceleration may be written as

$$\frac{v^2}{r} = \frac{qvB}{\gamma m}. \quad (4.108)$$

The momentum is

$$p = \gamma mv = qrB. \quad (4.109)$$

This expression is the same as the nonrelativistic result.

EXAMPLE 4-17
An electron has a radius of curvature of 1 m in a uniform magnetic field of 1 T. What is the momentum of the electron?

SOLUTION:
The momentum of the electron is

$$p = erB = e(1\,\text{m})(1\,\text{T}).$$

One T·m is equal to 1 V·s/m. Therefore,

$$p = erB = e(1\,\text{V} \cdot \text{s/m}) = 1\,\text{eV} \cdot \text{s/m}.$$

The momentum times the speed of light is

$$pc = (1\,\text{eV} \cdot \text{s/m})(3 \times 10^8\,\text{m/s}) = 300\,\text{MeV}.$$

The electron momentum is
$$p = 300\,\text{MeV/c}. \quad \blacksquare$$

In 1901 an experiment was performed by Walter Kaufmann that tested the relationship between momentum and speed. Kaufmann built a spectrometer similar to that used by J. J. Thomson (see Chapter 1). The source of electrons for this experiment was the beta decay of radioactive nuclei, as discovered earlier by Becquerel. Kaufmann simultaneously measured the speed of electrons and their radius of curvature in a magnetic field. From the relationship between momentum and radius of curvature (4.109), we expect that

$$r = \frac{p}{eB} = \frac{\gamma mv}{eB} = \frac{\gamma\beta mc}{eB}. \quad (4.110)$$

The radius of curvature should be proportional to $(\gamma\beta)$. Figure 4-12 shows the data of Kaufmann. These data were taken four years before the theory of special relativity was published by Einstein. From the work of Lorentz, Kaufmann knew about the gamma factor. Kaufmann used these data to deduce "an apparent mass of the electron," but what his data actually determined directly and conclusively was the *momentum* of the electron. These data were the first to show that the momentum of an electron depends on the speed as

$$p = \frac{mv}{\sqrt{1 - \frac{v^2}{c^2}}}. \qquad (4.111)$$

Transformation of Energy and Momentum

The energy and momentum of a particle depend on the reference frame in which they are evaluated. The Lorentz transformation relates the energy and momentum (E, p_x, p_y, p_z) in one frame to the energy and momentum (E', p_x', p_y', p_z') in a frame moving with a relative speed v in the x direction. The transformation of energy and momentum may be obtained by combining the velocity transforma-

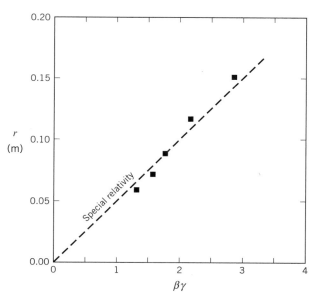

FIGURE 4-12 Measured speed dependence of the momentum of an electron.
These data were taken by W. Kaufmann in 1901, four years before Einstein published his famous paper on special relativity. Kaufmann simultaneously measured the electron speed and radius of curvature in a fixed magnetic field. From W. Kaufmann, *Göttingen Nachrichten*, 143 (1901).

tion (4.44, 4.48, and 4.49) with the definition of energy (4.95) and momentum (4.77). The result (see problem 47) is

$$E' = \gamma E - \beta \gamma p_x c, \qquad (4.112)$$

$$p_x' c = -\beta \gamma E + \gamma p_x c, \qquad (4.113)$$

$$p_y' c = p_y c, \qquad (4.114)$$

$$p_z' c = p_z c, \qquad (4.115)$$

where, as usual, $\gamma = (1 - v^2/c^2)^{-1/2}$ and $\beta = v/c$.

Consider a particle of mass m at rest in the frame S. In the frame S' that moves with a speed v in the x direction, the energy and momentum are given by

$$E' = \gamma m c^2, \qquad (4.116)$$

$$p_x' c = -\beta \gamma m c^2, \qquad (4.117)$$

$$p_y' c = 0, \qquad (4.118)$$

$$p_z' c = 0 \qquad (4.119)$$

This is often referred to as *boosting* a particle. The boosted energy is γmc^2 and the boosted momentum is γmv. These boosted quantities are just the energy and momentum of a particle with a speed v.

The inverse transformation is obtained by changing beta to minus beta:

$$E = \gamma E' + \beta \gamma p_x' c, \qquad (4.120)$$

$$p_x c = \beta \gamma E' + \gamma p_x' c, \qquad (4.121)$$

$$p_y c = p_y' c \qquad (4.122)$$

$$p_z c = p_z' c. \qquad (4.123)$$

EXAMPLE 4-18

Show that if we transform first in the x direction and then in the minus x direction we end up with the original energy and momentum.

SOLUTION:

The first transformation gives

$$E' = \gamma E - \beta \gamma p_x c,$$

and

$$p_x' c = -\beta \gamma E + \gamma p_x c,$$

with $p_y c$ and $p_z c$ unchanged. The inverse transformation gives for the energy

$$E'' = \gamma E' + \beta\gamma\, p_x'\, c$$
$$= \gamma(\gamma E - \beta\gamma\, p_x c) + \beta\gamma(-\beta\gamma E + \gamma\, p_x c)$$
$$= \gamma^2 E - \beta^2 \gamma^2 E = E,$$

and for the x component of momentum

$$p_x''\, c = \beta\gamma E' + \gamma p_x'\, c$$
$$= \beta\gamma(\gamma E - \beta\gamma p_x c) + \gamma(-\beta\gamma E + \gamma p_x c)$$
$$= \gamma^2 p_x c - \beta^2 \gamma^2 p_x c = p_x c.$$

The energy and momentum are unchanged if we transform first in the x direction and then in the minus x direction with the same speed. ∎

EXAMPLE 4-19

Show that the mass of the particle is unchanged by the Lorentz transformation.

SOLUTION:

The mass energy squared of the particle in the moving frame $(m'c^2)^2$ is

$$\left(m'c^2\right)^2 = E'^2 - \left(p_x'^2 c^2 + p_y'^2 c^2 + p_z'^2 c^2\right)$$
$$= \left(\gamma E - \beta\gamma p_x c\right)^2 - \left(-\beta\gamma E + \gamma p_x c\right)^2$$
$$\quad - p_y^2 c^2 - p_z^2 c^2$$
$$= \left(\gamma^2 - \beta^2 \gamma^2\right) E^2 - \left(\gamma^2 - \beta^2 \gamma^2\right)\left(p_x c\right)^2$$
$$\quad - p_y^2 c^2 - p_z^2 c^2$$
$$= E^2 - \left(p_x^2 c^2 + p_y^2 c^2 + p_z^2 c^2\right) = \left(mc^2\right)^2.$$

The transformation has left the mass unchanged. ∎

The Addition of Velocities

We have already seen the velocity addition rule in the context of the space–time coordinates. The Lorentz transformation contains the important result that no particle can travel faster than the speed of light. Let us look at this in terms of energy and momentum. Consider two electrons in the frame S that have speeds $v_1 = \beta_1 c$ and $v_2 = \beta_2 c$ in opposite directions. Our question is: *What is the speed ($v = \beta c$) of one electron in the rest frame of the other?* This speed is the relative speed of the two electrons.

EXAMPLE 4-20

Two electrons are directed toward each other. As measured in the laboratory frame, electron 1 has speed $v_1 = \beta_1 c$ and electron 2 has speed $v_2 = \beta_2 c$. What is the speed of one electron in the rest frame of the other? Give numerical answers for $\beta_1 = \beta_2 = 0.01$, 0.5, and 0.9.

SOLUTION:

Consider electron 1, which has a speed $v_1 = \beta_1 c$. In the laboratory frame, the energy and momentum of this electron are given by

$$E = \gamma_1 mc^2,$$

and

$$pc = \gamma_1 \beta_1 mc^2,$$

where m is the electron mass and

$$\gamma_1 = \frac{1}{\sqrt{1 - \beta_1^2}}.$$

To get the energy and momentum of electron 1 in the frame where electron 2 is at rest, we must make a Lorentz transformation with the speed $v_2 = \beta_2 c$ in the direction that "stops" electron 2 and speeds up electron 1. The energy and momentum of electron 1 are

$$E' = \gamma_2 E + \gamma_2 \beta_2 pc = \gamma_2 \gamma_1 mc^2 + \gamma_2 \beta_2 \gamma_1 \beta_1 mc^2,$$

and

$$p'c = \gamma_2 \beta_2 E + \gamma_2 pc = \gamma_2 \beta_2 \gamma_1 mc^2 + \gamma_2 \gamma_1 \beta_1 mc^2,$$

where

$$\gamma_2 = \frac{1}{\sqrt{1 - \beta_2^2}}.$$

The new speed of the electron (v') is given by

$$\frac{v'}{c} = \frac{p'c}{E'} = \frac{\gamma_2 \beta_2 \gamma_1 mc^2 + \gamma_2 \gamma_1 \beta_1 mc^2}{\gamma_2 \gamma_1 mc^2 + \gamma_2 \beta_2 \gamma_1 \beta_1 mc^2} = \frac{\beta_2 + \beta_1}{1 + \beta_1 \beta_2}.$$

Thus, the transformation contains the correction factor $(1 + \beta_1 \beta_2)^{-1}$. For $\beta_1 = \beta_2 = 0.01$, we have

$$\frac{v'}{c} = \frac{0.01 + 0.01}{1 + (0.01)(0.01)} = 0.02.$$

For $\beta_1 = \beta_2 = 0.5$,

$$\frac{v'}{c} = \frac{0.5 + 0.5}{1 + (0.5)(0.5)} = 0.8.$$

For $\beta_1 = \beta_2 = 0.9$,

$$\frac{v'}{c} = \frac{0.9 + 0.9}{1 + (0.9)(0.9)} = 0.9945. \qquad \blacksquare$$

Doppler Shift

The speed of light emitted from a moving source is c; however, the energy of the individual quanta (photons) depends on the velocity of the source. The shift in energy of a moving photon source compared to the source at rest is called the *Doppler shift*. A shift in energy also means a shift in wavelength and frequency. When the motion of the light source is toward the observer, the photon energy is higher, its frequency is higher, and its wavelength is shorter. This is referred to as a *blue shift*. If the motion of the source is away from the observer, the photon energy is lower, its frequency is lower, and its wavelength is longer. This is referred to as a *red shift*. The change in energy of a photon due to motion of the source (or observer) is easily calculated from the Lorentz transformation. Consider a source of photons with energy E, in the rest frame of the source. Since the photons are massless, their momentum is E/c. When these photons are measured in a frame that is moving toward the photon source with a speed $v = \beta c$, the observed energy is

$$E' = \gamma E + \beta \gamma E = E\gamma(1 + \beta) = E\sqrt{\frac{1+\beta}{1-\beta}}. \quad (4.124)$$

The ratio of photon energies is

$$\frac{E'}{E} = \sqrt{\frac{1+\beta}{1-\beta}}. \qquad (4.125)$$

In terms of the photon wavelength ($E = hc/\lambda$ and $E' = hc/\lambda'$), we have

$$\frac{\lambda'}{\lambda} = \sqrt{\frac{1-\beta}{1+\beta}}. \qquad (4.126)$$

This is the blue shift result.

If the photons are measured in a frame that is moving away from the photon source with a speed $v = \beta c$, the observed energy is

$$E' = \gamma E - \beta \gamma E = E\gamma(1 - \beta) = E\sqrt{\frac{1-\beta}{1+\beta}}. \quad (4.127)$$

The ratio of photon energies is

$$\frac{E'}{E} = \sqrt{\frac{1-\beta}{1+\beta}}. \qquad (4.128)$$

In terms of the photon wavelength ($E = hc/\lambda$ and $E' = hc/\lambda'$), we have

$$\frac{\lambda'}{\lambda} = \sqrt{\frac{1+\beta}{1-\beta}}. \qquad (4.129)$$

This is the red shift result. Figure 4-13 summarizes the Doppler shift. $\qquad \blacksquare$

EXAMPLE 4-21

A neighboring galaxy moves away from us with a relative speed, $v = 0.1\,c$. In the rest frame of the galaxy, photons from the L_α transition in hydrogen have a wavelength $\lambda = 122$ nm. Calculate the wavelength of the photons that are detected on earth.

SOLUTION:

In the rest frame of the galaxy the energy of the photons is

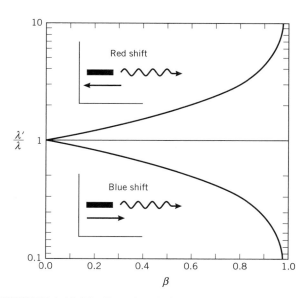

FIGURE 4-13 The Doppler shift.
The photons from a source at rest have a wavelength λ. When there is relative motion of the source toward the observer with speed v, the wavelength of the photons is observed to be blue-shifted to $\lambda' = \lambda[(1 - v/c)/(1 + v/c)]^{1/2}$. When there is relative motion of the source away from the observer with speed v, the wavelength of the photons is observed to be red shifted to $\lambda' = \lambda[(1 + v/c)/(1 - v/c)]^{1/2}$.

$$E = \frac{hc}{\lambda}.$$

The photons are Doppler red shifted. The observed energy of these photons is

$$E' = E\sqrt{\frac{1-\beta}{1+\beta}}.$$

The observed wavelength of the photons is

$$\lambda' = \frac{hc}{E'} = \frac{hc}{E}\sqrt{\frac{1+\beta}{1-\beta}} = (122\,\text{nm})\sqrt{\frac{1.1}{0.9}} = 135\,\text{nm}. \ \blacksquare$$

EXAMPLE 4-22

Calculate the speed of a galaxy relative to the earth if the Doppler red shift causes the photon wavelengths to be doubled.

SOLUTION:

In the rest frame of the galaxy the energy of the photons is

$$E = \frac{hc}{\lambda}.$$

The observed energy of these photons (E') is one-half as large because the wavelength is doubled.

$$E' = \frac{E}{2} = E\sqrt{\frac{1-\beta}{1+\beta}}.$$

Therefore

$$\sqrt{\frac{1-\beta}{1+\beta}} = \frac{1}{2}.$$

Solving for β, we get

$$\beta = \frac{3}{5} = 0.6.$$

The speed of the galaxy is

$$v = \beta c = (0.6)(3\times10^8\,\text{m/s}) = 1.8\times10^8\,\text{m/s}. \ \blacksquare$$

EXAMPLE 4-23

A π^0 particle ($m = 135\,\text{MeV}/c^2$) with an energy equal to 10 times its mass energy decays into two photons. One of the photons goes in the same direction as the π^0 particle and the other photon goes in the opposite direction. What are the resulting energies of the two photons?

SOLUTION:

In the rest frame of the π^0 particle, each photon has an energy

$$E = \frac{mc^2}{2} = \frac{135\,\text{MeV}}{2} = 67.5\,\text{MeV},$$

where m is the mass of the π^0 particle. In the frame where the π^0 particle is moving, both photons are Doppler shifted. The relativistic gamma factor of the pion is

$$\gamma = \frac{E}{mc^2} = 10.$$

The beta factor of the pion is

$$\beta = \sqrt{1 - \frac{1}{\gamma^2}} = \sqrt{0.99} = 0.995.$$

The energy of the blue shifted photon is

$$E' = \gamma E + \beta\gamma E = E\gamma(1+\beta)$$
$$= \left(\frac{135\,\text{MeV}}{2}\right)(10)(0.995) = 672\,\text{MeV}.$$

The energy of the red shifted photon is

$$E' = \gamma E - \beta\gamma E = E\gamma(1-\beta) = \frac{E}{\gamma(1+\beta)}$$
$$= \frac{135\,\text{MeV}}{(2)(10)(1.995)} = 3.38\,\text{MeV}.$$

Note that when beta is close to one, we want to avoid using factors of $(1-\beta)$ in getting a numerical answer. We do this by making use of the relationship: $\gamma^2 = [(1-\beta)(1+\beta)]^{-1}$. $\blacksquare$

* *Challenging*

4-4 FOUR-VECTORS

The second postulate of special relativity relates the space and time coordinates. The energy and momentum of a particle are related in the same fashion. We can make use of this relationship by defining space–time and energy–momentum vectors that each have four components. The main concept of a four-vector is that it is a quantity whose length is an invariant. The length of the space–time four-vector represents the physics that the speed of light is the same in all frames of reference. The length of the energy–momentum four-vector represents the physics that the mass of a particle is the same in all frames of reference. We shall have two benefits from the use of the four-vectors: (1) The Lorentz transformation will appear in a simple form that is very easy to remember, (2) many calculations are made easier with this simple notation.

The Space–Time Four-Vector

The zeroth component of the space–time four-vector (R_0) is defined to be the time coordinate times the speed of light:

$$R_0 \equiv ct. \tag{4.130}$$

The first, second, and third components of the four-vector (R_1, R_2, and R_3) are defined to be the space coordinates

$$R_1 \equiv x, \tag{4.131}$$

$$R_2 \equiv y, \tag{4.132}$$

and

$$R_3 \equiv z. \tag{4.133}$$

We write the four-vector as a column matrix,

$$\mathbf{R} = \begin{pmatrix} ct \\ x \\ y \\ z \end{pmatrix} = \begin{pmatrix} ct \\ \mathbf{x} \end{pmatrix}. \tag{4.134}$$

The dot product between two four-vectors,

$$\mathbf{R}_a = \begin{pmatrix} ct_a \\ \mathbf{x}_a \end{pmatrix}, \tag{4.135}$$

and

$$\mathbf{R}_b = \begin{pmatrix} ct_b \\ \mathbf{x}_b \end{pmatrix}, \tag{4.136}$$

is defined to be

$$\mathbf{R}_a \cdot \mathbf{R}_b \equiv ct_a ct_b - \mathbf{x}_a \cdot \mathbf{x}_b. \tag{4.137}$$

The length of the four-vector is

$$\sqrt{\mathbf{R} \cdot \mathbf{R}} = \sqrt{(ct)^2 - \left(x^2 + y^2 + z^2\right)^2}. \tag{4.138}$$

The length of the four-vector is an invariant; one obtains the same result in every inertial frame. This is a consequence of the fact that the speed of light is constant.

EXAMPLE 4-24
Show that the quantity $r_0{}^2 = \left[(ct)^2 - \left(x^2 + y^2 + z^2\right)\right]$ is unchanged by the Lorentz transformation.

SOLUTION:
Let

$$r_0{}'^2 = (ct')^2 - x'^2 - y'^2 - z'^2.$$

From the Lorentz transformation,

$$r_0{}'^2 = \left(\gamma ct - \frac{\gamma x v}{c}\right)^2 - (\gamma x - \gamma v t)^2 - y^2 - z^2$$

$$= \gamma^2 c^2 t^2 + \frac{\gamma^2 x^2 v^2}{c^2} - \left(\gamma^2 v^2 t^2 + \gamma^2 x^2\right) - y^2 - z^2$$

$$= \gamma^2 \left(1 - \frac{v^2}{c^2}\right) c^2 t^2 - \gamma^2 \left(1 - \frac{v^2}{c^2}\right) x^2 - y^2 - z^2.$$

Notice that

$$\gamma^2 \left(1 - \frac{v^2}{c^2}\right) = 1$$

by definition of γ. Therefore,

$$r_0{}'^2 = (ct)^2 - \left(x^2 + y^2 + z^2\right) = r_0^2.$$

The transformation has left the quantity r_0^2 unchanged. ∎

The Energy and Momentum Four-Vector

The energy and momentum of a particle form a four-vector, analogous to the space–time four-vector. The zeroth component of the energy–momentum four-vector (P_0) is defined to be the total energy:

$$P_0 \equiv E. \tag{4.139}$$

The first, second, and third components of the four-vector (P_1, P_2, and P_3) are defined to be the components of momentum times the speed of light:

$$P_1 \equiv p_x c, \tag{4.140}$$

$$P_2 \equiv p_y c, \tag{4.141}$$

and

$$P_3 \equiv p_z c. \tag{4.142}$$

The complete four-vector is

$$\mathbf{P} = \begin{pmatrix} E \\ p_x c \\ p_y c \\ p_z c \end{pmatrix} = \begin{pmatrix} E \\ \mathbf{p}c \end{pmatrix}. \tag{4.143}$$

The dot-product between two four-vectors,

$$\mathbf{P}_a = \begin{pmatrix} E_a \\ \mathbf{p}_a c \end{pmatrix} \tag{4.144}$$

and

$$\mathbf{P}_b = \begin{pmatrix} E_b \\ \mathbf{p}_b c \end{pmatrix}, \tag{4.145}$$

is

$$\mathbf{P}_a \cdot \mathbf{P}_b = E_a E_b - \mathbf{p}_a \cdot \mathbf{p}_b c^2. \tag{4.146}$$

The length of the four-vector is

$$\sqrt{\mathbf{P} \cdot \mathbf{P}} = \sqrt{E^2 - (pc)^2}. \tag{4.147}$$

The length of the energy–momentum four-vector is the particle mass energy. It is an invariant; one obtains the same result no matter what frame it is evaluated in.

The sum of two four-vectors $(\mathbf{P}_a + \mathbf{P}_b)$ is also a four-vector.

$$\mathbf{P}_a + \mathbf{P}_b = \begin{pmatrix} E_a + E_b \\ \mathbf{p}_a c + \mathbf{p}_b c \end{pmatrix}. \tag{4.148}$$

Its length is an invariant,

$$\begin{aligned}
(\mathbf{P}_a + \mathbf{P}_b)^2 &= (E_a + E_b)^2 - (\mathbf{p}_a c + \mathbf{p}_b c)^2 \\
&= (m_a c^2)^2 + (m_b c^2)^2 \\
&\quad + 2E_a E_b - 2\mathbf{p}_a \cdot \mathbf{p}_b c^2. \tag{4.149}
\end{aligned}$$

It may be evaluated in any frame with identical results.

The Lorentz Transformation as Matrix Multiplication

The four equations of the Lorentz transformation may be written as matrix multiplication. If we represent a four-vector as a (4×1) column vector, then the Lorentz transformation may be written as a (4×4) matrix times the column vector. The result of the matrix multiplication is a new (4×1) column vector whose components are the transformed four-vector. The space–time four-vector transforms as

$$\begin{pmatrix} ct' \\ x' \\ y' \\ z' \end{pmatrix} = \begin{pmatrix} \gamma & -\beta\gamma & 0 & 0 \\ -\beta\gamma & \gamma & 0 & 0 \\ 0 & 0 & 1 & 0 \\ 0 & 0 & 0 & 1 \end{pmatrix} \begin{pmatrix} ct \\ x \\ y \\ z \end{pmatrix}$$

$$= \begin{pmatrix} \gamma ct - \beta\gamma x \\ -\beta\gamma ct + \gamma x \\ y \\ z \end{pmatrix}. \tag{4.150}$$

The energy–momentum four-vector transforms as

$$\begin{pmatrix} E' \\ p_x' c \\ p_y' c \\ p_z' c \end{pmatrix} = \begin{pmatrix} \gamma & -\beta\gamma & 0 & 0 \\ -\beta\gamma & \gamma & 0 & 0 \\ 0 & 0 & 1 & 0 \\ 0 & 0 & 0 & 1 \end{pmatrix} \begin{pmatrix} E \\ p_x c \\ p_y c \\ p_z c \end{pmatrix}$$

$$= \begin{pmatrix} \gamma E - \beta\gamma p_x c \\ -\beta\gamma E + \gamma p_x c \\ p_y c \\ p_z c \end{pmatrix}. \tag{4.151}$$

The inverse transformation is obtained by changing the sign of beta.

Center-of-Mass Energy

Consider the case where a particle with mass m_a, momentum $\mathbf{p}_a$, and energy E_a collides with a particle of mass m_b at rest. If we take the projectile particle to be moving in the x direction, then the sum of the initial four vectors is

$$\begin{pmatrix} E_a \\ p_a c \\ 0 \\ 0 \end{pmatrix} + \begin{pmatrix} m_b c^2 \\ 0 \\ 0 \\ 0 \end{pmatrix} = \begin{pmatrix} E_a + m_b c^2 \\ p_a c \\ 0 \\ 0 \end{pmatrix}. \tag{4.152}$$

Now consider the collision in the *center-of-mass* frame, the frame where the sum of the momenta of the colliding particles is zero. We denote the momentum of each particle in the center-of-mass frame by $(p*)$, and the energies by $(E_a*$ and $E_b*)$. The sum of the four vectors in the center-of-mass frame is

$$\begin{pmatrix} E_a* \\ p*c \\ 0 \\ 0 \end{pmatrix} + \begin{pmatrix} E_b* \\ -p*c \\ 0 \\ 0 \end{pmatrix} = \begin{pmatrix} E_a*+E_b* \\ 0 \\ 0 \\ 0 \end{pmatrix}. \quad (4.153)$$

The total center-of-mass energy ($\sqrt{s}$) is

$$\begin{aligned} \sqrt{s} &= E_a*+E_b* \\ &= \sqrt{(p*c)^2+(m_ac^2)^2} \\ &\quad +\sqrt{(p*c)^2+(m_bc^2)^2}. \end{aligned} \quad (4.154)$$

Since the length of a four-vector is invariant, the total energy squared in the center-of-mass frame is

$$\begin{aligned} s &= (E_a*+E_b*)^2 = (E_a+m_bc^2)^2-(p_ac)^2 \\ &= E_a^2+(m_bc^2)^2+2E_am_bc^2-(p_ac)^2. \end{aligned} \quad (4.155)$$

Using

$$E_a^2 = (p_ac)^2+(m_ac^2)^2, \quad (4.156)$$

we have

$$s = (m_ac^2)^2+(m_bc^2)^2+2E_am_bc^2. \quad (4.157)$$

The total energy in the center-of-mass frame is

$$\sqrt{s} = \sqrt{(m_ac^2)^2+(m_bc^2)^2+2E_am_bc^2}. \quad (4.158)$$

We may use the expression for center-of-mass energy (4.154) to solve for the center-of-mass momentum:

$$\begin{aligned} p*c = \frac{1}{2\sqrt{s}}\Big[s&+(m_ac^2)^2+(m_bc^2)^2-2sm_ac^2 \\ &-2sm_bc^2-2m_ac^2m_bc^2 \Big]^{1/2}. \end{aligned} \quad (4.159)$$

For the center-of-mass energy large compared to the particle masses, we have

$$p*c \approx \frac{\sqrt{s}}{2} \approx \frac{\sqrt{2E_am_bc^2}}{2}. \quad (4.160)$$

EXAMPLE 4-25

Two protons each with energy of 500 GeV travel in the opposite direction and collide. Calculate the energy of one of the protons in the frame where the other proton is at rest.

SOLUTION:

The energy in the center-of-mass is

$$\sqrt{s} = 500\,\text{GeV} + 500\,\text{GeV} = 1000\,\text{GeV}.$$

We have

$$\sqrt{s} = \sqrt{m_ac^2+m_bc^2+2E_am_bc^2} \approx \sqrt{2E_am_bc^2}.$$

The proton energy is

$$E_a \approx \frac{s}{2m_bc^2} = \frac{(1000\,\text{GeV})^2}{(2)(0.94\,\text{GeV})} = 5.3\times10^5\,\text{GeV}. \blacksquare\ *$$

4-5 COMPTON SCATTERING

The photon interacts with all particles that have electric charge. The process of a photon scattering from a charged particle is called Compton scattering after Arthur Compton, who made the first measurements of photon–electron scattering in 1922. Compton studied the process

$$\gamma+e \rightarrow \gamma+e. \quad (4.161)$$

We shall consider the scattering of a photon from an electron at rest, the "laboratory frame." In the initial state (before the scattering) we let the photon have a momentum of p_1. In the final state (after the scattering) the photon has a momentum p_2 and scattering angle θ, and the electron has a momentum p_e and scattering angle ϕ, where both angles are measured with respect to the incident photon direction as shown in Figure 4-14.

Conservation of energy gives

$$p_1c+mc^2 = p_2c+\sqrt{p_e^2c^2+m^2c^4}. \quad (4.162)$$

Conservation of momentum gives

$$p_1c = p_2c\cos\theta+p_ec\cos\phi, \quad (4.163)$$

and

$$0 = -p_2c\sin\theta+p_ec\sin\phi, \quad (4.164)$$

p_1

(a) Before collision

p_2

θ

ϕ

p_e

(b) After collision

FIGURE 4-14 Definition of the variables in Compton scattering.
(a) A photon with momentum p_1 scatters from an electron (mass m) at rest. (b) After the scatter, the photon momentum is equal to p_2 and the electron momentum is equal to p_e. The scattering angle of the photon is θ and the scattering angle of the electron is ϕ.

where m is the electron mass. The electron energy and momentum are given by

$$\sqrt{p_e^2 c^2 + m^2 c^4} = p_1 c + mc^2 - p_2 c, \quad (4.165)$$

$$p_e c \cos\phi = p_1 c - p_2 c \cos\theta, \quad (4.166)$$

and

$$p_e c \sin\phi = p_2 c \sin\theta. \quad (4.167)$$

We may solve for the electron mass energy squared by squaring the electron energy and subtracting the square of the electron momentum:

$$\begin{aligned}
m^2 c^4 &= (p_1 c + mc^2 - p_2 c)^2 \\
&\quad -[(p_1 c - p_2 c \cos\theta)^2 + (p_2 c \sin\theta)^2] \\
&= p_1^2 c^2 + m^2 c^4 + p_2^2 c^2 + 2 p_1 c mc^2 - 2 p_2 c mc^2 \\
&\quad -2 p_1 c p_2 c - p_1^2 c^2 - p_2^2 c^2 \cos^2\theta \\
&\quad +2 p_1 c p_2 c \cos\theta - p_2^2 c^2 \sin^2\theta,
\end{aligned}$$

$$(4.168)$$

which reduces to

$$mc^2 (p_1 c - p_2 c) = p_1 c p_2 c (1 - \cos\theta). \quad (4.169)$$

Solving for $p_2 c$,

$$p_2 c = \frac{mc^2 p_1 c}{mc^2 + p_1 c(1 - \cos\theta)}. \quad (4.170)$$

This is the *Compton formula*, which gives the momentum of the scattered photon (p_2) in terms of the incident photon momentum (p_1), the scattering angle (θ), and the electron mass (m).

We now look at some limiting cases. When the incident photon energy is much smaller than the electron mass energy ($p_1 c \ll mc^2$), we have

$$p_2 c \approx p_1 c, \quad (4.171)$$

and the photon energy is unchanged by the scattering. This corresponds to a photon scattering from a "brick wall." Momentum is transferred but essentially no energy is transferred.

When the scattering angle is zero (forward scattering), we have

$$p_2 c = p_1 c, \quad (4.172)$$

and the photon energy is unchanged for any incident photon energy.

For $\theta = \pi$ (backward scattering), we have

$$p_2 c = \frac{mc^2 p_1 c}{mc^2 + 2 p_1 c}. \quad (4.173)$$

For backward scattering in the case of a large incident photon energy ($p_1 c \gg mc^2$), we have

$$p_2 c = \frac{mc^2}{2}. \quad (4.174)$$

The Compton scattering formula may be written in terms of the photon wavelengths before (λ_1) and after (λ_2) the scattering:

$$p_1 = \frac{h}{\lambda_1}, \quad (4.175)$$

and

$$p_2 = \frac{h}{\lambda_2}. \quad (4.176)$$

We have

$$mc^2 \left(\frac{hc}{\lambda_1} - \frac{hc}{\lambda_2} \right) = \frac{h^2 c^2 (1 - \cos\theta)}{\lambda_1 \lambda_2}, \quad (4.177)$$

or

$$(\lambda_2 - \lambda_1) = \frac{hc}{mc^2}(1 - \cos\theta). \qquad (4.178)$$

The change in wavelength of the photon depends only on the scattering angle θ. The quantity hc/mc^2 is called the *Compton wavelength* of the electron. The numerical value of the Compton wavelength (λ_C) is

$$\lambda_C = \frac{hc}{mc^2} = \frac{1240\,\text{eV}\cdot\text{nm}}{0.511\,\text{MeV}}$$
$$= 2.43 \times 10^{-6}\,\text{m}. \qquad (4.179)$$

The first data obtained by Compton are shown in Figure 4-15. Compton scattered energetic photons that he obtained from nuclear decays, and measured the wavelength at scattering angles of 0°, 45°, 90°, and 135°. The data show that the wavelength of the scattered photons depends linearly on $(1 - \cos\theta)$ and that the constant of proportionality is hc/mc^2.

4-6 DISCOVERY OF THE POSITRON

The *positron* was discovered in 1932 by Carl Anderson. Anderson observed 15 "positive electrons" in a sample of 1300 cosmic ray tracks photographed in a Wilson cloud chamber. The cloud chamber was placed in a magnetic field of 1.5 T. To establish the direction of motion of the particle, a 6-mm-thick lead plate was placed inside the chamber oriented such that the normal to the surface of the plate was perpendicular to the magnetic field. When a charged particle passes through the lead plate, it loses energy by electromagnetic interactions with the electrons in the lead atoms. The radius of curvature of the particle is therefore smaller after it passes through the plate. A cloud chamber photograph of the positron recorded by Anderson is shown in Figure 4-16. The radius of curvature is clearly smaller on the top portion of the photograph, indicating that the particle has traveled from bottom to top. The direction of the magnetic field, which is into the page in Figure 4-16, gives the sign of the electric charge to be positive. The momentum of the particle is determined by measurement of its radius of curvature and direction of travel. Before it enters the lead plate, the momentum of the particle is about 63 MeV/c. The momentum of the particle after it leaves the lead plate is about 23 MeV/c. The key to the identification of this and other similar tracks as being due to positrons is to establish that the particles have a positive charge and a small mass (much smaller than the proton mass).

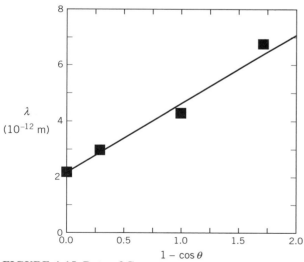

FIGURE 4-15 Data of Compton.
In his original experiment, Compton scattered photons of wavelength $\lambda = 2.2 \times 10^{-12}$ m from electrons and measured the wavelengths of radiation scattered at angles of 0°, 45°, 90°, and 135°. We plot his data in the form λ versus $(1-\cos\theta)$. The data fit a straight line with slope hc/mc^2. From A.H. Compton, *Phys. Rev.* **21**, 483 (1923).

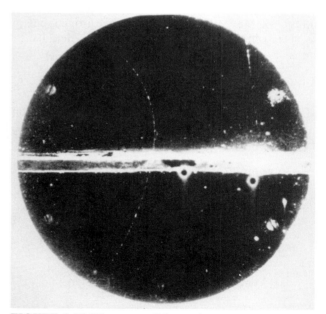

FIGURE 4-16 Discovery of the positron.
From C. Anderson, "The Positive Electron," *Phys. Rev.* **43**, 491 (1933).

EXAMPLE 4-26

Show that the change in momentum of the particle of Figure 4-16 is inconsistent with that of a proton. Find the maximum mass that is consistent with a change in momentum from 63 MeV/c to 23 MeV/c.

SOLUTION:

The momentum loss from passing through the lead plate is

$$\Delta p = 40 \text{ MeV}/c.$$

The minimum energy loss occurs when the particle is relativistic. Nonrelativistic particles lose more energy when passing through the plate because they spend more time experiencing the force of the atomic electrons. Since a relativistic particle has $E \approx pc$, the minimum energy loss (ΔE_{min}) is

$$\Delta E_{min} \approx 40 \text{ MeV}.$$

The kinetic energy of the particle, if it were a proton, would be

$$E_k = \frac{p^2}{2m} = \frac{(63 \text{ MeV}/c)^2}{(2)(940 \text{ MeV}/c)} \approx 2 \text{ MeV}.$$

The particle cannot be a proton because it does not have 40 MeV to lose! To be consistent with a change in momentum of $\Delta p = 40$ MeV/c, the particle must have a kinetic energy of at least 40 MeV. The kinetic energy of a particle of mass M is

$$E_k = \sqrt{(pc)^2 + (Mc^2)^2} - Mc^2.$$

The minimum kinetic energy is given by the expression

$$\sqrt{(pc)^2 + (Mc^2)^2} - Mc^2 = \Delta pc.$$

We solve this expression for the mass,

$$M = \frac{(pc)^2 - (\Delta pc)^2}{2\,\Delta pc}.$$

Numerically, we have

$$Mc^2 = \frac{(63 \text{ MeV})^2 - (40 \text{ MeV})^2}{(2)(40 \text{ MeV})} = 30 \text{ MeV}.$$

The maximum mass that is consistent with this change in momentum is 30 MeV/c^2. ∎

The precise value of the positron mass was measured in subsequent experiments. The mass of a particle mass is determined by measuring both its energy and its momentum. The positron mass is found to be *identical* to the electron mass:

$$m_{positron} = m_{electron}. \tag{4.180}$$

The positron (e^+) is the *antiparticle* of the electron (e^-), and vice versa. An electron–positron pair can annihilate by

$$e^+ + e^- + N \rightarrow \gamma + N, \tag{4.181}$$

where N is a nucleus. A pair can be created by

$$\gamma + N \rightarrow e^+ + e^- + N. \tag{4.182}$$

The processes of pair-production is shown in Figure 4-17.

CHAPTER 4: PHYSICS SUMMARY

- The speed of light in vacuum is

$$c = 3.00 \times 10^8 \text{ m/s},$$

  independent of the motion of the source.

- No particle can have a speed faster than the speed of light in vacuum (c).

- Space and time make a four–vector that transforms as

$$\begin{pmatrix} ct' \\ x' \\ y' \\ z' \end{pmatrix} = \begin{pmatrix} \gamma & -\beta\gamma & 0 & 0 \\ -\beta\gamma & \gamma & 0 & 0 \\ 0 & 0 & 1 & 0 \\ 0 & 0 & 0 & 1 \end{pmatrix} \begin{pmatrix} ct \\ x \\ y \\ z \end{pmatrix} = \begin{pmatrix} \gamma ct - \beta\gamma x \\ -\beta\gamma ct + \gamma x \\ y \\ z \end{pmatrix},$$

  when going from one frame to another frame with relative speed (v) in the x direction, where

$$\beta = \frac{v}{c},$$

  and

$$\gamma = \frac{1}{\sqrt{1 - \beta^2}}.$$

The inverse transformation is obtained by changing β to $-\beta$.

FIGURE 4-17 Electron-positron pair production.
The photograph was taken in a bubble chamber, a device similar to the cloud chamber except the chamber is filled with liquid hydrogen. The bubble chamber records the trajectories of charged particles in a magnetic field. A particle called a K^- meson (bottom of photo) enters the chamber and decays into a π^0 meson and a π^- meson. The π^0 decays into two photons each of which converts into an electron-positron pair on passing through a lead sheet. Lawrence Berkeley Laboratory/Science Photo Library/ Photo Researchers.

- A time interval measured in a frame where the clock is at rest is called the proper time. In all other frames of reference, the observed time interval is longer by a factor of γ, where the speed appearing in γ is the relative speed of the frame. Applied to the decay of a particle, the proper lifetime is the lifetime measured in the rest frame of the particle.

- A distance measured in a frame where the object is at rest is called the proper length. In all other frames of reference, the observed length is shorter by a factor of $1/\gamma$, where the speed appearing in γ is the relative speed of the frame.

- The total energy (E), momentum (p), and mass (m) of a particle are related by

$$E = \sqrt{\left(pc\right)^2 + \left(mc^2\right)^2}.$$

This quantity is observed to be conserved in all particle collisions.

- For the case where mass of a particle is not equal to zero, the momentum is given by

$$\mathbf{p} = \frac{m\mathbf{v}}{\sqrt{1 - \dfrac{v^2}{c^2}}} = m\mathbf{v}\gamma.$$

This quantity is observed to be conserved in particle collisions. An alternate expression for the energy in this case is

$$E = \gamma mc^2.$$

- The speed of a particle divided by the speed of light is

$$\frac{v}{c} = \frac{pc}{E}.$$

- The mass of a particle is the same in all frames of reference.

- Energy and momentum make a four-vector that transforms as

$$\begin{pmatrix} E' \\ p_x{}'c \\ p_y{}'c \\ p_z{}'c \end{pmatrix} = \begin{pmatrix} \gamma & -\beta\gamma & 0 & 0 \\ -\beta\gamma & \gamma & 0 & 0 \\ 0 & 0 & 1 & 0 \\ 0 & 0 & 0 & 1 \end{pmatrix} \begin{pmatrix} E \\ p_x c \\ p_y c \\ p_z c \end{pmatrix}$$

$$= \begin{pmatrix} \gamma E - \beta\gamma p_x c \\ -\beta\gamma E + \gamma p_x c \\ p_y c \\ p_z c \end{pmatrix}.$$

- For the circular motion of a charged particle in a magnetic field, the momentum, charge, radius of curvature, and magnetic field are related by

$$pc = erB.$$

- For the case of a photon scattering off a free electron at rest, the change in wavelength of the photon depends only on the scattering angle,

$$\left(\lambda_2 - \lambda_1 \right) = \frac{hc}{mc^2}\left(1 - \cos\theta \right).$$

- The positron is the antiparticle of the electron. The mass of the positron is identical to the electron mass. Electron–positron pairs are created when energetic photons interact with a nucleus,

$$\gamma + N \to e^+ + e^- + N.$$

REFERENCES AND SUGGESTIONS FOR FURTHER READING

C. Anderson, "The Positive Electron," *Phys. Rev.* **43**, 491 (1933).

A. H. Compton, "A Quantum Theory of the Scattering of X-Rays by Light Elements," *Phys. Rev.* **21**, 483 (1923).

A. H. Compton, "The Scattering of X-Ray Photons," *Am. J. Phys.* **14**, 80 (1946).

G. Gamow, *Mr. Tompkins in Wonderland,* Cambridge Univ. Press, London (1940).

G. Gamow, *Mr. Tompkins Explores the Atom,* Cambridge Univ. Press, London (1945).

J. C. Hafele and R. E. Keating, "Around-the-World Atomic Clocks: Predicted Relativistic Time Gains," *Science* **177**, 166 (1972).

J. C. Hafele and R. E. Keating, "Around-the-World Atomic Clocks: Observed Relativistic Time Gains," *Science* **177**, 168 (1972).

M. A. Handschy, "Re-examination of the 1887 Michelson-Morley Experiment," *Am. J. Phys.* **50**, 987 (1982).

W. Kaufmann, "Magnetic and Electric Deflectability of the Becquerel Rays and the Apparent Mass of the Electron," *Göttingen Nachrichten,* **143** (1901), trans. H. A. Boorse and L. Motz.

A. A. Michelson and E. M. Morley, "On the Relative Motion of the Earth and the Luminiferous Aether," *Phil. Mag.* **190**, 449 (1887).

L. Okun, "The Concept of Mass," *Phys. Today* **42**, No. 6, 31 (1989).

L. S. Swenson, "Michelson and Measurement," *Phys. Today* **40**, No. 5, 24 (1987).

E. F. Taylor and J. A. Wheeler, *Spacetime Physics,* Freeman, New York (1992).

QUESTIONS AND PROBLEMS

The foundations of special relativity

1. Consider two observers each in an inertial frame of reference. What two speed measurements do the two observers agree on?

2. In Newton's theory the gravitational force between two masses depends inversely on the square of their distance of separation. Show that this is inconsistent with the postulates of special relativity. (*Hint:* To calculate the forces between the two masses in Newton's theory, we need to know their relative positions at the same time.)

3. An airplane travels at a constant speed v for a distance of 3000 km as measured by a stationary observer. The pilot measures the flight time to be Δt and the stationary observer measures the flight time to be $\Delta t'$. (a) Which time interval is longer? (b) If $|\Delta t - \Delta t'| = 4$ ns, determine the speed of the airplane. (Time dilation has been directly measured using atomic clocks flown on commercial airplanes.)

4. Show that the phase shift in the Michelson-Morley experiment is given by

$$\Delta\phi = \frac{2Lv^2}{\lambda c^2}.$$

The relationship between space and time

5. An electron is shot with a relativistic speed v in the x direction and another is shot at a time Δt later in the y direction with the same speed. Does the relative speed of the two electrons depend on the time interval Δt?

6. In the laboratory frame a particle with a speed $v = 0.99c$ travels a distance of 1 mm before spontaneously decaying. What is the proper lifetime of the particle?

7. The tau lepton has a proper lifetime of 0.3 ps. What speed must tau particles have in order to travel an average distance of 1 mm?

8. (a) The muon has a proper lifetime of 2.2 μs. If a muon has a speed, $v = 0.99c$, what is the average distance that it travels before decaying? (b) The charged pion has a proper lifetime of 26 ns. If a pion has a speed, $v = 0.99c$, what is the average distance that it travels before decaying?

*9. *The twin paradox.* An astronaut is accelerated in his spaceship to a cruising speed v and travels from the earth to a faraway destination and back, a total distance d. The astronaut's twin stays at home on the earth. Assume that the time needed to reach cruising speed and the time needed to turn the spaceship around are negligible compared to the time taken to complete the trip. (a) Analyze the trip in the frame of the twin that stays home. Which twin has aged more and by how much? (b) Analyze the trip in the frame of the traveling twin. Which twin has aged more and by how much? (c) Which twin has actually aged more and by how much? (Hint: Make use of the first postulate of special relativity.)

10. Relativistic protons that have a certain speed v in the laboratory frame are selected by measuring the time that it takes the proton to travel between two detectors separated by a distance L. Each detector produces an electronic pulse of very short duration ($\Delta t \ll L/v$) when a proton passes through it. A coincidence circuit is made by delaying the pulse from the first detector by an amount L/v. The signals from the two detectors are fed into a logic circuit that produces an output pulse if the input pulses arrive at the same time. For input pulses that arrive at the same time as measured in the laboratory frame, calculate the time difference between arrival of the input pulses as measured in the rest frame of the proton.

11. An observer measures the velocity of two electrons and finds that one has a speed $c/2$ in the x direction and the other has a speed $c/2$ in the y direction. What is the relative speed of the two electrons?

The relationship between energy and momentum

12. What is the speed of a particle that has a momentum equal to its mass times the speed of light?

13. Determine the momentum and speed of a proton that has a kinetic energy of 1.00 GeV.

14. An electron has a kinetic energy of 1 MeV. What is the radius of curvature in a uniform magnetic field of 1 T, if the electron velocity is perpendicular to the field?

15. An electron and a proton have the same radius of curvature in a uniform magnetic field, and the electron speed is twice that of the proton. What is the momentum of each particle?

16. Calculate the radius of curvature of a 10-GeV electron in a magnetic field of 1 T.

17. Two protons and two neutrons bind together to form the nucleus of the helium atom (the alpha particle). The binding energy is 28.4 MeV. Calculate the mass of the alpha particle.

18. The carbon-14 nucleus decays into a nitrogen-14 nucleus plus an electron and a massless particle called the electron-antineutrino. Calculate the amount of energy released in the decay.

19. Lambda particles ($m = 1116$ MeV/c^2) are produced with momenta of 100 GeV/c as the result of energetic proton–proton collisions. The lambda decays into a proton and a pion with a lifetime of 10^{-10} s. In the frame in which the lambda has $p = 100$ GeV/c, which particle has a larger momentum on the average, the proton or the pion? Why?

*20. A Z^0 particle, which has a mass of about 91 GeV/c^2, is produced in proton–antiproton collisions, where each of the protons has an energy of 270 GeV. The proton and antiproton are composite objects, and the fundamental interaction for Z^0 production is the annihilation of a quark and an antiquark. If the quark has an energy of 30 GeV, calculate the energy of the antiquark. (Assume that the quark and antiquark are massless.)

21. An electron has a kinetic energy equal to its mass energy. Find the energy of a photon that has the same momentum as the electron.

22. A relativistic subatomic particle of mass m is moving away from a detector when it spontaneously decays, sending a photon toward the detector. The photon is observed to be red shifted by a factor of 100, that is, its energy is 1% of the energy of the photon measured in the frame where the decaying particle is at rest. What is the speed of the decaying particle?

23. Two electrons have a relative speed of 0.9c. Calculate the momentum of each electron in the center-of-mass frame, the frame where they have equal and opposite momenta.

24. A 10-MeV alpha particle ($mc^2 \approx 3700$ MeV) collides head on with an electron at rest. Find the speed of the electron after the collision.

25. A very energetic cosmic ray proton has a speed that differs from the speed of light by one part in 10^{24}. What is the energy of the proton?

26. A π^0 meson moving with speed v in the z direction decays into two photons. One of the photons travels in the z direction and the other travels in the minus z direction. (a) If one photon has an energy that is nine times that of the other photon, calculate the speed of the π^0 meson. (b) If the speed of the π^0 meson is $c/2$, determine the energies of the two photons.

27. In the frame S, a beam of photons has an energy E. In the frame S', which moves opposite the direction of the photon beam with speed v', the photons have

an energy of $E' = 10\ E$. Calculate the speed of the frame v'.

28. A π^0 meson is moving in the x direction when it spontaneously decays into two massless photons. One of the photons travels in the x direction and the other travels in the minus x direction. The difference in energy between the two photons is equal to the mass energy of the pion. Calculate the speed of the pion.

29. A Z^0 particle at rest decays into an electron and positron. The mass of the Z^0 particle is 91.2 GeV/c^2. Calculate the energy and momentum of the electron. By how much does the speed of the electron differ from c?

*30. Quarks were discovered inside the proton by scattering high energy (E_i) electrons from a hydrogen target and measuring the energy (E_f) and angle (θ) of the scattered electrons. The process is electron–quark elastic scattering,

$$e + q \rightarrow e + q.$$

In this process we may neglect the masses of both the quark and the electron. Let the variable x represent the proton momentum fraction of the quark that scatters, that is, if p^*_p is the momentum of the proton in the electron–quark center-of-mass system, then the quark momentum is $p^*_q = xp^*_p$. Derive an expression for x in terms E_i, E_f, θ, and the proton mass (M).

Four-vectors

31. A proton with a momentum of 200 GeV/c collides with a proton at rest. Calculate the total energy in the center-of-mass system.

*32. An experiment is designed to study pions produced by the interactions of relativistic protons with protons at rest. The experimenters wish to study relativistic pions produced at 90° with respect to the axis of the colliding protons in the proton–proton center-of-mass system. If the incident protons have momentum p ($pc >> mc^2$ where m is the proton mass), at what laboratory angle should the experimenters place their pion detector?

Compton scattering

33. A photon with kinetic energy equal to the mass energy of an electron collides with an electron at rest and scatters at an angle of $\pi/2$. Calculate the energy of the electron after the collision. (Give an exact answer in terms of the electron mass energy, mc^2).

34. A photon of energy E collides with an electron at rest.

Calculate the maximum amount of energy E_k that may be transferred to the electron. Make a graph of E_k versus E, labeling the scale in electronvolts.

35. Consider Compton scattering of photons with energy E by electrons with mass m and momentum p moving in a direction opposite the photon direction. (a) Give an exact expression relating the energy E' of backward scattered photons with the incident photon energy E, the incident electron momentum p and the electron mass m. (b) What is the value of E' for the special case of $p = 0$ and $E = 1$ MeV? (c) What is the value of E' for the special case of $p = E/c$?

36. (a) A photon with an energy of 10 GeV scatters with an electron at rest and scatters backwards. What is the energy of the electron after the collision? What is the energy of the photon after the collision? (b) A photon with an energy of 10 eV collides with an electron that has an energy of 10 GeV. The photon scatters backwards. What is the energy of the photon after the collision? What is the electron energy after the collision?

37. A 1-MeV photon collides with a proton at rest and is scattered at an angle of 45°. (a) Calculate the energy of the scattered photon. (b) Calculate the kinetic energy of the proton.

38. Prove that in Compton scattering of a photon with an electron at rest, the electron may not be scattered at an angle greater than $\pi/2$ in the laboratory frame.

*39. A beam of energetic photons is made by Compton scattering of a laser beam containing photons of energy 1 eV with electrons of energy 20 GeV. Calculate the maximum possible energy of the scattered photons.

*40. A photon of momentum p_1 scatters from a charged particle of mass m at rest and the particle gains an energy E_k from the collision. Calculate the angle (ϕ) that the momentum vector of the struck particle makes with respect to the photon direction.

*41. A photon with energy E collides with an electron at rest in the laboratory frame. (a) Calculate the speed of the electron in the center-of-mass frame. (b) Calculate the total energy in the center-of-mass frame. (c) The photon is scattered at 90° in the center-of-mass frame. Calculate the photon scattering angle in the laboratory frame.

Discovery of the positron

42. When a positron passes through a lead plate (see Figure 4-16), there is an electromagnetic force be-

tween the positron and the atomic electrons. Why can't a positron *gain* energy by passing through a lead plate?

43. A photon collides with an electron at rest creating an electron–positron pair,

$$\gamma + e^- \rightarrow e^+ + e^- + e^-.$$

Calculate the minimum photon energy for this process.

Additional problems

44. The relativistic Doppler shift causes the light from a distant galaxy to be blue shifted by 5%. What is the speed of the galaxy relative to the earth?

45. The *B* meson has a mass energy equal to about 5.3 GeV and a mean lifetime of about 1 ps. With what energy do *B* mesons need to be produced if they are to travel an average distance of 1 mm?

46. A well-collimated beam of π^+ mesons of energy *E* is directed through two counters separated by a distance

UPI/Bettmann Archive.

of 30 m. Some of the π^+ mesons decay in flight between the two counters so that the particle flux measured at the second counter is $1/e$ times that measured at the first counter. Determine the energy of the π^+ mesons. (The proper mean lifetime of the π^+ meson is 2.6×10^{-8} s; the mass of the π^+ meson is 140 MeV/c^2.)

***47.** Verify the validity of the Lorentz transformation for energy and momentum (4.112 to 4.115). (*Hint:* Consider a particle with energy *E* and momentum **p** in the frame S. Write the three velocity components in terms of *E* and **p**. Use the velocity transformation to determine the velocity components in a frame *S'* moving with relative speed *v* in the *x* direction. Calculate the velocity components in the frame *S'* in terms of *E'* and **p'** and show that they are identical to that obtained with the velocity transformation.)

48. A hydrogen atom in the ground state at rest absorbs a photon and makes a transition to the first excited state. Calculate the resulting speed of the atom.

49. An airplane with its running lights on is traveling at a speed of 300 m/s toward an observer. Calculate the Doppler shift of the light.

50. A galaxy is moving away from the earth with a speed of *c*/2. At what energy are the L$_\alpha$ photons in hydrogen (the $n = 2$ to $n = 1$ transition) expected to be observed, based on the relativistic motion of the source?

***51.** If a particle has a very large momentum, its trajectory in a magnetic field will be nearly a straight line. The maximum deviation of the particle trajectory from a straight line of length *L* that connects the beginning and end points of the particle path is called the *sagitta* (*s*). (a) Show that the momentum of a particle with charge *e* is given by

$$p = \frac{eBL^2}{8\,s}.$$

(b) Calculate the sagitta of 40-GeV electron that has a track length of 2 m in a magnetic field of 0.7 T. (c) For the electron of (b) what is the maximum allowed error in determination of the sagitta if the momentum is to be determined to a precision of 25%?

***52.** In 1955, a group led by Owen Chamberlain and Emilio Segrè working at the Berkeley Bevatron discovered the antiproton $(\bar{p})$ in the following reaction:

$$p + p \rightarrow p + p + p + \bar{p}.$$

If the initial state consists of a proton of energy E colliding with a proton at rest, calculate the minimum value of E for which the reaction may occur. (*Note:* The mass of the antiproton is identical to the mass of the proton).

53. Prove that the photoelectric effect cannot occur for a free electron, that is, that the process

$$\gamma + e^- \rightarrow e^-$$

is forbidden by conservation of energy and momentum.

*54. *The Bohr model.* Consider a scenario where the electron mass is nonzero but much smaller than 0.511 MeV/c^2 and where there is the possibility that an electron moves at relativistic speeds inside the hydrogen atom. Using the relativistic expression for momentum ($p = \gamma m v$) and the Bohr model quantization condition ($pr = n\hbar$), show that the speed of the electron in the ground state is equal to αc, independent of the electron mass.

55. *Positronium.* An electron and positron can form a bound state analogous to the hydrogen atom. (a) Use the Bohr model to show that the binding energy of positronium is equal to 6.8 eV. (b) Determine the Bohr radius.

In classical physics science started from the belief—or should we say illusion?—that we could describe the world or at least parts of the world without any reference to ourselves. This is actually possible to a large extent. We know that the city of London exists whether we see it or not. It may be said that classical physics is just that idealization in which we can speak about parts of the world without any reference to ourselves.

Werner Heisenberg

In classical physics, we speak of the position and momentum of a particle. We are not concerned with how the act of measuring the particle position may affect the particle momentum or vice versa. We assume that careful measurements may be made without disturbing the particle. This assumption holds true for macroscopic particles. For the electron this is not the case! It is not possible to accurately measure the position of an electron without significantly affecting its momentum, nor is it possible to accurately measure the momentum of the electron without significantly affecting its position. This has nothing to do with the technical quality of the apparatus, but rather with the measurement act itself. In order to know something about the position or momentum of the electron, we must necessarily interfere with the electron. We now explore this most interesting aspect of modern physics.

5-1 DEBROGLIE WAVELENGTH

The observation of the photoelectric effect and Compton scattering processes are convincing proof that electromagnetic waves have the properties of massless particles (photons). For electromagnetic radiation of a fixed wavelength (λ_{photon}), the energy of each photon is given by

$$E = \frac{hc}{\lambda_{photon}}. \tag{5.1}$$

From the theory of special relativity, the energy and momentum of the photons are related by

$$E = pc. \tag{5.2}$$

Combining these two expressions for the photon energy, we may write the photon wavelength as

$$\lambda_{photon} = \frac{h}{p}. \tag{5.3}$$

The wavelength of a photon is inversely proportional to its momentum. Radiation of wavelength λ_{photon} is composed of particles of momentum p. There is a particle–wave duality in the description of electromagnetic radiation.

Electron Wavelength and Momentum

In 1924, Louis deBroglie reasoned that if electromagnetic radiation can be interpreted as both particles and waves, then perhaps the electron, which had traditionally been interpreted as a particle, could also have a wave interpretation. If the electron has a wave behavior, then there are many questions to be answered, such as, *what is the wavelength of an electron?* The relationship between wavelength and momentum of an electron in an atom was already suggested by the Bohr condition on angular momentum quantization that successfully gives the ground state energy of the hydrogen atom. The quantization condition (3.85) is

$$L = pr = \frac{nh}{2\pi}, \tag{5.4}$$

where n is a positive integer ($1, 2, 3, \ldots$). If the electron has a wavelength $\lambda_{electron}$, then we may imagine that a stable circular orbit is possible if an integral number of wavelengths fit into the circumference of the orbit,

$$n\lambda_{electron} = 2\pi r, \tag{5.5}$$

This condition is called the *Bohr-Sommerfeld* quantization condition. We may combine the quantization conditions (5.4 and 5.5) to get

$$\lambda_{electron} = \frac{h}{p}. \tag{5.6}$$

The relationship between electron wavelength and momentum (5.6) is identical to that for the photon (5.3). Electromagnetic waves are equivalent to particles and the electron has a wave behavior. That is the way the world is. There is a particle–wave duality. For some situations we will find that a wave description is most convenient and for others we will be more comfortable with a particle description. They are equivalent.

We should make it clear that electrons do not actually move in circular orbits in atoms. The Bohr description of the hydrogen atom is only a model and we have yet to explain the origin of angular momentum quantization (5.4), except that it leads to the correct answer for the energy levels of hydrogen. The quantization condition is that the component of orbital angular momentum in some given direction is quantized in units of $\hbar$. We shall investigate this in much more detail.

DeBroglie hypothesized that *all particles* have a wave behavior with a universal relationship between the particle wavelength and momentum given by

$$\lambda = \frac{h}{p}. \tag{5.7}$$

This expression is called the *deBroglie* relationship and the wavelength that appears in the equation is called the *deBroglie wavelength*. The deBroglie wavelength is the wavelength of a particle that has a precise or *definite* momentum p. The momentum that appears in the deBroglie relationship (5.7) is the momentum that is conserved in collisions, that is, the relativistic momentum. The deBroglie relationship holds for all particles: photons, electrons, neutrons, quarks and baseballs. For radiation of definite wavelength, the momentum of a photon is precisely defined. The electron description is complicated because the electron is continuously interacting with photons. This interaction affects the electron momentum.

The wavelength of any particle with momentum p is given by

$$\lambda = \frac{h}{p}.$$

The deBroglie relationship tells us that for a given particle momentum, the observability of the wave properties (i.e., the size of λ) is dictated by the size of Planck's constant. (We seem to be finding that Planck's constant is appearing everywhere!) The universal relationship between particle wavelength and momentum discovered by deBroglie is plotted in Figure 5-1.

EXAMPLE 5-1

What is the deBroglie wavelength of a particle that has a momentum of 1 keV/c?

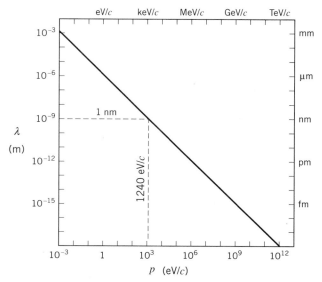

FIGURE 5-1 Wavelength as a function of momentum.
The wavelength of a particle is inversely proportional to the momentum of the particle, $\lambda = h/p$.

SOLUTION:

The deBroglie wavelength (5.7) is

$$\lambda = \frac{h}{p} = \frac{hc}{pc} = \frac{1240 \text{ eV} \cdot \text{nm}}{10^3 \text{ eV}} \approx 1 \text{ nm}. \quad \blacksquare$$

Calculating the Wavelength of Particles

The deBroglie relationship holds for all particles. Usually our physical intuition is such that when a wavelength is large (compared to some characteristic distance) we are comfortable with a wave description but when a wavelength is small we prefer a particle description. The wave nature that light exhibits when passing through a narrow slit is more easily observable when the wavelength is not too small compared to the size of the slit. The wave behavior of macroscopic objects is not observable because the wavelength is so small as to be far beyond experimental detection.

Wavelength of a Baseball

Your intuition should tell you that the wavelength of a baseball is small because it can be localized. When you look at a baseball you don't see it wiggling.

EXAMPLE 5-2

Calculate the wavelength of a baseball with a mass of 0.2 kg and a speed of 10 m/s.

SOLUTION:

The baseball is nonrelativistic ($v/c \ll 1$). The momentum is

$$p = mv = (0.2 \text{ kg})(10 \text{ m/s}) = 2 \text{ kg} \cdot \text{m/s}.$$

The wavelength of the baseball is

$$\lambda = \frac{h}{p} = \frac{6.6 \times 10^{-34} \text{ J} \cdot \text{s}}{2 \text{ kg} \cdot \text{m/s}} \approx 3 \times 10^{-34} \text{ m}.$$

This is indeed a very small wavelength. It is many orders of magnitude beyond the present capabilities of experimental detection (see Figure 1-13). $\quad \blacksquare$

The wavelength of the baseball is small because the momentum of the baseball is large. The momentum of the baseball is large because the mass of the baseball is large. *What if the speed of the baseball was small?* In the example above, we have taken the speed of the baseball to be 10 m/s, but what if the baseball was at rest, sitting on a table? What if the speed of the baseball was almost zero so that the speed times the mass is a small number and the wavelength is large compared to the wavelength of light. Then we might expect the baseball to look fuzzy. Are we not in trouble with the wave interpretation for the baseball, because a baseball sitting on a table appears to us to be localized and not to be wiggling? The answer is that the baseball is moving but we just don't detect it. The baseball is in thermal equilibrium with its surroundings and it is constantly absorbing and emitting radiation. It is impossible for the baseball to have zero momentum because momentum must be conserved when photons strike the baseball.

EXAMPLE 5-3
Estimate the deBroglie wavelength of a very *cold* baseball that has a speed of one atomic distance per thousand years.

SOLUTION:
One atomic distance is about 10^{-10} meters. One year is about 3×10^7 seconds. The speed of the baseball is

$$v \approx \frac{10^{-10} \text{ m}}{\left(10^3\right)\left(3 \times 10^7 \text{ s}\right)} \approx 3 \times 10^{-21} \text{ m/s}.$$

The momentum is

$$p = mv = \left(0.2 \text{ kg}\right)\left(3 \times 10^{-21} \text{ m/s}\right) = 6 \times 10^{-22} \text{ kg} \cdot \text{m/s}.$$

The wavelength of the cold baseball is

$$\lambda = \frac{h}{p} = \frac{6.6 \times 10^{-34} \text{ J} \cdot \text{s}}{6 \times 10^{-22} \text{ kg} \cdot \text{m/s}} \approx 10^{-12} \text{ m}.$$

The baseball has a tiny wavelength even if a ridiculously small speed is assumed, because the mass of the baseball is so large. ∎

Wavelength of an Electron in an Atom
How about an electron? The electron is the particle with the smallest nonzero mass. Therefore, the electron is the best candidate to exhibit observable wave properties. Let us calculate the wavelength of an electron in an atom. We shall learn how to calculate the energies of electrons in atoms with great precision, but we want to know the approximate answer immediately. We may use the Bohr model to estimate the electron momentum.

EXAMPLE 5-4
Estimate the wavelength of an electron in the hydrogen atom.

SOLUTION:
From the Bohr model, we know that the speed of an electron in the ground state of hydrogen (3.95) is given by

$$v = \alpha c = \frac{c}{137}.$$

The electron is nonrelativistic ($v/c \ll 1$), so the momentum (p) is

$$p = mv = \alpha mc.$$

The wavelength of the electron is

$$\lambda = \frac{h}{p} = \frac{hc}{pc} = \frac{hc}{\alpha mc^2}$$

$$\approx \frac{1240 \text{ eV} \cdot \text{nm}}{\left(\dfrac{1}{137}\right)\left(5.11 \times 10^6 \text{ eV}\right)} \approx 0.3 \text{ nm}. \quad ∎$$

The wavelength of an electron inside the hydrogen atom is equal to the size of the atom (1.8). In terms of the Bohr radius (3.90) the electron wavelength is

$$\lambda = 2\pi a_0, \tag{5.8}$$

which is equivalent to the Bohr-Sommerfeld quantization condition (5.5) for $n = 1$.

> The wavelength of an electron inside an atom is equal to the size of the atom.

It is not accidental that the wavelength of an electron in an atom is the size of the atom! The wave properties of the electron *give* the atom its size. The electrons in atoms are waves that fill up a space of roughly 10^{-29} m^3. The electrons inside an atom do not have a classical trajectory. The electron appears as a cloud with a size that defines the size of the atom. All atoms have nearly the same size because the size of the electron cloud is about the same for all atoms. The size of the electron cloud is about the same for all atoms because the outer electrons have approximately the same momenta for all atoms.

In estimating the wavelength of an electron inside an atom, we have taken the electron momentum from the Bohr model. The real question is: *What sets the scale of the momentum of the electron in the atom?* Since a larger momentum would make a smaller atom, *why doesn't the atom collapse?* The electron energy depends on three fundamental constants: the strength of the electromagnetic force (ke^2), the electron mass (m), and Planck's constant (h). Planck's constant appears because the electron wavelength determines the average distance between the electron and the nucleus, and this wavelength depends on Planck's constant. The average electron kinetic energy is one-half the potential energy but with the opposite sign:

$$\langle E_k \rangle = \frac{1}{2} \left\langle \frac{ke^2}{r} \right\rangle = \frac{\alpha}{2} \left\langle \frac{\hbar c}{r} \right\rangle. \qquad (5.9)$$

The quantity $< 1/r >$ is governed by the electron wavelength:

$$\left\langle \frac{1}{r} \right\rangle \approx \frac{2\pi}{\lambda} = \frac{p}{\hbar}. \qquad (5.10)$$

Combining these results we may write the average kinetic energy of the electron as

$$\langle E_k \rangle \approx \frac{\alpha pc}{2}. \qquad (5.11)$$

Since the kinetic energy is also equal to $p^2/2m$, we have an equation for the electron momentum:

$$\frac{p^2}{2m} \approx \frac{\alpha pc}{2}. \qquad (5.12)$$

Solving for the electron momentum, we get

$$p \approx \alpha mc. \qquad (5.13)$$

The wavelength of the electron (5.7) is

$$\lambda = \frac{h}{p} = \frac{hc}{pc} = \frac{hc}{\alpha mc^2} \approx 0.3 \, \text{nm}. \qquad (5.14)$$

Energy conservation determines the momentum of an electron in an atom. The momentum of the electron gives the wavelength of the electron. The wavelength of the electron gives the size of the atom. The atom does not collapse because this would correspond to a very tiny wavelength or equivalently a very large momentum. A large momentum violates energy conservation because potential energy in a collapsing atom varies as $-p$ but the kinetic energy varies as p^2.

Wavelength of a Neutron in a Nucleus
When a neutron is freed from a nucleus through a nuclear interaction, the order of magnitude of the neutron kinetic energy is 10 MeV. Since the neutron mass energy (mc^2) is

much greater than the kinetic energy, such a neutron is nonrelativistic and we may write its momentum as

$$p = \sqrt{2mE_k} = \sqrt{(2)(940 \, \text{MeV}/c^2)(10 \, \text{MeV})}$$
$$\approx 140 \, \text{MeV}/c. \qquad (5.15)$$

EXAMPLE 5-5
Calculate the wavelength of a 10-MeV neutron.

SOLUTION:
The neutron wavelength is

$$\lambda = \frac{h}{p} = \frac{hc}{pc} = \frac{1240 \, \text{MeV} \cdot \text{fm}}{140 \, \text{MeV}} \approx 9 \, \text{fm}.$$

The neutron wavelength is the same order of magnitude as the nuclear size. The wavelength of the neutron *gives* the nucleus its size! Using the Bohr-Sommerfeld condition (5.8), we may write the diameter of the nucleus ($d_{nucleus}$) in terms of the neutron wavelength:

$$d_{nucleus} = \frac{\lambda}{\pi} \approx \frac{9 \, \text{fm}}{\pi} \approx 3 \, \text{fm}. \qquad \blacksquare$$

Wavelength of a Quark Inside a Proton
We now consider a quark inside a proton. The quarks that make up a proton have small masses compared to the proton mass. The quarks have a large momentum, and so they are relativistic. The momentum distribution of quarks inside the proton has been measured by scattering electrons from the quarks. These scattering experiments will be discussed in Chapter 6. A typical fraction of a proton energy carried by an individual quark is 20%. Thus, the quark momentum is given by

$$pc \approx (0.2)m_p c^2$$
$$\approx (0.2)(940 \, \text{MeV}) \approx 200 \, \text{MeV}. \qquad (5.16)$$

EXAMPLE 5-6
Calculate the wavelength of a 200-MeV quark, assuming that the quark mass energy is much smaller than the kinetic energy.

SOLUTION:
The quark wavelength is

$$\lambda = \frac{hc}{pc} \approx \frac{1240 \, \text{MeV} \cdot \text{fm}}{200 \, \text{MeV}} \approx 6 \, \text{fm}.$$

The wavelength of a quark inside a proton is the same order of magnitude as the size of a proton. The wavelength of the quark *gives* the proton its size! The diameter of the proton (d_{proton}) is given by

$$d_{proton} = \frac{\lambda}{\pi} \approx \frac{6 \text{ fm}}{\pi} \approx 2 \text{ fm} \quad . \quad ■$$

Relationship Between Kinetic Energy and Wavelength

We can calculate the wavelength of a particle as a function of the kinetic energy of the particle. The kinetic energy of a particle is defined to be the total energy minus the rest energy:

$$E_k = E - mc^2$$
$$= \sqrt{(pc)^2 + (mc^2)^2} - mc^2. \quad (5.17)$$

Solving for the particle momentum, we have

$$pc = \sqrt{E_k^2 + 2mc^2 E_k} \quad . \quad (5.18)$$

(Notice that for the case $m = 0$, we have $pc = E_k$.) The wavelength is

$$\lambda = \frac{h}{p} = \frac{hc}{pc} = \frac{hc}{\sqrt{E_k^2 + 2mc^2 E_k}}. \quad (5.19)$$

In Figure 5-2, we plot the particle wavelength as a function of the particle kinetic energy for the photon ($mc^2 = 0$), the electron ($mc^2 = 0.511$ MeV), and the proton ($mc^2 = 938$ MeV). At kinetic energies that are small compared to the electron mass energy, the photon wavelength is much larger than the electron wavelength, which is much larger than the proton wavelength. At kinetic energies that are large compared to the mass energy of the proton, the wavelengths of the three particles are identical. The electron has no observed structure down to a wavelength of 10^{-18} m. The proton, however, is resolved into quarks at wavelengths smaller than about 1 fm. This exciting physics will be discussed in the next chapter.

5-2 MEASURING THE WAVE PROPERTIES OF THE ELECTRON

In 1927, the wave properties of the electron were directly observed in two independent experiments conducted by

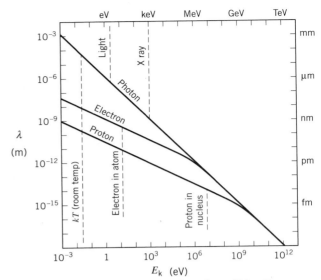

FIGURE 5-2 Wavelength as a function of kinetic energy.
For a given kinetic energy, the wavelength depends on the particle mass. For macroscopic objects, the wavelength is much smaller than the highest resolution experimentally achieved (10^{-18} m). For particles like the electron and proton, the wavelengths are comparatively large and their effects are observable.

Clinton Davisson and Lester Germer and by George Thomson and A. Reid. Both experiments involved the scattering of electrons from a crystalline target and the observation of interference effects.

The Davisson-Germer Experiment

Davisson and Germer did not plan their experiment with the goal of discovering the wave nature of the electron. Their goal, rather, was to measure the energy of electrons scattered from a metal surface. The classical picture of the electron at the time was that of a particle with a size r_c, where r_c is given by equating the electrical potential energy of a tiny sphere of charge with the mass energy of the electron:

$$\frac{ke^2}{r_c} = mc^2. \quad (5.20)$$

The magnitude of r_c, called the *classical electron radius,* was thought to be

$$r_c = \frac{ke^2}{mc^2} = \frac{1.44 \text{ MeV} \cdot \text{fm}}{0.511 \text{MeV}} \approx 3 \text{ fm} . \quad (5.21)$$

With the discovery of the nucleus by Rutherford and his collaborators and the subsequent development of the Bohr model, Davisson and Germer viewed electron scattering from a metal as the interaction between an electron and a single atomic nucleus. In describing the approach to the experiment, Davisson wrote:

We have been accustomed to think of the atom as rather like the solar system—a massive nuclear sun surrounded by planetary electrons moving in closed orbits. On this view the electron which strikes into a metal surface is like a comet plunging into a region rather densely packed with solar systems. There is a certain small probability, or at least there might seem to be, that the electron will strike into the atom in or near the surface of the metal, be swung about comet-wise, and sent flying out of the metal without loss of energy. The direction taken by such an electron as it leaves the metal should be a matter of private treaty between the electron and the individual atom. One does not see how the neighboring atoms could have any voice in the matter.

As sometimes happens in science, Davisson and Germer had a great stroke of luck. The glass vacuum tube of their apparatus broke, and in the course of assembling their nickel target into a new vacuum chamber, they heated and cooled the nickel to free it from surface contaminants that would degrade the vacuum. This annealing process caused the atoms to form a regular crystalline array. This crystal target was just what they needed to observe the interference effects of the electron waves. The scattering of electrons from a crystal is analogous to the scattering of x rays from a crystal.

The apparatus of Davisson and Germer is shown in Figure 5-3. An *electron gun* was made by heating a tungsten cathode and accelerating the ejected electrons through a potential difference (V). The electron kinetic energy or speed could be selected by choosing the voltage. A steady beam of electrons from the gun was directed at angle θ with respect to the surface of the nickel target.

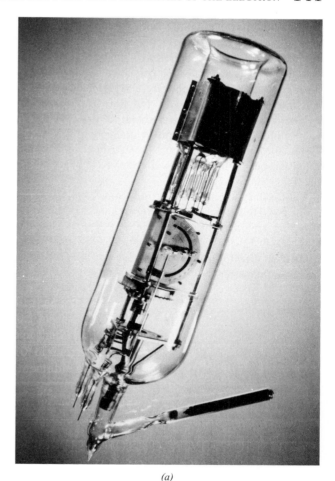

(a)

FIGURE 5-3 Experiment of Davisson and Germer discovering electron waves.
Electrons are directed at a nickel target and the intensity of the scattered electrons is measured as a function of the scattering angle. (*a*) Photograph courtesy of AT&T Bell Laboratory. (*b*) Schematic fashioned after C. J. Davisson, "Are Electrons Waves?," *Franklin Institute Journal* **205**, 597 (1928).

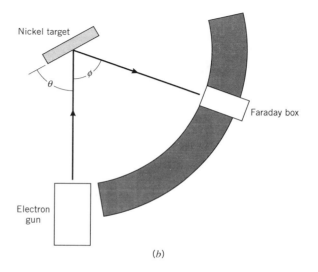

(b)

Electrons that were scattered from the nickel were collected in a small *Faraday box,* and the relative rate of charge collection was measured with a galvanometer. The Faraday box was movable along the arc of a circle so that the scattered electron intensity could be measured as a function of scattering angle. To their great surprise Davisson and Germer discovered that for certain combinations of electron acceleration voltages and scattering angles, the intensity of scattered electrons had a maximum. The beam of electrons was observed to be reflected just like a light wave!

The crystal geometry for electron scattering is shown in Figure 5-4. The incoming electron encounters layers planes of atoms separated by a distance d. Recall the Bragg condition (3.114) for constructive interference in the scattering of an x ray from a crystal where the spacing between the layers of atoms is d,

$$n\lambda = 2 d \sin\theta, \qquad (5.22)$$

where n is a positive integer. We may write the Bragg condition as

$$\frac{1}{\lambda} = \frac{n}{2 d \sin\theta}. \qquad (5.23)$$

Davisson and Germer noticed that the intensity of the scattered electrons had maxima in equally spaced intervals when plotted as a function of the square root of the acceleration voltage. These data are shown in Figure 5-5. Since the acceleration voltage is related to the electron kinetic energy (E_k) by

$$E_k = eV = \frac{1}{2} mv^2, \qquad (5.24)$$

we have

$$\sqrt{V} = v \sqrt{\frac{m}{2e}}. \qquad (5.25)$$

The square root of the acceleration voltage is directly proportional to the electron speed. If the physics of the x ray scattering is the same as the physics of the electron scattering, then the data of Figure 5-5 suggest that

$$\frac{1}{\lambda} \propto v, \qquad (5.26)$$

for a nonrelativistic electron.

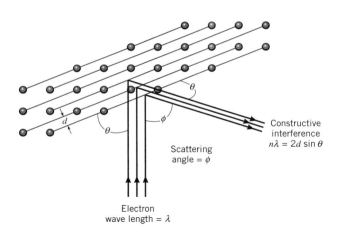

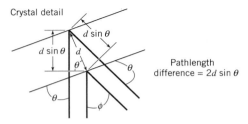

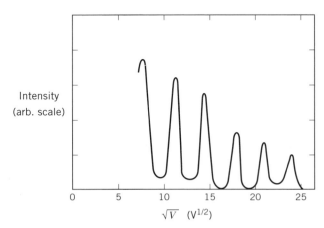

FIGURE 5-4 Crystal geometry for the Davisson-Germer experiment.
The incoming electron sees planes of atoms separated by a distance d. Constructive interference occurs when the difference in path lengths from scattering by successive planes is equal to an integer times the electron wavelength. This gives the Bragg condition, $n\lambda = 2d\sin\theta$.

FIGURE 5-5 Energy dependence of the scattering in the Davisson-Germer experiment.
For a fixed scattering angle, the intensity of the scattered electrons depends on the square root of the accelerating voltage, which is proportional to the electron speed. From C. J. Davisson, "Are Electrons Waves?," *Franklin Institute Journal* **205**, 597 (1928).

Davisson and Germer made a quantitative test of the relationship between electron speed and electron wavelength. They selected different electron speeds by choosing the acceleration voltage. They then measured the scattering angle where the intensity was maximum and used the Bragg condition to determine the electron wavelength. Their data are shown in Figure 5-6 as a plot of λ versus $V^{-1/2}$.

For nonrelativistic electrons (using $E_k = eV$) the deBroglie relationship (5.7) predicts

$$\lambda = \frac{hc}{\sqrt{2mc^2 eV}}$$

$$= \frac{1240 \text{ eV} \cdot \text{nm}}{\sqrt{(2)(5.11 \times 10^5 \text{ eV})(e)(V)}}$$

$$= \frac{1.23 \text{ nm} \cdot V^{1/2}}{\sqrt{V}}, \qquad (5.27)$$

in agreement with the data.

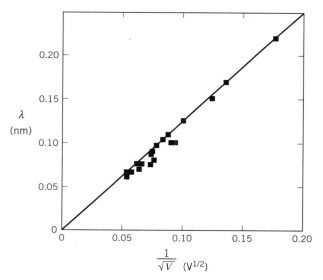

FIGURE 5-6 Testing the deBroglie equation for nonrelativistic electrons.
The electron wavelength is measured by measuring the scattering angle where the intensity maxima occur, using the Bragg condition, $n\lambda = 2d \sin\theta$. The wavelength is plotted as a function of the inverse square root of the accelerating voltage. The deBroglie equation predicts a linear relationship with a slope of 1.23 nm·$V^{1/2}$ (solid line). From C. J. Davisson, "Are Electrons Waves?," *Franklin Institute Journal* **205**, 597 (1928).

The Thomson-Reid Experiment

Thomson and Reid performed a simple important experiment that convincingly demonstrated the existence of electron waves. They passed a collimated beam of electrons through a thin target material of various types. The electron intensity was measured on a photographic plate placed at a distance of 0.1 m from the target as indicated in Figure 5-7. The atoms of the target form a regular crystal, resulting in an interference pattern in the detected electron intensity caused by the wave nature of the electron. The intensity is maximum at certain angles satisfying the Bragg condition (5.22). If many crystals are present, the interference pattern forms a continuous ring. The radii of the rings depend on the orientation of the crystals. Thomson and Reid measured the dependence of the radius of a given ring on the electron speed using nonrelativistic electrons with energies from 3.9 keV to 16.5 keV. They discovered that the radius of each ring (which is proportional to the electron wavelength from the Bragg condition) was inversely proportional to the electron speed in agreement with the result of Davisson and Germer (5.26).

Figure 5-8 shows a comparison of the diffraction pattern produced by x rays and electrons incident on an aluminum target. The order of magnitude of the wavelength of both the x rays and the electrons is 0.1 nm. The diffraction patterns are quantitatively identical in that the maxima occur when path lengths from the target to the detector differ by an integral number of wavelengths.

5-3 PROBABILITY AMPLITUDES
The Two-Slit Experiment

Consider an experiment where a beam of electromagnetic radiation is directed at a barrier that has two narrow openings or slits through which the photons may pass. The source of the radiation is a long distance away so that the incident radiation is perpendicular to the barrier. At some distance from the barrier is a detector that counts the photons that pass through the slits. The geometry of the two-slit experiment is shown in Figure 5-9. The two slits are separated by a distance d. The detection screen is placed at a distance z from the barrier. The number of photons striking the detection screen is measured as a function of the coordinate x along the screen.

On passing through the slit the rays diffract. If we follow the two possible paths of the waves, we expect maxima in the intensity when the path difference (ΔL) is equal to an integer number of wavelengths,

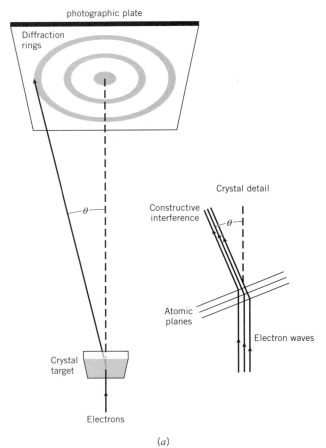

(a)

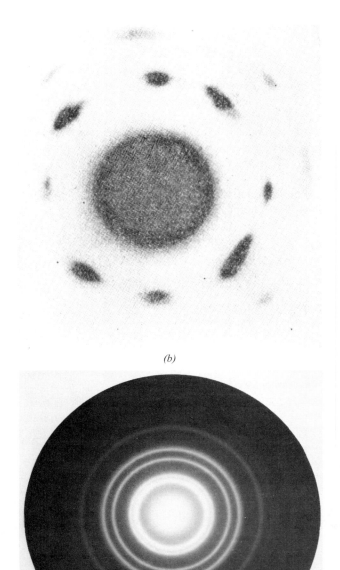

(b)

FIGURE 5-7 The experiment of Thomson and Reid discovering electron waves.
(a) Electrons are diffracted by a thin crystalline target and the scattering pattern is observed on a photographic plate. The intensity of the scattered electrons has maxima at certain scattering angles because of the interference of the electron waves. (b) Photographic plate made by Thomson using an aluminum target. From R. A. Millikan, *Electrons (+ and –), Protons, Photons, Neutrons, Mesotrons, and Cosmic Rays*, The University of Chicago Press (1947). (c) Photographic plate made by Thomson using a polycrystalline target. Courtesy of National Science Museum London.

(c)

$$\Delta L = n\lambda, \tag{5.28}$$

where n is a nonnegative integer $(0, 1, 2, \ldots)$. For example, at $x = 0$, we expect a maximum in the intensity since both particles have the same path length $(\Delta L = 0)$.

We expect minima in the intensities when the waves from the two paths interfere destructively. The destructive interference is greatest when the waves are out of phase by $180°$ or when the path difference is equal to

$$\Delta L = \left(n + \frac{1}{2}\right)\lambda, \tag{5.29}$$

where n is a nonnegative integer $(0, 1, 2, \ldots)$.

We can deduce the relationship between the path difference (ΔL), the distance between the slits (d), the distance to the detection screen (z), and the coordinate (x) along the detection screen. From the geometry shown in Figure 5-10, it is straightforward to show that (see problem 14) for small angles the distance (Δx) between adjacent maxima or minima is

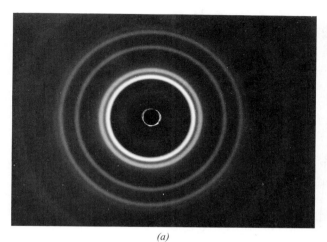

(a)

FIGURE 5-8 Diffraction pattern produced by (a) x rays and (b) electrons on passing through an aluminum crystalline target.
Courtesy of Educational Development Center, Massachusetts.

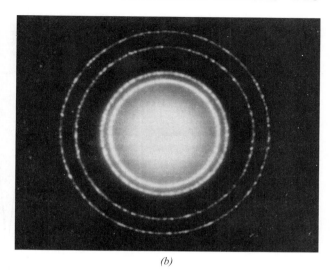

(b)

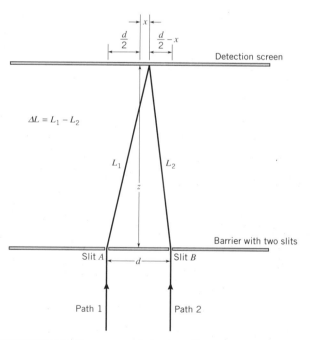

FIGURE 5-9 Geometry of the two-slit experiment.
Photons are directed toward a barrier having two narrow openings, *A* and *B*, separated by a distance *d*. The particles passing through the slits are detected on a screen placed at a distance *z*. The detection pattern is measured as a function of the coordinate *x*, as shown. The photons exhibit a wave behavior, and diffraction occurs when they pass through the slits. The intensity will be a maximum when the difference in path lengths (ΔL) is equal to an integer (*n*) times the wavelength (λ), $\Delta L = n\lambda$, and will be a minimum when $\Delta L = (n + 1/2)\lambda$.

$$\Delta x = \frac{z\lambda}{d}. \tag{5.30}$$

The condition that interference be observable is that detector resolution must be better than $z\lambda/d$. At some fixed distance (*z*), the expected intensity distribution, $I(x)$, depends on the distance between the slits, the wavelength of light, and also the width of the slits (*b*). The measured intensity distribution, for one choice of *b*, *d*, and λ, is shown in Figure 5-10*a*, compared with the predicted intensity distribution, Figure 5-10*b*.

Now consider conducting the experiment with massive particles. For macroscopic objects, the wavelength is infinitesimal and it is not possible to observe interference effects. Many oscillations of the intensity occur within a small distance (Δx), and the wave behavior of such particles is not observable. The intensity pattern detected through two slits (I_{12}) is the sum of the intensity patterns detected through the two slits individually (I_1 and I_2). There is no detectable interference:

$$I_{12} = I_1 + I_2. \tag{5.31}$$

If the projectile particles are electrons, the situation is different because wavelengths may be large enough to be observable. One experimental limitation is that the electron wavelength cannot be chosen to be very large because low-momentum electrons scatter easily and it is difficult to form them into a coherent beam. Another experimental difficulty is the challenge of making very tiny uniform

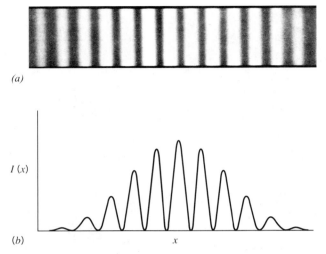

(a)

(b)

FIGURE 5-10 Results from the two-slit experiment using photons.
(*a*) Measured photon intensity as a function of position. From M. Cagnet, *Atlas of Optical Phenomena*, Springer-Verlag (1962). (*b*) Calculated photon intensity.

slits for the electrons to pass through. Finally, one must have a high-resolution technique for detecting the interference pattern.

The results of a two-slit experiment using electrons are shown in Figure 5-11. The energy of the electrons is 50 keV, corresponding to a wavelength of 5×10^{-12} m. The distance between the slits is 2×10^{-6} m. The observation plane is located at a distance of 0.35 m from the slits. For small angle scattering, the intensity pattern is expected to show a spacing between the maxima (5.30) of

$$\Delta x = \frac{\lambda z}{d} = \frac{\left(5 \times 10^{-12} \text{ m}\right)(0.35 \text{ m})}{2 \times 10^{-6} \text{ m}}$$
$$\approx 0.9 \times 10^{-6} \text{ m.} \tag{5.32}$$

This is confirmed by experiment. This is the same intensity pattern that is observed with electromagnetic radiation. Electrons are observed to behave like electromagnetic waves when passing through two slits.

We shall now investigate the cause of the interference in more detail. We have seen in Chapter 3 that a beam of electromagnetic waves is composed of individual quanta that we call photons. One might be tempted to speculate that the interference is connected with the fact that the waves contain so many photons or electrons that particles going through A and B arrive at the detector in or out of phase and interfere constructively or destructively. To test this we may decrease the intensity of the beam to such a small intensity that only a single photon or electron at a

time reaches the slits. Then surely the particle must go through either A or B. Now since we have one particle passing through one slit, we might expect the pattern to be that of the sum of the two single-slit patterns, that is, no interference. After counting up a large number of photons or electrons, we find the same interference pattern as with higher intensities. The interference does not in any way depend on the intensity of the photon or electron beam.

We take the experiment one step further. Imagine what would happen if we tried to determine which slit the photon or electron passes through. To do this we imagine that a tiny particle detector is placed near the slit. When the photon or electron passes through the slit it registers in our detector, thereby signaling which slit the particle has passed through. We then find that the interference pattern disappears. This is because the process of detecting the particle has necessarily scattered it and this scattering destroys the interference.

The Foundation of Quantum Mechanics

The two-slit experiment demonstrates a fundamental rule that is the cornerstone on which the quantum mechanical description of modern physics is built. The occurrence of every event, for example, a photon or electron arriving at the detector in a certain location, corresponds to the event probability (*P*). This probability may be represented by the square of a complex number, the *amplitude* (*A*) for the event,

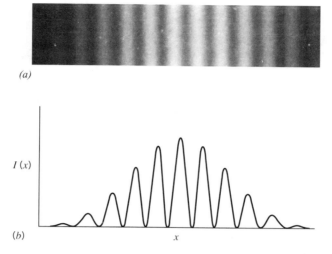

FIGURE 5-11 Results from the two-slit experiment using electrons.
(*a*) Measured electron intensity as a function of position. From C. Jönsson, Zeit. Phys. **161**, 454 (1961). (*b*) Calculated electron intensity.

$$P = |A|^2. \tag{5.33}$$

If the event can occur in more than one way, such as the particle having two slits that it could pass through, then there is an amplitude for each path involved in the event. The fundamental rule states: If an event can occur in more than one way, the event probability is the square of the sum of the probability amplitudes ($A_1 + A_2 + ...$),

$$P = |A_1 + A_2 + ...|^2. \tag{5.34}$$

For the special case where there are just two paths and two corresponding amplitudes (A_1 and A_2), then we may write

$$P = |A_1|^2 + |A_2|^2 + 2|A_1||A_2|\cos\delta. \tag{5.35}$$

where δ is the phase difference between the amplitudes A_1 and A_2. Note that the sign of the last term can be positive or negative. Thus, the probability is not equal to the sums of the squares of the amplitudes and we have interference.

If we determine which path is taken, this is equivalent to having only one path available in an event. The path can be different from event to event, but each event has only one path. The resultant probability is then the sum of the squares of the amplitudes and the interference is destroyed. Determining which path is taken has the same effect as choosing a path by closing off other options (such as closing one of the slits in the two-slit experiment). In that case, we see that knowing the path must necessarily destroy the interference.

How can this occur? The answer is that the process of determining which path was taken has altered the experiment. The experiment has been altered because the act of determining the path taken necessarily involves an additional photon interaction. This additional interaction destroys the interference, and the probability in this case is equal to the sums of the squares of the individual amplitudes. That's the way the quantum world works, add the amplitudes first and then square to get the probability.

> Any physical process may be represented by a complex number called the probability amplitude. The probability that the process occurs is the square of the amplitude. If a process can occur by more than one path, the probability amplitude is the sum of the individual amplitudes for each path and the net probability is the total amplitude squared.

There is a clear difference between classical physics and quantum physics. In classical physics if we have all the initial conditions, we can predict the outcome *exactly*. In quantum physics we can predict only the *probability* of various possible outcomes!

5-4 WAVE DESCRIPTION OF A PARTICLE

A plane wave may be represented by a sinusoidal function such as

$$F(x,t) = A\cos(kx - \omega t), \qquad (5\text{-}36)$$

which is a solution to the wave equation with speed v,

$$\frac{\partial^2 F}{\partial x^2} + \frac{\partial^2 F}{\partial y^2} + \frac{\partial^2 F}{\partial z^2} = \frac{1}{v^2}\frac{\partial^2 F}{\partial t^2}. \qquad (5.37)$$

The function $F(x,t)$ oscillates with time, and the frequency of oscillation is

$$f = \frac{\omega}{2\pi}. \qquad (5.38)$$

The wave number (k) is related to the wavelength (λ) by

$$k = \frac{2\pi}{\lambda}. \qquad (5.39)$$

The speed of propagation is

$$v = f\lambda = \frac{\omega}{k}. \qquad (5.40)$$

An equally valid function that can be used to represent the simple wave form is the sine function or the exponential function with an imaginary exponent

$$\begin{aligned}F(x,t) &= Ae^{i(kx-\omega t)}\\ &= A[\cos(kx - \omega t) + i\sin(kx - \omega t)].\end{aligned} (5.41)$$

Such a function is also a solution of the wave equation with the constraint $v = \omega/k$.

The simple wave form just described has one frequency and is called a pure harmonic (or plane wave). We may construct more complex wave forms by adding additional terms that are integer multiples of the fundamental frequency, ω:

$$F(x,t) = \sum_{n=-\infty}^{+\infty} A_n e^{in(kx-\omega t)}. \qquad (5.42)$$

This function is also a solution of the wave equation.

Useful Tool: Fourier Analysis

At a given time, any periodic wave form may be represented as

$$F(x) = \sum_{n=-\infty}^{+\infty} A_n e^{inkx}, \qquad (5.43)$$

where $k = 2\pi/\lambda$. By appropriate choice of the constants A_n, it is possible to represent any periodic function by a sum of simple wave forms. The technique of representing a function by a sum of harmonic functions is called the *Fourier series* representation. If the wave form has a relatively smooth shape, then an accurate representation is achieved with a small number of terms. (Functions which are sharply peaked require more terms.) The coefficients are obtained by multiplying each side of the Fourier sum (5.43) by e^{-imkx} and integrating over one period:

$$\int_0^{2\pi/k} dx\, F(x)\, e^{-imkx}$$
$$= \int_0^{2\pi/k} dx \sum_{n=-\infty}^{+\infty} A_n e^{i(n-m)kx}. \qquad (5.44)$$

The integral on the right-hand side is zero unless n is equal to m:

$$\int_0^{2\pi/k} dx\, F(x)\, e^{-imkx}$$
$$= \int_0^{2\pi/k} dx\, A_m = \frac{2\pi A_m}{k}. \qquad (5.45)$$

Therefore,

$$A_n = \frac{k}{2\pi}\int_0^{2\pi/k} dx\, F(x)\, e^{-inkx}. \qquad (5.46)$$

If the shape of the waveform is simple, then the constants (A_n) can be readily evaluated and the series converges rather rapidly. If the shape of the waveform is complicated, a computer may be used to evaluate the constants.

Packets of Waves

Now consider a waveform that is not periodic, but is nonzero only in some localized region of space. This type of waveform is called a *wave packet*. We may still represent this waveform as being composed of e^{ikx} terms provided that we treat k as a continuous variable. The distribution of wave numbers is given by the function $g(k)$. The function $\psi(x)$ may be written in terms of $g(k)$ as

$$\psi(x) = \frac{1}{\sqrt{2\pi}}\int_{-\infty}^{+\infty} dk\, g(k)\, e^{-ikx}. \qquad (5.47)$$

The function $g(k)$ may be obtained from $\psi(x)$ by the integral

$$g(k) = \frac{1}{\sqrt{2\pi}}\int_{-\infty}^{+\infty} dx\, \psi(x)\, e^{ikx}. \qquad (5.48)$$

There is a symmetry between the functions $g(k)$ and $\psi(x)$, and they are called the *Fourier integral transformations* of each other.

Wave packets may be used to represent particles. In this representation we define a function $\psi(x)$ that contains the information about the particle's position at a certain time. We allow the function $\psi(x)$ to be a complex number and define the probability of finding the particle per unit distance to be the square of $\psi(x)$:

$$\frac{dP}{dx} = |\psi(x)|^2. \qquad (5.49)$$

Since the probability of finding the particle somewhere is unity, the normalization condition on $\psi(x)$ is

$$\int_{-\infty}^{+\infty} dx\, |\psi(x)|^2 = 1. \qquad (5.50)$$

EXAMPLE 5-7

Construct a probability function $dP/dx = |\psi(x)|^2$ for a particle whose position is precisely known.

SOLUTION:

Let the particle be located at $x = a$. A normalized Gaussian function in the limit of zero width has just the required properties

$$\frac{dP}{dx} = |\psi(x)|^2 = \lim_{L \to 0}\left[\frac{1}{\sqrt{2\pi L^2}}\, e^{-(x-a)^2/2L^2}\right].$$

A function that is nonzero only at $x = a$ in such a manner that the integral is unity is called the Dirac delta function,

$$\delta(x-a) \equiv \lim_{L \to 0}\left[\frac{1}{\sqrt{2\pi L^2}}\, e^{-(x-a)^2/2L^2}\right]. \qquad \blacksquare$$

We now look at the wave number distribution $g(k)$ that corresponds to the function $\psi(x)$. The function $g(k)$ is obtained by taking the Fourier transform of $\psi(x)$:

$$g(k) = \frac{1}{\sqrt{2\pi}} \int_{-\infty}^{+\infty} dx\, \psi(x)\, e^{-ikx}. \qquad (5.51)$$

The square of $g(k)$ specifies the wave number probability distribution (dP/dk). The function $\psi(x)$, which is called the *wave function*, is the integral (5.47) of a distribution of wave numbers. The square of $\psi(x)$, which represents the probability per unit distance (dP/dx) of finding the particle, forms the wave packet.

EXAMPLE 5-8

What is the wave number distribution for a square wave packet, corresponding to equal probability for the particle to be found in the region $-L < x < L$?

SOLUTION:

We first construct the wave function $\psi(x)$. The particle has equal probability of being found in the interval $-L < x < L$, and so $\psi(x)$ is a constant

$$\psi(x) = C.$$

The normalization condition is

$$\int_{-\infty}^{+\infty} dx\, |\psi(x)|^2 = 1 = 2LC^2,$$

$$C = \frac{1}{\sqrt{2L}}.$$

The normalized wave function is

$$\psi(x) = \frac{1}{\sqrt{2L}}$$

for $-L < x < L$, and

$$\psi(x) = 0,$$

for $x < -L$ or $x > L$. The function $g(k)$ is

$$g(k) = \frac{1}{\sqrt{2\pi}} \int_{-\infty}^{+\infty} dx\, \psi(x)\, e^{-ikx}$$

$$= \frac{1}{\sqrt{4\pi L}} \int_{-L}^{+L} dx\, e^{-ikx} = \frac{1}{\sqrt{4\pi L}}\frac{2\sin(kL)}{k}$$

$$= \frac{1}{\sqrt{\pi L}}\frac{\sin(kL)}{k}.$$

The square of $g(k)$ is

$$|g(k)|^2 = \frac{1}{\pi L}\frac{\sin^2(kL)}{k^2}.$$

The functions $|\psi(x)|^2$ and $|g(k)|^2$ are plotted in Figure 5-12. $\qquad \blacksquare$

Group and Phase Velocity

Using the relationship between wavelength and wavenumber (5.39), we may write the deBroglie relationship (5.7) as

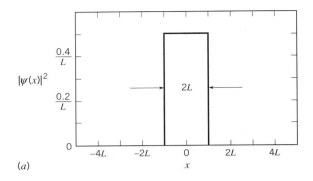

(a)

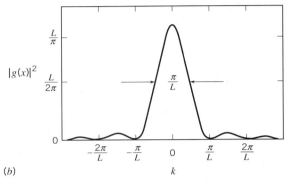

(b)

FIGURE 5-12 Square wave packet together with its wave number distribution.
(a) The square wave packet represents a particle that is confined to be in the region, $-L < x < L$. The wave function squared is normalized to unit area. (b) Distribution of wave numbers that correspond to the square wave packet. The wave number is localized in the region, $-\pi/L < k < \pi/L$. The wave packet in k space also has unit area. The product of the widths of the distributions, $\Delta x \Delta k \approx 1$ is of order unity, independent of L.

$$p = \frac{h}{\lambda} = \hbar k . \tag{5.52}$$

We may write an analogous deBroglie relationship between energy and frequency as

$$E = hf = \hbar \omega . \tag{5.53}$$

Consider a particle of mass m moving with a speed v. We may write the energy of the particle as

$$E = \frac{mc^2}{\sqrt{1 - \dfrac{v^2}{c^2}}} = \hbar \omega , \tag{5.54}$$

and the momentum as

$$p = \frac{mv}{\sqrt{1 - \dfrac{v^2}{c^2}}} = \hbar k . \tag{5.55}$$

In our expressions for energy (5.54) and momentum (5.55), both ω and k depend on the speed of the particle. The velocity of a single frequency component is called the *phase velocity*. The magnitude of the phase velocity (v_p) is defined by

$$v_p \equiv \frac{\omega}{k} . \tag{5.56}$$

From the de Broglie relationships (5.54 and 5.55), we have

$$v_p = \frac{E}{p} = \frac{c^2}{v} . \tag{5.57}$$

The velocity at which the wave packet propagates is called the *group velocity*. The magnitude of the group velocity (v_g) is defined by

$$v_g \equiv \frac{d\omega}{dk} . \tag{5.58}$$

From the de Broglie relationships (5.54) and (5.55), we have

$$v_g = \frac{d\omega}{dv}\frac{dv}{dk} = \left[\frac{mv}{h\left(1 - \dfrac{v^2}{c^2}\right)^{3/2}} \right] \left[\frac{h\left(1 - \dfrac{v^2}{c^2}\right)^{3/2}}{m} \right] = v . \tag{5.59}$$

The group velocity of the wave packet is equal to the particle velocity. The product of the magnitudes of the phase (5.57) and group (5.58) velocities is

$$v_p v_g = \left(\frac{c^2}{v} \right)(v) = c^2 . \tag{5.60}$$

The Uncertainty Principle

If a wave function $\psi(x)$ is more localized, a greater range of wave numbers are needed to compose $\psi(x)$. Both the particle wave function and the wave number distribution are localized. Consider a wave packet localized near $x = 0$ as shown in Figure 5-12a.

The size of the wave function is negligible except in the region $|x| < L$. The order of magnitude of the width (Δx) of the wave function is

$$\Delta x \approx L. \qquad (5.61)$$

The distribution of wave numbers may be approximated as

$$g(k) \approx \frac{1}{\sqrt{2\pi}} \int_{-L}^{+L} dx \, \psi(x) e^{-ikx}. \qquad (5.62)$$

If $kL \gg 1$, then e^{ikx} oscillates rapidly and $g(k)$ vanishes. If $kL \ll 1$, then $e^{ikx} \approx 1$ and the wave number distribution $g(k)$ is a constant. Therefore, the order of magnitude of the width (Δk) of the wave number distribution (see Figure 5-12b) is

$$\Delta k \approx \frac{1}{L}. \qquad (5.63)$$

The product $\Delta x \Delta k$ is independent of the size of the wave packet (L) and is of order unity,

$$\Delta x \Delta k \approx (L)\left(\frac{1}{L}\right) = 1. \qquad (5.64)$$

Using the deBroglie relationship $p_x = \hbar k$, the product of uncertainties (5.64) becomes

$$\Delta x \Delta p_x \approx \hbar. \qquad (5.65)$$

A particle is represented by a localized wave packet, which is composed of a distribution of wave numbers. The product of the spread in the position times the spread in the uncertainty has an intrinsic limit. The order of magnitude of this limit is given by Planck's constant. This is the *uncertainty principle*. The uncertainty principle is illustrated in Figure 5-13. A particle is localized so that its probability distribution (dP/dx) is zero except in some limited region of x. The length scale of the plot is chosen so that dP/dx has a width (Δx) that is approximately unity. The area under the curve is normalized to unity. A particle that is localized has a certain range of wave numbers or momenta (dP/dp_x). In units of Planck's constant divided by the same length scale for dP/dx, the width of the momentum distribution (Δp_x) is approximately unity. The plot dP/dp_x also has unit area. Consider what happens if the particle is localized to twice the precision of Figure 5-13a. The resulting probability distribution dP/dx is shown in Figure 5-13b. The width of the probability distribution dP/dx is now one-half the length compared to Figure 5-13a. The area under the curve is still unity. The corresponding momentum distribution dP/dp_x is twice as wide.

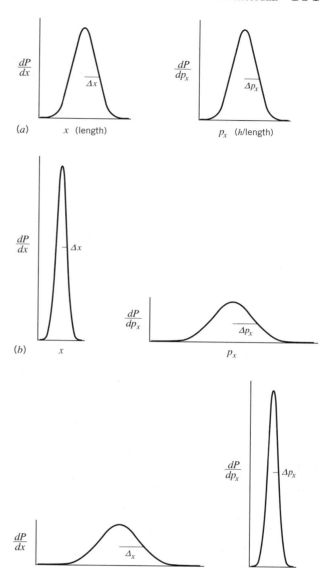

FIGURE 5-13 The uncertainty principle.
(a) A particle that is localized has a certain probability distribution (dP/dx) for being found at the position x. The width of the distribution is Δx. The unit of length in the plot is chosen so that Δx is approximately unity. The corresponding distribution of wave numbers gives the probability for finding the particle with a certain momentum (dP/dp_x). In units of Planck's constant divided by the length scale for dP/dx, the width of the momentum distribution is approximately unity. The plots dP/dx and dP/dp_x have unit area. (b) The particle is localized so that Δx is one-half the length compared to part (a). The corresponding momentum distribution dP/dp_x is twice as wide. (c) The particle is localized so that Δx is twice the length compared to part (a). The corresponding momentum distribution dP/dp_x is one-half as wide.

Now consider what happens if the particle is localized to only one-half of the precision of Figure 5-14a. The resulting probability distribution dP/dx is shown in Figure 5-13c. The corresponding momentum distribution dP/dp_x is one-half as wide. If the wave packet of the particle is twice as narrow, then the distribution of momenta is twice as large. Similarly, if the wave packet is twice as wide, then the distribution of momenta is two times smaller.

Note the two extremes. In a simple plane wave described by one value of k, the momentum is unique and the wave is not localized at all. If the wave packet is a delta function, then the opposite is true. An infinity of wave numbers are needed to describe the wave function, and in this case the particle is perfectly localized but there is no information about the particle momentum.

A wave packet that is very localized contains a large span of wave numbers. This has the profound consequence that there is a range of possible momenta of the particle.

Of special interest is the wave packet that has a Gaussian shape. Consider the Gaussian wave function

$$\psi(x) = Ce^{-x^2/2L^2}, \qquad (5.66)$$

where

$$\sigma_x = L. \qquad (5.67)$$

The distribution of wave numbers is

$$g(k) = \frac{C}{\sqrt{2\pi}} \int_{-\infty}^{+\infty} dx \, e^{-x^2/2L^2} \, e^{ikx}$$

$$= \frac{C}{\sqrt{2\pi}} \sqrt{\frac{\pi}{2L^2}} e^{-k^2L^2/2}. \qquad (5.68)$$

In this case, the distribution of wave numbers is also Gaussian. The standard deviation is

$$\sigma_k = \frac{1}{L}. \qquad (5.69)$$

The product of σ_x (5.67) and σ_k (5.69) is

$$\sigma_x \sigma_k = (L)\left(\frac{1}{L}\right) = 1. \qquad (5.70)$$

The probability of finding the particle per unit x as a function of x is represented by the square of the wave function:

$$\frac{dP}{dx} = |\psi(x)|^2. \qquad (5.71)$$

The standard deviation of dP/dx is

$$\sigma_x = \frac{L}{\sqrt{2}}. \qquad (5.72)$$

The corresponding probability of finding the particle per unit k as a function of k is represented by the square of the wave number distribution,

$$\frac{dP}{dk} = |g(k)|^2. \qquad (5.73)$$

The standard deviation of dP/dk is

$$\sigma_k = \frac{1}{\sqrt{2L}} \qquad (5.74)$$

The product of σ_x (5.72) and σ_k (5.74) is

$$\sigma_x \sigma_k = \frac{1}{2}. \qquad (5.75)$$

From the deBroglie relationship between wave number and momentum, we may write the standard deviation of the momentum distribution (σ_p) as

$$\sigma_x = \hbar \sigma_k, \qquad (5.76)$$

and thus,

$$\sigma_x \sigma_p = \frac{\hbar}{2}. \qquad (5.77)$$

The Gaussian form is unique in having this property. All other wave forms have a larger product of uncertainties in position (Δx) and uncertainties in momentum (Δp_x). Using $\Delta_x = \sigma_x$ and $\Delta p = \sigma_p$ we express this important result as

$$\Delta x \Delta p_x > \frac{\hbar}{2}. \qquad (5.78)$$

This is the precise statement of the uncertainty principle.

The uncertainty principle states the fundamental limitation on how precisely both the position and momentum of a particle can be simultaneously known,

$$\Delta x \Delta p_x > \frac{\hbar}{2}.$$

5-5 CONSEQUENCES OF THE UNCERTAINTY PRINCIPLE

Werner Heisenberg was the first to realize that in order to preserve the basic rules of quantum mechanics, there must be a fundamental limit to the accuracy with which one can simultaneously know both the position and the momentum of a particle. The product of these uncertainties must be larger than $\hbar/2$. This has nothing to do with the accuracy of the measurement apparatus, but rather with the quantum nature of matter and radiation. If we want to measure the position of a particle, we must scatter a photon or other particle such as an electron from the particle. This scattering will cause a momentum transfer to the particle. If we desire a more accurate measurement of position, we must use shorter wavelength photons. This will cause more momentum uncertainty from the scattering.

Note that there is no fundamental limit to how well we can measure the position of a particle. If a particle is localized to a great precision, however, we lose information on the momentum of the particle. Likewise, the momentum of a particle may be determined with great precision, but in that case we lose information on the position of the particle.

Localizing a Particle with a Slit

Consider a beam of particles with a definite momentum p_0 traveling in the y direction. The particles encounter a barrier with a single narrow horizontal slit (see Figure 1-2). Before the particles encounter the slit, they have no x component of momentum (p_x) and there is no uncertainty in p_x ($\Delta p_x = 0$). After passing through the slit, the particle is localized in the x direction to a precision,

$$\Delta x \approx d. \qquad (5.79)$$

The process of localizing the particle with the slit causes an uncertainty in the x component of momentum because the particle behaves like a wave and it diffracts when it passes through the slit. The uncertainty in momentum is

$$\Delta p_x \approx p_0 \Delta\theta \approx \frac{p_0 \lambda}{d} = \frac{h}{d}. \qquad (5.80)$$

The product of uncertainties is

$$\Delta p_x \Delta x \approx h, \qquad (5.81)$$

in agreement with the uncertainty principle (5.78).

Particle in a Box

One important consequence of the uncertainty principle is that a particle confined to occupy a limited region of space has a minimum average kinetic energy. This may be seen by observing that if the particle had zero average kinetic energy, then it would have a zero average momentum squared. But if the average momentum squared is zero, then the uncertainty in momentum is zero, in violation of the uncertainty principle!

The average kinetic energy is proportional to the average value of the square of the momentum $\left(\left\langle p^2 \right\rangle\right)$:

$$\left\langle E_k \right\rangle = \frac{\left\langle p^2 \right\rangle}{2m}. \qquad (5.82)$$

The uncertainty in momentum (Δp) is given by its root-mean-square deviation from the average momentum $\left(\left\langle p \right\rangle\right)$:

$$\Delta p = \sqrt{\left\langle p^2 \right\rangle - \left\langle p \right\rangle^2}. \qquad (5.83)$$

The average momentum is zero because it is equally likely to be positive or negative which gives

$$\left(\Delta p\right)^2 = \left\langle p^2 \right\rangle. \qquad (5.84)$$

Therefore,

$$\left\langle E_k \right\rangle = \frac{\left(\Delta p\right)^2}{2m}. \qquad (5.85)$$

Consider now that the particle is confined in the x direction such that the uncertainty in knowledge of the particle position is Δx. The uncertainty principle (5.78) gives us the uncertainty in momentum

$$\Delta p > \frac{\hbar}{2\Delta x}. \qquad (5.86)$$

Therefore, the particle has a minimum average kinetic energy given by

$$\left\langle E_k \right\rangle > \frac{\hbar^2}{8m\left(\Delta x\right)^2}. \qquad (5.87)$$

EXAMPLE 5-9

Estimate the minimum kinetic energy of a baseball confined inside a shoe box.

SOLUTION:

Estimate the size of the shoe box to be

$$\Delta x \approx 0.3\,\text{m},$$

and the mass of the baseball to be

$$m \approx 0.2\,\text{kg}.$$

The minimum kinetic energy is

$$\left\langle E_k \right\rangle_{\min} \approx \frac{\hbar^2}{8m(\Delta x)^2}$$

$$\approx \frac{\left(10^{-34}\,\text{J}\cdot\text{s}\right)^2}{(8)(0.2\,\text{kg})(0.3\,\text{m})^2}$$

$$\approx 10^{-67}\,\text{J} \approx 10^{-48}\,\text{eV}.$$

The uncertainty principle does not place a serious constraint on the kinetic energy of a massive object like a baseball. The situation is much different for an electron confined in an atom. ■

EXAMPLE 5-10

Estimate the minimum kinetic energy of an electron confined by the electromagnetic force to be inside an atom.

SOLUTION:

Estimate the uncertainty in the position of the electron to be the atomic radius

$$\Delta x \approx 0.1\,\text{nm}.$$

The mass of the electron is

$$m \approx 0.5\,\text{MeV/c}^2.$$

The minimum kinetic energy of the electron is

$$\left\langle E_k \right\rangle_{\min} \approx \frac{\hbar^2}{8m(\Delta x)^2}$$

$$\approx \frac{\hbar^2 c^2}{8mc^2(\Delta x)^2}$$

$$\approx \frac{(200\,\text{eV}\cdot\text{nm})^2}{(8)(5\times10^5\,\text{eV})(0.1\,\text{nm})^2} \approx 1\,\text{eV}. \quad ■$$

EXAMPLE 5-11

Estimate the minimum kinetic energy of a proton that is confined by the strong force to be inside a nucleus.

SOLUTION:

Estimate the uncertainty of the position of the proton to be the nuclear radius:

$$\Delta x \approx 2\,\text{fm}.$$

The uncertainty in momentum is

$$\Delta p = \frac{\hbar}{2\,\Delta x},$$

or

$$\Delta pc = \frac{\hbar c}{2\,\Delta x} \approx \frac{200\,\text{MeV}\cdot\text{fm}}{4\,\text{fm}} \approx 50\,\text{MeV}.$$

The mass energy of the proton is

$$mc^2 \approx 940\,\text{MeV},$$

so we see that the proton is not relativistic ($\Delta pc \ll mc^2$). The minimum kinetic energy of the proton is

$$\left\langle E_k \right\rangle_{\min} \approx \frac{(\Delta p)^2}{2m} = \frac{(\Delta pc)^2}{2mc^2}$$

$$\approx \frac{(50\,\text{MeV})^2}{(2)(940\,\text{MeV})} \approx 1\,\text{MeV}. \quad ■$$

EXAMPLE 5-12

Estimate the minimum kinetic energy of a light quark that is confined by the strong force to be inside a proton.

SOLUTION:

Estimate the uncertainty in the position of the quark to be the proton radius:

$$\Delta x \approx 1\,\text{fm}.$$

The uncertainty in momentum is

$$\Delta p = \frac{\hbar}{2\,\Delta x},$$

or

$$\Delta pc = \frac{\hbar c}{2\,\Delta x} \approx \frac{200\,\text{MeV}\cdot\text{fm}}{(2)(1\,\text{fm})} \approx 100\,\text{MeV}.$$

The mass energy of the light quark (see Figure 1-14) is

$$mc^2 \approx 5\,\text{MeV},$$

so we see that the quark is relativistic ($\Delta pc \gg mc^2$). The minimum kinetic energy of the quark is

$$\left\langle E_k \right\rangle_{\min} \approx \Delta pc \approx 100\,\text{MeV}. \quad ■$$

A particle that is confined to occupy a limited region of space has a minimum average kinetic energy given by the uncertainty principle, independent of the confining mechanism.

Particle in a Box in Three Dimensions

The previous examples have addressed the confinement of a particle in one dimension. The extension to three dimensions is straightforward. Consider a particle of mass m confined to a box of size L^3. The uncertainty in x position (Δx) is given by its root-mean-square deviation,

$$\Delta x = \sqrt{\left\langle x^2 \right\rangle - \left\langle x \right\rangle^2} = \sqrt{\left\langle x^2 \right\rangle} = \frac{L}{\sqrt{12}}. \quad (5.88)$$

The uncertainties in the y and z positions are identical. The minimum uncertainty in the x component of momentum is

$$\Delta p_x > \frac{\hbar}{2\,\Delta x} = \frac{\hbar\sqrt{12}}{2\,L}. \quad (5.89)$$

The minimum uncertainties in the y and z components of momentum are identical. The average kinetic energy is given by

$$\left\langle E_k \right\rangle = \frac{\left\langle p^2 \right\rangle}{2m} = \frac{\left\langle p_x^2 \right\rangle + \left\langle p_y^2 \right\rangle + \left\langle p_z^2 \right\rangle}{2m}. \quad (5.90)$$

The minimum value of average kinetic energy is

$$\left\langle E_k \right\rangle_{min} = \frac{3\left(\Delta p_x\right)^2}{2m} = \frac{9\hbar^2}{2mL^2}. \quad (5.91)$$

Measuring Energy in a Short Time

The energy of a nonrelativistic particle of mass m is

$$E = mc^2 + \frac{p^2}{2m}. \quad (5.92)$$

The uncertainty in the energy (ΔE) depends on the uncertainty in the momentum (Δp):

$$\Delta E = \frac{p\Delta p}{m}. \quad (5.93)$$

The speed of the particle is measured by measuring the displacement (Δx) in a short time (Δt). The momentum of a nonrelativistic particle is

$$p = m\frac{\Delta x}{\Delta t}. \quad (5.94)$$

The product of ΔE (5.93) and Δt, the time interval which was used to measured the momentum (5.94), is

$$\Delta E\Delta t = \left(p\frac{\Delta p}{m}\right)\left(\frac{m}{p}\Delta x\right) = \Delta p\Delta x. \quad (5.95)$$

Since the product of $\Delta x\Delta p$ is limited by the uncertainty principle, there is a fundamental limit on how accurately that we may know the energy of a particle if it is measured in a short time interval. This fundamental limit is

$$\Delta E\Delta t > \frac{\hbar}{2}. \quad (5.96)$$

This is an alternate from of the uncertainty principle.

The energy-time uncertainty relationship (5.96) also holds true for a relativistic particle. In the extreme relativistic case

$$E = pc, \quad (5.97)$$

and

$$\Delta E = c\Delta p. \quad (5.98)$$

Also

$$\Delta t = \frac{\Delta x}{c}, \quad (5.99)$$

so that again

$$\Delta E\Delta t = \left(c\Delta p\right)\left(\frac{\Delta x}{c}\right) = \Delta p\Delta x. \quad (5.100)$$

The energy resolution is limited when the energy is measured in a very short time because particles are continuously radiating and absorbing photons. A precise and repeatable determination of the energy can be performed only when a sufficiently long time interval is used such that the energy fluctuations average nearly to zero. This is a fundamental limit and is independent of the precision of the measurement apparatus. This places a fundamental limit (5.96) on the knowledge of the energy of a particle (ΔE) when it is determined in a finite length of time (Δt).

Particle Lifetimes and Natural Widths of Spectral Lines

For a particle that has a finite lifetime, the uncertainty principle places a fundamental limit on the knowledge of its energy for any given measurement. This does not mean that we cannot measure the energy to any arbitrary level of precision. We can do so by measuring the energy a large number of times and taking the average. We may write the uncertainty principle as a simple expression that relates a particle mean lifetime (τ) with its natural energy spread (Γ). The energy spread Γ is defined to be the full-width of the energy distribution at 1/2 its maximum value. With the identification (see Example 2-2)

$$\Delta t = \tau, \qquad (5.101)$$

and

$$2\,\Delta E = \Gamma, \qquad (5.102)$$

the uncertainty principle gives

$$\Gamma \tau = \hbar. \qquad (5.103)$$

If we know the particle lifetime, we may calculate the width of its energy uncertainty and vice versa. Some particle lifetimes and widths are given in Tables 5-1 and 5-2.

TABLE 5-1
MEASURED PARTICLE WIDTHS AND CORRESPONDING LIFETIMES. The particle width (Γ) is determined from experiment. The corresponding particle lifetime (τ) is calculated from the uncertainty principle.

Particle	Width (Γ)	Lifetime (τ)	Main Decay Mode
η	1.0 keV	6.6×10^{-19} s	$\gamma + \gamma$
Υ	43 keV	1.5×10^{-20} s	hadrons
J/ψ	60 keV	1×10^{-20} s	hadrons
ϕ	4.2 MeV	1.6×10^{-22} s	$K^+ + K^-$
ρ	153 MeV	4.3×10^{-24} s	$\pi^+ + \pi^-$
Z^0	2.5 GeV	2.6×10^{-25} s	hadrons

TABLE 5-2
MEASURED PARTICLE LIFETIMES AND CORRESPONDING WIDTHS. The particle lifetime (τ) is determined from experiment. The corresponding particle width (Γ) is calculated from the uncertainty principle.

Particle	Lifetime (τ)	Width (Γ)	Main Decay Mode
π^0	0.9×10^{-16} s	7 eV	$\gamma + \gamma$
π^-	2.60×10^{-8} s	2.53×10^{-8} eV	$\mu^- + \bar{\nu}_\mu$
μ^-	2.20×10^{-6} s	2.99×10^{-10} eV	$e^- + \bar{\nu}_e + \nu_\mu$
n	900 s	7×10^{-19} eV	$p + e^- + \bar{\nu}_e$

Measurement of the mass energy (E_0) of an unstable particle a large number of times determines the distribution dN/dE_0. This distribution is called the natural line shape of the particle and it follows a form called the *Breit-Wigner* or Lorentzian (see Figure 5-14). The form of the Breit-Wigner distribution is

$$\frac{dN}{dE_0} = \frac{C}{\left(E_0 - mc^2\right)^2 + \left(\dfrac{\Gamma}{2}\right)^2}. \qquad (5.104)$$

Near the maximum, the Breit-Wigner distribution is bell-shaped like the Gaussian but it has longer tails. We see directly that when E is equal to mc^2 plus or minus $\Gamma/2$, then the distribution is one-half its maximum value. The relationship between particle width and lifetime is shown in Figure 5-15.

EXAMPLE 5-13

A typical lifetime of an atomic transition is 10 ns. What is the minimum uncertainty in the energy of the photon that is emitted?

SOLUTION:

$$\Delta E = \frac{\hbar}{\tau} \approx \frac{6.6 \times 10^{-16} \text{ eV} \cdot \text{s}}{10^{-8} \text{ s}} \approx 7 \times 10^{-8} \text{ eV}. \qquad \blacksquare$$

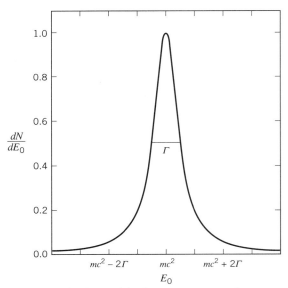

FIGURE 5-14 Measuring the mass energy of an unstable particle.

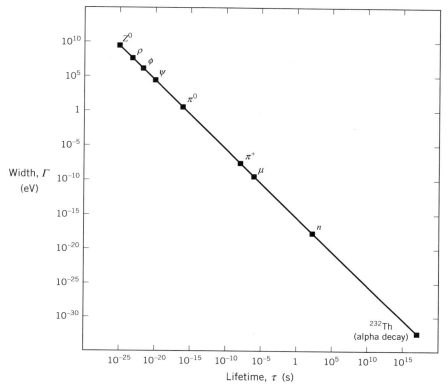

FIGURE 5-15 Particle width (Γ) as a function of lifetime (τ).
The relationship between the lifetime and width of all decaying particles from a very long-lived nucleus (^{232}Th, which decays by alpha particle emission) to the short-lived Z^0, is governed by the uncertainty principle, $\Gamma\tau = \hbar$.

EXAMPLE 5-14

The muon has a proper lifetime of 2.2 μs. What is the minimum uncertainty in the energy of the muon?

SOLUTION:

$$\Delta E = \frac{\hbar}{\tau} \approx \frac{6.6 \times 10^{-16} \text{ eV} \cdot \text{s}}{2.2 \times 10^{-6} \text{ s}} \approx 3 \times 10^{-10} \text{ eV}. \quad \blacksquare$$

EXAMPLE 5-15

The Z^0 particle has a natural line width of $\Gamma \approx 2.5$ GeV. What is the lifetime of the Z^0 particle?

SOLUTION:

$$\tau = \frac{\hbar}{\Gamma} \approx \frac{6.6 \times 10^{-16} \text{ eV} \cdot \text{s}}{2.5 \times 10^{9} \text{ eV}} \approx 3 \times 10^{-25} \text{ s}. \quad \blacksquare$$

Ranges of the Forces

We may use the uncertainty principle to estimate the range of a force. This form of the uncertainty principle states that a long observation time is required to measure the energy of a state to high precision. A fluctuation in the energy of a particle is unobservable if it occurs in a sufficiently small time. In the quantum description of the electromagnetic force operating by exchange of photons, the interacting particle can "lend" some energy to the photon that transmits the force, but there is a fundamental limit on the length of time that a given amount of energy can be "borrowed" by the photon. This maximum length of time is

$$\Delta t \approx \frac{\hbar}{2 \Delta E}. \quad (5.105)$$

The maximum distance (R) that the photon can travel in this time is

$$R \approx c\Delta t = \frac{\hbar c}{2\,\Delta E} \cdot \qquad (5.106)$$

Since the photon is massless, it can have an arbitrarily small amount of energy:

$$\Delta E = \frac{hc}{\lambda} \cdot \qquad (5.107)$$

The range of the photon existing on borrowed energy is

$$R \approx \frac{\lambda}{4\pi}, \qquad (5.108)$$

and that range extends to infinity for long wavelength photons. A virtual photon, which can exist a long distance from its parent charge, has a large wavelength and therefore transfers a very small momentum when captured by another distant charge. Thus, the strength of the electromagnetic force decreases with increasing distance. The range of the electromagnetic force is infinite because the photon has no mass.

The mechanism of the weak force is the exchange of a massive intermediate vector boson (W or Z^0). The masses of these particles are roughly 100 GeV/c^2. Therefore, the amount of energy that must be borrowed by the bosons for the exchange process is about 100 GeV. This produces an uncertainty in the mass energy of the particle that has emitted the boson. In order to remain within the energy constraints, this uncertainty can exist only for a time interval

$$\Delta t \approx \frac{\hbar}{2\,\Delta E} \cdot \qquad (5.109)$$

In this time interval the boson can travel a maximum distance of

$$\begin{aligned} R \approx c\Delta t &= \frac{\hbar c}{2\,\Delta E} \\ &\approx \frac{0.2\ \text{GeV} \cdot \text{fm}}{(2)(100\,\text{GeV})} = 10^{-18}\ \text{m}. \end{aligned} \qquad (5.110)$$

The range of the weak interaction is much smaller than the size of a proton.

What about the strong interaction? The strong interaction boson is the gluon and it too is massless. *Why, then, does the strong interaction not have an infinite range?* The range of the force between two quarks is infinite, in principle. In practice, however, the quarks are always observed to be bound into hadrons with no net strong charge. The strong interaction is so strong at large dis-

tances that quarks are permanently bound. If electrons and protons were permanently bound, the electromagnetic force would have a different behavior. There would be no free electric charges, just as there are no magnetic monopoles. The electromagnetic force would still have an infinite range. If we accelerate the electron inside the atom we could still make it radiate! *If we accelerate a quark inside the proton can we make it radiate gluons?* The answer is yes, but these gluons do not have an infinite range because they are carrying strong charge, unlike the photon, which is electrically neutral. It has taken many beautiful experiments to show this! Some of these data will be discussed in Chapters 17 and 18.

CHAPTER 5: PHYSICS SUMMARY

- The electron, proton, and neutron have observable wave properties analogous to that of the photon. The characteristic wavelength of any particle is inversely proportional to its momentum,

$$\lambda = \frac{h}{p} \cdot$$

- A physical process may be represented by a complex number called the probability amplitude. The probability that the process occurs is the square of the amplitude. If a process can occur by more than one path, then the probability amplitude is the sum of the individual amplitudes and the probability is the amplitude squared. The rule is *add amplitudes first, and then square to get the probability.*

- The uncertainty principle states the fundamental limitation on how precisely both the position and the momentum of a particle can be simultaneously known:

$$\Delta x \Delta p_x > \frac{\hbar}{2} \cdot$$

- A particle confined to occupy a limited region of space, such that the uncertainty in position is Δx, has a minimum average kinetic energy given by the uncertainty principle:

$$\langle E_k \rangle_{\min} \approx \frac{\hbar^2}{8\,m\,(\Delta x)^2} \cdot$$

- An alternate from of the uncertainty principle states the fundamental limitation on how precisely the energy of a particle can be known if it is measured in a short time:

$$\Delta E \Delta t > \frac{\hbar}{2}.$$

The natural width and lifetime of a particle are related by

$$\Gamma \tau > \hbar.$$

REFERENCES AND SUGGESTIONS FOR FURTHER READING

M. Berry, "The Geometric Phase," *Sci. Am.* **259**, No. 6, 46 (1988).

R. N. Bracewell, "The Fourier Transform," *Sci. Am.* **260**, No. 6, 86 (1989).

D. C. Cassidy, "Heisenberg, Uncertainty and the Quantum Revolution," *Sci. Am.* **266**, No. 5, 106 (1992).

C. J. Davisson, "Are Electrons Waves?," *Franklin Institute Journal* **205**, 597 (1928).

R. Feynman, R. Leighton, and M. Sands, *The Feynman Lectures on Physics,* Addison-Wesley (1963).

G. Gamow, "The Principle of Uncertainty," *Sci. Am.* **198**, No. 1, 51 (1958).

R. K. Geherenbeck, "Electron Diffraction: Fifty Years Ago," *Phys. Today* **31**, No. 1, 34 (1978).

W. Heisenberg, *Physics and Philosophy,* Harper (1958).

W. Heisenberg, *Encounters with Einstein, and Other Essays on People, Places and Particles,* Princeton University Press (1983).

C. Jönsson, Zeit. Phys. **161**, 454 (1961); "Electron Diffraction at Multiple Slits," *Amer. Jour. Phys.* **42**, 4 (1974), translated by D. Brandt and S. Hirschi.

I. G. Main, *Vibrations and Waves in Physics,* Cambridge University Press (1984).

D. Saxon, *Elementary Quantum Mechanics,* Holden-Day (1968).

A. Shimony, "The Reality of the Quantum World," *Sci. Am.* **258**, No. 1, 46 (1988).

V. Telegdi and V. Weisskopf, *Phys. Today* **45**, No. 7, 58 (1992).

G. P. Thomson and A. Reid, "Diffraction of Cathode Rays by a Thin Film," *Nature* **119**, 890 (1927).

A. Zeilinger et al., "Single- and Double-Slit Diffraction of Neutrons," *Rev. Mod. Phys.* **60**, 1067 (1988).

QUESTIONS AND PROBLEMS

DeBroglie wavelength

1. What is the wavelength of (a) a photon that has an energy of 10 eV, (b) an electron that has a kinetic energy of 6 MeV, (c) a neutron that has a momentum of 1 keV/c, and (d) a neutrino that has an energy of 1 GeV?

2. Determine the kinetic energy of (a) an electron that has a wavelength of 0.1 nm, (b) a photon that has a wavelength of 0.1 nm, and (c) an alpha particle that has a wavelength of 1 fm.

3. A particle with mass m and charge e is accelerated through a potential difference (V). What is the wavelength of the particle?

4. Estimate the wavelength of a neutron at room temperature.

5. Consider an x ray beam with an energy of 40 keV. What voltage must electrons be accelerated through if they are to have the same wavelength as the x rays?

6. Consider a microscopic object with a mass of 1 μg. What minimum speed must the particle have if its wavelength is to be smaller than the wavelength of light?

7. Electrons, neutrons, and photons each have wavelengths of 0.01 nm. Calculate their kinetic energies.

8. (a) Estimate the wavelength of a nitrogen molecule at room temperature. (b) Make a comparison of the wavelength of the gas molecule with the average distance between molecules at STP.

9. Calculate the wavelength of an electron that has a kinetic energy of (a) 10 eV, (b) 10 keV, (c) 0.5 MeV, (d) 1 GeV, and (e) 50 GeV. Are electrons with such energies found in nature?

10. Determine the range of speeds of electrons that have wavelengths equal to that of light.

11. An electron and a photon have the same wavelength and the total energy of the electron is equal to 1.0 MeV. Calculate the energy of the photon.

Measuring the wave properties of the electron

12. In the Davisson-Germer experiment, the scattering angle (ϕ) from nickel is measured to be 30 degrees when the electron kinetic energy is 83 eV (see Figures 5-3 and 5-4). Determine the scattering angle if the energy of the electrons is doubled.

13. In the Davisson-Germer experiment, the scattering angle (ϕ) from nickel is measured to be 50 degrees when the electron kinetic energy is 54 eV (see Figure 5-4 for the definition of the scattering angle). Determine the spacing of the atoms in the nickel crystal.

Probability amplitudes

14. (a) Show from the geometry of the two-slit experiment (see Figure 5-9) that the path length difference

(ΔL) to the detection screen from the two slits is

$$\Delta L = L_1 - L_2 = \frac{xd}{z}.$$

(b) Show that the distance (Δx) between adjacent maxima in the interference pattern is

$$\Delta x = \frac{z\lambda}{d}.$$

15. A two-slit experiment is performed using light as the incident particles. The distance between the slits is 1 millimeter and the photon detector is placed at a distance of 1 meter from the plane of the slits. Estimate the distance between the maxima in the resulting interference pattern. Does the light source need to be monochromatic?

16. In the two-slit experiment, what happens if the bombarding particles have some spread in energy, ΔE?

Wave description of a particle

17. A proton is located to a precision of 1 nm. What is the minimum uncertainty in the knowledge of the speed of the proton?

18. Imagine trying to beat the uncertainty principle with the following experiment. A beam of electrons is sent through a very narrow slit. The electrons diffract when they pass through the slit and emerge with a range of momenta in the x direction. A second very narrow slit of width Δx is placed at a large distance from the first slit to detect just those electrons that happen to have essentially no momentum in the x direction. The uncertainty in momentum of an electron that is directed into the second slit is $\Delta p_x \approx p\Delta x/z$. The product of uncertainties of x position and x component of momentum is $\Delta p_x \Delta x \approx p\Delta x^2/z$. Calculate the product of $\Delta p_x \Delta x$ for electrons that have an energy of 1 MeV, when the slit width is 0.1 μm and the distance to the second slit is 1 m. Compare your answer to Planck's constant. Why is the uncertainty principle not violated?

19. Consider the experiment of Jönsson, where 50-keV electrons were directed through a slit of width equal to 50 μm. Use the uncertainty principle to estimate the spread of the diffracted electron beam at a distance of 0.35 m from the slit.

Consequences of the uncertainty principle

20. Consider an electron and a proton each confined to a volume of 10^{-30} m³. Which particle has a larger minimum kinetic energy? Why?

21. Why is good mass resolution important in the search for a new long-lived particle?

22. Even a professional musician with "perfect pitch" will have trouble identifying the pitch of a note if the duration is too short. Why?

23. The tau lepton has a mass of about 1.8 GeV and a lifetime of about 0.3 picoseconds. What is the fractional width ($\Delta m/m$) of the tau particle?

24. A particle of mass m is confined to be in a small space. Estimate the minimum size of the confining space to make the particle relativistic ($v/c \approx 0.5$). Give numerical answers for (a) an electron, (b) a muon, and (c) a charm quark.

25. Estimate the minimum kinetic energy of an electron confined to be in a volume the size of the nucleus.

26. In the beta decay of a neutron a massless particle called the electron-antineutrino ($\overline{V}_e$) is produced. The decay process is

$$n \rightarrow p + e^- + \overline{v}_e.$$

(a) Make an estimate of the kinetic energy of an $\overline{V}_e$ confined to be inside a neutron. (The size of a neutron is the same as the size of a proton, i.e., about 1 fm.) (b) The order of magnitude of the energy of a $\overline{V}_e$ from beta decay is measured to be 1 MeV. What may we conclude from this?

27. If an amplifier is designed to respond to fast signals, then what must be true about its *bandwidth* or range of frequencies to which it is sensitive?

28. The K_s^0 meson decays into $\pi^+ + \pi^-$ or $\pi^0 + \pi^0$ with a lifetime of 8.9×10^{-9} s. What is the width of the K_s^0 meson?

29. A lower limit on the proton lifetime is measured to be 10^{33} years. What can be said about the width of the proton?

30. The width of the Δ particle is measured to be 115 MeV. What is the lifetime of the Δ?

Additional problems

*31. *Estimating the strength of the strong force.* (a) Use the uncertainty principle to estimate the kinetic energy of a proton that is confined in a nucleus that has a radius of 2 fm. (b) Consider the proton to behave as a classical harmonic oscillator (i.e., the force on the proton is proportional to its displacement) with a maximum displacement of 2 fm. Calculate the strength of the force at a displacement of 2 fm in MeV/fm. Compare your answer to the electric force between two protons separated by a distance of 2 fm. (c)

Calculate the maximum acceleration of the proton. How many "g" is this? ($g = 9.8$ m/s^2)

32. *Molecular vibrational energy.* The hydrogen molecule is a bound state of two protons and two electrons. The separation between the two protons is about 0.1 nm. The molecule has an additional degree of freedom that the atom does not have because the nuclei can vibrate along the axis that contains the two protons. Estimate the vibrational energy of the protons. (*Hint:* The protons are confined on a distance scale of 0.1 nm.)

33. The energy needed to remove one electron from potassium (the ionization energy) is about 4 eV. The ionization energy of argon is about 16 eV. (a) Estimate the root-mean-square speed of the outer electrons in potassium and argon. (*Hint:* What is the relationship between the total energy and the average kinetic energy?) (b) Use the deBroglie wavelengths to determine the relative sizes of the potassium and argon atoms.

34. (a) Using the Bohr model, show that the wavelength (λ) of an electron in the hydrogen atom is inversely proportional to the electron mass. (b) Show that the size of the atom is given by the electron wavelength, that is, both λ and the Bohr radius depend on the electron mass, electron charge, and Planck's constant in exactly the same manner.

35. The typical energy needed to ionize an atom is a few electronvolts. Use the uncertainty principle to estimate the size of an atom.

RUTHERFORD SCATTERING

Attention was drawn to the remarkable fact, first observed by Geiger and Marsden, that a small fraction of the swift α particles from radioactive substances were able to be deflected through an angle of more than 90° as the result of an encounter with a single atom... . In order to account for this large angle scattering of α particles, I supposed that the atom consisted of a positively charged nucleus of small dimensions in which practically all of the mass of the atom was concentrated. The nucleus was supposed to be surrounded by a distribution of electrons to make the atom electrically neutral, and extending to distances from the nucleus comparable to the ordinary accepted radius of the atom. Some of the swift α particles passed through the atoms in their path and entered the intense electric field in the neighborhood of the nucleus and were deflected from their rectilinear path. In order to suffer a deflection of more than a few degrees, the α particle has to pass very close to the nucleus, and it was assumed that the field of force in this region was not appreciably affected by the external electronic distribution. Supposing that the forces between the nucleus and the α particle are repulsive and follow the law of inverse squares, the α particle describes a hyperbolic orbit round the nucleus and its deflection can be simply calculated.

Ernest Rutherford

In 1909, Hans Geiger and Ernst Marsden observed that α particles from radioactive decays occasionally scatter at large angles (>90°) when passing through a thin layer of material. Scattering at large angles is possible only if a sufficiently large force is exerted on the α particle. Such a force can arise only if the α particle passes very close to the particle that causes the scattering, much closer than the size of an atom. This led Ernest Rutherford in 1912 to deduce that the positive charge of the atom must be concentrated in a nucleus. With that simple assumption, Rutherford was able to derive a detailed formula for the scattering, which was subsequently verified by more detailed experiments. This marked the beginning of an era of particle-scattering experiments that have led us to our present understanding of the fundamental structure of matter.

6-1 MEASURING STRUCTURE BY PARTICLE SCATTERING

The structure of matter is determined by scattering experiments. These scattering experiments have taken on a wide variety of forms. When we view a cell through a microscope, we scatter photons from the cell and observe the photons with our eye. Measurement of the scattered photons gives us an image of the cell. When light passes by an object, the light "bends" or diffracts. The amount of diffraction depends on the wavelength of the light. Therefore, the resolution of the image has a fundamental limitation that is set by the wavelength of the photons. Any type of particle may be used to probe the structure of matter. The key to achieving good resolution is to use particles of small wavelength.

At the beginning of the twentieth century, Ernest Rutherford designed an experiment to measure the structure of the atom using the α particle as a probe. The α particles were produced in the decay of radon. The α particle has a mass of about 3.7 GeV/c^2 and an electric charge of +2e. Rutherford determined that the α particle was a helium atom minus 2 electrons (doubly ionized helium). When Rutherford designed the experiment, he did not know the structure of an α particle, but he assumed that it was extremely small compared to the size of an

atom. As Rutherford and his associates soon discovered, the α particle is just the nucleus of the helium atom.

EXAMPLE 6-1

The α particles used in the original Rutherford experiment had a kinetic energy of $E_k = 5.5$ MeV, a typical energy of a nuclear decay. Estimate the spatial resolution that may be achieved by scattering of these α particles.

SOLUTION:

The spatial resolution limit of this experiment is determined by the wavelength of the α particle. The α particle mass energy is

$$m_\alpha c^2 \approx 3730 \text{ MeV} .$$

The α particle momentum is given by the expression

$$pc = \sqrt{\left(E_k + m_\alpha c^2\right)^2 - \left(m_\alpha c^2\right)^2}$$
$$= \sqrt{E_k^2 + 2 m_\alpha c^2 E_k} \approx \sqrt{2 m_\alpha c^2 E_k} .$$

The α particle wavelength (5.7) is

$$\lambda = \frac{h}{p} = \frac{hc}{pc} = \frac{hc}{\sqrt{2 m_\alpha c^2 E_k}}$$
$$\approx \frac{1240 \text{ MeV} \cdot \text{fm}}{\sqrt{(2)(3730 \text{ MeV})(5.5 \text{ MeV})}}$$
$$\approx 6 \text{ fm} = 6 \times 10^{-15} \text{ m} .$$

The α particle wavelength is much smaller than the size of the atom (about 10^{-10} m); it is about equal to the size of the nucleus. Nature had provided physicists with a tool well matched to the task at hand! ∎

The conceptual design of the Rutherford scattering experiment is indicated in Figure 6-1. A thin metal foil is bombarded with α particles and the angle θ of the scattered α particles is measured. The angular distribution, $dN/d\theta$, from the scattering of a large number of α particles gives us important details about the structure of the atom. At the time of the experiment, atoms were known to be electri-

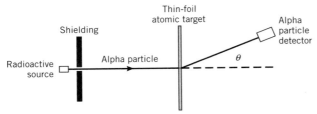

FIGURE 6-1 The Rutherford experiment.
Energetic α particles are passed through a thin metal foil and the angular scattering angle θ is measured.

cally neutral and contain electrons. Therefore, it was known that the atoms also contained positive charge, but there was no knowledge of how the positive charge was distributed in atoms. Consider the collision of an α particle with a single atom as it passes through the thin foil. We refer to the positive charge, however it is distributed in the atom, as the atomic scattering center. We shall neglect collisions of the α particle with electrons because the α particle mass (m_α) is so much greater than the electron mass (m),

$$m_\alpha \gg m. \tag{6.1}$$

EXAMPLE 6-2

Estimate the maximum kinetic energy that can be transferred to an electron from a collision with a 6-MeV α particle.

SOLUTION:

The maximum kinetic energy is transferred in a head-on collision. Let m_α be the mass of the α and m be the mass of the electron. Let v_α be the initial speed of the α and v be the recoil speed of the electron. The initial speed of the α particle is given by

$$\frac{1}{2} m_\alpha v_\alpha^{\,2} = E_k,$$

or

$$v_\alpha = \sqrt{\frac{2 E_k}{m_\alpha}}.$$

We know from momentum conservation that the speed of the electron after the collision is

$$v \approx 2 v_\alpha = 2 \sqrt{\frac{2 E_k}{m_\alpha}},$$

as may be verified by looking in the frame where the α is at rest. The increase in the kinetic energy of the electron (ΔE) is

$$\Delta E = \frac{1}{2} m v^2 = \frac{1}{2} m \left[2 \sqrt{\frac{2 E_k}{m_\alpha}} \right]^2$$

$$= \frac{4 m E_k}{m_\alpha} \approx \frac{(4)(0.511\,\text{MeV})(6\,\text{MeV})}{3730\,\text{MeV}} \approx 3\,\text{keV}.$$

Note that this energy is much larger than the typical energy of an outer electron in the atom (a few eV) but much smaller than the kinetic energy of the α particle (a few MeV). ∎

We now estimate the scattering angle of the α particle. Let the α particle be traveling with a speed v_α and come within a distance r of the atomic scattering center. Let q_1 be the charge of the α particle ($q_1 = 2e$) and let q_2 be the positive charge of the atom ($q_2 = Ze$). The change in momentum (Δp) of the α particle when it passes the atomic scattering center is the electric force acting for a time (Δt), or

$$\Delta p = F \Delta t \approx \left(\frac{k q_1 q_2}{r^2} \right) \left(\frac{2 r}{v_\alpha} \right), \tag{6.2}$$

where we have estimated Δt to be equal to $2r/v_\alpha$, the time that it takes the α particle to pass by the atomic scattering center. The scattering angle (θ_s) due to this single collision is approximately

$$\theta_s \approx \frac{\Delta p}{p} = \left(\frac{2 k q_1 q_2}{r v_\alpha} \right) \left(\frac{1}{m_\alpha v_\alpha} \right)$$

$$= \frac{k q_1 q_2}{r \left(\frac{1}{2} m_\alpha v_\alpha^{\,2} \right)} = \frac{k q_1 q_2}{r E_k}, \tag{6.3}$$

where E_k is the kinetic energy of the α particle. The important result of this estimate is that the scattering angle is inversely proportional to the distance (r) between the α and the atomic scattering center. If the scattering center is small in size, as indicated in Figure 6-2a, then r can be very small and large scattering angles are possible. If the atomic scattering center has a finite size (R), then there is a possibility that the α particle can penetrate the atomic scattering center, as indicated in Figure 6-2b. For a penetrating α particle, the force exerted is that due to a smaller charge. For a fixed value of r, the scattering angle is

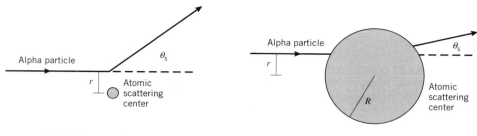

(a) Pointlike scattering center   (b) Finite size scattering center

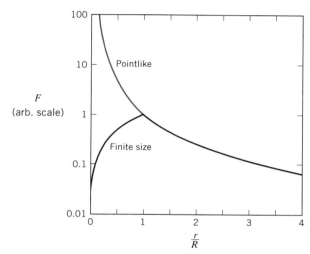

FIGURE 6-2 Force on an α particle as a function of distance (r) from the positive charge inside an atom.
(a) The α particle does not penetrate the positive charge of the atom. The α particle feels the full positive charge (Ze) of the atom. (b) The positive charge of the atom has a finite size and the α particle is able to penetrate the charge. The effective charge that scatters the α particle is only the fraction $(r/R)^3$, where R is the size of the charge. This results in a weaker force compared to (a) and the scattering angle is smaller at fixed r.

smaller for a penetrating α particle than for the case where the atomic scattering center is concentrated at a point.

EXAMPLE 6-3

Consider the charge Ze spread out uniformly in a sphere of radius R. Calculate the electric force at a distance r from the center of the sphere.

SOLUTION:

Gauss's law (B.1), is

$$\oiint d\mathbf{a} \cdot \mathbf{E} = 4\pi k q_{\text{tot}} \ .$$

For $r > R$, we have

$$4\pi r^2 E = 4\pi k Z e \ ,$$

or

$$E = \frac{Zke}{r^2} \ .$$

The electric force $(F_{r>R})$ on the charge e is

$$F_{r>R} = eE = \frac{Zke^2}{r^2} \ .$$

Now consider the case where $r < R$. Applying Gauss's law, we have

$$4\pi r^2 E = 4\pi k Z e \left(\frac{r}{R}\right)^3,$$

or

$$E = \frac{Z e k r}{R^3}.$$

The electric force ($F_{r<R}$) on the charge e is

$$F_{r<R} = eE = \frac{Z k e^2 r}{R^3}.$$

The force is maximum (F_{max}) when $r = R$:

$$F_{max} = \frac{Z k e^2}{R^2}. \qquad \blacksquare$$

If the positive charge in an atom were uniformly distributed, then the angular distribution of α particles should be sharply peaked in the forward direction.

EXAMPLE 6-4

Estimate the scattering angle for a single collision of a 6-MeV α particle with a platinum atom ($Z = 78$) if the positive charge was uniformly distributed in the atom.

SOLUTION:

The minimum value of R is the size of the atom. The scattering angle for a single collision is

$$\theta_s \approx \frac{k q_1 q_2}{R E_k} = \frac{(2)(78)(1.44 \text{ eV} \cdot \text{nm})}{(0.1 \text{ nm})(6 \times 10^6 \text{ eV})} = 4 \times 10^{-4}.$$

The scattering angle for a single collision is very small if the charge is spread out over the entire atom. Even a relatively large number of these collisions will not make a large net scattering angle. $\blacksquare$

A thin target foil contains a very large number of atomic layers, which each scatter the α particle through a small angle. The scattering from each layer can occur in any direction. The resulting distribution will be a Gaussian centered at zero. We can make an order of magnitude estimate of the width of the Gaussian.

EXAMPLE 6-5

Estimate the root-mean-square scattering angle of an α particle passing through a platinum foil of thickness 1 μm.

SOLUTION:

The α particle scatters a small amount as it passes each atom. Each scattering produces a random small change in the direction (angle) of the α particle. The resulting distribution is a Gaussian (see Chapter 2). The average scattering angle is zero and the root-mean-square is proportional to the square root of the number of scatters. The number of scatters is the foil thickness divided by the size of the atom:

$$N \approx \frac{10^{-6} \text{ m}}{10^{-10} \text{ m}} = 10^4.$$

Using θ_s from the previous example, the root-mean-square scattering angle (in radians) is

$$\theta_{rms} \approx \sqrt{10^4} \, \theta_s = (100)(4 \times 10^{-4}) = 0.04,$$

which corresponds to about 2°. Note that we have calculated the root-mean-square scattering angle by considering a very thin foil. The root-mean-square scattering angle increases with the square root of the thickness of the foil. If the foil is thick, the α particle also loses energy to the electrons (a small amount of energy times a large number of collisions). For these two reasons, Geiger and Marsden chose the thinnest foils they could obtain. $\blacksquare$

If the positive charge of the atom is concentrated in a tiny nucleus, the angular distribution of the scattered α particles is much different from the case of a uniform charge distribution. The reason for this is simple: the α particle can come much closer to this charge without penetrating the charge, and the force is much stronger at a shorter distance. A careful measurement of α particle scattering gives us detailed information about the structure of the atom.

Before discussing the results of the Rutherford experiment in more detail, we develop an important tool for the description of scattering experiments, the cross section.

6-2 DEFINITION OF CROSS SECTION

Consider the scattering of two hard spheres, a projectile bouncing off a target of radius R. A stream of projectile spheres is directed toward the target, as shown in Figure 6-3. The probability that a given projectile hits the target is called the scattering probability (P). This scattering probability is proportional to the cross-sectional area (σ) of the target:

$$P \propto \sigma = \pi R^2. \qquad (6.4)$$

The area σ is called the *total cross section* for the scattering process. Let the projectile particles cover an area a, where

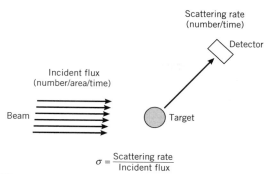

Scattering rate
(number/time)

Detector

Incident flux
(number/area/time)

Beam

Target

$$\sigma = \frac{\text{Scattering rate}}{\text{Incident flux}}$$

FIGURE 6-3 Definition of cross section.
Projectile particles are fired toward a target. The flux of projectile particles (number per area per time) covers an area larger than the target. The number of projectile particles that scatter per time is called the scattering rate. The cross section is the scattering rate divided by the incident flux. The cross section has units of area and its geometric interpretation is the cross-sectional area of the target that produces the scattering.

$a \gg R^2$, and be incident at the rate $\Delta N / \Delta t$. The incident flux (Φ_i) is

$$\Phi_i = \frac{\Delta N}{a \Delta t}. \qquad (6.5)$$

and the scattering rate (R_s) is

$$R_s = \left(\frac{\pi R^2}{a} \right) \left(\frac{\Delta N}{\Delta t} \right). \qquad (6.6)$$

The cross section is the scattering rate (6.6) divided by the incident flux (6.5),

$$\sigma = \frac{R_s}{\Phi_i} = \frac{\left(\dfrac{\pi R^2}{a} \right) \left(\dfrac{\Delta N}{\Delta t} \right)}{\left(\dfrac{\Delta N}{A \Delta t} \right)} = \pi R^2. \qquad (6.7)$$

Now consider the more general case of incoming particles of type A incident on target particles of type B with the reaction,

$$A + B \rightarrow \text{FINAL STATE}. \qquad (6.8)$$

The final state stands for any possible outcome that we wish to define. For example, the final state could be particle A having a momentum in some specified range, or it could be a state with one or more new particles created. The rate (number per time) that the final state is produced is called the *transition rate*. The probability for $A + B$ to

produce a certain final state or group of final states is specified by the interaction cross section σ which is defined to be

$$\sigma = \frac{\text{transition rate}: A + B \rightarrow \text{FINAL STATE}}{\Phi_i}. \qquad (6.9)$$

Cross section has the units of area. One barn (b), as in "hitting the broadside of a barn," is defined to be

$$1\,\text{b} \equiv 10^{-28}\,\text{m}^2. \qquad (6.10)$$

The cross section has the geometrical interpretation as the effective cross-sectional area of the target particle as seen by the incoming particle. Cross sections are a function of energy because the wavelengths of the interacting particles are energy-dependent.

> The cross section for any process, $A + B \rightarrow \text{FINAL STATE}$, is a unit of area that is proportional to its probability of occurrence and is defined to be the transition rate divided by the incident flux.

Consider the case of incoming protons striking a liquid hydrogen target. The target is made up of many protons and the fundamental interaction is a proton–proton interaction (not a proton–"target" interaction). Each incident proton has a chance to interact with target protons along the whole length of the target. In this case it is useful to calculate the cross section *per target proton*. The corresponding incident flux is equal to the rate at which the protons strike the target times the number of protons per area in the target. The number of protons per area in the target (N_p/a) depends on the length of the target (L),

$$\frac{N_p}{a} = \frac{N_A L \rho}{10^{-3}\,\text{kg}}, \qquad (6.11)$$

where N_A is Avogadro's number and ρ is the density of liquid hydrogen.

EXAMPLE 6-6
The total proton–proton strong interaction cross section is about 40 mb. Calculate the fraction of protons that scatter when a collimated beam of protons is sent through a liquid hydrogen target of length 0.3 meters. The density of liquid hydrogen is 70 kg/m³.

SOLUTION:
Let R_i be the rate of protons incident on the target and R_s be the scattering rate. The incident flux of protons is the incident rate times the number of protons per area in the

target. The number of protons per area (6.11) in the target is

$$\frac{N_p}{a} = \frac{N_A L\rho}{10^{-3} \text{ kg}}.$$

The incident flux is

$$\Phi_i = \frac{R_i N_p}{a} = \frac{R_i N_A L\rho}{10^{-3} \text{ kg}}.$$

The expression for the cross section *per proton* is the scattering rate divided by the incident flux:

$$\sigma = \frac{R_s}{\Phi_i} = \frac{R_s}{\left[\dfrac{R_i N_A L\rho}{10^{-3} \text{ kg}}\right]}.$$

The fraction of protons scattered is

$$\frac{R_s}{R_i} = \frac{N_A \sigma L\rho}{10^{-3} \text{ kg}}$$

$$= \frac{\left(6\times10^{23}\right)\left(40\times10^{-31} \text{ m}^2\right)\left(0.3 \text{ m}\right)\left(70 \text{ kg/m}^3\right)}{10^{-3} \text{ kg}}$$

$$\approx 0.05.$$

Note that the incident flux does not depend on the area of the beam, as long as it is smaller than the cross-sectional area of the target. The incident flux depends on the length of the target because each incident proton encounters target protons along the entire length of the target. ∎

Table 6-1 lists some cross sections for various processes. The sizes of the cross sections vary by many orders of magnitude because the interactions listed are governed by different forces. The first is a strong interaction, the

TABLE 6-1
EXAMPLES OF SOME PARTICLE CROSS SECTIONS AT A CENTER-OF-MASS ENERGY OF 10 GEV.

Process	σ (Approximate)
$p + p \rightarrow$ anything, by strong interaction	40 mb
$\gamma + p \rightarrow$ anything	100 µb
$e^+ + e^- \rightarrow \mu^+ + \mu^-$	1 nb
$\nu + N \rightarrow \mu^- +$ anything	1 pb

second two are electromagnetic interactions, and the last one is a weak interaction.

6-3 PROBING THE STRUCTURE OF THE ATOM

Discovery of the Nucleus

We now return to a discussion of the great discovery of Geiger and Marsden. When they scattered α particles from a thin platinum foil, they observed that an unexpectedly large fraction of the α particles were scattered at large angles. Even more surprising was the fact that α particles were occasionally scattered through angles greater than 90°. Some α particles were even observed to be scattered backwards! Rutherford was very excited about these experimental results which he described as:

> It was quite the most incredible event that has ever happened to me in my life. It was almost as if you fired a 15-inch shell at a piece of tissue paper and it came back and hit you.

What Rutherford did not know before the experiment, and what he discovered from the experiment was that the "15-inch shell" (the α) was hitting another "15-inch shell" (the nucleus). In 1911, Rutherford explained these data by correctly deducing that a large angle scatter could only be the result of a single collision and therefore that the atom has a dense charged nucleus.

To explain the large angle scattering data, Rutherford hypothesized that all the positive electric charge of the atom was concentrated in a very small volume. This was a revolutionary idea at the time. From his hypothesis, Rutherford applied Coulomb's law for the electric force and derived the form of the angular distribution of the scattered α particles. The angular distribution is called the *differential cross section* ($d\sigma/d\cos\theta$). (The integral of $d\sigma/d\cos\theta$ over all values of $\cos\theta$ gives the total cross section, σ. See problem 12.)

The form of the differential cross section is

$$\frac{d\sigma}{d\cos\theta} \propto \left(\frac{2Zke^2}{E_k}\right)^2 \frac{1}{\left(1-\cos\theta\right)^2}, \quad (6.12)$$

where $2e$ is the charge of the α particle, Ze is the charge of the nucleus and E_k is the α particle kinetic energy.

Note that $(ke^2/E_k)^2$ has dimensions of area. We can rewrite this result in terms of the dimensionless electromagnetic coupling ($\alpha = ke^2/\hbar c$) by multiplying and dividing by $(\hbar c)^2$. The result is

$$\frac{d\sigma}{d\cos\theta} \propto \alpha^2 \left(\frac{\hbar c}{E_k} \right)^2 \frac{1}{(1-\cos\theta)^2}. \quad (6.13)$$

We shall derive this important formula later in the chapter.

The differential cross section for the scattering of pointlike particles is proportional to the square of the fundamental interaction strength, inversely proportional to the kinetic energy squared and inversely proportional to $(1-\cos\theta)^2$,

$$\frac{d\sigma}{d\cos\theta} \propto \alpha^2 \left(\frac{\hbar c}{E_k} \right)^2 \frac{1}{(1-\cos\theta)^2}.$$

The theory of Rutherford scattering assumes that the nucleus is concentrated at a point, that is, the α particle does not penetrate the nucleus. If the α particle is energetic enough to penetrate the nucleus, then the Rutherford scattering formula does not hold true.

EXAMPLE 6-7
Estimate the kinetic energy of an α particle needed to penetrate the silver nucleus ($Z = 47$). The radius of the silver nucleus is about 5 fm.

SOLUTION:
The α particle can penetrate the nucleus when it has enough energy to overcome the Coulomb repulsion. This happens when

$$E_k = \frac{kq_1q_2}{R},$$

where $q_1 = 2e$, the α electric charge, $q_2 = 47e$, the silver nucleus electric charge, and $R = 5$ fm, the nuclear radius. Therefore,

$$E_k = \frac{(2)(47)(1.44 \text{ MeV} \cdot \text{fm})}{5 \text{ fm}} = 27 \text{ MeV}.$$

Such an α particle is not relativistic. ∎

In 1913, Geiger and Marsden carried out a second experiment to quantitatively test the validity of the Rutherford model. The objectives of the experiment were clearly stated in the introduction to the paper by Geiger and Marsden:

At the suggestion of Prof. Rutherford, we have carried out experiments to test the main conclusions of the above theory. The following points were investigated:

(1) Variation with scattering angle.

(2) Variation with thickness of the scattering material.

(3) Variation with atomic weight of the scattering material.

(4) Variation with velocity of incident α particles.

(5) The fraction of particles scattered through a definite angle.

The main difficulty of the experiments has arisen from the necessity of using a very intense and narrow source of α particles owing to the smallness of the scattering effect. All the measurements have been carried out by observing the scintillations due to the scattered α particles on a zinc-sulfide screen, and during the course of the experiments over 100,000 scintillations have been counted. It may be mentioned in anticipation that all the results of our investigation are in good agreement with the theoretical deductions of Prof. Rutherford, and afford strong evidence of the correctness of the underlying assumption that an atom contains a strong charge at the centre of all dimensions, small compared with the diameter of the atom.

The Apparatus
The source of α particles was purified radium contained in a thin-walled 1-mm diameter glass tube. The strength of the α source was huge by laboratory standards, about 0.1 Curie, or about 4 billion nuclear decays per second. Radium has a long decay chain. These decays are discussed in more detail in Chapter 11. The time scale of each decay is characterized by the half-life, the time it takes one-half of the particles to decay. The source was shielded with lead to reduce the background from electrons and photons, which are also produced in the chain of radioactive decays. The α particles were allowed to pass through a small diaphragm and were directed toward a thin foil that served as the nuclear target. Several different foils were mounted on a wheel so that they could be used as targets in successive measurements. The α particle detector was a small (10^{-6} m^2) zinc-sulfide screen mounted a few centimeters away from the target. On passing through the

detector, an α particle induces the emission of light. The screen was viewed with a microscope so that a flash of light was visible to the viewer when an α particle struck the screen. In order to observe scattered α particles at various angles, the screen and microscope assembly were allowed to rotate through the range from 5 to 150 degrees with respect to the α particle direction. The whole apparatus was placed in an evacuated container to eliminate the scattering and absorption of the α particles in air.

Angular Dependence

For the angular measurements, the main source of α particles was the decay of radon:

$$^{222}\mathrm{Rn} \rightarrow\ ^{218}\mathrm{Po} + \alpha. \qquad (6.14)$$

The energy of the α particle is 5.5 MeV and the half-life of the decay is 3.82 days. A typical set of angular measurements took about two days to complete, so the experimenters needed to make a correction for the decay of the source over the time in which the data were taken. The angular dependence of the α particle scattering was measured by rotating the detection screen plus microscope assembly. The scattering rate was by far the greatest at small angles. The maximum rate that could be accurately measured was roughly two scintillations per second. When the foil was removed, a small number of scintillations per time were observed due to scattering of the α particles from the walls of the container. The experimenters measured the scintillation rate with no foil present and subtracted this background to get the scattering rate in the foil. The background imposed a practical limit on the smallest scattering rate that could be accurately measured. This limit was determined to be about 0.1 scintillation per second. Thus, the experiment was sensitive to scattering rates in the range 0.1 to 2 per second. The rates could be adjusted to be in the sensitive range by choosing the distance from the foil to the detection screen. Since the lifetime of the source was 3.82 days, the large angle data were taken first when the radioactive source was "hottest" and the small angle data were taken after the source had substantially decayed.

Data of Geiger and Marsden for the angular dependence of α particles scattering from a silver foil are shown in Figure 6-4. The number N scattered at a given angle is proportional to the differential cross section. These data provide convincing proof that the angular dependence $(d\sigma/d\cos\theta)$ of the scattering cross section has the form

$$\frac{d\sigma}{d\cos\theta} \propto \frac{1}{\left(1 - \cos\theta\right)^2}. \qquad (6.15)$$

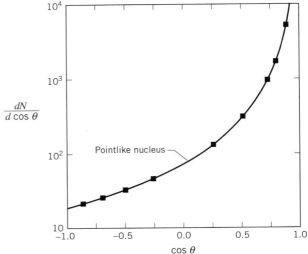

FIGURE 6-4 Angular distribution of α particles after passing through a thin silver foil.
The measured angular distribution is of the form $dN/d\cos\theta \propto 1/(1-\cos\theta)^2$, which can occur only if the scattering center is a "point." The data are from H. Geiger and E. Marsden, *Phil. Mag.* **25**, 604 (1913).

This result can hold true only if the scattering is due to a pointlike object. These data prove that the atom has a nucleus.

Variation with Foil Thickness

To study the variation of the scattering cross section with thickness, the experimenters had to take extra care to prepare a monoenergetic α source. For this set of measurements they did not need a long half-life and they used the decay

$$^{214}\mathrm{Bi} \rightarrow\ ^{210}\mathrm{Th} + \alpha. \qquad (6.16)$$

This radioactive material (^{214}Bi) was known at the time as *Radium C*. The α particle has an energy of 5.5 MeV and a half-life of about 20 minutes. If the large angle scatters are the result of a single close encounter of the α particle with the nucleus, then the scattering rate should be directly proportional to the target thickness. This would not be true for multiple scattering, where the rate would vary as the square-root of the thickness. Geiger and Marsden measured the scattering rate at a fixed angle near 25 degrees with targets composed of multiple foils. The results are shown in Figure 6-5. The data show that the scattering rate depends linearly on the target thickness, indicating that α particles emerge at large angles as the result of a single scatter.

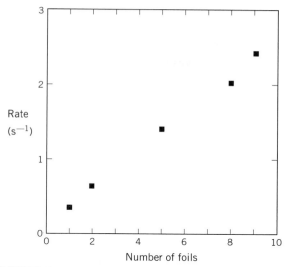

FIGURE 6-5 Alpha particle scattering rate at 25 degrees as a function of target thickness.
The target is made up of a number of gold foils. The scattering rate is proportional to the number of foils or to the thickness of the target. These data indicate that α particles emerge at large angles as the result of a single scatter. From H. Geiger and E. Marsden, *Phil. Mag.* **25**, 604 (1913).

Energy Dependence

To measure the energy dependence of the α scattering, the monoenergetic α source was used. The apparatus was modified so that sheets of mica could be inserted between the source and the target. When α particles pass through the mica, they lose energy by electromagnetic interaction with the electrons in the mica. The distance that an α particle can travel in a material before it loses all its kinetic energy and comes to rest is called the *range* of the particle. The range of α particles in mica had been measured by Geiger in a previous experiment. He found that the range was proportional to the cube of the speed of the α particle, or proportional to the kinetic energy to the 3/2 power. Geiger and Marsden slowed down the α particles with mica sheets to do their large angle scattering experiment at lower energies. They then determined the relative kinetic energies of the α particles by measuring the α particle ranges in mica. The results are shown in Figure 6-6. The energy units are arbitrary since the experimenters did not know the exact kinetic energy of the α particle from the radon decay (5.5 MeV). The data indicate that the scattering rate is inversely proportional to the square of the α particle kinetic energy,

$$\sigma \propto \frac{1}{E_k^2}. \qquad (6.17)$$

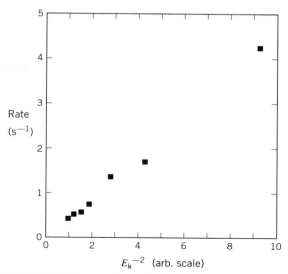

FIGURE 6-6 Energy dependence of the scattering.
From H. Geiger and E. Marsden, *Phil. Mag.* **25**, 604 (1913).

Z Dependence

Geiger and Marsden did not know the value of Z for the foils, but they did know the atomic mass numbers of the foils. The atomic mass number is the number of grams corresponding to one mole of atoms. The atomic mass number is also equal to the number of neutrons plus protons in the nucleus. The data showed that the scattering cross section varied approximately as the square of the atomic mass number (A). The reason the data show this effect is that the charge of the nucleus is roughly proportional to the atomic weight. In reality, Z is not exactly proportional to A because the fraction of neutrons increases with increasing A.

Geiger and Marsden determined the approximate value of Z by measuring the fraction of particles that were scattered at large angles. By making this measurement they could compare the theoretical cross section (which contains a factor Z^2) with their measured cross section. This comparison is shown in Figure 6-7.

Derivation of the Rutherford Cross Section

Consider an energetic α particle incident on a nucleus at rest. The trajectory of the α particle is indicated in Figure 6-8. If the nucleus would not scatter the α particle, the α particle would have a distance of closest approach equal to b. The variable b is called the *impact parameter*. The scattering angle θ is defined to be the angle of the outgoing

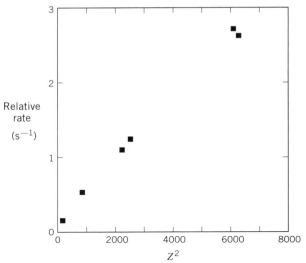

FIGURE 6-7 The Z dependence of the scattering cross section.
From H. Geiger and E. Marsden, *Phil. Mag.* **25**, 604 (1913).

α particle with respect to the incoming α particle. When b is large, the force between the α and the nucleus is small and the scattering angle is small. Conversely, a small impact parameter produces a large force and a large scattering angle. In terms of the impact parameter, the differential of the cross section ($d\sigma$) is equal to the area of a ring of radius b and thickness db:

$$d\sigma = 2\pi b \, db. \qquad (6.18)$$

The differential cross section ($d\sigma/db$) is

$$\frac{d\sigma}{db} = 2\pi b. \qquad (6.19)$$

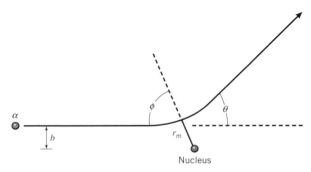

FIGURE 6-8 Definition of the variables for description of the Rutherford scattering experiment.
The incident α particle has an impact parameter b. The trajectory of the α particle is a hyperbola. The α particle has a distance of closest approach r_m with the nucleus and emerges at an angle θ.

For a hard sphere, the range of b is from 0 to R, the radius of the sphere. The cross section for this case is

$$\sigma = 2\pi \int_0^R db \, b = \pi R^2. \qquad (6.20)$$

The cross section for α scattering is much more interesting. A specific impact parameter will make a specific scattering angle. This is illustrated in Figure 6-9. When the impact parameter is between b and $b + \Delta b$, then the scattering angle is between θ and $\theta - \Delta\theta$. The differential cross section may be written

$$\frac{d\sigma}{d\cos\theta} = \frac{d\sigma}{db}\frac{db}{d\cos\theta} = 2\pi b \frac{db}{d\cos\theta}. \qquad (6.21)$$

We need to determine how b depends on $\cos\theta$ to arrive at an expression for the angular distribution of the scattered α particles.

We shall derive an expression for b by finding two independent expressions for the change in momentum Δp of the scattered α particle that involve b and θ. Let $\mathbf{p}_1$ be the momentum of the α particle before the scatter and $\mathbf{p}_2$ be the momentum after the scatter (see Figure 6-10). The square of the change in momentum is

$$(\Delta p)^2 = (\mathbf{p}_2 - \mathbf{p}_1)^2$$
$$= p_2^2 + p_1^2 - 2p_1 p_2 \cos\theta. \qquad (6.22)$$

Only the direction of the momentum changes, not its magnitude, provided that the mass of the nucleus is much

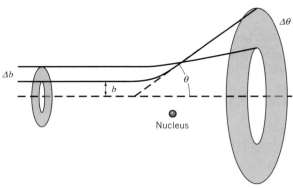

FIGURE 6-9 Relationship between impact parameter b and scattering angle θ.
When the impact parameter is between b and $b + \Delta b$, then the scattering angle is between θ and $\theta - \Delta\theta$. The cross section $\Delta\sigma$ for a particle detected between the rings defined by θ and $\theta - \Delta\theta$ is equal to the area between the rings defined by b and $b + \Delta b$, $\Delta\sigma = 2\pi b\Delta b$.

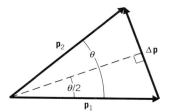

FIGURE 6-10 Momentum vector diagram of the scattered α particle.

larger than the mass of the α particle. The nucleus has momentum (Δp) transferred to it, but very little energy, like a ball bouncing off a brick wall. Since the magnitude of the momentum does not change, we write

$$p_2 = p_1 = p = m_\alpha v, \qquad (6.23)$$

where m_α is the mass of the α particle and v is the speed of the α particle. Therefore,

$$(\Delta p)^2 = 2p^2 - 2p^2 \cos\theta$$
$$= 2p^2(1 - \cos\theta), \qquad (6.24)$$

and we have one of our expressions for the momentum transfer,

$$\Delta p = m_\alpha v \sqrt{2(1 - \cos\theta)}. \qquad (6.25)$$

The momentum transfer is along a line that bisects the angle ($\pi - \theta$) as indicated in Figure 6-8. The magnitude of the force (F) on the α particle is

$$F = \frac{kq_1 q_2}{r^2}, \qquad (6.26)$$

where q_1 is the charge of the α particle ($2e$) and q_2 is the charge of the nucleus (Ze). The component of force in the direction of the momentum transfer is $F\cos\phi$. Therefore, the momentum transfer may be written as the time (t) integral of the force:

$$\Delta p = \int_{t_1}^{t_2} dt\, F \cos\phi = kq_1 q_2 \int_{t_1}^{t_2} dt\, \frac{\cos\phi}{r^2}. \quad (6.27)$$

In evaluating the integral we must keep in mind that both ϕ and r depend on t. At first sight the integral looks formidable, however, we shall be able to reduce it to an easy form by making use of angular momentum conservation. At any point along the path of the α particle, the component of velocity perpendicular to the direction of the force (v_T) is

$$v_T = r \frac{d\phi}{dt}. \qquad (6.28)$$

The angular momentum (L) of the α particle about the nucleus is

$$L = m_\alpha \left(r\frac{d\phi}{dt} \right) r = m_\alpha r^2 \frac{d\phi}{dt}. \qquad (6.29)$$

The angular momentum is a conserved quantity. When the α particle is a long distance from the nucleus, before the scatter, we have by definition of the impact parameter

$$L = m_\alpha vb. \qquad (6.30)$$

By conservation of angular momentum,

$$m_\alpha vb = m_\alpha r^2 \frac{d\phi}{dt}. \qquad (6.31)$$

So that

$$\frac{dt}{r^2} = \frac{d\phi}{vb}. \qquad (6.32)$$

Therefore, our expression for the momentum transfer (6.27) becomes

$$\Delta p = \frac{kq_1 q_2}{vb} \int_{-\phi_0}^{\phi_0} d\phi \cos\phi$$
$$= \frac{kq_1 q_2}{vb} [\sin\phi_0 - \sin(-\phi_0)]$$
$$= \left(\frac{2kq_1 q_2}{vb} \right) \sin\phi_0. \qquad (6.33)$$

Now we must convert the angle ϕ_0, which was a convenient variable for integrating the force, back to the scattering angle θ. The two angles are related by $2\phi_0 + \theta = \pi$, or

$$\phi_0 = \frac{\pi}{2} - \frac{\theta}{2}. \qquad (6.34)$$

The momentum transfer (6.33) is

$$\Delta p = \frac{2kq_1 q_2}{vb} \sin\left(\frac{\pi}{2} - \frac{\theta}{2} \right)$$
$$= \frac{2kq_1 q_2}{vb} \cos\left(\frac{\theta}{2} \right). \qquad (6.35)$$

Using the trigonometric identity

$$\cos\left(\frac{\theta}{2}\right) = \sqrt{\frac{1+\cos\theta}{2}}, \qquad (6.36)$$

we have

$$\Delta p = \frac{\sqrt{2}kq_1q_2}{vb}\sqrt{1+\cos\theta}. \qquad (6.37)$$

Now we equate the two expressions for the momentum transfer (6.25 and 6.37):

$$\Delta p = m_\alpha v\sqrt{2(1-\cos\theta)}$$

$$= \frac{\sqrt{2}kq_1q_2}{vb}\sqrt{1+\cos\theta}. \qquad (6.38)$$

This is our sought-after expression relating the impact parameter and the scattering angle. Solving for the impact parameter, we have

$$b = \frac{kq_1q_2}{m_\alpha v^2}\sqrt{\frac{1+\cos\theta}{1-\cos\theta}}. \qquad (6.39)$$

The square of the impact parameter is

$$b^2 = \left(\frac{kq_1q_2}{m_\alpha v^2}\right)^2\left(\frac{1+\cos\theta}{1-\cos\theta}\right). \qquad (6.40)$$

To calculate the cross section, we make the change of variables:

$$x \equiv \cos\theta, \qquad (6.41)$$

and the square of the impact parameter (6.40) becomes

$$b^2 = \left(\frac{kq_1q_2}{m_\alpha v^2}\right)^2\left(\frac{1+x}{1-x}\right). \qquad (6.42)$$

Differentiating b^2, we have

$$2b\,db = \left(\frac{kq_1q_2}{m_\alpha v^2}\right)^2\frac{(1-x+1+x)}{(1-x)^2}dx$$

$$= 2\left(\frac{kq_1q_2}{m_\alpha v^2}\right)^2\frac{dx}{(1-x)^2}. \qquad (6.43)$$

The differential cross section as a function of scattering angle is

$$d\sigma = 2\pi b\,db = 2\pi\left(\frac{kq_1q_2}{m_\alpha v^2}\right)^2\frac{dx}{(1-x)^2}, \qquad (6.44)$$

or

$$\frac{d\sigma}{dx} = 2\pi\left(\frac{kq_1q_2}{m_\alpha v^2}\right)^2\frac{1}{(1-x)^2}. \qquad (6.45)$$

Switching back to $\cos\theta$,

$$\frac{d\sigma}{d\cos\theta} = 2\pi\left(\frac{kq_1q_2}{m_\alpha v^2}\right)^2\frac{1}{(1-\cos\theta)^2}. \qquad (6.46)$$

We may write the kinetic energy of the incoming particle as $E_k = m_\alpha v^2/2$ to get

$$\frac{d\sigma}{d\cos\theta} = \frac{\pi}{2}\left(\frac{kq_1q_2}{E_k}\right)^2\frac{1}{(1-\cos\theta)^2}. \qquad (6.47)$$

If the electric charge of the projectile is $q_1 = ze$ and the electric charge of the nucleus is Ze, we have

$$\frac{d\sigma}{d\cos\theta} = \frac{\pi}{2}\left(\frac{zZke^2}{E_k}\right)^2\frac{1}{(1-\cos\theta)^2}. \qquad (6.48)$$

In terms of the electromagnetic coupling strength (α), we have

$$\frac{d\sigma}{d\cos\theta} = \frac{\pi}{2}z^2Z^2\alpha^2\left(\frac{\hbar c}{E_k}\right)^2\frac{1}{(1-\cos\theta)^2}. \qquad (6.49)$$

This is the *Rutherford formula*. It specifies the scattering rate as a function of angle for the electromagnetic interaction of two charged particles. For the special case of an α particle ($z = 2$) scattering from a nucleus, we get

$$\frac{d\sigma}{d\cos\theta} = 2\pi Z^2\alpha^2\left(\frac{\hbar c}{E_k}\right)^2\frac{1}{(1-\cos\theta)^2}. \qquad (6.50)$$

A Feynman diagram for alpha-nucleus scattering is shown in Figure 6-11. We examine each of the factors appearing the Rutherford formula with the aid of Figure 6-11:

1. The cross section is proportional to the electromagnetic coupling constant (α) squared. The electromagnetic force is mediated by photon exchange. The photon couples to the α particle and to the nucleus. Each coupling has a strength of α.

2. The cross section is inversely proportional to the square of the kinetic energy of the α particle. When

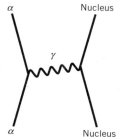

FIGURE 6-11 Feynman diagram for alpha-nucleus scattering.

the kinetic energy is small, the speed of the α particle is small and it spends more time near the nucleus experiencing the electric force. The constant $(\hbar c)$ divided by energy has units of length. This quantity squared has units of area. There is no other length scale in the physical process. Thus, we could have predicted on dimensional grounds that the cross section is proportional to $(\hbar c/E_k)^2$.

3. The angular dependence is $(1 - \cos\theta)^{-2}$. This angular factor comes from the fact that the force varies as r^{-2}. (Recall the elimination of the factor dt/r^2 in the expression for Δp.) This in turn comes from the fact that the quantum origin of the force is due to the exchange of massless particles (photons).

There is a singularity in the differential cross section at $\theta = 0$, where the cross section is infinite. This means that there is an infinitely large area (πb^2) that the α particle can be in and still get scattered. The total cross section is infinite because the electromagnetic force has an infinite range. Of course, the momentum transfer is very small when b is large.

The expression for the impact parameter is very useful for evaluating the cross section.

EXAMPLE 6-8

Calculate the cross section for a 12-MeV α particle scattering from a silver nucleus ($Z = 47$) at angles greater than (a) 90 degrees and (b) 10 degrees.

SOLUTION:

(a) We find the impact parameter (b_{max}) that corresponds to scattering at 90 degrees. Then impact parameters less than b_{max} produce scatters at angles greater than 90 degrees. The value of b_{max} is

$$b_{max} = \frac{kq_1q_2}{m_\alpha v^2}\sqrt{\frac{1 + \cos\theta}{1 - \cos\theta}} = \frac{kq_1q_2}{m_\alpha v^2}.$$

The cross section is

$$\sigma = \int_0^{b_{max}} db\,\frac{d\sigma}{db} = 2\pi \int_0^{b_{max}} db\,b$$

$$= \pi b_{max}^2 = \pi\left(\frac{kq_1q_2}{m_\alpha v^2}\right)^2.$$

Since the charge of the α is $2e$, we have

$$\sigma = \pi Z^2\left(\frac{ke^2}{E_k}\right)^2$$

$$= (3.14)(47)^2\left(\frac{1.44\ \text{Mev}\cdot\text{fm}}{12\ \text{MeV}}\right)^2$$

$$= 100\,(\text{fm})^2.$$

Since 1 $(\text{fm})^2 = 10^{-30}$ m$^2 = 10^{-2}$ b,

$$\sigma = 1.0\,\text{b}.$$

The silver nucleus is as big as a barn for this process.
(b) For scattering angles greater than 10 degrees, the cross section is larger by the factor

$$\frac{1 + \cos\theta}{1 - \cos\theta} \approx 130,$$

or

$$\sigma = 130\,\text{b}. \qquad \blacksquare$$

So far we have been the analyzing the scattering of a single particle from a single nucleus. In the Rutherford experiment the α particles were made to pass through a thin foil. For a given α particle, we cannot "choose" the impact parameter. When the α particle passes through the foil, it will have random sampling of impact parameters with many nuclei as it passes through the foil. If the α particle happens to have a small impact parameter, then a large angle scattering occurs. A good way to picture the α particle passing though the foil is to think of the nuclei in the foil as an incident flux of particles streaming in the general direction of the α. Consider a foil made of an element with atomic mass A and density ρ. The number of target nuclei per volume is

$$\frac{n}{V} = \frac{N_A\rho}{A(10^{-3}\ \text{kg})}. \qquad (6.51)$$

The number of target nuclei per cross-sectional area is the thickness (L) times n/V. If the incident rate of α particles is R_i, then the incident flux is

$$\Phi_i = \frac{R_i L n}{V} = \frac{R_i L N_A \rho}{A\left(10^{-3} \text{ kg}\right)}. \qquad (6.52)$$

The scattering cross section is

$$\sigma = \frac{R_s}{\Phi_i} = \frac{R_s}{\left[\dfrac{R_i L N_A \rho}{A\left(10^{-3} \text{ kg}\right)}\right]}$$

$$= \frac{R_s A\left(10^{-3} \text{ kg}\right)}{R_i L N_A \rho}. \qquad (6.53)$$

A measurement of the scattering rate is a measurement of the cross section. If the cross section is known, the scattering rate can be predicted.

EXAMPLE 6-9

A stream of 6-MeV α particles passes through a gold foil that has a thickness of 1 μm at a rate of 10^3 per second. Calculate the rate at which α particles are scattered at angles greater than 0.1 radian. The density of gold is 1.93 × 10^4 kg/m³.

SOLUTION:

The scattering rate (R_s) is

$$R_s = \frac{\sigma R_i N_A L \rho}{(197)\left(10^{-3} \text{ kg}\right)}.$$

We need to know the cross section that is given by the Rutherford formula (see Example 6-8):

$$\sigma = \pi Z^2 \left(\frac{ke^2}{E_k}\right)^2 \frac{1 + \cos\theta}{1 - \cos\theta}.$$

The cosine of 0.1 radian is

$$\cos(0.1) \approx 1 - \frac{(0.1)^2}{2} = 0.995 .$$

The angular factor is

$$\frac{1 + \cos\theta}{1 - \cos\theta} = \frac{1.995}{0.005} = 3.99 \times 10^2.$$

The cross section is

$$\sigma = (3.14)(79)^2 \left(\frac{1.44 \text{ Mev·fm}}{6 \text{ MeV}}\right)^2 \left(3.99 \times 10^2\right)$$

$$= 4.50 \times 10^5 \ (\text{fm})^2 = 4.50 \times 10^{-25} \text{ m}^2 .$$

The scattering rate is

$$R_s = \frac{\left(4.50 \times 10^{-25} \text{ m}^2\right)\left(10^3 \text{ s}^{-1}\right)\left(6 \times 10^{23}\right)\left(10^{-6} \text{ m}\right)}{0.197 \text{ kg}}$$

$$\times \left(1.93 \times 10^4 \text{ kg/m}^3\right)$$

$$= 26 \text{ s}^{-1} .$$

The fraction (f) scattered at angles greater than 0.1 radian is

$$f = \frac{\left(26 \text{ s}^{-1}\right)}{\left(1000 \text{ s}^{-1}\right)} = 0.026 . \qquad \blacksquare$$

6-4 PROBING THE STRUCTURE OF THE NUCLEUS

The distance of closest approach that the α particle makes with the nucleus depends on both the impact parameter and the kinetic energy of the α. For example, if the impact parameter is very small and the energy is small, the α particle gets repelled backwards and the distance of closest approach is much larger than the impact parameter. If the α particle kinetic energy is very large and the impact parameter is small, then the α particle can actually "strike" the nucleus if the nucleus has a finite size. By finite size, we mean that the charge distribution is not a point, but is spread out in space. In this case the α particle has a distance of closest approach that is smaller than the radial extent of the charge of the nucleus. The Rutherford formula for the angular distribution of the scattered α particle is no longer valid because the force is no longer given by kq_1q_2/r^2. The force is smaller because the effective charge that scatters the α particle is smaller than the total charge of the nucleus. The nucleus no longer appears as a pointlike particle to the incoming α particle. The breakdown of the Rutherford formula provides a powerful experimental technique for measuring the size of the nucleus.

We now derive an expression for the distance of closest approach for a non-relativistic α particle. Let the distance of closest approach be r_m and let the corresponding speed of the α particle be v_m. By conservation of energy (neglecting the nuclear recoil), we have

$$E_k = \frac{1}{2} m v_m^2 + \frac{2 Zke^2}{r_m}. \qquad (6.54)$$

Conservation of angular momentum relates the impact parameter b with r_m:

$$mvb = mv_m r_m . \qquad (6.55)$$

Therefore, the kinetic energy of the α particle (6.54) becomes

$$E_k = \frac{1}{2} m \left(\frac{vb}{r_m} \right)^2 + \frac{2 Zke^2}{r_m}$$

$$= \frac{E_k b^2}{r_m^2} + \frac{2 Zke^2}{r_m} , \qquad (6.56)$$

or

$$r_m^2 - \left(\frac{2 Zke^2}{E_k} \right) r_m - b^2 = 0 . \qquad (6.57)$$

The solution for r_m is

$$r_m = \frac{Zke^2}{E_k} + \frac{1}{2} \sqrt{ \left(\frac{2 Zke^2}{E_k} \right)^2 + 4 b^2 } . \qquad (6.58)$$

For $b = 0$, the distance of closest approach (6.58) is

$$r_m = \frac{2 Zke^2}{E_k} . \qquad (6.59)$$

EXAMPLE 6-10

Estimate the minimum kinetic energy of an α particle that can cause a deviation from the Rutherford formula in scattering from a gold nucleus ($Z = 79$). Approximate the gold nucleus as a closely packed group of 197 neutrons and protons, where the radius of each proton and neutron is equal to 1 fm.

SOLUTION:

The kinetic energy of the α particle must be greater than the potential energy at the surface (r_n) of the gold nucleus:

$$E_k = \frac{2 Zke^2}{r_n} .$$

We may estimate the size of the gold nucleus from

$$\frac{4}{3} \pi r_n^3 = (197) \left(\frac{4}{3} \pi \right) (1 \text{ fm})^3 ,$$

or

$$r_n = (197)^{1/3} \text{ fm} \approx 6 \text{ fm} .$$

The α kinetic energy needed to penetrate the nucleus is

$$E_k \approx \frac{(2)(79)(1.44 \text{ MeV} \cdot \text{fm})}{6 \text{ fm}} \approx 40 \text{ MeV} .$$

Note that this kinetic energy is much smaller than the mass energy of the α, so that the use of the nonrelativistic expression for kinetic energy is justified. ∎

Measuring the Size of the Nucleus

Electrons make excellent probes to study the electric charge distribution of the nucleus. The electron makes a better probe than the α particle for determining the charge distribution inside the nucleus because the electron behaves as a point charge (i.e., the electron has no detectable structure). A schematic of an electron scattering experiment is shown in Figure 6-12. A beam of electrons is incident on a target containing the nuclei to be studied. The electrons are scattered in all directions as a result of their electromagnetic interactions with the nuclei in the target. A small fraction of the scattered electrons enter a spectrometer, which is positioned at a scattering angle θ. The momentum of the scattered electron is measured in the spectrometer. The purpose of the spectrometer is to count the number of electrons that are elastically scattered at the angle θ. (Elastic scattering means that the energy of the electron does not change.) The spectrometer is mounted on a circular track so that a wide range of scattering angles may be studied. Since small scattering angles are the result of large impact parameter collisions, the nucleus appears as a pointlike object and the angular distribution, $d\sigma/d\cos\theta$, is to a good approximation given by the Rutherford formula with $z = 1$, $(d\sigma/d\cos\theta)_R$,

$$\frac{d\sigma}{d \cos\theta} \approx \left(\frac{d\sigma}{d \cos\theta} \right)_R$$

$$= \frac{\pi}{2} Z^2 \alpha^2 \left(\frac{\hbar c}{E_k} \right)^2 \frac{1}{(1 - \cos\theta)^2} . \qquad (6.60)$$

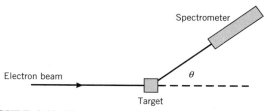

FIGURE 6-12 Electron–proton scattering experiment.

To get a more accurate prediction of the angular distribution, we would need to take into account relativistic and quantum effects and the recoil of the nucleus of mass (M). The result is

$$\left(\frac{d\sigma}{d\cos\theta}\right)_M$$

$$=\left(\frac{d\sigma}{d\cos\theta}\right)_R \frac{1+\cos\theta}{2\left[1+\dfrac{(1-\cos\theta)E_k}{Mc^2}\right]}. \quad (6.61)$$

This expression is called the Mott cross section (after Nevill Mott). The important part of this cross section is the Rutherford part, which contains the angular factor, $(1-\cos\theta)^{-2}$, and electron kinetic energy factor, E_k^{-2}. The correction to the Rutherford formula is of order unity except when the scattering angle is close to 180 degrees or when the incident electron has a very large energy. The factor of $(1+\cos\theta)/2$ is due to the effect of the electron magnetic moment which is a quantum mechanical effect. The electron magnetic moment is discussed further in Chapter 8. The other portion of the correction factor is due to the recoil of the nucleus.

When electrons are scattered from a pointlike particle, the cross section is given by the Mott formula. If the target particle has a charge distribution that is not a point, then the impact parameter may be so small that the electron penetrates the charge of the nucleus. In this case the force on the electron is smaller than the force due to a point charge and fewer electrons are scattered at large angles compared to the Mott formula. The amount of deviation from the Mott formula as a function of scattering angle or impact parameter gives us information on how the charge is distributed inside the nucleus. In the 1950s, electrons were used as probes to make high-resolution measurements of the nuclear size. The leader of this effort was Robert Hofstadter. In 1953, Hofstadter's group scattered electrons of 125 MeV from several different nuclei. Some of their data on electron–gold scattering is shown in Figure 6-13. For a pointlike nucleus, the angular distribution of scattered electrons would follow the Mott formula (6.61),

$$\frac{dN}{d\cos\theta} \propto \left(\frac{d\sigma}{d\cos\theta}\right)_M, \quad (6.62)$$

The data show a marked deviation from "pointlike" behavior of the nucleus. Since a given scattering angle is a given impact parameter, the amount of deviation from the pointlike cross section tells us the effective nuclear

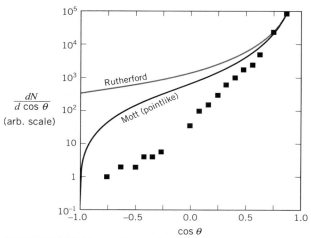

FIGURE 6-13 Determining the size of the nucleus. Electrons with a kinetic energy of 125 MeV are scattered from gold and the angular distribution of scattered electrons ($dN/d\cos\theta$) is measured. The data are inconsistent with a pointlike nucleus. The observed scattering rate is smaller than the rate expected from a pointlike nucleus because the electron penetrates the nucleus and experiences the force of a reduced charge. The amount of reduction from the pointlike scattering rate gives us the size of the gold nucleus, about 3×10^{-15} m. From R. Hofstadter et al., *Phys. Rev.* **92**, 978 (1953).

charge that scattered the electron, and therefore, how much the electron was able to penetrate the nucleus. The results of a large number of scattering measurements, like those shown in Figure 6-13, indicate that the charge distribution of the nucleus is spherical with an approximately constant density in the core. This is pictured in Figure 6-14. The size of the nuclear charge (R) is determined from the electron scattering experiments to be

$$R \approx (1.2 \text{ fm})\, A^{1/3}, \quad (6.63)$$

where A is the atomic mass number of the nucleus. The factor of $A^{1/3}$ is expected in a model, called the Fermi model (after Enrico Fermi), where the nucleus is considered to be a closely packed array of protons and neutrons. The neutrons have no electric charge, but their presence is part of the reason that the charges of the protons are spread out. The charge density distribution of the nucleus does not have a perfectly sharp edge. The details of the shape of the nuclear charge distributions are determined from the electron scattering experiments.

EXAMPLE 6-11

Electrons of 125 MeV are scattered from a gold nucleus. Calculate the deviation from the Rutherford formula at a

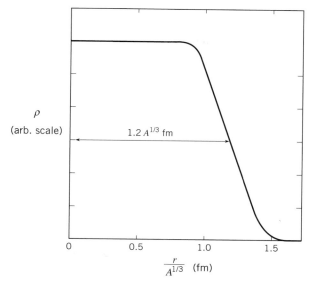

ρ
(arb. scale)

$1.2\,A^{1/3}$ fm

0   0.5   1.0   1.5

$\dfrac{r}{A^{1/3}}$ (fm)

FIGURE 6-14 Fermi model of the nucleus.
The electron–nucleus scattering data determine the charge distribution inside the nucleus. The data from many scattering experiments show that the charge density is spherically symmetric and has a characteristic size of $(1.2\text{ fm})A^{1/3}$. The edges of the nuclear charge distribution are not perfectly sharp.

scattering angle of 90 degrees due to the electron magnetic moment and nuclear recoil.

SOLUTION:

The effect of the electron magnetic moment is to introduce a factor of

$$\frac{1+\cos\theta}{2} = \frac{1}{2}.$$

The effect of the proton recoil is to introduce a factor of

$$\frac{1}{\left[1 + \dfrac{(1-\cos\theta)E_k}{Mc^2}\right]} \approx 1,$$

because

$$Mc^2 \gg 125\,\text{MeV}.$$

The net effect is a modification of the Rutherford formula by a factor of 2. This factor is much smaller than the deviation from the pointlike cross section that is observed in the data of Hofstadter. The deviation in the data is caused by a finite size of the nucleus. ∎

The Structure of the Nucleus

The nucleus is a bound state of neutrons and protons; the binding force is the strong interaction. (Chapter 11 is devoted to the properties of the nucleus.) It is possible to remove a neutron or a proton from a nucleus in a collision. The strong force has a short range. In a rough approximation, we may think of the nucleus as a closely packed "blob" of neutrons and protons. The neutrons and protons are very close together so that their wave-packets overlap. The nucleus is a complicated object when examined in detail! We proceed now to look at the simplest nucleus, that of hydrogen.

6-5 PROBING THE STRUCTURE OF THE PROTON

The proton does not behave like a pointlike object at distances shorter than about 1 femtometer. This was first observed by scattering electrons from a liquid hydrogen target. We have already remarked that the incoming particle must have sufficiently large energy in order to penetrate the target nucleus. For relativistic electrons, energy is inversely proportional to wavelength. The electron wavelength is a measure of the minimum possible distance of closest approach in a collision. Some diagrams for electron–proton scattering are shown in Figure 6-15. Electron–proton scattering is qualitatively different for three different sizes of the electron wavelength. When the electron wavelength is much larger than 1 femtometer (see Figure 6-15a), the proton appears as a point charge. When the wavelength is comparable to 1 femtometer (see Figure 6-15b), the electron is observed to penetrate the charge distribution of the proton. Finally, at wavelengths much smaller than 1 fm, the proton charge is resolved into individual quarks (see Figure 6-15c). We now discuss this in further detail.

Measuring the Proton Size

The structure of the proton was discovered and measured in great detail by electron scattering. Electron scattering is a powerful technique for determining the electric charge distribution inside the proton. If the proton were a point charge, then the cross section would be that given by the Mott formula (6.61).

If the proton is not a point charge, the structure of the proton charge distribution causes a deviation in the cross section. When an electron penetrates the proton, the force

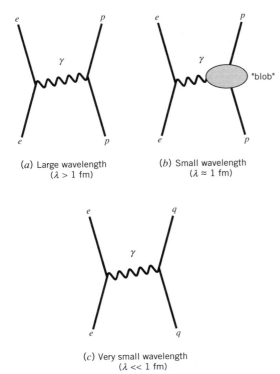

(a) Large wavelength
($\lambda > 1$ fm)

(b) Small wavelength
($\lambda \approx 1$ fm)

(c) Very small wavelength
($\lambda \ll 1$ fm)

FIGURE 6-15 Feynman diagrams for electron–proton scattering.
(*a*) When the electron wavelength is much larger than the proton size, the proton appears like a "point." (*b*) When the wavelength is comparable to the proton size, then the proton appears as a "blob" of charge. The spatial extent of the charge of the proton is observed to be finite with an exponential form. (*c*) When the wavelength is much smaller than the proton size, the proton is resolved into quarks. The quarks appear as point charges that carry a fraction *x* of the proton momentum.

on the electron is smaller, and the resulting scattering angle is smaller compared to the case of no penetration (see Figure 6-2). This cross section for a penetrating electron may be written

$$\left(\frac{d\sigma}{d\cos\theta}\right)_{\text{proton}} = \left(\frac{d\sigma}{d\cos\theta}\right)_{\text{M}} [F(\theta)]^2, \quad (6.64)$$

where the factor $[F(\theta)]^2$ is less than unity. The function F depends on the scattering angle because the impact parameter, which determines the value of the effective charge causing the scattering, depends on the scattering angle. The function F is called the proton *form factor*; it contains all the information about how the charge distribution inside the proton differs from a "point." For small-angle scattering,

$$F \approx 1. \quad (6.65)$$

In small-angle electron scattering, the proton size cannot be resolved, and the cross section follows the Mott formula (6.61). At slightly larger angles, the proton form factor differs from unity by an amount that is proportional to the average radius squared of the proton charge distribution:

$$F \approx 1 - C\langle r^2 \rangle. \quad (6.66)$$

The form factor is therefore related to the average of the radius squared of the charge distribution. The photon exchanged in the scattering has a wavelength small enough to resolve the proton size but too large to resolve the proton structure. The proton form factor was first measured by Robert Hofstadter. Data from Hofstadter is shown in Figure 6-16. Electrons with a kinetic energy of 550 MeV were scattered from protons and the angular distribution of scattered electrons ($dN/d\cos\theta$) was measured. The wavelength of a 550 MeV electron is

$$\lambda = \frac{hc}{pc} = \frac{1240 \text{ MeV·fm}}{550 \text{ MeV}} \approx 2 \text{ fm}. \quad (6.67)$$

The data are inconsistent with a pointlike proton according to the Mott formula. At large angles the measured

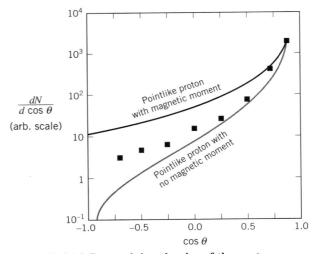

FIGURE 6-16 Determining the size of the proton.
Electrons with a kinetic energy of 550 MeV are scattered from protons and the angular distribution of scattered electrons ($dN/d\cos\theta$) is measured. The data are inconsistent with a pointlike proton. From E. E. Chambers and R. Hofstadter, *Phys. Rev.* **103**, 1454 (1956).

cross section is *larger* than that given by the Mott formula. *How can the cross section be larger than the Mott formula?*

Effect of the Proton Magnetic Moment

Since the measured electron–proton cross section at large angles is larger than the Mott formula, there must be a contribution to the electron–proton cross section that is not contained in the Mott formula. This contribution is due to the proton *magnetic moment*. The magnetic moment does not affect the angular distribution of particles with small speeds because the magnetic force is proportional to the speed. For relativistic electrons the effect of the proton magnetic moment is an important contribution to the scattering cross section. The effect of a magnetic moment is indicated in Figure 6-16. When the contribution to the magnetic moment is included, then the pointlike cross section is much larger than the measured rate. The data lie between that expected for a pointlike proton with no magnetic moment and a pointlike proton with the magnetic moment included. The observed scattering rate is *smaller* than the rate expected from a pointlike proton including magnetic moment because the electron penetrates the proton and experiences the force of a reduced charge and magnetic moment. Both the electric charge and the magnetic moment of the proton are spread out in space. The proton has a finite size. The amount of reduction from the pointlike scattering rate tells us how the charge and magnetic moment are distributed within the proton. The data indicate that both the charge and magnetic moment of the proton are exponentially distributed:

$$\rho = \rho_0 e^{-r/\delta}, \qquad (6.68)$$

where $\delta = 0.8 \times 10^{-15}$ m. In this context we say the size of the proton is about 1 fm.

Measuring the Proton Structure

We have so far only seen the tip of the iceberg concerning the physics contained in electron–proton scattering! The details of the proton structure are revealed by repeating the electron scattering experiment with higher resolution. This requires using a beam of electrons with higher momentum. A series of detailed measurements were made in the 1960s by a team of physicists led by Jerome Friedman, Henry Kendall and Richard Taylor. The experimenters used a beam of electrons with a momentum of 10 GeV/c. (The spectrometers used for these measurements are shown in color plate 15.) The wavelength of a 10-GeV electron is

$$\lambda = \frac{hc}{pc} = \frac{1240 \text{ MeV} \cdot \text{fm}}{10^4 \text{ MeV}} \approx 0.1 \text{ fm}. \qquad (6.69)$$

Thus, the experimenters had available to them a probe that was about 20 times finer than the electron probe of Hofstadter.

When a small wavelength electron collides with a proton, the charge distribution of the proton is measured with high resolution. The experimental result is remarkable. The charge distribution is found to be that of a number of pointlike objects. These pointlike particles are the constituent quarks. The proton charge is made up of a number of pointlike quarks that each carry a portion of the proton momentum. (The mass of the quarks is much smaller than the proton mass.) In the center-of-mass of the electron and quark, the scattering process is an elastic collision. The angular distribution of the scattered electrons when viewed in this frame is that due to the scattering of pointlike particles!

Electron–quark scattering is called "deep-inelastic" scattering and is written

$$e^- + p \rightarrow e^- + X, \qquad (6.70)$$

where X represents the target proton after it has been struck. The hadronic state X is in general complicated because energy can be converted to matter, but there remains one great simplicity of the interaction, that the electron remains an electron. The energy of the scattered electron E' and the scattering angle θ are measured (see Figure 6-17a). This is enough to uniquely determine the fraction of proton momentum (x) carried by the colliding quark. We now show how to calculate the momentum fraction (x) from the measured values of E' and θ. In the center-of-mass system (see Figure 6-17b), the electron and quark have momenta of equal magnitudes and opposite directions. Let β be the speed divided by c of the proton in the center-of-mass frame and as usual let $\gamma = (1 - \beta^2)^{-1/2}$. The proton momentum (times c) in the center-of-mass frame is given by the Lorentz transformation

$$p_p{}^* c = \gamma \beta M c^2, \qquad (6.71)$$

where M is the mass of the proton. The quark momentum (times c) is smaller than the proton momentum by the factor x:

$$p_q{}^* c = p_p{}^* c x = \gamma \beta M c^2 x. \qquad (6.72)$$

For energetic electrons we may neglect the electron mass. Then the electron momentum (times c) in the center-of-

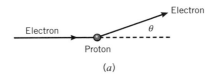

(a)

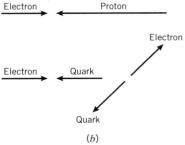

(b)

FIGURE 6-17 Electron–proton scattering.
(a) Laboratory system and (b) electron–quark center-of-mass system.

mass frame is

$$p_e{}^* c = \gamma E - \gamma \beta E . \qquad (6.73)$$

By definition of the center-of-mass frame, the electron and the quark have equal magnitudes of momentum:

$$p_e{}^* = p_q{}^* . \qquad (6.74)$$

Therefore,

$$\gamma E - \gamma \beta E = \gamma \beta M c^2 x . \qquad (6.75)$$

Solving for x, we get

$$x = \frac{E(1-\beta)}{\beta M c^2} . \qquad (6.76)$$

In the center-of-mass frame the electron energy is not changed after the scatter. The energy of the scattered electron in the center-of-mass frame is

$$E'{}^* = \gamma E' - \beta \gamma E' \cos\theta . \qquad (6.77)$$

Conservation of energy (neglecting the electron mass) gives

$$\gamma E - \gamma \beta E = \gamma E' - \beta \gamma E' \cos\theta . \qquad (6.78)$$

We may solve this expression for β,

$$\beta = \frac{E - E'}{E - E' \cos\theta} . \qquad (6.79)$$

We substitute β (6.79) into our expression for the proton momentum fraction (6.76) to get

$$x = \frac{E(1-\beta)}{\beta M c^2} = \frac{E E'(1-\cos\theta)}{(E-E') M c^2} . \qquad (6.80)$$

Thus, knowledge of the initial electron energy (E) and the proton mass (M) plus measurement of the electron scattering angle (θ) and the scattered electron energy (E') gives us the proton momentum fraction (x) carried by the quark that collided with the electron. Remarkable! The measurement of the scattered electron energies at a fixed scattering angle gives us the momentum distribution of quarks inside the proton. To an excellent approximation, all scattering angles and incident electron energies give the same x distribution of the quarks. (This last observation is called *scaling*. We shall return to the discussion of scaling in Chapter 17.)

The collective results of many scattering measurements are shown in Figure 6-18. The measured distribution, dP/dx, gives the probability that a single quark

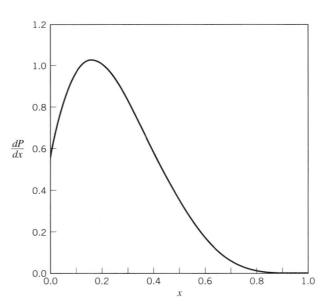

FIGURE 6-18 Structure of the proton.
When the impact parameter is much smaller than 1 femtometer, which is possible for very energetic electrons, then the proton charge is resolved into a distribution of pointlike charges that are called quarks. The scattering is inelastic and momentum is transferred from the electron to the quark that it collides with. The momentum and angular distributions of the scattered electrons gives us the momentum distributions of the quarks inside the proton.

carries a fraction x of the proton momentum. The physical interpretation of the electron–proton scattering is that the electron interacts with the proton by exchanging a photon with a quark inside the proton. The photon exchange occurs only if the quark has just the right momentum fraction x. The angular distribution of the scattered electrons in the electron–quark frame fits that of pointlike particles.

The Difference Between Size and Structure

It is important to make the distinction between the *size* of a particle and the *structure* of a particle. An electron with a momentum p has an approximate size R given by its deBroglie wavelength (5.7):

$$R \approx \lambda = \frac{h}{p}. \qquad (6.81)$$

By this we mean that the wavepacket of the electron actually has a characteristic size of h/p. The size of a particle is inversely proportional to its momentum as long as it is not a composite object. Scattering experiments show that the electron is not composite down to a distance of 10^{-18} meters. The electron has no structure that has been detected by experiment; it is a "pointlike" particle. We may picture an electron with momentum p as a "cloud" characterized by the approximate size h/p.

A proton behaves like a pointlike particle down to distances of about 1 femtometer. At low momentum, the approximate size of the proton is h/p. The momentum at which the proton wavelength is 1 fm is

$$p = \frac{h}{\lambda} = \frac{hc}{\lambda c} = \frac{1.24 \text{ GeV} \cdot \text{fm}}{(1 \text{ fm})c}$$
$$\approx 1 \text{ GeV}/c. \qquad (6.82)$$

At momenta larger than 1 GeV/c the proton has a smaller wavelength; however, its size remains about 1 fm. This is due to the fact that the proton is a composite object and its size is determined by the wavelengths of the quarks and gluons inside of it. A proton of high momentum will have some energetic quarks with small wavelengths, but it will also have some quarks that have a wavelength of about 1 fm. The proton size is given by the wavelengths of the least energetic quarks. We may make an analogy with a multielectron atom. The atom contains some electrons with small wavelengths (the inner ones), but it also has some electrons (the outer ones) that have a wavelength of

about 0.1 nm. The atomic size is determined by the wavelengths of the electrons with the smallest kinetic energy.

* *Challenging*

6-6 PROBING THE STRUCTURE OF THE QUARK

The deep-inelastic electron scattering data show that the quarks inside a proton appear as pointlike objects to an electron with a kinetic energy of 10 GeV. If we have a source of higher energy particles, then we can search for structure of the quarks by searching for a deviation from the pointlike cross section. If the energetic particle is another proton, then the fundamental scattering process is quark–quark scattering,

$$q + q \rightarrow q + q, \qquad (6.83)$$

and the quarks themselves may be used to probe the structure of quarks. In the 1980s, a detailed experiment was performed to search for possible structure within the quarks. The experiment was performed not by colliding two energetic protons, but rather by colliding protons and *antiprotons*. The antiproton is the *antiparticle* of the proton. The antiproton has the same mass as the proton, an electric charge of $(-e)$, and a structure identical to the proton except that it is made of pointlike *antiquarks* instead of quarks. Antiparticles are discussed in more detail in Chapter 17. The scattering experiment was performed with protons and antiprotons because, due to their opposite electric charge, they could be made to circulate in opposite directions with a single ring of magnets. In the experiment, the energy of the protons and antiprotons was chosen to be 315 GeV each. The process studied is quark–antiquark elastic scattering:

$$q + \bar{q} \rightarrow q + \bar{q}. \qquad (6.84)$$

The experiment is directly analogous to the original Rutherford experiment. The study of this process is greatly complicated by the fact that the quarks are confined inside the proton. When we try to pull a quark out of a proton, for example by striking the quark with another energetic particle, the quark experiences a potential energy barrier from the strong interaction that increases with distance. The quark can never penetrate this barrier, and what is observed experimentally is a sort of strong interaction "spark" in which several hadrons are created. The quark–quark scattering process is indicated in Figure 6-19. This strong interaction spark is somewhat analogous to an electric spark that is created if we make an extremely large electric field between two conductors. The process of a struck energetic quark (or antiquark) materializing as free hadrons is called quark *fragmentation*, which we may write as

$$q \rightarrow \text{hadrons}. \qquad (6.85)$$

The above reaction as we have written it violates all sorts of conservation laws including that of energy and momentum! Energy and momentum are conserved in the event as a whole because the recoiling quarks in the rest of the proton arrange themselves to accomplish just that. Recall that when an energetic photon converts itself into an electron–positron pair, energy and momentum are not exactly conserved in this subprocess and the nucleus must have some recoil momentum; however, essentially all of the photon energy appears in the electron and positron,

$$\gamma \rightarrow e^+ + e^-, \qquad (6.86)$$

and the vector sum of the electron plus positron momenta is very nearly equal to the photon momentum,

$$\mathbf{p}_{\text{photon}} \approx \mathbf{p}_{\text{electron}} + \mathbf{p}_{\text{positron}}. \qquad (6.87)$$

In the case of the struck quark, the situation is similar: Most of the energy of the quark appears in the hadrons and the vector sum of the momenta of the hadrons, is nearly equal to the quark momentum. The hadrons that arise from a struck quark are called a *jet*.

When quarks scatter with sufficiently large momentum transfer, then the jet of hadrons that they leave in a particle spectrometer becomes a measure of the quark momentum after the scatter ($\mathbf{p}_{\text{quark}}$),

$$q \rightarrow \text{jet}, \qquad (6.88)$$

$$\mathbf{p}_{\text{quark}} \approx \mathbf{p}_{\text{jet}}. \qquad (6.89)$$

Figure 6-20 shows the observed energy deposits in the detector from a large momentum transfer collision. The jet provides an experimental method for determination of the momentum of a scattered quark.

We have derived the Rutherford scattering cross section in the frame where the target particle is at rest. In the center of mass frame, the Rutherford cross section has the same form provided that we replace the kinetic energy squared by the total center-of-mass energy squared

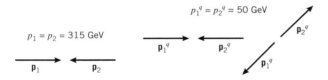

(a) Proton–antiproton system before scattering

(b) Quark–antiquark system

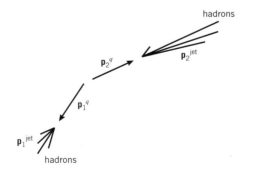

(c) Proton–antiproton system after scattering

FIGURE 6-19 The quark–quark scattering process.

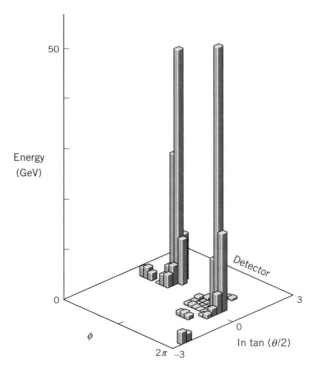

FIGURE 6-20 Measuring the angles of the scattered quarks.

From the UA1 experiment.

(s). The Rutherford cross section becomes

$$\frac{d\sigma}{d\cos\theta} \sim \alpha^2 \left(\frac{1}{s}\right)\frac{1}{(1-\cos\theta)^2}. \qquad (6.90)$$

Now we are ready to analyze the angular distribution of scattered quarks. The scattering cross section for two quarks is specified by the rules of the strong interaction. The fundamental physics for quarks scattering by strong interaction does not differ from the fundamental physics of charged particles scattering by the electromagnetic interaction. In each case the force is governed by the exchange of massless quanta by structureless particles. Figure 6-21 shows a Feynman diagram for quark–quark scattering. The two quarks exchange a massless particle called a *gluon*. The probability for the gluon exchange is proportional to the square of the strong interaction coupling parameter (α_s). The angular distribution for quark–quark scattering is given by

$$\frac{d\sigma}{d\cos\theta} \sim \alpha_s^2 \left(\frac{1}{s}\right)\frac{1}{(1-\cos\theta)^2}. \qquad (6.91)$$

The form of the strong interaction cross section (6.91) is identical to the form of the electromagnetic interaction cross section (6.90), except that we have replaced α by α_s. The numerical value of the strong interaction coupling parameter for the scattering of 100-GeV quarks is

$$\alpha_s \approx 0.1, \qquad (6.92)$$

compared to

$$\alpha = \frac{1}{137}. \qquad (6.93)$$

FIGURE 6-21 Feynman diagram for quark–antiquark scattering.
The quark and antiquark exchange a gluon. The probability for the exchange is proportional to the square of the strong interaction coupling parameter (α_s).

Let us examine the origin of each factor in cross section (6.91):

1. The strength of the coupling of a gluon to a quark is α_s for the amplitude squared, and this factor appears as the square because there are two such couplings, one to each of the scattered quarks.

2. The quarks have a size that is specified by their wavelength, and this wavelength is inversely proportional to the quark momentum, and energy and momentum are equivalent ($E \approx pc$) since the quarks are relativistic. The cross section is proportional to the square of the wavelength or inversely proportional to the square of the energy.

3. The angular distribution varies inversely as the square of the momentum transfer squared, and the momentum transfer squared is proportional to $(1-\cos\theta)$.

The quark–quark scattering experiment was performed to test the strong interaction version of the Rutherford formula. Agreement with the Rutherford formula is expected if quarks interact by the exchange of massless particles that mediate the strong interaction. A deviation from the Rutherford formula is expected if the quarks have internal structure when probed at short distances. The first high-energy quark–quark scattering data are shown in Figure 6-22. These data are from the UA1 experiment led by Carlo Rubbia. In this experiment, the UA1 physicists observed roughly the same number of quarks scattered at large angles as Geiger and Marsden observed α particles. The quark–quark scattering data agree beautifully with the pointlike cross section formula! These data show that the strong interaction of two quarks at short distances has the same fundamental quantum behavior as the electromagnetic interaction.

Whether or not a particle has structure also is not related to the mass of the particle. This may seem counterintuitive. For example, the bottom quark is pointlike in spite of the fact that it has a mass of about five times the proton mass! The question of why some pointlike particles are so much more massive than others is one of the most profound unanswered physics questions.

6-7 SUMMARY OF THE SCATTERING EXPERIMENTS

The results of the scattering experiments are summarized in Figure 6-23. The nucleus appears as a point like

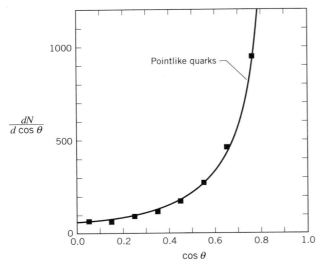

FIGURE 6-22 Angular distribution of scattered quarks and gluons from high energy proton–antiproton collisions measured in the center-of-mass frame of the colliding quarks and gluons.
The distribution fits the strong-interaction (QCD) Rutherford law, demonstrating that the quarks have no structure down to a distance scale of 10^{-18} m. From G. Arnison et al., *Phys. Lett.* **177B**, 244 (1986).

object to an incident charged particle with a kinetic energy of a few MeV. The angular distribution follows the Rutherford formula. When the energy of a charged particle is raised to 100 MeV, then the charged particle can penetrate the nucleus and the angular distribution deviates from the Rutherford formula. The measurement of this deviation determines the structure of the nucleus. The charge of the nucleus is made up of protons. Small angle electron scattering from a single proton follows the pointlike scattering cross section. Large angle "high-energy" electron scattering from a single proton shows a deviation from the pointlike scattering cross section, signaling that the proton is a composite object. Electron scattering at higher energies resolves the proton into its constituent quarks. The electron–quark scattering follows the pointlike scattering cross section. At even much higher energies, quark–antiquark scattering still is observed to follow the pointlike scattering cross section.

The use of the experimental technique pioneered by Rutherford to discover new structure has not been exhausted. Someday, quark–quark scattering experiments will be performed with sensitivity for detection of quark substructure well below the current limit of 10^{-18} meters. For this next exciting experiment, smaller wavelength,

higher energy quarks are needed than are currently available in the protons of today's accelerators.

CHAPTER 6: PHYSICS SUMMARY

- The nucleus was discovered by observing that α particles were occasionally scattered at very large angles (even backwards) when they passed through a thin foil. To explain this experimental fact, Rutherford deduced that the positive electric charge of the atom must be concentrated at a "point."

- The scattering cross section of a pointlike projectile particle of kinetic energy E_k and charge (ze) with a pointlike target particle with charge (Ze) by the electromagnetic interaction is given by the Rutherford formula

$$\frac{d\sigma}{d\cos\theta} = \frac{\pi}{2} z^2 Z^2 \alpha^2 \left(\frac{\hbar c}{E_k} \right)^2 \frac{1}{\left(1 - \cos\theta \right)^2} .$$

The Rutherford formula has been thoroughly tested by experiment.

- Rutherford scattering breaks down when the projectile particle is energetic enough to penetrate the nucleus. The nucleus is observed to be made up of a closely packed sphere of neutrons and protons held together by the strong interaction. The radius of the nucleus is determined from scattering experiments to be

$$R \approx (1.2 \,\text{fm}) A^{1/3} .$$

- When electrons with a momentum of about 1 GeV are scattered from protons or neutrons, a deviation in the Rutherford cross section is observed because the proton and neutron are not pointlike particles, but have a finite size. A quantitative measurement of the deviation from the Rutherford cross section shows that the size of the proton and neutron are about 1 fm.

- When electrons with a momentum of about 10 GeV are scattered from protons or neutrons, the detailed structure of the constituents is revealed. The scattering is due to the electromagnetic collision of the electron and a quark. The momentum distribution of the quarks inside the proton or neutron is measured by measuring the momentum transfer between the electron and the quark. In the electron–

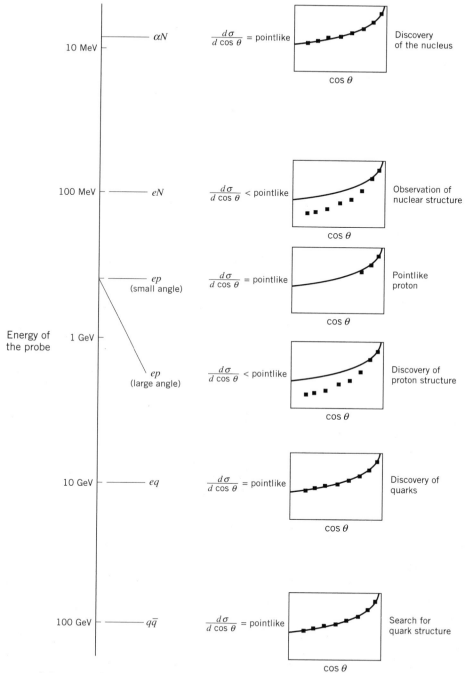

FIGURE 6-23 Summary of the scattering experiments.

quark center-of-mass system, the collision is elastic, and the angular distribution follows that expected for pointlike particles.

• When quarks with a momentum of about 50–100 GeV are scattered from each other, the quarks behave as free particles with regard to the protons that contain them. The quarks interact by the strong interaction, by the exchange of massless particles, analogous to the electromagnetic interaction. The angular distribution of the scattered quarks is expected to be of the Rutherford form, $(1-\cos\theta)^{-2}$, provided that the quarks have no internal structure. Quark–quark scattering data show that the quarks are pointlike down to a distance of 10^{-18} meters.

REFERENCES AND SUGGESTIONS FOR FURTHER READING

G. Arnison et al., "Angular Distributions for High-Mass Jet Pairs and a Limit of the Energy Scale of Compositeness for Quarks from the CERN $p\,\bar{p}$ Collider," *Phys. Lett.* **177B**, 244 (1986).

E. E. Chambers and R. Hofstadter, "Structure of the Proton," *Phys. Rev.* **103**, 1454 (1956).

J. I. Friedman, "Deep Inelastic Scattering: Comparisons with the Quark Model," *Rev. Mod. Phys.* **63**, 615 (1991).

H. Geiger and E. Marsden, "The Laws of Deflection of α Particles through Large Angles," *Phil. Mag.* **25**, 604 (1913).

R. Hofstadter, "The Atomic Nucleus," *Sci. Am.* **195**, No. 1, 55 (1956).

R. Hofstadter, "Structure of Nuclei and Nucleons," *Science* **136**, 1013 (1962).

R. Hofstadter, H. R. Fechter, and J. A. McIntyre, "High Energy Electron Scattering and Nuclear Structure Determinations," *Phys. Rev.* **92**, 978 (1953).

H. W. Kendall, "Deep Inelastic Scattering: Experiments on the Proton and the Observation of Scaling," *Rev. Mod. Phys.* **63**, 597 (1991).

H. W. Kendall and W. Panofsky, "The Structure of the Proton and the Neutron," *Sci. Am.* **224**, No. 6, 60 (1971).

E. Rutherford, "The Scattering of α and β Particles by Matter and the Structure of the Atom," *Phil. Mag.* **21**, 669 (1911).

E. Rutherford, "The Structure of the Atom," *Phil. Mag.* **27**, 488 (1914).

E. Rutherford and T. Royds, "The Nature of the α Particle from Radioactive Substances," *Phil. Mag.* **17**, 281 (1909).

R. E. Taylor, "Deep Inelastic Scattering: The Early Years," *Rev. Mod. Phys.* **63**, 573 (1991).

QUESTIONS AND PROBLEMS

Measuring structure by particle scattering

1. Why is the wavelength of an α particle that comes from a nuclear decay about equal to the size of a nucleus?

2. Explain how is it possible for an α particle to be scattered directly backwards ($\theta = \pi$) in the Rutherford experiment.

3. In the Rutherford experiment, is it possible to choose the impact parameter? Explain.

4. Macroscopic objects have extremely small wavelengths. Why do they not make good probes for measuring the structure of matter?

5. Could x rays have been used to discover the nucleus?

Definition of cross section

6. If the cross section is infinite, can the scattering rate also be infinite? Explain.

7. Experimenters wish to study the properties of a particle that has a production cross section in proton–proton collisions of 1 nb. They design a cylindrical liquid hydrogen target that has a diameter of 0.05 m and a length of 0.5 m. The proton beam has an intensity of 10^8 per second. If the apparatus can detect the rare particle with an efficiency of 10%, how long does it take to collect 10^6 events?

8. A collimated proton beam is sent into a liquid deuterium bubble chamber which has a thickness of 0.5 m. The density of liquid deuterium is 162 kg/m³. A sample of 10^5 pictures are analyzed and three rare events are found. (a) Calculate the production cross section for the process. (b) What is the 95% confidence level upper limit for the cross section?

9. The strong interaction has a short range, approximately 1 fm. Use this fact to estimate the cross section for the strong interaction of two energetic protons ($E \gg mc^2$). Compare your answer to 40 mb.

Probing the structure of the atom

10. When a particle has structure, why does a deviation from the Rutherford scattering formula at a fixed energy show up at large scattering angles rather than at small angles?

11. In the analysis of the Rutherford experiment, why can we treat the α as a particle scattering from a single nucleus and ignore the wave properties of the α? Make a comparison with the Davisson-Germer experiment.

12. Why is it convenient to write the differential cross section (6.12) as $d\sigma/d\cos\theta$ rather than $d\sigma/d\theta$? Show that if we write the differential cross section as $d\sigma/d\theta$ and integrate over all angles to get the total cross section that we get the same result as integrating $d\sigma/d\cos\theta$ over all values of $\cos\theta$.

13. Consider a beam of protons with momentum of 200 MeV/c scattering from aluminum nuclei. (a) Calculate the cross section for electromagnetic scattering at angles greater than 10°. (b) If the proton beam is sent into an aluminum foil 1 μm thick, what fraction are scattered to angles greater than 10°?

14. How important was it for Geiger and Marsden to evacuate their chamber? For α particles estimate the thickness of air that would have the same cross section as scattering in a gold foil of thickness 0.2 μm. (The density of gold is 1.9×10^4 kg/m³ and the density of air is 1.2 kg/m³ at atmospheric pressure and room temperature.)

15. A 10-MeV αparticle scatters from a silver nucleus at an angle of 90°. (a) Calculate the impact parameter. (b) Calculate the distance of closest approach.

16. Calculate the kinetic energy of an αparticle if the distance of closest approach to a gold nucleus is 10 fm when scattered at 90°.

17. Alpha particles of 5 MeV are scattered from a silver target 1 μm thick. If the scattering rate between 60 and 90 degrees is one per minute, what is the rate at which α particles strike the target?

18. Calculate the cross section for a 5-MeV αparticle to be scattered from platinum at an angle between 5 and 10°.

19. A beam of 5-MeV αparticles is directed into a silver foil of thickness 1 μm. The fraction of αparticles scattered at angles greater than a certain angle (θ) is measured to 10^{-3}. (a) What is the scattering cross section for this process? (b) What is the value of θ? (The density of silver is 1.05×10^4 kg/m³.)

Probing the structure of the nucleus

20. Protons of energy E_k are incident on a thin sheet of iron atoms at rest. Estimate the largest value of E_k for which the angular distribution of the scattered protons would be expected to follow the Rutherford angular distribution.

21. Alpha particles are scattered from a thin copper target. Estimate the minimum α particle kinetic energy that will cause a deviation from the Rutherford formula.

22. Consider a radioactive source that provides alpha particles with a kinetic energy of 6 MeV. Determine the approximate value of Z such that elements with nuclear charges greater than Z obey the Rutherford scattering law, and elements with nuclear charges smaller than Z show deviations.

Probing the structure of the proton

23. In an electron–proton scattering experiment, what factors limit the number of large angle scatters?

24. Electrons with an energy of 10 GeV are scattered from protons at rest at an angle of 30°. What is the maximum energy of the scattered electron?

25. If an electron with an energy of 10 GeV is scattered from a proton at rest and emerges with an energy of 5 GeV, what is the maximum scattering angle?

*26. An electron with an energy of 10 GeV scatters from a proton at rest and emerges at an angle of 5° with an energy of 5 GeV. What is the speed of the proton in the quark–electron center-of-mass frame?

Probing the structure of the quark

*27. The number of large angle scattering events observed by Geiger and Marsden in the discovery of the nucleus (Figure 6-4) was the same order of magnitude as the number of events observed by the UA1 Collaboration in the search for possible structure of quarks (Figure 6-22). Make an estimate of the total number of collisions needed to make the plots of (a) Figure 6-4 and (b) Figure 6-22.

*28. Consider the scattering of quarks at a center-of-mass energy of 100 GeV. Make an order of magnitude estimate of the probability that the quarks scatter at an angle between 45° and 90°.

Additional problems

29. (a) Calculate the maximum kinetic energy that can be transferred to a gold nucleus from a collision with a 6-MeV αparticle. (b) Calculate the maximum kinetic energy that can be transferred to an electron from a collision with a 6-MeV αparticle.

30. A beam of relativistic electrons with energy E is directed into a sheet of iron 0.01 meter thick. The fraction of scattered electrons with energy between E_1 and E_2 and angles between θ_1 and θ_2 is measured to be 10^{-6}. Calculate the scattering cross section per nucleon. (The density of iron is 7.87×10^3 kg/m³.)

31. A beam of αparticles is directed into a thin gold target. Scattered αparticles are detected in a small detector of cross-sectional area (a) that is placed a distance (R) facing the target at an angle (θ), ($a \ll R^2$). (a) Find an expression for the rate at which scattered αparticles strike the detector in terms of the rate at which the incident αparticles strike the foil (R_i), the kinetic energy of the αparticle (E_k), the thickness of the target (L), and the density (ρ), atomic number (Z), and atomic mass number (A) of the target. (b) Calcu-

late the detection rate in a 10^{-4} m^2 detector that is placed 0.1 m at an angle of 45 degrees from a 2 μm thick gold target, for 6-MeV α particles that are incident at a rate of 10^5 per second.

32. Consider the Rutherford scattering of projectile particles of charge q_1 with target particles of charge q_2. (a) Show that the cross section (σ) for scattering at angles between θ_1 and θ_2 ($\theta_2 > \theta_1$) is given by difference in the areas of two discs:

$$\sigma = \pi b_1^2 - \pi b_2^2,$$

where b_1 is the impact parameter corresponding to θ_1 and b_2 is the impact parameter corresponding to θ_2. Evaluate this expression using the relationship (6.40) between impact parameter squared and scattering angle. (b) Integrate the differential cross section (6.47)

$$\sigma = \int_{\cos \theta_1}^{\cos \theta_2} d\cos\theta \, \frac{d\sigma}{d\cos\theta},$$

to show explicitly that this gives an identical result for σ.

Being sufficiently versed in mathematics, I could not imagine how proper vibration frequencies could appear without boundary conditions. Later on I recognized that the more complicated form of the coefficients (i.e., the appearance of $V(x,y,z)$) takes charge, so to speak, of what is ordinarily brought about by boundary conditions, namely, the selection of definite values of E.

Erwin Schrödinger

We have seen that particles have wave properties. In this chapter we shall examine the wave equation for a nonrelativistic particle of mass m. The wave equation for a nonrelativistic particle is called the *Schrödinger equation*, after Erwin Schrödinger, who first published it in 1925. It is not possible to derive the Schrödinger equation, just as it is not possible to derive Maxwell's equations, but we can provide motivation for its form. We can then demonstrate the validity of the Schrödinger equation by comparing its tremendous predictive power with experiments.

Before we consider the wave equation for a nonrelativistic particle, we take note of two important properties of electromagnetic waves which will aid us in the interpretation of the Schödinger equation. The first property is that the speed of the wave is equal to the frequency times the wavelength:

$$c = f\lambda. \qquad (7.1)$$

Using the deBroglie relationships for energy and momentum, we may write the wave condition (7.1) as

$$c = \left(\frac{E}{h}\right)\left(\frac{h}{p}\right) = \frac{E}{p}. \qquad (7.2)$$

Thus, the electromagnetic wave equation contains the fundamental relationship between energy and momentum of a photon. The second property is that the energy per volume is proportional to the square of the electromagnetic field strength. In the electromagnetic wave equation, the components of the fields are the "wave functions." Since the energy of the waves is quantized, the probability per volume of finding a photon in the field is proportional to the square of the wave function.

7-1 FREE PARTICLE WAVE EQUATION

We begin by examining the wave equation for a free particle of fixed or *definite* total energy (E). We may represent a particle traveling in the x direction by the function

$$\Psi(x,t) = Ae^{ikx-i\omega t} = \psi(x)e^{-i\omega t}, \qquad (7.3)$$

where k is the wave number, ω is the frequency, and A is a normalization constant. The function $\Psi(x,t)$ is called the *time-dependent particle wave function*. The free-particle wave function (7.3) is called a *plane wave*. For a particle traveling in the minus-x direction the wave function is given by changing k to $-k$. The differential equation that

$\Psi(x,t)$ satisfies is called the time-dependent Schrödinger equation. We shall discuss the time-dependent Schrödinger equation at the end of this chapter.

The physics of this chapter is concerned with the spatial part of the wave function, $\psi(x)$, which is called the *time-independent* wave function. The wave function $\psi(x)$ satisfies the time-independent Schrödinger wave equation:

$$-\frac{\hbar^2}{2m}\frac{d^2\psi}{dx^2} = \left(E - mc^2\right)\psi(x). \qquad (7.4)$$

The solution of the Schrödinger equation (7.4) corresponding to motion in the x direction is $\psi(x) = Ae^{ikx}$. We verify that the wave function Ae^{ikx} satisfies the wave equation (7.4) by direct substitution, which gives

$$\left(-\frac{\hbar^2}{2m}\right)\left(-k^2\right)Ae^{ikx} = \left(E - mc^2\right)Ae^{ikx}. \qquad (7.5)$$

Solving for the particle energy, we have

$$E = mc^2 + \frac{\hbar^2 k^2}{2m}. \qquad (7.6)$$

Using the deBroglie relationship between momentum and wave number, $p = \hbar k$, we may write the particle energy (7.6) as

$$E = mc^2 + \frac{p^2}{2m}. \qquad (7.7)$$

The Schrödinger equation (7.4) for a free particle contains the fundamental relationship between energy and momentum (7.7) for a nonrelativistic particle analogous to the relationship between energy and momentum (7.2) for a photon.

For convenience and by convention the Schrödinger equation is written with the energy measured with the zero point defined to be the mass energy (as in the Bohr model). With the transformation, $E \leftarrow E - mc^2$, the Schrödinger equation becomes

$$-\frac{\hbar^2}{2m}\frac{d^2\psi}{dx^2} = E\psi(x). \qquad (7.8)$$

The solution for $\psi(x)$ is the same (7.4); however, our energy condition (7.7) becomes $E = p^2/2m$. The convenience of this notation will be appreciated when we consider a bound particle.

The interpretation of the wave function $\psi(x)$ was provided by Max Born in 1926. The square of $\psi(x)$, $|\psi(x)|^2$, is the *probability density*. The probability density gives the probability per unit x of finding the particle in any given region of space when an experiment is performed which is sensitive to an interaction of the particle. For example, consider an electron in a hydrogen atom. The electron does not have a classical trajectory but is spread-out in space. If we bombard the atom with photons of a sufficiently small wavelength, the probability that the electron interacts with the photon in a given region of space is given by $|\psi(x)|^2$.

The wave function $\psi(x)$ and its first derivative $d\psi/dx$ must be continuous so that the second derivative appearing in the Schrödinger equation exists. The continuity conditions are extremely important parts of the solution for $\psi(x)$.

7-2 PARTICLE IN A BOX

We now consider a particle that is absolutely confined to be in the interval $-L/2 < x < L/2$ as indicated in Figure 7-1. This problem is often referred to as a *particle in a box*. Since the particle may never be found outside the box, we immediately know that the square of the particle wave function is zero in this region,

$$|\psi(x)|^2 = 0 \tag{7.9}$$

for $x \le -L/2$ or $x \ge L/2$. This condition implies that the wave function is also zero in the same region:

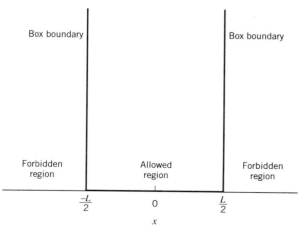

FIGURE 7-1 Particle in a box.
The particle is confined to be in the region $-L/2 < x < L/2$.

$$\psi(x) = 0. \tag{7.10}$$

Inside the box, $-L/2 < x < L/2$, $\psi(x)$ satisfies the free-particle Schrödinger equation (7.8) which we may write as

$$\frac{d^2\psi}{dx^2} = -\frac{2mE}{\hbar^2}\psi. \tag{7.11}$$

The general solution for $\psi(x)$ is a linear combination of sine and cosine terms. The wave function for a particle in a box differs from that of a free particle because of the boundary conditions imposed on $\psi(x)$. The boundary conditions are that $\psi(x)$ must be continuous at both $x = -L/2$ and $x = L/2$. Therefore,

$$\psi\left(-\frac{L}{2}\right) = \psi\left(\frac{L}{2}\right) = 0. \tag{7.12}$$

By choosing our coordinates such that $x = 0$ corresponds to the center of the box, we can make use of the symmetry of the problem. The particle is equally likely to be found on the left side of the box as on the right side. More specifically

$$|\psi(x)|^2 = |\psi(-x)|^2. \tag{7.13}$$

The symmetry condition (7.13) implies that either

$$\psi(x) = \psi(-x) \tag{7.14}$$

or

$$\psi(x) = -\psi(-x). \tag{7.15}$$

Therefore, the wave function may be a sine or cosine but not a mixture of the two. The wave function has a definite symmetry. One family of solutions may be written in the form

$$\psi(x) = A\cos kx, \tag{7.16}$$

where k is the wave number and A is the normalization constant. The boundary conditions (7.12) place a restriction on the allowed values of the wave number k:

$$k = \frac{n\pi}{L}, \tag{7.17}$$

where n is a positive odd integer (1, 3, 5, ...). We have found an infinite number of solutions, one for each allowed value of k. We may classify the solutions by assigning a serial number (n) to the wave function and the corresponding normalization constant and write the wave function as

$$\psi_n(x) = A_n \cos\left(\frac{n\pi x}{L}\right). \qquad (7.18)$$

where $n = 1, 3, 5\ldots$

The square of $\psi_n(x)$ gives the probability (P) of finding the particle per unit x. We write this condition as

$$\frac{dP}{dx} = |\psi_n(x)|^2. \qquad (7.19)$$

Since there is unit probability of finding the particle somewhere in the box, we have

$$\int_{-L/2}^{L/2} dx\, |\psi_n(x)|^2 = 1. \qquad (7.20)$$

The integral (7.20) is called the *normalization condition*. The normalization condition gives the constants A_n. For $n = 1$, we have

$$1 = \int_{-L/2}^{L/2} dx\, A_1^2 \cos^2\left(\frac{\pi x}{L}\right)$$

$$= A_1^2 \int_{-L/2}^{L/2} dx\, \sin^2\left(\frac{\pi x}{L}\right). \qquad (7.21)$$

The integral happens to be any easy one because

$$\int_{-L/2}^{L/2} dx\, \cos^2\left(\frac{\pi x}{L}\right)$$

$$= \frac{1}{2} \int_{-L/2}^{L/2} dx\left[\sin^2\left(\frac{\pi x}{L}\right) + \cos^2\left(\frac{\pi x}{L}\right)\right]$$

$$= \frac{1}{2} \int_{-L/2}^{L/2} dx(1) = \frac{L}{2}, \qquad (7.22)$$

which gives

$$A_1 = \sqrt{\frac{2}{L}}. \qquad (7.23)$$

The complete solution for the $n = 1$ wave function $\psi_1(x)$ is

$$\psi_1(x) = \sqrt{\frac{2}{L}} \cos\left(\frac{\pi x}{L}\right). \qquad (7.24)$$

Figure 7-2a shows a plot of the wave function $\psi_1(x)$. The probability of finding the particle per unit of x is given by the square of the wave function,

$$\frac{dP}{dx} = |\psi_1(x)|^2 = \frac{2}{L} \cos^2\left(\frac{\pi x}{L}\right). \qquad (7.25)$$

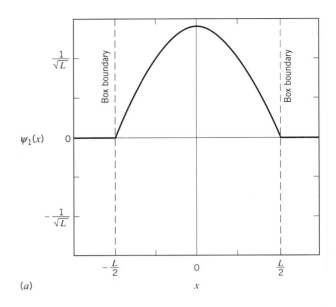

(a)

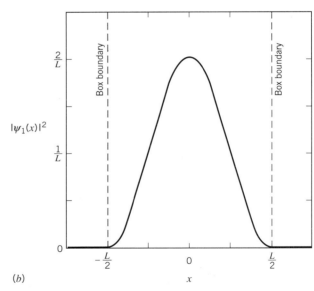

(b)

FIGURE 7-2 Particle in a box.
(a) The ground state ($n = 1$) wave function for a particle in a box. (b) The square of the ground state wave function.

This probability is shown in Figure 7-2b. The probability is maximum at the center of the box and zero at the boundaries.

The energy (E_1) corresponding to the wave function $\psi_1(x)$, is obtained by substituting the $n = 1$ wave function (7.24) into the Schrödinger equation (7.11). The result is

$$E_1 = \frac{\hbar^2 \pi^2}{2mL^2} = \frac{h^2}{8mL^2}. \qquad (7.26)$$

We shall see that all other solutions to the Schrödinger equation (7.11) for a particle in a box have *larger* energies. Accordingly, the state $\psi_1(x)$ is called the *ground state*.

A second family of solutions of the Schrödinger equation (7.11) may be written in the form

$$\psi(x) = B \sin kx, \qquad (7.27)$$

where k is the wave number and B is the normalization constant. The boundary conditions on $\psi(x)$ at the boundaries of the box give

$$\psi\left(-\frac{L}{2}\right) = \psi\left(\frac{L}{2}\right) = 0 = B \sin\left(\frac{kL}{2}\right). \qquad (7.28)$$

The boundary conditions are satisfied for

$$k = \frac{n\pi}{L}, \qquad (7.29)$$

where n is a positive even integer (2, 4, 6, ...). Using the boundary condition (7.28), we find that the wave functions (7.27) become

$$\psi_n(x) = B_n \sin\left(\frac{n\pi x}{L}\right), \qquad (7.30)$$

where $n = 2, 4, 6, \ldots$

EXAMPLE 7-1

Determine the value of the normalization constants A_n and B_n for all values of the quantum number n.

SOLUTION:

The normalization condition gives

$$\int_{-L/2}^{L/2} dx \, A_n^{\;2} \cos^2\left(\frac{n\pi x}{L}\right) = 1,$$

for n odd (1, 3, 5, ...). The value of the integral may be found in a handbook,

$$\int dx \cos^2(ax) = \frac{x}{2} + \frac{\sin(2ax)}{4a},$$

with $a = n\pi/L$. (This result may be verified by differentiation and simple trigonometry.) The normalization condition is

$$1 = \int_{-L/2}^{L/2} dx \, A_n^{\;2} \cos^2\left(\frac{n\pi x}{L}\right)$$

$$= A_n^{\;2}\left[\frac{L}{4} + \frac{L\sin(n\pi)}{4n\pi}\right] - A_n^{\;2}\left[-\frac{L}{4} + \frac{L\sin(-n\pi)}{4n\pi}\right]$$

$$= \frac{A_n^{\;2}L}{2},$$

or

$$A_n = \sqrt{\frac{2}{L}}.$$

For even values of the quantum number n (2, 4, 6, ...), the normalization condition gives

$$\int_{-L/2}^{L/2} dx \, B_n^{\;2} \sin^2\left(\frac{n\pi x}{L}\right) = 1.$$

The value of the integral is

$$\int dx \sin^2(ax) = \frac{x}{2} - \frac{\sin(2ax)}{4a},$$

with $a = n\pi/L$. The integral is

$$\int_{-L/2}^{L/2} dx \, B_n^2 \sin^2\left(\frac{n\pi x}{L}\right)$$

$$= B_n^{\;2}\left[\frac{L}{4} - \frac{L\sin(n\pi)}{4n\pi}\right] - B_n^{\;2}\left[-\frac{L}{4} - \frac{L\sin(-n\pi)}{4n\pi}\right]$$

$$= \frac{B_n^{\;2}L}{2},$$

which gives

$$B_n = \sqrt{\frac{2}{L}}.$$

Thus,

$$A_n = B_n = \sqrt{\frac{2}{L}},$$

for all values of n. In general, it is not the case that all the normalization constants are equal. ∎

Using the results of Example 7-1, we may summarize our particle-in-a-box wave functions as

$$\psi_n(x) = \sqrt{\frac{2}{L}} \cos\left(\frac{n\pi x}{L}\right) \qquad (7.31)$$

for n odd (1, 3, 5, ...) and

$$\psi_n(x) = \sqrt{\frac{2}{L}} \sin\left(\frac{n\pi x}{L}\right) \qquad (7.32)$$

for n even (2, 4, 6, ...). The integer n is called the *principal quantum number*. The wave functions and their squares for $n = 2$, 3, and 4 are shown in Figures 7-3, 7-4 and 7-5 respectively.

Energy Quantization

We may now solve for the energy of the particle by substituting our expression for $\psi_n(x)$ into the Schrödinger equation. There will be a different energy for each wave function so that we also assign a serial number to the energy, that is, E_1 corresponds to the energy of a particle with the wave function $\psi_1(x)$, E_2 to $\psi_2(x)$, and so on. The Schrödinger equations for each value of the quantum number n are

$$\frac{d^2\psi_n}{dx^2} = -\frac{2mE_n}{\hbar^2}\psi_n. \qquad (7.33)$$

The energy levels are obtained by substituting the expressions for $\psi_n(x)$ (7.31 and 7.32) into the Schrödinger equation (7.33). The result is

$$\left(\frac{n\pi}{L}\right)^2 \psi_n(x) = \frac{2mE_n}{\hbar^2}\psi_n(x). \qquad (7.34)$$

The allowed energies are

$$E_n = \frac{\hbar^2 n^2 \pi^2}{2mL^2} = \frac{n^2 h^2}{8mL^2}, \qquad (7.35)$$

where n is a positive integer (1, 2, 3, ...). We see that energy is quantized. The particle cannot have an arbitrary energy. The allowed energy levels are indicated in Figure 7-6. The energy quantization has come from the requirement that the wave function be equal to zero at the boundaries of the potential. This is how the integer n has appeared in the expression for the particle energy. The wave number corresponding to each value of the quantum number n may be written in terms of the energy,

$$k_n = \frac{\sqrt{2mE_n}}{\hbar}. \qquad (7.36)$$

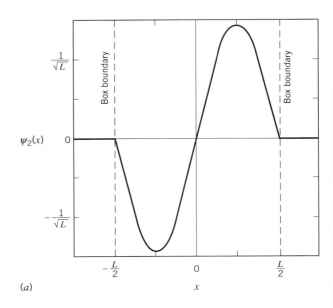

(a)

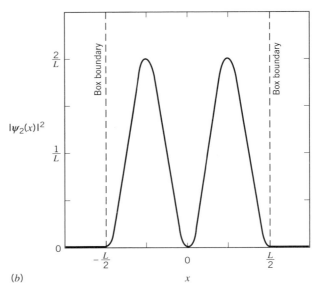

(b)

FIGURE 7-3 Particle in a box.
(a) The first excited state ($n = 2$) wave function for the infinite square-well. (b) The square of the $n = 2$ wave function.

EXAMPLE 7-2

What is the ground state energy of an electron that is confined to be in a "box" of size equal to the diameter of an atom (0.3 nm)?

SOLUTION:

The ground state energy is

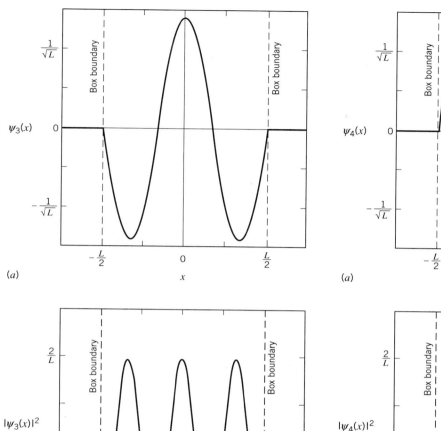

FIGURE 7-4 Particle in a box.
(a) The second excited state (n = 3) wave function for the infinite square-well. (b) The square of the n = 3 wave function.

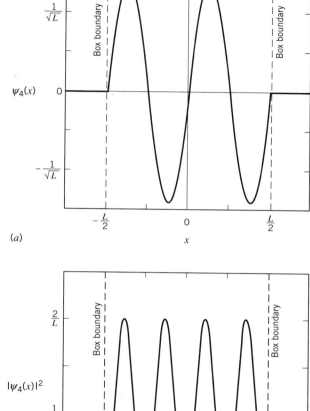

FIGURE 7-5 Particle in a box.
(a) The third excited state (n = 4) wave function for the infinite square-well. (b) The square of the n = 4 wave function.

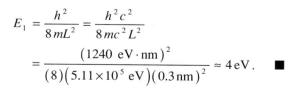

$$E_1 = \frac{h^2}{8mL^2} = \frac{h^2 c^2}{8mc^2 L^2}$$

$$= \frac{(1240 \ \text{eV} \cdot \text{nm})^2}{(8)(5.11 \times 10^5 \ \text{eV})(0.3 \, \text{nm})^2} \approx 4 \, \text{eV}. \quad \blacksquare$$

EXAMPLE 7-3

A particle confined to be in the region $-L/2 < x < L/2$ is in the ground state. Calculate the probability that the particle will be found in the region $-L/2 < x < -L/4$.

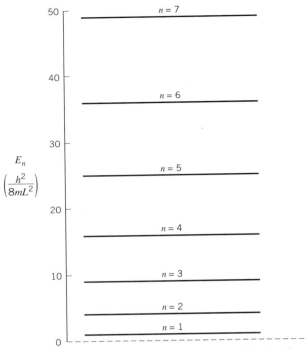

FIGURE 7-6 Allowed energy levels of a particle in a box of size L.
The energy is plotted in units of $h^2/8mL^2$. The minimum energy is $E_1 = h^2/8mL^2$. The energy of the nth level is $E_n = n^2 E_1$.

SOLUTION:
The particle wave function is

$$\psi_1(x) = \sqrt{\frac{2}{L}} \cos\left(\frac{\pi x}{L}\right).$$

The probability that the particle is found in the region $-L/2 < x < -L/4$ is

$$P = \int_{-L/2}^{-L/4} dx\, \psi^2 = \frac{2}{L} \int_{-L/2}^{-L/4} dx \cos^2\left(\frac{\pi x}{L}\right).$$

The value of the integral (see example 7-1) is

$$\int dx \cos^2(ax) = \frac{x}{2} + \frac{\sin(2ax)}{4a}.$$

Therefore,

$$P = \frac{2}{L}\left[-\frac{L}{8} + \frac{L}{4\pi}\sin\left(-\frac{\pi}{2}\right)\right] - \frac{2}{L}\left[-\frac{L}{4} + \frac{L}{8\pi}\sin(-\pi)\right]$$

$$= \frac{1}{4} - \frac{1}{2\pi} \approx 0.09. \qquad \blacksquare$$

Symmetry of the Wave Functions

The wave functions, $\psi_n(x)$, for the particle in a box possess a symmetry about the center of the box. For the ground state and all other odd values of the number n, the symmetry condition is

$$\psi_n(x) = \psi_n(-x). \qquad (7.37)$$

A wave function that does not change under a transformation of the coordinates

$$x \rightarrow -x, \qquad (7.38)$$

is called an *even* wave function.

For the first excited state and all other even values of the number n, the symmetry condition is

$$\psi_n(x) = -\psi_n(-x). \qquad (7.39)$$

A wave function that changes sign under transformation of the coordinates (7.38) is called an *odd* wave function.

The Standing Wave Interpretation

The particle in a box has a deBroglie wavelength (5.7)

$$\lambda = \frac{h}{p}. \qquad (7.40)$$

Confinement of the particle in the region $-L/2 < x < L/2$ may be interpreted as a standing wave (see Figure 3-2) in which an integer number of half-wavelengths must fit into the length L:

$$n\frac{\lambda}{2} = L. \qquad (7.41)$$

Using the deBroglie expression for the wavelength (7.40), the momentum may be written

$$p = \frac{h}{\lambda} = \frac{nh}{2L}. \qquad (7.42)$$

The kinetic energy for the standing wave (with n half-wavelengths) is

$$E_n = \frac{p^2}{2m} = \frac{n^2 h^2}{8mL^2}. \qquad (7.43)$$

The solution of the time-independent Schrödinger equation for a particle in a box is equivalent to standing waves with the wavelength given by the deBroglie relationship.

The standing wave interpretation may be made explicit by writing the wave functions as the sum of the spatial components of two traveling waves, one corresponding to motion in the x direction (e^{ikx}) and another corresponding to motion in the minus-x direction (e^{-ikx}). For the even solutions we have

$$\psi_n(x) = \sqrt{\frac{2}{L}}\cos\left(\frac{n\pi x}{L}\right)$$
$$= \sqrt{\frac{2}{L}}\left(\frac{e^{ikx}+e^{-ikx}}{2}\right), \qquad (7.44)$$

whereas for the odd solutions we have

$$\psi_n(x) = \sqrt{\frac{2}{L}}\sin\left(\frac{n\pi x}{L}\right)$$
$$= \sqrt{\frac{2}{L}}\left(\frac{e^{ikx}-e^{-ikx}}{2i}\right), \qquad (7.45)$$

where $k = n\pi/L$.

A particle with a definite energy, that is, a solution to the time-independent Schrödinger equation, may be interpreted as a standing wave with wavelength equal to Planck's constant divided by the particle momentum. The wave function for the nth state will have n "wiggles."

The Classical Limit

The kinetic energy of a particle in a box cannot have an arbitrary value. The correspondence principle states that when the quantum number n is very large, the fractional spacing between allowed energy values is small compared to experimental resolution and the kinetic energy appears as a continuous variable. When the principal quantum number n is large, the wave function oscillates many times per unit distance. Figure 7-7 shows the particle in a box wave function for $n = 20$. The classical regime is just the regime where the number of oscillations is large compared to the experimental resolution, Δx, of the position of the particle. Therefore the probability of finding the particle anywhere in the box is independent of x and the average value of the probability distribution, dP/dx, is equal to $1/L$. In the classical limit, the probability that the particle is in the interval Δx is

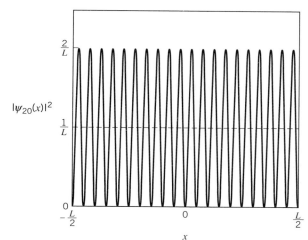

FIGURE 7-7 Probability distribution, $|\psi(x)|^2$, for a particle in a box for $n = 20$.
The number of oscillations in the probability distribution is equal to the value of the principle quantum number n.

$$\int_x^{x+\Delta x} dx\,\frac{dP}{dx} = \frac{1}{L}\int_x^{x+\Delta x} dx = \frac{\Delta x}{L}. \qquad (7.46)$$

EXAMPLE 7-4

Estimate the value of the quantum number n for a macroscopic particle, with a mass of 0.001 kg and a speed of 10^{-6} m/s, that is confined to be in a box of size 0.01 m.

SOLUTION:

The particle kinetic energy is

$$E_k = \frac{1}{2}mv^2.$$

The particle kinetic energy is quantized,

$$E = \frac{n^2 h^2}{8mL^2},$$

where n is a positive integer. Solving for n, we get

$$n = \sqrt{\frac{8mL^2E}{h^2}} = \sqrt{\frac{4m^2L^2v^2}{h^2}} = \frac{2mLv}{h},$$

or

$$n = \frac{(2)(10^{-3}\,\text{kg})(0.01\,\text{m})(10^{-6}\,\text{m/s})}{6.6\times10^{-34}\,\text{J}\cdot\text{s}} \approx 3\times10^{22}.$$

The quantum number n is large because the mass of the particle is large. There are 3×10^{22} oscillations of the wave function over a distance of 0.01 meter. ∎

Consistency with the Uncertainty Principle

According to the uncertainty principle, confinement of a particle implies that the particle must have a minimum kinetic energy. Our solution from the Schrödinger equation must be consistent with the uncertainty principle. The uncertainty in position (Δx) of a particle confined to be in a box of size L may be estimated to be (see Example 2-3)

$$\Delta x \approx \frac{L}{\sqrt{12}}. \qquad (7.47)$$

From the uncertainty principle, the uncertainty in momentum (Δp) is

$$\Delta p \approx \frac{\hbar}{2\,\Delta x} \approx \frac{\sqrt{12}\hbar}{2L}. \qquad (7.48)$$

The minimum kinetic energy according to the uncertainty principle is

$$\left\langle E_k^{min} \right\rangle = \frac{(\Delta p)^2}{2m} = \frac{3\hbar^2}{mL^2}. \qquad (7.49)$$

The exact answer from the solution to the Schrödinger equation,

$$E_1 = \frac{\hbar^2 \pi^2}{2mL^2}, \qquad (7.50)$$

is larger than the minimum required by the uncertainty principle by about 50%.

The particle energy is known precisely. Precise knowledge of the particle energy gives us precise knowledge of the particle momentum squared and precise knowledge of the absolute value of the momentum; however, we do not know the sign of the momentum! The standing wave is a linear combination of wave functions corresponding to motion in the x and minus-x directions. For a particle in the ground state in a box of size L, the uncertainty in momentum is

$$\Delta p = p = \frac{h}{\lambda} = \frac{h}{2L}, \qquad (7.51)$$

where we have used the fact that $\lambda = 2L$. The uncertainty in position is

$$\Delta x \approx \frac{L}{\sqrt{12}}. \qquad (7.52)$$

The product is

$$\Delta p \Delta x = \left(\frac{h}{2L} \right)\left(\frac{L}{\sqrt{12}} \right)$$
$$= \frac{h}{4\sqrt{3}} = \frac{\pi\hbar}{2\sqrt{3}} > \frac{\hbar}{2}. \qquad (7.53)$$

in agreement with the uncertainty principle.

7-3 FINITE SQUARE-WELL POTENTIAL

Suppose the particle is not free, but moves in the presence of a potential $V(x)$. The energy of the particle is

$$E = \frac{p^2}{2m} + V, \qquad (7.54)$$

where the potential may depend on x. The Schrödinger equation is

$$-\frac{\hbar^2}{2m} \frac{d^2\psi(x)}{dx^2} + V(x)\psi(x)$$
$$= E\psi(x). \qquad (7.55)$$

This is the time-independent Schrödinger equation, which describes a particle with definite energy E.

Consider a potential $V(x)$ that is zero if $-L/2 < x < L/2$ and a positive constant V_0 otherwise. This potential, called a *finite square-well*, is graphed in Figure 7-8. If the particle kinetic energy is greater than V_0, the particle is not confined to the well and energy is not quantized. We are interested in the situation where

$$E < V_0. \qquad (7.56)$$

Before we solve the Schrödinger equation (7.55), we examine what is known about the physics of the solution. If the potential V_0 is very large compared to the ground state energy, then we should have a solution similar to that of the particle in a box. There will be one important difference. In the particle in a box, the wave function is precisely equal to zero at the boundaries of the box. In the finite square-well with a large value of V_0, the wave function will be very small at the boundaries, but not identically zero. Since the integral of the square of the

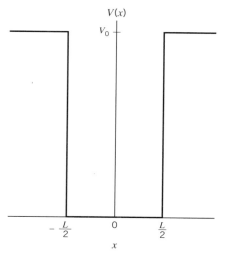

FIGURE 7-8 The finite square-well potential energy function $V(x)$.

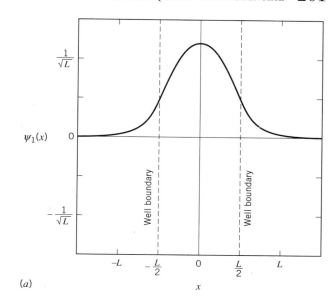

(a)

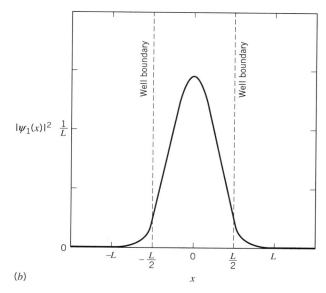

(b)

Figure 7-9 Finite square-well.
(a) The ground state ($n = 1$) wave function for the finite square-well. (b) The square of the $n = 1$ wave function.

wave function from $-\infty$ to $+\infty$ is unity, the wave function must go to zero at $x = -\infty$ and $x = +\infty$. Therefore, the wave function is identically zero at $x = -\infty$ and $x = +\infty$. The solutions to the finite square-well wave functions look similar to the particle in a box solutions, except that the wave functions will have a small extension beyond the well boundaries. The first four wave functions and their squares for the finite square-well are shown in Figures 7-9, 7-10, 7-11, and 7-12.

We now examine the quantitative solution for a particle confined by a finite square-well potential. Inside the well, the potential is zero and the Schrödinger equation is the same as for a particle in a box:

$$\frac{d^2\psi}{dx^2} = -\frac{2mE}{\hbar^2}\psi(x). \qquad (7.57)$$

The general solution is of the form

$$\psi(x) = A\cos kx + B\sin kx, \qquad (7.58)$$

where A and B are constants and the wave number and particle energy are related by

$$k = \frac{\sqrt{2mE}}{\hbar}, \qquad (7.59)$$

The wave function (7.58) is identical to the wave function for a particle in a box, except $\psi(x)$ is *not zero* at the well boundaries as it is for the box. Since the particle is bound, the boundary conditions on $\psi(x)$ are

$$\lim_{x \to \infty}\left[\psi(x)\right] = 0 \qquad (7.60)$$

and

$$\lim_{x \to -\infty}\left[\psi(x)\right] = 0. \qquad (7.61)$$

The boundary conditions (7.60) and (7.61) are necessary

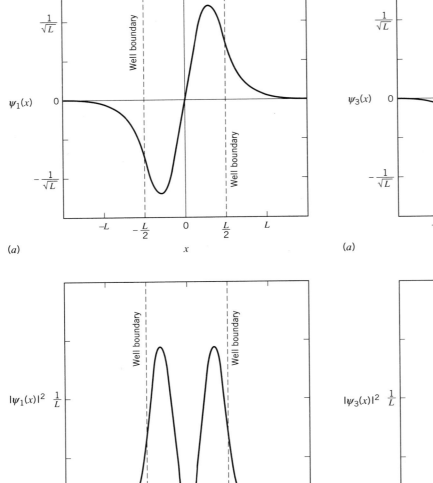

(a)

(b)

Figure 7-10 Finite square-well.
(a) The first excited state ($n = 2$) wave function for the finite square-well. (b) The square of the $n = 2$ wave function.

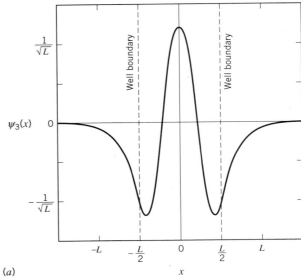

(a)

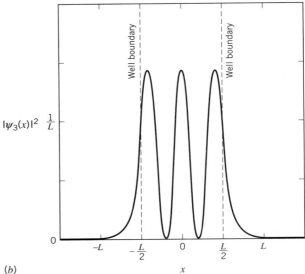

(b)

Figure 7-11 Finite square-well.
(a) The second excited state ($n = 3$) wave function for the finite square-well. (b) The square of the $n = 3$ wave function.

so that the integral of $|\psi|^2$ over all values of x from minus infinity to infinity is finite (namely, unity).

Outside the well, the Schrödinger equation (7.55) is

$$\frac{d^2\psi}{dx^2} = \frac{2m(V_0 - E)}{\hbar^2}\psi(x), \qquad (7.62)$$

which we may write as

$$\frac{d^2\psi}{dx^2} = \beta^2\psi(x) \qquad (7.63)$$

with

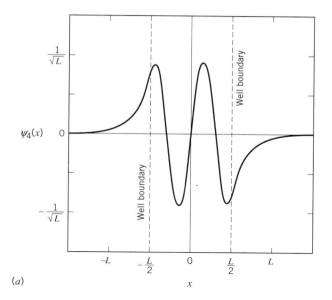

(a)

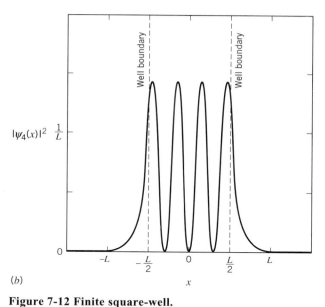

(b)

Figure 7-12 Finite square-well.
(a) The third excited state ($n = 4$) wave function for the finite square-well. (b) The square of the $n = 4$ wave function.

$$\beta \equiv \sqrt{\frac{2m(V_0 - E)}{\hbar^2}}. \qquad (7.64)$$

Since $V_0 - E$ is positive, β must be a real number. The solution of the Schrödinger equation is of the form

$$\psi(x) = Ce^{\beta x} + De^{-\beta x} \qquad (7.65)$$

where C and D are constants. Since $\psi(x) \to 0$ as $x \to +\infty$ and $x \to -\infty$, we have

$$\psi(x) = Ce^{\beta x} \qquad (7.66)$$

for $x < -L/2$, and

$$\psi(x) = De^{-\beta x} \qquad (7.67)$$

for $x > L/2$.

We must now determine the constants A, B, C, and D appearing in our expressions for the wave function (7.58, 7.66, and 7.67). To solve for these constants, we have four equations from the continuity of $\psi(x)$ and $d\psi/dx$ at $x = -L/2$ and $x = L/2$. We may greatly simplify the algebra by observing that a symmetric potential, $V(x) = V(-x)$, implies that the probability density $|\psi(x)|^2$ is also symmetric, $|\psi(x)|^2 = |\psi(-x)|^2$. Therefore, $\psi(x)$ must have a definite symmetry; $\psi(x)$ must be either an even or an odd function of x.

The Even Solutions

For the even solutions,

$$\psi(x) = \psi(-x). \qquad (7.68)$$

Since the sine is an odd function,

$$\sin(x) = -\sin(-x), \qquad (7.69)$$

the wave function inside the well cannot have a sine component. This means that the constant B must be equal to 0. Therefore,

$$\psi(x) = A\cos kx. \qquad (7.70)$$

The solution is similar to that of the infinite well except that now there is a small exponential part of the wave function that extends beyond the well boundary. Contrary to the classical result, there is a small but finite chance of finding the particle outside the well. One might be concerned that this would imply a negative kinetic energy since the total energy would be less than the potential energy; however, we are saved by the uncertainty principle! If the particle is outside the well boundary, then we have good information about its position because the wave function falls to zero exponentially in this region ($e^{-\beta x}$, for $x > 0$). If the particle is outside the boundary of the well, then it is confined to a region of size approximately $\Delta x \approx 1/\beta$. (Recall from Example 2-2, that the root-mean-square deviation of an exponential function, $e^{-\beta x}$, is equal to $1/\beta$.) By the uncertainty principle, the minimum average kinetic energy is

$$\left\langle E_K^{\min} \right\rangle \approx \frac{\hbar^2 \beta^2}{8m} = V_0 - E. \qquad (7.71)$$

The momentum of the particle plus its uncertainty is enough to get the particle over the barrier! The sum of the total particle energy and its minimum kinetic energy is

$$\left\langle E_K^{\min} \right\rangle + E \approx V_0 \qquad (7.72)$$

Therefore, it is not possible to measure a negative kinetic energy for the particle outside the box even though it has a finite probability of interacting there.

Finding the Energies of the Even Solutions

The wavelength of the particle is longer than in the case of a particle in a box, because the wave function now extends outside of the box. Therefore, the energy levels of the particle are lower in the finite square-well than for a particle in a box.

The boundary condition on the continuity of $\psi(x)$ at $x = L/2$ is

$$A\cos\left(\frac{kL}{2}\right) = De^{-\beta L/2}. \qquad (7.73)$$

The boundary condition on the continuity of $d\psi/dx$ at $x = L/2$ is

$$-kA\sin\left(\frac{kL}{2}\right) = -\beta De^{-\beta L/2}. \qquad (7.74)$$

The boundary conditions (7.73) and (7.74) may be combined to yield

$$\tan\left(\frac{kL}{2}\right) = \frac{\beta}{k}. \qquad (7.75)$$

The boundary conditions on the continuity of $\psi(x)$ and $d\psi/dx$ at $x = -L/2$ give no additional information, since we have already made use of the fact that $\psi(x)$ is symmetric. We may write the boundary condition (7.75) in terms of the particle energy by using the expressions for wave number (7.59) and β (7.64):

$$\tan\left(\frac{L\sqrt{2mE}}{2\hbar}\right) = \sqrt{\frac{(V_0 - E)}{E}}. \qquad (7.76)$$

This is a transcendental equation. It cannot be solved algebraically to give a closed-form solution for the energy

E. It can be solved on a computer and we can visualize the solution graphically. To simplify the notation we make the following change of variables,

$$\xi \equiv \frac{L\sqrt{2mE}}{2\hbar}, \qquad (7.77)$$

and

$$W \equiv \frac{mV_0 L^2}{2\hbar^2}. \qquad (7.78)$$

Then the energy in terms of ξ is

$$E = \frac{2\xi^2 \hbar^2}{mL^2}, \qquad (7.79)$$

and the transcendental equation (7.76) becomes

$$\tan\xi = \frac{\sqrt{W - \xi^2}}{\xi}. \qquad (7.80)$$

Figure 7-13 shows a graph of $\tan\xi$ and $(W-\xi^2)^{1/2}/\xi$ versus ξ. The allowed energy levels of the even states (7.79) are given by the values of ξ where the functions $\tan\xi$ and $(W-\xi^2)^{1/2}/\xi$ cross. The boundary conditions on $\psi(x)$ have led to energy quantization. The lowest energy state occurs at a value of ξ slightly below $\pi/2$. Note that if $\xi = \pi/2$, we get the ground state energy of the particle in

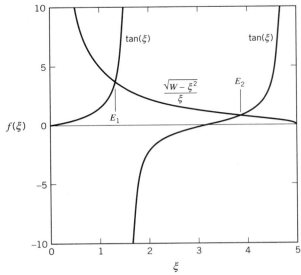

FIGURE 7-13 Graphical solution to the even solutions of the finite square-well.

a box. The second even state occurs near ξ equal to $3\pi/2$, corresponding to about nine times the energy of the ground state. There is not an infinite number of even states because the maximum allowed value of ξ is $W^{1/2}$. There must be at least one even state because even if the potential V_0 is very small compared to $\hbar^2/mL^2$, the function $(W-\xi^2)^{1/2}/\xi$ is nonzero and crosses $\tan \xi$ at least once because $\tan \xi$ goes to zero at $\xi = 0$. Note if $V_0 << \hbar^2/mL^2$, $W << 1$ and the energy is comparable to V_0.

When the height of the barrier (V_0) is much larger than the ground state energy, there is another method that we may use to estimate the ground state energy. The extension of the wave function outside of the barrier is exponentially attenuated by the length

$$\frac{1}{\beta} = \frac{\hbar}{\sqrt{2m\left(V_0 - E_1\right)}}. \qquad (7.81)$$

Therefore, the energy of the ground state is similar to the energy of the ground state of a particle in a box that is larger in size by an amount $2/\beta$. We estimate the energy as

$$E_1 \approx \frac{\hbar^2 \pi^2}{2m\left(L + \dfrac{2}{\beta}\right)^2}. \qquad (7.82)$$

Since β depends on E_1, we may estimate E_1 from the particle in a box of size L, use the relationship between β and E_1 (7.81) to get an estimate for β, and then make a revised estimate of the energy (7.82). The following two examples illustrate the procedure for estimating the ground state energy.

EXAMPLE 7-5
An electron is confined by a 0.2 nm finite square-well of strength 1 keV. What is the ground state energy?

SOLUTION:
The ground state energy for the particle in a box is

$$E_1^{box} = \frac{\hbar^2 \pi^2}{2mL^2} = \frac{(197\,\text{eV}\cdot\text{nm})^2 \, \pi^2}{(2)\left(5.11\times10^5 \,\text{eV}\right)(0.2\,\text{nm})^2}$$

$$= 9.4\,\text{eV}.$$

Estimating the energy for the finite square-well (E_1) to be 9.4 eV, the value of $1/\beta$ is

$$\frac{1}{\beta} = \frac{\hbar}{\sqrt{2m\left(V_0 - E_1\right)}}$$

$$= \frac{197\,\text{eV}\cdot\text{nm}}{\sqrt{(2)\left(5.11\times10^5\,\text{eV}\right)(1000\,\text{eV} - 9.4\,\text{eV})}}$$

$$= 0.00619\,\text{nm}.$$

The estimate of the ground state energy in the finite square-well is then refined to be

$$E_1 \approx \frac{\hbar^2 \pi^2}{2m\left(L + \dfrac{2}{\beta}\right)^2} = E_1^{box}\frac{L^2}{\left(L + \dfrac{2}{\beta}\right)^2}$$

$$\approx \frac{(9.4\,\text{eV})(0.20\,\text{nm})^2}{\left[0.20\,\text{nm} + (2)(0.0062\,\text{nm})\right]^2} \approx 8.3\,\text{eV}.$$

When we use this value of E_1 to recalculate $1/\beta$, we get $1/\beta = 0.00619$ nm, and the iteration technique has converged. This solution is in agreement with the exact solution obtained by numerical solution of the transcendental equation (7.76). ∎

EXAMPLE 7-6
An alpha particle is confined by a 3 fm finite square-well of strength 40 MeV. What is the ground state energy?

SOLUTION:
The mass energy of the alpha is 3730 MeV. The ground state energy for the particle in a box is

$$E_1^{box} = \frac{\hbar^2 \pi^2}{2mL^2} = \frac{(197\,\text{MeV}\cdot\text{fm})^2 \, \pi^2}{(2)(3730\,\text{MeV})(3\,\text{fm})^2} \approx 5.7\,\text{MeV}.$$

Estimating E_1 to be 5.7 MeV, the value of $1/\beta$ is

$$\frac{1}{\beta} = \frac{\hbar}{\sqrt{2m\left(V_0 - E_1\right)}}$$

$$= \frac{197\,\text{MeV}\cdot\text{fm}}{\sqrt{(2)(3730\,\text{MeV})(40\,\text{MeV} - 5.7\,\text{MeV})}}$$

$$\approx 0.39\,\text{fm}.$$

The estimate of the ground state energy in the finite square-well is

$$E_1 \approx \frac{\hbar^2 \pi^2}{2m\left(L+\dfrac{2}{\beta}\right)^2} = E_1^{\text{box}} \frac{L^2}{\left(L+\dfrac{2}{\beta}\right)^2}$$

$$= \frac{(5.7\,\text{MeV})(3\,\text{fm})^2}{\left[3\,\text{fm}+(2)(0.39\,\text{fm})\right]^2} \approx 3.6\,\text{MeV}.$$

When we use this value of E_1 to recalculate $1/\beta$, we get $1/\beta = 0.38$ fm, and the iteration technique has converged. ∎

The Odd Solutions

We may determine the energies of the odd solutions using the same technique used to find the energies of the even solutions. The only difference is that since the wave function inside the box is different (sine versus cosine), the boundary conditions lead to a different transcendental equation. For the odd solutions, the wave function inside the well (7.58) cannot have a cosine component, so that the constant A must be equal to 0. Therefore,

$$\psi(x) = B\sin kx. \qquad (7.83)$$

Finding the Energies of the Odd Solutions
The boundary condition on the continuity of $\psi(x)$ at $x = L/2$ is

$$B\sin\left(\frac{kL}{2}\right) = Ce^{-\beta L/2}. \qquad (7.84)$$

The boundary condition on the continuity of $d\psi/dx$ at $x = L/2$ is

$$kB\cos\left(\frac{kL}{2}\right) = -\beta Ce^{-\beta L/2}. \qquad (7.85)$$

The boundary conditions (7.84) and (7.85) give

$$\tan\left(\frac{kL}{2}\right) = -\frac{k}{\beta}. \qquad (7.86)$$

In terms of the particle energy, we have

$$\tan\left(\frac{L\sqrt{2mE}}{2\hbar}\right) = \sqrt{\frac{E}{(V_0 - E)}}. \qquad (7.87)$$

This is our transcendental equation for the odd states. In terms of the parameters ξ and W, we have

$$\tan\xi = \frac{\xi}{\sqrt{W-\xi^2}}. \qquad (7.88)$$

For $\xi = 0$, which satisfies the transcendental equation, we do not have a solution of the Schrödinger equation, since in this case $k = 0$ and $\psi(x)$ would be identically zero everywhere. The graphical solution is shown in Figure 7-14. It is possible that no odd bound states exist. The condition for at least one odd bound state to exist is given by the maximum value of the function $\xi/\sqrt{W-\xi^2}$.

The smallest energy corresponds to $\xi \approx \pi$. This function crosses at least one tangent line if

$$V_0 > \frac{\hbar^2}{2mL^2}. \qquad (7.89)$$

The energy levels of a particle in a finite square-well are summarized in Figure 7-15. The potential of the finite well is chosen to be

$$V_0 \approx \frac{12\hbar^2}{8mL^2}. \qquad (7.90)$$

There are three bound states. The energies of these states are slightly lower than the lowest three states of the infinite square-well.

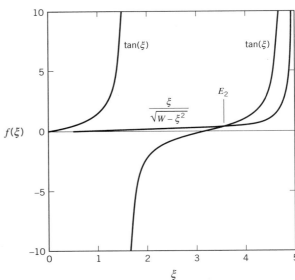

FIGURE 7-14 Graphical solution to the odd solutions of the finite square-well.
The value of $V_0 mL^2$ has been chosen so that there is one odd solution.

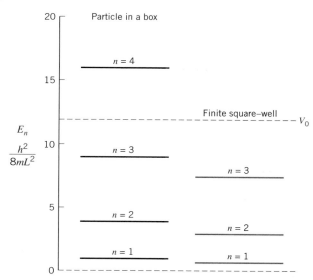

FIGURE 7-15 Energy levels of a particle in a finite square-well.

7-4 BARRIER PENETRATION

We have seen in the example of the finite square-well that the wave function extends into the classically forbidden region. In certain physical systems, these quantum tunneling phenomena are directly observable.

Alpha Decay

Certain heavy nuclei are observed to decay by emission of an alpha particle with a kinetic energy (E_k) of a few MeV. In an alpha decay, the daughter nucleus has two fewer protons and two fewer neutrons than the parent nucleus. The kinetic energy of the alpha particles are measured to be in a relatively small range from 4 to 9 MeV. The lifetimes of the alpha decay process vary dramatically from 10^{-7} seconds to 10^{10} years. For example, the decay

$$^{238}\text{U} \rightarrow \,^{234}\text{Th} + \alpha \qquad (7.91)$$

has $E_k = 4.18$ MeV and a half-life $t_{1/2} = 4.5 \times 10^9$ years, whereas the decay

$$^{212}\text{Po} \rightarrow \,^{208}\text{Pb} + \alpha \qquad (7.92)$$

has $E_k = 8.78$ MeV and $t_{1/2} = 0.3$ μs. (The half-life is defined as the time required for one-half the nuclei in a sample to decay.) *Why do similar decays yield alpha particle energies of the same order of magnitude while the*

lifetimes differ by more than 23 orders of magnitude? The answer is to be found in quantum tunneling.

Let us assume that the alpha particle can exist inside the nucleus. By this we mean that within the nucleus, which may be thought of as a dense conglomeration of protons and neutrons, there is a significant probability at any given time that two neutrons and two protons can find themselves bound together. This happens because the alpha particle is exceptionally stable; it has a binding energy near 7 MeV per nucleon. This is comparable to the binding energy per nucleon in a heavy nucleus (about 8 MeV per nucleon). The alpha particle is bound to the nucleus by the strong force, which has a very short range. We may approximate the potential energy of the alpha particle as an attractive square-well due to the strong interaction plus a repulsive Coulomb component outside the nucleus. The potential is shown in Figure 7-16.

EXAMPLE 7-7

Estimate the height of the Coulomb barrier for an alpha particle inside a uranium nucleus.

SOLUTION:

The charge of the alpha is $2e$ and the charge of the rest of the nucleus is $(Z-2)e$. When the alpha particle is at a distance R, the nuclear radius, it is just out of range of the strong force. The potential energy is

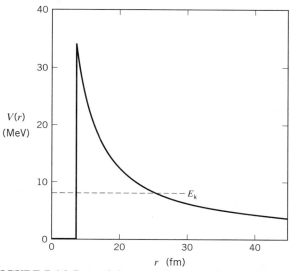

FIGURE 7-16 Potential energy for an alpha particle confined in the region of a nucleus.

$$V_0 = \frac{2(Z-2)ke^2}{R}.$$

We have seen in Chapter 6 that the size of the nucleus (6.63) is measured to be about

$$R \approx (1.2\,\text{fm})(238)^{1/3} \approx 7.4\,\text{fm}.$$

Therefore, the size of the barrier is approximately

$$V_0 \approx \frac{(2)(90)(1.44\,\text{MeV}\cdot\text{fm})}{7.4\,\text{fm}} \approx 35\,\text{MeV}. \quad \blacksquare$$

The size of the barrier (V_0) is several times larger than the kinetic energy of the alpha particle. The barrier provides the stability of the nucleus against alpha decay. Classically the alpha particle could never penetrate the barrier and the nucleus could not decay in this way because it is energetically forbidden.

If the alpha particle could exist far enough away from the nucleus, a few times the nuclear radius, then energy could be conserved with the alpha particle having escaped. This is what occurs in alpha decay. The quantum description of the alpha particle, given by its wave function, shows that it is concentrated in the region of the nucleus, but also extends outside the nucleus. There is a finite probability per unit time of finding the alpha particle outside the nucleus.

The magnitude of the wave function outside the nuclear well has an exponential form:

$$\psi(x) \approx A e^{-\beta x}, \tag{7.93}$$

where

$$\beta = \frac{\sqrt{2m(V_0 - E_k)}}{\hbar}. \tag{7.94}$$

Thus, the order of magnitude of the tunneling probability is

$$P \approx A^2 e^{-2\beta d}, \tag{7.95}$$

where d is the distance from the nucleus where the kinetic energy is equal to the potential energy. The value of d is given by energy conservation,

$$\frac{2(Z-2)ke^2}{d} = E_k. \tag{7.96}$$

or

$$d \approx \frac{2Zke^2}{E_k}. \tag{7.97}$$

The tunneling probability (7.95) is

$$P \approx A^2 e^{-2\beta d}$$
$$\approx A^2 e^{-4Zke^2 \sqrt{2m(V_0 - E_k)}/\hbar E_k}. \tag{7.98}$$

The decay half-life is inversely proportional to the tunneling probability. Taking the logarithm of the tunneling probability (7.98) and using the fact that $V_0 \gg E_k$, we have

$$\ln t_{1/2} \approx C_1 - \frac{C_2 Z}{E_k}. \tag{7.99}$$

where C_1 and C_2 are constants. A small change in alpha particle energy gives a huge change in tunneling probability because the energy appears in the exponential.

Figure 7-17 shows a plot of $t_{1/2}$ versus E_k for several heavy nuclei that decay by alpha particle emission. For each MeV increase in the alpha kinetic energy, the lifetime decreases by about a factor of 10^5. The four radioactive series, uranium, actinium, thorium, and neptunium, are indicated. The nuclei with the shortest lifetimes give the largest energy alpha particles in their decays.

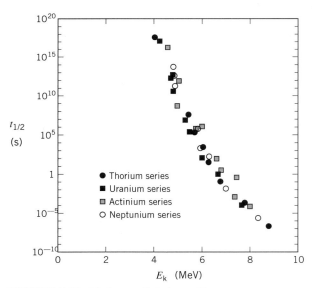

FIGURE 7-17 Alpha particle tunneling.
Several heavy nuclei have alpha particle decays with a wide range of lifetimes.

EXAMPLE 7-8

Estimate the number of times per second that an alpha particle strikes the potential barrier.

SOLUTION:

The typical kinetic energy of the alpha particle is

$$E_k \approx 6\,\text{MeV}.$$

The mass energy of the alpha particle is

$$mc^2 \approx 3730\,\text{MeV}.$$

The speed of the alpha particle is

$$v = c\sqrt{\frac{2E_k}{mc^2}}.$$

Taking $R \approx 5$ fm, the frequency with which the alpha particle strikes the barrier is

$$f \approx \frac{v}{2R} \approx \frac{\left(3\times10^8\,\text{m/s}\right)\sqrt{\dfrac{(2)(6\,\text{MeV})}{3730\,\text{MeV}}}}{10^{-14}\,\text{m}}$$

$$\approx 2\times10^{21}\,\text{s}^{-1}. \qquad \blacksquare$$

Scanning Tunneling Microscope

A scanning tunneling microscope (STM) consists of a sharp probe connected to a sensitive electrical circuit. The probe is scanned along the surface of the object to be studied. Electrons can tunnel from the object to the probe, and one obtains information about the location of the atoms from which the electrons tunneled. The resolution is given roughly by the tunneling distance (d),

$$d \approx \frac{1}{\beta} = \frac{\hbar}{\sqrt{2m(V_0 - E)}}. \qquad (7.100)$$

The binding energy of the electrons is given by the work function of the material so that $V_0 - E$ is a few electron-volts. The order of magnitude of d is

$$d \approx \frac{200\,\text{eV}\cdot\text{nm}}{\sqrt{(2)\left(0.5\times10^6\,\text{eV}\right)(4\,\text{eV})}} \approx 0.1\,\text{nm}. \qquad (7.101)$$

Thus, the STM can resolve single atoms (see color plate 2).

7-5 QUANTUM HARMONIC OSCILLATOR

The harmonic oscillator is an important problem in modern physics. (Recall the hypothesis of Planck in the explanation of the thermal radiation spectrum.) We begin our discussion of the quantum harmonic oscillator by examining the classical solution. The potential energy function of an oscillator is of the form

$$V(x) = \frac{1}{2}Kx^2, \qquad (7.102)$$

where K is a constant. This potential corresponds to a linear restoring force, a force that is proportional to the displacement and in the opposite direction:

$$F = -\frac{dV}{dx} = -Kx. \qquad (7.103)$$

The potential (7.102) is shown in Figure 7-18. The differential equation for the motion of the particle from Newton's second law is

$$m\frac{d^2x}{dt^2} = -Kx. \qquad (7.104)$$

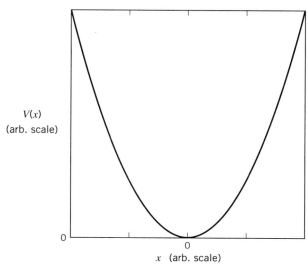

$V(x)$ (arb. scale)

x (arb. scale)

FIGURE 7-18 The harmonic oscillator potential energy function $V(x)$.
The potential energy function may be written $V(x) = Kx^2/2$, where K is a constant. This corresponds to a linear restoring force, $F = -Kx$.

One solution to the classical oscillator is

$$x = C \cos \omega t, \qquad (7.105)$$

where

$$m\omega^2 = K. \qquad (7.106)$$

Therefore, we may write the potential (7.102) in terms of the angular frequency with the result

$$V(x) = \frac{1}{2} m\omega^2 x^2. \qquad (7.107)$$

The Quantum Ground and First Excited States

Now consider the quantum solution. The Schrödinger equation with the potential (7.107) is

$$-\frac{\hbar^2}{2m} \frac{d^2\psi}{dx^2} + \frac{1}{2} m\omega^2 x^2 \psi(x) = E\psi(x), \qquad (7.108)$$

or

$$\frac{d^2\psi}{dx^2} = \frac{m^2\omega^2 x^2}{\hbar^2} \psi(x) - \frac{2mE}{\hbar^2} \psi(x). \qquad (7.109)$$

The wave function $\psi(x)$ must satisfy the boundary conditions $\psi(x) \rightarrow 0$ for $x \rightarrow +\infty$ and $x \rightarrow -\infty$. The solution must also be a function that gives itself back and x^2 times itself when we take the second derivative,

$$\frac{d^2\psi}{dx^2} = A\psi(x) + Bx^2 \psi(x), \qquad (7.110)$$

where A and B are constants. The simplest function that satisfies these conditions is the Gaussian. The harmonic oscillator ground state wave function is of the form

$$\psi_1(x) = Ce^{-\alpha x^2}. \qquad (7.111)$$

The first derivative of the wave function is

$$\frac{d\psi_1}{dx} = -2\alpha x \psi_1, \qquad (7.112)$$

and the second derivative is

$$\frac{d^2\psi_1}{dx^2} = 4\alpha^2 x^2 \psi_1 - 2\alpha \psi_1. \qquad (7.113)$$

Equating the x^2 terms in the expressions for $d^2\psi_1/dx^2$ (7.109 and 7.113), we have

$$4\alpha^2 = \frac{m^2\omega^2 x^2}{\hbar^2}, \qquad (7.114)$$

or

$$\alpha = \frac{m\omega}{2\hbar}. \qquad (7.115)$$

Equating the constant terms in the expressions for $d^2\psi_1/dx^2$ (7.109 and 7.113), we have

$$-2\alpha = -\frac{2mE_1}{\hbar^2}. \qquad (7.116)$$

Therefore, the ground state energy of the quantum oscillator is

$$E_1 = \frac{\alpha\hbar^2}{m} = \frac{\hbar\omega}{2}. \qquad (7.117)$$

Note that the ground state energy is not equal to zero.

EXAMPLE 7-9

Calculate the minimum energy of a pendulum that has a period of 1 second.

SOLUTION:

Classically, the minimum energy is zero. In the quantum solution, the minimum energy is not zero. The ground state energy is

$$E_1 = \frac{\hbar\omega}{2} = \frac{hf}{2}.$$

The frequency f is equal to the inverse of the period T:

$$E_1 = \frac{hf}{2} = \frac{h}{2T} = \frac{6.6 \times 10^{-34} \text{ J} \cdot \text{s}}{2 \text{ s}} = 3.3 \times 10^{-34} \text{ J}.$$

This is a negligible energy because the frequency is so small.   ■

EXAMPLE 7-10

Use the uncertainty principle to calculate the minimum energy of a quantum oscillator.

SOLUTION:

Let Δx be the uncertainty in x and Δp be the uncertainty in p. The energy of the oscillator must be at least as large as

$$E = \frac{(\Delta p)^2}{2m} + \frac{1}{2}m\omega^2 (\Delta x)^2.$$

From the uncertainty principle, we have

$$\Delta p \Delta x = \frac{\hbar}{2}.$$

The energy of the oscillator as a function of Δx is

$$E = \frac{\hbar^2}{8m(\Delta x)^2} + \frac{1}{2}m\omega^2 (\Delta x)^2.$$

We want to find the value of Δx where E is smallest. We set the derivative of E with respect to Δx equal to zero to find the minimum:

$$-\frac{\hbar^2}{4m(\Delta x)^3} + m\omega^2 (\Delta x) = 0.$$

Solving for Δx, we get

$$\Delta x = \sqrt{\frac{\hbar}{2m\omega}}.$$

The value of E for $\Delta x = (\hbar/2m\omega)^{1/2}$ is

$$E = \frac{\hbar^2}{8m(\Delta x)^2} + \frac{1}{2}m\omega^2 (\Delta x)^2 = \frac{\hbar\omega}{4} + \frac{\hbar\omega}{4} = \frac{\hbar\omega}{2}.$$

The minimum value of energy from the uncertainty principle happens in this case to be the actual minimum energy. The absolute limit, $\Delta p \Delta x = \hbar/2$, corresponds to a Gaussian wave function (see Chapter 5). This is the case for the ground state of a quantum harmonic oscillator. ∎

EXAMPLE 7-11

For the quantum oscillator in the ground state, calculate the probability that the particle is found beyond the maximum distance allowed by the classical solution.

SOLUTION:

The maximum displacement (x_m) in the classical solution is given when the energy of the particle is all potential energy and the kinetic energy is zero. This condition is

$$\frac{\hbar\omega}{2} = \frac{1}{2}m\omega^2 x_m^2,$$

or

$$x_m = \sqrt{\frac{\hbar}{m\omega}}.$$

The ground state wave function is

$$\psi(x) = Ce^{-m\omega x^2/2\hbar}.$$

The probability that $|x|$ is greater than x_m is given by the expression

$$P = \int_{-\infty}^{-x_m} dx |\psi(x)|^2 + \int_{x_m}^{+\infty} dx |\psi(x)|^2$$

$$= 2\int_{x_m}^{+\infty} dx |\psi(x)|^2.$$

The constant C is determined by the normalization condition

$$\int_{-\infty}^{+\infty} dx |\psi(x)|^2 = 1.$$

The wave function squared is a Gaussian with a standard deviation of

$$\sigma = \sqrt{\frac{\hbar}{2m\omega}} = \frac{x_m}{\sqrt{2}}.$$

Therefore the probability that we seek is equal to the fraction of the area under the Gaussian that is more than $2^{1/2}$ standard deviations from zero. (We know that the probability of being greater than one sigma away from zero is about 32% and the probability of being greater than two sigma from zero is about 5%). We may determine this number numerically. The answer is

$$P \approx 0.157. \qquad ∎$$

The first excited state solution is of the form

$$\psi_2(x) = Cxe^{-\alpha x^2}. \qquad (7.118)$$

The first derivative of the wave function is

$$\frac{d\psi_2}{dx} = Ce^{-\alpha x^2} - 2\alpha x^2 Ce^{-\alpha x^2}. \qquad (7.119)$$

The second derivative of the wave function is

$$\frac{d^2\psi_2}{dx^2} = -2\alpha xCe^{-\alpha x^2} - 4\alpha xCe^{-\alpha x^2}$$

$$-2\alpha x^2 (-2\alpha x)Ce^{-\alpha x^2}$$

$$= (-6\alpha + 4\alpha^2 x^2)\psi_2. \qquad (7.120)$$

The equation for α from the x^2 term is the same for the ground state, and the energy (E_2) is now three times larger because the constant term from the second derivative is three times larger. Therefore, the energy of the first excited state is

$$E_2 = \frac{3\hbar\omega}{2}. \qquad (7.121)$$

The energy difference (ΔE) between the first excited state and the ground state is

$$\Delta E = E_2 - E_1 = \frac{3\hbar\omega}{2} - \frac{\hbar\omega}{2} = \hbar\omega. \qquad (7.122)$$

* *Challenging*

The General Quantum Solution

The general solution of the harmonic oscillator is of the form

$$\psi_n = f_n(x)\, e^{-\alpha x^2}. \qquad (7.123)$$

We have found the first two solutions, $f_1 = C_1$ and $f_2 = C_2 x$. If we make the change of variables

$$y \equiv x\sqrt{\frac{m\omega}{2\hbar}} = x\sqrt{\alpha}, \qquad (7.124)$$

the harmonic oscillator wave function (7.123) becomes

$$\psi_n = f_n(y)\, e^{-y^2}. \qquad (7.125)$$

The Schrödinger equation becomes

$$\frac{d^2\psi}{dy^2} = 4y^2\psi - \frac{4E}{\hbar\omega}\psi. \qquad (7.126)$$

We now substitute our solution $\psi = f e^{-y^2}$ into the Schrödinger equation to get a differential equation for $f(y)$. The derivatives are

$$\frac{d\psi}{dy} = \frac{df}{dy} e^{-y^2} - 2yf e^{-y^2}, \qquad (7.127)$$

and

$$\frac{d^2\psi}{dy^2}$$

$$= \left(\frac{d^2 f}{dy^2} - 2y\frac{df}{dy} + 4y^2 f - 2f - 2y\frac{df}{dy} \right) e^{-y^2}. \qquad (7.128)$$

The Schrödinger equation gives a differential equation for $f(y)$:

$$\frac{d^2 f}{dy^2} - 4y\frac{df}{dy} + \left(\frac{4E}{\hbar\omega} - 2 \right) f = 0. \qquad (7.129)$$

This differential equation may be solved by the power series technique. In a power series solution, we expand $f(y)$ in powers of y and then determine the coefficients from the differential equation. Let the power series expansion for $f(y)$ be

$$f(y) = \sum_{k=0}^{\infty} C_k y^k. \qquad (7.130)$$

When we substitute this power series into the differential equation (7.129), we get

$$\sum_{k=0}^{\infty} y^k \left[C_{k+2}(k+2)(k+1) - 4kC_k + \left(\frac{4E}{\hbar\omega} - 2 \right) C_k \right]$$
$$= 0. \qquad (7.131)$$

The term in brackets must be zero for all values of the dummy summation variable k. This gives the following equations relating the coefficients: for $k = 0$

$$2C_2 + \left(\frac{4E}{\hbar\omega} - 2 \right) C_0 = 0; \qquad (7.132)$$

for $k = 1$,

$$6C_3 - 4C_1 + \left(\frac{4E}{\hbar\omega} - 2 \right) C_1 = 0; \qquad (7.133)$$

for $k = 2$,

$$12C_4 - 8C_2 + \left(\frac{4E}{\hbar\omega} - 2 \right) C_2 = 0; \qquad (7.134)$$

for $k = 3$,

$$20C_5 - 12C_3 + \left(\frac{4E}{\hbar\omega} - 2 \right) C_3 = 0; \qquad (7.135)$$

and so on. The solutions are: for $k = 0$, $C_0 = 1$, $C_1 = C_2 = C_3 \ldots = 0$, and $E = \hbar\omega/2$; for $k = 1$, $C_0 = 0$, $C_1 = 1$, $C_2 = C_3 = C_4 \ldots = 0$, and $E = 3\hbar\omega/2$; for $k = 2$, $C_0 = -1/4$, $C_1 = 0$, $C_2 = 1$, $C_3 = C_4 = C_5 \ldots = 0$, and $E = 5\hbar\omega/2$; for $k = 3$, $C_0 = 0$, $C_1 = -3/4$, $C_2 = 0$, $C_3 = 1$, $C_4 = C_5 = C_6 \ldots = 0$, and $E = 7\hbar\omega/2$; and so on.

The energy levels are given by

$$E_n = \left(n - \frac{1}{2} \right) \hbar\omega, \qquad (7.136)$$

where the quantum number n is a positive integer (1, 2, 3, ...). The energy levels are all equally spaced by $\hbar\omega$. The corresponding wave functions are

$$\psi_1 = Ce^{-y^2}, \qquad (7.137)$$

$$\psi_2 = Cye^{-y^2}, \qquad (7.138)$$

$$\psi_3 = C\left(y^2 - \frac{1}{4} \right)e^{-y^2}, \qquad (7.139)$$

$$\psi_4 = C\left(y^3 - \frac{3y}{4} \right)e^{-y^2}. \qquad (7.140)$$

and so on.   ✳

Figures 7-19, 7-20, 7-21 and 7-22 show the first four wave functions ($n = 1, 2, 3, 4$) and their squares for the quantum harmonic oscillator. Figure 7-23 shows the energy levels for the quantum harmonic oscillator.

The correspondence principle states that the quantum solution approaches the classical solution for large values of the quantum number n. For large n, the wave function squared oscillates many times per unit distance. Figure 7-24 shows the wave function of a quantum harmonic oscillator for $n = 61$ together with the classical probability. (Compare to Figure 7-7.)

7-6 SCHRÖDINGER EQUATION IN THREE DIMENSIONS

In the one-dimensional Schrödinger equation the term $(-\hbar^2/2m)d^2\psi/dx^2$ is identified with the kinetic energy of the particle, $p_x^2/2m$. In three dimensions the kinetic energy is $(p_x^2 + p_y^2 + p_z^2)/2m$ so that the Schrödinger equation contains the additional terms $(-\hbar^2/2m)\partial^2\psi/\partial y^2$ and $(-\hbar^2/2m)\partial^2\psi/\partial z^2$. The Schrödinger equation in three dimensions is

$$-\frac{\hbar^2}{2m}\left(\frac{\partial^2\psi}{\partial x^2} + \frac{\partial^2\psi}{\partial y^2} + \frac{\partial^2\psi}{\partial z^2} \right) + V(x,y,z)\psi(x,y,z)$$

$$= E\psi(x,y,z), \qquad (7.141)$$

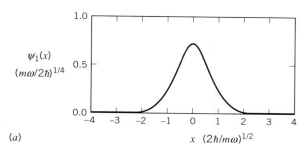

(a)

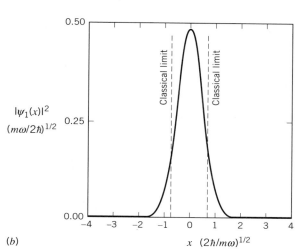

(b)

FIGURE 7-19 Quantum harmonic oscillator.
(a) The ground state ($n = 1$) wave function for the quantum harmonic oscillator. (b) The square of the $n = 1$ wave function.

where ψ and V depend on the coordinates x, y, and z. The square of the wave function gives the probability per volume for finding the particle. The normalization condition is

$$\int_{-\infty}^{+\infty}dx\int_{-\infty}^{+\infty}dy\int_{-\infty}^{+\infty}dz\,|\psi(x,y,z)|^2 = 1. \qquad (7.142)$$

There are boundary conditions to be satisfied for each coordinate x, y, and z. In order for the integral (7.142) to exist, $\psi(x,y,z)$ must go to zero for $x \to -\infty$, $x \to +\infty$, $y \to -\infty$, $y \to +\infty$, $z \to -\infty$, and $z \to +\infty$.

Separation of Variables

The Schrödinger equation may be solved for certain potentials by the technique of *separation of variables*. We shall illustrate the separation technique by solving the problem of a particle in a box, in three dimensions.

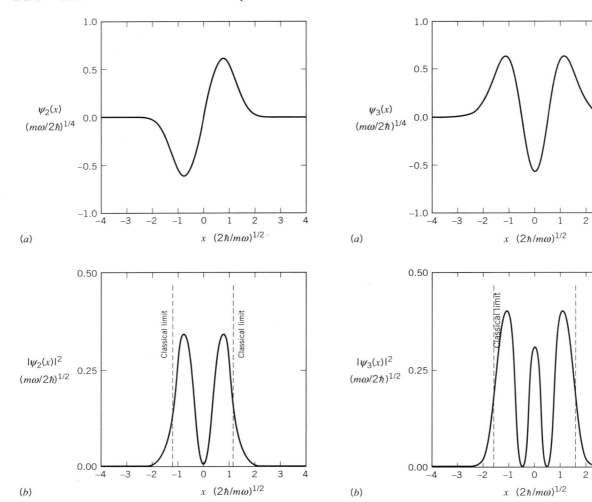

FIGURE 7-20 Quantum harmonic oscillator.
(a) The first excited state ($n = 2$) wave function for the quantum harmonic oscillator. (b) The square of the $n = 2$ wave function.

FIGURE 7-21 Quantum harmonic oscillator.
(a) The second excited state ($n = 3$) wave function for the quantum harmonic oscillator. (b) The square of the $n = 3$ wave function.

Consider a particle that is absolutely confined to be in the region $0 < x < L_1$, $0 < y < L_2$, and $0 < z < L_3$. The first step in the separation of variables is to write the wave function $\psi(x,y,z)$ as the product of three functions:

$$\psi(x, y, z) = F(x)G(y)H(z), \qquad (7.143)$$

where F depends only on x, G depends only on y, and H depends only on z. If we substitute $\psi(x,y,z)$ (7.143) into the Schrödinger equation (7.141), we get

or

$$-\frac{\hbar^2}{2m}\left[G(y)H(z)\frac{d^2F}{dx^2} + F(x)H(z)\frac{d^2G}{dy^2} \right.$$

$$\left. + F(x)G(y)\frac{d^2H}{dz^2} \right]$$

$$= E\,F(x)G(y)H(z), \qquad (7.144)$$

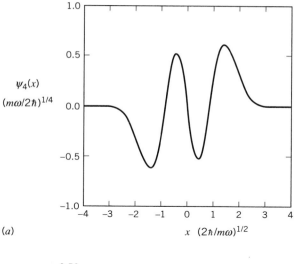

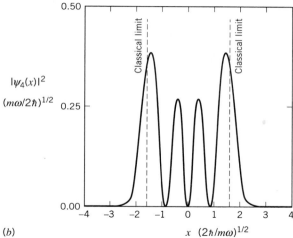

(a)

(b)

FIGURE 7-22 Quantum harmonic oscillator.
(a) The third excited state ($n = 4$) wave function for the quantum harmonic oscillator. (b) The square of the $n = 4$ wave function.

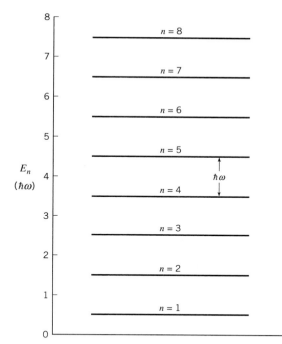

FIGURE 7-23 Energy levels for the quantum harmonic oscillator.

$$-\frac{\hbar^2}{2m}\left[\frac{1}{F}\frac{d^2F}{dx^2}+\frac{1}{G}\frac{d^2G}{dy^2}+\frac{1}{H}\frac{d^2H}{dz^2}\right]=E. \quad (7.145)$$

For a given solution, the energy is fixed. All of the x dependence is in the first term, all of the y dependence is in the second term, and all of the z dependence is in the last term. Since the equation holds for all values of x, y, and z, each of the three terms must be constant. Accordingly, we write

$$-\frac{\hbar^2}{2m}\frac{1}{F}\frac{d^2F}{dx^2}=C_x,$$

$$-\frac{\hbar^2}{2m}\frac{1}{G}\frac{d^2G}{dy^2}=C_y, \quad \begin{matrix}(7.146)\\(7.147)\end{matrix}$$

and

$$-\frac{\hbar^2}{2m}\frac{1}{H}\frac{d^2H}{dz^2}=C_z. \quad (7.148)$$

Each of the functions $F(x)$, $G(y)$, and $H(z)$ has a sinusoidal solution, analogous to the one-dimensional case. Let us examine the solution for $F(x)$. The boundary conditions are that $F(0) = 0$ and $F(L_1) = 0$. The solution is of the form

$$F(x)=C\sin\left(\frac{n_1\pi x}{L_1}\right), \quad (7.149)$$

where C is a constant and n_1 is a positive integer. The solutions for $G(y)$ and $H(z)$ are similar. The product of $F(x)$, $G(y)$, and $H(z)$ gives the total wave function $\psi(x,y,z)$:

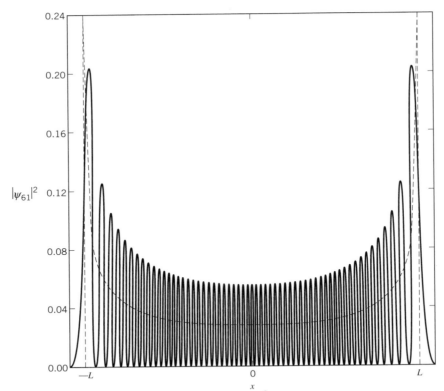

FIGURE 7-24 Probability density $|\psi|^2$ for a quantum harmonic oscillator with $n = 61$ compared with the classical probability distribution.
After K. Ziock, *Quantum Mechanics*, Wiley (1969).

$$\psi(x,y,z)$$

$$= A\sin\left(\frac{n_1\pi x}{L_1}\right)\sin\left(\frac{n_2\pi y}{L_2}\right)\sin\left(\frac{n_3\pi z}{L_3}\right), \quad (7.150)$$

where n_1, n_2, and n_3 are positive integers (1, 2, 3, …). The boundary conditions on each of the three coordinates x, y, and z has led to a quantum number. The energies are given by

$$E = \frac{\hbar^2\pi^2}{2m}\left[\left(\frac{n_1}{L_1}\right)^2 + \left(\frac{n_2}{L_2}\right)^2 + \left(\frac{n_3}{L_3}\right)^2\right]. \quad (7.151)$$

We assign subscripts (n_1,n_2,n_3) to both the wave function and its corresponding energy. The ground state energy is

$$E_{1,1,1} = \frac{\hbar^2\pi^2}{2m}\left[\left(\frac{1}{L_1}\right)^2 + \left(\frac{1}{L_2}\right)^2 + \left(\frac{1}{L_3}\right)^2\right]. \quad (7.152)$$

If the box is a cube, then $L_1 = L_2 = L_3 = L$ and the ground state energy is

$$E_{1,1,1} = \frac{3\hbar^2\pi^2}{2mL^2}. \quad (7.153)$$

For a cube, the first excited state is three-fold degenerate:

$$E_{2,1,1} = E_{1,2,1} = E_{1,1,2} = \frac{3\hbar^2\pi^2}{mL^2}. \quad (7.154)$$

The degeneracy is broken if the L_1, L_2, and L_3 have different lengths. The energy levels of a particle in a box in three dimensions are indicated in Figure 7-25.

* *Challenging*

7-7 TIME-DEPENDENT SCHRÖDINGER EQUATION

We have drawn two parallels between the wave equation for a nonrelativistic particle and the wave equation for a

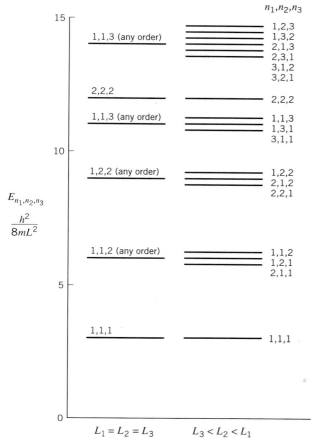

n_1, n_2, n_3

1,2,3
1,3,2
2,1,3
2,3,1
3,1,2
3,2,1

2,2,2

1,1,3
1,3,1
3,1,1

1,2,2
2,1,2
2,2,1

1,1,2
1,2,1
2,1,1

1,1,1

E_{n_1,n_2,n_3}

$\dfrac{h^2}{8mL^2}$

$L_1 = L_2 = L_3$ $L_3 < L_2 < L_1$

FIGURE 7-25 Allowed energy levels for a particle in a box in three dimensions.
If the box is a cube, $L_1 = L_2 = L_3 = L$, then the energy levels are degenerate for certain values of the quantum numbers n_1, n_2 and n_3. The degeneracy is broken if the sides of the box have different lengths.

photon: in each case (1) the square of the wave function represents the probability of interacting with the particle and (2) the wave equation gives a relationship between the energy and momentum of the particle. There is a major difference, however, between the electromagnetic wave equation and the Schrödinger equation. The electromagnetic wave equation

$$\frac{\partial^2 F}{\partial x^2} + \frac{\partial^2 F}{\partial y^2} + \frac{\partial^2 F}{\partial z^2} = \frac{1}{c^2}\frac{\partial^2 F}{\partial t^2}, \quad (7.155)$$

represents a traveling wave whose amplitude varies with time:

$$F(\mathbf{x},t) = Ce^{i(\mathbf{k}\cdot\mathbf{x}-\omega t)}. \quad (7.156)$$

In contrast the Schrödinger equation (7.55) represents a standing wave, that is, the solution $\psi(x)$ is independent of time.

The time dependence of the wave function is given by the *time-dependent* Schrödinger equation. For a free particle the time-dependent Schrödinger equation is

$$-\frac{\hbar^2}{2m}\left(\frac{\partial^2 \Psi}{\partial x^2} + \frac{\partial^2 \Psi}{\partial y^2} + \frac{\partial^2 \Psi}{\partial z^2}\right) = i\hbar\frac{\partial \Psi}{\partial t}, \quad (7.157)$$

where $\Psi(\mathbf{x},t)$ is the time-dependent wave function. It is not possible to derive this equation. If the particle has a fixed momentum ($\mathbf{p} = \hbar\mathbf{k}$), then it can be represented by the traveling wave

$$\Psi(\mathbf{x},t) = Ce^{i(\mathbf{k}\cdot\mathbf{x}-\omega t)-imc^2 t/\hbar}$$

$$= \psi(\mathbf{x})e^{-i\omega t - imc^2 t/\hbar}, \quad (7.158)$$

where

$$\psi(\mathbf{x}) = Ce^{i\mathbf{k}\cdot\mathbf{x}}. \quad (7.159)$$

Substituting the wave function (7.158) into the Schrödinger equation (7.157) gives

$$\frac{\hbar^2 k^2}{2m} + mc^2 = \hbar\omega, \quad (7.160)$$

which is a statement of energy conservation for a non-relativistic particle if we identify the particle energy and momentum with the deBroglie relationships ($p = \hbar k$ and $E = \hbar\omega$). By convention the mass factor is usually left out of the wave function (7.158) so that energy is measured relative to the mass energy and the deBroglie condition becomes ($\hbar\omega = \hbar^2 k^2/2m$).

For a particle that is not free but is moving in a potential $V(x,t)$, the time-dependent Schrödinger equation becomes

$$-\frac{\hbar^2}{2m}\left(\frac{\partial^2 \Psi}{\partial x^2} + \frac{\partial^2 \Psi}{\partial y^2} + \frac{\partial^2 \Psi}{\partial z^2}\right) + V(\mathbf{x},t)\Psi(\mathbf{x},t)$$

$$= i\hbar\frac{\partial \Psi}{\partial t}. \quad (7.161)$$

If a particle is represented by a wavepacket, a distribution of wavenumbers is necessary for the description of the particle in accordance with the uncertainty principle. Each wavenumber component will propagate at its own speed (ω/k) and the wavepacket will spread out in time.

If the potential V does not depend on time, then we may solve the time-dependent Schrödinger equation (7.161) by making a separation of variables and writing

$$\Psi(\mathbf{x}, t) = \psi(\mathbf{x})e^{-iEt/\hbar}, \qquad (7.162)$$

where E is the particle energy (kinetic plus potential). Substituting the wave function (7.162) into the Schrödinger equation (7.161) gives us the time-independent Schrödinger equation:

$$-\frac{\hbar^2}{2m}\left(\frac{\partial^2\psi}{\partial x^2} + \frac{\partial^2\psi}{\partial y^2} + \frac{\partial^2\psi}{\partial z^2}\right) + V(\mathbf{x})\psi(\mathbf{x})$$
$$= E\psi(\mathbf{x}). \qquad (7.163)$$

*

CHAPTER 7: PHYSICS SUMMARY

- For a particle with a mass m moving in a potential $V(x)$, the time-independent Schrödinger equation gives the wave function for the particle with a definite energy E:

$$-\frac{\hbar^2}{2m}\frac{d^2\psi}{dx^2} + V(x)\psi(x) = E\psi(x).$$

- The absolute square of $\psi(x)$ is the probability of finding the particle per unit of x as a function of x:

$$\frac{dP}{dx} = |\psi(x)|^2.$$

The boundary condition on $\psi(x)$ is that $\psi(x) \to 0$ as $x \to +\infty$ and $x \to -\infty$, so that

$$\int_{-\infty}^{+\infty} dx\,|\psi(x)|^2 = 1.$$

$\psi(x)$ is continuous, and $d\psi/dx$ is continuous, unless $V(x)$ is infinite.

- If the potential energy function $V(x)$ is symmetric, then the wave functions possess a definite symmetry. The ground state is always an even function.
- The energy levels of a particle in a box (one dimension) are given by

$$E_n = \frac{\hbar^2 n^2 \pi^2}{2mL^2} = \frac{n^2 h^2}{8mL^2},$$

where n is a positive integer (1, 2, 3 ...).

- The energy levels of a quantum harmonic oscillator are given by

$$E_n = \left(n - \frac{1}{2}\right)\hbar\omega,$$

where n is a positive integer (1, 2, 3 ...).

- The Schrödinger equation in three dimensions is

$$-\frac{\hbar^2}{2m}\left(\frac{\partial^2\psi}{\partial x^2} + \frac{\partial^2\psi}{\partial y^2} + \frac{\partial^2\psi}{\partial z^2}\right) + V(x,y,z)\psi(x,y,z)$$
$$= E\psi(x,y,z).$$

The solution is obtained by the technique of separation of variables:

$$\psi(x,y,z) = F(x)G(y)H(z).$$

The boundary condition on each of the components of the wave function $F(x)$, $G(y)$, and $H(z)$ produce three quantum numbers n_1, n_2, and n_3. The energy levels of a particle in a box of size (L_1, L_2, L_3) are given by

$$E_{n_1, n_2, n_3} = \frac{\hbar^2 \pi^2}{2m}\left[\left(\frac{n_1}{L_1}\right)^2 + \left(\frac{n_2}{L_2}\right)^2 + \left(\frac{n_3}{L_3}\right)^2\right].$$

REFERENCES AND SUGGESTIONS FOR FURTHER READING

G. Binnig and H. Rohrer, "Scanning Tunneling Microscopy—from Birth to Adolescence," *Rev. Mod. Phys.* **59**, 615 (1987).

G. Binnig and H. Rohrer, "The Scanning Tunneling Microscope," *Sci. Am.* **253**, No. 2. 50 (1985).

W. E. Boyce and R. C. DiPrima, *Elementary Differential Equations and Boundary Value Problems*, Wiley (1986).

E. Ruska, "The Development of the Electron Microscope and of Electron Microscopy," *Rev. Mod. Phys.* **59**, 627 (1987).

D. Saxon, *Elementary Quantum Mechanics*, Holden Day (1968).

E. Schrödinger, *Four Lectures on Wave Mechanics*, Blackie and Sons (1928).

E. Schrödinger, "What is Matter?" *Sci. Amer.*, **189**, No. 3, 52 (1953).

H. Wickramasinghe, "Scanned-Probe Microscopes," *Sci. Am.* **261**, No. 4, 98 (1989).

QUESTIONS AND PROBLEMS

Free particle wave equation

1. A free electron has a kinetic energy of 100 eV. What is the wave function of the electron?

Particle in a box

2. (a) For a particle in a box of size L, under what circumstance may the particle be found near the edge of the box with significant probability? (b) What is the maximum probability that the particle can have for being within Δx of one of the walls of the box?

3. In the Bohr model of the hydrogen atom, the average kinetic energy of the electron in the first excited state is four times smaller than in the ground state. For a particle in a box, the kinetic energy of the first excited state is four times larger than the ground state. Explain.

4. A particle is confined to be in the region $-L/2 < x < L/2$. (a) What is the probability that the particle is found in the region $0 < x < L/2$? Does this probability depend on the quantum number n? (b) If the particle is in the ground state, calculate the probability that the particle is found in the central half of the box, $-L/4 < x < L/4$. How does this probability change if the particle is in a higher energy state?

5. A particle is confined to be in the region $-L/2 < x < L/2$. (a) Calculate the average value of x^2 as a function of the quantum number n. Show that for very large values of n, the root-mean-square value of x approaches the classical value, $L/(12)^{1/2}$.

6. A proton is confined in a box of size 2 fm. Calculate the energies of the ground and first excited states.

7. For a particle with principle quantum n in a box of size L, what is the average value of momentum? What is the average value of momentum squared?

8. Consider an electron in a box. If the energy of the first excited state is 10 eV larger than the energy of the ground state, what is the size of the box?

9. A particle is confined to be in the region $-L/2 < x < L/2$. The particle is in the ground state. (a) What are the average position and momentum of the particle? (b) What is the root-mean-square deviation (Δx) of the position of the particle from the average position? (c) What is the root-mean-square deviation (Δp) of the momentum of the particle from the average momentum? (d) Show that the product $\Delta x \Delta p$ is in agreement with the uncertainty principle.

The finite square-well potential

10. For a particle confined in a finite square-well with many bound states, make a qualitative sketch of the wave function of the $n = 5$ state.

11. For a particle in a finite square-well, how does wavelength of a particle in the ground state depend on the size of the potential V_0?

12. Consider a finite square-well for which the size of the potential is $V_0 = \varepsilon h^2/8mL^2$, where $\varepsilon < 1$. Show that one and only one bound state exists. What is the approximate value of the energy of the bound state?

13. An electron is confined by a finite square-well of height 10 eV. If the ground state energy is 1 eV, what is the width of the well?

Barrier penetration

14. If the size of the nucleus was smaller, how would the alpha decay probability change?

15. An alpha particle from the decay of a heavy nucleus has a kinetic energy of 4 MeV. Use Figure 7-17 to make a rough estimate of the lifetime of the nucleus.

16. An electron in a copper wire behaves like a free particle inside the wire. Estimate the distance that the electron wave function extends beyond the surface of the wire.

17. An electron with a kinetic energy of 3 eV is incident on a square barrier of height 4 eV and width 0.1 nm. Estimate the probability that the electron penetrates the barrier.

18. Consider a marble of mass m in a shoe box. The kinetic energy of the marble is less than the barrier presented by the wall of the box, *i.e.*,

$$\frac{1}{2}mv^2 \ll mgz,$$

where v is the speed of the marble and z is the height of the box. Make a rough estimate of the probability that the marble can be found outside the box at any given time. Are you convinced that the marble cannot tunnel outside the box, in agreement with classical physics?

Quantum harmonic oscillator

19. For the quantum harmonic oscillator, how does the average value of the coordinate x depend on the energy? How does the root-mean-square value of x depend on the energy?

20. For the quantum harmonic oscillator in the ground state, calculate the average value of x and the root-mean-square value of x.

21. The ground state wave function of a quantum harmonic oscillator is a Gaussian. Show that the standard deviation of this Gaussian corresponds to the classical turning point, the position where the kinetic energy is zero in the classical solution.

22. Show by direct substitution into the Schrödinger equation that the wave function

$$\psi(x) = C\left(\alpha x^2 - \frac{1}{4}\right)e^{-\alpha x^2},$$

where $\alpha = m\omega/2\hbar$ and C is a normalization constant, is a solution of the harmonic oscillator. Calculate the corresponding energy.

23. Show by direct substitution into the Schrödinger equation that the wave function

$$\psi(x) = C\left(\alpha^{3/2} x^3 - \frac{3}{4}\sqrt{\alpha}\, x\right)e^{-\alpha x^2},$$

where $\alpha = m\omega/2\hbar$ and C is a normalization constant, is a solution of the harmonic oscillator. Calculate the corresponding energy.

24. Consider a particle of mass m subject to the potential, $V(x) = m\omega^2 x^2/2$ (the quantum harmonic oscillator). The particle is in the ground state. (a) What can be said about the uncertainty in the position (Δx) of the particle? (b) What can be said about the uncertainty in the momentum (Δp) of the particle? (c) If the particle is an electron and $\hbar\omega = 10$ eV, what are the numerical values of Δx and Δp?

25. Consider a quantum harmonic oscillator with the particle in the first excited state. (a) Calculate the classical turning point. (b) Make an estimate of the probability that the particle is found in the classically forbidden region. Compare your answer to that found for the ground state (see Example 7-11).

Schrödinger equation in three dimensions

26. (a) Calculate the minimum momentum of an electron that is confined to be in a cube of volume 10^{-29} m³. (b) Calculate the ground state energy.

27. A particle of mass m is confined inside a box. Two sides of the box have equal dimensions (L), while the third side is half as large ($L/2$). Calculate the energies, quantum numbers, and degeneracy of the states that have the five lowest unique energies.

28. Consider an electron in a three-dimensional cubical box of size L. What value of L corresponds to a ground state energy of 13.6 eV?

Time-dependent Schrödinger equation

29. (a) Show that the functions $A\cos(kx-\omega t)$ and $B\sin(kx-\omega t)$, where A and B are constants are not solutions of the time-dependent Schrödinger equation. (b) Show that the function $Ce^{i(kx-\omega t)}$ where C is a constant is a solution.

30. Show that $\psi(x)e^{-iEt/\hbar}$ is a solution of the time-dependent Schrödinger equation provided that $\psi(x)$ satisfies the time-independent Schrödinger equation.

Additional problems

31. Consider a nonrelativistic particle of mass m that is confined by the following potential: $V(x) = \infty$ for $x < 0$ or $x > L$, and $V(x) = -V_0$ for $0 \leq x \leq L$. (a) Determine the allowed energies of the particle. (b) If the particle is an electron, $L = 0.1$ nm and $V_0 = 10$ eV, how many states have a negative energy?

32. Consider a particle in a box in one dimension where boundaries of the box are chosen to be at $x = 0$ and $x = L$. Show that the wave functions are

$$\psi_n(x) = \sqrt{\frac{2}{L}}\sin\left(\frac{n\pi x}{L}\right),$$

where n is a positive integer. Calculate the energy levels.

33. Estimate the ground state energy of an electron if the electron were to be confined inside a nucleus.

*34. Consider two wave functions for a particle in a box, ψ_n and ψ_m, where n and m are the quantum numbers of the states. Show that

$$\int_{-\infty}^{+\infty} dx\, \psi_n^* \psi_m = 0,$$

if $n \neq m$.

*35. Approximate the strong interaction potential of a neutron in a nucleus as a one-dimensional finite square-well of size V_0. Use your knowledge of the size of the nucleus and the measured binding energy of the deuteron (2.2 MeV) to estimate the magnitude of V_0 for the deuteron. (The deuteron the nucleus of the isotope of hydrogen with $A=2$.)

It is very important that this problem should receive further experimental and theoretical attention. When an accuracy of comparison of 0.1 megacycle per second has been reached, it will mean that the energy separations of the 2s and 2p states of hydrogen agree with theory to a precision of a few parts in 10^9 of their binding energy or that the exponent in Coulomb's law of force is 2 with comparable accuracy.

Willis E. Lamb, Jr.

In this chapter we examine the Schrödinger equation for an electron in the hydrogen atom. The electron is bound to the proton by the Coulomb potential. We shall refer to the position of the electron using a coordinate system whose origin is at the center of the proton. The Coulomb potential is

$$V(x,y,z) = -\frac{ke^2}{\sqrt{x^2 + y^2 + z^2}}, \qquad (8.1)$$

and the Schrödinger equation is

$$-\frac{\hbar^2}{2m}\left(\frac{\partial^2 \psi}{\partial x^2} + \frac{\partial^2 \psi}{\partial y^2} + \frac{\partial^2 \psi}{\partial z^2}\right) - \frac{ke^2}{\sqrt{x^2 + y^2 + z^2}}\psi(x,y,z)$$
$$= E\psi(x,y,z). \qquad (8.2)$$

This is the wave equation for the electron in the hydrogen atom. The square of the wave function specifies the electron probability density (probability per volume of finding the electron at any given position). The mass m is the electron mass. Since the mass of the proton is much greater than the mass of the electron, we ignore the motion of the proton. If we wish to account for the proton motion, we can replace the electron mass with the reduced mass (3.111) as discussed in the context of the Bohr model. The solutions of the Schrödinger equation (8.2) give states of fixed electron energy E. By convention we have ignored the mass energy so that E is a negative number.

A shorthand notation is used to write the second derivatives appearing in the Schrödinger equation:

$$\nabla^2\psi = \frac{\partial^2 \psi}{\partial x^2} + \frac{\partial^2 \psi}{\partial y^2} + \frac{\partial^2 \psi}{\partial z^2}. \qquad (8.3)$$

In vector calculus (see Appendix C) the symbol ∇^2 represents the divergence of the gradient (often designated as the *Laplacian*.). With this notation the Schrödinger equation (8.2) takes the compact form

$$-\frac{\hbar^2}{2m}\nabla^2\psi - \frac{ke^2}{\sqrt{x^2 + y^2 + z^2}}\psi = E\psi. \qquad (8.4)$$

In Cartesian coordinates (x,y,z), the Schrödinger equation is not separable, that is, we cannot make the substitution $\psi = F(x)G(y)H(z)$ and rearrange the terms so that each variable is separated. If we change to spherical coordinates (r,θ,ϕ), the Schrödinger equation is separable. In spherical coordinates, the potential energy has a simple form,

$$V(r) = -\frac{ke^2}{r}, \qquad (8.5)$$

(see Figure 8-1) and the Schrödinger equation (8.4) is written

$$-\frac{\hbar^2}{2m}\nabla^2\psi - \frac{ke^2}{r}\psi(r,\theta,\phi) = E\psi(r,\theta,\phi). \qquad (8.6)$$

The expression $\nabla^2\psi(r,\theta,\phi)$ is given by

$$\nabla^2\psi = \frac{1}{r^2 \sin\theta}\left[\sin\theta\frac{\partial}{\partial r}\left(r^2\frac{\partial \psi}{\partial r}\right) + \frac{\partial}{\partial\theta}\left(\sin\theta\frac{\partial \psi}{\partial\theta}\right)\right.$$
$$\left. + \frac{1}{\sin\theta}\frac{\partial^2 \psi}{\partial\phi^2}\right] \qquad (8.7)$$

(see Appendix E). As usual, the boundary conditions on ψ are that $|\psi|^2$ must be integrable. In spherical coordinates this implies that the wave function must vanish at large values of r,

$$\lim_{r\to\infty}\psi = 0. \qquad (8.8)$$

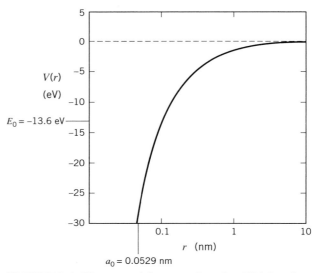

FIGURE 8-1 The potential energy function $V(r)$ for the hydrogen atom.
The potential energy as a function of the electron–proton separation is $V(r) = (-1.44 \text{ eV·nm})/r$. In the Bohr model of the hydrogen atom, the ground state energy is $E_0 = -13.6$ eV and the radius of the ground state orbit is $a_0 = 0.0529$ nm.

8-1 THE GROUND STATE SOLUTION

Since $V(r)$ is spherically symmetric, there are solutions to the Schrödinger equation for the hydrogen atom that are spherically symmetric, that is, solutions for which ψ does not depend on θ or ϕ. For the spherically symmetric solutions we may write the wave function as

$$\psi = f(r). \qquad (8.9)$$

The Schrödinger equation for the hydrogen atom (8.6) becomes

$$-\frac{\hbar^2}{2mr^2}\frac{d}{dr}\left(r^2\frac{df}{dr}\right) - \frac{ke^2}{r}f(r) = Ef(r), \qquad (8.10)$$

or

$$\frac{d^2f}{dr^2} + \frac{2}{r}\frac{df}{dr} + \frac{2mke^2}{r\hbar^2}f + \frac{2mE}{\hbar^2}f = 0. \qquad (8.11)$$

It is instructive to examine the form of the wave function $f(r)$ for large r. At large r the second and third terms of the differential equation (8.11) are small so that

$$\frac{d^2f}{dr^2} + \frac{2mE}{\hbar^2}f \sim 0. \qquad (8.12)$$

Since E is negative, the solution at large r is an exponential

$$f(r) \sim Ce^{-r\sqrt{-2mE}/\hbar}. \qquad (8.13)$$

The exponential is a good candidate solution for $f(r)$ in the Schrödinger equation (8.11) because it goes to zero at $r = \infty$, and it reproduces itself when we take the first and second derivatives. Let us try a solution that is of the form

$$f(r) = Ce^{-r/\delta}, \qquad (8.14)$$

where C and δ are constants, and see if it works. The derivatives of $f(r)$ are

$$\frac{df}{dr} = -\frac{C}{\delta}e^{-r/\delta}, \qquad (8.15)$$

and

$$\frac{d^2f}{dr^2} = \frac{C}{\delta^2}e^{-r/\delta}. \qquad (8.16)$$

The Schrödinger equation (8.11) gives

$$\left(\frac{1}{\delta^2} - \frac{2}{\delta r} + \frac{2mke^2}{\hbar^2 r} + \frac{2mE}{\hbar^2}\right)Ce^{-r/\delta}$$
$$= 0. \qquad (8.17)$$

The solution requires that

$$\left(\frac{1}{r}\right)\left(\frac{2mke^2}{\hbar^2} - \frac{2}{\delta}\right) + \left(\frac{2mE}{\hbar^2} + \frac{1}{\delta^2}\right) = 0, \qquad (8.18)$$

for all values of r. The $1/r$ terms and the constant terms must both be equal to zero. From the $1/r$ terms we get

$$\delta = \frac{\hbar^2}{mke^2}. \qquad (8.19)$$

The constant δ is the Bohr radius (3.90),

$$\delta = a_0 = 5.29 \times 10^{-11} \text{ m}. \qquad (8.20)$$

From the constant terms we get

$$E = -\frac{\hbar^2}{2m\delta^2} = -\frac{m\left(ke^2\right)^2}{2\hbar^2}$$
$$= -\frac{\alpha^2 mc^2}{2} = -13.6 \text{ eV} \quad . \qquad (8.21)$$

We have determined the ground state solution of the hydrogen atom! We may summarize what we have discovered as follows: the function $\psi(r) = Ce^{-r/a_0}$ is one solution of Schrödinger's equation for the hydrogen atom if and only if the energy corresponding to this wave function is equal to $-\alpha^2mc^2/2$.

EXAMPLE 8-1

Determine the normalization constant C for the ground state wave function.

SOLUTION:

The normalization condition is

$$\iiint_V dV |\psi|^2 = \int_0^{2\pi} d\phi \int_{-1}^{1} d\cos\theta \int_0^{\infty} dr\, r^2 |\psi|^2$$
$$= 4\pi \int_0^{\infty} dr\, r^2 |\psi|^2 = 1,$$

or

$$4\pi C^2 \int_0^{\infty} dr\, r^2 e^{-2r/a_0} = 1.$$

We evaluate the integral by parts twice, noting that the integrated part vanishes each time:

$$\int_0^\infty dr\, r^2 e^{-2r/a_0}$$

$$= \int_0^\infty dr\, (2r)\left(\frac{a_0}{2}\right)e^{-2r/a_0} = \int_0^\infty dr\, (2)\left(\frac{a_0}{2}\right)^2 e^{-2r/a_0}$$

$$= \left[-(2)\left(\frac{a_0}{2}\right)^3 e^{-2r/a_0}\right]_{r=0}^{r=\infty} = \frac{a_0^3}{4}.$$

Therefore,

$$C = \frac{1}{\sqrt{\pi}\, a_0^{3/2}}. \qquad \blacksquare$$

Using the normalization constant calculated in the Example 8-1, we write the wave function for the ground state of hydrogen as

$$\psi(r) = \frac{1}{\sqrt{\pi}a_0^{3/2}}e^{-r/a_0}, \qquad (8.22)$$

The wave function (8.22) is maximum at $r = 0$. At first sight this might sound alarming, for at $r = 0$, the electron is inside the proton! At small r, there is very little volume so there is actually very little probability of finding the electron there. The probability per unit of r of finding the electron is called the radial probability density (dP/dr) and is given by

$$\frac{dP}{dr} = 4\pi r^2 \left|\psi\right|^2 = \frac{4r^2}{a_0^3}e^{-2r/a_0}. \qquad (8.23)$$

The units of dP/dr are inverse distance and the integral of dP/dr over some range of r is the probability that the electron is found in that range of r. A convenient way to plot the radial probability is to plot $a_0\,dP/dr$ versus r/a_0 as shown in Figure 8-2. The radial probability density is zero both at $r = 0$ and at $r = \infty$. The radial probability density is a maximum when $r = a_0$, indicating that the electron is most likely to be found in the vicinity of the Bohr radius.

EXAMPLE 8-2
Estimate the probability that the electron will be found inside the proton for a hydrogen atom in the ground state.

SOLUTION:
Estimate the proton radius (R) to be

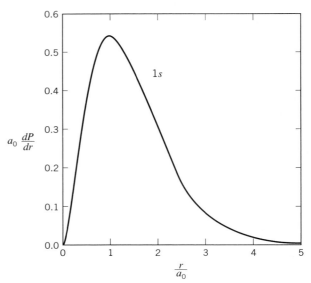

FIGURE 8-2 Probability of finding (interacting with) the electron as a function of radius for a hydrogen atom in the ground state.

The ground state is referred to as the $1s$ state. The wave function is $\psi = Ce^{-r/a_0}$ and the radial probability density is $dP/dr = 4\pi r^2|\psi|^2$. The area under the curve is normalized to unity.

$$R \approx 1\,\text{fm}.$$

The probability for the electron to be inside the proton is

$$P = \frac{4}{a_0^3}\int_0^R dr\, r^2 e^{-2r/a_0}.$$

Since $R \ll a_0$,

$$e^{-2r/a_0} \approx 1,$$

and

$$P \approx \frac{4R^3}{3a_0^3} = \frac{(4)\left(10^{-15}\,\text{m}\right)^3}{(3)\left(0.5\times10^{-10}\,\text{m}\right)^3} \approx 10^{-14}.$$

The electron is not very likely to be found inside the proton simply because the volume of the proton is so small compared with the cube of the Bohr radius. $\blacksquare$

EXAMPLE 8-3
Show that the electron is most likely to be found near $r = a_0$ for a hydrogen atom in the ground state.

SOLUTION:
The radial probability density is

$$\frac{dP}{dr} = Cr^2 e^{-2r/a_0}.$$

The value of radius (r_m), where the radial probability density is maximum, is found by differentiating dP/dr and setting it equal to zero,

$$\left[\frac{d}{dr}\left(\frac{dP}{dr}\right) \right]_{r=r_m} = 0$$

$$= \left[2rCe^{-2r/a_0} - \frac{2r^2}{a_0}Ce^{-2r/a_0} \right]_{r=r_m},$$

or

$$r_m = a_0.$$

The radial probability density is maximum at the Bohr radius. ∎

EXAMPLE 8-4

Calculate the probability that an electron in the ground state of the hydrogen atom will be found in the interval $0.9\, a_0 < r < 1.1\, a_0$.

SOLUTION:

The probability is given by the expression

$$P = \frac{4}{a_0^3} \int_{0.9\,a_0}^{1.1\,a_0} dr\, r^2 e^{-2r/a_0}.$$

Since the interval of r is small ($0.2a_0$), we can approximate the answer by

$$P \approx \left(\frac{4}{a_0^3}\right)(0.2\,a_0)a_0^2 e^{-2} \approx 0.11.$$

To get the exact answer, we integrate by parts. The result, which can be verified by consulting a handbook, is

$$\int_{0.9\,a_0}^{1.1\,a_0} dr\, r^2 e^{-2r/a_0}$$

$$= \left[e^{-2r/a_0}\left(-\frac{a_0 r^2}{2} - \frac{a_0^2 r}{2} - \frac{a_0^3}{4} \right) \right]_{r=0.9\,a_0}^{r=1.1\,a_0}.$$

The probability is

$$P = e^{-2.2}\left[-(2)(1.1)^2 - (2)(1.1) - 1 \right]$$
$$- e^{-1.8}\left[-(2)(0.9)^2 - (2)(0.9) - 1 \right]$$
$$\approx 0.108.$$ ∎

EXAMPLE 8-5

Calculate the average distance between the electron and the proton in the ground state of the hydrogen atom.

SOLUTION:

The average value of r is given by the expression

$$\langle r \rangle = \int_0^\infty dr\, r\,\frac{dP}{dr} = \frac{4}{a_0^3}\int_0^\infty dr\, r^3 e^{-2r/a_0}.$$

The integral may be done by parts. The result is

$$\langle r \rangle$$

$$= \frac{4}{a_0^3}\left[e^{-2r/a_0}\left(-\frac{a_0 r^3}{2} - \frac{3a_0^2 r^2}{4} - \frac{3a_0^3 r}{4} - \frac{3a_0^4}{8} \right) \right]_{r=0}^{r=\infty}$$

$$= \frac{3a_0}{2}.$$ ∎

EXAMPLE 8-6

The probability density for an electron in the ground state of hydrogen extends to a distance greater than four times the Bohr radius (see Figure 8-2). Show that this is consistent with the uncertainty principle.

SOLUTION:

The potential energy of the electron at a distance of $4a_0$ is

$$V = -\frac{ke^2}{4a_0} = -\frac{1.44\text{ eV}\cdot\text{nm}}{(4)(0.053\,\text{nm})} = -6.8\text{ eV}.$$

Since the total energy of the electron is -13.6 eV, an electron at a distance of $4a_0$ corresponds to a kinetic energy of -6.8 eV, a negative value! If an electron is localized to $r > 4a_0$, its position is well known because the square of the wave function decreases exponentially with increasing r,

$$\psi^2 = C^2 e^{-2r/a_0},$$

that is, we know that the electron must be close to $r = 4a_0$. Since the root-mean-square deviation of an exponential function is the reciprocal of the exponential factor, we may estimate the uncertainty in position to be

$$\Delta r \approx \frac{a_0}{2}.$$

From the uncertainty principle

$$\Delta p \Delta r > \frac{\hbar}{2}.$$

We estimate the minimum kinetic energy to be

$$E_k \approx \frac{(\Delta p)^2}{2m} \approx \frac{\hbar^2}{8m(\Delta r)^2} \approx \frac{\hbar^2 c^2}{2mc^2 a_0^2}$$

$$\approx \frac{(197\,\text{eV}\cdot\text{nm})^2}{(10^6\,\text{eV})(0.0529\,\text{nm})^2} \approx 14\,\text{eV}.$$

The uncertainty principle explains why the electron probability density may extend beyond $4a_0$. If an electron is known to be at such a radial distance, then its position is well known and the electron has a minimum average kinetic energy. A negative value of the kinetic energy can never be measured! Note that this is true no matter what distance we choose. At large r, the electron is always localized to $a_0/2$ and the minimum average kinetic energy is always about 14 eV.

8-2 SEPARATION OF VARIABLES

The general solution for the wave function of the hydrogen atom is obtained by the technique of separation of variables. We can write the wave function as the product of three functions, one that depends only on the radial coordinate, $R(r)$, one that depends only on the colatitude angle, $P(\theta)$, and one that depends only on the azimuthal angle, $F(\phi)$:

$$\psi(r,\theta,\phi) = R(r)P(\theta)F(\phi). \quad (8.24)$$

The boundary conditions on each of these functions $R(r)$, $P(\theta)$, and $F(\phi)$ will give a quantum number, analogous to the case of the particle in a box. The relationship between the three quantum numbers for the hydrogen atom is more complicated than for a particle in a box because the boundary conditions on the hydrogen atom wave functions are more complex. With the change of variables (8.24), the Schrödinger equation (8.6) becomes

$$-\frac{\hbar^2}{2m}\left[\frac{P(\theta)F(\phi)}{r^2} \frac{d}{dr}\left(r^2 \frac{dR}{dr} \right) \right.$$

$$+ \frac{R(r)F(\phi)}{r^2 \sin\theta} \frac{d}{d\theta}\left(\sin\theta \frac{dP}{d\theta} \right)$$

$$\left. + \frac{R(r)P(\theta)}{r^2 \sin^2\theta} \frac{d^2 F}{d\phi^2} \right]$$

$$- \frac{ke^2}{r} R(r)P(\theta)F(\phi)$$

$$= E\,R(r)P(\theta)F(\phi). \quad (8.25)$$

Multiplying both sides by $-2mr^2/\hbar^2 RPF$ gives

$$\frac{1}{R}\frac{d}{dr}\left(r^2 \frac{dR}{dr} \right) + \frac{1}{P\sin\theta}\frac{d}{d\theta}\left(\sin\theta \frac{dP}{d\theta} \right)$$

$$+ \frac{1}{F\sin^2\theta}\frac{d^2 F}{d\phi^2} + \frac{2mEr^2}{\hbar^2} + \frac{2mke^2 r}{\hbar^2} = 0, \quad (8.26)$$

or

$$\left[\frac{1}{R}\frac{d}{dr}\left(r^2 \frac{dR}{dr} \right) + \frac{2mEr^2}{\hbar^2} + \frac{2mke^2 r}{\hbar^2} \right]$$

$$+ \left[\frac{1}{P\sin\theta}\frac{d}{d\theta}\left(\sin\theta \frac{dP}{d\theta} \right) + \frac{1}{F\sin^2\theta}\frac{d^2 F}{d\phi^2} \right] = 0.$$

$$(8.27)$$

The radial part of the equation has been separated, and therefore we may set it equal to a constant (C_r),

$$\frac{1}{R}\frac{d}{dr}\left(r^2 \frac{dR}{dr} \right) + \frac{2mEr^2}{\hbar^2} + \frac{2mke^2 r}{\hbar^2} \equiv C_r. \quad (8.28)$$

The constant C_r is constrained by the boundary conditions on $\psi(r,\theta,\phi)$. For the special case of spherical symmetry, $C_r = 0$.

The differential equation for the angular part is

$$C_r + \frac{1}{P\sin\theta}\frac{d}{d\theta}\left(\sin\theta \frac{dP}{d\theta} \right) + \frac{1}{F\sin^2\theta}\frac{d^2 F}{d\phi^2} = 0. \quad (8.29)$$

This equation is not separated yet because the last term contains both θ and ϕ. We may rearrange the terms to get

$$\frac{1}{F}\frac{d^2 F}{d\phi^2} = -C_r \sin^2\theta - \frac{\sin\theta}{P}\frac{d}{d\theta}\left(\sin\theta \frac{dP}{d\theta} \right). \quad (8.30)$$

The angular part of the Schrödinger equation has now been separated. We define the separation constant (C_ϕ) by

$$\frac{1}{F}\frac{d^2F}{d\phi^2} \equiv C_\phi. \qquad (8.31)$$

Therefore, the equation for $P(\theta)$ is

$$C_\phi = -C_r \sin^2\theta - \frac{\sin\theta}{P}\frac{d}{d\theta}\left(\sin\theta\frac{dP}{d\theta}\right), \quad (8.32)$$

We have separated the variables in the Schrödinger equation for the hydrogen atom. We must now solve the differential equations for $R(r)$ (8.28), $P(\theta)$ (8.32), and $F(\phi)$ (8.31). The differential equations are coupled because the solution for $F(\phi)$ places a restriction on the constant C_ϕ, which appears in the equation for $P(\theta)$, and the solution for $P(\theta)$ places a restriction on the constant C_r, which appears in the equation for $R(r)$.

8-3 THREE QUANTUM NUMBERS

The mathematical solution to the differential equations for $R(r)$ (8.28), $P(\theta)$ (8.32), and $F(\phi)$ (8.31) is discussed in Appendix H. The solution gives three quantum numbers, which we call n, ℓ, and m_ℓ. The allowed values of n, ℓ, and m_ℓ are determined by the solution. The quantum number n, which arises within the solution to the radial equation, is called the *principal* quantum number. The allowed values of n are

$$n = 1, \ 2, \ 3... \qquad (8.33)$$

The quantum number ℓ is called the *orbital angular momentum* quantum number and is related to the separation constant C_r by

$$C_r = \ell(\ell+1). \qquad (8.34)$$

The allowed values of ℓ are

$$\ell = 0, \ 1, \ 2,... \ n-1. \qquad (8.35)$$

The quantum number m_ℓ is called the *magnetic* quantum number and is related to the separation constant C_ϕ by

$$C_\phi = -m_\ell^2. \qquad (8.36)$$

The allowed values of m_ℓ are

$$m_\ell = -\ell, \ -\ell+1, \ -\ell+2,... \ \ell-2, \ \ell-1, \ \ell. \quad (8.37)$$

The radial portion of the wave function depends on both n and ℓ, the theta part depends on ℓ and m_ℓ, and the phi part depends on m_ℓ. Therefore, we assign serial numbers (n,ℓ,m_ℓ) to ψ and write the hydrogen atom wave functions as

$$\psi_{n,\ell,m_\ell}(r,\theta,\phi) = R_{n,\ell}(r)P_{\ell,m_\ell}(\theta)e^{im_\ell\phi}. \quad (8.38)$$

The atomic states are named with a number-letter code according to the value of the quantum numbers n and ℓ. This labeling is called *spectroscopic notation*. The number corresponds to n. The spectroscopic letter code is s for $\ell = 0$, p for $\ell = 1$, d for $\ell = 2$, and f for $\ell = 3$. The letters have arisen from the early classification of associated spectral lines as being *sharp*, *principal*, *diffuse*, and *fundamental*.

The first few wave functions are summarized in Table 8-1. For $n=1$ there is only one possible value of ℓ, which is zero (the 1s state). The 1s state is the ground state. The ground state is spherically symmetric, that is, there is no angular dependence. The value of m_ℓ is zero. The 1s radial probability distribution is given in Figure 8-2.

For $n = 2$, there are two possibilities for ℓ: 0 and 1 (the 2s and 2p states). The 2s state is spherically symmetric ($\ell = 0$ and $m_\ell = 0$). For the 2p state there are three possibilities for m_ℓ: −1, 0, and 1. We see from the form of the wave function that its square does not depend on m_ℓ, that is, there is no phi dependence to the electron probability distribution. The wave function squared does depend on theta, but the average over all angles depends only on $R_{n,\ell}$ so that the radial probability distribution does not depend on m_ℓ. The radial probability distributions for the 2s and 2p states are shown in Figure 8-3.

For $n = 3$, there are three possibilities for ℓ: 0, 1, and 2 (the 3s, 3p, and 3d states). The radial probability density for the 3s, 3p, and 3d states are shown in Figure 8-4.

Figures 8-2 to 8-4 show the quantitative behavior of the radial probability distributions. It is also instructive to examine plots of $|\psi|^2$ in three dimensions, that is, without averaging over all angles. Figure 8-5 shows some sketches of $|\psi|^2$.

The Principal Quantum Number, *n*

The quantum number n appears in the solution of the radial equation (8.28). The radial equation with the allowed values of the separation constant (8.34) is

$$\frac{1}{R}\frac{d}{dr}\left(r^2\frac{dR}{dr}\right) + \frac{2mEr^2}{\hbar^2} + \frac{2mke^2r}{\hbar^2} = \ell(\ell+1). \quad (8.39)$$

TABLE 8-1
HYDROGEN ATOM WAVE FUNCTIONS.

Spectroscopic Notation	$\psi_{n,\ell,m_\ell}(r,\theta,\phi)$
1s	$\psi_{1,0,0} = \dfrac{1}{\sqrt{\pi}\, a_0^{3/2}}\, e^{-r/a_0}$
2s	$\psi_{2,0,0} = \dfrac{1}{4\sqrt{2\pi}\, a_0^{3/2}}\left(2 - \dfrac{r}{a_0}\right) e^{-r/2a_0}$
2p ($m_\ell = 0$)	$\psi_{2,1,0} = \dfrac{1}{4\sqrt{2\pi}\, a_0^{3/2}}\left(\dfrac{r}{a_0}\right) e^{-r/2a_0}\, \cos\theta$
2p ($m_\ell = +1$)	$\psi_{2,1,1} = \dfrac{1}{8\sqrt{\pi}\, a_0^{3/2}}\left(\dfrac{r}{a_0}\right) e^{-r/2a_0}\, \sin\theta\, e^{i\phi}$
2p ($m_\ell = -1$)	$\psi_{2,1,-1} = \dfrac{1}{8\sqrt{\pi}\, a_0^{3/2}}\left(\dfrac{r}{a_0}\right) e^{-r/2a_0}\, \sin\theta\, e^{-i\phi}$
3s	$\psi_{3,0,0} = \dfrac{1}{81\sqrt{3\pi}\, a_0^{3/2}}\left(27 - \dfrac{18r}{a_0} + \dfrac{2r^2}{a_0^2}\right) e^{-r/3a_0}$
3p ($m_\ell = 0$)	$\psi_{3,1,0} = \dfrac{2}{81\sqrt{2\pi}\, a_0^{3/2}}\left(\dfrac{6r}{a_0} - \dfrac{r^2}{a_0^2}\right) e^{-r/3a_0}\, \cos\theta$
3p ($m_\ell = +1$)	$\psi_{3,1,1} = \dfrac{1}{81\sqrt{\pi}\, a_0^{3/2}}\left(\dfrac{6r}{a_0} - \dfrac{r^2}{a_0^2}\right) e^{-r/3a_0}\, \sin\theta\, e^{i\phi}$
3p ($m_\ell = -1$)	$\psi_{3,1,-1} = \dfrac{1}{81\sqrt{\pi}\, a_0^{3/2}}\left(\dfrac{6r}{a_0} - \dfrac{r^2}{a_0^2}\right) e^{-r/3a_0}\, \sin\theta\, e^{-i\phi}$
3d ($m_\ell = 0$)	$\psi_{3,2,0} = \dfrac{1}{81\sqrt{6\pi}\, a_0^{3/2}}\left(\dfrac{r^2}{a_0^2}\right) e^{-r/3a_0}\, (3\cos^2\theta - 1)$
3d ($m_\ell = +1$)	$\psi_{3,2,1} = \dfrac{1}{81\sqrt{\pi}\, a_0^{3/2}}\left(\dfrac{r^2}{a_0^2}\right) e^{-r/3a_0}\, \sin\theta\cos\theta\, e^{i\phi}$
3d ($m_\ell = -1$)	$\psi_{3,2,-1} = \dfrac{1}{81\sqrt{\pi}\, a_0^{3/2}}\left(\dfrac{r^2}{a_0^2}\right) e^{-r/3a_0}\, \sin\theta\cos\theta\, e^{-i\phi}$
3d ($m_\ell = +2$)	$\psi_{3,2,2} = \dfrac{1}{162\sqrt{\pi}\, a_0^{3/2}}\left(\dfrac{r^2}{a_0^2}\right) e^{-r/3a_0}\, \sin^2\theta\, e^{2i\phi}$
3d ($m_\ell = -2$)	$\psi_{3,2,-2} = \dfrac{1}{162\sqrt{\pi}\, a_0^{3/2}}\left(\dfrac{r^2}{a_0^2}\right) e^{-r/3a_0}\, \sin^2\theta\, e^{-2i\phi}$

The boundary condition that the wave function $R(r)$ go to zero at infinity causes the integer n to appear (compare to the particle in a box, the finite square-well, and the harmonic oscillator). The solutions to the radial equation (see Table 8-1 and Appendix H) are polynomials in r times an exponential, e^{-r/na_0}, where n is the principal quantum number. The radial equation (8.39) contains a term with d^2R/dr^2. This second derivative with respect to r produces a term proportional to $1/n^2$. This leads to solutions for the electron energy proportional to $1/n^2$. The energy levels in the hydrogen atom are quantized. We give each energy level a serial number n and write

$$E_n = -\frac{\alpha^2 mc^2}{2n^2} = -\frac{13.6\,\text{eV}}{n^2}, \qquad (8.40)$$

where n is a positive integer $(1, 2, 3, \ldots)$. According to the Schrödinger equation, the energy levels do not depend on the values of the quantum numbers ℓ and m_ℓ. For example, the $2s$ and $2p$ states have the same energy even though they have different radial probability distributions. Figure 8-6 shows the energy levels of an electron in the hydrogen atom.

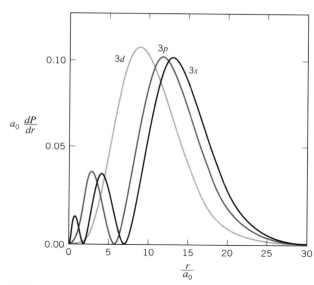

FIGURE 8-4 Probability of finding the electron as a function of radius for a hydrogen atom in the second excited state.
There are three possible states: $n = 3$ and $\ell = 0$ called the $3s$ state, $n = 3$ and $\ell = 1$ called the $3p$ state, or $n = 3$ and $\ell = 2$ called the $3d$ state. The area under each curve is normalized to unity.

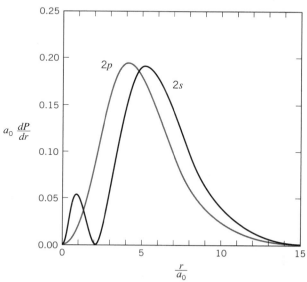

FIGURE 8-3 Probability of finding the electron as a function of radius for a hydrogen atom in the first excited state.
There are two possible states: $n = 2$ and $\ell = 0$ called the $2s$ state, or $n = 2$ and $\ell = 1$ called the $2p$ state. The area under each curve is normalized to unity.

EXAMPLE 8-7
Use the wave functions in Table 8-1 to calculate the energies of the $2s$ and $2p$ states.

SOLUTION:
The radial wave function for the $2s$ state is

$$R_{2,0} = C_0\left(2 - \frac{r}{a_0}\right)e^{-r/2a_0},$$

where C_0 is a constant. The derivatives are

$$\frac{dR_{2,0}}{dr} = C_0\left(-\frac{1}{a_0}\right)e^{-r/2a_0}$$

$$+ C_0\left(2 - \frac{r}{a_0}\right)\left(-\frac{1}{2a_0}\right)e^{-r/2a_0}$$

$$= C_0\left(\frac{r}{2a_0^2} - \frac{2}{a_0}\right)e^{-r/2a_0},$$

and

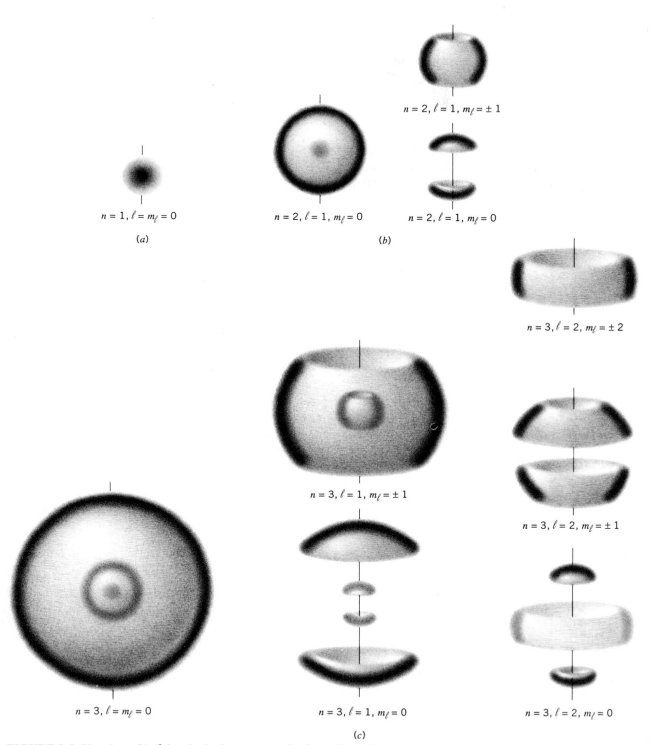

$n = 1, \ell = m_\ell = 0$

(a)

$n = 2, \ell = 1, m_\ell = 0$

$n = 2, \ell = 1, m_\ell = \pm 1$

$n = 2, \ell = 1, m_\ell = 0$

(b)

$n = 3, \ell = m_\ell = 0$

$n = 3, \ell = 1, m_\ell = \pm 1$

$n = 3, \ell = 1, m_\ell = 0$

$n = 3, \ell = 2, m_\ell = \pm 2$

$n = 3, \ell = 2, m_\ell = \pm 1$

$n = 3, \ell = 2, m_\ell = 0$

(c)

FIGURE 8-5 Sketches of $|\psi|^2$ for the hydrogen atom in three dimensions:
(a) the $n = 1$ state, (b) the $n = 2$ states, and (c) the $n = 3$ states. After R. Eisberg and R. Resnick, *Quantum Physics of Atoms, Molecules, Solids, Nuclei, and Particles*, Wiley (copyright © 1985).

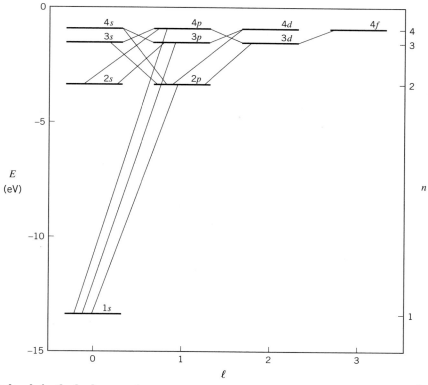

FIGURE 8-6 Energy levels in the hydrogen atom.
The favored transitions are those in which ℓ changes by ± 1.

$$\frac{d}{dr}\left(r^2 \frac{dR_{2,0}}{dr} \right)$$

$$= C_0 \left(\frac{3r^2}{2a_0{}^2} - \frac{4r}{a_0} \right) e^{-r/2a_0}$$

$$+ C_0 \left(\frac{r^3}{2a_0{}^2} - \frac{2r^2}{a_0} \right) \left(\frac{-1}{2a_0} \right) e^{-r/2a_0}$$

$$= C_0 \left(-\frac{4r}{a_0} + \frac{5r^2}{2a_0{}^2} - \frac{r^3}{4a_0{}^3} \right) e^{-r/2a_0} .$$

The radial equation (8.39) with $\ell = 0$ is

$$\frac{d}{dr}\left(r^2 \frac{dR}{dr} \right) = -\frac{2mke^2 r}{\hbar^2} R - \frac{2mEr^2}{\hbar^2} R .$$

Therefore, the radial equation gives

$$C_0 \left(-\frac{4r}{a_0} + \frac{5r^2}{2a_0{}^2} - \frac{r^3}{4a_0{}^3} \right) e^{-r/2a_0}$$

$$= -\frac{2mke^2}{\hbar^2} C_0 \left(2r - \frac{r^2}{a_0} \right) e^{-r/2a_0}$$

$$- \frac{2mE_2}{\hbar^2} C_0 \left(2r^2 - \frac{r^3}{a_0} \right) e^{-r/2a_0} .$$

Comparison of the r terms gives the Bohr radius,

$$a_0 = \frac{\hbar^2}{mke^2} .$$

Comparison of the r^3 terms gives the electron energy,

$$E_2 = -\frac{\hbar^2}{8ma_0{}^2} = -\frac{1}{8}mc^2\alpha^2 .$$

Comparison of the r^2 terms gives no extra information but shows that the solution works. The radial wave function for the $2p$ state is

$$R_{2,1} = C_1 \frac{r}{a_0} e^{-r/2a_0},$$

where C_1 is a constant. The derivatives are

$$\frac{dR_{2,1}}{dr} = C_1 \left(\frac{1}{a_0} \right) e^{-r/2a_0} + C_1 \left(\frac{r}{a_0} \right) \left(-\frac{1}{2a_0} \right) e^{-r/2a_0}$$

$$= C_1 \left(\frac{1}{a_0} - \frac{r}{2a_0^2} \right) e^{-r/2a_0},$$

and

$$\frac{d}{dr} \left(r^2 \frac{dR_{2,1}}{dr} \right)$$

$$= C_1 \left(\frac{2r}{a_0} - \frac{3r^2}{2a_0^2} \right) e^{-r/2a_0}$$

$$+ C_1 \left(\frac{r^2}{a_0} - \frac{r^3}{2a_0^2} \right) \left(\frac{-1}{2a_0} \right) e^{-r/2a_0}$$

$$= C_1 \left(\frac{2r}{a_0} - \frac{2r^2}{a_0^2} + \frac{r^3}{4a_0^3} \right) e^{-r/2a_0}.$$

The radial equation (8.39) with $\ell=1$ is

$$\frac{d}{dr} \left(r^2 \frac{dR}{dr} \right) = -\frac{2mke^2 r}{\hbar^2} R - \frac{2mEr^2}{\hbar^2} R + 2R.$$

Therefore, the radial equation gives

$$C_1 \left(\frac{2r}{a_0} - \frac{2r^2}{a_0^2} + \frac{r^3}{4a_0^3} \right) e^{-r/2a_0}$$

$$= -\frac{2mke^2}{\hbar^2} \left(C_1 \frac{r^2}{a_0} \right) e^{-r/2a_0}$$

$$- \frac{2mE_2}{\hbar^2} \left(C_1 \frac{r^3}{a_0} \right) e^{-r/2a_0} + 2C_1 \frac{r}{a_0} e^{-r/2a_0}.$$

The r terms cancel, the r^2 terms give the Bohr radius, and the r^3 terms give the energy:

$$E_2 = -\frac{\hbar^2}{8ma_0^2} = -\frac{1}{8} mc^2 \alpha^2.$$

The energies of the $2s$ and $2p$ states are identical. The energy does not depend on the quantum number ℓ. ∎

The Orbital Angular Momentum Quantum Number, ℓ

The quantum number ℓ has appeared in the solution for the angular dependence $P(\theta)$. The separation constant C_r is equal to the product of two consecutive nonnegative integers that we write as $\ell(\ell+1)$. This separation constant also appears in the radial equation (8.39) which we may write as

$$-\frac{\hbar^2}{2mr^2} \frac{d}{dr} \left(r^2 \frac{dR}{dr} \right)$$

$$+ \left[\frac{\ell(\ell+1)\hbar^2}{2mr^2} - \frac{ke^2}{r} \right] R = ER. \quad (8.41)$$

The solution for radial motion should depend only on the coordinate r. The total energy E that appears in the radial equation is the kinetic plus the potential energy. The kinetic energy contains both radial motion ($p_r^2/2m$) and orbital motion ($L^2/2mr^2$), where L is the *orbital angular momentum*. We may write the electron energy as

$$E = \frac{p^2}{2m} + V = \left(\frac{p_r^2}{2m} + \frac{L^2}{2mr^2} \right) - \frac{ke^2}{r}, \quad (8.42)$$

and the radial equation (8.41) becomes

$$-\frac{\hbar^2}{2mr^2} \frac{d}{dr} \left(r^2 \frac{dR}{dr} \right)$$

$$+ \left[\frac{\ell(\ell+1)\hbar^2}{2mr^2} - \frac{p_r^2}{2m} - \frac{L^2}{2mr^2} \right] R = 0. (8.43)$$

Thus, we see that we have a *pure* radial equation if

$$\frac{L^2}{2mr^2} = \frac{\ell(\ell+1)\hbar^2}{2mr^2}, \quad (8.44)$$

or

$$L^2 = \ell(\ell+1)\hbar^2. \quad (8.45)$$

Orbital angular momentum is quantized.

The quantum number ℓ specifies the orbital angular momentum of the electron. Zero orbital angular momentum corresponds to spherical symmetry since the separation constant (8.34) is zero and there is no angular dependence. There is a limit on how large ℓ may be for a given value of n. The solution requires that $\ell < n$. For example, if $n = 1$ then $\ell = 0$, that is, there is no $1p$ state. If ℓ is not equal to zero, the wave function vanishes at the origin (see Table 8-1 and Figure 8-5) and the wave function is proportional to r^ℓ.

The Magnetic Quantum Number, m_ℓ

The quantum number m_ℓ has appeared in the solution for $F(\phi)$. The separation constant C_ϕ is equal to minus one times the square of an integer (m_ℓ). The minimum value of m_ℓ is $-\ell$ and the maximum value is ℓ.

For the $2p$ wave functions, there are three possibilities for m_ℓ: -1, 0, and $+1$. For $m_\ell = 0$, the wave function (see Table 8-1) is proportional to $\cos\theta$ and the electron is most likely to be found near the z axis. We can see from Figure 8-5 that the *average* of the z component of angular momentum (L_z) is zero when $m_\ell = 0$. It turns out that the z component of angular momentum is *exactly* zero when $m_\ell = 0$. For $m_\ell = +1$ or $m_\ell = -1$, the wave function is proportional to $\sin\theta$ and the electron is most likely to be found near the x-y plane. The exact value of L_z is $\hbar$ for $m_\ell = +1$ and $-\hbar$ for $m_\ell = -1$. We summarize these results with

$$L_z = m_\ell \hbar, \qquad (8.46)$$

where m_ℓ may take on integer values from $-\ell$ to ℓ. The total number of states for a given value of ℓ is $2\ell + 1$. The z direction is special in spherical coordinates because the polar angle is measured with respect to the z axis. All of the wave functions squared are symmetric about the z axis, that is, $|\psi|^2$ does not depend on ϕ. The quantization condition (8.46) is similar to that used in the Bohr model, where the electron is assumed to have a circular trajectory. In this case, the z direction is the direction perpendicular to the plane of the orbit. In the Bohr model, the minimum value of $|L_z|$ is $\hbar$, whereas in the Schrödinger formulation (8.46) it is zero.

If both the total angular momentum and the z component of angular momentum are quantized, then the angular momentum cannot point in an arbitrary direction in space. Angular momentum quantization is observable when there is a magnetic field that selects a special direction in space. The magnetic field causes a shift in the atomic energy level by an amount that is proportional to the quantum number m_ℓ. We shall discuss the reason for this in the next section. Thus, m_ℓ is known as the *magnetic* quantum number. Angular momentum quantization is illustrated in Figure 8-7, where we plot L_z versus $(L_x^2 + L_y^2)^{1/2}$. (Note that $L_x^2 + L_y^2 = L^2 - L_z^2$.)

The uncertainty principle tells us that we may not simultaneously know the exact values of any two components of the angular momentum. For example, if L_x and L_y were known exactly, then we would know the exact trajectory, position, and momentum of the particle in the x-y plane. If the z component of angular momentum (8.46) is known exactly, then nothing is known about the azimuthal angle. Both the magnitude of the electron orbital angular momentum L and the z component L_z may be specified exactly. The direction of $\mathbf{L}$ is not known; however, $\mathbf{L}$ is constrained to lie on a cone (Figure 8-8).

The solution to the Schrödinger equation gives three quantum numbers for an electron in the hydrogen atom. The primary quantum number n gives the electron energy,

$$E_n = -\frac{13.6 \text{ eV}}{n^2},$$

where $n = 1, 2, 3, \ldots$ The orbital quantum number ℓ gives the electron orbital angular momentum,

$$L = \sqrt{\ell(\ell+1)}\, \hbar,$$

where $\ell = 0, 1, 2, 3, \ldots n-1$. The magnetic quantum number m_ℓ gives the z component of orbital angular momentum,

$$L_z = m_\ell \hbar,$$

where $m_\ell = -\ell, -\ell+1, \ldots 0, \ldots \ell-1, \ell$.

---

* *Challenging*

Operators in Quantum Mechanics

The Schrödinger equation may be written as

$$\frac{1}{2m}\left(\mathbf{p}^{\text{opp}}\right)^2 \psi + V\psi = E\psi, \qquad (8.47)$$

where the symbol $\mathbf{p}^{\text{opp}}$ is defined to be

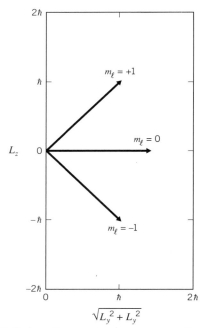

FIGURE 8-7 Angular momentum quantization for $\ell=1$.
When the orbital quantum number is one, the orbital angular momentum is $L = \sqrt{2}\,\hbar$. There are only three possibilities for the z component of orbital angular momentum: $-\hbar$, 0, or $\hbar$. Thus, the total angular momentum vector cannot point in any arbitrary direction. This effect is observable when the atom is placed in an external magnetic field, which serves to define a specific direction.

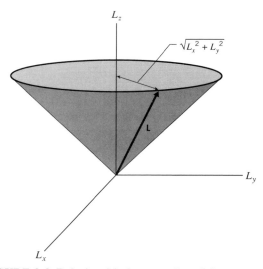

FIGURE 8-8 Relationship between L and L_z.
Both the magnitude of the electron orbital angular momentum and the z component may be specified exactly. The exact direction of **L** is not known but is constrained to lie on a cone.

$$\mathbf{p}^{\mathrm{opp}} \equiv -i\hbar \mathbf{\nabla}. \qquad (8.48)$$

The symbol $\mathbf{p}^{\mathrm{opp}}$ is called the momentum *operator*. In the Schrödinger equation (8.47) the square of an operator means that the operation is applied twice.

The operator for the z component of angular momentum $\mathbf{L}_z^{\mathrm{opp}}$ is

$$\mathbf{L}_z^{\mathrm{opp}} = x\mathbf{p}_y - y\mathbf{p}_x = -i\hbar\left(x\frac{\partial}{\partial y} - y\frac{\partial}{\partial x}\right). \quad (8.49)$$

From the definition of spherical coordinates (see problem 15), we get

$$\frac{\partial}{\partial x} = \frac{\partial r}{\partial x}\frac{\partial}{\partial r} + \frac{\partial \theta}{\partial x}\frac{\partial}{\partial \theta} + \frac{\partial \phi}{\partial x}\frac{\partial}{\partial \phi}$$
$$= \sin\theta\cos\phi\frac{\partial}{\partial r} + \frac{1}{r}\cos\theta\cos\phi\frac{\partial}{\partial \theta} - \frac{\sin\phi}{r\sin\theta}\frac{\partial}{\partial \phi}, \quad (8.50)$$

and

$$\frac{\partial}{\partial y} = \frac{\partial r}{\partial y}\frac{\partial}{\partial r} + \frac{\partial \theta}{\partial y}\frac{\partial}{\partial \theta} + \frac{\partial \phi}{\partial y}\frac{\partial}{\partial \phi}$$
$$= \sin\theta\sin\phi\frac{\partial}{\partial r} + \frac{1}{r}\cos\theta\sin\phi\frac{\partial}{\partial \theta} + \frac{\cos\phi}{r\sin\theta}\frac{\partial}{\partial \phi}. \quad (8.51)$$

Therefore, $\mathbf{L}_z^{\mathrm{opp}}$ (8.49) may be written

$$\mathbf{L}_z^{\mathrm{opp}} = -i\hbar\frac{\partial}{\partial \phi}. \qquad (8.52)$$

Applying $\mathbf{L}_z^{\mathrm{opp}}$ to the hydrogen atom wave functions determines the value of $\mathbf{L}_z$,

$$\mathbf{L}_z^{\mathrm{opp}}\psi_{n,\ell,m_\ell} = -i\hbar\frac{\partial \psi_{n,\ell,m_\ell}}{\partial \phi}$$
$$= (-i\hbar)(im_\ell)\psi_{n,\ell,m_\ell}$$
$$= m_\ell\hbar\psi_{n,\ell,m_\ell}. \qquad (8.53)$$

The hydrogen atom wave functions have a definite angular momentum given by

$$L_z = m_\ell\hbar. \qquad (8.54)$$

Similarly, the operators for the other components of angular momentum (see problem 15) are

$$\mathbf{L}_x{}^{\mathrm{opp}} = -i\hbar\left(y\frac{\partial}{\partial z} - z\frac{\partial}{\partial y}\right)$$

$$= i\hbar\left(\sin\phi\,\frac{\partial}{\partial\theta} + \frac{\cos\theta\cos\phi}{\sin\theta}\frac{\partial}{\partial\phi}\right), \quad (8.55)$$

and

$$\mathbf{L}_y{}^{\mathrm{opp}} = -i\hbar\left(z\frac{\partial}{\partial x} - x\frac{\partial}{\partial z}\right)$$

$$= i\hbar\left(-\cos\phi\,\frac{\partial}{\partial\theta} + \frac{\cos\theta\sin\phi}{\sin\theta}\frac{\partial}{\partial\phi}\right). \quad (8.56)$$

The total angular momentum operator $\mathbf{L}^{\mathrm{opp}}$ may be constructed from the component operators, $\mathbf{L}_x{}^{\mathrm{opp}}$, $\mathbf{L}_y{}^{\mathrm{opp}}$, and $\mathbf{L}_z{}^{\mathrm{opp}}$:

$$\left(\mathbf{L}^{\mathrm{opp}}\right)^2 = \left(\mathbf{L}_x{}^{\mathrm{opp}}\right)^2 + \left(\mathbf{L}_y{}^{\mathrm{opp}}\right)^2 + \left(\mathbf{L}_z{}^{\mathrm{opp}}\right)^2, \quad (8.57)$$

which reduces to (see problem 16)

$$\left(\mathbf{L}^{\mathrm{opp}}\right)^2$$

$$= -\hbar^2\left[\frac{1}{\sin\theta}\frac{\partial}{\partial\theta}\left(\sin\theta\frac{\partial}{\partial\theta}\right) + \frac{1}{\sin^2\theta}\frac{\partial^2}{\partial\phi^2}\right]. (8.58)$$

The angular momentum operator applied twice gives the angular portion of the $\mathbf{\nabla}^2$ operator times $\hbar$. Therefore, the solution to the Schrödinger equation gives

$$\left(\mathbf{L}^{\mathrm{opp}}\right)^2\psi_{n,\ell,m_\ell}$$

$$= -\hbar^2\left[\frac{1}{\sin\theta}\frac{\partial}{\partial\theta}\left(\sin\theta\frac{\partial}{\partial\theta}\right) + \frac{1}{\sin^2\theta}\frac{\partial^2}{\partial\phi^2}\right]\psi_{n,\ell,m_\ell}$$

$$= \ell(\ell+1)\hbar^2\psi_{n,\ell,m_\ell}. \quad (8.59)$$

Thus, the value of L^2 is $\ell(\ell+1)\hbar^2$.

EXAMPLE 8-8

The average value of an observable quantity $\langle O\rangle$ whose corresponding operator is $\mathbf{O}^{\mathrm{opp}}$ is given by the expression

$$\langle O\rangle = \iiint_V dV\,\psi^*\,\mathbf{O}^{\mathrm{opp}}\,\psi.$$

Calculate the average kinetic energy of an electron in the ground state of the hydrogen atom.

SOLUTION:

The average kinetic energy is given by

$$\left\langle\frac{p^2}{2m}\right\rangle = \frac{1}{2m}\iiint_V dV\,\psi^*\left(\mathbf{p}^{\mathrm{opp}}\right)^2\psi$$

$$= \frac{1}{2m}\iiint_V dV\,\psi^*\left(-i\hbar\nabla\right)^2\psi$$

$$= -\frac{\hbar^2}{2m}\iiint_V dV\,\psi^*\nabla^2\psi.$$

For the ground state,

$$\psi = Ce^{-r/a_0}$$

and

$$\nabla^2\psi = \frac{1}{r^2}\frac{d}{dr}\left(r^2\frac{d}{dr}Ce^{-r/a_0}\right)$$

$$= C\left(\frac{1}{a_0{}^2} - \frac{2}{a_0 r}\right)e^{-r/a_0}.$$

where $c^2 = 1/\pi a_0{}^3$. The average kinetic energy is

$$\left\langle\frac{p^2}{2m}\right\rangle = -\frac{\hbar^2}{2m}\iiint_V dV\left[\frac{C^2 e^{-2r/a_0}}{a_0{}^2} - \frac{2C^2 e^{-2r/a_0}}{a_0 r}\right]$$

$$= \left(-\frac{\hbar^2}{2ma_0{}^2}\right)\left(\frac{1}{\pi a_0{}^3}\right)4\pi\int_0^\infty dr\,r^2 e^{-2r/a_0}$$

$$+ \left(\frac{\pi\hbar^2}{ma_0}\right)\left(\frac{1}{\pi a_0{}^3}\right)4\pi\int_0^\infty dr\,re^{-2r/a_0}$$

$$= -\frac{\hbar^2}{2ma_0{}^2} + \frac{\hbar^2}{ma_0{}^2}$$

$$= \frac{\hbar^2}{2ma_0{}^2} = \frac{1}{2}\alpha^2 mc^2 = 13.6\,\mathrm{eV}.$$

The *average* value of the kinetic energy plus the *average* value of the potential energy is equal to the energy of the state

$$\left\langle\frac{p^2}{2m}\right\rangle + \left\langle-\frac{ke^2}{r}\right\rangle = E_n. \quad\blacksquare$$

*

8-4 INTRINSIC ANGULAR MOMENTUM

The Schrödinger equation in three dimensions has given us three quantum numbers (n, ℓ, and m_ℓ) for the description of an electron in a hydrogen atom. In spite of the great success of the Schrödinger equation in explaining much of the structure of the hydrogen atom, experiments tell us that the Schrödinger theory of the hydrogen atom is not complete! To appreciate where the Schrödinger theory fails, we proceed with a more detailed discussion of the angular momentum of the electron in the hydrogen atom.

Electron Angular Momentum and Magnetic Moment

The electron in the hydrogen atom does not have a classical trajectory; the wave properties of the electron are inconsistent with such an interpretation. The electron is described by a probability density given by the square of the wave function. The probability density is peaked near $n^2 a_0$ for the nth state. If there is no angular (θ) dependence to the probability density, then there is no net orbital angular momentum of the electron, $L = 0$. If there is a θ dependence of the probability density, then there is a nonzero value of orbital angular momentum. For example, if the probability density is maximum at $z = 0$ and $(x^2+y^2)^{1/2}$ = constant, corresponding to a ring in the x-y plane, then there is a net angular momentum in the z direction (either positive or negative). This is the case for the $2p$ state when $m_\ell = +1$ or -1 (see Figure 8-5). In the classical limit, the probability density is a delta function corresponding to an electron traveling in a circle either clockwise or counterclockwise with an orbital angular momentum, $L = mvr$. In the quantum interpretation, all we can say is that at any given time the electron can interact or be found near the ring with a fixed energy E and a fixed orbital angular momentum L. The electron is moving with a characteristic speed $v \approx \alpha c$ and the result of this motion constitutes a net orbital angular momentum.

A moving charge corresponds to a current and a magnetic moment. Consider a loop of current (I) enclosing an area (A) in the x-y plane. The classical magnetic moment is

$$\mu = IA. \qquad (8.60)$$

The orientation of the magnetic moment vector is perpendicular to the loop. For a circulating charge (q) the time-averaged current is

$$I = \frac{q}{T}, \qquad (8.61)$$

where T is the period of revolution. Kepler's law states that the area swept out per time is a constant,

$$\frac{A}{T} = \frac{L}{2m}. \qquad (8.62)$$

For a displacement $d\mathbf{r}$ along the path (see Figure 8-9),

$$dA = \frac{1}{2}|\mathbf{r} \times d\mathbf{r}|, \qquad (8.63)$$

so

$$\frac{dA}{dt} = \frac{1}{2}\left|\mathbf{r} \times \frac{d\mathbf{r}}{dt}\right| = \frac{\left|\mathbf{r} \times \left(m\dfrac{d\mathbf{r}}{dt}\right)\right|}{2m} = \frac{L}{2m}, \qquad (8.64)$$

and

$$\frac{A}{T} = \frac{L}{2m}. \qquad (8.65)$$

This is easily verified for a circular orbit where $L = mvr$, $T = 2\pi r/v$ and $A = \pi r^2$. The magnetic moment of a moving charge is

$$\mu = IA = \left(\frac{q}{T}\right)\left(\frac{TL}{2m}\right) = \frac{q}{2m}L. \qquad (8.66)$$

The direction of μ is along the direction of L,

$$\boldsymbol{\mu} = \frac{q}{2m}\mathbf{L}. \qquad (8.67)$$

By writing the magnetic moment in terms of orbital angular momentum, we have circumvented the need to know the exact trajectory of the charge. This is just the situation of our knowledge of the motion of an electron in the atom: the orbital angular momentum is well-defined but the exact motion of the electron is not well-defined. Note that for electrons $q = -e$ so that the direction of the magnetic moment is opposite the direction of the orbital angular momentum:

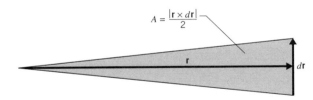

FIGURE 8-9 Kepler's law.

$$\boldsymbol{\mu} = -\frac{e}{2m}\mathbf{L}. \qquad (8.68)$$

An electron with orbital angular momentum L is a tiny electromagnet.

The ratio $-e/2m$ is commonly called the electron *gyromagnetic* ratio. The gyromagnetic ratio is the magnetic moment divided by the orbital angular momentum. The smallest unit of angular momentum is $\hbar$. The *Bohr magneton* (μ_B) is defined as

$$\mu_B \equiv \frac{e\hbar}{2m} = 9.274 \times 10^{-24} \text{ A} \cdot \text{m}^2$$
$$= 5.788 \times 10^{-5} \text{ eV/T}. \qquad (8.69)$$

Since the magnetic moment is proportional to the orbital angular momentum, the direction of the magnetic moment vector $\boldsymbol{\mu}$ is restricted (see Figure 8-7). Therefore, we may write the z component of $\boldsymbol{\mu}$ as

$$\mu_z = -\frac{e}{2m}L_z = -\mu_B m_\ell. \qquad (8.70)$$

A magnet placed in a magnetic field experiences a torque (τ),

$$\boldsymbol{\tau} = \frac{d\mathbf{L}}{dt} = \boldsymbol{\mu} \times \mathbf{B} = \frac{q}{2m}\mathbf{L} \times \mathbf{B}, \qquad (8.71)$$

which tends to align $\boldsymbol{\mu}$ with $\mathbf{B}$. This is illustrated in Figure 8-10. This results in a precession when $\boldsymbol{\mu}$ is not parallel $\mathbf{B}$ because the change in angular momentum is perpendicular to both $\mathbf{L}$ and $\mathbf{B}$. This is analogous to a spinning top in a gravitational field. The frequency of precession ω_L, called the *Larmor* frequency (see problem 35), is

$$\omega_L = \frac{eB}{2m}. \qquad (8.72)$$

The potential energy of the magnetic dipole in an external magnetic field is

$$V = -\boldsymbol{\mu} \cdot \mathbf{B}. \qquad (8.73)$$

The energy of an electron in an external field (B_z) is shifted by

$$\Delta E = \mu_z B_z = \hbar \omega_L m_\ell. \qquad (8.74)$$

The Stern-Gerlach Experiment

According to the analysis of the preceding section, an atom in the s state ($\ell = 0$ and $m_\ell = 0$), has no energy shift

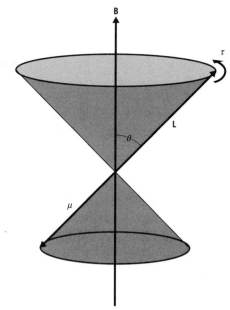

FIGURE 8-10 An electron with orbital angular momentum L in a magnetic field (B).

when placed in an external magnetic field. In 1921 Otto Stern and Walter Gerlach performed an experiment that proved this not to be correct. Stern and Gerlach chose to experiment with the silver atom because it could be readily detected by photographic techniques. The silver atom behaves like the hydrogen atom; a single outer electron moves in a Coulomb potential generated by a charge e due to 47 protons and 46 inner electrons. The orbital angular momentum of the outer electron in the ground state is zero. The structure of multielectron atoms is discussed in the next chapter.

Stern and Gerlach directed a beam of silver atoms into a region where the magnetic field was nonuniform, as indicated in Figure 8-11. The force on the atom is in the direction of the magnetic field gradient and has a magnitude (F_z),

$$F_z = -\mu_z \frac{\Delta B}{\Delta z}. \qquad (8.75)$$

After passing through the nonuniform field, the beam of silver atoms was observed to be split in two, as shown in Figure 8-12. The Stern-Gerlach experiment shows that the silver atom has two possible values for the z component of the magnetic moment (μ_z).

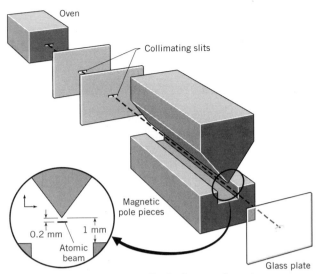

FIGURE 8-11 The Stern-Gerlach experiment.
A beam of atoms enters a region where the magnetic field varies in the direction transverse to the velocity (**v**) of the atoms. There is a force on the atoms that depends on the orientation of the magnetic moment (μ). The number of paths that the atoms may follow through the magnetic field region is equal to the number of possible orientations of μ . For the case μ = 0, there is no force on the atoms. For atoms with a single electron in the $\ell = 0$ state, two possible deflections are observed corresponding to two possible orientations of μ . This magnetic moment is generated by the intrinsic angular momentum of the electron. After R. L. Sproull and W. A. Phillips, *Modern Physics: The Quantum Physics of Atoms, Solids, and Nuclei*, Wiley (copyright © 1985).

Intrinsic Angular Momentum

Recall that the total number of states for a given value of the orbital angular momentum quantum number ℓ is given by the number of possibilities for the quantum number m_ℓ, which is $2\ell + 1$. The Stern-Gerlach experiment shows that atoms in the s state ($\ell = 0$) are split into *two states* in a magnetic field. In 1925 Samuel A. Goudsmit and George E. Uhlenbeck postulated that the electron had an *intrinsic* angular momentum. The intrinsic angular momentum causes the electron to behave like a tiny magnet even when the orbital angular momentum is zero. We assign the electron an additional quantum number (*s*), called the *intrinsic angular momentum* quantum number,

$$s = \frac{1}{2}, \qquad (8.76)$$

corresponding to the

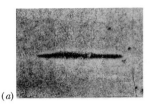

(a)

(b)

FIGURE 8-12 Results from the Stern-Gerlach experiment.
(a) A beam of silver atoms from a hot oven is deposited on a piece of glass and detected by photographic techniques. (b) The beam passes through a region of nonuniform magnetic field before reaching the detector. The beam is split into two due to interaction of the magnetic moment of the atom with the magnetic field. The magnetic moment of the silver atom is caused by the intrinsic angular momentum of the electron. (Photos courtesy of Dr. Vera Rubin.)

$$2s + 1 = 2 \qquad (8.77)$$

states. The intrinsic angular momentum quantum number is commonly referred to as the *spin* quantum number. The term "spin" is somewhat misleading since the electron is not really "spinning" like a top. Intrinsic angular momentum is a purely quantum mechanical effect with no classical analogy. The magnitude of the intrinsic angular momentum is

$$S = \sqrt{s(s+1)}\,\hbar = \frac{\sqrt{3}}{2}\hbar\,, \qquad (8.78)$$

analogous to the orbital angular momentum L (8.45). The z component of the intrinsic angular momentum is

$$S_z = m_s\hbar\,, \qquad (8.79)$$

where $m_s = -1/2$ or $1/2$, analogous to L_z (8.46).

Corresponding to the intrinsic angular momentum, the electron has an intrinsic magnetic moment,

$$\boldsymbol{\mu}_s = -\frac{e}{2m}g\,\mathbf{S}\,, \qquad (8.80)$$

where the factor g accounts for the intrinsic charge-to-mass ratio of the electron. (The g factor for orbital motion is 1.) The *g* factor is measured to be very nearly equal to 2:

$$g \approx 2, \tag{8.81}$$

The small deviation from 2 is discussed further in Section 8-9. The intrinsic magnetic moment (8.80) is a fundamental property of the electron analogous to the electric charge.

The total magnetic moment of an electron in an atom is

$$\boldsymbol{\mu} = -\frac{e}{2m}(\mathbf{L} + g\mathbf{S}) \approx -\frac{e}{2m}(\mathbf{L} + 2\mathbf{S}). \tag{8.82}$$

In the Stern-Gerlach experiment, the silver atom behaves like an atom with one electron because net orbital angular momentum of the inner electrons is zero and the net intrinsic angular momentum of the inner electrons is zero. The quantum numbers $\mathbf{L}$ and $\mathbf{S}$ of the atom are given by the quantum numbers of the outer electron. In the ground state, the orbital angular momentum of the outer electron is zero ($L = 0$). The magnetic moment of the atom is given by the magnetic moment of the outer electron

$$\boldsymbol{\mu} = -\frac{e}{2m}g\,\mathbf{S} \approx -\frac{e}{m}\mathbf{S}. \tag{8.83}$$

The intrinsic angular momentum vector ($\mathbf{S}$) specifies the magnetic moment. The vector $\mathbf{S}$ can point in one of two directions, as indicated in Figure 8-13:

$$\mu_z = -\frac{e}{m}m_s\hbar. \tag{8.84}$$

Since m_s can be $+1/2$ or $-1/2$, the value of μ_z can be $-\mu_B$ or $+\mu_B$. There are two possibilities for the direction of the force, and the beam of silver atoms is split into two when it passes through the region of nonuniform magnetic field.

The electron has an internal degree of freedom, specified by the quantum numbers $s = 1/2$ and m_s, not predicted by the Schrödinger equation. The electron has an intrinsic angular momentum

$$S = \sqrt{s(s+1)}\,\hbar = \frac{\sqrt{3}}{2}\,\hbar,$$

and *z* component

$$S_z = m_s\hbar,$$

where $m_s = -1/2$ or $+1/2$.

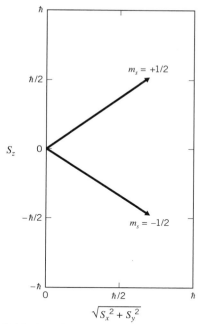

FIGURE 8-13 Angular momentum quantization for an atom with one outer electron in an $\ell = 0$ state.
The total angular momentum is not zero because the electron has an intrinsic angular momentum. The *z* component (S_z) of the intrinsic angular momentum can be $\hbar/2$ or $-\hbar/2$. The total angular momentum (S) is $\sqrt{3}\hbar/2$.

8-5 TOTAL ANGULAR MOMENTUM

The total angular momentum ($\mathbf{J}$) is the vector sum of the orbital and intrinsic parts,

$$\mathbf{J} = \mathbf{L} + \mathbf{S}. \tag{8.85}$$

The total angular momentum ($\mathbf{J}$) is conserved in all interactions, and not necessarily $\mathbf{L}$ or $\mathbf{S}$ individually. The total angular momentum has the value

$$J = \sqrt{j(j+1)}\,\hbar, \tag{8.86}$$

where the possible values of *j* are $|\ell-s|$, $|\ell-s|+1,\ldots |\ell+s|$. The *z* component of angular momentum has the value

$$J_z = m_j\hbar, \tag{8.87}$$

where the possible values of m_j are $-j$, $-j+1$, $-j+2,\ldots j$.

In the spectroscopic notation for the atomic states, we often label the value of the total angular momentum quantum number (*j*) with a subscript following the letter code for orbital angular momentum. Thus, the two 2*p* states of hydrogen (see Example 8-10) are called $2p_{1/2}$ (for $j = 1/2$) and $2p_{3/2}$ (for $j = 3/2$).

EXAMPLE 8-9

What is the total angular momentum (J) and z component (J_z) for an electron in the ground state of hydrogen?

SOLUTION:

The orbital angular momentum quantum number is

$$\ell = 0.$$

The intrinsic angular momentum quantum number is

$$s = \frac{1}{2}.$$

There is only one possibility for the total angular momentum quantum number:

$$j = \ell + s = \frac{1}{2}.$$

There are two possible values for the quantum number m_j: $-1/2$ and $1/2$. The total electron angular momentum is

$$J = \sqrt{j(j+1)}\,\hbar = \frac{\sqrt{3}}{2}\hbar.$$

The possible values of the z component of angular momentum are $-\hbar/2$ or $\hbar/2$. ∎

The observed total angular momentum of the electron in the ground state of hydrogen is

$$J = \frac{\sqrt{3}}{2}\hbar. \tag{8.88}$$

This value of total angular momentum is between that predicted by the Bohr model ($J = \hbar$) and the Schrödinger equation ($J = 0$). The relativistic formulation of the wave equation of the electron was made in 1927 by Paul A. M. Dirac. The *Dirac equation* correctly predicts the angular momentum of the electron (see Table 8-2).

EXAMPLE 8-10

What are the possible values for the total angular momentum (J) for an electron in a p state?

SOLUTION:

First we determine the possible values of the total angular momentum quantum number (j). The orbital quantum number is

TABLE 8-2
TOTAL ANGULAR MOMENTUM OF THE ELECTRON IN THE GROUND STATE OF HYDROGEN.

Theory	*Angular Momentum*
Bohr Model	$\hbar$
Schrödinger Equation	0
Dirac Equation	$\dfrac{\sqrt{3}}{2}\hbar$

$$\ell = 1.$$

The intrinsic angular momentum quantum number is

$$s = \frac{1}{2}.$$

The possible values of j are $1/2$ or $3/2$. Thus,

$$J = \sqrt{\frac{1}{2}\left(\frac{1}{2}+1\right)}\,\hbar = \frac{\sqrt{3}}{2}\hbar,$$

or

$$J = \sqrt{\frac{3}{2}\left(\frac{3}{2}+1\right)}\,\hbar = \frac{\sqrt{15}}{2}\hbar. \quad ∎$$

The intrinsic angular momentum vector **S** and the orbital angular momentum vector **L** are shown in Figure 8-14.

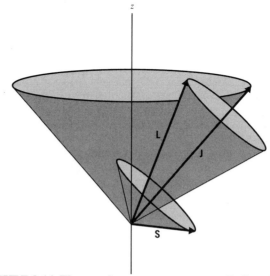

FIGURE 8-14 The angular momentum vectors L, S, and J.

The vectors **L** and **S** precess about **J** such that the resultant vector **J** has a fixed magnitude and a fixed z component. The vector **J** precesses about the z axis. The addition of angular momentum, $\mathbf{J} = \mathbf{L} + \mathbf{S}$, for $\ell = 1$ and $s = 1/2$ is shown in Figure 8-15.

8-6 THE SPIN–ORBITAL INTERACTION: FINE STRUCTURE

If the orbital angular momentum is not zero, then the orbital motion of the electron creates an internal magnetic field ($\mathbf{B}_{int}$). In the rest frame of the electron, the proton appears as a cloud with an angular momentum (**L**). The internal magnetic field is oriented in the same direction as the orbital angular momentum. The portion of electron

magnetic moment (μ_s) due to the electron intrinsic angular momentum (**S**) interacts with this internal magnetic field, causing an energy shift (ΔE). The size of this energy shift is

$$\Delta E = -\mathbf{\mu}_s \cdot \mathbf{B}_{int}$$

$$= \frac{e}{2m} g\mathbf{S} \cdot \mathbf{B}_{int} = C\,\mathbf{S} \cdot \mathbf{L}, \qquad (8.89)$$

where C is a positive constant. The amount of energy shift depends on the orientation of the electron intrinsic angular momentum vector (**S**) with respect to **L**. If **S** is aligned with **L**, then the energy shift is positive, whereas if **S** is antialigned with **L**, then the energy shift is negative. Consider the $2p$ states of hydrogen. The energy shift is negative for the case where $\mathbf{S} \cdot \mathbf{L}$ is negative, which cor-

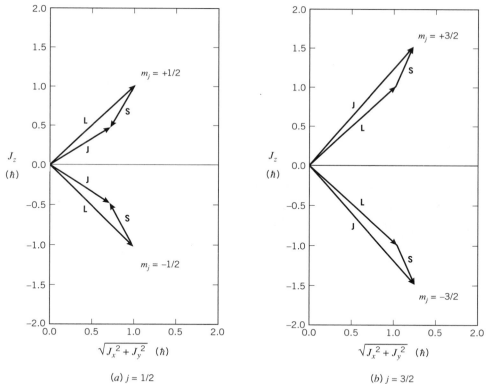

(a) $j = 1/2$ (b) $j = 3/2$

FIGURE 8-15 The total angular momentum, J=L+S, for a p state electron.
The total angular momentum is the vector sum of the orbital (**L**) and the intrinsic (**S**) angular momentum. For a p state, $\ell = 1$, corresponding to an orbital angular momentum of $\sqrt{2}\hbar$. The electron spin quantum number is 1/2, corresponding to an intrinsic angular momentum of $\sqrt{3}\hbar/2$. The two possibilities for the total angular momentum quantum number are (a) $j = 1/2$ or (b) 3/2 corresponding to $J = \sqrt{3}\hbar/2$ or $\sqrt{15}\hbar/2$. The smaller value of J corresponds to **L** and **S** being approximately antialigned, and the larger value of J corresponds to **L** and **S** being approximately aligned.

responds to the smaller value of **J** ($j = 1/2$). The energy shift is positive for the case where $\mathbf{S} \cdot \mathbf{L}$ is positive, which corresponds to the larger value of **J** ($j = 3/2$). Thus, the $2p$ states of hydrogen are split into two states, with the $2p_{3/2}$ state having a larger energy than the $2p_{1/2}$ state. Note that there is no fine-structure splitting of the $s_{1/2}$ state because $L = 0$. The fine structure splitting of hydrogen is shown in Figure 8-16.

The amount of fine-structure splitting is tiny compared to the $2p$ energy level (-3.4 eV). We may make an estimate of the fine-structure energy splitting by using the Bohr model of the atom. The magnetic field due to a current I is given by Ampère's law

$$\oint_L d\mathbf{l} \cdot \mathbf{B} = \frac{4\pi k I}{c^2}. \tag{8.90}$$

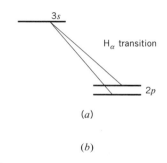

$3s$

H_α transition

$2p$

(a)

(b)

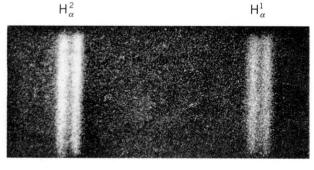

H_α^2 H_α^1

656.1   656.2   656.3

λ (nm)

FIGURE 8-16 The fine structure of hydrogen.
(a) The $2p$ state of hydrogen is split by the spin-orbital interaction. This results in the splitting of the H_α ($3s \rightarrow 2p$) transition into two lines. The order of magnitude of the energy splitting is 1 eV times α^2. *(b)* Observation of fine-structure splitting in hydrogen (H_α^1) and deuterium (H_α^2). After H. E. White, *Introduction to Atomic Spectra*, McGraw-Hill (1934). Photo by Lewis and Spedding.

The magnetic field at the center of a current loop of radius r is

$$B = \left(\frac{4\pi k}{c^2} \right)\left(\frac{I}{2r} \right). \tag{8.91}$$

The current is equal to the charge times the frequency of revolution

$$I = \frac{ev}{2\pi r}, \tag{8.92}$$

where v is the electron speed. The field at the center of the current loop is

$$B = \left(\frac{4\pi k}{c^2} \right)\left(\frac{I}{2r} \right)$$
$$= \left(\frac{4\pi k}{c^2} \right)\left(\frac{ev}{4\pi r^2} \right) = \frac{kev}{c^2 r^2}. \tag{8.93}$$

An estimate of the energy shift is

$$\Delta E \approx \frac{e}{2m} g\mathbf{S} \cdot \mathbf{B} \approx \left(\frac{e}{2m} \right)(2)\left(\frac{\hbar}{2} \right)\left(\frac{kev}{c^2 r^2} \right)$$
$$= \alpha \left(\frac{v}{c} \right)\left(\frac{\hbar^2}{2mr^2} \right), \tag{8.94}$$

For an electron in the $n = 2$ Bohr orbit, the speed is given by

$$\frac{v}{c} = \frac{\alpha}{2}, \tag{8.95}$$

and the radius of the orbit is

$$r = 4a_0 = \frac{4\hbar c}{\alpha mc^2}. \tag{8.96}$$

Thus, we may write the estimate of the energy shift as

$$\Delta E \approx \alpha \left(\frac{\alpha}{2} \right)\left(\frac{\hbar^2 \alpha^2 m^2 c^4}{32 m\hbar^2 c^2} \right) \approx \frac{\alpha^4 mc^2}{64}. \tag{8.97}$$

The energy shift is proportional to α^4, whereas the energy level without the shift is proportional to α^2. For this reason, the constant α has historically been referred to as the *fine-structure constant*. The energy shift is ΔE for the $2p_{3/2}$ state and $-\Delta E$ for the $2p_{1/2}$ state. The estimate of the energy splitting between the $2p_{1/2}$ and $2p_{3/2}$ states is

$$E_{2p_{3/2}} - E_{2p_{1/2}} \approx 2\,\Delta E \approx (2)\frac{5.1\times10^5\text{ eV}}{(64)(137)^4}$$

$$\approx 4.5\times10^{-5}\text{ eV}. \qquad (8.98)$$

* *Challenging*

Closer Look at the Energy Shifts

We have made an estimate of the energy splitting (8.98) between the $2p$ states of hydrogen using the Bohr model. We now reexamine this fine-structure splitting in light of the interpretation of **J** as the total angular momentum. The total angular momentum is a *constant of the motion* in the hydrogen atom. If there was no spin–orbital interaction, then the intrinsic angular momentum **S** and the orbital angular momentum **L** would also both be constants of the motion. The presence of the spin–orbital interaction destroys this because there is a torque acting on both **S** and **L** that changes their directions!

In the rest frame of the electron, the magnetic field due to the motion of the proton may be written in terms of the orbital angular momentum:

$$\mathbf{B} = -\frac{\mathbf{v}\times\mathbf{E}}{c^2} = \frac{ke\mathbf{L}}{mc^2 r^3}, \qquad (8.99)$$

so that the energy shift (8.89) for a level with principle quantum number n may be written in terms of the average value of $\mathbf{S}\cdot\mathbf{L}$:

$$\Delta E_n = -\langle\mu\cdot\mathbf{B}\rangle = \left\langle\left(\frac{e}{2m}\right)(2\mathbf{S})\cdot\left(\frac{ke\mathbf{L}}{mc^2 r^3}\right)\right\rangle$$

$$= \frac{ke^2}{m^2 c^2}\left\langle\frac{\mathbf{S}\cdot\mathbf{L}}{r^3}\right\rangle. \qquad (8.100)$$

Our expression for ΔE_n is not exact because there are relativistic corrections. We shall first proceed with the evaluation of ΔE_n and return later to these relativistic corrections.

The spin–orbital interaction causes a change or *perturbation* in the wave functions; however, the interaction causes such a small shift in the energy levels (8.98) that we may use the unperturbed wave functions to calculate the average of $\mathbf{S}\cdot\mathbf{L}/r^3$. Therefore,

$$\left\langle\frac{\mathbf{S}\cdot\mathbf{L}}{r^3}\right\rangle = \iiint_V dV\,\psi^*\,\frac{\mathbf{S}\cdot\mathbf{L}}{r^3}\,\psi. \qquad (8.101)$$

Since all of the radial dependence of the wave function is contained in $R_{n,\ell}(r)$ and the angular momentum part $\mathbf{S}\cdot\mathbf{L}$ does not depend on r, we have

$$\left\langle\frac{\mathbf{S}\cdot\mathbf{L}}{r^3}\right\rangle = \left[\int_0^\infty dr\,r^2 R_{n,\ell}{}^2(r)\frac{1}{r^3}\right]$$

$$\times\left[\int_0^\pi d\theta\sin\theta\int_0^{2\pi}d\phi\,P_{\ell,m_l}(\theta)e^{-im_l\phi}\right.$$

$$\left.\mathbf{S}\cdot\mathbf{L}P_{\ell,m_l}(\theta)e^{im_l\phi}\right]. \qquad (8.102)$$

The average of $\mathbf{S}\cdot\mathbf{L}/r^3$ is $1/r^3$ averaged over the radial part of the wave function and $\mathbf{S}\cdot\mathbf{L}$ averaged over the angular part. The average value of $1/r^3$ depends on the quantum numbers n and ℓ. The evaluation of the average of $1/r^3$ is straightforward although somewhat tedious. The result is

$$\left\langle\frac{1}{r^3}\right\rangle = \frac{2}{a_0{}^3 n^3 \ell(\ell+1)(2\ell+1)}. \qquad (8.103)$$

To compute the average of $\mathbf{S}\cdot\mathbf{L}$ over the angular part we write out the square of the total angular momentum vector:

$$J^2 = \mathbf{J}\cdot\mathbf{J} = (\mathbf{L}+\mathbf{S})\cdot(\mathbf{L}+\mathbf{S})$$

$$= \mathbf{L}\cdot\mathbf{L}+\mathbf{S}\cdot\mathbf{S}+2\mathbf{L}\cdot\mathbf{S}$$

$$= L^2 + S^2 + 2\mathbf{L}\cdot\mathbf{S}. \qquad (8.104)$$

Therefore,

$$\langle\mathbf{S}\cdot\mathbf{L}\rangle = \frac{1}{2}\langle J^2 - L^2 - S^2\rangle. \qquad (8.105)$$

The average value of J^2, L^2, and S^2 are determined by the quantum numbers j, ℓ, and s so that

$$\langle\mathbf{S}\cdot\mathbf{L}\rangle$$

$$= \frac{j(j+1)\hbar^2 - \ell(\ell+1)\hbar^2 - s(s+1)\hbar^2}{2}. \qquad (8.106)$$

The value of s is 1/2 and there are two possible values for j: $\ell+1/2$ or $\ell-1/2$ unless $\ell = 0$ in which case $j = 1/2$. Therefore, the two possible values of $\langle\mathbf{S}\cdot\mathbf{L}\rangle$ are for $j = \ell + 1/2$,

$$\langle \mathbf{S} \cdot \mathbf{L} \rangle = \frac{\ell \hbar^2}{2} , \qquad (8.107)$$

and for $j = \ell - 1/2$,

$$\langle \mathbf{S} \cdot \mathbf{L} \rangle = -\frac{(\ell+1)\hbar^2}{2} \qquad (8.108)$$

provided that ℓ is not zero. Putting these results together, for $j = \ell + 1/2$ we have

$$\left\langle \frac{\mathbf{S} \cdot \mathbf{L}}{r^3} \right\rangle = \left[\frac{2}{a_0{}^3 n^3 \ell(\ell+1)(2\ell+1)} \right] \left[\frac{\ell \hbar^2}{2} \right]$$

$$= \frac{\hbar^2}{a_0{}^3 n^3 (\ell+1)(2\ell+1)}, \qquad (8.109)$$

and for $j = \ell - 1/2$ we have

$$\left\langle \frac{\mathbf{S} \cdot \mathbf{L}}{r^3} \right\rangle$$

$$= \left[\frac{2}{a_0{}^3 n^3 \ell(\ell+1)(2\ell+1)} \right] \left[-\frac{(\ell+1)\hbar^2}{2} \right]$$

$$= -\frac{\hbar^2}{a_0{}^3 n^3 \ell(2\ell+1)}, \qquad (8.110)$$

provided that ℓ is not zero. Using $E_n = -mc^2\alpha^2/2n^2$, the energy shift (8.100) for $j = \ell + 1/2$ is

$$\Delta E_n = \frac{ke^2 \hbar^2}{m^2 c^2 a_0{}^3 n^3 (\ell+1)(2\ell+1)}$$

$$= \frac{\alpha^4 mc^2}{n^3 (\ell+1)(2\ell+1)}$$

$$= -\frac{2 E_n \alpha^2}{n(\ell+1)(2\ell+1)}, \qquad (8.111)$$

and for $j = \ell - 1/2$ is

$$\Delta E_n = -\frac{ke^2 \hbar^2}{m^2 c^2 a_0{}^3 n^3 \ell(2\ell+1)}$$

$$= -\frac{\alpha^4 mc^2}{n^3 \ell(2\ell+1)} = \frac{2 E_n \alpha^2}{n\ell(2\ell+1)}, \qquad (8.112)$$

provided that ℓ is not zero.

We now discuss the relativistic corrections to the energy shifts. The first is due to the fact that the orbital time calculated in the rest frame of the electron is not equal to the orbital time calculated in the rest frame of the proton due to time dilation. The angular frequency in the rest frame of the electron is L/mr^2. The angular frequency in the rest frame of the proton is $L/\gamma mr^2$, where $\gamma = (1 - v^2/c^2)^{-1/2}$. The difference in the two angular frequencies (ω_T) is

$$\omega_T = \frac{L}{\gamma mr^2} - \frac{L}{mr^2} = \frac{L}{mr^2}\left(\frac{1}{\sqrt{1 - \dfrac{v^2}{c^2}}} - 1 \right)$$

$$\approx \frac{Lv^2}{2mr^2 c^2} = \frac{ke^2 L}{2m^2 c^2 r^3}, \qquad (8.113)$$

where we have used $mv^2/r = ke^2/r^2$ to eliminate v. Therefore, the angular momentum vector precesses at the frequency ω_T, a phenomenon called *Thomas precession*. Thomas precession causes an energy shift that has the opposite sign of ΔE_n (8.100). The magnitude of the shift can be obtained by examining the Larmor precession frequency of $\mathbf{S}$ in an external field $\mathbf{B}$, which is

$$\omega_L = -g\frac{eB}{2m} = -\frac{eB}{m} = -\frac{ke^2 L}{m^2 c^2 r^3}. \qquad (8.114)$$

We see that ω_T is one-half as large as ω_L so that the energy shift due to Thomas precession (ΔE_T) is

$$\Delta E_T = -\frac{\Delta E_n}{2}. \qquad (8.115)$$

A second type of relativistic correction arises because we have approximated the kinetic energy as $p^2/2m$. The expression for the kinetic energy is

$$E_k = \sqrt{(mc^2)^2 + (pc)^2} - mc^2$$

$$\approx \frac{(pc)^2}{2mc^2} - \frac{(pc)^4}{8(mc^2)^3}. \qquad (8.116)$$

Since the order of magnitude of the electron speed is αc, the correction will be proportional to α^4, that is, it is the same order of magnitude as ΔE_n. We need to calculate the average value of the correction to the kinetic energy, which may be written

$$\Delta E_{\mathrm r} = \left\langle \frac{(pc)^4}{8(mc^2)^3} \right\rangle$$

$$= \frac{1}{2mc^2} \left\langle \left(\frac{p^2}{2m}\right)^2 \right\rangle = \frac{\left\langle (E_n - V)^2 \right\rangle}{2mc^2}$$

$$= \frac{E_n^2 + \left\langle V^2 \right\rangle - 2E_n \left\langle V \right\rangle}{2mc^2}. \qquad (8.117)$$

This calculation is straightforward:

$$\Delta E_{\mathrm r} = \frac{E_n^2}{2mc^2}$$

$$+ \frac{1}{2mc^2} \iiint_V dV \psi^* \left(-\frac{ke^2}{r}\right)^2 \psi$$

$$- \frac{E_n}{mc^2} \iiint_V dV \psi^* \left(-\frac{ke^2}{r}\right) \psi. \qquad (8.118)$$

The result depends on n and ℓ:

$$\Delta E_{\mathrm r} = -\frac{\alpha^2 E_n}{4n^2} \left(3 - \frac{8n}{2\ell+1}\right). \qquad (8.119)$$

The complete first-order corrections to the energy level shifts are

$$\Delta E_{\mathrm{tot}} = \Delta E_n + \Delta E_{\mathrm T} + \Delta E_{\mathrm r}$$

$$= -\frac{\alpha^2 E_n}{4n^2} \left(3 - \frac{8n}{2j+1}\right), \qquad (8.120)$$

which is valid for both $j = \ell+1/2$ and $j = \ell-1/2$ (see problem 36). For the ground state of hydrogen, $n = 1$ and $j = 1/2$ and the energy shift is

$$\Delta E_{1s_{1/2}} = -\frac{\alpha^2 (13.6\,\mathrm{eV})}{4} = -1.8 \times 10^{-4}\ \mathrm{eV}. \qquad (8.121)$$

For the first excited state, $n = 2$ and $j = 1/2$ or $j = 3/2$. The energy shifts are

$$\Delta E_{2s_{1/2}} = \Delta E_{2p_{1/2}} = -\frac{5\alpha^2 (13.6\,\mathrm{eV})}{64}$$

$$= -5.6 \times 10^{-5}\ \mathrm{eV}, \qquad (8.122)$$

and

$$\Delta E_{2p_{3/2}} = -\frac{\alpha^2 (13.6\,\mathrm{eV})}{64} = -1.1 \times 10^{-5}\ \mathrm{eV}. \qquad (8.123)$$

The energy splitting between the $2p_{3/2}$ and $2p_{1/2}$ states is

$$\Delta E_{2p_{3/2}} - \Delta E_{2p_{1/2}} = 4.5 \times 10^{-5}\ \mathrm{eV}. \qquad (8.124)$$

These energy shifts are summarized in Figure 8-17. *****

Hyperfine Structure

The proton also has an intrinsic angular momentum. We have seen the contribution of the proton intrinsic angular momentum to the electron–proton scattering cross section (see Chapter 6). There is also an interaction of the electron intrinsic angular momentum with the proton intrinsic angular momentum in the hydrogen atom. This interaction produces a *hyperfine structure* in the ground state of hydrogen ($1s_{1/2}$ state) because the energy of the state in which the electron has $m_j = 1/2$ has a slightly higher energy than the state in which the electron has $m_j = -1/2$. When hydrogen makes the transition between the two states, the wavelength of the emitted radiation is 0.21 meter. Detection of radiation at a wavelength of 0.21 meter is used as a signature for the presence of hydrogen in outer space.

8-7 ATOMIC TRANSITIONS AND SELECTION RULES

The photon has an intrinsic angular momentum characterized by the quantum number,

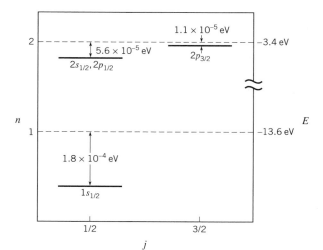

FIGURE 8-17 The fine structure of hydrogen.

$$s = 1. \qquad (8.125)$$

There are only two possibilities for the quantum number m_s: $m_s = +1$ or $m_s = -1$, corresponding to the two photon polarizations. (The value $m_s = 0$ is not allowed by special relativity; this is related to the fact that there exists no frame where the photon is at rest.) Angular momentum conservation places restrictions on the allowed atomic transitions involving the absorption or emission of a photon. These restrictions may be expressed in terms of the allowed changes in the quantum numbers of the electron. The restrictions are called *selection rules*. If a selection rule is not satisfied, then the transition is either forbidden or suppressed.

The selection rule on the electron orbital angular momentum is that ℓ must change by one unit:

$$\Delta \ell = \pm 1. \qquad (8.126)$$

The change in the magnetic quantum number must be

$$\Delta m_\ell = 0 \quad \text{or} \quad \pm 1. \qquad (8.127)$$

In addition, the electron intrinsic angular momentum magnetic quantum number (m_s) must not change

$$\Delta m_s = 0, \qquad (8.128)$$

and the change in the electron total angular momentum quantum number (j) must be

$$\Delta j = 0 \quad \text{or} \quad \pm 1, \qquad (8.129)$$

with the restriction that the transition $j = 0 \rightarrow j = 0$ is forbidden.

8-8 THE ZEEMAN EFFECT

If a hydrogen atom is placed in an external magnetic field, the magnetic moment of the electron in the atom will interact with the field causing an energy shift:

$$\Delta E = -\boldsymbol{\mu} \cdot \mathbf{B}. \qquad (8.130)$$

When the external magnetic field is much larger than the fields inside the atom, this energy shift (8.130) will dominate over the spin–orbital interaction. As usual, we choose the z direction to be direction of the field, so that

$$\Delta E = -\mu_z B_z = \frac{e}{2m} \left(L_z + 2 S_z \right) B_z$$

$$= \frac{e\hbar}{2m} \left(m_\ell + 2 m_s \right) B_z. \qquad (8.131)$$

The splitting of energy levels in an external magnetic field was first observed by Pieter Zeeman in 1896 and is known as the *Zeeman effect*. The Zeeman effect for strong fields was studied by F. Paschen and E. Back and is referred to as the *Paschen-Back effect*. The Paschen-Back effect for hydrogen is illustrated in Figure 8-18. The p states are split into five states, and the s states are split into two states. The $s \rightarrow p$ transitions are split into two triplets of lines according to the selection rule $\Delta m_s = 0$. The triplet from the hydrogen H_α transition measured by Paschen and Back is shown in Figure 8-18c. The Zeeman effect is discussed further in Chapter 9.

EXAMPLE 8-11

Into how many states does the $3d$ state of hydrogen divide in a strong external magnetic field?

SOLUTION:

The value of the orbital angular momentum quantum number is

$$\ell = 2.$$

The possible values of m_ℓ are -2, -1, 0, 1, and 2. The possible values of $2m_s$ are -1, and 1. The possible values of $(m_\ell + 2m_s)$ are $-3, -2, -1, 0, 1, 2,$ and 3. Therefore, the $3d$ state is split into 7 states in the strong external magnetic field. ∎

EXAMPLE 8-12

Calculate the magnetic field strength needed to make an upward fractional energy shift ($\Delta E/E$) of 10^{-4} on the $3d_{5/2}$ state of hydrogen.

SOLUTION:

The energy of the $3d$ state is

$$E_3 = \frac{-13.6\,\text{eV}}{9}.$$

The energy shift is

$$\Delta E = \frac{\left(10^{-4} \right)\left(13.6\,\text{eV} \right)}{9}.$$

The quantum numbers of the electron are $\ell = 2$, $s = 1/2$ and $j = 5/2$. The maximum shift will occur when $m_j = 5/2$ corresponding to $m_\ell = 2$ and $m_s = 1/2$, so

$$m_\ell + 2 m_s = 3.$$

The magnetic field is

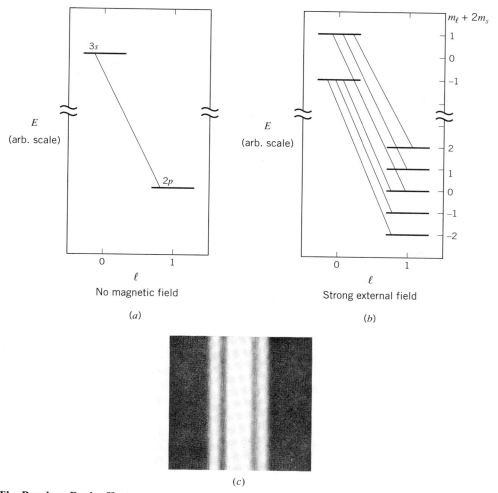

FIGURE 8-18 The Paschen-Back effect.

(*a*) The 2*p* and 3*s* states of hydrogen with no external magnetic field, neglecting fine structure. (*b*) In a strong external magnetic field (compared to the internal magnetic field), the energy shift of each state is proportional to $m_\ell + 2m_s$ which splits the the 2*p* state into 5 states and the 3*s* state into 2 states. For the $3s \rightarrow 2p$ transitions (H_α) , there is a selection rule $\Delta m_s = 0$ corresponding to the 6 transitions shown. There are only 3 unique photon energies. (*c*) First observation of the hydrogen H_α transition in a strong magnetic field showing 3 photon lines. From F. Paschen and E. Back *Ann. d. Phys.* **39**, 897 (1912).

* *Challenging*

8-9 THE LAMB SHIFT

The electron in a hydrogen atom interacts with itself by emission and absorption of photons as indicated in Figure 8-19. The effect of this self-interaction is to cause a "smearing" of the electron over a distance scale of approximately 0.1 fm. This effect is observable when

$$B = \frac{\Delta E}{3\mu_B} = \frac{\left(10^{-4}\right)\left(\dfrac{3.6\,\text{eV}}{9}\right)}{(3)\left(5.79 \times 10^{-5}\ \text{eV/T}\right)} \approx 0.23\,\text{T}. \quad \blacksquare$$

the electron is very close to the proton. In this case, the "smeared" electron has a slightly weaker attraction to the proton compared to an "unsmeared" electron. This means that a state where the electron has a higher probability of being near the proton will have a slightly higher energy. (It will take slightly less energy to remove the electron because the force is weaker.) Therefore, the hydrogen $2s_{1/2}$ state has a slightly higher energy than the $2p_{1/2}$ state (see Figure 8-3). This energy shift due to the self-interaction of the electron is called the *Lamb shift*, after Willis Lamb, who first accurately measured it in 1951. The magnitude of the Lamb shift can be calculated to many significant figures, and comparison with measurement of the energy level differences provides the most powerful quantitative test of the quantum theory of electrodynamics. It proves that the picture of the electron as continually emitting and absorbing photons is the correct one!

An extremely stringent test of quantum electrodynamics comes from the accurate calculation of the difference in energy of the $2s_{1/2}$ and $2p_{1/2}$ levels in the hydrogen atom. Part of the energy difference arises from the self interaction of the electron causing a deviation of g from 2. We have seen that the spin–orbital correction to the energy depends on the total angular momentum

quantum number (j). In the $2s_{1/2}$ state, $j = 1/2$ comes from the electron intrinsic angular momentum, whereas in the $2p_{1/2}$ state, $j = 1/2$ comes from one unit of orbital angular momentum minus one-half unit of intrinsic angular momentum. If g is exactly equal to 2, then the z component of electron magnetic moment due to intrinsic angular momentum is exactly equal to the z component of electron magnetic moment due to one unit of orbital angular momentum. For $g = 2$ the energies of the states of fixed n depend only on the total angular momentum quantum number j and not the addition of ℓ and s producing j.

The first-order correction to g (in powers of α) is calculated in quantum electrodynamics to be

$$g \approx 2 + \frac{\alpha}{\pi}, \tag{8.133}$$

or

$$\frac{g-2}{2} \approx \frac{\alpha}{2\pi}. \tag{8.134}$$

Thus, g is slightly smaller than 2 and the $2p_{1/2}$ state is slightly smaller in energy than the $2s_{1/2}$ state.

The difference in energy levels was first measured by Willis E. Lamb. The energy shift is indicated in Figure 8-20. In his experiment, Lamb formed a beam of hydrogen atoms in the $2s_{1/2}$ state. These atoms cannot make transitions to the $1s_{1/2}$ state because of the selection rule, $\Delta\ell = \pm 1$. (The electromagnetic classical analogy is that a spherically symmetric charge and current distribution cannot radiate.) The atoms can make transitions to the $2p$ levels, and from there to the $1s$ ground state. Lamb put his hydrogen atoms in a magnetic field so that the energy levels were split by the Zeeman effect. He then exposed the atoms to radio waves of a fixed frequency (2395 MHz) and varied the magnetic field while measuring the transitions

$$2s \rightarrow 2p \rightarrow 1s. \tag{8.135}$$

Lamb used his data to determine the zero magnetic field energy separation of the $2s_{1/2}$ and $2p_{1/2}$ states. Expressed as the frequency of a transition photon, this splitting is

$$f = 1057\,\text{MHz}, \tag{8.136}$$

or

$$\Delta E = hf = \left(4.136 \times 10^{-15}\,\text{eV}\cdot\text{s}\right)\left(1.057 \times 10^{9}\,\text{s}^{-1}\right)$$

$$= 4.372 \times 10^{-6}\,\text{eV}. \tag{8.137}$$

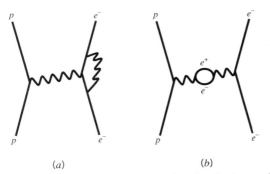

(a)   (b)

FIGURE 8-19 Feynman diagrams for the electron self-interaction.
Two diagrams (*a*) and (*b*) with amplitude of order α^2 are shown. The electron interacts with itself by emitting and absorbing photons, causing the electron g factor to differ slightly from 2. Interations of the type shown in diagram (*b*) cause a small amount of charge screening of the proton from the electron, resulting in a slightly weaker force when the electron is in the immediate vicinity of the proton. Because the wave function of the $2s$ state is finite at $r = 0$ compared to the wave function of the $2p$ state, which vanishes at $r = 0$, the force on an electron in the $2s$ state is smaller than an electron in the $2p$ state, and the energy of the $2s$ state is slightly larger than the energy of the $2p$ state.

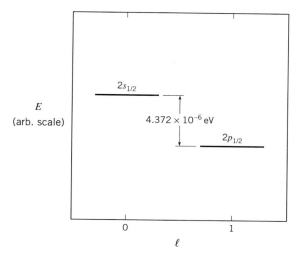

FIGURE 8-20 The Lamb shift in hydrogen.
The spin–orbital interaction gives corrections to the energy levels in the hydrogen atom that depend only on the quantum numbers n and j. The $2s_{1/2}$ and the $2p_{1/2}$ states of hydrogen have the same energy after applying this correction. The gyromagnetic factor (g) of the electron is slightly greater than 2 so that a component of spin angular momentum of $g\hbar/2$ contributes more to the electron magnetic moment than a component of orbital angular momentum $\hbar$. This causes the $2s_{1/2}$ state to have a slightly higher energy than the $2p_{1/2}$ state. The factor g can be calculated to many decimal places so that the measurement of the energy difference of the $2s_{1/2}$ and $2p_{1/2}$ states provides a precision test of quantum electrodynamics.

This energy difference is about 1/10 as large as the energy splitting between the $2p$ states due to the spin–orbital interaction. The Lamb shift has been measured to many decimal places, and comparison with higher order calculations of quantum electrodynamics shows perfect agreement!   *

CHAPTER 8: PHYSICS SUMMARY

- The solution to the Schrödinger equation with a Coulomb potential energy gives a quantitatively correct description of the energy levels and electron probability distributions of the hydrogen atom (apart from small corrections due to the effects of intrinsic angular momentum and relativity).

- The boundary conditions on the hydrogen atom wave functions lead to quantization of energy (E), angular momentum (L), and the projection of angular momentum in any given direction (L_z). The quantization conditions are:

$$E_n = -\frac{13.6\,\text{eV}}{n^2},$$

where n is a positive integer (1, 2, 3, …),

$$L = \sqrt{\ell(\ell+1)}\,\hbar,$$

where ℓ is an integer less than n (0, 1, 2, 3, … $n-1$), and

$$L_z = m_\ell\,\hbar,$$

where m_ℓ is an integer from $-\ell$ to ℓ.

- A complete description of the physics of the hydrogen atom must include the intrinsic angular momentum of the electron, which has a fixed value of

$$S = \frac{\sqrt{3}\,\hbar}{2},$$

corresponding to a spin quantum number $s = 1/2$. The component of intrinsic angular momentum in any given direction (S_z) is quantized ($S_z = \hbar/2$ or $S_z = -\hbar/2$) corresponding to $m_s = 1/2$ or $m_s = -1/2$.

- The total angular momentum of an atomic electron is the vector sum of its orbital angular momentum and its intrinsic angular momentum:

$$\mathbf{J} = \mathbf{L} + \mathbf{S}.$$

- The notation used for atomic states is nl_j, where n and j are the principal and total angular momentum quantum numbers and l is a letter (s, p, d, or f) that represents the value of the orbital angular momentum quantum number: s for $\ell = 0$, p for $\ell = 1$, d for $\ell = 2$, and f for $\ell = 3$.

- The magnetic moment of an electron in an atom is

$$\boldsymbol{\mu} = -\frac{e}{2m}(\mathbf{L} + g\mathbf{S}),$$

where the gyromagnetic factor (g) of the electron is very nearly equal to 2.

- There is a contribution to the fine-structure splitting of p states in atoms caused by the interaction of the electron spin with the internal magnetic field due to the electron orbital angular momentum. The order of magnitude of the splitting is α^2 times the energy of the state.

REFERENCES AND SUGGESTIONS FOR FURTHER READING

R. Eisberg, *Fundamentals of Modern Physics,* Wiley (1961).

R. Eisberg and R. Resnick, *Quantum Physics of Atoms, Molecules, Solids, Nuclei and Particles,* Wiley (1985).

S. Gasiorowicz, *Quantum Physics,* Wiley (1974).

P. Kusch, "The Electron Dipole Moment — A Case History," *Phys. Today* **19**, No. 2, 23 (1966).

W. E. Lamb, "Fine Structure of the Hydrogen Atom," *Science* **123**, 439 (1956).

R. Leighton, *Principles of Modern Physics,* McGraw-Hill 1959).

A. Pais, "George Unlenbeck and the Discovery of Electron Spin," *Phys. Today* **42**, No. 12, 34 (1989).

L. Pauling and E. B. Wilson, *Introduction to Quantum Mechanics,* McGraw-Hill (1935).

F. K. Richtmyer, E. H. Kennard and J. N. Cooper, *Introduction to Modern Physics,* McGraw-Hill (1969).

R. L. Sproull and W. A. Phillips, *Modern Physics,* Wiley (1980).

QUESTIONS AND PROBLEMS

The ground state solution

1. For a hydrogen atom in the ground state, what is the probability of finding the electron *exactly* at the Bohr radius?

2. (a) Use the radial probability distribution (Figure 8-2) to *estimate* the average kinetic energy of an electron in the ground state of hydrogen. (b) Use the radial probability distribution to estimate the uncertainty in position Δr of the electron. (c) Show that these results are in agreement with the uncertainty principle.

3. For the ground state of hydrogen, calculate the probability of finding the electron at a distance less than the Bohr radius.

4. In the Bohr model, the radii of the ground and first excited state orbits are a_0 and $4a_0$, respectively. For the ground state of hydrogen, calculate the probability of finding the electron in the interval $a_0 < r < 4a_0$.

5. For the ground state of hydrogen, the electron is most likely to be found near the Bohr radius (a_0). (a) Use the graph of the radial probability distribution (Figure 8-2) to make a quick estimate of the root-mean-square deviation of r from a_0. (b) Calculate the root-mean-square deviation of r from a_0.

*6. (a) For the ground state of hydrogen, determine the radial distance for which the probability of finding the electron at a radius less than this distance is 90%. Give a numerical answer of two significant figures. (b) Repeat the calculation for a 99% probability.

Separation of variables

7. The electron energy appears in the radial equation but not in the angular equations. What physical fact does this explain?

8. Why does the boundary condition on r lead to nodes in the radial probability distribution?

Three quantum numbers

9. Estimate the probability that the electron will be found inside the proton for (a) the 2s state and (b) the 2p state. Compare your answers to the result (Example 8-2) for the 1s state.

10. How can an electron in the 2s or 2p state of hydrogen have very nearly the same energy when their spatial distributions are significantly different?

11. (a) What are the possible values of L and L_z for the 3p state? (b) For the 3d state?

12. A hydrogen atom is in an excited state, $n = 5$. (a) What are the possible values of the quantum numbers ℓ and m_ℓ? (b) What are the possible values of the orbital angular momentum L?

13. (a) Show that the function

$$\psi = Cre^{-r/2\delta} \cos\theta$$

is a solution of the hydrogen atom, where $\delta = \hbar^2/mke^2$. (b) Determine the energy of the state. (c) What is the value of the angular momentum L?

*14. (a) Show by direct substitution that the 3s wave function (see Table 8-1) is a solution of the Schrödinger equation with an energy $-\alpha^2mc^2/18$. (b) Show that the 3p wave functions give the same energy.

*15. (a) From the definition of spherical coordinates, $x = r \sin\theta \cos\phi$, $y = r \sin\theta \sin\phi$, and $z = r \cos\theta$, use the chain rule for derivatives to show that

$$dx = \sin\theta \cos\phi \, dr + r \cos\theta \cos\phi \, d\theta - r \sin\theta \sin\phi \, d\phi,$$

$$dy = \sin\theta \sin\phi \, dr + r \cos\theta \sin\phi \, d\theta + r \sin\theta \cos\phi \, d\phi,$$

and

$$dz = \cos\theta \, dr - r \sin\theta \, d\theta.$$

(b) Show that the solutions for dr, $d\theta$, and $d\phi$ are

$$dr = \sin\theta \cos\phi \, dx + \sin\theta \sin\phi \, dy + \cos\theta \, dz,$$

$$d\theta = \frac{1}{r} \cos\theta \cos\phi \, dx + \frac{1}{r} \cos\theta \sin\phi \, dy - \frac{1}{r} \sin\theta \, dz,$$

and

$$d\phi = -\frac{\sin\phi}{r\sin\theta}\,dx + \frac{\cos\phi}{r\sin\theta}\,dy\,.$$

(c) Verify the expressions for the orbital angular momentum operators $\mathbf{L}_x^{\text{opp}}$ (8.55), $\mathbf{L}_y^{\text{opp}}$ (8.56), and $\mathbf{L}_z^{\text{opp}}$ (8.52).

*16. Verify the expression for the total orbital angular momentum operator (8.58).

Intrinsic angular momentum

17. What would be the difficulty of performing the Stern-Gerlach experiment with electrons?

18. In the Stern-Gerlach experiment, nonrelativistic silver atoms with a kinetic energy of E_k are sent through a nonuniform magnetic field that has a gradient (dB/dz) in the direction perpendicular to the initial trajectory of the silver atoms. (a) If the atoms pass a distance L through the magnetic field, show that the beam is split by an amount

$$x = \frac{\mu_B L^2}{4E_k}\frac{dB}{dz},$$

where μ_B is the Bohr magneton. (b) If the silver atoms come from an oven at a temperature of 1000 K, and they travel a distance of 0.05 m through the magnetic field, calculate the magnetic field gradient needed to make a splitting of 0.001 m. (*Hint:* The average kinetic energy of a particle escaping from an oven at temperature T is $2kT$, see problem 36 in Chapter 2.)

Total angular momentum

19. What are the possible values for the total angular momentum (J) for an electron in a d state?

20. Can the orbital and intrinsic angular momentum vectors ever be exactly aligned? Why or why not?

21. (a) Show that the total number of states with total angular momentum quantum number j is $2j + 1$. (b) Write the quantum numbers of the $n = 2$ states of hydrogen in terms of n, ℓ, m_ℓ and m_s and in terms of n, ℓ, j and m_j. (c) If $m_j = 3/2$, what are the possible values of ℓ, m_ℓ and m_s? (d) If $m_j = 1/2$, what are the possible values of ℓ, m_ℓ and m_s?

The spin–orbital interaction: fine structure

22. Consider the transitions $3d \rightarrow 2p$ in hydrogen. Calculate the possible energies of the emitted photons, taking into account the spin–orbital interaction. How many photon lines are there?

23. Use the Bohr model to estimate the spin–orbital splitting of the $3p$ states.

Atomic transitions and selection rules

24. Consider a hydrogen atom in the state $3d_{3/2}$. Find all the allowed states of lower energy to which the atom can make a transition while satisfying the selection rules.

The Zeeman effect

25. Consider a hydrogen atom in the $n = 3$ state in a strong external magnetic field (B). (a) Make a diagram of the energy levels for all possible values of the quantum numbers m_ℓ and m_s. (b) Calculate all the possible photon energies observable for the transitions from the $n = 3$ to $n = 1$ state.

26. Calculate the energy levels of the $2p$ states of hydrogen in an external magnetic field of 5 tesla.

27. An electron in the hydrogen atom makes the transition $3d \rightarrow 2p$. (a) Calculate the energy of the emitted photon. (b) The transition occurs in a strong external magnetic field. Make an energy diagram showing the splitting of the $3d$ and $2p$ states. (c) An experimenter measures photons from the transition $3d \rightarrow 2p$ in a magnetic field with a resolution insufficient to resolve the splitting but good enough to observe a broadening of the photon "line." What value of B will produce an photon energy spread of $\Delta E/E = 10^{-4}$?

The Lamb shift

28. Why did Willis Lamb perform his experiment in a magnetic field when his objective was to determine the "zero field" separation of the $2s_{1/2}$ and $2p_{1/2}$ states?

Additional problems

29. Consider a model of an electron as a tiny uniform sphere of size 10^{-18} meter corresponding to the experimental limit on possible electron structure. Suppose that the electron intrinsic angular momentum is due to a spinning motion of the sphere with an angular frequency ω. Calculate the value of ω. Why is this a "bad" model of electron spin? (Useful information: Recall that the moment of inertia of a sphere spinning about an axis passing through its center is $I = 2mr^2/5$.)

30. Use the expression $L^2/2mr^2$ to make an estimate of the amount of kinetic energy due to orbital motion of an electron in the $2p$ state of hydrogen.

31. (a) For a hydrogen atom in the $2p$ state, determine the radius near which the electron is most likely to be found. (b) Make an estimate of the probability of finding the electron within $\pm 10\%$ of the value determined in part (a).

32. Show that the average value of the potential energy is equal to $-ke^2/n^2a_0$.

33. Calculate the average distance between the electron and the proton in the hydrogen atom for (a) the $2s$ state and (b) the $2p$ state.

34. What is the relationship between the most probable radial position of an electron and the width of the radial probability distribution? Explain in the context of the uncertainty principle. (*Hint*: What is the relationship between average kinetic energy and principal quantum number n?)

35. *The Larmor frequency.* (a) Show from the relationship between **B**, **μ**, and **L** (see Figure 8-10) that

$$dL = L \sin\theta \; d\phi,$$

where θ is the angle between **B** and **μ**, and ϕ is the azimuthal angle of **L**. (b) Use the definition of torque to show that

$$\frac{dL}{dt} = \frac{eLB \sin\theta}{2m}.$$

(c) Show that

$$\omega_L \equiv \frac{d\phi}{dt} = \frac{eB}{2m}.$$

36. Show that the sum of the hydrogen fine-structure energy shifts ΔE_n (8.111 and 8.112), ΔE_T (8.115) and ΔE_r (8.119) may be written in the compact form,

$$\Delta E_{tot} = -\frac{\alpha^2 E_n}{4n^2}\left(3 - \frac{8n}{2j+1}\right),$$

valid for both possible values of j.

37. (a) Use the uncertainty principle and the wave function plotted in Figure 8-2 to make an estimate of the minimum kinetic energy of an electron in the $n = 1$ state of hydrogen. (b) Use Figure 8-3 to estimate of the minimum kinetic energy of an electron in the $n = 2$ state.

38. The radial dependence of the wave function for an electron in the hydrogen atom at large r is

$$\psi \sim e^{-r/4a_0}.$$

Determine the energy of the electron.

39. Consider a one electron atom with a nucleus of charge Ze. Show that the wave functions (see Table 8-1) are the hydrogen atom wave functions with the Bohr radius replaced by a_0/Z.

*40. The average value of r for an electron in the hydrogen atom is given by

$$\langle r \rangle = n^2 a_0 \left[\frac{3}{2} - \frac{\ell(\ell+1)}{2n^2} \right].$$

(a) Verify that this formula holds for the $1s$, $2s$ and $2p$ states. (b) Use this result to compare the average radii for the $2s$ and $2p$ states and the $3s$, $3p$, and $3d$ states.

41. Show that the average value of $1/r$ does not depend on the orbital angular momentum.

The fundamental idea can be stated in the following way: The complicated numbers of electrons in closed subgroups are reduced to the simple number one if the division of the groups by giving the values of the 4 quantum numbers of an electron is carried so far that every degeneracy is removed. An entirely non-degenerate energy level is already "closed," if it is occupied by a single electron; states in contradiction with this postulate have to be excluded.

Wolfgang Pauli

Atoms with more than one electron are significantly more complicated than the hydrogen atom because the electrons interact with each other as well as with the nucleus. In spite of this difficulty, much of the physics of multielectron atoms may be qualitatively understood with the assumption that each electron in the atom is described by an independent wave function.

9-1 INDEPENDENT PARTICLE APPROXIMATION

Consider a neutral atom with Z electrons. The potential energy of a given electron is position dependent, not only because of the $1/r$ dependence of the Coulomb potential, but also because of the presence of the other $Z - 1$ electrons. When the given electron is far away from both the nucleus and all the other electrons ($r \gg a_0$), then the potential energy is $-ke^2/r$, as in the hydrogen atom. In the other extreme where the given electron is closer to the nucleus than all the other electrons ($r \ll a_0$), the potential energy is $-Zke^2/r$. For intermediate distances ($r \approx a_0$) the potential energy depends on the location of the other electrons but it is bounded by $-ke^2/r$ and $-Zke^2/r$. This is illustrated in Figure 9-1.

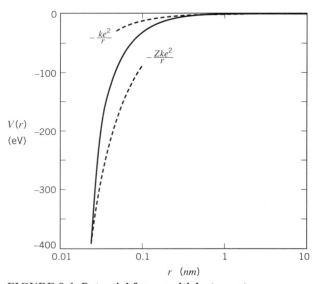

FIGURE 9-1 Potential for a multielectron atom.
Each electron moves in a Coulomb potential that is bounded by the hydrogen atom potential at large distances and that of a one-electron atom with nuclear charge Z at short distances.

The *independent particle approximation* assumes that each electron may be described by the potential of Figure 9-1. The same technique is used in calculating the orbits of planets about the sun, that is, to a good approximation the gravitation interaction between the planets may be ignored. The independent particle approximation is far better for the planets than for electrons in the atom because the solar mass is so much larger than the sum of the planetary masses, whereas the same is not true for the charges in the atom, that is, the sum of the electron charges equals the nuclear charge. The independent particle approximation allows us to determine an independent wave function for each electron. Each wave function is similar to a hydrogen wave function and is described by a set of four quantum numbers (n, ℓ, m_ℓ, and m_s). The central question before us is: *What are the values of these quantum numbers for each electron for an atom in the ground state*?

9-2 THE PAULI EXCLUSION PRINCIPLE

Photons are emitted by atoms when the electrons make transitions between allowed energy levels. Energy and angular momentum is conserved in these transitions. Since the energy and angular momentum of the initial and final states are given by the quantum numbers of the electrons, the existence of certain photon energies as well as the absence of others gives us much information on the quantum numbers of the electrons in the atoms. In 1925, by a detailed examination of atomic spectra, especially those in which the lines were split by a magnetic field, Wolfgang Pauli discovered an empirical law governing the quantum numbers of the electrons in an atom. The empirical law of Pauli was motivated by the observation that the number of *states* available to an electron of an atom in the first column of the periodic table (hydrogen, lithium, sodium, potassium) was *identical* to the increase in the number of *electrons* in the corresponding element in the last column (helium, neon, argon, krypton). For example, the ground state of hydrogen is twofold degenerate and the helium atom has two electrons; the ground state of lithium is eightfold degenerate and neon has eight more electrons than helium, and so on. Pauli's empirical law is that no two electrons in an atom may have an identical set of four quantum numbers (n, ℓ, m_ℓ, and m_s). An atomic energy level is *full* if it is occupied by a *single electron*. There is no room for a second electron. This rule is called the *Pauli exclusion principle*.

An atomic state may accommodate only a single electron. No two electrons in an atom may have an identical set of four quantum numbers (n, ℓ, m_ℓ and m_s).

Symmetry of the Wave Function

The Pauli exclusion principle is a quantum phenomenon, which is ultimately connected with the symmetry of the electron wave functions. An atom with two electrons may be described with a *total wave function*, $\psi(\mathbf{r}_1, \mathbf{r}_2)$, where the vectors $\mathbf{r}_1$ and $\mathbf{r}_2$ denote the coordinates of the two electrons. The Schrödinger equation for the two-electron atom is

$$-\frac{\hbar^2}{2m^2}\nabla_1^2 \psi(\mathbf{r}_1, \mathbf{r}_2) - \frac{\hbar^2}{2m^2}\nabla_2^2 \psi(\mathbf{r}_1, \mathbf{r}_2)$$
$$+ V(\mathbf{r}_1, \mathbf{r}_2)\psi(\mathbf{r}_1, \mathbf{r}_2)$$
$$= E\psi(\mathbf{r}_1, \mathbf{r}_2), \qquad (9.1)$$

where ∇_1 and ∇_2 denote differentiation with respect to $\mathbf{r}_1$ and $\mathbf{r}_2$, respectively. The electron probability density is given by the square of the wave function,

$$\frac{dP}{dV} = \left| \psi(\mathbf{r}_1, \mathbf{r}_2) \right|^2. \qquad (9.2)$$

Consider what happens if we exchange the positions of the two electrons. Since the electrons are identical, the probability density is unchanged:

$$\left| \psi(\mathbf{r}_1, \mathbf{r}_2) \right|^2 = \left| \psi(\mathbf{r}_2, \mathbf{r}_1) \right|^2. \qquad (9.3)$$

The wave function squared (9.3) gives two possibilities for the total wave function, a *symmetric* solution (ψ_S),

$$\psi_S(\mathbf{r}_1, \mathbf{r}_2) = \psi_S(\mathbf{r}_2, \mathbf{r}_1), \qquad (9.4)$$

or an *antisymmetric* solution (ψ_A),

$$\psi_A(\mathbf{r}_1, \mathbf{r}_2) = -\psi_A(\mathbf{r}_2, \mathbf{r}_1). \qquad (9.5)$$

We now construct the total wave function from individual wave functions of the two electrons. Let **a** and **b** represent the four quantum numbers of the two electrons. The *symmetric wave function* (9.4) may be written

$$\psi_S(\mathbf{r}_1, \mathbf{r}_2)$$
$$= \frac{1}{\sqrt{2}}\left[\psi_a(\mathbf{r}_1)\psi_b(\mathbf{r}_2) + \psi_b(\mathbf{r}_1)\psi_a(\mathbf{r}_2) \right], \qquad (9.6)$$

where we have included a normalization factor, $1/\sqrt{2}$. Similarly, the antisymmetric wave function (9-5) may be written

$$\psi_A(\mathbf{r}_1, \mathbf{r}_2)$$
$$= \frac{1}{\sqrt{2}}\left[\psi_a(\mathbf{r}_1)\psi_b(\mathbf{r}_2) - \psi_b(\mathbf{r}_1)\psi_a(\mathbf{r}_2) \right]. \qquad (9.7)$$

The antisymmetric wave function is *identically zero* if the two electrons have the same set of quantum numbers, $\mathbf{a} = \mathbf{b}$. The antisymmetric wave function (9.7) gives the observed behavior of an electron pair, that is, it is consistent with the Pauli exclusion principle. We shall return to a discussion of symmetric and antisymmetric wave functions in Chapter 12.

9-3 SHELL STRUCTURE AND THE PERIODIC TABLE

The independent particle approximation together with the Pauli exclusion principle results in a *shell structure* of the atom that is well established by experiment. The energy and position of each electron are most affected by the principal quantum number n. More than one electron can have the same value of n, but the Pauli exclusion principle determines the maximum number of electrons (N_n) that can have the same n to be

$$N_n = 2\sum_{\ell=0}^{n-1}(2\ell+1)$$
$$= 2(1+3+5+\dots n-1) = 2n^2. \qquad (9.8)$$

The electrons in an atom that have the same value of n form a shell. For each value of n, the electron position and energy are affected somewhat by the value of the orbital angular momentum quantum number ℓ. The electrons having a given allowed value of ℓ form a *subshell* with the shell.

For $n = 1$ there are two unique states, those with quantum numbers (n, ℓ, m_ℓ, m_s) equal to $(1, 0, 0, -1/2)$ and $(1, 0, 0, 1/2)$. These two electrons form the *K shell*. The electron configuration of the K shell is abbreviated $1s^2$, which stands for two electrons with $n = 1$ and $\ell = 0$.

For $n = 2$ there are eight unique states, those with quantum numbers $(2, 0, 0, -1/2)$, $(2, 0, 0, 1/2)$, $(2, 1, -1, -1/2)$, $(2, 1, -1, 1/2)$, $(2, 1, 0, -1/2)$, $(2, 1, 0, 1/2)$, $(2, 1, 1, -1/2)$, and $(2, 1, 1, 1/2)$. These eight electrons form the *L shell*. The electron configuration of the L shell is abbreviated $2s^2 2p^6$, which stands for two electrons with $n = 2$ and $\ell = 0$, and six electrons with $n = 2$ and $\ell = 1$.

Similarly, for $n = 3$ there are 18 unique states that form the *M shell*. The electron configuration of the *M* shell is $3s^2 3p^6 3d^{10}$. For $n = 4$ there are 32 unique states that form the *N shell*. The electron configuration of the *N* shell is $4s^2 4p^6 4d^{10} 4f^{14}$. The *N* shell corresponds to the shell with the largest value of n that can be completely filled in any atom. Atoms with large values of Z may have electrons in the $n = 5$, 6, and 7 states, but these shells can never be full because the nucleus of an atom with so many electrons would not be stable. The electron shell notation is summarized in Table 9-1.

EXAMPLE 9-1

How many atomic states are there with principal quantum number n less than or equal to 7?

SOLUTION:

The number of electronic states with $n \leq 7$ is

$$\sum_{n=1}^{7} 2n^2 = 2(1 + 4 + 9 + 16 + 25 + 36 + 49) = 280 \ .$$

No atom has this many electrons because a nucleus with so many protons is not stable. Nuclei have been artificially created with values of Z up to 109. Nuclei with such large Z are short lived. For example, the lifetime of element number 109 is about 5 ms. ∎

The Pauli exclusion principle and the resulting shell structure for an atom with a large number of electrons ($Z \gg 1$) is qualitatively illustrated in Figure 9-2. Each electron occupies the lowest energy state available. Only one electron is allowed in each quantum state. The energy of each state is strongly dependent on the principal quantum number n and to a lesser degree on the orbital angular momentum quantum number ℓ.

TABLE 9-1
ELECTRON SHELLS.

n	Shell	Max. No. of e	Configuration
1	K	2	$1s^2$
2	L	8	$2s^2 2p^6$
3	M	18	$3s^2 3p^6 3d^{10}$
4	N	32	$4s^2 4p^6 4d^{10} 4f^{14}$

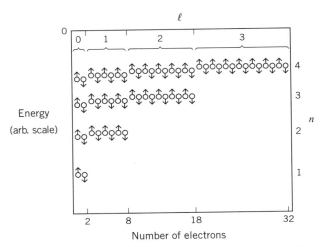

FIGURE 9-2 Pauli exclusion principle for an atom with a very large number of electrons ($Z \gg 1$).
Each electron resides in the lowest energy state available and only one electron is allowed in each quantum state. The energy of each state is strongly dependent on the principle quantum number n and somewhat dependent on the orbital angular momentum quantum number ℓ.

The validity and limitation of the independent particle approximation and the existence of shells is illustrated in Figure 9-3, which shows the electron radial probability density calculated for the krypton atom. If we measure the position of an electron in a multielectron atom, we cannot determine with absolute certainty to which shell the electron "belongs." The wave functions of the electrons overlap; however, the shell structure of the atom is evident. On the average, we know the approximate positions of the $n = 1$, $n = 2$, and $n = 3$ electrons.

Energy Dependence on the Quantum Number ℓ

In the independent particle approximation, the potential energy for an innermost electron in a multielectron atom is bounded from below by $-Zke^2/r$. The Pauli exclusion principle allows two electrons, those with opposite values of m_s, to occupy the innermost states. Therefore, a better approximation for the potential energy of an innermost electron is $-(Z-1)ke^2/r$. In fact, this relationship was found empirically by Moseley (see Figure 3-24). For an outermost electron the potential is bounded from above by $-ke^2/r$. For an electron between these two extremes, the exact form of potential energy is much more complicated. The potential energy depends on the position (r) of the

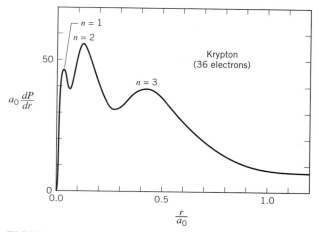

FIGURE 9-3 Shell structure of the krypton atom.
The probability of finding an electron versus distance from the nucleus, as calculated from the Schrödinger equation, is plotted for the Kr atom. The Kr atom has 36 electrons in the configuration $1s^2 2s^2 2p^6 3s^2 3p^6 4s^2 3d^{10} 4p^6$. All subshells are filled and the electron probability distribution is spherically symmetric. The bumps in the radial probability distribution are due to the filled K ($n = 1$), L ($n = 2$) and M ($n = 3$) shells. The 8 N shell ($n = 4$) electrons are most likely to be found at large distances. After R. Leighton, *Principles of Modern Physics,* McGraw-Hill (1959).

electron. Since the wave function squared depends on the orbital angular momentum, the potential energy depends on the quantum number ℓ.

The lowest energy levels in a multielectron atom are the $1s$ states. The $1s$ states have the lowest energy because electrons in these states are closer to the nucleus than in any other states. (Lower total energy implies a larger kinetic energy and a smaller wavelength.) More energy is required to remove an electron from the $1s$ state than from any other state. The energy level of an electron in the $1s$ state (E_{1s}) can be approximated by ignoring the presence of the other electrons except the other electron in the $1s$ state:

$$E_{1s} \approx (Z-1)^2 (-13.6\,\text{eV}). \qquad (9.9)$$

The exact energy differs from this simple expression because the electron will be repelled by the other electrons in the atom.

The next lowest energy levels are the states with $n = 2$. There are two possible values of the orbital angular momentum quantum number, zero or one, corresponding to $2s$ and $2p$ states. In the hydrogen atom, the $2s$ and $2p$ states have the same energy ($E_2 = -3.4$ eV) except for corrections of order $\alpha^2 E_2$. This is not true in a multi-electron atom. In a multielectron atom, the effect of the

other electrons depends on position. To a reasonable approximation, an electron in the $n = 2$ state feels an effective nuclear charge of $Z - 2$ due to the Z protons plus the two electrons in the $1s$ states. The energy of the $2s$ state (E_{2s}) may be approximated as

$$E_{2s} \approx (Z-2)^2 (-3.4\,\text{eV}). \qquad (9.10)$$

This *shielding effect* of the two $1s$ electrons on the nucleus is smaller for a $2s$ electron than for a $2p$ electron because the $2s$ state has a slightly greater probability to be found closer to the nucleus than an electron in the $2p$ state. Thus, the energy levels of the $2s$ states are lower than the energy levels of the $2p$ states:

$$E_{2s} < E_{2p}. \qquad (9.11)$$

EXAMPLE 9-2

Make a rough estimate of the energy levels of the $2s$ and $2p$ states in the lithium atom ($Z = 3$).

SOLUTION:

The lithium atom has three electrons. By the Pauli exclusion principle, two of the electrons are in the $1s$ state (with opposite spin quantum numbers). The third electron must be in the $n = 2$ state. For the $2p$ state, we estimate that the energy is approximately equal to the energy of the $2p$ state in hydrogen:

$$E_{2p} \approx \frac{-13.6\,\text{eV}}{4} = -3.4\,\text{eV}.$$

For the $2s$ state, we need to estimate the fraction of the probability density that is less than the Bohr radius. From Figure 8-3, we see that this fraction is approximately

$$f \approx (0.05)(0.5) = 0.025 \ .$$

For an orbit close to the nucleus, the energy would be

$$E_{1s} \approx (Z-1)^2 (-13.6\,\text{eV})$$
$$= (4)(-13.6\,\text{eV}) \approx -54\,\text{eV}.$$

Therefore, we estimate the energy of the $2s$ state as

$$E_{2s} \approx f E_{1s} + (1-f) E_{2p}$$
$$\approx (0.025)(-54\,\text{eV}) + (0.975)(-3.4\,\text{eV})$$
$$\approx -5\,\text{eV}.$$

We predict the energy of the $2s$ state to be smaller than the energy of the $2p$ state in lithium. Results from measurement of the energy levels in lithium are

$$E_{2p} = -3.6\,\text{eV}$$

and

$$E_{2s} = -5.4\,\text{eV}. \qquad\blacksquare$$

The next lowest energy levels are the states with $n = 3$. There are three possible values of the orbital angular momentum quantum number, $\ell = 0$, $\ell = 1$ or $\ell = 2$ corresponding to $3s$ states, $3p$ states, and $3d$ states. From Figure 8-4, we see that both the $3s$ and the $3p$ wave functions have a small component close to the nucleus, with the $3s$ having a larger component than the $3p$ at radii less than the Bohr radius. By the same reasoning as above, states with lower values of orbital angular momentum have lower energy:

$$E_{3s} < E_{3p} < E_{3d}. \qquad (9.12)$$

Since the energy depends on both n and ℓ, it is possible that states with large values of ℓ have larger energy than states with larger n and smaller ℓ. For example, in the potassium atom ($Z = 19$), the $3d$ state has a slightly *larger* energy than the $4s$ state. Therefore, the $4s$ states are filled before the $3d$ states. We must remember, however, that the energy levels depend on Z. For large Z, the energies of the $3d$ states are *smaller* than the energies of the $4s$ states. Therefore, we should make a distinction between the order in which the energy levels are filled, and the ordering of the energy levels for filled shells.

By comparison of the wave functions, we arrive at the following general rule for the order in which the energy levels are filled in multielectron atoms:

$$< E_{1s} < E_{2s} < E_{2p} < E_{3s} < E_{3p} < E_{4s}$$
$$< E_{3d} < E_{4p} < E_{5s} < E_{4d} < E_{5p} < E_{6s}$$
$$< E_{4f} < E_{5d} < E_{6p} < E_{7s} < E_{5f} < E_{6d}. \qquad (9.13)$$

There are a few minor exceptions to this general rule. All of these exceptions occur when d states have energies approximately equal to that of neighboring s or f states. Figure 9-4 indicates the filling order on a plot of quantum numbers, n versus ℓ.

The filling order (9.13) indicates that when the $3d$ states are all unoccupied, $E_{4s} < E_{3d}$. The ordering of the energy levels changes when the states are occupied. When a subshell is full, the energy of states with larger values of n is always larger, independent of ℓ. For example, the $4s$ states are occupied before the $3d$ states, but when the $3d$ states are fully occupied, $E_{3d} < E_{4s}$. (In some cases, when the $3d$ states are only *partially* occupied, $E_{3d} < E_{4s}$.) In the copper atom ($Z = 29$), the $4s$ state has a larger energy than

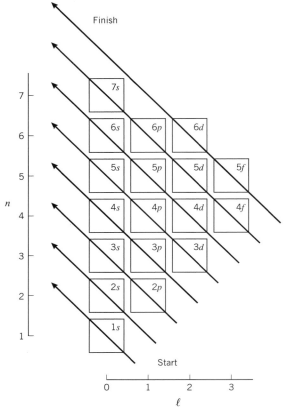

FIGURE 9-4 The filling of atomic states.
In a plot of n versus ℓ, if we start at $n = 1$ and $\ell = 0$, and follow the diagonal lines as indicated, we obtain the filling order of the energy levels: $1s$, $2s$, $2p$, $3s$, $3p$, $4s$, $3d$, $4p$, $5s$, $4d$, $5p$, $6s$, $4f$, $5d$, $6p$, $7s$, $5f$, $6d$. There are a few minor exceptions to this simple ordering in a few elements. There can be 2 electrons in each s state, 6 electrons in each p state, 10 electrons in each d state, and 14 electrons in each f state. The indicated states can accommodate 112 electrons, all with unique values of (n, ℓ, m_ℓ, m_s).

the $3d$ state. For filled subshells, the order of the energy levels is

$$E_{1s} < E_{2s} < E_{2p} < E_{3s} < E_{3p} < E_{3d}$$
$$< E_{4s} < E_{4p} < E_{4d} < E_{4f} < E_{5s} < E_{5p}$$
$$< E_{5d} < E_{5f} < E_{6s} < E_{6p} < E_{6d} < E_{7s}. \qquad (9.14)$$

> For any given value of the principal quantum number n, the states with smaller values of orbital angular momentum have lower energies.

Ionization Energy

The energy levels in a multielectron atom are quantized and *well-defined* even though the wave functions in our independent particle approximation overlap. The electron shell structure of the elements given by the Pauli exclusion principle is revealed by measuring the energy of the largest energy electron in each type of atom. This largest energy is determined by measuring the amount of energy necessary to remove one electron or singly *ionize* the atom. Figure 9-5 shows a plot of the energy level of the most energetic electron (the negative of the ionization energy) as a function of the atomic number Z. The electron with the largest total energy is the outermost electron, since the square of its wave function extends to larger values of r than the other electrons. Notice the regularity in the plot. The atoms with the electrons of the highest energy are the atoms with a single electron in an s state. The atoms with an outer electron of the lowest energy are the atoms with fully occupied p states and an empty s state in the next shell.

When an atom has a completely filled subshell, the electron probability distribution is spherically symmetric. Atoms that have a spherically symmetric electron probability distribution tend to be chemically inert if the energy difference between the highest energy occupied state and the lowest energy unoccupied state is significant, that is, greater than a few electronvolts. This energy gap is especially significant for the elements helium, neon, argon, krypton, xenon, and radon. The properties of an atom depend on the quantum numbers of the outermost (largest energy) electrons because these are the electrons that interact with neighboring atoms.

The Periodic Table of the Elements

A plot of n versus ℓ of the elements is shown in Figure 9-6. The atomic number Z of each element is indicated inside each box. Each horizontal row corresponds to a shell, and each column corresponds to a value of ℓ. Two electrons can have $\ell = 0$, 6 can have $\ell = 1$, 10 can have $\ell = 2$, and 14 can have $\ell = 3$. The electron configuration of any element

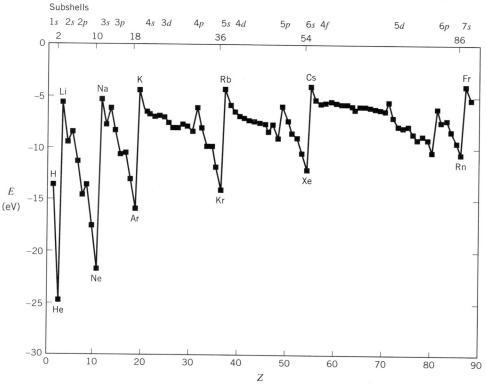

FIGURE 9-5 Energy level of the highest energy electron in a neutral atom as a function of the atomic number, Z.
The energy is highest for those atoms that have a single electron in the s state and lowest for those atoms that have a filled p subshell and no electrons in the next highest s state.

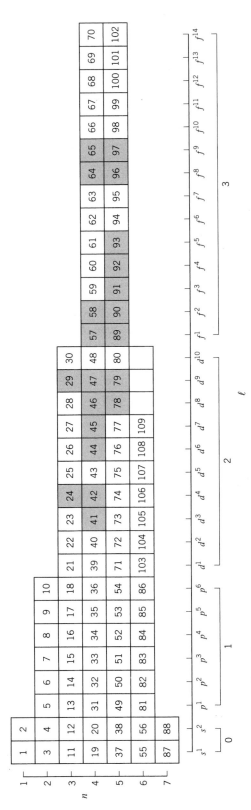

FIGURE 9-6 Table of the elements.

The atomic number of each element is indicated inside each box. Each horizontal row corresponds to a shell. Each column corresponds to a value of ℓ. The filling order of the elements follows Figure 9-4, except for the shaded elements where an s state fills before a d state or a d state fills before an f state.

is just the electron configuration of the preceding element plus one additional electron in the next available subshell. (The filling order of the subshells is given in Figure 9-4.) A few exceptions to this rule occur when the s and d or d and f subshells are very close together.

We can modify the table of the elements (Figure 9-6) by placing each element that has a similar *outer* electron configuration in the same column. This is shown in Figure 9-7. Each column then represents a group of elements with similar chemical properties. (An exception is He which is

an inert gas because of the relatively large energy gap between the $1s$ and $2s$ states.

9-4 THE COUPLING OF ANGULAR MOMENTA

The energy levels of electrons in atoms depend strongly on the principal quantum number n. The angular momentum of the electrons also affects the energy levels. We have

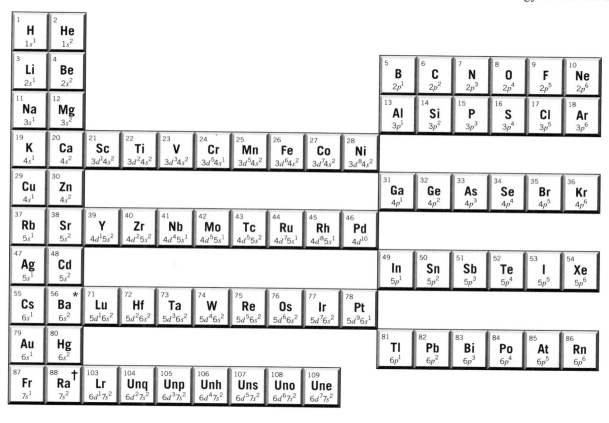

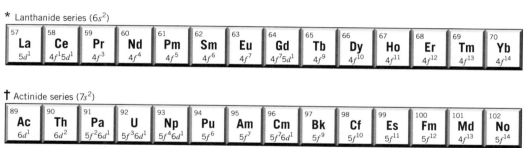

FIGURE 9-7 Outer electron configurations of the elements.
The inner electron configuration consists of filled subshells.

already qualitatively discussed the effect of the orbital angular momentum **L** on the energy of electron subshells (s, p, d, etc.) in the context of the independent particle approximation. The intrinsic angular momentum **S** also affects the energy levels. We may classify the corrections to the energy levels as: *orbital–orbital*, *spin–spin*, and *spin–orbital*. The orbital–orbital interactions are obviously important because of the Coulomb repulsion between the electrons whose probability density depends on **L**. The spin–spin interactions are also very important. This is not due to the interaction between the magnetic moments of the electrons, which is very weak, but because of the Pauli exclusion principle. For atoms where Z is not too large, the spin–spin and orbital–orbital interactions are the most important, that is, the spin–spin and orbital–orbital interactions have the largest effect on the energies. The spin–orbital interactions have a much smaller effect on the energy levels (recall that the size of the spin–orbital interaction in hydrogen is proportional to $\alpha^4 mc^2$).

In the independent particle approximation, the energy of each state is roughly determined by n and ℓ. It is convenient to characterize an atom with more than one outer electron by a total intrinsic angular momentum quantum number (*spin–spin coupling*) and a total orbital angular momentum quantum number (*orbital–orbital coupling*).

Spin–Spin Coupling

Consider an atom with two outer electrons. The intrinsic angular momentum quantum number of each electron (s_1 and s_2) is equal to one-half:

$$s_1 = s_2 = \frac{1}{2}. \qquad (9.15)$$

We may form the total intrinsic angular momentum vector (**S**) by adding the intrinsic angular momentum vectors of each electron (**S**$_1$ and **S**$_2$),

$$\mathbf{S} = \mathbf{S}_1 + \mathbf{S}_2. \qquad (9.16)$$

This addition (9.16) must respect the quantization of angular momentum, analogous to the addition of **L**+**S** to get the total angular momentum in the hydrogen atom. The quantization condition is that the resulting total intrinsic angular momentum quantum number (s) can be zero or one. The magnitude of the intrinsic angular momentum (S) is related to the intrinsic angular momentum quantum number (s) by

$$S^2 = s(s+1)\hbar^2. \qquad (9.17)$$

The energy depends on the value of the quantum number s. If two electrons have parallel spins ($s = 1$), then they tend to be further apart than if the spins are antiparallel ($s = 0$). This is a consequence of the Pauli exclusion principle which does not allow the four quantum numbers of the two electrons to be identical. (The total wave function of the two electrons must be antisymmetric; if the spin part is symmetric as is the case for parallel spins, then the space part is antisymmetric. An antisymmetric space part means that the electrons are further apart, on the average.) If the electrons are further apart, then the contribution to the energy from their Coulomb repulsion is reduced and the energy is lower. The order of magnitude of this effect is 1 eV. Therefore, in multielectron atoms it is nearly always observed that states with larger values of s have lower energy. For example, in the ground state of carbon the two $2p$ electrons have parallel spins ($s = 1$).

If an atom has three outer electrons, the situation is similar. Then the total intrinsic angular momentum is the vector sum of three terms:

$$\mathbf{S} = \mathbf{S}_1 + \mathbf{S}_2 + \mathbf{S}_3. \qquad (9.18)$$

There are two possibilities for the total intrinsic angular momentum quantum number, $s = 1/2$ or $s = 3/2$. The states with $s = 3/2$ tend to have lower energies then the states with $s = 1/2$.

Orbital–Orbital Coupling

The energy levels of multielectron atoms also depend on the total orbital angular momentum quantum number. For the case of two electrons with orbital angular momentum **L**$_1$ and **L**$_2$, we may form the total orbital angular momentum vector (**L**) by vector addition,

$$\mathbf{L} = \mathbf{L}_1 + \mathbf{L}_2. \qquad (9.19)$$

If ℓ_1 and ℓ_2 are the orbital angular momentum quantum numbers of the two electrons, then the possible values for the total angular momentum quantum number (ℓ) are

$$\ell = |\ell_1 - \ell_2|, |\ell_1 - \ell_2| + 1, \ldots \ell_1 + \ell_2. \qquad (9.20)$$

The magnitude of the total orbital angular momentum vector is related to the total orbital angular momentum quantum number by

$$L^2 = \ell(\ell+1)\hbar^2. \qquad (9.21)$$

The states with the largest value of ℓ are the states in which the electrons are furthest apart, on the average. (Recall that larger values of ℓ have probability densities that extend to larger distances.) In multielectron atoms it is nearly always observed that states with larger values of ℓ have lower energy.

Spin–Orbital Coupling

The effect of the spin–orbital interaction is accounted for by adding together the total orbital angular momentum ($\mathbf{L}$) and the total intrinsic angular momentum ($\mathbf{S}$) to get the total angular momentum ($\mathbf{J}$):

$$\mathbf{J} = \mathbf{L} + \mathbf{S}. \qquad (9.22)$$

The total angular momentum quantum number (j) may take on the values

$$j = |\ell - s|, |\ell - s| + 1, \dots \ell + s. \qquad (9.23)$$

The spin–orbital interaction makes a small correction to the energy compared to the spin–spin and orbital–orbital effects, provided that Z is not very large. States with a lower value of total angular momentum quantum number (j) have a lower energy. The spin–orbital coupling is often referred to as *LS coupling* or Russell-Saunders coupling after Henry Norris Russell and F. A. Saunders who applied the spin–orbital interaction theory to the interpretation of stellar spectra.

The effect of the spin–spin, orbital–orbital, and spin–orbital interactions on the lowest energy states in the carbon atom is qualitatively illustrated in Figure 9-8.

Carbon has two outer electrons in the $2p$ states. The intrinsic angular momentum quantum number of each electron is equal to 1/2. The total intrinsic angular momentum quantum number (s) may be zero or one. The state with $s = 1$ has a lower energy. The state with $s = 1$ must have $\ell = 1$, because it must be antisymmetric under the exchange of two electrons in accordance with the Pauli exclusion principle. In determining the exchange symmetry, we must take into account all the quantum numbers of the electron. It is useful to consider the total wave function to be the product of a space part and a spin part. Even values of ℓ make symmetric space wave functions, and odd values of ℓ make antisymmetric space wave functions. The state $s = 1$ corresponds to spins aligned, that is, exchanging intrinsic angular momenta does not change the sign of the wave function. The total wave function of the electron pair is given by the product of the symmetries of the space part and the spin part

$$\psi_{\text{A}} = \left(\psi_{\text{space}} \right)_{\text{A}} \left(\psi_{\text{spin}} \right)_{\text{S}}. \qquad (9.24)$$

Therefore, under electron exchange,

$$\psi_{\text{total}} \rightarrow (-1)^{\ell + s + 1} \psi_{\text{total}}. \qquad (9.25)$$

Since there is only one allowed value of ℓ for the $s = 1$ state ($\ell = 1$), there is no orbital–orbital splitting. Adding $s = 1$ and $\ell = 1$ gives three possible values of the total angular momentum quantum number j (0, 1, and 2) so that the

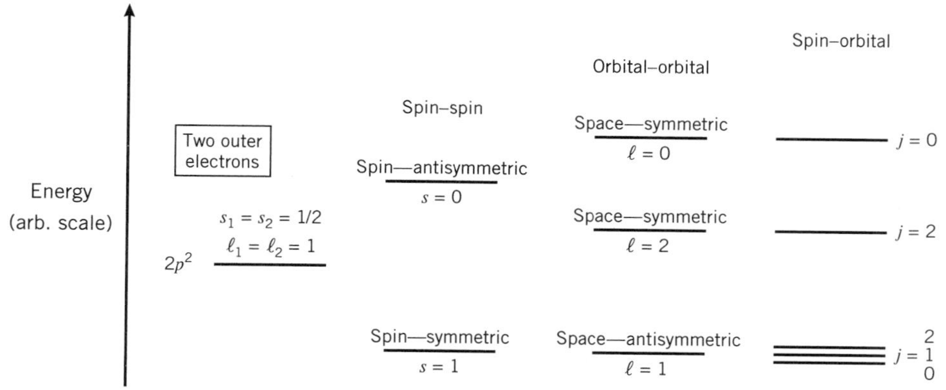

FIGURE 9-8 Coupling of angular momenta in the carbon atom.
The spin–spin interaction splits the states into two groups with higher spin states having lower energy. The orbital–orbital interaction splits the $s = 0$ states into two states with the state of higher total orbital angular momentum having a lower energy. The spin–orbital interaction further splits the states with $s = 1$ into a triplet of states. The order of magnitude of the spin–spin and orbital–orbital splitting is an electronvolt. The spin–orbital splitting is much smaller and is exaggerated in the diagram.

spin–orbital interaction splits the state into three states. The spin–orbital interaction is very weak compared to the spin–spin interaction and the splitting of the states ($j = 0,1,2$) is exaggerated in Figure 9-8. The lowest lying states form a *triplet*. The state with $s = 0$ must have $\ell = 0$ or 2, so that the total wave function (9.25) is antisymmetric. Thus, we have

$$\psi_A = \left(\psi_{\text{space}} \right)_S \left(\psi_{\text{spin}} \right)_A. \tag{9.26}$$

The orbital–orbital interaction splits these states into two states with the $\ell = 2$ having the lower energy. The states with $s = 0$ have no spin–orbital interaction, and there is no further energy splitting.

JJ Coupling

The order of magnitude of the size of the spin–orbital interaction in hydrogen is $\alpha^4 mc^2$. For other atoms the size of the spin–orbital interaction is obtained by replacing ke^2 with Zke^2, resulting in a correction of size $Z^4 \alpha^4 mc^2$. Therefore, when Z is very large, the spin–orbital interaction becomes larger than the spin–spin or orbital–orbital interactions. In the treatment of atoms with large Z, it is more logical to treat the spin–orbital interaction first and then make smaller corrections due to the spin–spin and orbital–orbital interactions.

We first form total angular momentum vectors for each electron, ($\mathbf{J}_1$ and $\mathbf{J}_2$), where

$$\mathbf{J}_1 = \mathbf{L}_1 + \mathbf{S}_1, \tag{9.27}$$

and

$$\mathbf{J}_2 = \mathbf{L}_2 + \mathbf{S}_2. \tag{9.28}$$

The two total angular momentum quantum numbers are j_1 and j_2. The values of j_1 and j_2 will have a large effect on the total energy. We then sum the two total angular momentum vectors to get the net value

$$\mathbf{J} = \mathbf{J}_1 + \mathbf{J}_2. \tag{9.29}$$

This treatment is called *JJ coupling*. The difference between *JJ* coupling and *LS* coupling is the order in which angular momenta are added and used at each stage to make corrections to the energy levels.

9-5 EXCITED STATES OF ATOMS

Hydrogenlike Atoms

The simplest atoms are those with a single outer electron. An example is the lithium atom with three electrons in the configuration $1s^2 2s^1$. The excited states of the lithium atom in which the $2s$ electron occupies a state with larger energy are shown in Figure 9-9. In this diagram two electrons always occupy the $1s$ states. The excited states are labeled by the value of quantum numbers n and ℓ of the outer electron. The energy diagram is similar to that of hydrogen (see Figure 8-6) except that there is an ℓ dependence of the energies. The allowed transitions that satisfy the selection rule $\Delta \ell = 1$ are indicated on the diagram.

All atoms that have a single outer electron in an s state have a spectrum that is similar to hydrogen. These atoms are lithium, sodium, potassium, copper, rubidium, silver, cesium, gold, and francium (see Figure 9-7). The energy levels of the sodium atom are shown in Figure 9-10.

The Helium Atom

Singly ionized helium (He$^+$) has energy levels similar to hydrogen, except in helium the electrical force is twice as strong because there are two protons attracting the electron. The only difference is that we must replace ke^2 by Zke^2, where $Z = 2$. The energy levels of He$^+$ are given by

$$E_n = \frac{-m \left(Zke^2 \right)^2}{2 \hbar^2 n^2} = \frac{(4)(-13.6 \, \text{eV})}{n^2}$$

$$= \frac{-54.4 \, \text{eV}}{n^2}. \tag{9.30}$$

Four times more energy is required to remove the electron from the ground state of ionized helium than is required to remove the electron from the ground state of hydrogen. All of the energy levels of He$^+$ are lower than hydrogen by a factor of $Z^2 = 4$.

The neutral helium atom has two electrons with the configuration $1s^2$. Consider what happens when a second electron is added to the helium ion. The "second" electron experiences a potential due to both the nucleus and the "first" electron. The "first" electron also experiences a potential due to both the nucleus and the "second" electron so that the energy levels of both of the electrons are affected by the presence of the other electron. The two

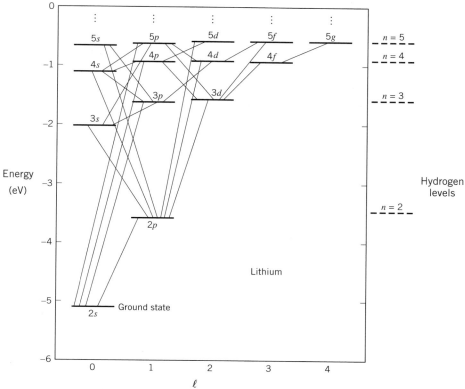

FIGURE 9-9 Energy level diagram for the lithium atom ($Z = 3$).

In this diagram, two of the electrons occupy the $1s$ states. The quantum numbers of the other electron are indicated.

electrons are identical particles; they are indistinguishable. The energy levels can be obtained by solving the Schrödinger equation. The description of the atom contains two coordinates, $\mathbf{r}_1$ and $\mathbf{r}_2$, corresponding to the positions of the two electrons. The potential energy is of the form,

$$V(\mathbf{r}_1, \mathbf{r}_2) = -\frac{ke^2}{r_1} - \frac{ke^2}{r_2} + \frac{ke^2}{|\mathbf{r}_2 - \mathbf{r}_1|}. \quad (9.31)$$

The last term corresponds to the electrical repulsion of the two electrons (V_{ee}). Because of the complicated form of the potential, the Schrödinger equation can no longer be solved in closed form, but we can solve for the energies using an iterative approximation technique for which many algorithms have been invented. Consider the energy of the two electrons together, the energy of the pair. The ground state energy (E_g) of the pair is the amount of energy

required to remove both electrons from the atom. If we neglect the electron–electron interaction, then the energy required to remove both electrons from the atom is two times 54.4 eV, or about 109 eV/c; however, the repulsion of the two electrons will make it slightly easier to remove both electrons. (Note that the term V_{ee} is positive.) The simplest estimate of the ground state energy may be made by using the hydrogen atom wave function to calculate the average value of V_{ee}. The ground state energy may then be approximated by

$$E_g \approx -54.4\,\text{eV} - 54.4\,\text{eV} + \langle V_{ee} \rangle. \quad (9.32)$$

Such a technique is called perturbation theory. The above approximation gives an answer for helium accurate to about 5%. We can make a second-order correction by using the new energy to correct the wave function and then compute a new correction to the energy. The final answer is

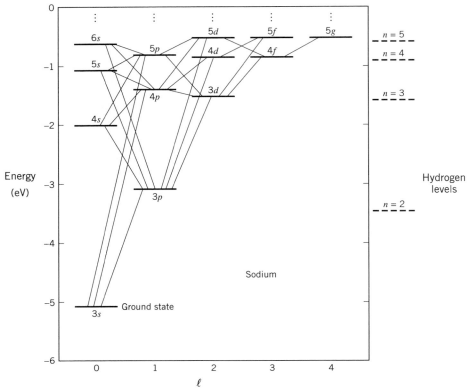

FIGURE 9-10 Energy level diagram for the sodium atom ($Z = 11$).
In this diagram, 10 of the electrons are in the states $1s^2 2s^2 2p^6$. The quantum numbers of the other electron are indicated. Notice the similarity between sodium and lithium.

$$E_g = -79 \, \text{eV} . \tag{9.33}$$

Experiment confirms that an energy of 79 eV is required to remove both electrons from helium. We also know that it will take 54.4 eV to remove the electron from He$^+$. Therefore, the amount of energy needed to remove one electron from helium (E_1) is

$$E_1 = 79 \, \text{eV} - 54.4 \, \text{eV} = 24.6 \, \text{eV} . \tag{9.34}$$

This energy, often expressed as a voltage by dividing by the electron charge (e), is called the first ionization potential. It is the minimum amount of energy needed to remove one electron from the atom.

In specifying the energy levels of the helium atom, we must specify the quantum numbers of both electrons. We may divide the energy levels of the helium atom into two classes, depending on the total intrinsic angular momentum quantum number s, which is the sum of the spin quantum numbers of the two electrons. There are two possibilities for the total intrinsic angular momentum quantum number, zero or one. The $s = 0$ states are called *singlet* states because there is only one possible value for m_s: $m_s = 0$. The $s = 1$ states are called *triplet* states because there are three possible value for m_s: $m_s = -1$, 0, or 1.

The energy level diagram for helium, where one of the electrons is in the $1s$ state, is shown in Figure 9-11. The quantum numbers of the second electron are indicated. The energy levels of the singlet ($s = 0$) states are plotted on the left and the triplet ($s = 1$) states on the right. There is only one possibility for the electron spin quantum numbers in the ground state because the two electrons cannot all have the same quantum numbers. Therefore, the spins must be *antialigned* and the ground state has $s = 0$. The electron configuration of the ground state is $1s^2$, meaning that there are two electrons in the $1s$ state. The ground state is a singlet state. The energy of the ground state is −24.6 eV, which means that if both electrons are in the $1s$ level

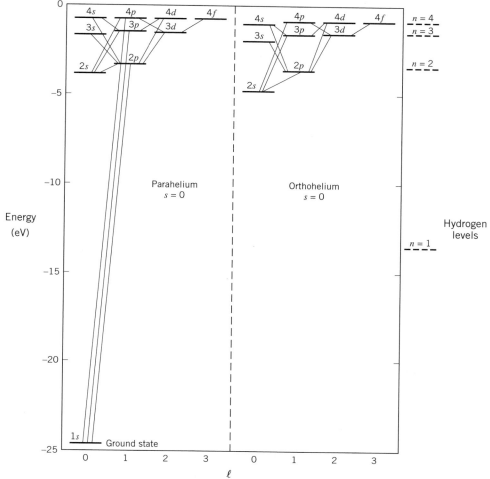

FIGURE 9-11 Energy level diagram for the helium atom ($Z = 2$).
In this diagram, one of the electrons is always in the $1s$ state. The quantum numbers of the other electron are indicated. There are two possibilities for the electron spins, aligned (parahelium) or opposite (orthohelium). The selection rules are $\Delta\ell = 1$ and $\Delta s = 0$.

(with opposite spins) then 24.6 eV of energy is needed to remove *one* of the electrons to infinity.

The $2p$ state has an electron configuration $1s^12p^1$, meaning that there is one electron in the $1s$ state and one in the $1p$ state. The energy level of this state is −3.4 eV, which means that an energy of

$$\Delta E = (-3.4\,\text{eV}) - (-24.6\,\text{eV}) = 21.2\,\text{eV} \quad (9.35)$$

is needed to move one of the electrons from the $1s$ state to the $2p$ state.

As in the hydrogen atom, the transitions between energy levels obey the selection rules $\Delta\ell = 1$ and $\Delta s = 0$. The

selection rules do not absolutely forbid the corresponding transitions to occur, but they strongly suppress them. The selection rule that prohibits the spin from changing prevents transitions between the singlet and triplet states. The collection of singlet states is called *parahelium*, and the collection of triplet states is called *orthohelium*. The allowed transitions for the helium atom are indicated in Figure 9-11.

The Carbon Atom

The carbon atom has a ground state electron configuration of $1s^22s^22p^2$. The total intrinsic angular momentum of the atom may be in a state $s = 0$ if the outer two electrons have antiparallel spin or $s = 1$ if they have parallel spin. The transition rule $\Delta s = 0$ will suppress transitions between the states of different spin. Therefore, as in the case of the helium, it is convenient to separate these states into singlet ($s = 0$) and triplet ($s = 1$) states.

When one of the $2p$ electrons is excited, we may abbreviate the resulting excited states $1s^2 2s^2 2p^1 3s^1$, $1s^2 2s^2 2p^1 3p^1$, $1s^2 2s^2 2p^1 3d^1$, ... by the notation of the excited electron, $3s$, $3p$, $3d$, ... Another possibility is an excitation of one of the $2s$ electrons to the state $1s^2 2s^1 2p^3$. We abbreviate these states as $2p^3$. The energy level diagram for the carbon atom is shown in Figure 9-12. The horizontal axis is the *total* orbital angular momentum quantum number ℓ of the two outer electrons. For example, the triplet $2p$ state must have $\ell = 1$ and the singlet $2p$ states must have $\ell = 0$ or $\ell = 2$ in order to satisfy the Pauli exclusion principle (see Figure 9-8). The values of the total angular momentum quantum number j are indicated in Figure 9-12.

The Nitrogen Atom

The nitrogen atom has a ground state electron configuration of $1s^2 2s^2 2p^3$. The total intrinsic angular momentum of the atom may be in a state $s = 1/2$ if two of the outer electrons have spins antialigned, or $s = 3/2$ if all spins are aligned. The transition rule $\Delta s = 0$ will suppress transitions between the states of different spin. Again, it is convenient to separate these two spin states. The states with $s = 1/2$ are called *doublet* states because there are two possible value of m_s ($-1/2$ or $1/2$). The states with $s = 3/2$ are called *quartet* states because there are four possible values of m_s ($-3/2$, $-1/2$, $1/2$, or $3/2$). The energy level diagram for the nitrogen atom is shown in Figure 9-13. When one of the $2p$ electrons is excited, the states are labeled by the quantum numbers of the excited electron. Another possibility is an excitation of one of the $2s$ electrons to the state $1s^2 2s^1 2p^4$. We abbreviate these states as $2p^4$. The horizontal axis is the total orbital angular momentum of the three outer electrons.

The Width of Spectral Lines

When an electron in an atom makes a transition between two energy levels (E_2 and E_1) in an atom by photon emission, the energy of the photon is given by

$$E_{\text{photon}} = E_2 - E_1. \qquad (9.36)$$

The photon energy is not perfectly sharp. If we measure a large number of transitions with extremely good energy resolution, the distribution of photon energies will not be a delta function. The photon energy distribution will be a Gaussian-like curve with a width that is governed by several effects.

The uncertainty principle sets a lower limit on the width (ΔE) of the photon energy distribution

$$\Delta E > \frac{\hbar}{\tau}, \qquad (9.37)$$

where τ is the lifetime of the state. Since the atom is moving when it emits the photon, the photon energy will be Doppler shifted. In addition, the photon energy is affected by atomic collisions. Since these collisions are pressure dependent, the photon energy width depends on the pressure. A very strong electric field can also cause an energy broadening.

9-6 ATOMS IN AN EXTERNAL MAGNETIC FIELD

If we neglect electron–electron interactions and the spin–orbit interaction, then the s plus p states of an atomic shell have a degeneracy of eight. There are eight states with the same energy that depends only on the principal quantum number n. This degeneracy is due to the symmetry of the electric potential. The electron–electron interaction removes the degeneracy between the s and p states, with the s states getting a lower energy because electrons in the s state penetrate the inner electron shield more than the p states. On a finer scale, the interaction of electron spin with the internal magnetic field of the atom removes some of the degeneracy (six) of the p states by splitting them into two states. The degeneracy of the electron energy levels is completely removed if the atom is placed in an external magnetic field (see Figure 9-14). There are two possibilities that give qualitatively different results depending on the strength of the external magnetic field compared to the internal field of the atom.

Weak Magnetic Field

If the external magnetic field is weaker than the internal magnetic field of the atom, then the splitting caused by the external field is smaller than the splitting caused by the spin–orbital interaction. The shift in energy levels is given by finding the component of the magnetic moment in the

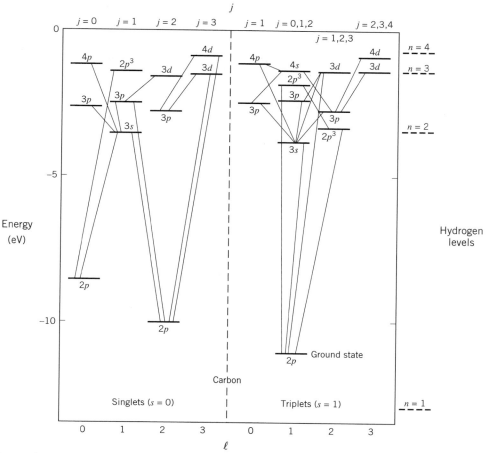

FIGURE 9-12 Energy level diagram for the carbon atom ($Z = 6$).
For most of the states in this diagram, five of the electrons are in the states $1s^2 2s^2 2p^1$. The quantum numbers of the other electron are indicated as $2p$, $3s$, $3p$, ... The other states where the electron configuration is $1s^2 2s^1 2p^3$ are labeled $2p^3$.

direction of B

$$\Delta E = -\boldsymbol{\mu} \cdot \mathbf{B} = \frac{e}{2m}(\mathbf{L} + 2\mathbf{S}) \cdot \mathbf{B}. \quad (9.38)$$

We cannot evaluate this expression directly because the directions of $\mathbf{L}$ and $\mathbf{S}$ are changing as these vectors precess about the total angular momentum vector ($\mathbf{J}$):

$$\mathbf{J} = \mathbf{L} + \mathbf{S}. \quad (9.39)$$

We can evaluate the dot product by evaluating the component of $\boldsymbol{\mu}$ in the direction of $\mathbf{J}$ and then multiplying by the component of $\mathbf{B}$ in the direction of $\mathbf{J}$:

$$\Delta E = \left(\frac{e}{2m}\right)\left[\frac{(\mathbf{L} + 2\mathbf{S}) \cdot (\mathbf{J})}{J}\right]\left(\frac{\mathbf{J} \cdot \mathbf{B}}{J}\right). \quad (9.40)$$

Taking B to be in the z direction, we have

$$\Delta E = \left(\frac{e}{2m}\right)\left[\frac{(\mathbf{L} + 2\mathbf{S}) \cdot (\mathbf{L} + \mathbf{S})J_z B}{J^2}\right]. \quad (9.41)$$

The component of total angular momentum in the z direction is

$$J_z = m_j \hbar. \quad (9.42)$$

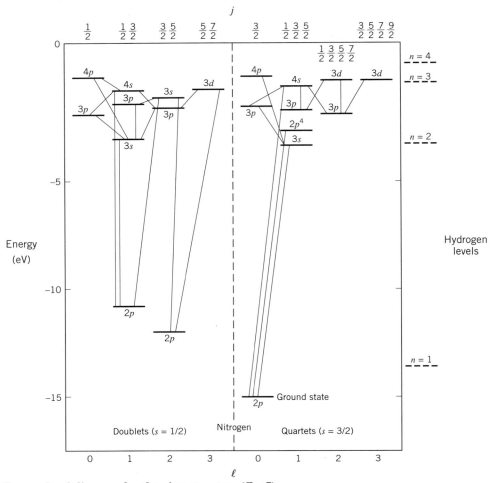

FIGURE 9-13 Energy level diagram for the nitrogen atom ($Z = 7$).
For most of the states in this diagram, six of the electrons are in the states $1s^2 2s^2 2p^2$. The quantum numbers of the other electron are indicated as $2p$, $3s$, $3p$,... . The state with an electron configuration of $1s^2 2s^1 2p^4$ is labeled $2p^4$.

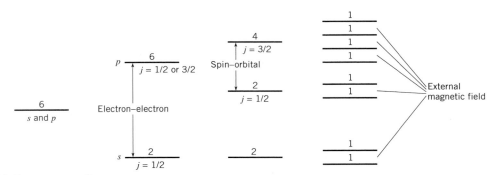

FIGURE 9-14 Degeneracy of states.

The shift in energy levels is

$$\Delta E = \left(\frac{e}{2m}\right)\left[\frac{\left(L^2 + 2S^2 + 3\mathbf{L}\cdot\mathbf{S}\right)m_j\hbar B}{J^2}\right]. \quad (9.43)$$

Using

$$3\mathbf{L}\cdot\mathbf{S} = \frac{3}{2}\left(J^2 - L^2 - S^2\right). \quad (9.44)$$

we get

$$\Delta E = \left(\frac{m_j e\hbar B}{2m}\right)\left[\frac{\left(3J^2 - L^2 + S^2\right)}{2J^2}\right]. \quad (9.45)$$

The amount of shift due to the external field is proportional to the z component of the total angular momentum of the electron. The factor,

$$g_L \equiv \frac{3J^2 - L^2 + S^2}{2J^2}$$

$$= 1 + \frac{J^2 - L^2 + S^2}{2J^2}, \quad (9.46)$$

is called the *Landé factor*.

The energy shift is

$$\Delta E = \frac{g_L m_j e\hbar B}{2m} = g_L m_j \mu_B B, \quad (9.47)$$

where g_L may be evaluated as

$$g_L = 1 + \frac{j(j+1) + s(s+1) - \ell(\ell+1)}{2j(j+1)}. \quad (9.48)$$

Figure 9-15 shows an energy level diagram for the splitting of s and p states in a weak external magnetic field. Ten unique photon energies are observed when an electron makes the transition $p \to s$ in the magnetic field, obeying the selection rule $\Delta m_j = -1$, 0, or 1. Figure 9-15 shows the measured photon lines from sodium.

EXAMPLE 9-3

Calculate the Zeeman splitting for the $3s_{1/2}$, $3p_{1/2}$, and $3p_{3/2}$ states of sodium in an external magnetic field of 1.0 T.

SOLUTION:

For the $3s_{1/2}$ states, the Landé factor is

$$g_L = 1 + \frac{\left(\frac{1}{2}\right)\left(\frac{3}{2}\right) + \left(\frac{1}{2}\right)\left(\frac{3}{2}\right) - 0}{2\left(\frac{1}{2}\right)\left(\frac{3}{2}\right)} = 2,$$

and

$$m_j = \frac{1}{2} \quad \text{or} \quad m_j = -\frac{1}{2}.$$

The energy shifts are

$$\Delta E = g_L m_j \mu_B B = (2)\left(\frac{1}{2}\right)\left(5.8 \times 10^{-5} \text{ eV/T}\right)(1.0 \text{ T})$$

$$= 5.8 \times 10^{-5} \text{ eV},$$

and

$$\Delta E = -5.8 \times 10^{-5} \text{ eV}.$$

The splitting between the $3s_{1/2}$ states is 1.2×10^{-4} eV. For the $3p_{1/2}$ states, the Landé factor is

$$g_L = 1 + \frac{\left(\frac{1}{2}\right)\left(\frac{3}{2}\right) + \left(\frac{1}{2}\right)\left(\frac{3}{2}\right) - (1)(2)}{(2)\left(\frac{1}{2}\right)\left(\frac{3}{2}\right)} = \frac{2}{3}.$$

The energy shifts are

$$\Delta E = g_L m_j \mu_B B$$

$$= \left(\frac{2}{3}\right)\left(\frac{1}{2}\right)\left(5.8 \times 10^{-5} \text{ eV/T}\right)(1.0 \text{ T})$$

$$= 1.9 \times 10^{-5} \text{ eV},$$

and

$$\Delta E = -1.9 \times 10^{-5} \text{ eV}.$$

The splitting between the $3p_{1/2}$ states is 3.9×10^{-5} eV. For the $3p_{3/2}$ state, the Landé factor is

$$g_L = 1 + \frac{\left(\frac{3}{2}\right)\left(\frac{5}{2}\right) + \left(\frac{1}{2}\right)\left(\frac{3}{2}\right) - (1)(2)}{(2)\left(\frac{3}{2}\right)\left(\frac{5}{2}\right)} = \frac{4}{3},$$

and

$$m_j = \frac{3}{2}, \quad m_j = \frac{1}{2}, \quad m_j = -\frac{1}{2} \quad \text{or} \quad m_j = -\frac{3}{2}.$$

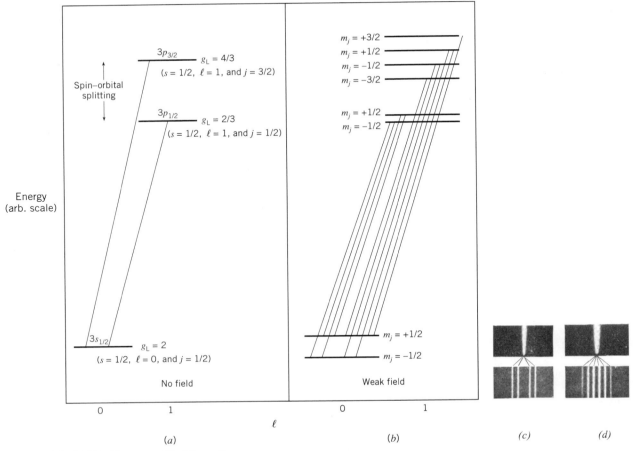

FIGURE 9-15 The Zeeman effect in sodium.
(*a*) The 3*s* and 3*p* states of sodium with no external magnetic field. The *p* states are split by the spin-orbit interaction. (*b*) In a weak external magnetic field (compared to the internal magnetic field), the energy shift of each state is proportional to g_L and m_j. The $3p_{1/2} \rightarrow 3s_{1/2}$ transitions result in 4 photon lines and the $3p_{3/2} \rightarrow 3s_{1/2}$ transitions result in 6 photon lines. (*c*) Observation of the $3p_{1/2} \rightarrow 3s_{1/2}$ transitions with no magnetic field (top) and weak external field (bottom). (*d*) Observation of the $3p_{3/2} \rightarrow 3s_{1/2}$ transitions with no magnetic field (top) and weak external field (bottom). Photographs from H. E. White, *Introduction to Atomic Spectra*, McGraw-Hill (1934).

The energy shifts are

$$\Delta E = g_L m_j \mu_B B$$

$$= \left(\frac{4}{3}\right)\left(\frac{3}{2}\right)\left(5.8 \times 10^{-5} \text{ eV/T}\right)\left(1.0 \text{ T}\right)$$

$$\approx 1.2 \times 10^{-4} \text{ eV},$$

$$\Delta E = \left(\frac{4}{3}\right)\left(\frac{1}{2}\right)\left(5.8 \times 10^{-5} \text{ eV/T}\right)\left(1 \text{ T}\right)$$

$$\approx 3.9 \times 10^{-5} \text{ eV},$$

$$\Delta E \approx -3.9 \times 10^{-5} \text{ eV},$$

and

$$\Delta E \approx -1.2 \times 10^{-4} \text{ eV}.$$

The splitting between the $3p_{3/2}$ states is 2.3×10^{-4} eV. ∎

Strong Magnetic Field

If the external magnetic field is much stronger than the internal magnetic field of the atom, then the Zeeman splitting dominates over splitting due to the spin–orbital interaction. The splitting of energy levels in a strong external magnetic field, the Paschen-Back effect, was illustrated for hydrogen in Figure 8-18. The energy shift is proportional to the magnetic moment, which is proportional to $(m_\ell + 2m_s)$,

$$\Delta E = \frac{(m_\ell + 2m_s)e\hbar B}{2m}. \tag{9.49}$$

The p levels are split into five states and the s levels are split into two states. There are only three unique photon energies in a $p \rightarrow s$ transition because there are only three possible values of the change in the sum $(m_\ell + 2m_s)$:

$$\Delta(m_\ell + 2m_s) = -1, \ 0 \text{ or } 1. \tag{9.50}$$

This results in a triplet of lines (see Figure 8-18c).

CHAPTER 9: PHYSICS SUMMARY

- The qualitative description of multielectron atoms is greatly simplified assuming that each electron moves independently in a potential that is bounded from above by $-ke^2/r$ and bounded from below by $-Zke^2/r$.

- The Pauli exclusion principle states that each electron in an atom must have a unique set of quantum numbers (n, ℓ, m_ℓ, m_s).

- The independent particle approximation and the Pauli exclusion principle leads to atomic shell structure. The K shell holds the 2 lowest energy, innermost electrons ($1s^2$). The L shell holds 8 electrons ($2s^2 2p^6$), the M shell 18 electrons ($3s^2 3p^6 3d^{10}$), and the N shell 32 electrons ($4s^2 4p^6 4d^{10} 4f^{14}$).

- The energy of electrons within a shell depends on the orbital angular momentum ℓ and is smallest for states with the smallest values of ℓ.

- The properties of atoms are governed by the quantum numbers of the outer electrons.

- For multielectron atoms with small Z, the most important corrections to the energy levels in the independent particle approximation are due to spin–spin and orbital–orbital interactions of the electrons. Of less importance is the spin–orbital interaction. For large Z, the spin–orbital corrections are the most important.

- The splitting of atoms in a weak magnetic field is proportional to $m_j g_L$, where

$$g_L = 1 + \frac{j(j+1) + s(s+1) - \ell(\ell+1)}{2j(j+1)}.$$

The splitting of energy levels in a strong external magnetic field is proportional to $(m_\ell + 2m_s)$.

REFERENCES AND SUGGESTIONS FOR FURTHER READING

E. U. Condon, "The Atom," *Phys. Today* **5**, No. 1, 4 (1952).

E. U. Condon and J. E. Mack, "An Interpretation of Pauli's Exclusion Principle," *Phys. Rev.* **35**, 579 (1930).

G. Gamow, "The Exclusion Principle," *Sci. Am.* 201, No. 1, 74 (1959).

G. Herzberg, *Atomic Spectra and Atomic Structure*, Prentice-Hall (1937).

H. E. White, *Introduction to Atomic Spectra*, McGraw-Hill (1934).

P. Zeeman, *On the Influence of Magnetism on the Nature of Light Emitted by a Substance*, Amsterdam (1897), reprinted in the *Am. J. Phys.* **16**, 216 (1948).

Also see references for Chapter 8.

QUESTIONS AND PROBLEMS

Independent particle approximation

1. Estimate the energy of an innermost electron in (a) the sodium atom, (b) the silver atom, and (c) the uranium atom.

2. Consider the potential energy $V(r)$ of an electron in an atom with atomic number Z. Approximate the charge distribution of the other $Z-1$ electrons to be distributed like the probability density for the ground state of hydrogen. Calculate the potential energy $V(r)$ and verify that the result conforms to Figure 9-1.

The Pauli exclusion principle

3. Consider the total wave function ψ for three electrons. Give an expression for ψ as a function of the individual electron wave functions $\psi_a(\mathbf{r}_1)$, $\psi_b(\mathbf{r}_2)$, and $\psi_c(\mathbf{r}_3)$ where **a**, **b**, and **c** represent the quantum numbers of each electron and $\mathbf{r}_1$, $\mathbf{r}_2$, and $\mathbf{r}_3$ represent the coordinates.

4. (a) List the possible values of the quantum numbers n, ℓ, m_ℓ, and m_s for a $2p$ state. (b) If an atom has two $2p$ electrons, how many states are there?

Shell structure and the periodic table

5. If there were a stable element 113, what would be your guess of the quantum numbers n and ℓ of the most energetic electron? Why might such an element not exist?

6. Following Example 9-2, use Figure 8-4 to make a rough estimate of the energy levels of the $3s$, $3p$, and $3d$ states of lithium. Compare your estimate with the exact answer (Figure 9-9).

The coupling of angular momenta

7. Determine the possible values of the total angular momentum, J, for an outer electron in the scandium atom ($Z = 21$). Express your answer in units of $\hbar$.

8. What are the possible values of the total angular momentum of a pair of $3d$ electrons?

9. What are the possible quantum numbers for an atom with an outer electron configuration of $2p^1 3d^1$?

10. An atom has two electrons in the d subshell. What are the possible values of the z component of the total angular momentum?

11. An atom has a single outer electron in the $n = 3$ state. What are the possible values of the total angular momentum quantum number (j)?

Excited states of atoms

12. In an energy level diagram for a multielectron atom, why must we specify the quantum numbers of every electron in the atom to define an energy level?

13. What is the highest energy photon that can be emitted from the helium atom?

14. How do the energy levels of singly ionized lithium compare to the energy levels of helium?

15. Make an energy level diagram for the helium atom showing the states where one electron is in the $1s$ state and the other electron is in the $5p$ state. Indicate the allowed transitions to lower energy states.

16. Make a qualitative sketch of the energy levels in the silicon atom ($Z = 14$).

17. Make a qualitative sketch of the energy levels in the arsenic atom ($Z = 33$).

18. Make an estimate of the amount of energy needed to remove all three electrons from the lithium atom.

Atoms in an external magnetic field

19. In a Stern-Gerlach type of experiment, for what elements (if any) would you expect to see no splitting, a threefold splitting, or a fourfold splitting?

20. A $3d_{5/2}$ state with an unsplit energy of E is placed in a weak magnetic field. Determine the number of states

that level is split into and make a sketch of the resulting energy levels.

21. How many states is a $2p$ state split into in a strong external magnetic field? Calculate the energy splitting in an external field of 1 tesla.

22. (a) What are the possible values of the quantum numbers n, ℓ and j of outermost electron in the potassium atom ($Z = 19$) in the ground state and first excited state which satisfies the selection rule $\Delta \ell = \pm 1$. (b) Which of these states is split by the spin–orbit interaction? Make an energy diagram of the states indicating the splitting. Label each state by its quantum numbers. (c) Consider the addition of an external magnetic field. Make an energy diagram indicating the additional splitting of each state. Label each state by its quantum numbers.

23. Before the discovery of electron intrinsic angular momentum, atoms were considered to exhibit a "normal" Zeeman effect if a transition was split into an odd number of lines in an external magnetic field. Which elements would you expect to have a normal Zeeman effect? Give three specific examples of transitions that are split into three lines.

Additional problems

24. Which of the following elements have their ground state split by the spin–orbital interaction: Li, B, Na, Al, K, and Ga?

*25. Sodium atoms are placed in magnetic field of 1.5 T. (a) Calculate the Zeeman splitting of the ground state. (b) At what temperature are 1/3 of the atoms in the higher energy state? (c) At what temperature are 49% of the atoms in the higher energy state?

26. The spectroscopic notation for the ground state of hydrogen is $1s_{1/2}$. Give the corresponding spectroscopic notation for the ground states of K, Al, Y, and Au.

*27. *Hund's rules*. The rules for finding the the ground state quantum numbers of an atom with more than one electron in a subshell are: 1. The total intrinsic angular momentum quantum number is the maximum value consistent with the Pauli exclusion principle, and 2. The total orbital angular momentum quantum number is the maximum total m_ℓ consistent with the Pauli exclusion principle. (a) Show that Hund's rules give the correct ground state quantum numbers for carbon (see Figure 9-12) and nitrogen (see Figure 9-13). (b) Find the ground state quantum numbers for iron.

28. In an excited state of Zr, one of the $4d$ electrons is in the $5p$ state. (a) What are the possible values for the

total intrinsic and orbital angular momentum quantum numbers? (b) Determine the total angular momentum quantum number of the state with lowest energy.

29. Find the the angle between **L** and **S** for (a) $p_{1/2}$ and (b) $p_{3/2}$ states.

*30. (a) Show that the magnetic moment of an atom for which LS coupling holds is

$$\mu = \sqrt{j(j+1)}\, g_L \mu_B.$$

(b) Calculate the magnetic moment for a $p_{3/2}$ state.

31. A sample of sodium atoms is placed in an external magnetic field of 1.0 tesla. Calculate the resulting shifts in energy levels for the $3s$, $3p_{1/2}$ and $3p_{3/2}$ states.

As a subject becomes older, and better understood, it is usually possible to present its fundamentals in a simpler form than at the outset. This is true of the rotational distortion of molecular spectral terms... . The energy of a molecule whose center of gravity is at rest is approximately separable into the internal energy of the electrons relative to "clamped nuclei," the energy of molecular vibration, and the rotational energy of a rigid body whose constants are obtained by regarding the molecule as built out of an ensemble of rigidly connected nuclei. This fact seems reasonably obvious physically, and also can be demonstrated by a systematic application of perturbation theory.

John H. Van Vleck

Various combinations of atoms are observed to attract each other to form molecules. The number of atoms in a molecule ranges from two in a simple diatomic molecule to roughly 10^{10} in molecules of deoxyribonucleic acid (DNA)! The properties of molecules are dictated by their electron configurations. The fundamental physics of electrons in molecules is identical to electrons in atoms; however, the detailed form of the electromagnetic potential is different. In this chapter we shall apply what we have learned about the quantum mechanics of electrons in atoms together with the quantum mechanics of vibrations and rotations in order to understand the fundamental interaction of two atoms in a molecule.

Molecular bonding is caused by electromagnetic interactions of electrons and protons in atoms. A molecular bond occurs when it is energetically favorable for two or more atoms to be close to each other, that is, energy is required to separate the atoms. The molecular *bond length* is defined to be the distance between the nuclei of the atoms. A molecule is a configuration of atoms that are sufficiently close together such that the probability densities of the outer electrons overlap. This causes the energy levels to be perturbed because the electrons of a given atom interact with the electrons and nuclei of the neighboring atoms. If the total energy of all the electrons in the atoms is smaller when the atoms are close together, a molecular bond occurs. In the language of special relativity, the mass of the molecule is smaller than the sum of the masses of all the atoms that make up the molecule:

$$m_{molecule} < \sum m_{atom} . \qquad (10.1)$$

The difference in rest energies (E_b),

$$E_b = \sum m_{atom} c^2 - m_{molecule} c^2, \qquad (10.2)$$

is a measure of the energy needed to break the molecular bonds, the molecular binding energy. The order of magnitude for the strongest type of molecular bonds is a few electronvolts (see Table 10-1). (Some types of molecular bonds, however, are significantly weaker.) The molecular bond is responsible for the attraction of molecules in a liquid or solid state; it keeps you from falling apart! The energy of the molecular bonds determines many physical properties of materials such as the melting point.

Molecular bonds are caused by the arrangement of electrons into a lower energy configuration in the molecule as compared to the separated atoms. Molecular bonds are often classified according to the type of electron arrangement that occurs. The types of molecular bonds are listed in Table 10-2. A classification of the bonding types

TABLE 10-1
BOND LENGTHS AND STRENGTHS.

Molecule	Formula	Bond	Length (nm)	Strength (eV)
Hydrogen	H_2	H–H	0.075	4.5
Nitrogen	N_2	N–N	0.11	9.8
Oxygen	O_2	O–O	0.12	5.2
Fluorine	F_2	F–F	0.14	1.6
Chlorine	Cl_2	Cl–Cl	0.20	2.5
Carbon monoxide	CO	C–O	0.11	11.1
Methane	CH_4	$H–CH_3$	0.11	4.5
Water	H_2O	H–OH	0.096	5.2

serves to qualitatively explain why specific combinations of atoms bond together into molecules. In only a few ideal cases does a single type of bond exist. Most molecules are a mixture of two or more bonding types. (A good general picture of a molecule is that the electron energy levels are shifted because of the presence of the neighboring atoms.)

> The typical size of a bond length is the atomic radius ($d_{atom}/2$) resulting in a typical molecular binding energy of a few electronvolts.

10-1 THE HYDROGEN MOLECULE

The H_2^+ Molecule

The simplest molecule is the hydrogen ion (H_2^+), a bound state of two protons and one electron. The total potential

TABLE 10-2
CLASSIFICATION OF MOLECULAR BONDS.

Type	Physical Mechanism	Example
Ionic	Electron transfer	NaCl
Covalent	Electron sharing (highly localized)	H_2
Hydrogen	Hydrogen atom link	DNA
Van der Waals	Dipole attraction	Water
Metallic	Electrons shared by many atoms (weakly localized)	Metal crystal

energy V of the molecule, the energy required to separate all three particles, may be analyzed as a function of the distance r between the two protons. There are two contributions to V, the repulsion of the two protons and the attraction of the electron to each of the protons. The repulsion of the two protons is simply ke^2/r. The contribution to V from the attraction of the electron is -13.6 eV when the protons are far apart, that is, the electron is near one of the protons and the effect of the other proton is negligible. If the protons are very close together, then the contribution to V from the attraction of the electron is four times greater (-54.4 eV), corresponding to the ionization energy of He^+. The contributions to V are shown in Figure 10-1. The potential energy V has a minimum of -16.2 eV near $r = 0.1$ nm. More energy is required to separate all three particles when the two protons are 0.1 nm apart than for any other separation distance. Therefore, a proton plus a hydrogen atom make a molecular bond with a bond length of 0.1 nm. The strength of the bond between the proton and the hydrogen atom is the energy required to separate the H_2^+ into $H + p$. This binding energy may be written

$$E_b = m_H c^2 + m_p c^2 - m_{H^+} c^2 \approx 2.6 \text{ eV.} \quad (10.3)$$

In the H_2^+ molecule the electron probability density is symmetric, that is, the electron is equally likely be found near either proton. We illustrate this by examining the potential energy of the electron as a function of the electron position z along the axis containing the two protons. This electron potential energy is shown in Figure 10-2 for three different proton separations. The origin is taken to be at $z = 0$ and the location of the protons is (a) ± 0.15 nm, (b) ± 0.015 nm, and (c) ± 0.05 nm. Figure 10-2c corresponds to the bond length of the H_2^+ molecule. The electron probability distribution is symmetric about $z = 0$. The electron probability density is maximum at $z = 0$, when the electron is near both protons. (An antisymmetric total wave function, which has a node at $z = 0$ has a higher energy and does not make a bound state.)

The H_2 Molecule

The neutral H_2 molecule has two protons and two electrons. The H_2 molecule is similar in many respects to the H_2^+ molecule except for one important difference: The two electrons must obey the Pauli exclusion principle. Therefore, in the independent particle approximation, the ground state is twofold degenerate, as in the hydrogen atom, and each electron can occupy the ground state provided that the two electrons have opposite values of m_s (spins antialigned). The spin component of the total wave function is therefore antisymmetric. The total wave function must be antisymmetric according to the Pauli exclusion principle. Therefore, the spatial part of the wave function must be symmetric. A symmetric wave function implies that the electrons have a relatively large probability of being found in the region between the two protons.

We may analyze the total potential energy $V(r)$ as a function of distance between the two protons. The two extremes are (1) the two protons are far apart, forming two hydrogen atoms, and (2) the two protons have essentially no separation. In the first case the amount of energy required to separate both electrons from both protons is twice the ionization energy for hydrogen, or 27.2 eV. In the second case, the amount of energy is 79 eV, corresponding to the energy (9.33) needed to doubly ionize helium. The form of $V(r)$ for the neutral hydrogen molecule is similar that of the hydrogen molecular ion (Figure 10-1). The two protons each share both electrons forming a molecule. The molecular bond length is about 0.075 nm and the molecular binding energy is about 4.5 eV. Total energy of the molecule is lower than in the case of two separated atoms. The bond length is shorter and the binding energy is larger in H_2 than for H_2^+ because in H_2 there are two electrons pulling the protons together (i.e., 79 eV > 54 eV).

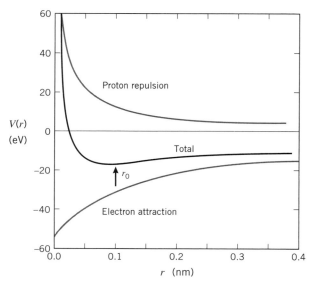

FIGURE 10-1 Potential energy of two protons and one electron as a function of the distance (r) between the two protons.

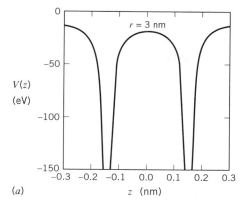

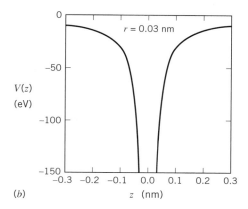

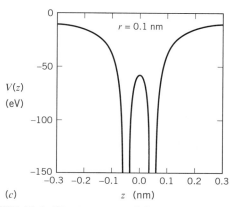

FIGURE 10-2. Electron potential energy along the axis (z) of two protons.
The proton separation is (a) 0.3 nm, (b) 0.03 nm, and (c) 0.1 nm.

10-2 THE SODIUM-CHLORIDE MOLECULE

The NaCl molecule is an example of the ionic bond. Sodium ($Z = 11$) has the electron configuration $1s^2 2s^2 2p^6 3s^1$.

Chlorine ($Z = 17$) has the electron configuration $1s^2 2s^2 2p^6 3s^2 3p^5$. Sodium donates its outer electron (becoming Na$^+$) to chlorine (becoming Cl$^-$). The Na$^+$ and Cl$^-$ then attract each other by Coulomb's law.

The key to the physics of the ionic bond is that the sodium gives up its outer electron because it is energetically favorable for it to do so. To remove the outer electron from Na requires 5.1 eV (see Figure 9-10). (*Note:* In hydrogen, $n = 3$, the ionization energy is 13.6 eV/9 ≈ 1.5 eV. In sodium the ionization energy is larger because the $3s$ electron penetrates the cloud of the other 10 electrons and therefore feels the force of a larger charge.) The energy released when an electron is added to chlorine, called the electron affinity, is about 3.6 eV. The energy difference between the neutral atoms and the separated ions is

$$\Delta E = 5.1\,\text{eV} - 3.6\,\text{eV} = 1.5\,\text{eV} . \quad (10.4)$$

This energy (and more) is supplied by the potential energy of the Na$^+$ and Cl$^-$ attraction. The total energy, $E(r)$, of the molecule is the sum of the contributions due to ionization, electron affinity, Coulomb attraction, and the Pauli exclusion principle. We may write the molecular energy as

$$E(r) = 5.1\,\text{eV} - 3.6\,\text{eV} - \frac{ke^2}{r} + C\frac{e^{-ar}}{r}, \quad (10.5)$$

where the last term represents the effect of the Pauli exclusion principle (with C and a constants). In Figure 10-3 we plot this total energy (10.5) versus the separation distance r between the Na$^+$ and Cl$^-$ ions. (The constants C and a have been chosen to give appropriate behavior of the potential at short distances.) The minimum value of the total energy is the binding energy (E_b) of the molecule, the amount of energy needed to separate the neutral atoms to infinity. At $r \approx 1$ nm we have $E = 0$. As r decreases, the Coulomb energy decreases as $-1/r$. At very small r, the energy increases because electrons are forced to go into higher orbits by the exclusion principle. The minimum energy, $E_b = 3.6$ eV, occurs at $r = 0.24$ nm.

Thus far, we have been discussing the bonding in a *single* NaCl molecule. In a solid, the NaCl molecules form a regular array or *crystal,* and each atom interacts with all of its neighbors. The structure of the NaCl salt crystal is called *face-centered cubic* because the basic cell has atoms on all corners and all faces of a cube. The entire crystal can be built up from spatial translations of the basic cell. In the face-centered-cubic lattice, each sodium atom is surrounded by six chlorine atoms at equal distance. Also, each chlorine atom is surrounded by six sodium

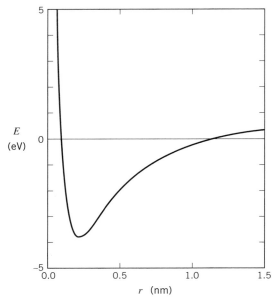

FIGURE 10-3 Molecular potential.
Total energy (ionization + affinity + Coulomb attraction + Pauli exclusion principle) versus separation distance r between Na^+ and Cl^-.

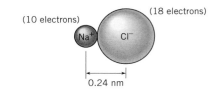

(a) Single molecule

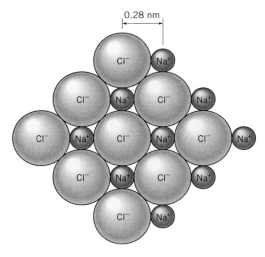

(b) Salt crystal

FIGURE 10-4 Sketch of the electron distributions in (a) the NaCl molecule, and (b) the NaCl crystal.
This crystal lattice structure is called face-centered cubic.

atoms all at equal distance. When the molecules are packed into a crystal, the resulting binding energy is $E_b = 6.4$ eV with $r_0 = 0.28$ nm.

In the NaCl crystal, one can verify that the ionic bonding mechanism is dominant by studying the electron distribution by x-ray diffraction. One finds that the electron distributions are spherically symmetric in both the sodium and chlorine, corresponding to closed subshells, except for small distortions near the boundaries of the atoms. Ten electrons are observed to be clustered around each sodium nucleus and 18 electrons are observed to be clustered around each chlorine nucleus.

10-3 MOLECULAR VIBRATIONS AND ROTATIONS

The energy levels of electrons in molecules are quantized so that molecules have electronic excitations similar to atoms. In addition to electronic excitations, molecules have degrees of freedom not available to single atoms. These additional degrees of freedom arise from the fact that the nuclei in a molecule may have a relative motion even though the center of mass of the molecule is fixed. One of these degrees of freedom is *vibrational* motion of

the nuclei. We shall analyze the vibrational motion of the nuclei as a quantum harmonic oscillator.

Vibrational Levels

Consider a diatomic molecule. A small displacement in the distance between the nuclei will cause a change in energy, which may be described by a potential energy function similar to that of Figure 10-3. We may expand the potential about the minimum ($r = r_0$) to get (see Appendix F)

$$V(r) \approx V(r_0) + \frac{1}{2}(r - r_0)^2 \left[\frac{\partial^2 V}{\partial r^2}\right]_{r=r_0}. \quad (10.6)$$

Therefore, the molecular potential energy near its minimum may be approximated by a quadratic function:

$$V(x) = \frac{1}{2}Kx^2, \quad (10.7)$$

where x is the displacement from the equilibrium bond length. We have seen that the quantum solution of the quadratic potential leads to quantized energy levels in the simple harmonic oscillator. Thus, the vibrational energy levels will have quantized energies

$$E_n \approx \left(n - \frac{1}{2} \right) \hbar \omega, \qquad (10.8)$$

where n is a positive integer (1, 2, 3, …) and ω is the vibrational angular frequency. The minimum vibrational energy is equal to $\hbar\omega/2$. Recall that the reason that the minimum energy of vibration is not zero is that the nuclei are confined. The vibrational frequency is related to the force constant K by (see problem 35)

$$\omega = \sqrt{\frac{K}{M}}, \qquad (10.9)$$

where M is the reduced mass of the vibrating molecule ($M = M_1 M_2/(M_1 + M_2)$ for a diatomic molecule composed of atoms with masses M_1 and M_2).

We now estimate the order of magnitude of the vibrational energies. The energy scale of the vibrations (E_v) is given by

$$E_v \approx \hbar\omega = \hbar \sqrt{\frac{K}{M}}. \qquad (10.10)$$

The value of K depends on the form of the electromagnetic potential that confines the atom. The distance between nuclei in a molecule is about equal to the diameter of an atom (d_{atom}). Therefore, the distance scale that confines the atom in a molecule is d_{atom}. The order of magnitude electronic energy scale (E_e) of the atom is given by the molecular potential energy (10.7) at $x \approx d_{atom}/2$:

$$E_e \approx \frac{1}{2} K \left(\frac{d_{atom}}{2} \right)^2 = \frac{1}{8} K d_{atom}^2. \qquad (10.11)$$

We may also relate the electronic energy scale directly with the size of the atom. Equating the electronic energy scale with the ground state energy (7.26) for a particle in a box of size d_{atom}, we have

$$E_e \approx \frac{h^2}{8 m d_{atom}^2}, \qquad (10.12)$$

where m is the electron mass. Numerically, the electronic energy scale (taking $d_{atom} \approx 0.3$ nm) is

$$E_e \approx \frac{(1240 \, eV \cdot nm)^2}{(8)(5 \times 10^5 \, eV)(0.3 \, nm)^2}$$
$$\approx 4 \, eV. \qquad (10.13)$$

Combining the two expressions for the electronic energy scale (10.11 and 10.12), we have

$$\frac{1}{8} K d_{atom}^2 \approx \frac{h^2}{8 m d_{atom}^2}. \qquad (10.14)$$

Solving for the force constant K, we have

$$K \approx \frac{h^2}{m d_{atom}^4}. \qquad (10.15)$$

The energy scale of the vibrational excitations (10.10) is

$$E_v = \hbar \sqrt{\frac{K}{M}} = \frac{h^2}{2\pi d_{atom}^2 \sqrt{mM}}. \qquad (10.16)$$

We now compare the vibrational energy scale (10.16) with the electronic energy scale (10.12). The ratio is

$$\frac{E_v}{E_e} \approx \frac{\left(\dfrac{h^2}{2\pi d_{atom}^2 \sqrt{mM}} \right)}{\left(\dfrac{h^2}{8 m d_{atom}^2} \right)} \approx \sqrt{\frac{m}{M}}. \qquad (10.17)$$

The vibration energy is smaller than the electronic energy by the approximate ratio $\sqrt{m/M}$. The order of magnitude of the vibrational energy scale is

$$E_v \approx E_e \sqrt{\frac{m}{M}} \approx (4 \, eV) \sqrt{\frac{5 \times 10^5 \, eV}{10^9 \, eV}}$$
$$\approx 0.1 \, eV. \qquad (10.18)$$

EXAMPLE 10-1

The measured vibrational frequency for the oxygen molecule (O_2) is $f = \omega/2\pi = 4.7 \times 10^{13}$ Hz. Calculate the energy difference between vibrational states in the oxygen molecule and estimate the value of the force constant K.

SOLUTION:

The vibrational excitation energy is

$$E_v = hf$$
$$= \left(4.14 \times 10^{-15}\ \text{eV} \cdot \text{s}\right)\left(4.7 \times 10^{13}\ \text{s}^{-1}\right)$$
$$\approx 0.2\ \text{eV}.$$

The reduced mass (M) of the oxygen molecule is

$$M = \frac{(16)(16)}{16+16}\left(940\ \text{MeV}\right) \approx 7.5 \times 10^9\ \text{eV}.$$

The force constant is

$$K = M\omega^2 = \frac{Mc^2\hbar^2\omega^2}{\hbar^2c^2} = \frac{Mc^2E_v^2}{\hbar^2c^2}$$
$$\approx \frac{\left(7.5 \times 10^9\ \text{eV}\right)\left(0.2\ \text{eV}\right)^2}{\left(200\ \text{eV} \cdot \text{nm}\right)^2}$$
$$\approx 8 \times 10^3\ \text{eV/nm}^2.$$

In Newtons per meter, the force constant is

$$K \approx \left(8 \times 10^3\ \text{eV/nm}^2\right)\left(\frac{1.6 \times 10^{-19}\ \text{J}}{\text{eV}}\right)$$
$$\times \left(\frac{10^9\ \text{nm}}{\text{m}}\right)^2\left(\frac{\text{N} \cdot \text{m}}{\text{J}}\right)$$
$$\approx 1000\ \text{N/m}.$$

The value of K is the same order of magnitude as a common mechanical spring. ∎

EXAMPLE 10-2

To excite the hydrogen-chloride molecule (HCl) from the ground state to the first vibrational state requires an energy of 0.36 eV. Calculate the vibrational frequency of the HCl molecule.

SOLUTION:

The vibrational frequency of the HCl molecule is given by

$$f = \frac{E}{h} = \frac{0.36\ \text{eV}}{4.14 \times 10^{-15}\ \text{eV} \cdot \text{s}} \approx 8.7 \times 10^{13}\ \text{s}^{-1}.$$

This is a typical value for a molecular vibrational frequency. ∎

The vibrational energy levels are not equally spaced as found for the harmonic oscillator because the molecular potential is not exactly quadratic (see Figure 10-3). This results in the higher energy excited states being closer together than the ground and first excited states. The vibrational frequency is defined to be the frequency of radiation needed to cause a transition from the ground state to the first excited state. The energy levels may be approximated by

$$E_n \approx \left(n - \frac{1}{2}\right)\hbar\omega - C\left(n - \frac{1}{2}\right)^2\hbar\omega, \quad (10.19)$$

where n is a positive integer $(1, 2, 3, \ldots)$ and C is a constant. For a fixed electronic level, there is a selection rule on the change in the quantum number n, which is

$$\Delta n = \pm 1. \quad (10.20)$$

The change in vibrational levels is only to a neighboring level. Thus, the molecule radiates at the same frequency with which it vibrates. Figure 10-5 shows a diagram of molecular vibrational energy levels. The measured vibrational frequencies of some molecules are given in Table 10-3.

> The order of magnitude of the energy of a molecular vibrational excitation is 10^{-1} eV. This corresponds to energies in the infrared region of the electromagnetic spectrum.

Measuring the Vibrational Levels

The vibrational levels may be measured by a variety of experimental methods. One powerful technique is that of

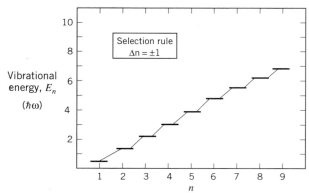

FIGURE 10-5 Vibrational energy levels of a molecule. The lower energy levels are proportional to $n - 1/2$, where n is the molecular vibrational quantum number. Higher energy levels are more closely spaced. The selection rules for transitions between levels are $\Delta n = \pm 1$. The order of magnitude of the photon energy from a typical transition is 10^{-1} eV.

TABLE 10-3
VIBRATIONAL FREQUENCIES OF SOME DIATOMIC MOLECULES.

Molecule	$f\,(10^{13}\ \text{Hz})$
H_2	13
N_2	7.0
CO	6.4
NO	5.7
O_2	4.7

photoelectron spectroscopy. In photoelectron spectroscopy a photon of known energy is absorbed by a molecule, which ejects an electron, leaving the molecular ion in a state of vibrational excitation. The time scale for the ejection of the electron is short compared to the period ($T = 1/f$) of a molecular vibration. The molecular ion may be left in one of several possible vibrational states. The energy of the ejected electron is measured in a spectrometer. By conservation of energy, the incoming photon energy (E_γ) is equal to the sum of the ejected electron kinetic energy (E_k), the electron ionization energy (E_{ion}), and the energy that goes into exciting the molecular ion out of the ground state (ΔE):

$$E_\gamma = E_k + E_{ion} + \Delta E. \qquad (10.21)$$

If the molecular ion is left in the nth vibrational excited state, then $\Delta E = E_n - E_1$, where E_n is the energy (10.19) of the nth excited state and E_1 is the energy of the ground state.

One common photon source uses an electrical discharge in helium gas. In this process, helium atoms are excited by collisions with electrons. When the helium atoms make transitions to lower energy states, photons of many energies are produced (see Figure 9-11). The largest number of photons come from the singlet $2p \rightarrow 1s$ transition at a wavelength of 58.4 nanometers in the far ultraviolet. The energy of these photons is 21.22 eV. Photons arising from transitions to the $2s$ state have energies limited to about 4 eV. Therefore, when atoms or molecules having an ionization energy greater than 4 eV are exposed to the helium light source, an electron can be emitted only by collision with a 21.22-eV photon.

Hydrogen
The energy levels of H_2^+ may be studied by directing the 21.22-eV photons from the helium lamp into a sample of hydrogen molecules. Electrons are ejected by the process

$$\gamma + H_2 \rightarrow H_2^+ + e^-. \qquad (10.22)$$

The kinetic energy of the electron is measured in a spectrometer. The H_2^+ gets essentially zero kinetic energy because of its large mass compared to the electron mass. By conservation of energy (10.21), the molecular excitation energy is

$$\Delta E = (21.22\,\text{eV}) - E_k - E_{ion}. \qquad (10.23)$$

The resulting electron kinetic energy spectrum of the electron is shown in Figure 10-6. Fifteen peaks are observed in the spectrum, corresponding to the ground state and fourteen vibrational excited states of H_2^+. The ground state ($\Delta E = 0$) occurs at $E_k = 5.79$ eV corresponding to $E_{ion} = 15.43$ eV. (An energy of 15.43 eV is required to remove one electron from H_2.) The first excited state occurs at $\Delta E = 0.28$ eV and $E_k = 5.51$ eV. The energy difference between the ground state and first vibrational excited state is $\Delta E = 0.28$ eV, corresponding to an H_2^+ vibrational frequency of 6.8×10^{13} s^{-1}. For excitation energies greater than 2.6 eV, the H_2^+ molecule dissociates. Recall that the binding energy (10.3) of the H_2^+ molecule is 2.6 eV.

Water
Molecules with more than two atoms have more complicated vibrational spectra than diatomic molecules because there is more than one mode of vibration. Consider the water molecule (H_2O). Figure 10-7 shows the photoelectron spectrum of water, again using the 21.22-eV helium photon. The reaction is

$$\gamma + H_2O \rightarrow H_2O^+ + e^-. \qquad (10.24)$$

Three bands are observed with rich structure. The structure is caused by vibrational excitations of the water molecule. The two principle modes of excitation are stretching and bending of the H–O bonds. The regular structure of the vibrational levels in the middle band near $E_k = 6$ eV is caused by the H–O bending mode.

Rotational Levels

Molecules may also rotate about an axis through the center of mass. For a diatomic molecule rotating with angular frequency ω, the kinetic energy due to rotation (E_r) is

$$E_r = \frac{1}{2} I \omega^2, \qquad (10.25)$$

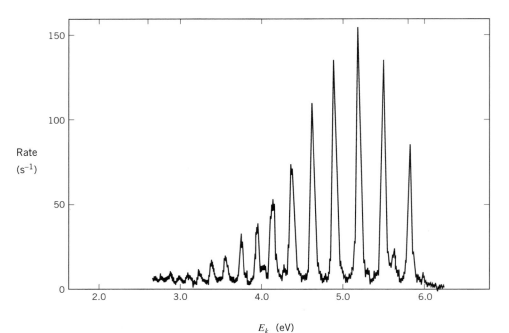

E_k (eV)

FIGURE 10-6 Photoelectron spectrum of H_2^+.
The spectrum is made by directing photons of energy 21.22 eV from a helium lamp onto a sample of hydrogen gas. The peaks in the electron kinetic energy spectrum are due to the H_2^+ molecule being left in 15 different vibrational states. From D. W. Turner, C. Baker, A. D. Baker, and C. R. Brundle, *Molecular Photoelectron Spectroscopy,* Wiley-Interscience (1970). Reprinted by permission of John Wiley & Sons, Inc.

where I is the moment of inertia. We may express the moment of inertia in terms of the atomic masses (M_1 and M_2) and the atomic distances to the rotation axis (r_1 and r_2) as shown in Figure 10-8,

$$I = M_1 r_1^2 + M_2 r_2^2. \qquad (10.26)$$

In terms of the angular momentum (L), the rotational kinetic energy is

$$E_r = \frac{L^2}{2I}. \qquad (10.27)$$

The angular momentum is quantized. The quantization condition may be written in an identical fashion to that in an atom:

$$L^2 = \ell(\ell+1)\hbar^2, \qquad (10.28)$$

where ℓ is a nonnegative integer ($0, 1, 2, 3, \ldots$). The energy due to rotation is given by

$$E_r = \frac{\ell(\ell+1)\hbar^2}{2\left(M_1 r_1^2 + M_2 r_2^2\right)}. \qquad (10.29)$$

Taking $M_1 = M_2 = M$ and $r_1 = r_2 = d_{atom}/2$, we may estimate the order of magnitude of a rotational excitation energy to be

$$E_r \approx \frac{\hbar^2}{M d_{atom}^2} = \frac{h^2}{4\pi^2 M d_{atom}^2}. \qquad (10.30)$$

Comparing the order of magnitude of the rotational energy scale (10.30) with the electronic energy scale (10.12), we have

$$\frac{E_r}{E_e} \approx \frac{m}{M}. \qquad (10.31)$$

The typical order of magnitude of a rotational excitation energy is

$$E_r = E_e \frac{m}{M} = (4 \text{ eV}) \frac{\left(5 \times 10^5 \text{ eV}\right)}{10^9 \text{ eV}} \qquad (10.32)$$

$$\approx 10^{-3} \text{ eV}.$$

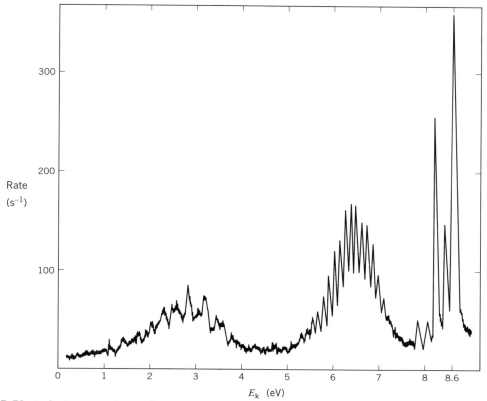

FIGURE 10-7 Photoelectron spectrum of water.
From D. W. Turner, C. Baker, A. D. Baker, and C. R. Brundle, *Molecular Photoelectron Spectroscopy,* Wiley-Interscience (1970). Reprinted by permission of John Wiley & Sons, Inc.

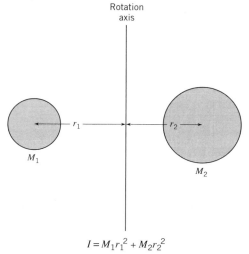

FIGURE 10-8 Rotation of a diatomic molecule.

Thus, the energy scale of the rotational modes of excitation is about four orders of magnitude smaller than the electronic and about two orders of magnitude smaller than the vibrational. The results are summarized in Table 10-4.

TABLE 10-4
COMPARISON OF THE ELECTRONIC, VIBRATIONAL AND ROTATIONAL MOLECULAR ENERGY LEVELS. In the energy estimates m is the electron mass, m is the molecular mass, and d_{atom} is the diameter of an atom.

Excitation	Energy Estimate	Typical Energy
Electronic	$\dfrac{h^2}{8md_{atom}^2}$	4 eV
Vibrational	$\dfrac{h^2}{2\pi d_{atom}^2 \sqrt{mM}}$	10^{-1} eV
Rotational	$\dfrac{h^2}{4\pi^2 Md_{atom}^2}$	10^{-3} eV

> The order of magnitude of the energy of a molecular rotational excitation is 10^{-3} eV.

The molecular rotational energy levels of a molecule are shown in Figure 10-9. Conservation of angular momentum imposes the selection rule, $\Delta\ell = \pm 1$, so that transitions occur only between neighboring levels. When a molecule goes from level $\ell+1$ to level ℓ, the energy of the emitted photon is

$$E_\gamma = \frac{(\ell+1)(\ell+2)\hbar^2}{2I} - \frac{\ell(\ell+1)\hbar^2}{2I}$$
$$= \frac{(\ell+1)\hbar^2}{I}. \qquad (10.33)$$

A pure rotational spectrum, where the electronic and vibrational levels do not change, consists of a number of equally spaced lines with energies of $\hbar^2/I$, $2\hbar^2/I$, $3\hbar^2/I$, and so on.

EXAMPLE 10-3
Calculate the energy needed to excite an oxygen molecule (O_2) from the ground state to the first rotational level.

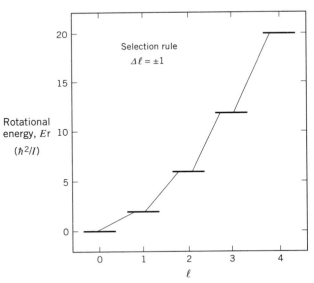

FIGURE 10-9 Rotational energy levels of a molecule. The energy units are the reduced Planck's constant squared divided by the moment of inertia of the molecule. The energy levels are proportional to $\ell(\ell + 1)$, where ℓ is the molecular angular momentum quantum number. The selection rules for transitions between levels are $\Delta\ell = \pm 1$. The order of magnitude of the photon energy from a typical transition is 10^{-3} eV.

SOLUTION:
The bond length of oxygen is 0.12 nm. Taking $r_1 = r_2 = r$, we have

$$r + r = 0.12 \, \text{nm}.$$

The mass of the oxygen atom is

$$M = \left(931.5 \, \text{MeV}/c^2\right)(16) = 14.9 \, \text{GeV}/c^2.$$

From the expression for the rotational energy levels (10.29) the excitation energy (ΔE) to go from $\ell = 0$ to $\ell = 1$ is

$$\Delta E = \frac{1(1+1)\hbar^2}{4Mr^2} - \frac{0(0+1)\hbar^2}{4Mr^2} = \frac{\hbar^2}{2Mr^2},$$

or

$$E_r = \frac{(\hbar c)^2}{2Mc^2 r^2}$$
$$= \frac{(200 \, \text{eV} \cdot \text{nm})^2}{(2)(1.49 \times 10^{10} \, \text{eV})(0.060 \, \text{nm})^2}$$
$$\approx 3.7 \times 10^{-4} \, \text{eV}. \qquad \blacksquare$$

EXAMPLE 10-4
Make an estimate of the energy difference between the ground state and the first rotational excited state of the NaCl molecule.

SOLUTION:
The bond length is 0.24 nm. Let r_1 be the distance to the rotation axis of the Na nucleus, mass M_1, and r_2 be the distance for the Cl nucleus, mass M_2. Then

$$M_1 r_1 = M_2 r_2,$$

and

$$r_1 + r_2 = 0.24 \, \text{nm}.$$

From these two equations we have

$$r_1 = \frac{M_2}{M_1 + M_2}(0.24 \, \text{nm}),$$

and

$$r_2 = \frac{M_1}{M_1 + M_2}(0.24 \, \text{nm}).$$

The moment of inertia is

$$I = M_1 r_1^2 + M_2 r_2^2 = \frac{M_1 M_2}{M_1 + M_2}(0.24 \text{ nm})^2.$$

The energy difference between the levels is

$$\Delta E = \frac{1(1+1)\hbar^2}{2I} - 0 = \frac{\hbar^2}{I} = \frac{\hbar^2(M_1 + M_2)}{M_1 M_2 (0.24 \text{ nm})^2}.$$

The mass of the Na atom is

$$M_1 = (931.5 \text{ MeV}/c^2)(23) = 21.4 \text{ GeV}/c^2.$$

The mass of the Cl atom is

$$M_2 = (931.5 \text{ MeV}/c^2)(35) = 32.6 \text{ GeV}/c^2.$$

The energy difference between the levels is

$$\Delta E = \frac{(200 \text{ eV} \cdot \text{nm})^2 (21.4 \text{ GeV} + 32.6 \text{ GeV})}{(21.4 \text{ GeV})(32.6 \text{ GeV})(10^9 \text{ eV})(0.24 \text{ nm})^2}$$

$$= 5.4 \times 10^{-5} \text{ eV}. \qquad \blacksquare$$

10-4 MOLECULAR SPECTRA

Vibrational–Rotational Spectra

For a given electronic level (E_e), we may write the energy of a molecular state as the sum of electronic, vibrational, and rotational terms. Two quantum numbers are needed to specify the energy level, the vibrational quantum number (n) and the rotational quantum number (ℓ). We may write the molecular energy as

$$E_{n,\ell} \approx E_e + \left(n - \frac{1}{2}\right)\hbar\omega + \frac{\ell(\ell+1)\hbar^2}{2I}, \quad (10.34)$$

where we have neglected the quadratic correction term to the vibrational energy. There is a selection rule for molecular transitions:

$$\Delta\ell = \pm 1. \qquad (10.35)$$

For molecular absorption from the $n = 1$ state, the initial energy is

$$E_{1,\ell} \approx E_e + \frac{\hbar\omega}{2} + \frac{\ell(\ell+1)\hbar^2}{2I}. \qquad (10.36)$$

After absorbing the photon, the molecule is in the first vibrational excited state ($n = 2$). There are two possibilities

for the rotational contribution to the final energy, because the quantum number ℓ can increase by one or decrease by one (unless $\ell = 0$). If ℓ increases by one, then the molecular energy is

$$E_{2,\ell+1} \approx E_e + \frac{3\hbar\omega}{2} + \frac{(\ell+1)(\ell+2)\hbar^2}{2I}. \quad (10.37)$$

The corresponding photon energy (E_γ) is

$$E_\gamma = E_{2,\ell+1} - E_{1,\ell} \approx \hbar\omega + \frac{(\ell+1)\hbar^2}{I}, \quad (10.38)$$

where ℓ is a nonnegative integer (0, 1, 2,...). These transitions are called the *R branch*. If ℓ decreases by one, then the molecular energy is

$$E_{2,\ell-1} \approx E_e + \frac{3\hbar\omega}{2} + \frac{(\ell-1)(\ell)\hbar^2}{2I}. \quad (10.39)$$

The corresponding photon energy (E_γ) is

$$E_\gamma = E_{2,\ell-1} - E_{1,\ell} \approx \hbar\omega - \frac{\ell\hbar^2}{I}, \quad (10.40)$$

where ℓ is a positive integer (1, 2, 3,...). These transitions are called the *P branch*.

Thus, the absorption spectrum consists of a number of approximately equally spaced lines with an energy difference (ΔE) of

$$\Delta E = \frac{\hbar^2}{I}, \qquad (10.41)$$

and there is a double-sized gap of energy $2\Delta E$ that occurs at the energy $\hbar\omega$. The location of this center gap can readily be used to find the value of the vibrational frequency, ω. A diagram of allowed vibrational–rotational transitions is shown in Figure 10-10. Figure 10-11 shows photon lines emitted from the HCl molecule when the photons make transitions from $n = 2$ to $n = 1$. The vibrational energy (about 0.358 eV) can be quickly detected by the location of the gap. Note also that the lines become closer together for large values of the quantum number ℓ. This is due to the fact that the vibrational frequency of the molecule changes slightly because the molecule becomes distorted by the rotation.

Electronic–Vibrational Spectra

The total energy (10.34) of a molecular state is composed of three terms: electronic, vibrational, and rotational. The

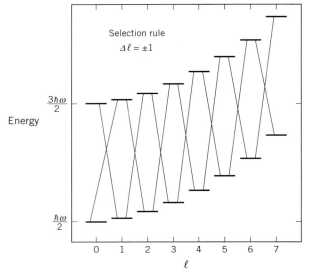

FIGURE 10-10 Vibrational–rotational transitions between the levels $n = 1$ and $n = 2$.
The rotational levels are shown for $\ell = 0$ to 7. The selection rule is $\Delta\ell = \pm 1$. The photon energies are $\Delta E = \hbar\omega - 2\ell E_0$, where $\ell = 1, 2, 3, \ldots$.

energy levels are dominated by the electronic term, which has a typical scale of a few electronvolts (see Table 10-4). The electronic levels have a fine structure due to the molecular vibrations on a scale of 10^{-1} electronvolts. Each vibrational level has a superfine structure due to molecular rotations on the scale of 10^{-3} eV (see Figure 10-12).

When an electronic transition takes place, the selection rules are modified because there are more ways in which

angular momentum can be conserved. We must now also consider the angular momentum of the electron and in particular the component of angular momentum along the molecular axis (the z component of angular momentum). This component of angular momentum we may write as

$$J_z = m_j\hbar. \qquad (10.42)$$

(The ground state of almost all molecules has $m_j = 0$.) The total angular momentum (J) is the vector sum of the electronic angular momentum and the rotational angular momentum of the molecule. As usual, we may write

$$J^2 = j(j+1)\hbar^2. \qquad (10.43)$$

The rotational energy levels of the molecule (E_j) may be written

$$E_j \approx \frac{j(j+1)\hbar^2}{2I} + \frac{m_j{}^2\hbar^2}{2I_e}, \qquad (10.44)$$

where I_e is the *electron* moment of inertia. Since the electron has a small mass, $I_e \ll I$. The total energy of the molecule is

$$E_{tot} \approx E_e + \left(n - \frac{1}{2}\right)\hbar\omega$$
$$+ \frac{j(j+1)\hbar^2}{2I} + \frac{m_j{}^2\hbar^2}{2I_e}. \qquad (10.45)$$

For transitions between two levels, the photon energy is given by the difference between the two levels, which will

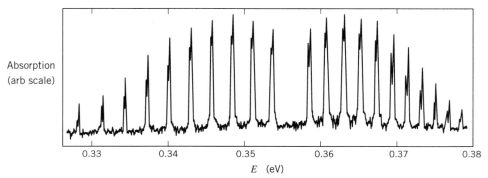

FIGURE 10-11 Vibrational–rotational photon lines from the HCl molecule.
The transitions are from the vibrational first excited state ($n = 2$) to the vibrational ground state ($n = 1$). The transitions obey the angular momentum selection rule, $\Delta\ell = \pm 1$. Each photon line is split in two due to the presence of both ^{35}Cl and ^{37}Cl. After R. L. Sproull and W. A. Phillips, *Modern Physics: The Quantum Physics of Atoms, Solids, and Nuclei*, Wiley (copyright © 1985).

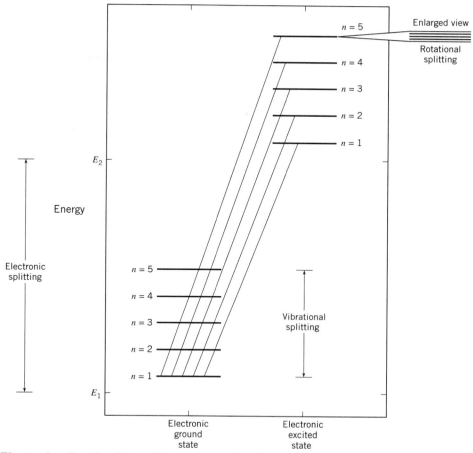

FIGURE 10-12 Electronic–vibrational transitions between the electronic ground state and an electronic excited state.
The vibrational levels are shown for $n = 1$ to 5. Only the transitions to $n = 1$ in the electronic ground state are shown, for clarity. There is no selection rule on n when the transition occurs between two electronic levels. The photon energies are $\Delta E = (E_2 - E_1) + (n-1)\hbar\omega$, where $n = 1, 2, 3, \ldots$. On a finer scale, each line is split into a series of lines due to rotational excitations.

depend on the electron quantum numbers, the vibrational quantum number, the total angular momentum quantum number, and the quantum number for the z component of angular momentum. There is no longer any restriction on the change in the vibrational quantum number. Thus, transitions that are forbidden within a single vibrational band, such as $n = 3$ to $n = 1$, are allowed if the electron level also changes. The selection rule for the z component of angular momentum is

$$\Delta m_j = \pm 1, 0. \tag{10.46}$$

The selection rule of the rotational quantum number j is also modified to

$$\Delta j = \pm 1, 0 \tag{10.47}$$

except when $m_j = \Delta m_j = 0$, in which case

$$\Delta j = \pm 1. \tag{10.48}$$

Photon lines from some of the electronic–vibrational–rotational transitions from the cyanogen molecule (CN) are shown in Figure 10-13. For this portion of the spectra, the change in electronic and vibrational levels is fixed; the multiple photon lines are caused by the various rotational transitions indicated. The line with longest wavelength is referred to as the *band head*. For the spectra of Figure 10-13, the band head is at $\lambda = 388.34$ nm.

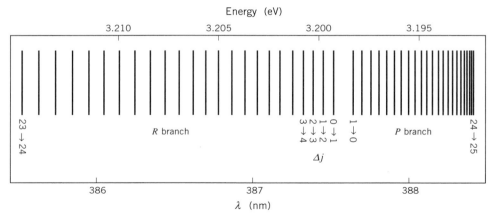

FIGURE 10-13 Vibrational–rotational–electronic spectrum from the CN molecule in the near ultraviolet.
The band head is at 388.34 nanometers. The initial and final rotational levels are indicated. The wavelengths have been measured by H. S. Uhler and R. A. Patterson, *Astrophys. J.* **42**, 434 (1915).

10-5 ABSORPTION FROM ζ OPHIUCI

There is an interesting bit of physics connected with the molecular excitation of cyanogen (CN) molecules. Cyanogen molecules are found to be present in a gaseous interstellar cloud in the constellation Ophiuchus, "The Serpent Bearer." Ophiuchus is located on the celestial equator between the constellations Libra and Aquila. The Ophiuchus cloud lies between us and the star ζ Ophiuci, which illuminates it from behind. Analysis of the Fraunhofer absorption lines in the spectrum of ζ Ophiuci shows the presence of the CN molecule in the interstellar space. A photon line in the ultraviolet region is observed at a wavelength near $\lambda = 387.5$ nm, corresponding to absorption from the $j = 0$ state (see Figure 10-14a). A second strong absorption line is observed with a slightly larger wavelength ($\Delta\lambda = 0.061$ nm), corresponding to absorption from the $j = 1$ state. The second line was not expected. The presence of the second line requires some physical mechanism for the CN molecules to be in the first rotational excited state.

EXAMPLE 10-5

The photon line from the $j = 1$ state was observed at an intensity of 1/4 that of the photon line from the $j = 0$ state. Estimate the temperature of the CN molecules.

SOLUTION:

The relative populations of the $j = 0$ and $j = 1$ states are given by the Maxwell-Boltzmann distribution and the density of states. The energy difference between the two states is

$$\Delta E = \frac{hc}{\lambda_2} - \frac{hc}{\lambda_1} = \frac{hc\Delta\lambda}{\lambda^2},$$

where

$$\lambda = 387.5 \, \text{nm}.$$

The energy difference is

$$\Delta E = \frac{hc\Delta\lambda}{\lambda^2} = \frac{(1240 \, \text{eV} \cdot \text{nm})(0.061 \, \text{nm})}{(387.5 \, \text{nm})^2}$$

$$\approx 5.0 \times 10^{-4} \, \text{eV}.$$

Since the density of states is proportional to $(2j + 1)$, there are three times as many states for $j = 1$ as for $j = 0$. The ratio of the number of molecules in the electronic ground state with $j = 1$ (n_1) to the number with $j = 0$ (n_0) is measured to be 1/4. We may write this ratio as

$$\frac{n_1}{n_0} = \frac{(2 \times 1 + 1)}{(2 \times 0 + 1)}\left(\frac{e^{-E_1/kT}}{e^{-E_0/kT}}\right) = 3e^{-\Delta E/kT} = \frac{1}{4},$$

where E_1 are the energies of the two states and $\Delta E = E_1 - E_0$. Therefore,

$$\frac{\Delta E}{kT} = \ln(12).$$

The estimated temperature is

$$T = \frac{\Delta E}{k \ln(12)}$$

$$\approx \frac{\left(5.0 \times 10^{-4}\,\text{eV}\right)}{\ln(12)} \left(\frac{300\,\text{K}}{0.026\,\text{eV}}\right)$$

$$\approx 2.3\,\text{K}. \qquad \blacksquare$$

The presence of the second line in the spectrum from ζ Ophiuci implies an excitation temperature of about 2.3 kelvin (see Figure 10-14). At the time of this discovery (1940), there was no known source of thermal radiation in interstellar space at such a low temperature. Gerhard Herzberg, one of the world's experts on molecular spectroscopy, summed up the understanding of the physics with the following statement:

In high dispersion spectra of distant stars a few exceedingly sharp lines occur which have been shown to be due to absorption in interstellar space.... While some of these lines were readily identified... several others remained unidentified until it was realized... that they are due to interstellar molecules in their lowest rotational levels.... From the intensity ratio of the [CN] lines with [ℓ = 0 and ℓ = 1] a rotational temperature of 2.3 K follows, which has of course a very restricted meaning.

We now know, of course, that the CN molecules are excited into the rotational excited state by the cosmic background radiation discovered by Penzias and Wilson in 1965. The effects of the cosmic background radiation had been observed 25 years earlier! Even the temperature of the radiation had been correctly estimated. After the observation of Penzias and Wilson, molecular spectra from several other stars were analyzed. Collectively these data all showed a cosmic radiation temperature of about 2.7 kelvin. These data were historically an important part of the first determination of the cosmic background radiation spectrum, before the COBE satellite mission, because the radiation near the peak of the cosmic spectrum is strongly absorbed by the atmosphere. The COBE satellite data are discussed in Chapter 19.

10-6 ABSORPTION SPECTRUM OF WATER

As a final topic in this chapter, we examine the absorption spectrum of water. The spectrum is shown in Figure 10-15 over a wide range of photon wavelengths from about 1 km to 10^{-15} m. The general feature of the absorption spectrum of water is that all wavelengths are absorbed except for a very narrow window corresponding to visible radiation. *How can this be so?* We now examine the reason for the absorption in each wavelength region. In the photon wavelength region of 1 m to 10^{-2} m, the absorption occurs as rotational modes of the water molecule are excited. At photon wavelengths from about 10^{-2} m to 10^{-6} m, vibrational modes of the water molecule are excited. At photon wavelengths from about 10^{-6} m to 10^{-9} m, electronic modes of the water molecule are excited. At photon wavelengths from about 10^{-9} m to 10^{-11} m, the photoelectric effect dominates. The photon absorption results in an electron being ejected from an atom. At photon wavelengths from about 10^{-11} m to 10^{-13} m, Compton scattering dominates. When an energetic photon Compton scatters from an electron in the molecule, the photon loses energy. The lower energy photon can be absorbed by one of the processes previously discussed. Finally, for the shortest wavelengths ($\lambda < 10^{-13}$ m), the photons can interact with the oxygen nuclei and convert their energy into electron–positron pairs.

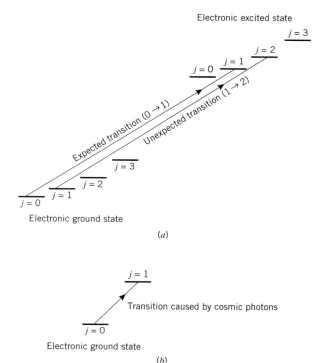

FIGURE 10-14 Energy levels of the CN molecule.
(*a*) Observed transitions from ζ Ophiuci, and (*b*) transition caused by cosmic photons.

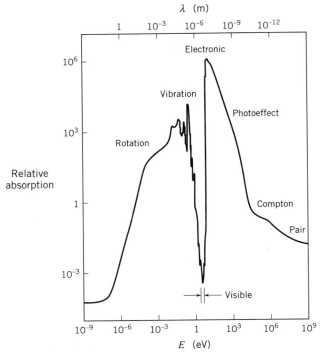

FIGURE 10-15 Absorption spectrum of water.
Adapted from J. D. Jackson, *Classical Electrodynamics,* Wiley (1975).

The net result is that water absorbs radiation of all wavelengths except those wavelengths corresponding to the visible region (0.4–0.7 μm). Water does not absorb radiation in this region because there is no physical mechanism for it to do so. Photons in the visible wavelength region have energies of about 2 to 3 eV. This is too much energy to cause a vibrational excitation and too little energy to cause an electronic transition. Since there is water everywhere on our planet, the existence of this tiny absorption window is crucial for the evolution of life.

CHAPTER 10: PHYSICS SUMMARY

- Certain combinations of atoms form molecules because the electrons in the molecule have lower energies than they do when the atoms are separated. The typical bond length is $d_{atom}/2$. The typical binding energy of the strongest type of molecular bonds is a few electronvolts.

- Molecules have vibrational degrees of freedom that are not present in single atoms. The electromag-

netic potential energy of a molecule (as a function of the separation between the nuclei) has the shape of a quadratic near its minimum. Treating the molecule as a quantum harmonic oscillator, we estimate the molecular vibrational energy scale to be

$$E_v = \frac{h^2}{2\pi d_{atom}^2 \sqrt{mM}},$$

where m is the electron mass and M is the reduced mass of the molecule. The order of magnitude of a molecular vibrational excitation energy is 10^{-1} eV, corresponding to radiation in the infrared region.

- Molecules have rotational degrees of freedom that are not present in single atoms. The angular momentum of a rotating molecule is quantized, which leads to an estimate of the molecular vibrational energy scale:

$$E_r \approx \frac{h^2}{4\pi^2 M d_{atom}^2}.$$

where m is the electron mass and M is the reduced mass of the molecule. The order of magnitude of a molecular rotational excitation energy is 10^{-3} eV, corresponding to radiation in the microwave region.

- The absorption spectrum of water has a narrow window in the visible region between vibrational and electronic excitations.

REFERENCES AND SUGGESTIONS FOR FURTHER READING

Jun-ichi Aihara, "Why Aromatic Compounds Are Stable," *Sci. Am.* **266**, No. 3, 62 (1992).

F. H. C. Crick, "The Structure of the Hereditary Material," *Sci.Am.* **191**, No. 4, 54 (1954).

R. F. Curl and R. E. Smalley, "Fullerenes," *Sci. Am.* **265**, No. 4, 54 (1991).

Handbook of Chemistry and Physics, published each year by the Chemical Rubber Company.

S. Harache and J.-M. Raimond, "Cavity Quantum Electrodynamics," *Sci. Am.* **268**, No. 4, 54 (1993).

G. Herzberg, *Molecular Spectra and Molecular Structure,* D. Van Nostrand (1950).

B. Mahan, *University Chemistry,* Addison-Wesley (1969).

J. L. Schnapf and D. A. Baylor, "How Photoreceptor Cells Respond to Light," *Sci. Am.* **256**, No. 4, 40 (1987).

E. Schrödinger, *What Is Life,* Cambridge Univ. Press (1944).

E. Schrödinger, *Mind and Matter,* Cambridge Univ. Press (1967).

L. Stryer, "The Molecules of Visual Excitation," *Sci. Am.* **257**, No. 1, 42 (1987).

D. W. Turner, C. Baker, A. D. Baker, and C. R. Brundle, *Molecular Photoelectron Spectroscopy,* Wiley-Interscience (1970).

J. H. Van Vleck, "The Coupling of Angular Momentum Vectors in Molecules," *Rev. Mod. Phys.* **23**, 213 (1951).

J. Walker, "The Amateur Scientist," *Sci. Am.* **256**, No. 2, 134 (1987).

R. A. Weinberg, "The Molecules of Life," *Sci. Am.* **253**, No. 4, 48 (1985).

D. C. Youvan and B. L. Marrs, "Molecular Mechanisms of Photosynthesis," *Sci. Am.* **256**, 47 (June 1987).

QUESTIONS AND PROBLEMS

The hydrogen molecule

1. Why don't two helium atoms bond together to make a molecule like H_2, N_2, or O_2?

2. An experimenter provides 4.5 eV of energy in order to dissociate an H_2 molecule into two H atoms. Where has the energy gone?

3. Why is the strength of a covalent or ionic molecular bond the same order of magnitude as the energy needed to remove an outer electron from an atom?

4. How do you expect the molecular excitations of the H_2 molecule to compare with those of the H_2^+ molecule?

5. The formation of two water molecules by burning two hydrogen molecules releases an energy of 5.7 eV:

$$2H_2 + O_2 \rightarrow 2H_2O + (5.7 \text{ eV}) .$$

Use this result and the data in Table 10-1 to determine the amount of energy needed to break both bonds in the water molecule.

6. (a) The binding energy for each of the carbon atoms in CH_4 is nearly identical: 4.5 eV for $H\text{-}CH_3$, 4.8 eV for $H\text{-}CH_2$, 4.4 eV for H-CH, and 3.5 eV for H-C. Why is this true? (b) Hydrogen atoms and carbon atoms are combined to make 1 kg of methane. Calculate the mass of the uncombined atoms.

7. Does a longer bond length imply a smaller molecular binding energy? Why or why not?

The sodium-chloride molecule

8. Why is the bond length greater in the sodium chloride salt crystal than in an isolated sodium chloride molecule? How can the binding energy per molecule be larger in the crystal if the bond length is larger?

9. Which atom will make the strongest bond with the boron atom? Make a rough estimate of the strength of the bond. Look up the data on boron bond strengths in a handbook to check your estimate.

10. The ionization energy of lithium is 5.4 eV and the electron affinity of fluorine is 3.4 eV. The bond length of the LiF molecule is 0.16 nm. Make an estimate of the molecular binding energy.

11. The ionization energy of potassium is 4.3 eV and the electron affinity of iodine is 3.1 eV. The binding energy of the KI molecule is 3.3 eV. Make an estimate of the molecular bond length.

Molecular vibrations and rotations

12. Why is more energy required to induce a vibrational molecular excitation than to induce a rotational excitation?

13. Consider the $\ell = 1$ to $\ell = 0$ transition in the HCl molecule. If the sample contains two isotopes of chlorine, ^{35}Cl and ^{37}Cl, the photon line is split into two. Which isotope produces a photon of larger energy? Calculate the energy difference between the photons from the two isotopes.

14. The force constant (K) in the HCl molecule is determined to be 480 N/m. Calculate the vibrational frequency of the molecule.

15. The vibrational frequency of the CO molecule is 6.42×10^{13} Hz. Estimate the amplitude of the molecular vibrations. Compare your answer to the bond length.

16. For hydrogen gas at room temperature, make an estimate of the fraction of molecules that are in the first vibrational excited state.

17. The vibrational frequency of the carbon monoxide (CO) molecule is 6.42×10^{13} Hz. Assuming that the molecule is a quantum harmonic oscillator, make an estimate of the fraction of the molecules that are in the first vibrational excited state at room temperature.

18. The vibrational frequency of the carbon monoxide (CO) molecule is 2.04×10^{13} Hz. Assuming that the molecule is a quantum harmonic oscillator, calculate the force constant (K). Make an order-of-magnitude estimate of the force constant of an ordinary spring of coiled metal. How does this compare to the molecule?

19. The lithium chloride molecule (LiCl) has a bond length of about 0.26 nm. Calculate the wavelength of

radiation needed to excite the molecule from the ground state to the first rotational excited state.

20. The carbon monoxide (CO) molecule absorbs radiation at a wavelength of 2.61 millimeters, corresponding to the rotational excitation from $\ell = 0$ to $\ell = 1$. Calculate the CO bond length.

21. The bond length of the N_2 molecule is 0.11 nm. Determine the wavelength of radiation corresponding to transitions between the $\ell = 1$ and $\ell = 2$ rotational states.

22. (a) For a gas of O_2 molecules at room temperature, how many molecules are in the rotational ground state ($\ell = 0$) compared to the first excited state ($\ell = 1$)? (b) What temperature would be needed in order to decrease the population ratio by 1%?

Molecular spectra

23. Why does the removal of an electron from a molecular bonding orbital by photoabsorption cause the molecular ion to be left in an excited vibrational state?

24. The photon spectrum from the CaCl molecule has a relatively strong line near $\lambda = 8$ millimeters. Estimate the bond length of the molecule. (What assumption do you need to make?)

25. The CO molecule absorbs radiation at a wavelength of 0.652 millimeters. What transition is taking place?

Absorption from z Ophiuci

26. Consider an experiment to measure the absorption of cosmic background radiation by cyanogen molecules on the earth. What are the difficulties?

Absorption spectrum of water

27. Why are clouds white?

28. When radar was first developed for the detection of aircraft, the presence of water in the atmosphere imposed a technical limit on the resolution of the radar. Why was this so? How do you think that this limit was eventually overcome?

Additional problems

29. How is it that N_2 and CO can have different bond strengths if both molecules have 14 electrons? Why is the N_2 bond slightly weaker? (*Hint:* Examine the energy levels of the outer electron in the atoms C, N, and O.)

*30. Consider a system of molecules in thermal equilibrium at a temperature T. The number of molecules in any given rotational state (N_ℓ) is given by the density of states times the Maxwell-Boltzmann

factor. (a) Show that

$$N_\ell = A(2\ell+1)e^{-\ell(\ell+1)\hbar^2/2IkT}$$

where I is the moment of inertia of the molecule and A is a normalization constant. (b) Show that the rotational level with the largest number of molecules is given by the value of ℓ which is the closest to the quantity ℓ_{max} given by

$$\ell_{max} = \frac{\sqrt{IkT}}{\hbar} - \frac{1}{2}$$

(c) Use this result to estimate the temperature at which the HCL spectrum of Figure 10-11 was taken.

31. The water molecule is symmetric in the sense that both hydrogens are bound to the oxygen with equal strength, however, more energy is required to break the first hydrogen bond (either one) than to break the second bond. Why? How much more energy is required to break the first bond?

32. In the combustion of one molecule of octane (C_8H_{18}), the carbon atoms end up in carbon dioxide (CO_2) and the hydrogen atoms end up in water (H_2O). The molecular binding energies are 102 eV for C_8H_{18}, 16.5 eV for CO_2, 5.1 eV for O_2, and 9.5 eV for H_2O. Calculate the energy released in the burning of 1 kilogram of octane.

33. Chemists often give a molecular bond strength as the number of kilocalories (kcal) needed to dissociate one mole of the bonds (1 cal = 4.18 J). Calculate the size of a typical bond in kcal/mole.

34. Two carbon atoms can make a single(C–C), double (C=C), or triple bond (C≡C) with each other. The vibrational frequency of molecules with the double carbon bond is about 5×10^{13} Hz. Make an estimate of the vibrational frequencies of molecules that have a single carbon bond and a triple carbon bond.

*35. Consider a classical harmonic oscillator with masses M_1 and M_2 separated by a distance x. (a) Show that for vibrations along the axis of the masses, the total kinetic energy of the particles may be written

$$E_k = \frac{1}{2}\mu\left(\frac{dx}{dt}\right)^2,$$

where μ is the reduced mass, that is. $\mu = M_1M_2/(M_1+M_2)$. (b) Use this result to verify the relationship (10.9) between the frequency (ω), force constant (K) and reduced mass,

$$\omega = \sqrt{\frac{K}{\mu}} \,.$$

36. Show that the molecular rotational energy may be written as

$$E_r = \frac{\ell(\ell+1)\hbar^2}{2\mu r_0^2},$$

where r_0 is the bond length and μ is the reduced mass.

*37. (a) Use the Bohr model to arrive at an expression for the molecular rotational levels of a diatomic molecule. (b) Calculate the photon energies corresponding to the five lowest molecular rotational transitions in H_2, in both the Bohr model and the exact quantum mechanical result. (c) Show that for large quantum numbers (>>1), the two results agree.

CHAPTER

11

THE NUCLEUS

In disentangling the problems of the atom, one of the major steps has been the recognition that it is useful to speak of individual orbits of the electrons in the atoms. This, to be sure is only an approximation, in fact a crude approximation, but still it provides a quite invaluable starting point for the study of complex atoms containing large numbers of electrons.

When physicists became reasonably certain that the nucleus was constructed of protons and neutrons, questions were raised concerning the orbital behavior of these particles. Could nuclear structure be interpreted on the general pattern of atomic structure by attributing to the various neutrons and to the various protons within the nucleus something like individual orbits and individual states?...

Consider one nucleon in the nucleus traveling along its orbit among the other nucleons. If the collision mean free path is λ this nucleon would collide with the other neutrons and protons in the nucleus and its orbit would be lost after it had gone the distance of its free path.... Now it is a very difficult problem to decide the length of the mean free path, but if one takes somewhat literally the strength of the interactions between the neutron and other components of the nucleus, one is led to a value that seems discouragingly short....

In spite of this argument, evidence has been accumulating for the last few years... to the effect that orbits do exist. The best known feature of this evidence has been the discovery of the so-called magic numbers. They are the numbers 2, 8, 20, 50, 82, 126. When a nucleus contains a number of either neutrons or protons equal to one of the magic numbers, it is particularly stable....

It would appear that for some reason the mean free path must be longer than is given by a somewhat crude estimate of its length. One possible reason for this may be the Pauli principle, according to which collisions between two particles may be forbidden when, after the collision, one of the particles would go to an occupied state.

Enrico Fermi

We might well refer to the period from 1910 to 1930 as the "golden age" of atomic physics (see Table 11-1). During this period, the basics of atomic physics were understood for the first time with the development of quantum mechanics. Atoms were described by the energies and angular momenta of electrons. The fundamental structure of matter was established at an energy scale from a few eV to 50 keV (x ray energies).

The period from about 1930 to 1950 may well be referred to as the "golden age" of nuclear physics. During this period, the study of the nucleus was the frontier of fundamental science. The first modern accelerator technology was developed and the energy frontier was pushed to 100 MeV. The basic constituents of the nucleus were established to be neutrons and protons. Since the size of the nucleus was known to be a few femtometers, physicists were led to conclude that a new "strong" force was responsible for the interaction between protons and neutrons. The strength of this force relative to the strength of the electromagnetic force is responsible for the small size of the nucleus compared to the atomic size. Together with the discovery of the nuclear constituents came new fundamental questions. One major difficulty was that the force that gives the nucleus its small size also causes a large interaction probability between protons and neutrons, yet protons and neutrons were observed to have stable configurations ("nuclear orbits") inside nuclei. Another major challenge was to understand why certain nuclei would spontaneously decay, whereas others were stable.

11-1 DISCOVERY OF THE NEUTRON

By 1932, physicists had measured both the atomic mass number A and the nuclear charge Ze of the elements. Since the nucleus was found to contain nearly all the mass of the atom and to have the electric charge of Z protons, it was natural to expect that the nuclear mass was due to Z protons ($A = Z$). For all nuclei (except a single proton), however, it was found that $A > Z$. The discovery of the nucleus had left physicists with an interesting problem: *Why are A and Z different?* Since protons were known to be very massive compared to electrons, one plausible scenario was that the nucleus is a bound state of A protons and $A - Z$ electrons, which would give the nucleus the observed mass and charge. This interesting possibility was qualitatively supported by the observation that electrons were emitted in certain types of nuclear decays, called *beta* decays. Recall that beta decays were discovered a decade before the nucleus was discovered!

If the nucleus decays into an electron plus other particles, is it not natural to assume that the electron is present inside the nucleus? An important clue comes from the energies of the emitted electrons, which are typically an MeV. The kinetic energy of the beta decay electrons imposed a very serious problem for a nucleus made of

TABLE 11-1
FRONTIERS IN PHYSICS.

Period	Energy	Matter	Composition
1910–30	keV	Atoms	Electrons + nucleus
1930–50	MeV	Nucleus	Protons + neutrons
1950–80	GeV	Hadrons	Quarks
1980–present	TeV	Quarks + leptons	???

electrons and protons because an electron confined in such a small space as the nuclear size has a small wavelength and large kinetic energy. If we approximate the nucleus as a box with a size of a few fm, then the maximum wavelength of an electron confined inside the nucleus is about 5 fm. The momentum of such an electron is

$$p = \frac{h}{\lambda}, \qquad (11.1)$$

or

$$pc = \frac{hc}{\lambda} = \frac{1240 \text{ MeV} \cdot \text{fm}}{5 \text{ fm}} \approx 250 \text{ MeV}. \quad (11.2)$$

Such an electron is relativistic; the electron kinetic energy is

$$E_k \approx pc \approx 250 \text{ MeV}. \qquad (11.3)$$

This energy is far greater than the observed electron energies from nuclear decays. Another problem with electrons bound inside the nucleus is that the observed force between an electron and a nucleus is not strong enough to bind the electron. In other words, protons and neutrons experience the strong force but electrons do not. The electromagnetic potential energy between an electron and a gold nucleus at a distance of 5 fm is

$$V = \frac{Zke^2}{r} = \frac{(79)(1.44 \text{ MeV} \cdot \text{fm})}{(5 \text{ fm})}$$
$$\approx 23 \text{ MeV}. \qquad (11.4)$$

This energy is not sufficient to bind the electron in a volume of nuclear size because, as we have seen, the kinetic energy of an electron confined to such a small space is an order of magnitude greater:

$$250 \text{ MeV} \gg 23 \text{ MeV}. \qquad (11.5)$$

Of course, the electromagnetic force does bind electrons and nuclei into atoms, but the strength of the force results in an atomic size of about 0.1 nm, or about 10^5 times the nuclear size.

The solution to the nuclear puzzle of why A and Z are not identical was provided by James Chadwick in 1932 with the discovery of the neutron. (The year 1932 was a "banner" year for physics, as Carl Anderson also discovered the positron and John Cockroft and Ernest Walton made the first artificial disintegration of the nucleus.)

A new type of radiation particle (Y) was discovered by bombarding a beryllium foil with alpha particles. The reaction was

$$\alpha + {}^9\text{Be} \rightarrow (\text{new nucleus}) + Y. \qquad (11.6)$$

The superscript specifies the atomic mass number of beryllium ($A = 9$). The atomic number of beryllium is $Z = 4$. The alpha particle (the helium nucleus) has $A = 4$ and $Z = 2$. This reaction had been studied by several physicists, and it was known that the Y particle was electrically neutral and that it could penetrate a few centimeters of lead. Irène Curie and Frédéric Joliot scattered Y particles from a paraffin target. Paraffin contains hydrogen atoms that are bound by only a few electronvolts; paraffin is a source of protons that are essentially free. Protons were observed to be freed from the paraffin in the reaction,

$$Y + p \rightarrow Y + p. \qquad (11.7)$$

The kinetic energy of the protons was measured to be 5.7 MeV. If the Y particle was a photon, then a large photon energy (about 50 MeV) is needed to produce protons with an energy of 5.7 MeV.

EXAMPLE 11-1

If the Y particle is the photon, calculate the photon kinetic energy required to produce 5.7-MeV protons.

SOLUTION:

The process is Compton scattering—not the usual scattering of an energetic photon from an *electron*, but rather the scattering of an energetic photon from a *proton*. The photon interacts with all charged particles! We can use the Compton formula (4.170) with m replaced by the proton mass to get an exact answer. We may simplify the analysis, however, by observing that the scattered proton is nonrelativistic since its kinetic energy is much smaller than its mass energy (5.7 MeV $\ll$ 940 MeV). Therefore, a significant momentum is transferred to the proton but not too much kinetic energy, similar to a photon scattering from a "brick wall." The maximum momentum is transferred from the photon to the proton when the scattering is backwards. If the initial photon momentum is p_x, the change in momentum of the photon is

$$\Delta p \approx 2 p_x.$$

The kinetic energy of the recoiling proton is

$$E_k = \frac{\Delta p^2}{2M} = \frac{2 p_x^2}{M},$$

where M is the proton mass. Solving for p_x, we have

$$p_x = \sqrt{\frac{ME_k}{2}}.$$

The initial photon energy (E) is

$$E = p_x c = \sqrt{\frac{Mc^2 E_k}{2}}$$

$$= \sqrt{\frac{(940 \text{ MeV})(5.7 \text{ MeV})}{2}}$$

$$\approx 50 \text{ MeV}. \qquad \blacksquare$$

The exact energy of the Y particles was not known; however, Chadwick did not believe that the Y particles could be produced with a kinetic energy as large as 50 MeV, because the typical energy scale of nuclear decays had been established to be a few MeV.

Part of the great contribution of Chadwick was to show that the particle Y was *not* the photon. To establish the identity of the mysterious Y particle, Chadwick scattered the Y particles from several different targets: hydrogen, helium, nitrogen, oxygen, and argon. The experimental arrangement is indicated in Figure 11-1. Chadwick measured the kinetic energy of the recoiling nucleus and then compared his result to that expected from Compton scattering (see Example 11-1) for each target. The results were found to be clearly inconsistent with Compton scattering. Furthermore, Chadwick observed that the scattering *rate* for Y particles was greater than that of Compton scattering by several orders of magnitude.

The data of Chadwick showed that the Y particle is more efficient than a massless photon at transferring *energy* to the target particle. The scattering data could be explained if the Y particle had a mass that is approximately

equal to the proton mass and an interaction with the proton that was significantly stronger than the photon–proton interaction. For example, if the Y particle has the same mass as the proton, then a Y particle kinetic energy of 5.7 MeV is sufficient to scatter protons with a kinetic energy of 5.7 MeV. Chadwick had demonstrated that the Y particle had a mass approximately equal to the proton mass. The Y particle is called the *neutron* (n), from the Italian *neutrone* meaning "the large neutral one."

By the preceding analysis, Chadwick knew that the neutron mass (m_n) was approximately equal to the proton mass (m_p). Chadwick made an accurate determination of the neutron mass by analysis of the reaction,

$$\alpha + {}^{11}\text{B} \rightarrow {}^{14}\text{N} + n, \qquad (11.8)$$

and applying conservation of energy and momentum. Chadwick chose a boron target because the masses of boron (m_B) and nitrogen (m_N) were well known at the time of the experiment. By conservation of energy,

$$\frac{1}{2}m_\alpha v_\alpha^2 + m_\alpha c^2 + m_B c^2$$
$$= \frac{1}{2}m_N v_N^2 + m_N c^2 + \frac{1}{2}m_n v_n^2 + m_n c^2, \quad (11.9)$$

where v_α is the speed of the incoming alpha and v_N and v_n are the speeds of the outgoing nitrogen and neutron. (All the particles are nonrelativistic.) Conservation of momentum gives

$$m_\alpha \mathbf{v}_\alpha = m_n \mathbf{v}_n + m_N \mathbf{v}_N. \qquad (11.10)$$

Since the neutron mass is much smaller than the nitrogen mass, $m_n \ll m_N$, the neutron gets nearly all the kinetic energy of the final state by conservation of momentum. The nitrogen nucleus gets just enough kinetic energy to conserve momentum. Therefore, energy conservation (11.9) gives

$$\frac{1}{2}m_n v_n^2 + m_n c^2$$
$$\approx \frac{1}{2}m_\alpha v_\alpha^2 + m_\alpha c^2 + m_B c^2 - m_N c^2. \quad (11.11)$$

Solving for the mass energy of the neutron,

$$m_n c^2 \approx \frac{\frac{1}{2}m_\alpha v_\alpha^2 + m_\alpha c^2 + m_B c^2 - m_N c^2}{1 + \frac{v_n^2}{2c^2}}. \qquad (11.12)$$

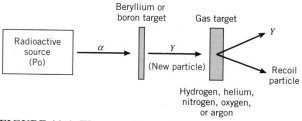

FIGURE 11-1 The experiment of Chadwick.
A new particle (Y) is produced when alpha particles from the radioactive decay of polonium hit a beryllium or boron target. The Y particle is observed to transfer its energy to various gas particles, indicating it is massive. The mass of the Y particle is determined by conservation of energy in the boron target from the reaction $\alpha + {}^{11}\text{B} \rightarrow {}^{14}\text{N} + Y$. The Y particle is called the neutron.

All quantities on the right-hand side were known by Chadwick except the neutron speed. Chadwick measured the neutron speed by allowing the neutron to collide with a proton and then measuring the resulting proton speed, applying momentum conservation with $m_n \approx m_p$. Chadwick's result for the neutron mass energy was 938 MeV with an experimental error of 1.8 MeV. The experimental error includes Chadwick's errors due to his knowledge of the alpha, boron, and nitrogen masses.

More refined measurements show that the mass energy of the neutron is

$$m_n c^2 = 939.57 \text{ MeV}. \qquad (11.13)$$

The mass of the neutron is slightly greater than the mass of the proton. The difference in the neutron and proton mass energies is

$$m_n c^2 - m_p c^2 = 1.29 \text{ MeV}. \qquad (11.14)$$

11-2 BASIC PROPERTIES OF THE NUCLEUS

The nucleus is a bound state of protons and neutrons. Protons and neutrons are collectively called *nucleons*. Before embarking on the description of the nucleus, we make a few remarks about the nature of the strong force between nucleons.

The Force Between Nucleons

The neutron and the proton have nearly identical masses. In other fundamental respects, the neutron is similar to the proton except that the neutron has zero electric charge. For example, the proton and neutron have identical intrinsic angular momentum quantum numbers:

$$s = \frac{1}{2}. \qquad (11.15)$$

The neutron and proton are observed to have identical strong interactions. The strong force between any two nucleons is observed to be *attractive*. This strong force does not reflect the most fundamental aspect of the strong interaction because the nucleons are composite objects (see Chapter 6). The strong attraction between a neutron and a proton is analogous to the electromagnetic attraction of two neutral hydrogen atoms to form the H_2 molecule. In the molecule, the fundamental electromagnetic attraction is between the charges, electrons and protons. The mol-

ecule is formed as sort of a residual effect; when the atoms are close together, they feel the composite charges of each other, which results in a net attraction. The approximate range of the residual electromagnetic force between atoms is just the size of the atom. In a nucleus, the fundamental strong interaction is between quarks. The quarks have the property (strong charge which is also called color) that causes the force, but the neutron and proton have no *net* color. The neutron and proton are *color neutrals* just like the hydrogen atom is electrically neutral. When a neutron and a proton are close together, they feel the composite quarks, resulting in a net attraction. The *residual* strong interaction between two nucleons has a short range because the nucleons have no net color. The approximate range of the residual strong force between nucleons is just the size of the nucleon (about 1 fm). Nucleons inside a nucleus feel the residual strong force only from their immediate neighbors. The fundamental aspects of the strong interaction between quarks is discussed in Chapter 18.

The Deuteron

The deuteron is a bound state of one neutron and one proton. When we combine two particles with $s = 1/2$, there are two possibilities for the resulting intrinsic angular momentum quantum number: zero or one. The strong interaction between a neutron and a proton depends on the orientation of the intrinsic angular momentum vectors, that is, the strong interaction is *spin dependent*. The strong force is spin dependent because it is mediated by gluon exchange and the gluons have intrinsic angular momentum ($s = 1$). If the intrinsic angular momentum vectors of a proton and neutron are not parallel, then the force is not strong enough to hold the neutron and the proton together. The state with antiparallel spins is not stable. A bound state of two neutrons, which by the Pauli exclusion principle would have to have opposite intrinsic angular momentum vectors, does not exist in nature. (The bound state of two protons also does not exist in nature; besides the Pauli exclusion principle, the electromagnetic repulsive force also prevents the binding of two protons.) For a neutron and proton with spins parallel, the force is strong enough to hold the neutron and proton together in a bound state. The bound state has

$$s = 1, \qquad (11.16)$$

and is called the deuteron. The binding energy of the deuteron is 2.22 MeV. The proton (m_p), neutron (m_n), and deuterium (m_d) masses are related by

$$m_d c^2 + 2.22 \text{ MeV} = m_p c^2 + m_n c^2 . \quad (11.17)$$

The atom made up of a bound state of a deuteron and an electron is called deuterium. Deuterium exists on earth with a natural abundance of about 1.5×10^{-4} times that of hydrogen. The deuterons that are found on earth (inside deuterium atoms) were made in nuclear reactions in the early universe.

Nuclear Stability

The nuclear states are summarized in Appendix J. Some nuclei are stable against all decays, whereas others decay spontaneously. The decays of nuclei are discussed in more detail in the next section. There are a total of 254 stable nuclei found in nature and many more unstable ones. Figure 11-2 shows a plot of the number of protons versus the number of neutrons for the stable nuclei. The unstable nuclei tend to have values of A and Z very close to that of a stable nucleus. From this plot, we see that for small values of Z ($Z < 20$) the numbers of neutrons and protons tend to be equal. At larger values of Z, however, there are more neutrons than protons in the nucleus.

The stability of a nucleus increases with increasing A because there are more nucleons that are all attracting each other. There are two additional factors, however, that *reduce* the stability of the nucleus: (1) the Pauli exclusion principle and (2) the Coulomb repulsion of the protons. We may think of the nucleus as neutrons and protons in a box. The nucleons will occupy quantized energy levels similar to the electrons in an atom. The neutrons and protons are observed to obey the Pauli exclusion principle, that is, no two protons and no two neutrons can have an identical set of quantum numbers. (The Pauli exclusion principle applies to all particles with $s = 1/2$.) For example, the ^{12}C nucleus has six protons and six neutrons. This configuration has a lower energy and is therefore more stable than seven protons and five neutrons (^{12}N) or five protons and seven neutrons (^{12}B). The Pauli exclusion principle favors a nucleus in which the number of neutrons and protons are equal. If this was the only factor, the stable nuclei would lie along the 45 degree line of Figure 11-2. This is the case for small Z; however, at large Z there is a significant deviation from equal numbers of neutrons and protons. A nucleus with $A - Z > Z$ is more stable than a nucleus with $A = Z$ because in the later case a greater fraction of the nucleons repel each other by the electromagnetic force. At small values of Z the proton repulsion is unimportant compared to the Pauli exclusion principle

but at large Z it becomes important. The heaviest stable nucleus is ^{209}Bi, which has 83 protons and 126 neutrons.

Nuclear Binding Energy

The sum of the masses of the individual components of a nucleus, the protons and neutrons, is greater than the mass of the nucleus. This excess energy is called the nuclear binding energy. The amount of binding energy is a function of the atomic mass number. The binding energy per nucleon versus A is shown in Figure 11-3. The binding energy per nucleon rises very rapidly with increasing A and reaches a broad maximum of about 8.5 MeV at about $A = 55$. For larger values of A, the binding energy per nucleon drops gradually to about 7 MeV for very high values of A. Note that the alpha particle ($A = 4$) has an exceptionally large binding energy (about 7 MeV per nucleon) compared to other light nuclei.

EXAMPLE 11-2
The mass of the alpha particle (m_α) is 3727.41 MeV/c^2. Calculate the binding energy and binding energy per nucleon of the alpha particle.

SOLUTION:
The alpha particle is a bound state of two neutrons and two protons. The binding energy of the alpha particle is

$$E_b = 2 m_p c^2 + 2 m_n c^2 - m_\alpha c^2 ,$$

or

$$E_b = (2)(938.27 \text{ MeV}) + (2)(939.57 \text{ MeV})$$
$$- 3727.41 \text{ MeV}$$
$$\approx 28.3 \text{ MeV} .$$

The binding energy per nucleon of the alpha particle is

$$\frac{E_b}{A} \approx \frac{28.3 \text{ MeV}}{4} \approx 7.1 \text{ MeV} . \qquad \blacksquare$$

EXAMPLE 11-3
The mass of the ^{238}U nucleus (m_U) is 221697.7 MeV/c^2. Calculate the binding energy and binding energy per nucleon in ^{238}U.

SOLUTION:
The ^{238}U nucleus has 92 protons and 146 neutrons. The total binding energy is

$$E_b = 92 m_p c^2 + 146 m_n c^2 - m_U c^2 ,$$

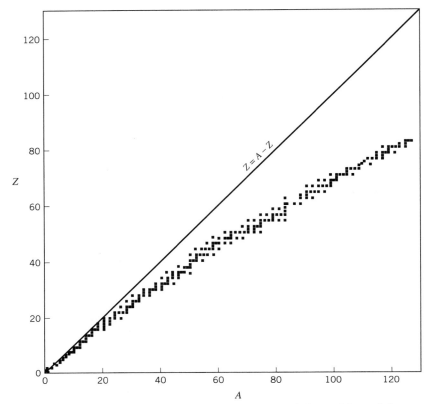

FIGURE 11-2 Number of protons (Z) versus number of neutrons ($A - Z$) for stable nuclei.

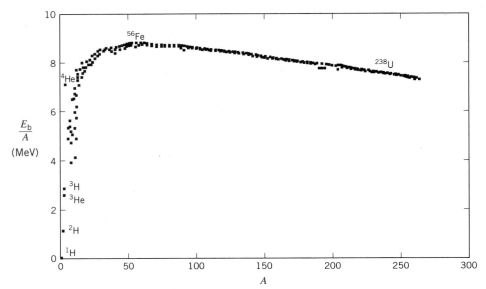

FIGURE 11-3. Binding energy per nucleon versus atomic mass number (A).

or

$$E_b = (92)(938.3 \text{ MeV})$$
$$+ (146)(939.6 \text{ MeV}) - 22,1697.7 \text{ MeV}$$
$$\approx 1810 \text{ MeV}.$$

The binding energy per nucleon is

$$\frac{E_b}{A} \approx \frac{1810 \text{ MeV}}{238} \approx 7.6 \text{ MeV}. \qquad \blacksquare$$

11-3 NUCLEAR MODELS

Liquid Drop Model

Scattering experiments show that nuclei have constant densities (see Figure 6-14). This has led to a model of the nucleus as a very dense incompressible spherical *liquid drop*. In the liquid drop model, the nuclear binding energy as a function of atomic mass number A (see Figure 11-3) is semiempirically fit to the form

$$E_b \approx C_1 A - C_2 A^{2/3}, \qquad (11.18)$$

where C_1 and C_2 are positive constants. The first term in the binding energy (11.18) is proportional to the number of nucleons (A) because each nucleon feels the presence of all the other nucleons. The second term accounts for the fact that the nucleons on the surface of the liquid drop have fewer neighbors. (The surface area of the drop is proportional to $A^{2/3}$.) We find that this expression (11.18) gives a fairly good representation of the observed nuclear binding energies (Figure 11-3) with $C_1 \approx 16$ MeV and $C_2 \approx 18$ MeV.

The calculation of the binding energy (11.18) may be refined by taking into account the Coulomb energy of proton repulsion and the effect of the Pauli exclusion principle. The reduction of the binding energy due to the Coulomb repulsion ($\Delta E_b^{\text{Coulomb}}$) is proportional to Z^2 and inversely proportional to the nuclear radius ($A^{1/3}$),

$$\Delta E_b^{\text{Coulomb}} \approx -\frac{C_3 Z^2}{A^{1/3}}. \qquad (11.19)$$

The constant C_3 is found from the data on binding energies to be about 0.711 MeV. The Pauli exclusion principle favors $A = 2Z$, and the effect of the exclusion principle is less important at large values of A. Accordingly, the reduction of the binding energy ($\Delta E_b^{\text{Pauli}}$) may be written in the form

$$\Delta E_b^{\text{Pauli}} \approx -\frac{C_4 (A - 2Z)^2}{A}. \qquad (11.20)$$

The constant C_4 is found from the data on binding energies to be about 23.7 MeV. Finally, a fifth correction, also due to the Pauli exclusion principle, may be added to the binding energy, which accounts for the fact that nuclei with an even numbers of both neutrons and protons (*even–even* nuclei) are observed to be especially stable, while those with odd numbers of neutrons and protons (*odd–odd* nuclei) tend to be unstable. The last term is not needed if Z is even and $A - Z$ is odd or vice versa (*even–odd* nuclei).

The resulting empirical formula for the nuclear binding energy is for even–odd nuclei,

$$E_b^{\text{even–odd}} = (15.75 \text{ MeV}) A - (17.8 \text{ MeV}) A^{2/3}$$
$$- \frac{(0.711 \text{ MeV}) Z^2}{A^{1/3}}$$
$$- \frac{(23.7 \text{ MeV})(A - 2Z)^2}{A}; \qquad (11.21)$$

for even–even nuclei,

$$E_b^{\text{even–even}} = E_b^{\text{even–odd}} + \frac{11.18 \text{ MeV}}{\sqrt{A}}; \qquad (11.22)$$

and for odd–odd nuclei,

$$E_b^{\text{odd–odd}} = E_b^{\text{even–odd}} - \frac{11.18 \text{ MeV}}{\sqrt{A}}. \qquad (11.23)$$

This parameterization of the binding energy is called the *Weizsaecker formula*.

EXAMPLE 11-4

Calculate the binding energy of ^{56}Fe and compare the result with the liquid drop model.

SOLUTION:

The binding energy is

$$E_b = Z m_p c^2 + (A - Z) m_n c^2 - M c^2,$$

where M is the nuclear mass, $A = 56$, and $Z = 26$. Consulting Appendix J to find the nuclear mass, we have

$$E_b = (26)(938.27 \text{ MeV}) + (30)(939.57 \text{ MeV})$$
$$- (52090.2 \text{ MeV}) = 492 \text{ MeV}.$$

From leading terms of the Weizsaecker formula we have

$$E_b \approx (15.75 \text{ MeV})(56) - (17.8 \text{ MeV})(56)^{2/3}$$
$$= 621.5 \text{ MeV}.$$

The Coulomb correction is

$$\Delta E_b^{\text{Coulomb}} \approx -\frac{(0.711 \text{ MeV}) Z^2}{A^{1/3}}$$
$$= -\frac{(0.711 \text{ MeV})(26)^2}{(56)^{1/3}}$$
$$= -125.6 \text{ MeV}.$$

The Pauli exclusion principle correction is

$$\Delta E_b^{\text{Pauli}} \approx -\frac{(23.7 \text{ MeV})(A - 2Z)^2}{A}$$
$$= -\frac{(23.7 \text{ MeV})(4)^2}{56}$$
$$= -6.8 \text{ MeV}.$$

The even–even correction is

$$\Delta E_b^{\text{even-even}} \approx \frac{11.18 \text{ MeV}}{\sqrt{A}} = \frac{11.18 \text{ MeV}}{\sqrt{56}} = 1.5 \text{ MeV}.$$

The calculated binding energy is

$$E_b = 621.5 \text{ MeV} - 125.6 \text{ MeV} - 6.8 \text{ MeV} + 1.5 \text{ MeV}$$
$$= 491 \text{ MeV},$$

which deviates from the actual value (492 MeV) by about 0.2%. ∎

Shell Structure

There are more stable nuclei with even numbers of neutrons and protons than with odd numbers. Furthermore, when the number of neutrons is

$$2, 8, 20, 28, 50, 82, \text{ or } 126, \qquad (11.24)$$

the nucleus is observed to be particularly stable. These numbers are referred to as *magic numbers*. The magic numbers also apply to protons, except that there is no element number 126; when the number of protons in a nucleus is equal to a magic number, then the nucleus is also very stable. Nuclei in which the number of neutrons and

the number of protons are *both* equal to magic numbers $\left({}^4_2\text{He}, {}^{16}_8\text{O}, {}^{40}_{20}\text{Ca}, {}^{48}_{20}\text{Ca}, {}^{208}_{82}\text{Pb} \right)$ tend to be particularly stable.

The existence of the magic numbers is a signature for a *shell structure* of the nucleus analogous to the atomic shell structure. Recall that atoms that have the most stable electron configurations are those that have a number of electrons (2, 10, 18, 36, 54, or 86) corresponding to certain filled subshells. In the nuclear shell model, each nucleon has an independent wave function analogous to our description of electrons in atoms. We now examine the origin of the magic numbers. The total angular momentum (**J**) of a single neutron or proton is given by the sum of its spin (**S**) and orbital (**L**) parts:

$$\mathbf{J} = \mathbf{L} + \mathbf{S}. \qquad (11.25)$$

The spin quantum number s is equal to 1/2. The orbital quantum number ℓ is equal to a nonnegative integer (0, 1, 2, …), and as in atomic physics these states are called s, p, d, and so on. The possible values of the total angular momentum quantum number (j) are given by the addition rule for angular momentum. For s states,

$$j = \frac{1}{2}. \qquad (11.26)$$

For p states,

$$j = \frac{1}{2} \quad \text{or} \quad \frac{3}{2}. \qquad (11.27)$$

For d states,

$$j = \frac{3}{2} \quad \text{or} \quad \frac{5}{2}, \qquad (11.28)$$

and so on.

In nuclear physics, the principle quantum number n is defined so that the energy levels (in increasing energy) for s states are called $1s, 2s, 3s, …$; the energy levels for p states are $1p, 2p, 3p$, and so on. In atomic systems there is no $1p$ state. We have defined the quantum numbers in atomic physics so that the states with nearly the same energy (e.g., $2s$ and $2p$) have the same value of the quantum number n. There is no corresponding symmetry for the nucleus. For this reason, the states with the lowest energy for each value of the quantum number ℓ are labeled starting with $n = 1$.

In atoms we found that states with lower values of orbital angular momentum have lower energy. The same is true for the nucleus. In addition, there is a strong spin dependence to the force between nucleons. The spin dependence results in nuclear fine-structure splitting that is quite large for large values of ℓ. For a given value of ℓ, the energy is lower when **L** and **S** are parallel compared to the case where **L** and **S** are antiparallel. Thus, the larger of the two values of the quantum number j has the lower energy. (Recall that the deuteron has proton and neutron spins parallel.) The energy levels for the shell model are indicated in Figure 11-4.

The liquid drop model assumes that the nucleons are strongly coupled to each other in the nucleus. In contrast, the shell model assumes that each nucleon moves as an independent particle. Both models have their virtues and deficiencies. More sophisticated nuclear models have been made by combining the features of the liquid drop and shell models.

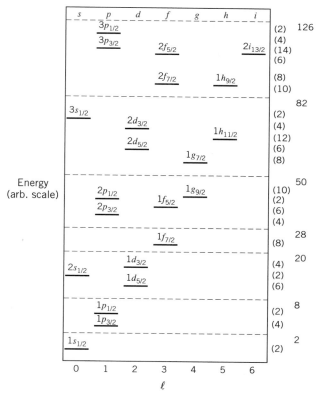

FIGURE 11-4 Schematic diagram of the nuclear shell model.
The numbers in parenthesis indicate the number of nucleons per subshell. The numbers on the far left indicate the total number of nucleons in filled subshells.

11-4 RADIOACTIVE DECAYS

Mean Lifetime and Half-Life

The first nuclear disintegrations, those of a uranium compound, were discovered by Becquerel in 1896. Radioactive decay is a random process. Any given particle has a certain probability per unit time of spontaneous decay. The decay probability is independent of the previous life of the particle. If $N(t)$ is the number of particles in a sample as a function of time, then the decay rate $(-dN/dt)$ is proportional to N,

$$-\frac{dN}{dt} = \lambda N. \qquad (11.29)$$

The constant of proportionality (λ) has dimensions of inverse time. If we begin with N_0 particles, then the number of particles as a function of time is

$$N(t) = N_0 e^{-\lambda t}. \qquad (11.30)$$

The number of particles decreases exponentially.

When we speak of a single particle, we usually refer to its mean lifetime. The mean lifetime (τ) of a particle is

$$\tau = \frac{1}{\lambda}, \qquad (11.31)$$

as was shown in Example 2-1. For a large sample of particles, $1/e$ of them (about 37.8%) will have *not decayed* after a time τ. In nuclear physics, lifetimes are usually specified by the half-life ($t_{1/2}$) the time after which one-half of the sample has decayed. The half-life is shorter than the mean-life, τ and $t_{1/2}$ are related by

$$e^{-t_{1/2}/\tau} = \frac{1}{2}, \qquad (11.32)$$

or

$$t_{1/2} = \tau \ln 2. \qquad (11.33)$$

The exponential decay law for ^{222}Rn is shown in Figure 11-5.

Alpha Decay

Alpha particles were first classified as the decay products of natural radioactivity that could not penetrate matter. The α particle, or helium nucleus (^{4}He), has an exceptionally large binding energy per nucleon:

$$E_b = 28 \text{ MeV}, \qquad (11.34)$$

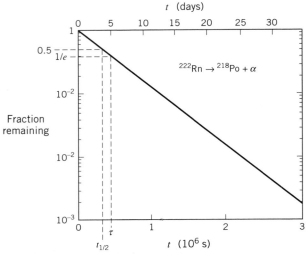

FIGURE 11-5 The exponential decay law.
Radon-222, part of the uranium series, decays by alpha emission with a half-life of 3.82 days. For an initial sample of ^{222}Rn at $t = 0$, we plot the fraction that has not decayed as a function of time.

or 7 MeV per nucleon. The large binding energy accounts for the very high stability of the alpha particle. We have considered α decay already as an example of barrier penetration. Quantum mechanical tunneling explains the small variation in α energies and the large range of lifetimes (see Figure 7-17). There is a finite probability per unit time that the four nucleons will find themselves in the form of an α particle outside of the nucleus.

The number of neutrons and protons are both conserved in α decay; they are simply rearranged. In an α decay, the Z of the decaying or *parent* nucleus is lowered by 2 and the value of A is lowered by 4. The energy released in α decay (Q) appears as kinetic energy of the α particle and the resulting *daughter* nucleus. If M_P is the mass of the parent nucleus, M_D is the mass of the daughter nucleus, and m_α is the mass of the alpha particle, then the energy released in the decay is

$$Q = (M_P - M_D - m_\alpha)c^2. \quad (11.35)$$

Not all nuclei can decay by alpha emission because of conservation of energy. Alpha decay occurs if and only if $Q > 0$. Since the decay has only two particles in the final state, the α particle and the daughter nucleus have equal and opposite momentum (p). By energy conservation,

$$Q = \frac{p^2}{2M_D} + \frac{p^2}{2m_\alpha}, \quad (11.36)$$

or

$$Q = \frac{p^2}{2m_\alpha}\left(1 + \frac{m_\alpha}{M_D}\right)$$
$$= \frac{p^2}{2m_\alpha}\left(1 + \frac{4}{A-4}\right). \quad (11.37)$$

Solving for the kinetic energy of the alpha particle, we have

$$\frac{p^2}{2m_\alpha} = \frac{Q(A-4)}{A}. \quad (11.38)$$

EXAMPLE 11-5

Calculate the kinetic energy of an alpha decay from ^{238}U.

SOLUTION:

The decay product of ^{238}U is ^{234}Th. The decay process is

$$^{238}\text{U} \rightarrow ^{234}\text{Th} + \alpha.$$

The energy released in the decay is

$$Q = M_P c^2 - M_D c^2 - m_\alpha c^2.$$

Numerically, we have

$$Q = 221{,}697.68 \text{ MeV} - 217{,}965.99 \text{ MeV}$$
$$-3727.41 \text{ MeV}$$
$$\approx 4.3 \text{ MeV}.$$

The kinetic energy (11.38) of the alpha particle is

$$E = \frac{Q(A-4)}{A} \approx \frac{(4.3 \text{ MeV})(234)}{238} \approx 4.2 \text{ MeV}. \quad \blacksquare$$

Figure 11-6 shows a photograph taken by C. F. Powell and G. Occhialini of the decay sequence

$$^{228}\text{Th} \rightarrow ^{224}\text{Ra} + \alpha, \quad (t_{1/2} = 1.9 \text{ y}); \quad (11.39)$$

$$^{224}\text{Ra} \rightarrow ^{220}\text{Rn} + \alpha, \quad (t_{1/2} = 3.6 \text{ d}); \quad (11.40)$$

$$^{220}\text{Rn} \rightarrow ^{216}\text{Po} + \alpha, \quad (t_{1/2} = 54.5 \text{ s}); \quad (11.41)$$

and

$$^{216}\text{Po} \rightarrow ^{212}\text{Pb} + \alpha, \quad (t_{1/2} = 0.16 \text{ s}). \quad (11.42)$$

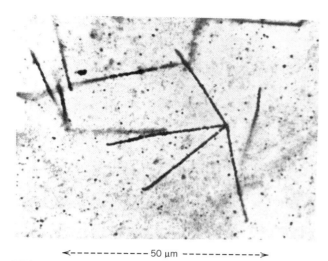

<----------- 50 μm ----------->

FIGURE 11-6 Alpha particles from radioactive decays observed in emulsion.

The four alpha particles are produced in the sequential spontaneous decays of ^{228}Th, ^{224}Ra, ^{220}Rn and ^{216}Po. From C. F. Powell and G. Occhialini, *Nuclear Physics in Photographs*, Oxford University Press (1947). Reprinted by permission of Oxford University Press.

**TABLE 11-2
THE FOUR RADIOACTIVE SERIES.**

Series	Parent	Lifetime	First Decay	End Product
Thorium	^{232}Th	1.40×10^{10} y	$^{232}\text{Th} \rightarrow {}^{228}\text{Ra} + \alpha$	^{208}Pb
Neptunium	^{237}Np	2.14×10^{6} y	$^{237}\text{Np} \rightarrow {}^{233}\text{Pa} + \alpha$	^{209}Bi
Uranium	^{238}U	4.17×10^{9} y	$^{238}\text{U} \rightarrow {}^{234}\text{Th} + \alpha$	^{206}Pb
Actinium	^{235}U	7.04×10^{8} y	$^{235}\text{U} \rightarrow {}^{231}\text{Th} + \alpha$	^{207}Pb

The picture was made by mixing a sample of ^{228}Th into a photographic emulsion. The parent nucleus (^{228}Th) has a long lifetime so that enough of it must be introduced into the emulsion so that some will decay in a few days. The decay products all have much shorter lifetimes and thus, are "guaranteed" to decay in a relatively short time period. All four alpha particles are observed as heavily ionizing tracks originating from the same point because the nuclei do not move.

Alpha decays decrease the atomic mass number A by four. Therefore, the nuclei that are products of a chain of alpha decays will have atomic mass numbers that differ by four times an integer. The decay of ^{232}Th produces nuclei with $A = 228, 224, 220,\dots$; ^{238}U produces nuclei with $A = 234, 230, 226\dots$; ^{237}Th produces nuclei with $A = 233, 229, 225,\dots$; and ^{235}U produces nuclei with $A = 231, 227, 223\dots$ There are four and only four parent nuclei that give unique chains or *series* of radioactive decay. The radioactive series are listed in Table 11-2. The decays of Figure 11-6 are part of the thorium series.

Beta Decay

The beta particle is the electron. Nuclear decays in which an electron is emitted are called *beta-minus* (β^-) decays. In

β^- decay, a neutron is converted into a proton. The simplest example of beta-minus decay is the decay of a free neutron. The neutron mass is larger than the proton mass by about 1.3 MeV. Free neutrons are observed to decay with a mean lifetime of about 1000 seconds. The electron momentum distribution from beta decay is shown in Figure 11-7. In the rest frame of the neutron, the proton and electron do not have opposite momentum and they are not monoenergetic as would be expected for a simple two-body decay ($n \rightarrow p + e$). This was the first indication of the existence of the *neutrino*. The neutrino has zero electric charge, zero mass (< 7 eV/c^2), and an intrinsic angular momentum quantum number $s = 1/2$. The neutrino is the

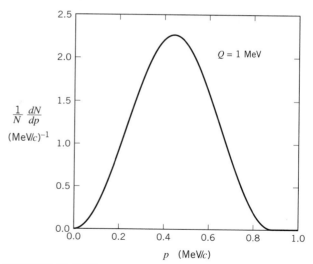

FIGURE 11-7 Electron momentum distribution from beta decay.

The shape of the spectrum is characteristic of a final state with three particles. The spectrum is plotted for a reaction Q value of 1 MeV, corresponding to a maximum electron momentum of about 0.85 MeV/c.

weak interaction partner of the electron. The particle that is emitted in neutron decay is called the *electron antineutrino* ($\bar{v}_e$); the process of beta decay produces one particle and one antiparticle. The neutron beta decay process is

$$n \rightarrow p + e^- + \bar{v}_e. \qquad (11.43)$$

Beta decay is an example of the weak interaction.

A neutron inside the alpha particle cannot decay by conservation of energy. The decay process would be $\alpha \rightarrow$ $^4\text{Li} + e^- + \bar{v}_e$. Such a process is forbidden by energy conservation because the alpha mass is smaller than the ^4Li mass. (The alpha mass is smaller than the ^4Li mass because the binding energy of the alpha is so large.)

In many nuclei, the binding energies are such that beta decay is possible resulting in an unstable nucleus. For example, a neutron can decay inside the nitrogen nucleus,

$$^{16}\text{N} \rightarrow {}^{16}\text{O} + e^- + \bar{v}_e. \qquad (11.44)$$

The mass of the ^{16}N nucleus is 14,906.1 MeV/c^2, the mass of the ^{16}O nucleus is 14,895.2 MeV/c^2, and the electron mass is 0.5 MeV/c^2. The mass of ^{16}N is greater than the sum of the ^{16}O mass plus the electron mass. The decay occurs with a half-life of about 7.13 seconds. In β^- decay the value of Z increases by one and A is unchanged.

A second type of beta decay occurs in which a proton is converted to a neutron (the inverse of β^- decay). This type of beta decay is called *beta-plus* decay (β^+) because a positively charged particle called the *positron* is emitted. The positron (e^+) is the *antiparticle* of the electron. The mass of the positron is identical to the mass of the electron. The β^+ decay process is

$$p \rightarrow n + e^+ + v_e. \qquad (11.45)$$

This process cannot occur for a free proton because of energy conservation ($m_p < m_n$). The β^+ decay (11.45) can occur inside some nuclei. The β^+ decay transforms the nucleus by increasing Z by one while leaving A unchanged. The condition for β^+ decay to occur is that the binding energy of the transformed nucleus must exceed the binding energy of the original nucleus by at least $m_n c^2 - m_p c^2 + mc^2$ (where m is the electron mass). Some examples of β^+ decay are

$$^{11}\text{C} \rightarrow {}^{11}\text{Be} + e^+ + v_e, \qquad (11.46)$$

$$^{23}\text{Mg} \rightarrow {}^{23}\text{Na} + e^+ + v_e, \qquad (11.47)$$

and

$$^{41}\text{Sc} \rightarrow {}^{41}\text{Ca} + e^+ + v_e. \qquad (11.48)$$

The Discovery of the Neutrino

The need for the neutrino was recognized by Pauli in 1931, from theoretical analysis of the beta decay process. The interaction of neutrinos with matter is so weak that it took 25 years to detect their interactions! Neutrino detection is possible only if an extremely large flux is available. The first direct observation of the neutrino interacting with matter was made by Frederick Reines, Clyde Cowan, Jr., and collaborators in 1956. Their neutrino source was the nuclear reactor at the Savannah River Plant in South Carolina. A nuclear reactor is a source of electron antineutrinos from β^- decays.

In 1958, Reines and Cowan repeated their neutrino experiment with greatly improved sensitivity. Figure 11-8 shows the experimental arrangement of Reines and Cowan. The process they observed was

$$\bar{v}_e + p \rightarrow n + e^+. \qquad (11.49)$$

The target protons were provided by a 1400 liter liquid detector loaded with a cadmium compound. The region around the neutrino target was instrumented with liquid *scintillator*. The scintillator emits light in response to an energetic gamma ray interaction. The scintillation light is detected by photomultipliers.

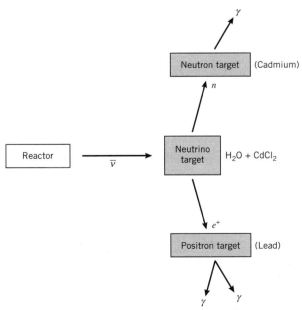

Figure 11-8 The experiment of Reines and Cowan.

The presence of the positron from the antineutrino interaction (11.49) was detected by its annihilation with an electron

$$e^+ + e^- \rightarrow \gamma + \gamma. \qquad (11.50)$$

The presence of the neutron was detected by its absorption in cadmium

$$n + {}^{108}\text{Cd} \rightarrow {}^{109}\text{Cd}* \rightarrow {}^{109}\text{Cd} + \gamma. \qquad (11.51)$$

The experimental signature of the $\bar{\nu}_e$ interaction is the scintillator light from the three gamma rays. (The γ from the neutron interaction is produced about 10^{-5} s after the e^+e^- annihilation.) The experimenters proved that the scintillation signals were caused by electron antineutrinos from the reactor by making measurements with the reactor both on and off.

EXAMPLE 11-6

The $\bar{\nu}_e$ flux passing through a 1.4 ton detector is about 10^{15} m^{-2}s^{-1} and the measured neutrino interaction is 10^{-2} s^{-1}. Estimate the neutrino cross section per target proton.

SOLUTION:

The total cross section for interaction in the detector is the interaction rate divided by the incident flux:

$$\sigma_T = \frac{R_i}{\Phi_i}.$$

The cross section per proton is obtained by dividing by the number of protons in the target,

$$\sigma = \frac{\sigma_T}{N_p} = \frac{R_i}{\Phi_i N_p}.$$

The number of protons in the target is

$$N_p = \frac{MN_A}{(2)(10^{-3} \text{ kg})},$$

where M is the mass of the target and we have assumed that 1/2 of the target mass is contained in protons. Therefore,

$$\sigma \approx \frac{R_i (2)(10^{-3} \text{ kg})}{\Phi_i MN_A}$$

$$\approx \frac{(10^{-2} \text{ s}^{-1})(2)(10^{-3} \text{ kg})}{(10^{15} \text{ m}^{-2}\text{s}^{-1})(1.4 \times 10^3 \text{ kg})(6 \times 10^{23})}$$

$$\approx 10^{-47} \text{ m}^2.$$

The interaction of the neutrino with matter is indeed "weak." ∎

* *Challenging*

Fermi's Golden Rule

The dynamics of beta decay was first worked out by Fermi. The number of beta decays per unit time is given by the *Golden Rule*, which states that the transition rate (W) is proportional to the density of states (ρ) times the square of a matrix element ($|M|^2$),

$$W = \frac{2\pi}{\hbar} \rho |\mathcal{M}|^2. \qquad (11.52)$$

The matrix element contains the physics of the specific interaction. For the purpose of calculating a transition rate for any given process, $|\mathcal{M}|^2$ is just a number evaluated from the integral of the wave function times its complex conjugate with the interaction potential sandwiched in between. Often, $|\mathcal{M}|^2$ is too complicated to calculate and must be determined from experiment. The most important part of $|\mathcal{M}|^2$ is that it contains a factor of the coupling strength of the force squared (α_w^2 for the weak force).

> Fermi's Golden Rule: The transition rate is proportional to the density of final states times a matrix element squared:
>
> $$W = \frac{2\pi}{\hbar} \rho |\mathcal{M}|^2.$$

For one particle, the density of states (2.90) is

$$\rho = \frac{dn_s}{dE} = \frac{dn_s}{dp} \frac{dp}{dE}, \qquad (11.53)$$

and

$$\frac{dn_s}{dp} = \frac{4\pi p^2}{h^3}. \qquad (11.54)$$

(The density of states is discussed further in Chapter 12.) For beta decay we must include both the electron and the neutrino,

$$\frac{d^2 n_s}{dp_e \, dp_\nu} = \frac{16\pi^2 p_e^2 p_\nu^2}{h^6}. \qquad (11.55)$$

Since we may neglect the kinetic energy of the nucleus, conservation of energy gives

$$E = cp_v + E_e, \qquad (11.56)$$

where E is the total energy available in the decay, cp_v is the neutrino energy, and E_e is the electron energy;

$$dp_v = \frac{dE}{c}. \qquad (11.57)$$

The density of states factor is

$$\rho = \frac{d^2 n_s}{dE dp_e} = \frac{16\pi^2}{c^3 h^6} p_e^2 (E - E_e)^2. \qquad (11.58)$$

The transition rate is

$$\begin{aligned} W &= \frac{2\pi}{\hbar} \rho |\mathcal{M}|^2 \\ &= \frac{2\pi}{\hbar} \frac{16\pi^2}{c^3 h^6} p_e^2 (E - E_e)^2 |\mathcal{M}|^2. \end{aligned} \qquad (11.59)$$

The square root of the transition rate divided by the electron momentum is a linear function of the electron energy:

$$\frac{\sqrt{W}}{p_e} \propto (E - E_e). \qquad (11.60)$$

A plot of $\sqrt{W}/p$ versus E_e is called a *Kurie plot* after Franz Kurie, J. R. Richardson, and H. C. Paxton, who first plotted the beta decay data in this manner in 1936. A Kurie plot for promethium-147 is shown in Figure 11-9. Kurie plots have been made for many different beta decays. The Kurie plots prove that the decay rate is proportional to the density of states (11.58). These data show that the Golden Rule works for beta decay.

Parity Violation

Parity is a transformation that changes the algebraic sign of the coordinate system. Under the parity transformation

$$x' = -x, \qquad (11.61)$$

$$y' = -y, \qquad (11.62)$$

and

$$z' = -z. \qquad (11.63)$$

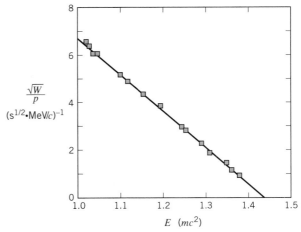

FIGURE 11-9 Kurie plot of ^{147}Pm.
The energy spectrum of the electrons from beta decay is measured. In a plot of the square root of the decay rate divided by the electron momentum versus electron energy, the data lie on a straight line as predicted by Fermi's Golden Rule. After L. M. Langer et al., *Phys. Rev.* **77**, 798 (1950).

The parity transformation changes a right-handed coordinate system into a left-handed coordinate system, as indicated in Figure 11-10. If we apply the parity transformation twice to any vector, we return to the original vector.

The electromagnetic and strong interactions are invariant under the parity transformation. In the early 1950s, it was widely assumed that the weak interaction was also invariant under parity. This turned out not to be true! In 1956, Tsung Dao Lee and Chen Ning Yang predicted the nonconservation of parity in the weak interaction. In 1957, a classic experiment was performed by C. S. Wu and her collaborators that demonstrated parity violation in the weak interaction. In the Wu experiment, ^{60}Co nuclei were aligned in a magnetic field (Figure 11-11). The cobalt undergoes beta decay

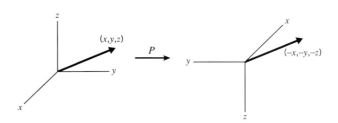

FIGURE 11-10 The parity transformation.
Under the parity transformation, a right-handed system is changed into a left-handed system.

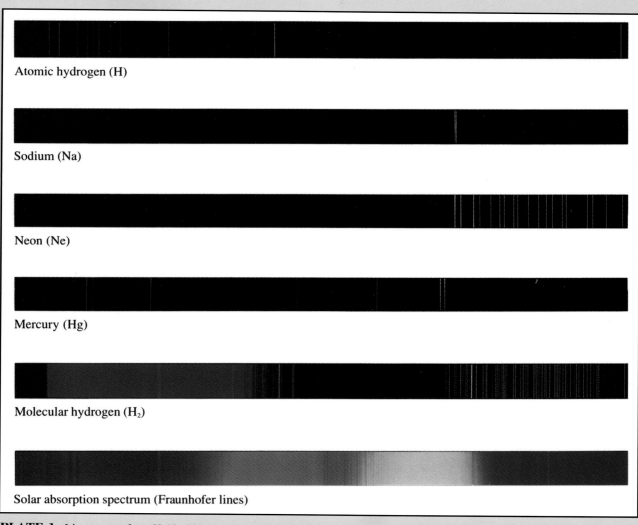

Atomic hydrogen (H)

Sodium (Na)

Neon (Ne)

Mercury (Hg)

Molecular hydrogen (H$_2$)

Solar absorption spectrum (Fraunhofer lines)

PLATE 1 Line spectra from H, Na, Ne, Hg, H$_2$, and the continuous solar spectrum. *Courtesy of Bausch and Lomb.*

PLATE 2 Image of a DNA molecule taken with a scanning tunneling microscope. *Courtesy of John D. Baldeschwieler, Caltech, Nature **346**, 294 (1990).*

PLATE 3 First laser built by T.H. Maiman. *Courtesy of Hughes Research Laboratory.*

PLATE 4 Helium-neon laser beam used to monitor flow of blood, useful in diagnosing eye diseases. *Alexander Tsiaras/Photo Researchers.*

PLATE 6   Triple exposure of a hologram from 3 angles. *Chuck O'Rear/Westlight.*

PLATE 5   Making a hologram.   *Courtesy Newport Corporation.*

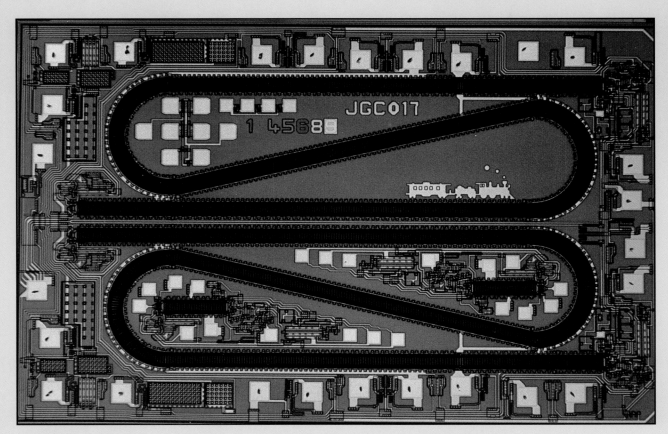

PLATE 7   Integrated circuit, 2 channel shift register.   *Courtesy of LeCroy Corporation, Chestnut Ridge, NY.*

PLATE 8 Fastbus module, 96 channel ADC. *Courtesy of LeCroy Corporation, Chestnut Ridge, NY.*

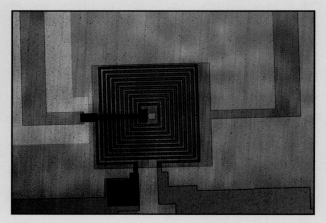

PLATE 10 High-T_c squid. *Courtesy of Conductus, Inc., Sunnyvale,California.*

PLATE 9 Levitation of a goldfish with a high-T_c super-conducting magnet. *Courtesy of Mr. Shoji Tanaka, International Superconductivity Technology Center, Tokyo, Japan.*

PLATE 11 Cockcroft-Walton accelerator at Fermilab. *Courtesy Fermilab Visual Media Services.*

PLATE 12 Super Proton Synchroton. *Courtesy CERN, Geneva, Switzerland.*

PLATE 13 Antiproton Accumulator Ring. *Courtesy CERN, Geneva, Switzerland.*

PLATE 15 Electron spectrometers at End Station A where the quark structure of the proton was first observed. *Courtesy Stanford Linear Accelerator Center*.

PLATE 14 RF cavity for the LEP accelerator. *Courtesy CERN, Geneva, Switzerland*.

PLATE 16 UA1 detector used to discover the W and Z^0 particles in proton-antiproton collisions. *Courtesy CERN, Geneva, Switzerland*.

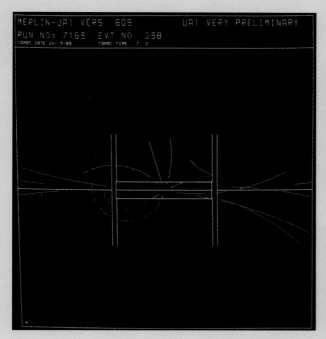

PLATE 17 Electronic image from the UA1 drift chamber of the ionization trails left by charged particles produced in a proton-antiproton collision. *Courtesy CERN, Geneva, Switzerland.*

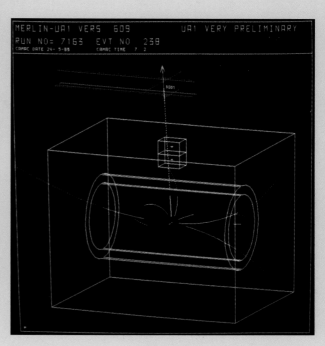

PLATE 18 Decay of a *W* particle into a muon and neutrino observed in the UA1 detector. *Courtesy CERN, Geneva, Switzerland.*

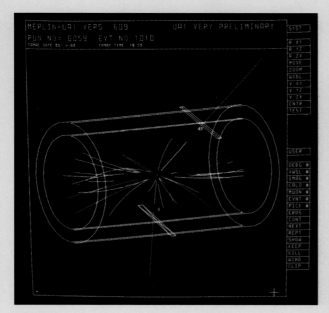

PLATE 19 Decay of a Z^0 particle into electron and positron observed in the UA1 detector. *Courtesy CERN, Geneva, Switzerland.*

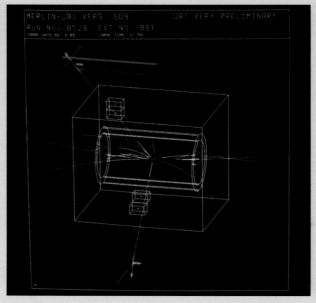

PLATE 20 Decay of a Z^0 particle into muons observed in the UA1 detector. *Courtesy of CERN, Geneva, Switzerland.*

PLATE 21 Aleph detector at LEP. *Courtesy Aleph Collaboration and CERN, Geneva, Switzerland.*

PLATE 23 Spiral galaxy M51. *Courtesy James Wray.*

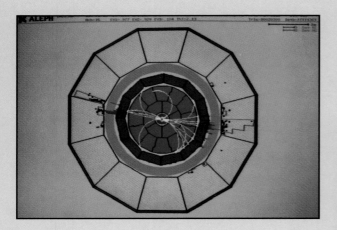

PLATE 22 Decay of a Z^0 particle into quarks observed in the Aleph detector. *Courtesy Aleph Collaboration and CERN, Geneva, Switzerland.*

PLATE 24 Ultraviolet image of the spiral galaxy M74 (color coded to indicate the intensity) taken from the space shuttle Columbia on December 7, 1990. *Courtesy NASA.*

$${}^{60}\text{Co} \rightarrow {}^{60}\text{Ni} + e^- + \overline{\nu}_e, \qquad (11.64)$$

with a half-life of 10.5 minutes. The directions of the emitted electrons were measured. If parity were to be conserved in this decay, equal numbers of electrons would be emitted in directions parallel and antiparallel to the magnetic field. Since parity is not conserved in the weak interaction, more electrons are emitted in the direction opposite the magnetic field and, therefore, opposite to the direction of the nuclear spin.

A handy tool for analysis of weak decays is the helicity (H) of a particle defined as the component of intrinsic angular momentum (**S**) in the direction of the velocity vector (**v**):

$$H = \frac{\mathbf{S} \cdot \mathbf{v}}{Sv}. \qquad (11.65)$$

The helicity is frame dependent. Consider an electron with its spin parallel to the velocity vector ($H = +1$) in Frame A as shown in Figure 11-12. In Frame B, which is moving with a velocity **V** in the same direction as **v**

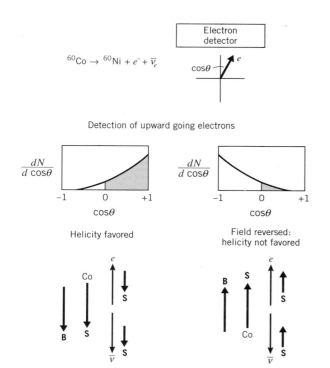

FIGURE 11-11 The discovery of parity violation in the weak interaction.
The ^{60}Co nuclei decay by the weak interaction into ^{60}Ni nuclei, an electron, and an antineutrino. The experiment was first performed by C. S. Wu et al.

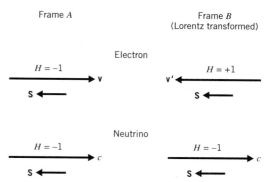

FIGURE 11-12 Helicity.

($V > v$), the spin is antiparallel to the velocity vector ($H = -1$). Now consider the case of a neutrino. Since the neutrino is massless, its speed is always equal to the speed of light, c. When we transform with any velocity **V** ($V < c$) in the direction of the neutrino, the speed of the neutrino is still c. There is no frame in which the neutrino is moving in the opposite direction. We cannot stop a neutrino with a Lorentz transformation!

Now comes an amazing piece of physics. The neutrino is defined to be the particle emitted together with the positron in β^+ decays. All neutrinos observed in nature are found to have negative helicity. The intrinsic angular momentum of the neutrino always points in a direction opposite the velocity. Such a particle is called a *left-handed* particle. Nobody has ever detected the presence of a *right-handed* neutrino, where the spin points in the same direction as the velocity. The antineutrino is defined to be the particle emitted together with the electron in β^- decays. All antineutrinos observed in nature are found to have positive helicity; the antineutrinos are right-handed.

Electron Capture

Nuclei that undergo β^+ decay can also decay by *electron capture*. The electron capture process is

$$p + e^- \rightarrow n + \nu_e, \qquad (11.66)$$

occurring inside the nucleus. The emitted neutrino is monoenergetic. Electron capture may occur when β^+ decay is forbidden by energy conservation because a smaller difference in binding energy is needed for electron capture than for β^+ decay by twice the electron mass energy (about 1 MeV). Usually, the inner (K shell) electrons are captured, and the process is referred to as K capture. An example of electron capture is

$$^7\text{Be} + e^- \rightarrow {}^7\text{Li} + \nu_e, \qquad (11.67)$$

which occurs with a half-life of 53 days.

When an electron is captured, it leaves a vacancy or *hole* in the atom. The hole will be filled by an electron making a transition from a higher energy level, usually with the emission of an x ray. Occasionally, it is observed that the transition causes an electron to be ejected from the atom. Such an ejected electron is called an *Auger* electron. Auger electrons are emitted with a well-defined energy typically in the keV energy region.

Gamma Decay

Often a nucleus is left in an excited state after a decay. The nuclear excited state is analogous to the atomic excited state. The excited nucleus (N*) may decay to the ground state by emission of a photon without changing the value of Z or A:

$$^Z_A\text{N*} \rightarrow {}^Z_A\text{Ni} + \gamma. \qquad (11.68)$$

This process is the nuclear analogy of photon emission in atomic systems. The typical energy scale of the photons from nuclear decay is an MeV. Gamma decay is an electromagnetic process.

Table 11-3 lists some commonly used radioactive sources.

The Radioactive Series

The four series of radioactive decays (thorium, actinium, neptunium, and uranium) are shown in Figure 11-13. All of the series contain beta decays as well as alpha decays. The reason for this is that alpha decays change both A and Z by 2 units. The lighter nuclei are more stable when they have a smaller neutron fraction than the heavier nuclei (by the Pauli exclusion principle). The β^- decays change neutrons into protons as needed to provide the nuclear stability.

All of the heavy elements were created by collapsing stars in approximately equal numbers. Thorium, uranium, and actinium are all present in the earth; however, there is no neptunium. All of the neptunium that was present in the earth has decayed; the half-life of ^{237}Np is

$$t_{1/2} = 2.14 \times 10^6 \text{ y}, \qquad (11.69)$$

and the age of the earth is much larger than this, about 4.6×10^9 years.

The three radioactive series (thorium, uranium, and actinium) account for most of the activity found in nature.

TABLE 11-3
PROPERTIES OF SOME COMMONLY USED RADIOACTIVE SOURCES.

Source	half-life (y)	Type	Radiation	Energy (keV)
^{22}Na	2.60	β^+, EC	Positron	546 (max)
			Gamma	511
			Gamma	1275
^{55}Fe	2.73	EC	Gamma	5.89
			Gamma	6.49
			Gamma	14.4
^{60}Co	5.27	β^-	Electron	318 (max)
			Gamma	1173
			Gamma	1332
^{90}Sr	28.5	β^-	Electron	546 (max)
		β^-	Electron	2284 (max, from ^{90}Y)
^{207}Bi	32.2	EC	Gamma	70
			Gamma	1064
			Gamma	1770
			Electron	976 (Auger)
			Electron	482 (Auger)
			Electron	1048 (Auger)
^{241}Am	432	α	Alpha	5486
			Alpha	5443
			Gamma	60
			Gamma	18
			Gamma	14

Handbook of Chemistry and Physics.

In addition, a small amount of radioactive material, such as ^{14}C, is continuously produced by cosmic rays. There are a few additional radioactive elements that are not pro-

FIGURE 11-13 The four radioactive series.
(*a*) The thorium series begins with ^{232}Th which has a half-life of about fourteen billion years. By a sequence of alpha and beta decays, the stable element ^{208}Pb is produced. (*b*) The actinium series begins with ^{235}U which has a half-life of about seven-hundred million years, and ends with ^{207}Pb. (*c*)The neptunium series begins with ^{237}Np which has a half-life of about two million years, and ends with ^{209}Bi. (*d*) The uranium series begins with ^{238}U which has a half-life of about five billion years, and ends with ^{206}Pb.

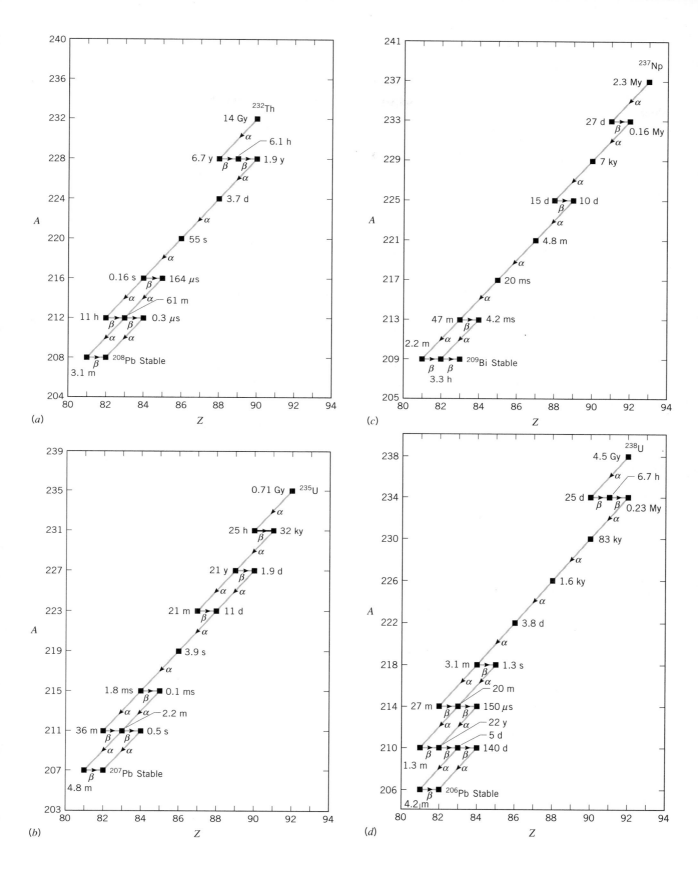

duced by cosmic rays and are present in the earth. These elements are all lighter than lead, the endpoint of the three naturally occurring radioactive series. These elements must also have lifetimes longer than about a billion years or they would have all decayed. An example is ^{40}K, which has a half-life of about 1.3×10^9 years.

Radioactive Dating

Carbon Dating

Our atmosphere is continuously bombarded by energetic cosmic ray protons. These protons produce neutrons by the reaction

$$p + \text{nucleus} \rightarrow n + X. \tag{11.70}$$

The flux of neutrons that is generated is about $2 \times 10^4 \ \text{m}^{-2}\text{s}^{-1}$. The neutrons continuously produce ^{14}C in the earth's atmosphere through the process

$$n + {}^{14}\text{N} \rightarrow p + {}^{14}\text{C}. \tag{11.71}$$

The ^{14}C isotope is not stable and decays with a half-life of 5730 years by β^- decay,

$$^{14}\text{C} \rightarrow {}^{14}\text{N} + e^- + \bar{\nu}_e. \tag{11.72}$$

The chemistry of ^{14}C is identical to that of ^{12}C. Radioactive ^{14}C forms carbon dioxide just like ^{12}C does. Since the lifetime of the radioactive ^{14}C is relatively long, it is distributed throughout the earth's atmosphere and enters all living objects: plants, animals, and people. All living things have been activated with ^{14}C by the cosmic radiation. The ^{14}C also ends up in the ocean in inorganic forms of dissolved carbon dioxide, bicarbonate, and carbonate. The fraction of carbon that is in the form of ^{14}C is measured to be

$$f = 1.3 \times 10^{-12}. \tag{11.73}$$

All living things are radioactive! The activity of a radioactive sample is the number of decays per second.

EXAMPLE 11-7

Calculate the activity in a living sample that contains 1 kilogram of carbon.

SOLUTION:

The activity (R) of the sample is

$$R = \frac{N}{\tau} = \frac{N \ln 2}{t_{1/2}},$$

where N is the number of ^{14}C atoms in the sample, τ is the ^{14}C mean lifetime, and $t_{1/2}$ is the ^{14}C half-life. The number of ^{14}C atoms in the sample is f times the total number of carbon atoms in the sample:

$$N = \frac{f N_A (1 \ \text{kg})}{(12)(10^{-3} \ \text{kg})}.$$

The activity is

$$R = \frac{(1.3 \times 10^{-12})(6.02 \times 10^{23})(1 \ \text{kg})(\ln 2)}{(12)(10^{-3} \ \text{kg})(5730 \ \text{y})(3.15 \times 10^7 \ \text{s/y})} = 250 \ \text{s}^{-1}.$$

The activity of a living sample due to radioactive ^{14}C decays is 250 decays per second per kilogram of carbon. ∎

In 1947, Willard F. Libby developed the method of carbon dating. Libby determined that all living objects have a constant fraction of radioactive carbon, but that when they die the ^{14}C is not replenished and decays exponentially with its 5730-year half-life. By determining the ^{14}C activity of a once-living object, one can determine how long it has been dead. This technique assumes that the ^{14}C fraction (11.73) has remained constant (i.e., the cosmic ray flux has not changed). Figure 11-14 shows the

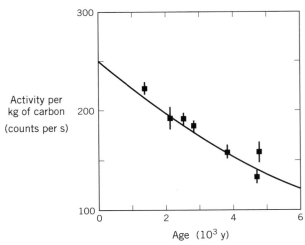

FIGURE 11-14 Activity of ^{14}C (disintegrations per second per kilogram of carbon) for several samples of known age.
The data are from W. Libby, *Radiocarbon Dating* Univ. of Chicago Press (1952).

measured ^{14}C activity for several samples of known age. Carbon dating by measuring ^{14}C activity is limited to objects less than 25,000 years old (a few half-lifes), because the ^{14}C activity is too low in older objects.

A modern version of carbon dating is made possible by use of a mass spectrometer. In a mass spectrometer, one can measure the ratio of ^{14}C to ^{12}C atoms before the ^{14}C atoms decay. This allows a much more sensitive measurement of objects older than 20,000 years, and allows extension of the carbon dating technique to ages of 50,000 years. Another advantage of the mass spectrometer technique is that objects may be dated by using samples that are one thousand times smaller than that needed for measuring ^{14}C activity.

EXAMPLE 11-8

The activity in a piece of bone is measured to be 100 decays per second per kilogram. What is the age of the bone?

SOLUTION:

Let T be the age of the bone. The activity is

$$R = R_0 e^{-T/\tau},$$

where R_0 is the initial decay rate in the bone (250 s^{-1}). The age of the bone is

$$T = -\tau \ln\left(\frac{R}{N_0}\right) = \tau \ln\left(\frac{R_0}{R}\right) = \frac{t_{1/2}}{\ln 2}\ln\left(\frac{R_0}{R}\right)$$

$$= \frac{5730\,\text{y}}{\ln 2}\ln\left(\frac{250}{100}\right) \approx 7600\,\text{y}. \qquad \blacksquare$$

Age of the Earth

As already mentioned, the material of the earth once contained all heavy elements in roughly equal proportions. As the earth has aged, the radioactive elements have decayed. The absence of the neptunium series (half-life of 2×10^6 years) in the earth indicates that the earth is much older than a few million years. The existence of the thorium series (half-life of 1.4×10^{10} years) in the earth indicates that the earth is less than one-hundred billion years old because the thorium has not yet all decayed. The age of the earth may be determined by *uranium dating*. In uranium dating, one measures the ratio of ^{238}U (half-life of 4.17×10^9 y) to ^{206}Pb (stable). Measurements of this ratio on the earth, in meteorites, and on the moon shows that the age of the solar system is about 4.6×10^9 years.

11-5 NUCLEAR REACTIONS

Nuclear Fusion

An examination of the atomic mass number (A) dependence of the binding energy of Figure 11-3 shows that for light elements the binding energy per nucleon is increasing with increasing A. Each additional nucleon contributes to the binding energy of the other nucleons. Therefore, if we fuse two light nuclei together, they are more tightly bound than the two original nuclei.

> If two light nuclei are combined into a heavier nucleus (fusion), the sum of the masses of the two lighter nuclei is greater than the mass of the heavy nucleus. Therefore, energy is released when the heavier nucleus is formed.

In the fusion of deuterium and tritium,

$$d + t \rightarrow \alpha + n, \qquad (11.74)$$

(equivalently written ^{2}H + ^{3}H $\rightarrow$ ^{4}He + n), an energy of 17.6 MeV is released. This energy appears in the kinetic energy of the alpha particle and the neutron. Table 11-4 lists some fusion reactions and their Q values.

TABLE 11-4
SELECTED FUSION REACTIONS AND THEIR Q VALUES.

Reaction	Q (MeV)
$p + p \rightarrow d + e^+ + \nu_e$	1.4
$He^3 + \alpha \rightarrow {}^7Be + \gamma$	1.6
$d + d \rightarrow t + p$	3.3
$d + d \rightarrow {}^3He + n$	4.0
$p + d \rightarrow {}^3He + \gamma$	5.5
$C^{12} + p \rightarrow {}^{13}N + \gamma$	7.6
$^3He + {}^3He \rightarrow \alpha + p + p$	12.9
$^7Li + p \rightarrow \alpha + \alpha$	17.3
$d + t \rightarrow \alpha + n$	17.6
$^3He + d \rightarrow \alpha + p$	18.3

EXAMPLE 11-9
Calculate the energy released in deuterium-tritium fusion.

SOLUTION:
The energy released is

$$Q = m_d c^2 + m_t c^2 - m_\alpha c^2 - m_n c^2 ,$$

where the masses correspond to deuterium, tritium, alpha, and neutron. Numerically, we have

$$Q = 1875.63 \text{ MeV} + 2808.94 \text{ MeV} - 3727.41 \text{ MeV}$$
$$- 939.57 \text{ MeV}$$
$$= 17.6 \text{ MeV}. \qquad \blacksquare$$

Energy Production in Stars
The process by which energy is produced in stars is nuclear fusion. Sir Arthur Stanley Eddington was the first to realize that the main source of solar energy was the combining of four nucleons into an alpha particle. In 1938, Hans Bethe determined the sequence of reactions that powers most stars, the carbon cycle. The nuclear reactions that make up the carbon cycle are

$$p + {}^{12}\text{C} \rightarrow {}^{13}\text{N} + \gamma , \qquad (11.75)$$

$$^{13}\text{N} \rightarrow {}^{13}\text{C} + e^+ + \nu_e , \qquad (11.76)$$

$$p + {}^{13}\text{C} \rightarrow {}^{14}\text{N} + \gamma , \qquad (11.77)$$

$$p + {}^{14}\text{N} \rightarrow {}^{15}\text{O} + \gamma , \qquad (11.78)$$

$$^{15}\text{O} \rightarrow {}^{15}\text{N} + e^+ + \nu_e , \qquad (11.79)$$

and

$$p + {}^{15}\text{N} \rightarrow {}^{12}\text{C} + \alpha . \qquad (11.80)$$

In these reactions, carbon is a catalyst for the burning of four protons to make an alpha particle. The net reaction is $4p \rightarrow \alpha + 2e^+ + 2\nu_e + 3\gamma$. The amount of energy released in these reactions is about 25 MeV. The collision cross sections are extremely small, but the collision rate in stars is enormous. The carbon cycle is the dominate source of energy in stars that have internal temperatures greater than about 10^8 kelvin.

A second set of proton-burning reactions in stars is the proton cycle. The nuclear reactions of the proton cycle are

$$p + p \rightarrow d + e^+ + \nu_e , \qquad (11.81)$$

$$p + d \rightarrow {}^{3}\text{He} + \gamma , \qquad (11.82)$$

and

$$^{3}\text{He} + {}^{3}\text{He} \rightarrow p + p + \alpha . \qquad (11.83)$$

The net reaction is the same as for the carbon cycle. An alternate sequence can include

$$^{3}\text{He} + \alpha \rightarrow {}^{7}\text{Be} + \gamma , \qquad (11.84)$$

$$^{7}\text{Be} + e^- \rightarrow {}^{7}\text{Li} + \nu_e , \qquad (11.85)$$

$$^{7}\text{Li} + p \rightarrow {}^{8}\text{Be} + \gamma , \qquad (11.86)$$

$$^{8}\text{Be} \rightarrow \alpha + \alpha , \qquad (11.87)$$

and

$$^{7}\text{Be} + p \rightarrow {}^{8}\text{B} + \gamma , \qquad (11.88)$$

$$^{8}\text{B} + e^- \rightarrow {}^{8}\text{Be} + \nu_e , \qquad (11.89)$$

$$^{8}\text{Be} \rightarrow \alpha + \alpha . \qquad (11.90)$$

At temperatures below about 10^8 kelvin the proton cycle dominates over the carbon cycle. The temperature of the solar core is about 1.5×10^7 kelvin, and fusion in the sun is dominated by the proton–proton cycle.

EXAMPLE 11-10
The age of the sun is roughly 10 billion years. The power output of the sun is about 4×10^{26} watts. Show that the energy source of the sun cannot be of gravitational origin.

SOLUTION:
The gravitational binding energy (E_b) of the sun depends on how the mass is distributed. For an object of mass M and radius R, the order of magnitude of the energy is (Chap. 1, prob. 46)

$$E_b \approx \frac{GM^2}{R} .$$

For the sun, this energy is

$$E_b \approx \frac{\left(7 \times 10^{-11} \text{ m}^3 \cdot \text{kg}^{-1} \cdot \text{s}^{-2}\right)\left(2 \times 10^{30} \text{ kg}\right)^2}{7 \times 10^8 \text{ m}}$$
$$\approx 4 \times 10^{41} \text{ J}.$$

If the sun were contracting, then energy would be produced. Consider a small change in solar radius (ΔR). The energy produced (ΔE_b) by this change in radius is

$$\Delta E_b \approx \frac{GM^2 \Delta R}{R^2}.$$

If the solar energy were due to this gravitational origin, then the fractional change in radius *per second* would be

$$\frac{\Delta R}{R} \approx \frac{4 \times 10^{26} \text{ J}}{E_b} = \frac{4 \times 10^{26} \text{ J}}{4 \times 10^{41} \text{ J}} = 10^{-15}.$$

The solar radius would change by 10% in a mere 10^{14} seconds (about 3 million years). Furthermore, protons would be burned faster as the sun collapsed. The energy source of the sun cannot be of gravitational origin because its age is about 10 billion years. ∎

EXAMPLE 11-11

Show that the proton fusion provides the necessary energy to account for the solar luminosity and age.

SOLUTION:

Four protons combine to produce 25 MeV of solar energy. The rate (R) that protons are consumed is related to the power (P) output of the sun by

$$R = \frac{4P}{25 \text{ MeV}} = \frac{(4)(4 \times 10^{26} \text{ J/s})}{(25 \text{ MeV})(1.6 \times 10^{-13} \text{ J/MeV})}$$

$$\approx 4 \times 10^{38} \text{ s}^{-1}.$$

Assuming that the mass of the sun is dominated by protons, the number of protons (N) in the sun is about

$$N = \frac{2 \times 10^{30} \text{ kg}}{1.7 \times 10^{-27} \text{ kg}} \approx 10^{57}.$$

If the sun were to shine at its present rate, it would last for a time (T)

$$T = \frac{N}{R} = \frac{10^{57}}{4 \times 10^{38} \text{ s}^{-1}} \approx 3 \times 10^{18} \text{ s},$$

or about 100 billion years. Proton fusion can account for a long solar lifetime ($T \gg 10$ billion years). The sun will not last this long, however. As the protons are consumed, there is a reduction in radiation pressure that causes a gravitational contraction. The contraction causes an increase in temperature and an increase in the fusion

rate. The total time that the sun is expected to shine is roughly 15 billion years. (The evolution of the stars is discussed in Chapter 18.) ∎

Detecting the Neutrinos from the Sun

A detailed model has been made of the numerous nuclear processes occurring inside the sun. This standard solar model assumes that the source of energy production inside the sun is nuclear fusion and that the isotopic abundances are given by nuclear reactions. The solar model is "tuned" to match the observed solar luminosity. The solar model gives a prediction of the number of neutrinos produced as a function of their energy.

Table 11-5 lists the main processes inside the sun that produce neutrinos. Figure 11-15 shows the neutrino flux from the sun expected at the earth. These neutrinos are all electron neutrinos (ν_e). The flux is enormous, nearly 10^{15} m^{-2}s^{-1}! The size of this flux makes solar neutrino detection possible despite the weakness of the neutrino interaction. The neutrino interaction probability per nucleon is energy dependent. The rate of neutrinos from the sun is expressed in solar neutrino units (SNU). One SNU is defined to be the quantity of solar neutrinos that cause an interaction rate of one per second in a target of 10^{36} nucleons.

The pioneering solar neutrino detector was built in 1970 by Raymond Davis and collaborators. It consists of a large tank containing 615 tons of the cleaning fluid perchloroethylene located in the Homestake gold mine in South Dakota. The tank contains about 2.3×10^{30} chlorine atoms. Neutrinos interact by the process

$$\nu_e + {}^{37}\text{Cl} \rightarrow e^- + {}^{37}\text{Ar}. \tag{11.91}$$

TABLE 11-5
SUMMARY OF THE MAIN PROCESSES PRODUCING NEUTRINOS IN THE SUN.
The two-body final states produce a neutrino of fixed energy (E) and the three-body states produce a β spectrum with a maximum energy (E_{max}).

Process	Neutrino Energy
$p + p \rightarrow d + e^+ + \nu_e$	$E_{max} = 0.25$ MeV
$p + e^- + p \rightarrow d + \nu_e$	$E = 1.44$ MeV
${}^7\text{Be} + e^- \rightarrow {}^7\text{Li} + \nu_e$	$E = 0.86$ MeV
${}^7\text{Be} + e^- \rightarrow {}^7\text{Li*} + \nu_e$	$E = 0.38$ MeV
${}^8\text{B} \rightarrow {}^8\text{Be} + e^+ + \nu_e$	$E_{max} = 14.06$ MeV
${}^3\text{He} + p \rightarrow \alpha + e^+ + \nu_e$	$E_{max} = 18.77$ MeV

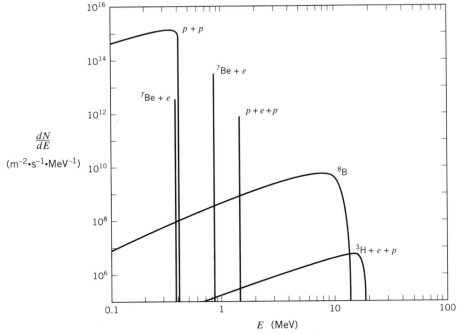

FIGURE 11-15 Calculated neutrino flux from the sun incident on the earth.
The neutrinos are produced by nuclear fusion inside the sun. The spectrum consists of β spectra from three-body final states and lines from two-body final states. The height of the lines is plotted so that the scale gives the flux in $m^{-2} \cdot s^{-1}$. Flux calculations are reported by J. N. Bahcall and R. K. Ulrich, "Solar Models, Neutrino Experiments, and Helioseismology," *Rev. Mod. Phys.* **60**, 297 (1988).

The production rate is about one argon atom every two days.

A few times per year the ^{37}Ar is removed by flushing the perchloroethylene with helium. The ^{37}Ar-He gas mixture is put in proportional ionization counters where the number of radioactive decays are counted. The ^{37}Ar decays by electron capture, the inverse reaction of its production:

$$e^- + {}^{37}\text{Ar} \rightarrow \nu_e + {}^{37}\text{Cl}. \quad (11.92)$$

The half-life of the ^{37}Ar is 35.1 days. The electron capture leaves a vacancy in the K shell of the chlorine atom. The electronic transition that fills the K shell vacancy produces an Auger electron with an energy of 2.8 keV. The number of argon decays gives the number of interacting solar neutrinos. The energy threshold for activation of the chlorine is 0.81 MeV. Examination of Figure 11-15 shows that the Homestake detector is sensitive mainly to the boron neutrinos. The measured solar neutrino rate (R) from the Homestake detector, averaged over 20 years is

$$R \approx 2 \text{ SNU}. \quad (11.93)$$

The prediction of the standard solar model is about 8 SNU. This discrepancy, known since the 1970s, is called the *solar neutrino puzzle,* and has been the subject of many theoretical and experimental investigations. It is tempting to speculate that some new physics has affected the neutrino flux; however, the simplest solution of the solar neutrino puzzle would be that the solar model is incorrect.

EXAMPLE 11-12
Estimate the flux of solar neutrinos at the surface of the earth.

SOLUTION:
Each 25 MeV produced by the proton–proton cycle produces two neutrinos. The total rate (R) of neutrino production is

$$R = \frac{(2)(4 \times 10^{26} \text{ J/s})}{(25 \text{ MeV})(1.6 \times 10^{-13} \text{ J/MeV})} \approx 2 \times 10^{38} \text{ s}^{-1}.$$

The distance (D) to the sun is

$$D = 1.5 \times 10^{11} \text{ m}.$$

The neutrinos are isotropic. The flux (f) of neutrinos on the earth is

$$f = \frac{R}{4\pi D^2} \approx \frac{2 \times 10^{38} \text{ s}^{-1}}{(4\pi)(1.5 \times 10^{11} \text{ m})^2} \approx 7 \times 10^{14} \text{ m}^{-2}\text{s}^{-1}.$$

∎

Nuclear Fission

An examination of the atomic mass number (A) dependence of the binding energy (Figure 11-3) shows that for heavy elements the binding energy per nucleon is increasing with decreasing A. Therefore, if we break apart a heavy nucleus into two lighter nuclei, a process called *fission*, the binding energy per nucleon of the end products will be greater than the binding energy per nucleon of the parent nucleus. There will be energy released in the fission process. Nuclear fission may be induced by a variety of particles, but the most effective is by slow neutrons which have a large cross section for interaction with a heavy nucleus.

> If a heavy nucleus is split into two lighter nuclei (fission), the sum of the masses of the two lighter nuclei is smaller than the mass of the heavy nucleus. Therefore energy is released when a heavy nucleus is broken apart .

In the fission of a heavy nucleus such as uranium or plutonium by slow neutrons, the nuclei that are produced tend to be of unequal mass. The lighter nucleus has an atomic mass number in the range 88–100 and the heavier nucleus has an atomic mass number in the range 130–142. The fission-product nuclei are neutron-rich (see Figure 11-2) and are unstable against beta-decay. Nuclear fission was first observed in 1939 by Otto Hahn and F. Strassmann by bombarding uranium with neutrons. The correct interpretation was provided by Lise Meitner and O. Frisch. The fission signature was the observation of the decays of the elements with atomic number 140:

$$^{140}_{54}\text{Xe} \rightarrow {}^{140}_{55}\text{Cs} + e^- + \bar{\nu}_e, \tag{11.94}$$

$$^{140}_{55}\text{Cs} \rightarrow {}^{140}_{56}\text{Ba} + e^- + \bar{\nu}_e, \tag{11.95}$$

$$^{140}_{56}\text{Ba} \rightarrow {}^{140}_{57}\text{La} + e^- + \bar{\nu}_e, \tag{11.96}$$

and

$$^{140}_{57}\text{La} \rightarrow {}^{140}_{58}\text{Ce} + e^- + \bar{\nu}_e. \tag{11.97}$$

Figure 11-16 shows the fission of ^{235}U observed in photographic emulsions. Fission of a single ^{235}U nucleus produces about 200 MeV of kinetic energy. The kinetic energy appears in the fission fragments as well as neutrinos, electrons, gamma rays and neutrons.

The number of neutrons produced in fission is distributed according to Poisson statistics. The mean number of neutrons produced per fission is 2.4 for ^{235}U and 2.9 for ^{239}Pu. The neutron energy distribution of the fission neutrons is peaked at about 1 MeV. Since more than one fission neutron is produced on the average, it is possible to generate a nuclear chain reaction, resulting in a violent explosion.

In some of the heaviest nuclei, such as ^{254}Cf, fission can occur spontaneously.

11-6 NUCLEAR SPIN

The spin quantum number of a nucleus is the net result of the addition of the spin quantum numbers of the neutrons and protons. This addition must be done according to the rules of quantum mechanics.

EXAMPLE 11-13
Determine the spins of the ^{3}He and ^{4}He nuclei.

SOLUTION:
The ^{3}He nucleus has two protons and one neutron. By the rules of addition of angular momenta, the spin of three nucleons could be 1/2 or 3/2. By the Pauli exclusion principle, the two protons must have opposite spin. (Two protons that have parallel spins must be in a higher energy state.) Therefore, the spin of the ^{3}He nucleus is 1/2.

The ^{4}He nucleus has two protons and two neutrons. By the rules of addition of angular momenta, the spin of four nucleons could be 0, 1, or 2. By the Pauli exclusion principle, the two protons must have opposite spin and the two neutrons must have opposite spin. Therefore, the spin of the ^{4}He nucleus is 0. ∎

Magnetic Moments of the Proton and Neutron

The intrinsic angular momentum of the proton ($s = 1/2$) gives the proton a magnetic moment. (The actual source of

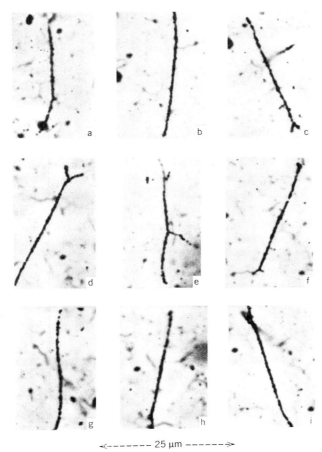

FIGURE 11-16 **Uranium fission observed in emulsion.**
Uranium is introduced into the emulsion which is bombarded by
slow neutrons. The fission products travel in opposite directions
appearing as a single "track." From C. F. Powell and G. Occhialini,
Nuclear Physics in Photographs, Oxford University Press (1947).
Reprinted by permission of Oxford University Press.

the intrinsic angular momentum of the proton is the
constituent quarks.) Analogous to the definition of the
Bohr magneton for the electron, we may define the *nuclear
magneton* (μ_N) to be the proton charge times $\hbar$ divided by
twice the mass,

$$\mu_N = \frac{e\hbar}{2m_p} = 3.15 \times 10^{-8} \text{ eV/T}. \quad (11.98)$$

We then write the proton magnetic moment (μ_p) as

$$\mu_p = g\mu_N, \quad (11.99)$$

where g is the *nuclear g* factor. The g factor tells us how
the actual magnetic moment of the proton differs from the

simple ratio ($e\hbar/2m_p$). The g factor of the proton is measured to be

$$g = 2.79. \quad (11.100)$$

Recall that for the electron, the g factor is very nearly equal
to 2. The g factor of the proton is significantly larger than
2 because the proton has structure.

The spin quantum number of the neutron is also 1/2.
The neutron has zero charge and we might expect the
neutron magnetic moment to be zero, however, the neutron has structure analogous to the proton. This results in
a relatively large magnetic moment (μ_n).

$$\mu_n = (-1.91)\mu_N. \quad (11.101)$$

The effect of the nuclear magnetic moment is to cause
a shift in energy level in a magnetic field. This is illustrated
in Figure 11-17a. The additional potential energy (V) due
to the interaction of the magnetic moment with the magnetic field is

$$V = -\boldsymbol{\mu} \cdot \mathbf{B}. \quad (11.102)$$

EXAMPLE 11-14

A sample of protons (for example, the hydrogen nuclei in
water) at room temperature is placed in a magnetic field of
1 tesla. Estimate the ratio of protons in the higher energy
state to those in the lower energy state.

SOLUTION:

The lower energy state is $m_s = 1/2$ and the higher energy
state is $m_s = -1/2$. The component of the magnetic moment
in the direction of the field (the z direction) is

$$\mu_z = g\mu_N m_s.$$

The energy difference (ΔE) between the two states is

$$\Delta E = g\mu_N B\left(\frac{1}{2}\right) - g\mu_N B\left(-\frac{1}{2}\right) = g\mu_N B$$

$$= (2.79)(3.15 \times 10^{-8} \text{ eV/T})(1\,\text{T})$$

$$= 8.79 \times 10^{-8} \text{ eV}.$$

The relative population of the nuclear spin states is given
by the Maxwell-Boltzmann factor:

$$\frac{N_{1/2}}{N_{-1/2}} = e^{\Delta E / kT}.$$

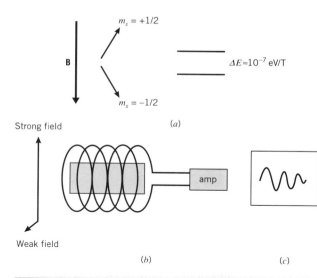

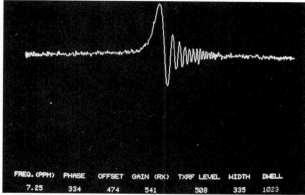

FIGURE 11-17 Nuclear magnetic resonance.
(*a*) Energy levels of a proton are split in an external magnetic field due to an interaction of the proton intrinsic angular momentum with the field. (*b*) Schematic of apparatus for observing transitions between the 2 states. (*c*) Nuclear magnetic resonance signal. (J. Prince/Photo Researchers)

The value of $\Delta E/kT$ is approximately

$$\frac{\Delta E}{kT} = \frac{8.79 \times 10^{-8} \text{ eV}}{\left(\frac{1}{40} \text{ eV}\right)} = 3.5 \times 10^{-6}.$$

Therefore,

$$\frac{N_{1/2}}{N_{-1/2}} = e^{\Delta E/kT} \approx 1 + \frac{\Delta E}{kT} = 1 + 3.5 \times 10^{-6}. \quad \blacksquare$$

Nuclear Magnetic Resonance

Recall the analysis of the precession of the electron spin. If a proton is placed in a magnetic field, it experiences a torque, which tends to align μ with **B**, resulting in a precession because μ is proportional to **B**. The frequency of precession is

$$\omega = \frac{2\mu_p B}{\hbar}. \quad (11.103)$$

The precession of a proton (or other nucleus) in a magnetic field may be detected by a technique called *nuclear magnetic resonance* (NMR).

EXAMPLE 11-15

Calculate the frequency of precession of a proton in a magnetic field of 10^{-4} T.

SOLUTION:
The angular frequency of precession is

$$
\begin{aligned}
\omega &= \frac{2\mu_p B}{\hbar} \\
&= \frac{(2)(2.79)(3.15 \times 10^{-8} \text{ eV/T})(10^{-4} \text{ T})}{6.58 \times 10^{-16} \text{ eV} \cdot \text{s}} \\
&= 2.67 \times 10^4 \text{ s}^{-1}. \quad \blacksquare
\end{aligned}
$$

The NMR technique is illustrated in Figure 11-17. A sample containing protons (hydrogen nuclei) is placed in a strong magnetic field. This field causes a small imbalance in the nuclear spin populations; there will be a tiny excess of protons with their spins aligned with the magnetic field. A second weak magnetic field is oriented perpendicular to the strong field. If the strong magnetic field is suddenly switched off, then the proton spins precess around the direction of the weak field. If a coil is placed around the proton sample, the nuclear precession induces an oscillating voltage in the coil. The voltage, although small, can be amplified and detected in an external circuit. The frequency of the voltage oscillation is equal to the frequency of precession.

Nuclear magnetic resonance has many uses. The precession frequency may be used to accurately determine the strength of a magnetic field. Nuclear magnetic resonance is also a powerful tool for measuring molecular structure because the electrons in the atoms of the molecules create internal magnetic fields. The presence of these internal magnetic fields and therefore information on the location of the atoms may be detected by nuclear magnetic resonance. Finally, magnetic resonance is an extremely useful medical tool for imaging the body.

11-7 THE MÖSSBAUER EFFECT

Consider an atomic transition from a state with energy E_2 to a state with energy E_1. We usually write the energy of the emitted photon (E_{photon}) as

$$E_{photon} = E_2 - E_1. \qquad (11.104)$$

The foregoing equation is an excellent approximation but it is not *exact* because momentum must be conserved. The atom must recoil with the same magnitude of momentum as the photon. The recoil momentum (p) of the atom is

$$p = \frac{E_{photon}}{c}. \qquad (11.105)$$

The kinetic energy of the atom is

$$E_k = \frac{p^2}{2M} = \frac{(pc)^2}{2Mc^2}, \qquad (11.106)$$

where M is the atomic mass. The energy scale of a photon emitted from an electronic transition between outer energy levels is electronvolts. The mass energy of the atom is much larger than the photon energy ($Mc^2 \gg pc$). This means that even for a light atom like hydrogen, the recoil kinetic energy of the atom is small because the mass is large. For an iron atom ($Mc^2 \approx 50$ GeV), the recoil kinetic energy is

$$E_k = \frac{(1\,eV/c)^2}{100\,GeV} \approx 10^{-11}\,eV. \qquad (11.107)$$

Since the atom gains this energy, the photon energy is smaller than $E_2 - E_1$ by an amount E_k. We usually can neglect such a small energy correction! Notice, however, that because the photon energy is slightly smaller than $E_2 - E_1$, we might expect that the emitted photon could not be readily reabsorbed by the same type of atom because the photon energy is a bit too small. The emitted photons *are* observed to be reabsorbed, however. This phenomenon is called *resonance absorption*. Resonance absorption occurs because of the natural width of the photon energy distribution, corresponding to the finite lifetime of the atomic state. The natural width and atomic lifetime are related by the uncertainty principle. The typical lifetime of an atomic transition is 10^{-8} seconds, corresponding to an energy width (Γ) of

$$\Gamma = \frac{\hbar}{\tau} = \frac{6.6 \times 10^{-16}\,eV \cdot s}{10^{-8}\,s} \approx 10^{-7}\,eV. \qquad (11.108)$$

The natural width is much larger than the recoil energy:

$$\Gamma \gg E_k. \qquad (11.109)$$

The natural spread in photon energies allows atoms to reabsorb the radiation.

Now consider the case of photons emitted from an excited nucleus (gamma transitions). The nuclear lifetimes are of the same order of magnitude as the atomic lifetimes because the same force, electromagnetism, is responsible for the decays. The photon energies from nuclear transitions, however, are about 10^5 times greater than photons from atomic transitions between outer electron levels. The recoil energy of the nucleus, again taking the iron mass, is

$$E_k = \frac{(pc)^2}{2Mc^2} = \frac{(10^5\,eV)^2}{100\,GeV} \approx 10^{-1}\,eV. \qquad (11.110)$$

In nuclear systems, the kinetic energy of the recoil nuclei are typically much greater than the width:

$$E_k \gg \Gamma. \qquad (11.111)$$

This means that photons emitted in nuclear decay ordinarily do not have enough energy to be reabsorbed by the same type of nucleus. Resonance absorption cannot readily occur in nuclei.

The phenomenon of resonance absorption is illustrated in Figure 11-18. A transition occurs between two energy levels causing a photon of energy (E) to be emitted. The average emitted photon energy is smaller than the energy difference between the energy levels (E_0), but the photon has an energy width given by the uncertainty principle. For a photon to be absorbed, it must have an energy to allow for the recoil. The absorption spectrum has the same shape as the emission spectrum. The amount of overlap determines the amount of resonance absorption. For atomic systems, the overlap is complete. For nuclear systems, the overlap is zero.

In 1957, Rudolf Mössbauer discovered that nuclear transitions may occur with negligible nuclear recoil if the decaying nucleus is embedded in a crystal lattice. In this case the entire lattice takes up the recoil momentum. Since the mass of the crystal (M_c) can be huge, the crystal as a whole conserves momentum and gains essentially zero kinetic energy. The kinetic energy gained by the recoiling crystal is

$$E_k = \frac{(pc)^2}{2M_c c^2} \approx 0. \qquad (11.112)$$

This means that like the atomic system, there can be complete overlap between the emission and absorption energy spectra, and resonance absorption can occur. The nuclear system first studied by Mössbauer was ^{191}Ir. The nuclear energy levels of ^{191}Ir are indicated in Figure 11-19.

EXAMPLE 11-16

Estimate the size of the crystal necessary to cause nuclear resonance absorption in ^{191}Ir. The lifetime of the ^{191}Ir* state is 1.9×10^{-10}s and the emitted photon has an energy of 129 keV.

SOLUTION:

From the uncertainty principle, the width of the ^{191}Ir* state is

$$\Gamma = \frac{\hbar}{\tau} = \frac{6.6 \times 10^{-16} \text{ eV} \cdot \text{s}}{1.3 \times 10^{-10} \text{ s}} \approx 3.5 \times 10^{-6} \text{ eV}.$$

A significant amount of resonance absorption will occur when the recoil kinetic energy is

$$E_k \approx \Gamma.$$

The recoil kinetic energy may be found from

$$E_k = \frac{(pc)^2}{2M_c c^2} \approx \Gamma,$$

so the mass of the crystal is given by

$$M_c c^2 = \frac{(pc)^2}{2\Gamma} = \frac{(129 \text{ keV})^2}{(2)(3.5 \times 10^{-6} \text{ eV})}$$

$$\approx 2 \times 10^{15} \text{ eV} = 2 \times 10^6 \text{ GeV}.$$

The mass energy of ^{191}Ir is about 178 GeV. Therefore, the number of atoms in the crystal necessary to cause resonance absorption is

$$N \approx \frac{2 \times 10^6 \text{ GeV}}{178 \text{ GeV}} \approx 10^4.$$

Only a small crystal is necessary. ■

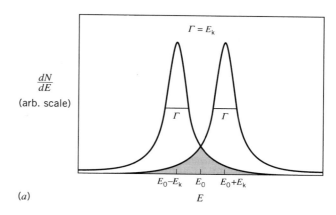

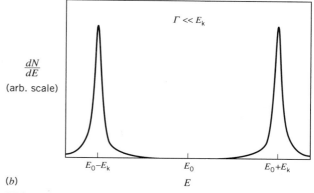

(a)

(b)

FIGURE 11-18 Resonance absorption.
A transition occurs between two energy levels, causing a photon of energy (E) to be emitted. The photon energy is slightly different from the difference between the energy levels due to the recoil energy (E_k) and the natural width due to the uncertainty principle. (a) Photon energy distribution for the case $\Gamma = E_k$. The resonance absorption probability is proportional to the amount of overlap of the two curves. (b) Photon energy distribution for the case $\Gamma \ll E_k$.

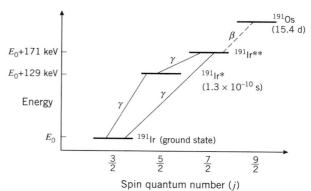

FIGURE 11-19 Nuclear energy levels for ^{191}Ir.

The experimental setup of Mössbauer is shown in Figure 11-20. The radioactive source is ^{191}Os, which beta decays into ^{191}Ir:

$$^{191}\text{Os} \rightarrow \,^{191}\text{Ir}^* + e^- + \bar{v}_e, \qquad (11.113)$$

and the excited iridium decays by gamma emission

$$^{191}\text{Ir}^* \rightarrow \,^{191}\text{Ir} + \gamma. \qquad (11.114)$$

The energy of the photon is 129 keV. The half-life of the ^{191}Ir* decay is 1.3×10^{-10}s, corresponding to a natural fractional width of

$$\frac{\Gamma}{E} = 4 \times 10^{-11}. \qquad (11.115)$$

Normally, the energy shift due to the motion of the nucleus will be larger than the natural width by several orders of magnitude. Mössbauer discovered that if the ^{191}Os is embedded in a crystal, then a small fraction of the decays have a negligible nuclear recoil. The recoilless photon emission produces an extremely narrow photon "line." The photons from the decay of ^{191}Ir* were detected by Mössbauer by resonance absorption in an iridium absorber. The process in the absorber is

$$^{191}\text{Ir} + \gamma \rightarrow \,^{191}\text{Ir}^*. \qquad (11.116)$$

The absorber must also be a crystal for resonance absorption to occur (see Figure 11-18).

In the Mössbauer experiment, the width of the photon source may be measured by giving the source some motion so that the line is broadened by the Doppler effect.

The absorption fraction is then measured as a function of the speed of the source. When the photon has the wrong energy, its absorption probability is reduced. The data of Mössbauer are shown in Figure 11-21. A speed of only a few centimeters per second is sufficient to destroy the absorption.

EXAMPLE 11-17

Calculate the speed of the source necessary to destroy the resonance absorption in ^{191}Ir.

SOLUTION:
The natural line width is

$$\Gamma = \frac{\hbar}{\tau} = \frac{6.6 \times 10^{-16} \text{ eV} \cdot \text{s}}{1.3 \times 10^{-10} \text{ s}} = 3.5 \times 10^{-6} \text{ eV}.$$

The fractional width is

$$\frac{\Gamma}{E} = \frac{3.5 \times 10^{-6} \text{ eV}}{129 \text{ keV}} = 2.7 \times 10^{-11}.$$

The motion of the decaying ^{191}Ir* nucleus causes a Doppler broadening of the photon. Consider motion of the ^{191}Ir* in the direction of the emitted photon. The Doppler formula for the Lorentz boosted photon energy (E') is

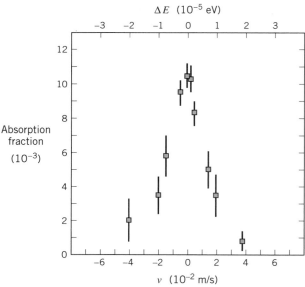

FIGURE 11-21 Recoilless gamma emission and absorption from ^{191}Ir.
From R. L. Mössbauer, *Naturwissenschaften* **45**, 538 (1958).

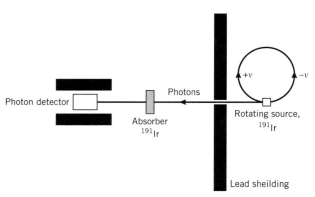

FIGURE 11-20 Experimental arrangement of Mössbauer.

$$E' = E\sqrt{\frac{1+\beta}{1-\beta}} = E\gamma(1+\beta) \approx E(1+\beta).$$

The fractional change in the photon energy is

$$\frac{E'-E}{E} = \beta.$$

The resonance absorption is destroyed when β is equal to a few times Γ/E. For $\beta = \Gamma/E$, we have

$$\beta = \frac{v}{c} = 2.7\times10^{-11}.$$

This corresponds to a speed of

$$v = (3\times10^8 \text{ m/s})(2.7\times10^{-11}) \approx 10^{-2} \text{ m/s}.$$

Speeds of centimeters per second destroy the resonance absorption! ■

The unprecedented resolution of the Mössbauer technique has many interesting applications in physics. Perhaps the most ingenious of these is contained in the work of Robert V. Pound and Glen A. Rebka, Jr. The resolution in the original experiment of Mössbauer is limited by the natural width of the ^{191}Ir* state due to its short lifetime. Pound and Rebka studied the Mössbauer effect in ^{57}Fe, which has an excited state with a much longer lifetime. An energy level diagram of ^{57}Fe is shown in Figure 11-22. The 14.4-keV photon originates from the decay of the ^{57}Fe* state, which has a lifetime of about 10^{-7}s. This lifetime corresponds to a natural width of about $\Gamma \approx 10^{-8}$ eV. The fractional width is

$$\frac{\Gamma}{E} \approx 10^{-12}. \tag{11.117}$$

Pound and Rebka realized that if they could achieve this resolution, then they could measure the change in energy of the 14.4-keV photon caused by its gravitational interaction with the earth! This experiment provided an important fundamental test of the theory of general relativity.

In 1959, Pound and Rebka set out to fabricate a high-resolution Mössbauer source appropriate for detection of the gravitational interaction of the photon. In this section we shall describe the ^{57}Fe source; the results of the gravitational experiment will be discussed in Chapter 19. The key to high resolution is to make a source and absorber that have identical energy levels. In the case of ^{57}Fe, a shift in energy levels caused by the intrinsic angular momentum of the nuclei is enough to cause significant broadening of the photon line observed by resonant absorption. The experimenters made an ^{57}Fe source by electroplating ^{57}Co atoms onto one face of a thin sheet of iron. The source was then heated to 1220 K for an hour. The heat treatment caused diffusion of the cobalt into the iron to a mean distance of about 300 nanometers or about 1000 atomic distances. The absorber consisted of a thin (about 14 μm) sheet of iron that was also annealed. The heat treatment was discovered to be a crucial step for achieving high resolution. The source was cemented on a moving-coil magnetic transducer (a "loudspeaker") that was driven at 10 Hz. The resulting Mössbauer spectrum obtained by Pound and Rebka is shown in Figure 11-23. The resonant absorption is a maximum of about 17.5% (compare to Figure 11-21). The line shape fits that of a Breit-Wigner with a full width of

$$\Gamma = 1.6\times10^{-8} \text{ eV}, \tag{11.118}$$

or

$$\frac{\Gamma}{E} = 1.1\times10^{-12}. \tag{11.119}$$

The contribution to the width of the photon line from finite lifetime of the ^{57}Fe* state is twice the natural width, or about 9×10^{-8} eV. Thus, the experimental resolution is only about a factor of 2 greater than the contribution from the natural width.

The resolution of the Pound-Rebka apparatus was good enough to observe hyperfine splitting, as shown in Figure 11-23b. Hyperfine splitting is caused by the interaction of the magnetic moment of the nucleus with the internal

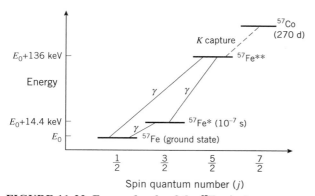

FIGURE 11-22 Energy levels of the ^{57}Fe nucleus.

magnetic field caused by the electrons. The intrinsic angular momentum quantum number of ^{57}Fe* is $j = 3/2$. Thus, ^{57}Fe* is expected to be split into four states by the nuclear hyperfine interaction. The intrinsic angular momentum quantum number of ^{57}Fe is $j = 1/2$. The ground state, ^{57}Fe, is expected to be split into two states by the nuclear hyperfine interaction. Thus, six photon transitions may be expected to be present from

$$^{57}\text{Fe}^* \rightarrow {}^{57}\text{Fe} + \gamma. \qquad (11.120)$$

The observed spectrum can be quite complicated because it depends on how the energy levels in the source and absorber overlap. The data of Figure 11-23b show that the

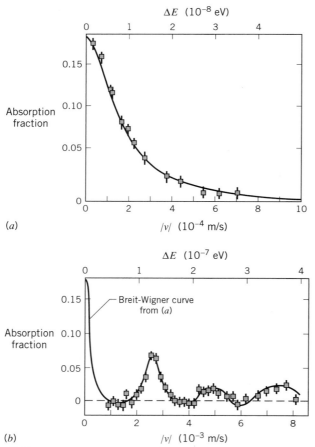

(a)

(b)

FIGURE 11-23 Mössbauer effect in ^{57}Fe.
(a) Principal line. (b) Satellite lines are observed due to the interaction of the nuclear magnetic moment with the internal magnetic field generated by electrons. From R. V. Pound and G. A. Rebka Jr., *Phys. Rev. Lett.* **3**, 554 (1959).

splitting between the states is about 1.3×10^{-7} eV. This technique provided unprecedented resolution for the study of nuclear structure.

11-8 PASSAGE OF RADIATION THROUGH MATTER

When radiation passes through matter, it deposits energy. The amount of energy deposited depends on the type of radiation, the energy of the radiating particles, and the type of absorbing matter.

Ionization Loss

Charged particles lose energy in matter by their electromagnetic interactions with atomic electrons. This process is called energy loss by ionization. This change in energy per unit distance traversed is called dE/dx. The ionization loss depends on the speed of the particle as well as properties of the medium. At small speeds, dE/dx is inversely proportional to the square of the speed. At relativistic speeds, dE/dx passes through a minimum and then grows logarithmically. The reason for the logarithmic growth is that the Lorentz contraction of the x coordinate causes the incoming particle to see a greater density of atomic electrons. The formula for dE/dx, which is discussed in more detail in Chapter 16, is called the *Bethe-Bloch* equation:

$$\frac{dE}{dx} = -\frac{CZ\rho}{A\beta^2}\left[\ln\left(\frac{2m\gamma^2\beta^2c^2}{I}\right) - \beta^2\right], \quad (11.121)$$

where $C = 0.0307$ MeV·m²/kg. A useful approximation of the ionization loss is

$$-\left(\frac{dE}{dx}\right)_{\text{min}} \approx \left(0.2\,\text{MeV}\cdot\text{m}^2/\text{kg}\right)\rho. \quad (11.122)$$

EXAMPLE 11-18
A 10-GeV cosmic ray muon passes through a person, from head to toe. How much energy does the muon deposit inside the person?

SOLUTION:
The muon is relativistic ($\gamma = E/mc^2 \approx 100$) and is approximately minimum ionizing. Estimate the density of the person to be that of water

$$\rho \approx 10^3 \text{ kg/m}^3.$$

The rate of energy loss per distance traveled by the muon inside the person is

$$-\left(\frac{dE}{dx}\right) \approx \left(0.2\,\text{MeV}\cdot\text{m}^2/\text{kg}\right)\rho$$
$$\approx \left(0.2\,\text{MeV}\cdot\text{m}^2/\text{kg}\right)\left(10^3\,\text{kg/m}^3\right)$$
$$= 200\,\text{MeV/m}.$$

Estimate the height, Δx, of the person to be

$$\Delta x \approx 1.70\,\text{m}.$$

The energy deposited in the person by the muon is

$$\Delta E \approx -\frac{dE}{dx}\Delta x \approx \left(200\,\text{MeV/m}\right)\left(1.70\,\text{m}\right)$$
$$\approx 340\,\text{MeV}. \qquad \blacksquare$$

Radiation Units and Doses

The SI unit of radioactivity or *activity* is the becquerel (Bq), named after the discoverer of nuclear disintegrations:

$$1\,\text{Bq} = 1\,\text{decay/s}. \qquad (11.123)$$

The activity of radioactive sources is commonly stated in units of the curie (Ci),

$$1\,\text{Ci} = 3.7\times10^{10}\,\text{Bq}. \qquad (11.124)$$

One curie is the approximate activity of one gram of radium. The activity includes the decay not only of radium but also that of the daughter nuclei. The heavily ionizing tracks produced by alpha particles from radium and its daughter decays are illustrated in Figure 11-24. The experimenters put a tiny speck of radium on a photographic plate and developed it a few days later. The activity of 1 kg of natural uranium is about 1.3×10^{13} Bq and the activity of 1 kg natural thorium is about 4.11×10^{12} Bq.

The energy absorbed per mass is called the *radiation absorbed dose*. The SI unit is the gray (Gy):

$$1\,\text{Gy} = 1\,\text{J/kg} = 6.24\times10^{12}\,\text{MeV/kg}. \quad (11.125)$$

Some types of energy deposits do more biological damage then others. For example, a 5-MeV alpha particle does more biological damage than a 5-MeV electron. We take this into account by defining the *radiation dose equivalent*

as the radiation absorbed dose multiplied by a damage factor (Q), which depends on the type of radiation. The more dangerous the radiation is to your health, the higher the Q. Values of Q are given in Table 11-6. The SI unit of the radiation dose equivalent is the sievert (Sv):

$$1\,\text{Sv} = \left(1\,\text{Gy}\right)\times Q. \qquad (11.126)$$

The average annual background radiation received by a person on the earth is in the range of 0.4 to 4 millisieverts (mSv). The maximum recommended occupational dose is 50 mSv per year. A lethal dose is about 3 sieverts.

EXAMPLE 11-19

A person works for 4 hours in a room that contains an unshielded 10-mCi ^{60}Co source. The person keeps a certain distance from the source. At what distance is the estimated radiation dose in 4 hours equal to one-hundredth of the maximum permitted annual occupational dose?

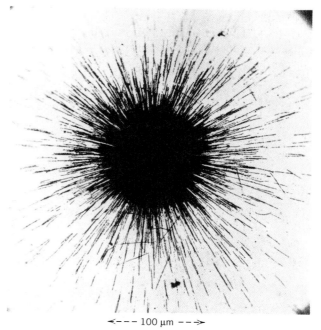

$\longleftarrow --- 100\,\mu\text{m} --- \longrightarrow$

FIGURE 11-24 Alpha particles from radium observed in emulsion.

A "speck" of radium, too small to be seen under a microscope, was inserted into emulsion by first depositing the radium on a needle and then shaking the needle over the photographic plate. The emulsion was developed after a few days. From C. F. Powell and G. Occhialini, *Nuclear Physics in Photographs*, Oxford University Press (1947). Reprinted by permission of Oxford University Press.

TABLE 11-6
VALUES OF Q FOR VARIOUS TYPES OF RADIATION.

Radiation	Q
Gamma	1
Electron	1
Protons (10 MeV)	1
Protons (1 GeV)	2
Neutrons (thermal)	3
Neutrons (fast)	10
Alpha	20

SOLUTION:

Cobalt-60 is a beta emitter. The decay is

$$^{60}\text{Co} \rightarrow {}^{60}\text{Ni} + e^- + \bar{\nu}_e .$$

(^{60}Ni is stable.) The typical kinetic energy of the electron is 1 MeV. The angular distribution of the electrons is isotropic. The activity of the source is

$$10\,\text{mCi} = \left(10^{-2}\,\text{Ci}\right)\left(3.7 \times 10^{10}\,\text{Bq/Ci}\right) = 3.7 \times 10^8\,\text{Bq} .$$

The total energy that appears in electrons in 4 hours is

$$E \approx \left(3.7 \times 10^8\,\text{s}^{-1}\right)\left(4\,\text{h}\right)\left(\frac{3600\,\text{s}}{1\,\text{h}}\right)\left(1\,\text{MeV}\right)$$

$$= 5.3 \times 10^{12}\,\text{MeV} .$$

If all of this energy were absorbed by the person of mass 60 kg, the radiation absorbed dose would be

$$\frac{E}{60\,\text{kg}} = \left(\frac{5.3 \times 10^{12}\,\text{MeV}}{60\,\text{kg}}\right)\left(\frac{1\,\text{Gy}}{6.24 \times 10^{12}\,\text{MeV}/\text{kg}}\right)$$

$$= 0.014\,\text{Gy} .$$

Since the Q value of beta radiation is unity, the maximum radiation dose is 0.014 Sv. To get one-hundredth of the maximum annual dose, the fraction of the electrons absorbed would be

$$f = \frac{5 \times 10^{-4}\,\text{Sv}}{0.014\,\text{Sv}} = 3.6 \times 10^{-2} .$$

At a distance R, a person with a cross-sectional area (A) absorbs the fraction

$$f = \frac{A}{4\pi R^2} ,$$

or

$$R = \sqrt{\frac{A}{4\pi f}} .$$

Let us estimate the cross-sectional area of the person to be $A \approx 0.5\,\text{m}^2$. The distance that the person must keep from the source to limit the dose to 0.5 mSv is

$$R = \sqrt{\frac{0.5\,\text{m}^2}{4\pi\left(3.6 \times 10^{-2}\right)}} \approx 1\,\text{m} .$$

Notice that the radiation dose varies as the inverse square of the distance to the source. The person is protected by the distance to the source because of the "solid angle" factor. A person 10 m away gets 100 times less radiation (5 μSv). At the other extreme, a person just next to the source for 4 hours will absorb roughly one-half of all the radiation, or about 7 mSv. ∎

Radon gas can be a particularly dangerous health hazard. Radon is produced naturally on earth as part of the uranium series. The fraction of radon in air is about one part in 10^{21}, on the average. This corresponds to a radiation dose to the lungs of about 0.1 mSv. Radon becomes a health hazard when it accumulates in mines or in the basements of buildings. Part of the radon decay chain is

$$^{222}\text{Ra} \rightarrow {}^{218}\text{Po} + \alpha , \qquad (11.127)$$

$$^{218}\text{Po} \rightarrow {}^{214}\text{Pb} + \alpha , \qquad (11.128)$$

$$^{214}\text{Pb} \rightarrow {}^{214}\text{Bi} + e^- + \bar{\nu}_e , \quad (11.129)$$

$$^{214}\text{Bi} \rightarrow {}^{214}\text{Po} + e^- + \bar{\nu}_e , \quad (11.130)$$

and

$$^{214}\text{Po} \rightarrow {}^{210}\text{Pb} + \alpha . \qquad (11.131)$$

The daughter decay products, which are normally solids, get stuck to dust particles in the air. The half-lifes of ^{218}Po, ^{214}Pb, ^{214}Bi, and ^{214}Po are all relatively short. Lead-210 has a half-life of 22.3 years. The three alpha particles dominate the radiation damage from radon to your lungs because of their large Q factors. Radon is considered to be a health hazard if its presence in air causes an activity greater than about 150 Bq/m^3.

EXAMPLE 11-20

A person works for an hour in a basement that contains radon gas. The activity of the air in the basement is 300 Bq/m^3. Estimate the radiation dose equivalent to the lungs that the person receives during this time.

SOLUTION:

Each radon decay produces five disintegrations. The ^{222}Ra decay rate (R) is

$$R = \frac{300 \, \text{Bq/m}^3}{5} = 60 \, \text{Bq/m}^3.$$

The half-life ($\tau_{1/2}$) of ^{222}Ra is 3.92 days. The number of ^{222}Ra per volume (n) is

$$
\begin{aligned}
n &= \frac{R\tau_{1/2}}{\ln 2} \\
&= \frac{\left(60 \, \text{Bq/m}^3\right)\left(3.92 \, \text{d}\right)\left(24 \, \text{h/d}\right)\left(3600 \, \text{s/h}\right)}{0.693} \\
&= 2.9 \times 10^7 \, \text{m}^{-3}.
\end{aligned}
$$

The total number density of air molecules is

$$V = \frac{6.02 \times 10^{23}}{0.0224 \, \text{m}^3} = 2.7 \times 10^{25} \, \text{m}^{-3}.$$

The fraction of radon in the air is about one part in 10^{18} or 1000 times the normal level. The annual dose to the lungs at this rate would be

$$\left(10^3\right)\left(0.1 \, \text{mSv}\right) = 100 \, \text{mSv}.$$

In one hour, the dose is

$$\frac{\left(100 \, \text{mSv}\right)\left(3600 \, \text{s}\right)}{3.16 \times 10^7 \, \text{s}} \approx 10 \, \mu\text{Sv}.$$

A prolonged exposure at this rate is hazardous because the radiation dose is concentrated in the lungs. ∎

EXAMPLE 11-21

A detector for a particle physics experiment is fashioned out of a thin wafer of silicon. The silicon detector is designed to operate for 10^7 seconds in a flux of energetic charged particles of 10^{10} per square meter per second. Make an estimate of the radiation absorbed dose that the silicon will get in the experiment.

SOLUTION:

Let A and Δx be the cross-sectional area and thickness of the silicon wafer and let ρ be the density of silicon. The energy deposited in the wafer in 10^7 seconds is

$$E = \left(0.2 \, \text{MeV} \cdot \text{m}^2/\text{kg}\right)\rho\Delta x\left(10^7 \, \text{s}\right)\left(10^{10} \, \text{m}^{-2}\right)A.$$

The mass of the wafer is

$$M = \rho\Delta x A.$$

The radiation absorbed dose is

$$
\begin{aligned}
\frac{E}{M} &\approx \left(0.2 \, \text{MeV} \cdot \text{m}^2/\text{kg}\right)\left(10^7 \, \text{s}\right)\left(10^{10} \, \text{m}^{-2}\right) \\
&= 2 \times 10^{16} \, \text{MeV/kg},
\end{aligned}
$$

or

$$
\begin{aligned}
\frac{E}{M} &\approx \left(2 \times 10^{16} \, \text{MeV/kg}\right)\left(\frac{1 \, \text{Gy}}{6.24 \times 10^{12} \, \text{MeV/kg}}\right) \\
&= 3.2 \, \text{kGy}.
\end{aligned}
$$

CHAPTER 11: PHYSICS SUMMARY

- The nucleus is made up neutrons and protons (nucleons) bound together by the strong interaction. Neutrons and protons have an identical strong interaction behavior. The spin quantum number of both the neutron and proton is

$$s = \frac{1}{2}.$$

- For small values of Z, the most stable nuclear configurations are those with approximately equal numbers of neutrons and protons. The reason for this is that the neutrons and protons each obey the Pauli exclusion principle. At large values of Z, the stable nuclei have more neutrons than protons. This is due to the electromagnetic repulsion of the protons.

- Nuclei with a magic number (2, 8, 20, 28, 50, 82, or 126) of neutrons or protons are observed to be especially stable. These magic numbers can be accounted for in the shell model of the nucleus, where each nucleon is assumed to move in a stable orbit as an independent particle.

- Fusion is the combining of two elements lighter than iron to make a single nucleus, resulting in an increase in the binding energy per nucleon. Fission is the splitting of an element heavier than iron, resulting in an increase in the binding energy per nucleon.

- The three types of radioactive decays are alpha, beta, and gamma. The alpha decay process is

$$_Z^A P \rightarrow _{Z-2}^{A-4} D + \alpha$$

where P and D are the parent and daughter nuclei. There are two types of beta decays, beta-minus

$$_Z^A P \rightarrow _{Z+1}^{A} D + e^- + \bar{\nu}_e$$

and beta-plus

$$_Z^A P \rightarrow _{Z-1}^{A} D + e^+ + \nu_e .$$

Gamma decays are transitions between an excited nuclear state ($P*$) and a lower energy state (P):

$$_Z^A P* \rightarrow _Z^A P + \gamma .$$

- The Mössbauer effect occurs when a radioactive nucleus is embedded in a crystal. Nuclear gamma decays can occur without nuclear recoil, resulting in extremely narrow line widths.

- One becquerel is the unit of radioactivity:

$$1\,\text{Bq} = 1\,\text{decay/s} .$$

One curie is the approximate activity from one gram of radium:

$$1\,\text{Ci} = 3.7 \times 10^{10}\ \text{Bq} .$$

- The radiation absorbed dose is the energy absorbed per mass.

$$1\,\text{Gy} = 1\,\text{J/kg} = 6.24 \times 10^{12}\ \text{MeV/kg} .$$

- The radiation dose equivalent is the radiation absorbed dose multiplied by a factor (Q) that depends on the type of radiation,

$$1\,\text{Sv} = (1\,\text{Gy}) \times Q .$$

Radiation that is more harmful to your health has a higher Q factor. The Q factor ranges from unity for electrons to about 20 for alpha particles.

REFERENCES AND SUGGESTIONS FOR FURTHER READING

F. Ajzenberg-Selove and E. K. Warburton, "Nuclear Spectroscopy," *Phys. Today* **36**, No. 11, 26 (1983).

P. Armbruster and G. Münzenberg, "Creating Superheavy Elements," *Sci. Am.* **260**, No. 5, 66 (1989).

John N. Bahcall, "The Solar Neutrino Problem," *Sci. Amer.* **262**, No. 5, 54 (1990).

J. M. Blatt and V. Weisskopf, *Theoretical Nuclear Physics*, Wiley (1952).

J. Chadwick, "The Existence of a Neutron," *Proc. of the Royal Soc.* **136A,** 692 (1932).

E. Fermi, "The Nucleus" *Phys. Today* **5**, No. 3, 6 (1952).

H. Frauenfelder and E. M. Henley, *Subatomic Physics,* Prentice-Hall (1991).

R. E. M. Hedges and J. A. J. Gowlett, "Radiocarbon Dating by Accelerator Mass Spectrometry," *Sci. Am.* **254**, No. 1, 100 (1986).

J. Hans D. Jensen, "The History of the Theory of Structure of the Atomic Nucleus," *Sci.* **147**, 1419 (1965).

A. D. Krisch, "Collisions Between Spinning Protons," *Sci. Am.* **257**, No. 2, 42 (1987).

W. F. Libby, *Radioactive Carbon Dating*, Univ. of Chicago Press (1952).

M. G. Mayer, "The Structure of the Nucleus," *Sci. Am.* **184**, No. 3, 22 (1951).

M. Goeppert Mayer and J. H. D. Jensen, *Elementary Theory of Nuclear Shell Structure,* Wiley (1955).

A. C. Melissinos, *Experiments in Modern Physics,* Academic Press (1966).

M. K. Moe and S. P. Rosen, "Double-Beta Decay," *Sci. Am.* **261**, No. 5, 48 (1989).

E. Segrè, "The Discovery of Nuclear Fission," *Phys. Today* **42**, No. 7, 38 (1990).

E. Segrè, *Nuclei and Particles*, Benjamin (1977).

C. P. Slichter, *Prinicples of Magnetic Resonance,* Springer (1978).

L. Wolfenstein and E. W. Beier, "Neutrino Oscillations and Solar Neutrinos," *Phys. Today* **42**, No. 7, 28 (1990).

QUESTIONS AND PROBLEMS

Discovery of the neutron

1. Recall that in the discovery of the neutron, Chadwick ruled out Compton scattering partly on the basis of the observed scattering rate. Why is the neutron–proton cross section much greater than the photon–proton cross section?

2. In the decay of the neutron,

$$n \rightarrow p + e^- + \bar{\nu}_e,$$

what can be said about the relative wavelengths of the three particles?

3. An unknown particle scatters from the nitrogen nucleus and the nucleus is measured to recoil with a maximum kinetic energy of 1.4 MeV. (a) Calculate the energy of the incident particle if the particle is a photon. (b) Calculate the energy of the incident particle if the particle is a neutron.

Basic properties of the nucleus

4. Why does the binding energy per nucleon increase with increasing atomic mass number (A) for small values of A? Why does the binding energy per nucleon decrease with increasing atomic mass number for large values of A?

5. Why does carbon-12 have a greater binding energy than nitrogen-12?

6. The binding energy per nucleon of ^{235}U is 7.59 MeV. (a) Calculate the mass of the uranium nucleus. (b) Calculate the mass of the ^{235}U atom.

7. Calculate the binding energy per nucleon for all the stable isotopes of manganese, iron, and cobalt. Which nucleus has the largest binding energy per nucleon?

8. Which nucleus do you expect will have a larger binding energy, tritium (^{3}H) or ^{3}He? Explain. Calculate the binding energy of each.

9. Which nucleus do you expect to have a larger binding energy, ^{64}Ni or ^{64}Zn? Explain. Calculate the binding energy of each.

10. Which nucleus do you expect to have the largest binding energy, ^{12}Be, ^{12}B, ^{12}C, or ^{12}N? Explain. Calculate the binding energy of each.

Nuclear models

11. Calculate the binding energies of ^{55}Fe, ^{57}Co, and ^{58}Ni. Compare the actual binding energies to the Weizsaecker formula of the liquid drop model.

Radioactive decays

12. Why are there β^- decays in the four radioactive series? Why are there no β^+ decays?

13. Why do two of the radioactive series start with uranium isotopes?

14. Why is the lifetime of ^{237}Np so much shorter than the lifetimes of leading nuclei in the other three series?

15. If the activity of a substance drops by a factor of 32 in 5 seconds, what is the radioactive half-life?

16. Can a given nucleus have both β^- and a β^+ decay modes?

17. Why are the photons from nuclear gamma transitions so much larger in energy than photons from atomic transitions?

18. A piece of bone found at the Palace of Knossos contains 100 grams of carbon and has a ^{14}C activity of 1000 per minute. How long has this chap been dead?

19. (a) What is the initial ^{238}U decay rate from one gram of ^{238}U? (b) What is the initial ^{234}Th decay rate from the same sample?

Nuclear reactions

20. (a) A neutron is added to ^{16}O to make the stable isotope ^{17}O. Calculate the binding energy of the neutron. (b) Another neutron is added to ^{17}O to make the stable isotope ^{18}O. Calculate the binding energy of the neutron.

21. Iron-55 decays by electron capture. (a) What are its end products? Is the resulting nucleus stable? (b) The capture process results in an atomic x ray. Make an estimate of the energy of the x ray. (c) Estimate the radiation dose equivalent (sieverts) that a person receives in 1 hour working 3 meters from a iron-55 source that has an activity of 5 mCi.

22. For each of the following reactions, identify the particle "X":

(a) ^{23}Mg $\rightarrow e^- + \nu_e + X$, (b) $X \rightarrow ^{186}$Os $+ \alpha$,

(c) ^{11}C $\rightarrow e^+ + \nu_e + X$, (d) ^{12}N $\rightarrow ^{12}$C $+ \nu_e + X$,

and (e) $p + X \rightarrow ^2$H $+ e^+ + \nu_e$.

23. Calculate the kinetic energy of an alpha particle from the decay of ^{239}Pu.

24. The energy (Q) released in beta decay is given by

$$Q = M_P c^2 - M_D c^2 - mc^2,$$

where M_P and M_D are the masses of the parent and daughter *nuclei,* and m is the electron or positron mass. If one uses the atomic masses (M_P' and M_D') then show that

$$Q = M_P'c^2 - M_D'c$$

for β^- decay, and

$$Q = M_P'c^2 - M_D'c^2 - 2mc^2$$

for β^+ decay.

25. Why does the proton cycle dominate over the carbon cycle at lower temperatures?

Nuclear spin

26. Show that the decay $n \rightarrow pe$ cannot conserve angular momentum.

27. Make a prediction of the spin quantum numbers of the following nuclei (a) ^{12}C, (b) ^{14}N, (c) ^{16}O, and (d) ^{19}F.

The Mössbauer effect

28. Estimate the size of the crystal necessary to cause resonant absorption in ^{57}Fe.

29. Calculate the source speed that corresponds to change in energy of (a) 10^{-5} eV for ^{191}Ir, and (b) 10^{-8} eV for ^{57}Fe.

Passage of radiation through matter

30. Is a 1-curie radioactive source dangerous to handle? Why or why not?

31. A piece of earth one-half meter thick with an area of one square kilometer contains roughly 1 gram of radium, corresponding to an activity of 1 curie. Does this radiation constitute a health hazard?

32. Which is more of a radiation safety problem, radioactive waste with a long half-life or radioactive waste with a short half-life? Why?

33. Can a person absorb a lethal dose of radiation without having a significant change in temperature? Make a rough calculation to support your answer.

34. A person works at a distance of 10 meters from a ^{60}Co source that is not shielded. Estimate the maximum strength of the source that does not present a radiation hazard to the worker.

35. A person manages to eat 1 milligram of radium. Is this hazardous to the health? Estimate the radiation dose received in one day.

36. Estimate the amount of radium, that if eaten, would double the amount of radiation that a person normally receives from environmental background.

37. A basement room has a size of 250 cubic meters. What mass of ^{222}Rn in the air will make the room a health-hazard?

38. A physicist wishes to test a transistor for its capability to operate under intense radiation by charged particles. The physicist brings the transistor to an accelerator and places it in an energetic electron beam. About how many electrons does the physicist need to send through the transistor to give it a radiation dose of 10 kilograys?

39. Make an estimate of the radiation dose equivalent received by a person from cosmic rays during one year.

40. The recommended permissible annual oral intake of ^{226}Ra is about $0.1\ \mu Ci$. How many ^{226}Ra atoms is it safe to eat per year?

Additional problems

41. The giant "star" shown in Figure 11-24 was made by depositing specks of radium, "in many cases too small to be seen under the microscope," on a photographic plate. (a) Make a rough estimate of the number of radium atoms in a "speck." (b) How many radium nuclei decay from the speck in one day?

42. In the β^+ decay of ^8B an energy of 11.15 MeV is released. (a) Calculate the mass of the resulting ^8C nucleus. (b) What is the binding energy of ^8C? (c) The ^8C nucleus is extremely unstable resulting in the following decays:

$$^8C \rightarrow {}^7B + p,$$

$$^7B \rightarrow {}^6Be + p,$$

and

$$^6Be \rightarrow {}^4He + p + p.$$

How much energy is released?

Let us suppose that only antisymmetrical states occur in nature... *two particles cannot occupy the same state...* . It leads to a special statistics, which was first studied by Fermi, so we shall call particles for which only antisymmetrical states occur in nature *fermions*. Let us suppose now that only symmetrical states occur in nature... . It is now possible for two or more of the states... to be the same, so that two or more particles can be in the same state. In spite of this, the statistics of the particles is not the same as the usual statistics of the classical theory. The new statistics was first studied by Bose, so we shall call particles for which only symmetrical states occur in nature *bosons*. We can see the difference of Bose statistics from usual statistics by considering a special case—that of only two particles and only two independent states a and b for a particle. According to classical mechanics, if the assembly of two particles is in thermodynamic equilibrium at high temperature, each particle will be equally likely to be in either state. There is a probability of 1/4 for both particles being in state a, a probability 1/4 for both particles being in state b, and a probability 1/2 for one particle being in each state. In the quantum theory there are three independent symmetrical states for the particles... . For thermodynamic equilibrium at high temperature these three states are equally probable... so that there is a probability 1/3 of both particles being in state a, a probability 1/3 of both particles being in state b, and a probability 1/3 of one particle being in each state. Thus *with Bose statistics the probability of two particles being in the same state is greater than with classical statistics*. Bose statistics differ from classical statistics in the opposite direction to Fermi statistics, for which the probability of two particles being in the same state is zero. In building up a theory of atoms... to get agreement with experiment one must assume that two electrons are never in the same state. This rule is known as *Pauli's exclusion principle*. It shows us that electrons are fermions. Planck's law of radiation shows us that photons are bosons, as only the Bose statistics for photons will lead to Planck's law. Similarly, for each of the other kinds of particle known in physics, there is experimental evidence to show that either they are fermions or bosons.

Paul A. M. Dirac

In the statistical description of an ideal gas at a temperature T, we have used the Maxwell-Boltzmann distribution function

$$f_{MB}(E) = Ce^{-E/kT}, \qquad (12.1)$$

to specify the probability that a molecule is in a state with energy E. The Maxwell-Boltzmann distribution follows from the *second law of thermodynamics,* which states that any division of a total amount of energy (E_{tot}) among N particles is equally probable to occur in nature. This leads to an exponential form of the probability distribution, as discussed in Chapter 2. Two basic assumptions are made in the derivation $f_{MB}(E)$:

1. In the calculation of the number of ways of dividing up E_{tot}, it is assumed that all particles are *distinguishable*, that is, if we interchange the energies of any two particles, we have a different state, and

2. Any number of particles are allowed to have a given energy E.

The Maxwell-Boltzmann distribution has widespread application because there are many physical systems where these two assumptions are valid.

12-1 PARTICLE DISTINGUISHABILITY

The Maxwell-Boltzmann distribution applies to distinguishable particles. Particles are distinguishable if they are not *identical*. For example, an electron is distinguishable from a proton in the hydrogen atom. Identical particles are distinguishable if we can resolve them in a scattering experiment. Our ability to distinguish identical particles depends on their separation distance compared to their

deBroglie wavelengths. Figure 12-1a shows the probability densities for two particles whose separation is large compared to their deBroglie wavelengths. We may say with certainty that one of the particles is near x_1 and the other is near x_2. The particles are distinguishable by their positions.

There are also many physical systems where the two Maxwell-Boltzmann assumptions do not hold. Two identical particles are *indistinguishable* if they are close enough together such that their probability densities overlap significantly. Figure 12-1b shows the wave function of two particles whose deBroglie wavelengths are large compared to their separation. In this case the particles are clearly indistinguishable. The degree of indistinguishability is governed by the amount of overlap of the probability densities. For electrons in an atom, the electron probability densities overlap significantly and we cannot distinguish them from one another. This is a violation of assumption 1. Furthermore, there is a definite restriction on the number of electrons allowed to be in any given shell. This is a violation of assumption 2. The Maxwell-Boltzmann distribution does not apply to electrons in atoms.

In order for the second law of thermodynamics to have general application, we must take into account the indistinguishability of particles when counting the number of possible states. We must also introduce new rules for determining how many particles may occupy a given state. The particles found in nature may be divided into two classes: *fermions* and *bosons*. The classification is based on the symmetry of the total wave function of a system of identical particles. The total wave function for a system of identical fermions is antisymmetric, that is, the wave function (9.7) changes sign with the exchange of any two particles. This is the case for electrons. The total wave function for a system of identical bosons is symmetric, that is, the wave function (9.6) does not change with the

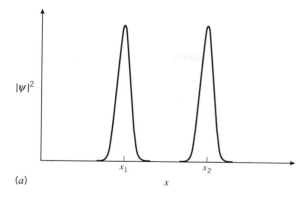

(a)

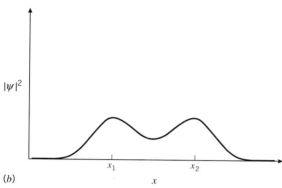

(b)

FIGURE 12-1 Distinguishability of two identical particles.
Probability densities for (a) two particles that are separated by a distance larger than the sizes of their individual wave packets, and (b) two particles with overlapping wave packets. In (a) the two particles are distinguished by their position, but in (b) one can no longer distinguish between the two particles with complete certainty.

exchange of any two particles. This is the case for photons. There is a fundamental connection between the symmetry and intrinsic angular momentum quantum number. Fermions have half-integer spin quantum numbers (1/2, 3/2, 5/2, …) and bosons have integer spin quantum numbers (0, 1, 2, …). Some examples of fermions and bosons are given in Table 12-1.

The number of particles allowed in each state is determined by the symmetry of the total wave function. There are two possibilities:

1. For fermions only one particle is allowed in each state.

2. For bosons any number of particles are allowed in each state.

TABLE 12-1
EXAMPLES OF FERMIONS AND BOSONS.

	Particle	Spin
Fermions	Electron	1/2
	Muon	1/2
	Proton	1/2
	Neutron	1/2
	Ω	3/2
	Quarks	1/2
	^{3}He	1/2
	^{55}Fe	3/2
Bosons	Alpha	0
	Photon	1
	Pion	0
	Deuteron	1
	Gluon	1
	W	1
	Z^0	1

Rule 1 applies to electrons, and is a statement of the Pauli exclusion principle. Rule 2 applies to photons. Application of these rules results in two *quantum distribution functions*, one for fermions and another for bosons. In the limit of small wavelengths (large energies), both quantum distributions reduce to the Maxwell-Boltzmann distribution.

Two particles are distinguishable when the distance between them (ℓ) is large compared to their deBroglie wavelengths (λ),

$$\ell \gg \lambda = \frac{h}{p}. \qquad (12.2)$$

Consider the case of a gas of oxygen molecules at a temperature T. Do the wave functions of the oxygen molecules significantly overlap? The relationship between temperature and average kinetic energy (2.30) of the gas particles gives

$$\langle E_k \rangle = \frac{\langle p^2 \rangle}{2m} = \frac{3}{2}kT. \qquad (12.3)$$

The root-mean-square momentum,

$$\sqrt{\langle p^2 \rangle} = \sqrt{3mkT}, \qquad (12.4)$$

gives the typical momentum of a molecule (see Figure 2-9). From this typical momentum (12.4) the deBroglie relationship (5.7) gives

$$\lambda = \frac{h}{\sqrt{3mkT}}. \qquad (12.5)$$

Therefore, the condition that the oxygen molecule wave functions do not overlap is that

$$\ell \gg \frac{h}{\sqrt{3mkT}}. \qquad (12.6)$$

EXAMPLE 12-1

Show that oxygen molecules in a gas at STP are distinguishable particles.

SOLUTION:

We need to show that the wavelength is much smaller than the average distance between molecules. In Example 2-10, we calculated the average distance between molecules to be

$$\ell \approx 3 \, \text{nm}.$$

The wavelength of the molecule is

$$\lambda = \frac{h}{\sqrt{3mkT}} = \frac{hc}{\sqrt{3mc^2 kT}}.$$

Using

$$mc^2 \approx 30 \, \text{GeV}$$

and

$$kT \approx \frac{1}{40} \text{eV},$$

we have

$$\lambda \approx \frac{1240 \text{ eV} \cdot \text{nm}}{\sqrt{(3)(3 \times 10^{10} \text{ eV})\left(\frac{1}{40} \text{eV}\right)}} \approx 0.03 \text{ nm}.$$

Thus,

$$\lambda \ll \ell.$$

The wave functions of the oxygen molecules do not overlap. The oxygen molecules are distinguishable, and the Maxwell-Boltzmann distribution function applies to their statistical description. ∎

12-2 BOSONS

Symmetric Wave Functions

Consider the total wave function for two *distinguishable* particles, for example, a proton and a neutron. The total wave function (ψ_D) may be written

$$\psi_D = \psi_a(\mathbf{r}_1) \, \psi_b(\mathbf{r}_2), \qquad (12.7)$$

where $\mathbf{r}_1$ and $\mathbf{r}_2$ represent the coordinates of the particles and the subscript (**a** or **b**) represents the complete set of quantum numbers of the particle (mass, intrinsic angular momentum, etc.). With the usual interpretation of $\psi_D^* \psi_D$ as the probability density, the normalization condition on the total wave function is

$$\int d^3\mathbf{r}_1 \int d^3\mathbf{r}_2 \, \psi_D^* \psi_D = 2, \qquad (12.8)$$

corresponding to two particles, where the limits of integration cover all possible values of the position vectors $\mathbf{r}_1$ and $\mathbf{r}_2$.

Now consider a state of two *identical indistinguishable bosons*. The total wave function for the two bosons is symmetric under particle exchange. The symmetric total wave function (ψ_S) may be written

$$\psi_S(\mathbf{r}_1, \mathbf{r}_2)$$
$$= \frac{1}{\sqrt{2}} \left[\psi_a(\mathbf{r}_1) \, \psi_b(\mathbf{r}_2) + \psi_b(\mathbf{r}_1) \, \psi_a(\mathbf{r}_2) \right], \qquad (12.9)$$

where the factor of $1/\sqrt{2}$ is included for normalization. The wave function (12.9) satisfies the symmetry condition, $\psi_S(\mathbf{r}_1,\mathbf{r}_2)=\psi_S(\mathbf{r}_2,\mathbf{r}_1)$.

> The total wave function for two identical bosons is symmetric under particle exchange.

We may construct a symmetric total wave function for any number of identical bosons. Consider a system of three identical bosons. Three coordinates ($\mathbf{r}_1$, $\mathbf{r}_2$, and $\mathbf{r}_3$)

and three sets of quantum numbers (**a**, **b**, and **c**) are needed to specify the total wave function. There are 3! or six permutations for the pairing of the quantum numbers with the coordinates. The total wave function is

$$\psi_S(\mathbf{r}_1, \mathbf{r}_2, \mathbf{r}_3) = \frac{1}{\sqrt{3!}}[\, \psi_a(\mathbf{r}_1)\, \psi_b(\mathbf{r}_2)\, \psi_c(\mathbf{r}_3)$$
$$+ \psi_a(\mathbf{r}_1)\, \psi_c(\mathbf{r}_2)\, \psi_b(\mathbf{r}_3)$$
$$+ \psi_b(\mathbf{r}_1)\, \psi_c(\mathbf{r}_2)\, \psi_a(\mathbf{r}_3)$$
$$+ \psi_b(\mathbf{r}_1)\, \psi_a(\mathbf{r}_2)\, \psi_c(\mathbf{r}_3)$$
$$+ \psi_c(\mathbf{r}_1)\, \psi_a(\mathbf{r}_2)\, \psi_b(\mathbf{r}_3)$$
$$+ \psi_c(\mathbf{r}_1)\, \psi_b(\mathbf{r}_2)\, \psi_a(\mathbf{r}_3)\,].$$

$$(12.10)$$

The factor of $1/\sqrt{3!}$ is included for normalization.

We now compare the probability that two identical bosons occupy the same quantum state with the probability that two distinguishable particles occupy the same state. For distinguishable particles, the probability density with **a** equal to **b** is

$$\psi_D{}^* \psi_D = \psi_a{}^2(\mathbf{r}_1)\, \psi_a{}^2(\mathbf{r}_2). \quad (12.11)$$

The probability density for the two identical bosons is

$$\psi_S{}^* \psi_S = \frac{1}{2}[\psi_a(\mathbf{r}_1)\psi_a(\mathbf{r}_2) + \psi_a(\mathbf{r}_1)\psi_a(\mathbf{r}_2)]^2$$
$$= 2\psi_a{}^2(\mathbf{r}_1)\psi_a{}^2(\mathbf{r}_2) = 2\psi_D{}^*\psi_D.$$

$$(12.12)$$

The probability for two identical bosons to be in the same quantum state is twice as large as that for two distinguishable particles. Now consider the case of three particles in the same quantum state. For distinguishable particles, we have

$$\psi_D{}^*\psi_D = \psi_a{}^2(\mathbf{r}_1)\, \psi_a{}^2(\mathbf{r}_2)\, \psi_a{}^2(\mathbf{r}_3), \quad (12.13)$$

while for identical bosons we have

$$\psi_S{}^*\psi_S = 6\,\psi_a{}^2(\mathbf{r}_1)\, \psi_a{}^2(\mathbf{r}_2)\, \psi_a{}^2(\mathbf{r}_3)$$
$$= 6\,\psi_D{}^*\psi_D. \quad (12.14)$$

The probability for three identical bosons to be in the same quantum state is six times larger than that for three distinguishable particles. For n identical bosons, the prob-

ability is $n!$ times greater that they are in the same quantum state compared to the case of distinguishable particles! If there are already n identical bosons in some particular state, then the probability of adding one more boson is enhanced by a factor of $(n+1)$ by the quantum mechanics of identical bosons. For example, if $W_{a\to b}$ is the transition rate from state **a** to state **b** for *distinguishable* particles, then the corresponding transition rate for *identical bosons* ($W_{a\to b}^{\text{boson}}$) is

$$W_{a\to b}^{\text{boson}} = (n_b + 1)W_{a\to b}, \quad (12.15)$$

where n_b is the number of bosons in state **b**. Identical bosons prefer to be in the same quantum state.

Bose-Einstein Distribution Function

Consider a system of distinguishable particles with two allowed states (**a** and **b**) with energies E_a and E_b. Let the number of particles in state **a** be n_a and in state **b** be n_b. In thermal equilibrium, the ratio n_a/n_b is given by the Maxwell-Boltzmann distribution,

$$\frac{n_a}{n_b} = \frac{e^{-E_a/kT}}{e^{-E_b/kT}}. \quad (12.16)$$

The particles make transitions between the two states. The number of particles in state **b** is proportional to the probability per time that a particle makes a transition from state **a** to state **b** ($W_{a\to b}$). Similarly, the number of particles in state **a** is proportional to the transition rate from state **b** to state **a** ($W_{b\to a}$). Therefore, we may write the ratio n_a/n_b (12.16) as

$$\frac{n_a}{n_b} = \frac{W_{b\to a}}{W_{a\to b}} = \frac{e^{-E_a/kT}}{e^{-E_b/kT}}. \quad (12.17)$$

For a similar system of identical bosons, the transition rates $W_{a\to b}^{\text{boson}}$ and $W_{b\to a}^{\text{boson}}$ are enhanced according to the number of bosons that already occupy each state:

$$W_{a\to b}^{\text{boson}} = (n_b + 1)W_{a\to b}, \quad (12.18)$$

and

$$W_{b\to a}^{\text{boson}} = (n_a + 1)W_{b\to a}. \quad (12.19)$$

For a system of bosons, the ratio n_a/n_b is the ratio of boson transition rates,

$$\frac{n_a}{n_b} = \frac{W_{b \to a}^{boson}}{W_{a \to b}^{boson}} = \frac{(n_a + 1) W_{b \to a}}{(n_b + 1) W_{a \to b}}$$

$$= \frac{(n_a + 1)}{(n_b + 1)} \frac{e^{-E_a/kT}}{e^{-E_b/kT}}. \qquad (12.20)$$

Rearranging the terms in the ratio n_a/n_b (12.20) gives

$$\frac{(n_a + 1) e^{-E_a/kT}}{n_a} = \frac{(n_b + 1) e^{-E_b/kT}}{n_b}. \qquad (12.21)$$

The left-hand side does not depend on state **b** (the number of particles in state **b** or the energy of state **b**), and the right-hand side does not depend on state **a**. Therefore, we set each side of the above expression (12.21) equal to a temperature-dependent parameter, $A(T)$:

$$\frac{(n_a + 1) e^{-E_a/kT}}{n_a} = A(T). \qquad (12.22)$$

Solving for n_a gives

$$n_a = \frac{1}{(A e^{E_a/kT} - 1)}. \qquad (12.23)$$

This expression is called the *Bose-Einstein* distribution function (f_{BE}) and is generally written,

$$f_{BE} = \frac{1}{A e^{E/kT} - 1}. \qquad (12.24)$$

The Bose-Einstein distribution function, which applies to any system of identical bosons, is analogous to the Maxwell-Boltzmann distribution function; it specifies the probability that a state with energy E is occupied.

An important application of the Bose-Einstein distribution function is a *gas of photons*. We have already encountered the concept of the photon gas in our description of thermal radiation. For the photon gas

$$A = 1. \qquad (12.25)$$

The Bose-Einstein distribution function (12.24) becomes

$$f_{BE}(E) = \frac{1}{e^{E/kT} - 1}. \qquad (12.26)$$

Photons are not subject to the Pauli exclusion principle. It is energetically favorable for all photons to be in the lowest energy state. The distribution function (12.26) that

the photons obey becomes infinite at $E = 0$ and reduces to the Maxwell-Boltzmann at large energies. The Bose-Einstein distribution function appears in the thermal radiation formula (3.32 with $E = hc/\lambda$) as part of the factor specifying the average energy per photon. The "−1" in the denominator of f_{BE} (12.26) has an important effect at energies comparable to kT (large photon wavelengths), and it has negligible effect at energies much larger than kT (small photon wavelengths).

12-3 FERMIONS

Antisymmetric Wave Functions

A state of two identical fermions satisfies the Pauli exclusion principle; the two particles cannot occupy the same state. We may represent the two wave functions as $\psi_a(\mathbf{r}_1)$ and $\psi_b(\mathbf{r}_2)$. The exclusion principle requires that

$$\mathbf{a} \neq \mathbf{b}. \qquad (12.27)$$

The total wave function for two identical fermions is antisymmetric under particle exchange. The antisymmetric total wave function (ψ_A) is

$$\psi_A(\mathbf{r}_1, \mathbf{r}_2)$$
$$= \frac{1}{\sqrt{2}} \left[\psi_a(\mathbf{r}_1) \psi_b(\mathbf{r}_2) - \psi_b(\mathbf{r}_1) \psi_a(\mathbf{r}_2) \right]. \qquad (12.28)$$

The wave function ψ_A is antisymmetric because if we switch particle coordinates $(\mathbf{r}_1)$ and $(\mathbf{r}_2)$, then $\psi_A(\mathbf{r}_1, \mathbf{r}_2) = -\psi_A(\mathbf{r}_2, \mathbf{r}_1)$. The antisymmetric total wave function (12.28) satisfies the Pauli exclusion principle because if we assign two identical fermions the same set of quantum numbers ($\mathbf{a} = \mathbf{b}$), then ψ_A is identically zero.

The total wave function for two particles may be written as the product of a space part and a spin part. Consider the case where the spin part is symmetric (corresponding to spins aligned). Then the space part of the wave function is symmetric for a pair of bosons and antisymmetric for a pair of fermions. (If the spin part is antisymmetric, corresponding to spins antialigned, then the reverse is true.) The space part of the wave function for two bosons and two fermions (one dimension) is sketched in Figure 12-2. The wave function for the boson pair is symmetric while the wave function for the fermion pair is antisymmetric. The two wave functions have a different amplitude squared because of interference.

We may construct an antisymmetric total wave function for any number of identical fermions. Consider a system of three identical fermions with the coordinates ($\mathbf{r}_1$, $\mathbf{r}_2$, and $\mathbf{r}_3$) and quantum numbers ($\mathbf{a}$, $\mathbf{b}$, and $\mathbf{c}$). The antisymmetric wave function is

$$
\begin{aligned}
\psi_A(\mathbf{r}_1, \mathbf{r}_2, \mathbf{r}_3) = \frac{1}{\sqrt{3!}} \big[& \psi_a(\mathbf{r}_1)\,\psi_b(\mathbf{r}_2)\,\psi_c(\mathbf{r}_3) \\
& - \psi_a(\mathbf{r}_1)\,\psi_c(\mathbf{r}_2)\,\psi_b(\mathbf{r}_3) \\
& + \psi_b(\mathbf{r}_1)\,\psi_c(\mathbf{r}_2)\,\psi_a(\mathbf{r}_3) \\
& - \psi_b(\mathbf{r}_1)\,\psi_a(\mathbf{r}_2)\,\psi_c(\mathbf{r}_3) \\
& + \psi_c(\mathbf{r}_1)\,\psi_a(\mathbf{r}_2)\,\psi_b(\mathbf{r}_3) \\
& - \psi_c(\mathbf{r}_1)\,\psi_b(\mathbf{r}_2)\,\psi_a(\mathbf{r}_3) \big].
\end{aligned}
$$

$$(12.29)$$

The factor of $1/\sqrt{3!}$ is included for normalization.

> The total wave function for two identical fermions is antisymmetric under particle exchange.

Fermi-Dirac Distribution Function

For a system of identical fermions, there are only two possibilities for the number of particles in any state: zero or one. Consider the transitions between two states ($\mathbf{a}$ and $\mathbf{b}$). For identical fermions, the possible values of n_a and n_b are zero or one. The transition rate $W_{a \to b}^{\text{fermion}}$ is zero if $n_b = 1$. If $n_b = 0$, the transition rate is identical to that of distinguishable particles ($W_{a \to b}$). Therefore, we may write

$$W_{a \to b}^{\text{fermion}} = (1 - n_b) W_{a \to b}, \qquad (12.30)$$

Similarly,

$$W_{b \to a}^{\text{fermion}} = (1 - n_a) W_{b \to a}. \qquad (12.31)$$

Using the Maxwell-Boltzmann distribution to evaluate n_a/n_b (12.16), we arrive at the following expression for fermions analogous to the boson expression (12.21):

$$\frac{(1 - n_a)\,e^{-E_a/kT}}{n_a} = \frac{(1 - n_b)\,e^{-E_b/kT}}{n_b}. \qquad (12.32)$$

We set each side of the foregoing expression (12.32) equal to a temperature-dependent parameter $A(T)$:

$$\frac{(1 - n_a)\,e^{-E_a/kT}}{n_a} = A(T). \qquad (12.33)$$

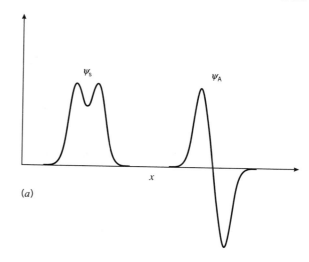

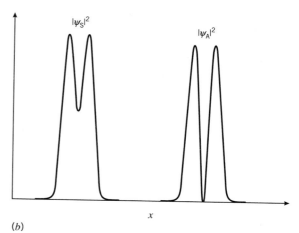

FIGURE 12-2 Space wave function for two particles in one dimension.
Sketch of the (a) amplitude and (b) amplitude squared. The two wave functions have a different amplitude squared because of interference.

Solving for n_a gives

$$n_a = \frac{1}{A\,e^{E_a/kT} + 1}. \qquad (12.34)$$

This expression is called the *Fermi-Dirac* distribution function (f_{FD}). The temperature-dependent parameter $A(T)$ is conventionally written as

$$A = e^{-E_F/kT}, \qquad (12.35)$$

where E_F is called the *Fermi energy*. (The Fermi energy is discussed in greater detail later in Chapter 14.) With the parameter A written in terms of the Fermi energy, the Fermi-Dirac distribution function (12.34) becomes

$$f_{FD} = \frac{1}{e^{(E-E_F)/kT} + 1},$$ (12.36)

The Fermi-Dirac distribution function reduces to the Maxwell-Boltzmann at large energies. The "+1" in the denominator of f_{FD} (12.36) has an important effect at energies less than the Fermi energy (large electron wavelengths), and it has negligible effect at energies greater than the Fermi energy (small electron wavelengths).

12-4 SCATTERING OF IDENTICAL FERMIONS AND BOSONS

Consider the scattering of distinguishable nuclei ³He and ⁴He. The electromagnetic properties of ³He and ⁴He are identical because they both have an electric charge of $2e$. The chemistry of the ³He and ⁴He atoms is identical. If we scatter ³He and ⁴He nuclei, they will interact by the electromagnetic interaction. Consider a Rutherford-type scattering experiment in the center-of-mass system. (We choose a scattering energy not so large that the nuclei come close enough to interact by the strong interaction.) We define $P_{3,4}$ is to be the probability for scattering into some suitably small interval about 90°. The two particle types are distinguishable in a spectrometer because of their different masses. There are two possible trajectories of the ³He and ⁴He particles, as shown in Figure 12-3a. The probability ($P_{3,4}$) may be expressed as the sum of the squares of a probability amplitude (A) for either of the two possible scatterings. There is no interference because the particles are distinguishable:

$$P_{3,4} = A^2 + A^2 = 2A^2.$$ (12.37)

Now examine the scattering of two identical particles, ³He and ³He. The ³He nucleus has a spin quantum number of 1/2. In the ground state the protons spins are anti-aligned and the single neutron does not have its spin "paired." The ³He particle is a fermion with $s = 1/2$. The total wave function for two ³He particles must be antisymmetric because the two ³He particles are not allowed to be in the same state by the Pauli exclusion principle. What happens if we try to follow the trajectories of the two ³He particles as they scatter? We find that in order to scatter at large angles, the particles must come close enough together that their wave functions significantly overlap, and we cannot tell which ³He particle scatters into a given

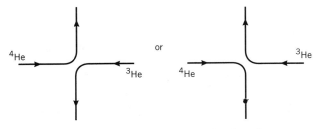

(a) Distinguishable particles, $P_{3,4} = A^2 + A^2 = 2A^2$

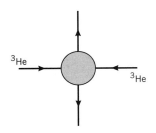

(b) Identical fermions, $P_{3,3} = (A - A)^2 = 0$

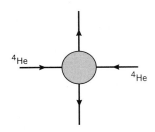

(c) Identical bosons, $P_{3,3} = (A + A)^2 = 4A^2$

FIGURE 12-3 Quantum interference in particle scattering.
Scattering of (a) distinguishable particles, (b) identical fermions, and (c) identical bosons.

detector. The actual process of scattering is pictured as a "blob" in Figure 12-3b. Inside the "blob," the notion of a classical trajectory loses its meaning and rules of quantum mechanics must be used to get the scattering probability. There are two possible paths that the ³He particles can take. By the rules of quantum mechanics we must add the amplitudes for the two processes and then square the result to get the scattering probability. Consider the path where the ³He particle that is incident from the left-hand side in Figure 12-3b scatters "up" and let the amplitude for this process be A. The amplitude for this ³He particle to scatter "down" may be obtained by exchanging the two ³He particles. Since the total wave function is antisymmetric under particle exchange, the amplitude for this process must be equal to $-A$. The scattering probability ($P_{3,3}$) becomes

$$P_{3,3} = (A - A)^2 = 0. \qquad (12.38)$$

The two ^{3}He particles cannot scatter at 90°! This is a most remarkable prediction of quantum mechanics and it is verified by experiment. If we insist on following the trajectory of one of the particles through the region of the "blob," then the act of following the particle path causes an interaction with the particle that disturbs the scattering process and the quantum interference is destroyed. In this case the particles can scatter at 90°! We then get the same scattering probability as if the particles were distinguishable; we have made them distinguishable by following the trajectory of one of the particles.

Now examine the scattering of the two identical particles, ^{4}He and ^{4}He (see Figure 12-3c). In the ground state the proton spins are antialigned and the neutron spins are antialigned. The ^{4}He particle is a boson with $s = 0$. The total wave function for two ^{4}He particles must be symmetric, and the amplitudes for the two paths have the same sign. The scattering probability ($P_{4,4}$) is

$$P_{4,4} = (A + A)^2 = 4A^2. \qquad (12.39)$$

The scattering probability is twice as large as in the case of distinguishable particles. The amplitudes have interfered constructively.

12-5 COMPARISON OF THE DISTRIBUTION FUNCTIONS

The three distribution functions all contain an exponential term that dominates at large energy. At large enough energies, the particle wavelengths are small compared to the distance between particles, and quantum mechanics is not needed for a statistical description of the occupation probability. (Quantum mechanics still may be needed to describe what energy states are allowed.) The distribution function gives us the occupation probability for the allowed states. The distribution functions are

$$f_{MB} = \frac{1}{Ae^{E/kT}}, \qquad (12.40)$$

$$f_{FD} = \frac{1}{Ae^{E/kT} + 1}, \qquad (12.41)$$

and

$$f_{BE} = \frac{1}{Ae^{E/kT} - 1}. \qquad (12.42)$$

The rather innocent looking "+1" in the Fermi-Dirac distribution and the "−1" in the Bose-Einstein distribution are needed to specify the quantum physics of a system of fermions or bosons at low energies. The three distribution functions are plotted in Figure 12-4. Their properties are summarized in Table 12-2.

12-6 DENSITY OF STATES

Along with the distribution functions, we need to evaluate the number of possible energy states (N) per volume (V)

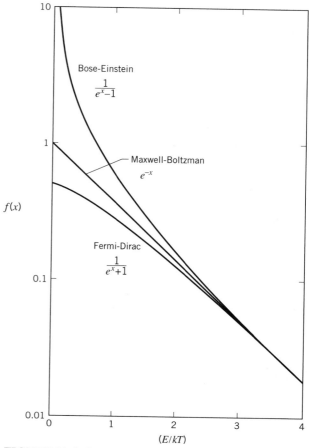

FIGURE 12-4 Comparison of the distribution functions. The distribution functions are plotted in terms of the dimensionless variable $x = E/kT$. The constant A is chosen to be unity. All three distribution functions are identical at large values of x.

TABLE 12-2
COMPARISON OF THE DISTRIBUTION FUNCTIONS.

	Maxwell-Boltzmann	Bose-Einstein	Fermi-Dirac
Distribution	$f_{MB} = \dfrac{1}{Ae^{E/kT}}$	$f_{BE} = \dfrac{1}{Ae^{E/kT} - 1}$	$f_{FD} = \dfrac{1}{Ae^{E/kT} + 1}$
Particle type	Distinguishable	Identical bosons	Identical fermions
Spin	Any	0, 1, 2, ...	1/2, 3/2, 5/2, ...
Maximum number of particles per state	No limit	No limit	1
Example	Ideal gas	Thermal radiation	Electrons in metal

for specific physical systems. A system might consist of molecules of an ideal gas, electrons in a metal, or black-body radiation. The number of states is energy dependent because of the wave nature of particles. There are more ways for a smaller wavelength particle to fit into a box of fixed size than there are for a larger wavelength particle. Since a smaller wavelength corresponds to a larger energy, the number of states $N(E)$ increases with increasing energy. We write the number of possible energy states per volume (n_s) as

$$n_s(E) = \frac{N(E)}{V}. \tag{12.43}$$

The density of states (ρ) is defined to be the derivative of $n_s(E)$ with respect to E:

$$\rho(E) = \frac{dn_s}{dE}. \tag{12.44}$$

This derivative is a positive number since $n_s(E)$ increases with E. The density of states is the number of states per volume per unit energy. The density of states function specifies which states are allowed but does not tell us which states are occupied.

Electrons

For electrons in a metal we can treat each electron as a nonrelativistic particle in a box. Recall the solution of the three-dimensional infinite square-well. The particle energies (7.151) are given by

$$E = \frac{h^2\left(n_1^2 + n_2^2 + n_3^2\right)}{8mL^2}. \tag{12.45}$$

In (n_1, n_2, n_3) space, we define a radius vector (R),

$$R \equiv \sqrt{n_1^2 + n_2^2 + n_3^2}. \tag{12.46}$$

We may express the energy in terms of R:

$$E = \frac{h^2 R^2}{8mL^2}, \tag{12.47}$$

or

$$R = \frac{2\sqrt{2mE}\,L}{h}. \tag{12.48}$$

Within a radius R, the number of states is one-eighth of the volume of a sphere because n_1, n_2, and n_3 are positive. There is an additional factor of two to accommodate the two possible electron spin states. The number of states (N) is

$$N = (2)\left(\frac{1}{8}\right)\left(\frac{4}{3}\right)\pi R^3$$
$$= \left(\frac{8\pi}{3}\right)(2mE)^{3/2}\left(\frac{L}{h}\right)^3. \tag{12.49}$$

The number of states per volume is

$$n_s = \frac{N}{L^3} = \left(\frac{8\pi}{3}\right)\frac{(2mE)^{3/2}}{h^3}. \tag{12.50}$$

The density of states for nonrelativistic electrons is

$$\rho = \frac{dn_s}{dE} = \frac{4\pi(2m)^{3/2}}{h^3}\sqrt{E}. \tag{12.51}$$

The density of states is proportional to the square root of the electron energy. Recall from Chapter 2 that the density of states for an ideal gas (2.93) was also determined to be proportional to the square root of the energy.

Photons

For photons, we may calculate the density of states in a similar fashion. We have done this already in Chapter 3 by counting the number of standing waves in a box of size L. The difference between photons and nonrelativistic electrons is the relationship between energy and momentum (or energy and wavelength). For photons, we have

$$E = pc = \frac{hc}{\lambda}. \tag{12.52}$$

The wave equation for the photon gives

$$\left(\frac{n_1 \pi}{L}\right)^2 + \left(\frac{n_2 \pi}{L}\right)^2 + \left(\frac{n_3 \pi}{L}\right)^2 = \left(\frac{2\pi}{\lambda}\right)^2. \tag{12.53}$$

Therefore,

$$n_1^2 + n_2^2 + n_3^2 = \left(\frac{2L}{\lambda}\right)^2. \tag{12.54}$$

Using the relationship between wavelength and energy, we have

$$n_1^2 + n_2^2 + n_3^2 = \left(\frac{2LE}{hc}\right)^2. \tag{12.55}$$

As before, we use

$$R^2 = n_1^2 + n_2^2 + n_3^2, \tag{12.56}$$

so that

$$R^2 = \left(\frac{2LE}{hc}\right)^2, \tag{12.57}$$

or

$$R = \frac{2LE}{hc}. \tag{12.58}$$

The number of states (standing waves in the box) is

$$N = (2)\left(\frac{1}{8}\right)\left(\frac{4}{3}\right)\pi R^3$$

$$= \left(\frac{\pi}{3}\right)\left(\frac{2LE}{hc}\right)^3, \tag{12.59}$$

where the factor of 2 accounts for the two possible polarizations of the photon. The number of states per volume is

$$n_s = \frac{N}{L^3} = \left(\frac{\pi}{3}\right)(8)\frac{E^3}{(hc)^3}. \tag{12.60}$$

The density of states for photons is

$$\rho = \frac{dn_s}{dE} = \frac{8\pi}{(hc)^3}E^2. \tag{12.61}$$

The density of states for photons is proportional to the energy squared. Note that in the determination of the density of states for both the electron and the photon, the key to the physics is the wave nature of the particles. The density of states increases much faster with energy for photons compared to nonrelativistic electrons because of the energy dependence of the wavelength.

Phase Space

We may gain some physical insight on the meaning of density of states by considering the six-dimensional space composed of the three spatial coordinates and the three momentum components: (x, y, z, p_x, p_y, p_z). This space is called *phase space*. The volume of phase space (V_{PS}) is an integral over all six coordinates:

$$V_{PS} = \int dx \int dy \int dz \int dp_x \int dp_y \int dp_z. \tag{12.62}$$

The limits of the integration are given by the boundaries of the physical system. We may write the phase space volume as a product of the ordinary volume (V) and a volume in momentum space (V_p).

$$V_{PS} = V V_p. \tag{12.63}$$

Because the wavelength and the momentum are dependent on one another, there is a fixed volume of phase space that a particle occupies. Since

$$p\lambda = h, \tag{12.64}$$

this fixed volume of phase space is h^3. The number of states (N) as a function of momentum may be written as the total phase space volume times the number of internal states (g) due to intrinsic angular momentum, divided by the phase space volume per state (h^3):

$$N(p) = g\frac{V_{PS}}{h^3} = g\frac{V V_p}{h^3}. \tag{12.65}$$

Since the volume in momentum space for a momentum p is $4\pi p^3/3$, the number of states per volume is

$$n_s(p) = \frac{N(p)}{V} = g\frac{V_p}{h^3} = \frac{4}{3}\pi g \frac{p^3}{h^3}. \quad (12.66)$$

This is pictured in Figure 12-5. The density of states is obtained by differentiating n_s (12.66) with respect to E. In carrying out the differentiation, we make use of the fundamental relationship between energy (E), mass (m), and momentum of any particle, $E^2 = (mc^2)^2 + (pc)^2$. The result is

$$
\begin{aligned}
\rho &= \frac{dn_s}{dE} = \frac{dn_s}{dp}\frac{dp}{dE} \\
&= \frac{4\pi g p^2}{h^3}\frac{d}{dE}\left(\frac{1}{c}\sqrt{E^2 - \left(mc^2\right)^2}\right) \\
&= \frac{4\pi g p^2}{h^3}\frac{E}{pc^2} = \frac{4\pi g p^2}{h^3 v}, \quad (12.67)
\end{aligned}
$$

where $v = pc^2/E$ is the speed of the particle. Thus, the density of states is proportional to the momentum squared divided by the speed.

EXAMPLE 12-2
Calculate the density of states for a nonrelativistic electron "gas."

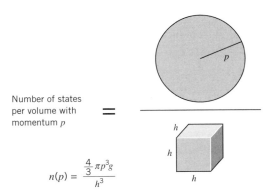

Number of states per volume with momentum p $=$

$$n(p) = \frac{\frac{4}{3}\pi p^3 g}{h^3}$$

FIGURE 12-5 Determination of the number of states per volume with momentum p.
Each state occupies a volume in phase space (x, y, z, p_x, p_y, p_z) of h^3. The momentum part of the phase space volume for a momentum p is $4\pi p^3/3$. The factor g accounts for the number of possible spin orientations. The ratio $4\pi p^3 g/3$ divided by h^3 is the number of states per volume with momentum p.

SOLUTION:
The relationship between momentum, energy, and speed is

$$p = mv = \sqrt{2mE}.$$

The spin quantum number of the electron is $s = 1/2$. The degeneracy is

$$g = 2s + 1 = 2.$$

The density of states (12.67) is

$$\rho = \frac{4\pi g p^2}{h^3 v} = \left(\frac{4\pi}{h^3}\right)(2)\frac{2mE}{\sqrt{\frac{2E}{m}}} = \frac{4\pi(2m)^{3/2}}{h^3}\sqrt{E},$$

in agreement with (12.51). ∎

EXAMPLE 12-3
Calculate the density of states for a photon gas.

SOLUTION:
The relationship between kinetic energy and momentum is

$$p = \frac{E}{c}.$$

There are two possible polarization states of the photon:

$$g = 2.$$

The density of states (12.67) is

$$\rho = \frac{4\pi g p^2}{h^3 v} = \left(\frac{4\pi}{h^3}\right)(2)\frac{\left(\frac{E}{c}\right)^2}{c} = \frac{8\pi}{h^3 c^3}E^2,$$

in agreement with (12.61). ∎

As an example of density of states, consider the routine chore of parallel parking an automobile into a "tight spot." Everyone who has ever tried this experiment will agree that it is easier to drive a car out of a parking place than to drive the car into the parking place. There is a simple reason for this. To get into the parking spot, the number of possible final positions of the car is relatively small. An "acceptable" parking position does not leave the car a meter from the curb. We say that the density of states of the parked car is small or the "phase space" is small. In getting the car out of the spot, there are many more places for the car to go. The density of states is large or the "phase space" is large. It is easier to get the car out than to get the car in!

12-7 EXAMPLES OF QUANTUM DISTRIBUTIONS

Photon Gas: The Thermal Radiation Spectrum

Thermal radiation is an example of a "gas" of photons. The photon spectrum is described statistically by the Bose-Einstein distribution together with the density of states. The density of states for a gas of photons is shown in Figure 12-6. The number of photons per volume per energy (dn/dE), the density of *occupied* states, is the density of states times the Bose-Einstein factor:

$$\frac{dn}{dE} = \rho(E) f_{BE}(E)$$

$$= \left(\frac{8\pi E^2}{c^3 h^3}\right)\left(\frac{1}{e^{E/kT}-1}\right). \quad (12.68)$$

The energy per volume is equal to the photon energy (E) times dn/dE,

$$\frac{du}{dE} = E\frac{dn}{dE} = \left(\frac{8\pi}{c^3 h^3}\right)\left(\frac{E^3}{e^{E/kT}-1}\right). \quad (12.69)$$

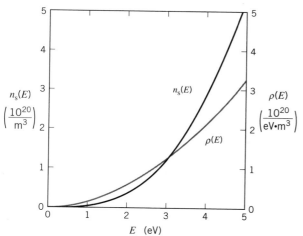

FIGURE 12-6 Number of states per volume (n_s) and density of states ρ as a function of energy for a gas of photons.
The units of n_s are 10^{20} per m³ and the units of ρ are 10^{20} per eV per m³.

This is the universal thermal radiation spectrum that we have arrived at in Chapter 3.

Electron Gas: Theory of Metals

In a metal, the outermost electrons from each atom are free to move, resulting in good heat and electrical conductivity. (Conductivity is discussed in more detail in Chapter 14.) For example, the sodium atom ($Z = 11$) has an electron configuration ($1s^2 2s^2 2p^6 3s^1$) in which the K shell and L shell are full and there is a single electron in the $3s$ state. The $3s$ electron is weakly bound and is free to move from atom to atom in the metal.

We see from the Fermi-Dirac distribution,

$$f_{FD} = \frac{1}{e^{(E-E_F)/kT}+1}, \quad (12.70)$$

that the probability of a state being occupied is equal to 1/2 when $E = E_F$. For the case $T = 0$ K, $f_{FD} = 1$ when $E < E_F$, and $f_{FD} = 0$ when $E > E_F$. States with energy less than the Fermi energy are all filled, and states with energy greater than the Fermi energy are all empty. The Fermi-Dirac distribution function is shown in Figure 12-7. A typical value for the Fermi energy in a metal is a few electronvolts. We shall learn how to calculate Fermi energies in Chapter 14. From this energy scale, we know immediately that the order-of-magnitude speed of an electron in a metal is $c\alpha$. This is true even at very low temperatures. The Pauli exclusion principle prevents the electrons from all having low energies.

> The Fermi energy is the boundary between occupied and unoccupied states. For $kT \ll E_F$, states below E_F are occupied and states above E_F are empty.

EXAMPLE 12-4
Estimate the speed of an electron in a metal that has a kinetic energy near the Fermi energy.

SOLUTION:
The electron kinetic energy is

$$E_k = \frac{1}{2}mv^2.$$

Taking

$$E_k = E_F \approx 5\,\text{eV},$$

and solving for the electron speed we have

$$v = \sqrt{\frac{2E_F}{m}} = c\sqrt{\frac{2E_F}{mc^2}}$$

$$\approx (3\times10^8 \text{ m/s})\sqrt{\frac{(2)(5\text{eV})}{5\times10^5 \text{ eV}}} \approx 10^6 \text{ m/s}.$$

The speed of an electron in a metal is much larger than would be expected if the electrons obeyed Maxwell Boltzmann statistics. ∎

EXAMPLE 12-5

Estimate the root-mean-square speed of an electron in a metal at $T = 300$ K and $T = 3$ K, if the electron were to obey Maxwell-Boltzmann statistics.

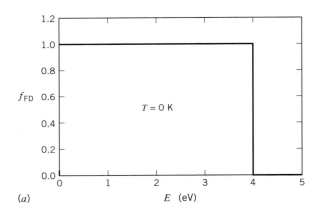

(a)

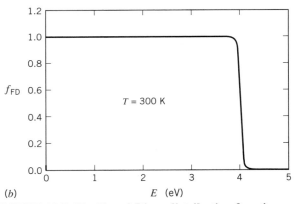

(b)

FIGURE 12-7 The Fermi-Dirac distribution function. The Fermi energy is taken to be 4 eV. (a) For $T = 0$ K, all states with energies lower than E_F are filled and all states higher than E_F are empty. (b) For $T = 300$ K, some of the electrons with energies near E_F are excited to energies just above E_F.

SOLUTION:
The average kinetic energy of the electron is given by

$$\langle E_k \rangle = \frac{3}{2}kT.$$

The root-mean-square speed is

$$v_{rms} = \sqrt{\frac{3kT}{m}} = c\sqrt{\frac{3kT}{mc^2}}.$$

For $T = 300$ K,

$$v_{rms} \approx (3\times10^8 \text{ m/s})\sqrt{\frac{(3)\left(\frac{1}{40}\text{ eV}\right)}{5\times10^5 \text{ eV}}} \approx 10^5 \text{ m/s}.$$

For $T = 3$ K, the root-mean-square speed would be smaller by a factor of $(300 \text{ K}/3 \text{ K})^{1/2} = 10$,

$$v_{rms} \approx 10^4 \text{ m/s}. \qquad ∎$$

The density of states for electrons in a metal (see Example 12-2) is

$$\rho(E) = \frac{dn_s}{dE} = \frac{4\pi(2m)^{3/2}}{h^3}\sqrt{E}. \qquad (12.71)$$

The density of states for electrons is shown in Figure 12-8. The number of electrons per volume per unit energy (dn/dE) is

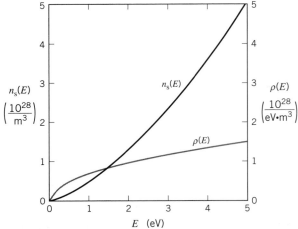

FIGURE 12-8 Number of states per volume (n) and density of states (ρ) as a fraction of kinetic energy for a gas of nonrelativistic electrons.
The units of n_s are 10^{28} per m³ and the units of ρ are 10^{28} per eV per m³.

$$\frac{dn}{dE} = \rho(E)\frac{1}{e^{(E-E_F)/kT}+1}, \qquad (12.72)$$

or

$$\frac{dn}{dE} = 4\pi(2m)^{3/2}\frac{\sqrt{E}}{h^3}\frac{1}{e^{(E-E_F)/kT}+1}. \quad (12.73)$$

CHAPTER 12: PHYSICS SUMMARY

- In a system of bosons at temperature T, the Bose-Einstein distribution function specifies the probability that a state with a given energy is occupied:

$$f_{BE} = \frac{1}{e^{E/kT}-1}.$$

  The Bose-Einstein distribution function applies to the photon ($A = 1$) and appears in the thermal radiation formula. There is no limit to the number of photons that can be in the same quantum state.

- In a system of fermions at temperature T, the Fermi-Dirac distribution function specifies the probability that a state with a given energy is occupied:

$$f_{FD} = \frac{1}{e^{(E-E_F)/kT}+1}.$$

  The Fermi-Dirac distribution function applies to the electron. The maximum number of electrons that can be in any quantum state is one.

- A typical value of the Fermi energy for electrons in a metal is a few eV. At low temperatures ($kT \ll E_F$) states with energy less than the Fermi energy are all filled and states with energy greater than the Fermi energy are all empty.

- For particles with small enough wavelengths (large energies) so that the wave functions do not overlap, the particles are distinguishable and the distributions (f_{MB}, f_{BE}, and f_{FD}) are identical, decreasing exponentially with increasing energy.

- The density of states, the number of available states per volume per unit energy, is given by

$$\rho = \frac{dn_s}{dE} = \frac{4\pi g p^2}{h^3 v},$$

  where g is the number of internal ("spin") states, p is the particle momentum, and v is the particle

speed. For electrons in a metal, ρ is proportional to $E^{1/2}$. For photons, ρ is proportional to E^2.

- The number of particles per volume per unit energy is equal to the density of states times the occupation probability (f_{BE} for bosons or f_{FD} for fermions).

REFERENCES AND SUGGESTIONS FOR FURTHER READING

P. A. M. Dirac, *The Principles of Quantum Mechanics,* Oxford Univ. Press (1958).

P. A. M. Dirac, "The Evolution of the Physicists Picture of Nature," *Sci. Am.* **208**, No. 5, 45 (1963).

R. J. Donnelly, "Superfluid Turbulence," *Sci. Am.* **259**, No. 5, 100 (1988).

R. Eisberg and R. Resnick, *Quantum Physics of Atoms, Molecules, Solids, Nuclei, and Particles,* Wiley (1985).

K. Huang, *Statistical Mechanics,* Wiley (1963).

F. Laloë and J. H. Freed, "The Effects of Spin in Gases," *Sci. Am.* **258**, No. 4, 94 (1988).

O. V. Lounasmaa and G. Pickett, "The ³He Superfluids," *Sci. Am.* **262**, No. 6, 104 (1990).

R. C. Tolman, *The Principles of Statistical Mechanics,* Oxford University Press (1938).

F. Wilczek, "Anyons," *Sci. Am.* **264**, No. 5, 58 (1991).

QUESTIONS AND PROBLEMS

Particle distinguishability

1. Estimate the average separation distance and the deBroglie wavelength for a gas of helium atoms at atmospheric pressure and a temperature just above liquification. Can the system be described by Maxwell-Boltzmann statistics?

2. Protons in a plasma fusion reactor have a temperature of 3×10^7 K. (a) What is the average kinetic energy of the protons? (b) The density of protons in the plasma is 10^{21} m^{-3}. Can we describe the system with Maxwell-Boltzmann statistics?

Bosons

3. Write down the total wave function for a system of four identical bosons.

4. For what energy is the Bose-Einstein factor equal to 10 at room temperature?

Fermions

5. Consider a system of 6 spin-1/2 fermions sharing 10 units of energy. Determine the energy distribution

function df/dE. Make an estimate of the Fermi energy.

6. Write down the total wave function for a system of four identical fermions.

7. In the Chapter 2 discussion of thermionic emission of electrons from metals, we used the Maxwell-Boltzmann distribution. Why is this valid?

Scattering of identical fermions and bosons

8. Make an analogy between the helium-scattering experiment and the double-slit experiment.

9. Consider the scattering of two ^{3}He nuclei or two ^{4}He nuclei at small angles. Is there quantum interference? Explain!

Comparison of the distribution functions

10. Consider a system of five particles sharing 10 units of energy analogous to the discussion of Chapter 2 (see Figures 2-10 and 2-11). Determine the energy distribution function df/dE if the particles are (a) distinguishable, (b) identical spin-zero bosons, and (c) identical spin-1/2 fermions.

11. For a gas of particles at fixed temperature, which particle type, fermion or boson, causes the greatest pressure? Why?

Density of states

12. For electrons in a metal, why is the density of states at the Fermi energy approximately equal to the density of free electrons divided by the Fermi energy?

13. Which system has more states available at 1 eV, a gas of electrons or a gas of photons? Why?

14. (a) Why are there no states of zero energy for a distribution of thermal photons? (b) Why are there no states of zero energy for a distribution of electrons in a metal?

Examples of quantum distributions

15. (a) Use the density of states for a photon gas and the Bose-Einstein distribution function to obtain an expression for the energy per volume per unit wavelength $(du/d\lambda)$ of radiation in a cavity at a temperature T. (b) Show that your answer is equivalent to the thermal radiation formula (3.32) by calculating the power per area per unit wavelength $(dR/d\lambda)$ radiated by the cavity walls.

16. Deduce the expression for the density of states for a gas of relativistic electrons.

Additional problems

17. What is the relationship between the size of the electron cloud in a single free atom and the spacing between those atoms in a chunk of metal?

18. Is a red marble a boson or a fermion? Explain!

19. Consider two particles in an infinite square-well in one dimension with one particle in the ground state and one particle in the first excited state. (a) Write down the total wave function (space part) for the two particles if they are distinguishable particles. (b) Write down the total wave function (space part) for the two particles if they are identical bosons in a space symmetric state. (c) Write down the total wave function (space part) for the two particles if they are identical fermions in a space antisymmetric state.

*20. *Bose-Einstein condensation.* (a) Treating a system of helium atoms as a boson gas, give an expression for the number of particles per volume (n) as a function of temperature. Take the parameter A in f_{BE} to be unity. (b) The density of liquid helium at the boiling point is 125 kg/m^3. Estimate the temperature (T_c) where the calculated value of n is equal to that of the liquid.

$$\left(Hint : \int_0^\infty dx \sqrt{x}\left(e^x - 1\right)^{-1} \approx 2.32. \right)$$

(c) When the temperature of the liquid helium is reduced below T_c, a fraction of the helium atoms condense into the ground state forming what is called a *superfluid*. The superfluid density may be specified by adding a constant term (n_{sup}) to the particle density. Show that

$$n_{sup} = n\left[1 - \left(\frac{T}{T_c}\right)^{3/2} \right].$$

21. Give an expression for the number of particles per volume per unit energy for a gas of (a) nonrelativistic α particles, (b) relativistic α particles, (c) nonrelativistic ^{3}He particles, and (d) relativistic ^{3}He particles.

22. A beam of thermal neutrons ($T = 300$K) emerges from an open port in a nuclear reactor with a flux of 10^{13}m^{-2}s^{-1}. Estimate the density of neutrons in the beam. Can the energy distribution of the neutrons be described with Maxwell-Boltzmann statistics?

MASERS AND
LASERS

Although the word maser was coined as an acronym for microwave amplification by stimulated emission of radiation, it is now being used more widely for amplification by atomic or molecular processes in the radio frequency, infra-red and optical regions. Perhaps it might now be interpreted to mean molecular amplification by stimulated emission of radiation. One can show quite generally that atoms or molecules in thermal equilibrium cannot coherently amplify an electromagnetic wave. Consider normal thermal radiation. Within a given frequency interval there is a maximum value, that for a black body, of the radiation intensity from any material at a certain temperature. A coherent wave can be represented by an exceedingly high temperature and, if it were amplified, the resulting amplified wave would be at a still higher temperature. Hence, coherent (e.g., without change of frequency) amplification cannot be produced by a system which is in thermal equilibrium. One must therefore look for systems which are not in thermal equilibrium if amplification is to be achieved.

Charles H. Townes

Consider a system of atoms or molecules where the population of an excited state is larger than the ground state. Incident radiation that has an energy equal to the difference in energy between the excited state and the ground state can induce or *stimulate* radiative transitions to the ground state. The process is

$$(\text{atom})^* + \gamma \rightarrow (\text{atom}) + \gamma + \gamma, \qquad (13.1)$$

where (atom)* represents an excited atomic state. The stimulated radiation is *coherent*; it has the same wavelength, direction, and polarization as the incident radiation. The amplitude squared for two coherent waves is *four* times the amplitude squared for one wave, that is, $(A + A)^2 = 4A^2$. The stimulated emission process may be used to amplify the radiation. This is pictured in Figure 13-1. A single photon interacts with an atom, resulting in two photons, these two photons each interact with atoms resulting in four photons, and so on. At each stage the amplitude is doubled and the square of the amplitude increases by a factor of four.

Such amplification was first achieved in the microwave region by Charles H. Townes and his collaborators in 1954. Microwaves are the portion of the electromagnetic spectrum between infrared and radio waves. The invention of Townes is called a *maser*, which stands for microwave amplification by stimulated emission of radiation. The same principle may be used to amplify radiation in the infrared or optical region. A maser operating in the optical region is called a *laser*, for light amplification by stimulated emission of radiation.

> *Maser* stands for microwave amplification by stimulated emission of radiation, and *laser* stands for light amplification by stimulated emission of radiation.

The operation of the maser or laser is based on the concept of stimulated emission (13.1). We have learned that the probability for a boson to make a transition to a state that already contains n identical bosons is proportional to $n + 1$. The more bosons that exist in a state, the greater the probability for adding another. Bosons tend to be in the same state. Stimulated emission occurs when a photon of just the right wavelength induces an atom or molecule to make a transition, resulting in coherent photons. In the first section of this chapter, we make an overview of the photon–atom interaction processes. We shall then proceed to discuss the stimulated emission process in more detail.

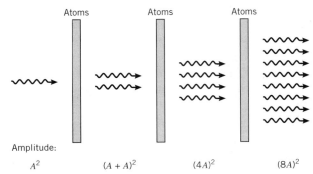

FIGURE 13-1 Photon amplification.
Photons interact with atoms in the excited state and stimulate the emission of additional coherent photons.

13-1 SUMMARY OF PHOTON–ATOM INTERACTIONS

Our description of modern physics has involved, to a large extent, the understanding of the interaction between radiation and matter. The processes that are observed to occur when photons interact with atoms are summarized in Figure 13-2. These processes are

(a) *Absorption:* An atom may absorb a photon (γ) with one of the electrons in the atom moving to a higher energy level. We may write the process as

$$(\text{atom}) + \gamma \rightarrow (\text{atom})^*. \qquad (13.2)$$

This is a resonant process. The transition probability has a sharp maximum when the photon energy is equal to the difference in energy levels of the atom. We have seen that photon absorption occurs in molecules through excitation of rotational and vibrational modes as well as electronic excitations.

(b) *Spontaneous emission:* An atom in an excited state can spontaneously decay into a lower energy state with the emission of a photon:

$$(\text{atom})^* \rightarrow (\text{atom}) + \gamma. \qquad (13.3)$$

Spontaneous emission is the inverse process of absorption.

(c) *Stimulated emission:* A photon may also induce an atomic transition from a higher energy state to a lower

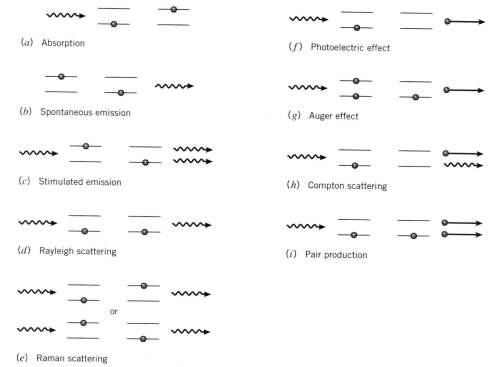

FIGURE 13-2 Summary of the photon-atom interaction.
(*a*) Absorption, (*b*) spontaneous emission, (*c*) stimulated emission, (*d*) Rayleigh scattering, (*e*) Raman scattering, (*f*) photoelectric effect, (*g*) Auger effect, (*h*) Compton scattering and (*i*) pair production.

energy state (13.1). Like the absorption process, stimulated emission is a resonant process. The transition probability for stimulated emission has a sharp maximum when the photon has an energy that is equal to the energy difference between the atomic states. The final state contains the incident photon plus an additional photon from the electron transition. The two photons are coherent.

(d) *Rayleigh scattering:* A photon may scatter elastically with an electron in the atom, leaving the electron energy unchanged:

$$(\text{atom}) + \gamma \rightarrow (\text{atom}) + \gamma . \qquad (13.4)$$

The cross section for this process varies as the inverse fourth power of the photon wavelength. When we look at the sky in the daytime, we are seeing light from the sun that is scattered by the atmosphere of the earth (i.e, we are not looking directly at the source). The shorter wavelength light scatters more, and this, coupled with the wavelength sensitivity of the eye, is why the sky is blue. When we view the sun at sunset or sunrise, we are looking directly at the

source. The light that reaches us is the light that scatters the least (longer wavelengths) and the sunset appears orange-red. The effect is enhanced when the sun is near the horizon because light reaching us passes a long distance though the atmosphere, many times the atmospheric thickness.

(e) *Raman scattering:* A photon may scatter from an atom, and in the process the atom may end up in an excited state. The process is

$$(\text{atom}) + \gamma \rightarrow (\text{atom})* + \gamma . \qquad (13.5)$$

The energy of the scattered photon is correspondingly reduced. The inverse process is also possible:

$$(\text{atom})* + \gamma \rightarrow (\text{atom}) + \gamma . \qquad (13.6)$$

In this case the energy of the photon is increased.

(f) *Photoelectric effect:* An atom may absorb a photon that has sufficient energy to free the electron from the atom. In this case the electron emerges with a kinetic energy equal to the incident photon energy minus the electron binding energy:

$$(\text{atom}) + \gamma \rightarrow (\text{atom minus } e^-) + e^-. \quad (13.7)$$

(g) *Auger effect:* If a photon excites an inner electron from an atom, then the atom may return to the ground state with the emission of an electron. (The Auger effect can also occur when an inner electron is captured by a nucleus.) The process is

$$(\text{atom}) + \gamma \rightarrow (\text{atom})^* \rightarrow (\text{atom minus } e^-) + e^-. \quad (13.8)$$

The atomic excitation energy has appeared as kinetic energy of an ejected electron.

(h) *Compton scattering:* An energetic photon ($E \gg 1$ eV) may scatter off an electron in an atom. Kinetic energy is transferred from the photon to the electron:

$$(\text{atom}) + \gamma \rightarrow (\text{atom minus } e^-) + e^- + \gamma. \quad (13.9)$$

In this case the photon energy is so large that the electron may be considered as a free particle and the rest of the atom plays no role in the scattering.

(i) *Pair production:* An energetic photon ($E > 1$ MeV) may interact with the nucleus of an atom and produce an electron–positron pair,

$$(\text{atom}) + \gamma \rightarrow (\text{atom}) + e^+ + e^-. \quad (13.10)$$

13-2 STIMULATED EMISSION OF RADIATION

We now examine the processes of spontaneous emission and stimulated emission more closely. The two processes are very similar. In the case of spontaneous emission, the excited atomic state decays without external stimulus, as in the case of natural radioactive decay. The energy of the emitted photon is given by the energy difference between the atomic states. In the process of stimulated emission, the atom is induced to make the same transition. The external stimulus is a photon that has an energy equal to the energy difference between the atomic states.

We have discussed spontaneous decays on several occasions. The stimulated decay is something new. We shall find it useful to think of the atom as a quantum oscillator. Stimulated emission occurs when photons of the correct frequency cause the atom to oscillate and radiate. The source of these photons is an external electromagnetic field. The field shakes the atom and causes the atom to radiate!

How then does the spontaneous decay occur? The answer lies in the quantization condition that applies to the electromagnetic field. Like other quantized systems, the ground state of the electromagnetic field is not empty or zero but contains a small number of photons that cause the atom to vibrate and radiate even when there is no external field applied. This should not seem too surprising. We know that there is no such thing as an electron without photons bumping into it! Part of the source of these photons is the electron itself. The atom is shaking all by itself, even when no external electromagnetic field is present. *Why doesn't an atom in the ground state decay?* Since the electron is confined, it has a minimum kinetic energy given by the uncertainty principle. The atom cannot decay from the ground state because there is no state of lower energy to decay into.

The process of stimulated emission is also closely related to the absorption process. We should like to know how the absorption probability is related to the stimulated emission probability. Einstein (again!) was the first person to understand that the two probabilities are identical. We shall now demonstrate that this is the case.

Consider two atomic states S_1 and S_2, where the energy of the state S_1 is smaller than the energy of the state S_2. For simplicity, let the degeneracy of the states be unity. When atoms make transitions between the two states, there are three main processes to consider: absorption, spontaneous emission, and stimulated emission. The other two processes that can induce transitions, Raman scattering and Rayleigh scattering occur, with significantly smaller probability because they are nonresonant processes.

For atoms and radiation in thermal equilibrium, transitions occur between S_1 and S_2. The upward transition rate, the number of atoms going from S_1 to S_2 per unit time ($W_{1 \rightarrow 2}$), is equal to the downward transition rate, the number of atoms going from S_2 to S_1 per unit time ($W_{2 \rightarrow 1}$). The upward transitions occur only by absorption, whereas the downward transitions occur by both spontaneous and stimulated emission. The equilibrium condition may be written,

$$W_{1 \rightarrow 2} = W_{2 \rightarrow 1} = W_{\text{spon}, 2 \rightarrow 1} + W_{\text{stim}, 2 \rightarrow 1}, \quad (13.11)$$

where ($W_{\text{spon}, 2 \rightarrow 1}$) and ($W_{\text{stim}, 2 \rightarrow 1}$) are the spontaneous and stimulated transition rates.

The transition rate for spontaneous emission is proportional to the number of atoms in S_2 (N_2). The transition rate may be written

$$W_{\text{spon},2\to1} = AN_2, \qquad (13.12)$$

where the factor A specifies the temperature-independent spontaneous emission probability. This factor is called the *Einstein A coefficient*.

The transition rate for absorption is proportional to the number of atoms in S_1 (N_1). The absorption rate is also proportional to the number of photons with the correct wavelength (λ) to be absorbed. If the energies of S_1 and S_2 are E_1 and E_2, the condition for resonant absorption is

$$E_2 - E_1 = \frac{hc}{\lambda}. \qquad (13.13)$$

The absorption rate is written

$$W_{1\to2} = B_{12} N_1 \frac{du}{d\lambda}. \qquad (13.14)$$

The factor B_{12} gives the temperature-independent absorption probability. This factor is called the *Einstein B coefficient*. The factor $du/d\lambda$ gives the photon density, the number of photons per volume per unit wavelength. The photon density is given by the thermal radiation formula of Planck:

$$\frac{du}{d\lambda} = \frac{c}{4} \frac{dR}{d\lambda} = \frac{\pi hc^3}{2\lambda^5} \frac{1}{e^{hc/\lambda kT} - 1}. \qquad (13.15)$$

The transition rate for stimulated emission is proportional to both N_2 and the photon energy density ($du/d\lambda$). The transition rate for stimulated emission is written

$$W_{\text{stim},2\to1} = B_{21} N_2 \frac{du}{d\lambda}, \qquad (13.16)$$

where a second Einstein B coefficient (B_{21}) gives the temperature-independent stimulated emission probability.

Einstein demonstrated that equality of the absorption and stimulated emission probabilities ($B_{12} = B_{21}$) leads to the universal radiation formula discovered by Planck. Taking the Planck distribution as an experimental fact, we use the same argument to prove that absorption and stimulated emission occur with equal probability.

In terms of the Einstein A and B coefficients, the rate equation (13.11) gives

$$B_{12} N_1 \frac{du}{d\lambda} = AN_2 + B_{21} N_2 \frac{du}{d\lambda}. \qquad (13.17)$$

Solving for the photon density ($du/d\lambda$), we have

$$\frac{du}{d\lambda} = \frac{A}{B_{12}\dfrac{N_1}{N_2} - B_{21}}. \qquad (13.18)$$

The ratio of the number of atoms in S_1 to the number in S_2 is given by the Boltzmann factor:

$$\frac{N_1}{N_2} = \frac{e^{-E_1/kT}}{e^{-E_2/kT}} = e^{(E_2-E_1)/kT}, \qquad (13.19)$$

and the photon density (13.18) becomes

$$\frac{du}{d\lambda} = \frac{A}{B_{12}e^{(E_2-E_1)/kT} - B_{21}}$$

$$= \frac{\left(\dfrac{A}{B_{21}}\right)}{\dfrac{B_{12}}{B_{21}}e^{(E_2-E_1)/kT} - 1}. \qquad (13.20)$$

We now compare the photon density (13.20) with the universal thermal spectrum obtained by Planck (13.15). The photon density becomes infinite as the temperature (T) approaches infinity. Therefore *at large values of temperature* we have

$$B_{12} = B_{21}. \qquad (13.21)$$

Since the Einstein coefficients (A, B_{12}, and B_{21}) *do not depend on temperature*, the equality of B_{12} and B_{21} holds at all temperatures. The two Einstein B coefficients are identical; the absorption probability is equal to the stimulated emission probability. We may simplify our notation by defining the constant (B) as

$$B \equiv B_{12} = B_{21}. \qquad (13.22)$$

(Note that we have assumed a degeneracy of unity. For degeneracies g_1 and g_2 of states S_1 and S_2, we have $B_{12}g_1 = B_{21}g_2$.) The photon energy density (13.20) is

$$\frac{du}{d\lambda} = \frac{\left(\dfrac{A}{B}\right)}{e^{(E_2-E_1)/kT} - 1}. \qquad (13.23)$$

Therefore, the ratio of the Einstein coefficients is

$$\frac{A}{B} = \frac{\pi h c^3}{2 \lambda^5}. \tag{13.24}$$

We now compare the stimulated and spontaneous transition rates:

$$\frac{W_{\text{stim}, 2\to1}}{W_{\text{spon}, 2\to1}} = \frac{BN_2 \dfrac{du}{d\lambda}}{AN_2}$$

$$= \frac{B}{A} \frac{du}{d\lambda}$$

$$= \frac{1}{e^{(E_2 - E_1)/kT} - 1}. \tag{13.25}$$

If the energy difference between the atomic levels is much larger than the energy of a typical radiation photon in thermal equilibrium, $E_2 - E_1 \gg kT$, then spontaneous emission occurs with a much greater probability than stimulated emission. There are simply not many photons that can stimulate the transition.

If the typical energy of a radiation photon is much larger than the energy difference between the atomic levels, $kT \gg E_2 - E_1$, then stimulated emission dominates over spontaneous emission. In this situation, the ratio of the total emission rate, spontaneous plus stimulated, to the absorption rate is

$$\frac{W_{2\to1}}{W_{1\to2}} = \frac{\left(AN_2 + B\dfrac{du}{d\lambda}N_2\right)}{B\dfrac{du}{d\lambda}N_1}$$

$$= \left(1 + \frac{A}{B\dfrac{du}{d\lambda}}\right)\frac{N_2}{N_1} \approx \frac{N_2}{N_1}. \tag{13.26}$$

Since kT is large compared to $E_2 - E_1$, we have

$$\frac{N_2}{N_1} \approx 1, \tag{13.27}$$

and

$$\frac{W_{2\to1}}{W_{1\to2}} \approx 1. \tag{13.28}$$

The absorption rate is equal to the emission rate for $kT \gg E_2 - E_1$.

In a system with two states S_1 and S_2, the probability per occupied state for resonant photon absorption $(S_1 + \gamma \to S_2)$ is identical to the probability for stimulated emission $(S_2 + \gamma \to S_1 + \gamma + \gamma)$.

EXAMPLE 13-1

Make an estimate of the ratio of stimulated emission probability to spontaneous emission probability at room temperature for systems that radiate in (a) the microwave region, (b) the visible region, and (c) the x-ray region.

SOLUTION:

The ratio of stimulated emission probability to spontaneous emission probability (13.25) is

$$\frac{W_{\text{stim}, 2\to1}}{W_{\text{spon}, 2\to1}} = \frac{1}{e^{(E_2 - E_1)/kT} - 1}.$$

(a) For microwave radiation,

$$\lambda \approx 1 \text{ mm},$$

$$E_2 - E_1 = \frac{hc}{\lambda} = \frac{1240 \text{ eV} \cdot \text{nm}}{10^6 \text{ nm}} \approx 10^{-3} \text{ eV},$$

$$\frac{E_2 - E_1}{kT} \approx \frac{10^{-3} \text{ eV}}{\left(\dfrac{1}{40} \text{ eV}\right)} = 4 \times 10^{-2},$$

and

$$\frac{W_{\text{stim}, 2\to1}}{W_{\text{spon}, 2\to1}} = \frac{1}{e^{(E_2 - E_1)/kT} - 1}$$

$$\approx \frac{kT}{E_2 - E_1} = 25.$$

(b) For visible radiation,

$$E_2 - E_1 \approx 2 \text{ eV},$$

$$\frac{E_2 - E_1}{kT} \approx \frac{2 \text{ eV}}{\left(\dfrac{1}{40} \text{ eV}\right)} = 80,$$

and

$$\frac{W_{stim,\,2\to1}}{W_{spon,\,2\to1}} = \frac{1}{e^{(E_2-E_1)/kT} - 1}$$

$$\approx e^{-80} \approx 10^{-35}.$$

(c) For x rays,

$$E_2 - E_1 \approx 25 \text{ keV},$$

$$\frac{E_2 - E_1}{kT} \approx \frac{25 \times 10^3 \text{ eV}}{\left(\frac{1}{40} \text{ eV}\right)} = 10^6,$$

and

$$\frac{W_{stim,\,2\to1}}{W_{spon,\,2\to1}} = \frac{1}{e^{(E_2-E_1)/kT} - 1}$$

$$\approx e^{-10^6} \approx 10^{-430000}. \quad\blacksquare$$

13-3 AMPLIFICATION OF RADIATION

In this section we give the general conditions necessary in order to cause amplification by stimulated emission. In order for amplification to occur, the number of photons from stimulated emission must be very large compared to the case of thermal equilibrium.

Population Inversion

Consider a molecular or atomic system where amplification is to be achieved by stimulated emission between the energy levels E_2 and E_1. Let the population of the energy level E_2 be N_2 and the population of the energy level E_1 be N_1. A photon with energy (E_2-E_1) is required to stimulate the transition; however, a photon of the same energy can also be absorbed. Since the stimulated emission and absorption probabilities per occupied state are equal (13.21), amplification can occur only if

$$N_2 > N_1. \quad (13.29)$$

Obviously, this is a not an equilibrium condition. A condition in which a higher energy state has a larger probability of being occupied than a lower energy state is called a *population inversion*.

In order to achieve a population inversion, energy must be supplied to the system to destroy the thermal equilib-

rium. There are a variety of techniques for providing this energy. The technique that is used depends on the exact energy levels of the particular molecular or atomic system. A necessary and important feature of the system is that the lifetime of the higher energy state must be long compared to the time that it takes to populate the state and initiate the stimulated transitions.

Containment

In order to achieve photon amplification, there must be some method for containment of the photons so that they can stimulate further transitions. The technique used to contain the radiation varies with the wavelength and the type of atom that is radiating. For amplification in the microwave region, a cavity length is tuned so that it supports standing waves of the appropriate wavelength. For amplification in the visible region, the walls of a cavity are made reflecting. Some examples are discussed in the next sections.

Coherence

Photons from stimulated transitions are coherent; each photon has the same energy, direction, polarization, and phase. A large number of photons compose a single plane wave whose amplitude is proportional to the number of radiating atoms. The intensity of the wave is proportional to the amplitude squared or to the square of the number of radiating atoms. In contrast, ordinary *noncoherent* radiation is composed of many random phases. The amplitude of noncoherent radiation is proportional to the square root of the number of radiating atoms (compare to the random walk problem). The intensity of noncoherent radiation is proportional to the number of radiating atoms. Thus, very large intensities are obtainable with coherent radiation compared to noncoherent radiation.

> Maser or laser operation requires (1) a molecular or atomic system with a long-lived intermediate state, so that a population inversion can be achieved, (2) an energy source to populate the excited state, and (3) containment of the photons from stimulated emission so that they can stimulate further transitions.

13-4 THE AMMONIA MASER

The first microwave amplification by stimulated emission was observed with the ammonia molecule (NH_3). The ammonia molecule has many states of electronic–vibrational–rotational excitation. For any given state there is an additional degree of freedom, which depends on the orientation of the nitrogen atom. The ammonia molecule is shown in Figure 13-3. The three hydrogen atoms form a plane. The nitrogen molecule may be on either side of this plane, as indicated in Figure 13-3a. The potential energy of the nitrogen molecule as a function the distance from the plane is indicated in Figure 13-3b. The solution to the Schrödinger equation for this potential has a symmetric and an antisymmetric solution. The energy difference between the two states (ΔE) is about

$$\Delta E \approx 10^{-4} \text{ eV}, \tag{13.30}$$

as indicated in Figure 13-3c. This energy difference is so small that the two states are equally populated at room temperature ($kT \gg \Delta E$). Transitions between the two states occur with the absorption and emission of photons of frequency

$$f = 2.39 \times 10^{10} \text{ Hz}, \tag{13.31}$$

corresponding to a wavelength of

$$\lambda = 0.0126 \text{ m}. \tag{13.32}$$

This wavelength is in the microwave region.

The ammonia molecule has an electric dipole moment whose orientation is given by the position of the nitrogen atom. When an electric dipole passes through a region that contains a nonuniform electric field, a force acts on the dipole. The direction of the force depends on the orientation of the dipole. (Molecules in the higher energy state are deflected in the direction of decreasing electric field.) Thus, a nonuniform electric field may be used to select one of the ammonia states.

The experimental arrangement of Gordon, Zeiger, and Townes is shown in Figure 13-4. A beam of ammonia molecules is sent into a region that has a cylindrical electric quadrupole field. When the ammonia molecules pass through the quadrupole field, there is an outward (defocusing) radial force on those molecules in the lower energy state and an inward (focusing) force on those molecules in the higher energy state. Only those ammonia molecules in the higher energy state are focused onto a small opening in a copper cavity. Thus, the quadrupole field is used to create a population inversion of the ammo-

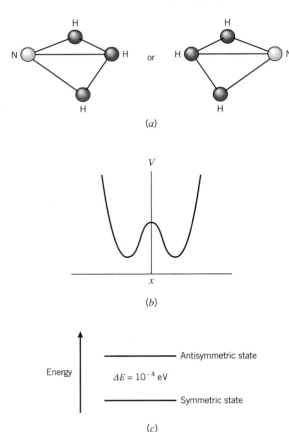

(a)

(b)

(c)

FIGURE 13-3 The ammonia molecule.
(a) The nitrogen atom may be on either side of the plane of hydrogen atoms. (b) The potential energy V(x) is symmetric, where x is the distance of the nitrogen atom from the hydrogen plane. (c) The energy difference between the symmetric and antisymmetric states is about 10^{-4} eV.

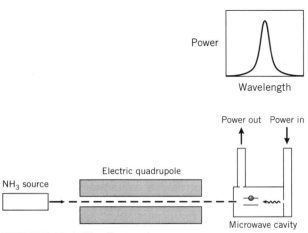

FIGURE 13-4 The first maser.
After J. P. Gordon, H. J. Zeiger, and C. H. Townes, *Phys. Rev.* **95**, 282 (1954).

nia molecules that enter the cavity. Microwave power of variable frequency is transmitted through the cavity. When the frequency of radiation in the cavity corresponds to the energy difference between the two states of the ammonia molecule (2.39×10^8 Hz), stimulated emission occurs. The experimenters observed that self-sustained oscillations occurred in the cavity because the microwaves from stimulated emission cause additional transitions. Thus, the ammonia maser could be used as an accurate clock. The experimenters also observed that the microwave power could be amplified with this device.

The ammonia maser is not a practical amplifier because it cannot be tuned; it can work at only one frequency. The ammonia maser was the predecessor of other masers that could be tuned. Microwave amplification may be achieved by stimulating transitions between energy levels that are split by the Zeeman effect. The frequency of operation can then be tuned by adjusting the magnetic field. A maser was used by Penzias and Wilson to discover the cosmic background radiation.

13-5 AMPLIFICATION AT INFRARED AND OPTICAL WAVELENGTHS

Design Considerations

In a maser, the size of the radiation cavity is comparable to the wavelength of the radiation. For infrared or optical wavelengths, this is no longer true. In 1958, Arthur Schawlow and Townes outlined a method for extension of the maser technique to infrared and optical wavelengths. Schawlow and Townes proposed that stimulated emission at infrared and optical wavelengths could occur in a resonant cavity that has a characteristic dimension of centimeters. The radiation could be contained by making the ends of the cavity highly reflecting. The radiation in a maser cavity may be thought of as a standing wave; the radiation in a laser cavity may be thought of as a plane wave that is reflected at the ends of the cavity. Schawlow and Townes proposed several possible laser sources including a crystal of aluminum oxide (Al_2O_3), or sapphire, in the shape of a rod. Sapphire was suggested because of its chemical inertness and excellent transparency.

A laser cavity has *multimodes* because its size is much larger than the wavelength of the radiation. There are many possible standing waves in the cavity. An important condition for the operation of the laser is that the rate of radiation energy loss due to absorption by the walls of the cavity not exceed the rate of energy produced by the stimulated emission. We may write the radiation intensity (I) as a function of time as

$$I(t) = I_0 e^{-t/t_c} \qquad (13.33)$$

where I_0 is the initial intensity and t_c is the lifetime of the radiation in the cavity. Consider a cubical cavity of length L and let the reflection coefficient at the walls of the cavity be a. The time that it takes to make n reflections is

$$t = \frac{nL}{c}. \qquad (13.34)$$

After n reflections, the radiation intensity is

$$I_0 e^{-nL/ct_c} = I_0 a^n. \qquad (13.35)$$

The lifetime of the radiation in the cavity is

$$t_c = \frac{L}{c \ln\left(\dfrac{1}{a}\right)} \approx \frac{L}{c(1-a)}, \qquad (13.36)$$

where we have used the approximation,

$$\ln\left(\frac{1}{a}\right) \approx 1 - a, \qquad (13.37)$$

valid for $a \approx 1$.

The rate of stimulated emission is proportional to difference in the number of atoms in the excited state and the ground state ($N_2 - N_1$). The laser condition is

$$N_2 - N_1 = \frac{N_m \tau}{t_c}. \qquad (13.38)$$

where (N_m) is the number of modes in a small wavelength interval ($\Delta\lambda$) corresponding to the line width, and (τ) is the lifetime of the excited state. The power that must be supplied to maintain the population difference, called the *pumping* power, is

$$P = \frac{N_m hc}{\lambda t_c}. \qquad (13.39)$$

The following example was used to illustrate the feasibility of constructing an infrared maser.

EXAMPLE 13-2

Basic parameters of an infrared maser. The device considered by Schawlow and Townes, which was a cubical cavity of gas with a volume of 10^{-6} m, reflection coefficient of $a = 0.98$, and wavelength of $\lambda = 1$ μm. (a) Estimate the width of the spectral line ($\Delta\lambda/\lambda$) if it is dominated by Doppler broadening. (b) Estimate the number of modes of oscillation in the wavelength interval $\Delta\lambda$. (c) Estimate the lifetime (t_c) of the radiation in the cavity. (d) Estimate the size of the population inversion needed so that stimulated emission dominates over cavity losses. (e) Estimate the minimum pumping power that must be supplied so that stimulated emission dominates over cavity loss.

SOLUTION:

(a) The width of the line is

$$\frac{\Delta\lambda}{\lambda} \approx \sqrt{\frac{2\,kT}{Mc^2}},$$

where M is the mass of the gas molecule. If we take the mass energy of the gas molecule to be $100\,\mathrm{GeV}$, or $10^{11}\,\mathrm{eV}$, we have

$$\frac{\Delta\lambda}{\lambda} \approx \sqrt{\frac{0.05\ \mathrm{eV}}{10^{11}\ \mathrm{eV}}} \approx 10^{-6}.$$

(b) The number of modes of oscillation in the wavelength interval ($\Delta\lambda$) is given by the density of states factor. In a volume V, the number of modes with wavelengths between λ and $\lambda + \Delta\lambda$ is given by

$$\frac{N_m}{V} = \frac{8\pi}{\lambda^3}\frac{\Delta\lambda}{\lambda},$$

or

$$N_m = \frac{8\pi}{\lambda^3}\frac{\Delta\lambda}{\lambda}V \approx \frac{(8)(\pi)\left(10^{-6}\right)\left(10^{-2}\ \mathrm{m}\right)^3}{\left(10^{-6}\ \mathrm{m}\right)^3}$$

$$\approx 3\times 10^7.$$

(c) The lifetime of the radiation in the cavity is given by

$$t_c = \frac{L}{(1-a)c} = \frac{0.01\ \mathrm{m}}{(1-0.98)\left(3\times 10^8\ \mathrm{m/s}\right)} \approx 2\times 10^{-9}\ \mathrm{s}.$$

(d) The rate of stimulated emissions is proportional to the difference in the number of atoms in the excited state and the ground state (N_2-N_1). The laser condition is

$$N_2 - N_1 = \frac{N_m\tau}{t_c},$$

where τ is the lifetime of the excited state against spontaneous decay. A typical value of a spontaneous decay lifetime is

$$\tau \approx 10^{-8}\ \mathrm{s},$$

which gives a population inversion of

$$N_2 - N_1 = \frac{\left(3\times 10^7\right)\left(10^{-8}\ \mathrm{s}\right)}{2\times 10^{-9}\ \mathrm{s}} \approx 2\times 10^8.$$

(e) The minimum pumping power that must be provided to overcome cavity losses is

$$P = \frac{\left(N_2 - N_1\right)hc}{\tau\lambda} \approx \frac{\left(2\times 10^8\right)(1240\ \mathrm{eV\text{-}nm})}{\left(10^{-8}\ \mathrm{s}\right)\left(10^3\ \mathrm{nm}\right)}$$

$$\approx 2\times 10^{16}\ \mathrm{eV/s} \approx 1\ \mathrm{mW}. \qquad\blacksquare$$

Since the number of modes (N_m) varies inversely as the third power of the wavelength (for fixed $\Delta\lambda/\lambda$), the laser pumping power (13.39) varies as the inverse fourth power of the wavelength. Thus, enormous pumping powers are required for the operation of lasers at small wavelengths. In Example 13-2 we have estimated that a laser with a volume of $10^{-6}\ \mathrm{m}^3$ operating at $\lambda = 1$ μm requires a pumping power of 1 mW. Figure 13-5 shows the required pumping power for other lasers of the same size as a function of the laser wavelength.

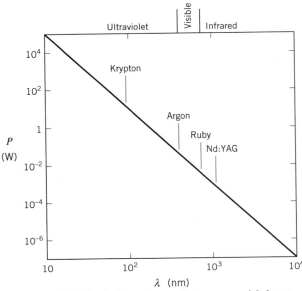

FIGURE 13-5 Variation of pumping power with laser wavelength for lasers with a cavity volume of $10^{-6}\,\mathrm{m}^3$.

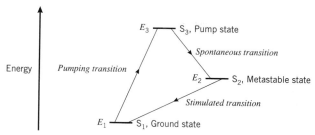

FIGURE 13-6 Energy-level diagram for a three-level laser.

Configuration of the Energy Levels

Three-Level Laser

Consider three energy states in an atom as shown in Figure 13-6. In thermal equilibrium at room temperature, essentially all of the atoms are in the ground state. Suppose that the state of intermediate energy (S_2) has a long lifetime. This situation occurs if spontaneous transitions from the intermediate state are suppressed by a selection rule. Such a long-lived state is called a *metastable* state. An example of a metastable state is the 2s state of helium. The atomic transition to the 1s state is suppressed by the orbital angular momentum selection rule, $\Delta\ell = \pm 1$. In order to excite an electron from the ground state to the metastable state, without violating the selection rule, we may excite the electron to a higher energy state (S_3) that has an allowed spontaneous decay to the metastable state:

$$S_1 + \gamma \rightarrow S_3 \rightarrow S_2 + \gamma. \qquad (13.40)$$

The radiation that is used to excite the atoms out of the ground state is called *pumping* radiation. In the three-level system of Figure 13-6, radiation quanta with energy ($E_3 - E_1$) must be supplied to the system in order to excite atoms from the ground state into the pump state. The pump state is very short-lived, with a decay by spontaneous emission into the long-lived metastable state. The pumping of atoms into the excited state can also be achieved electrically by creating a current of electrons that collide with the electrons in the atoms and excite them. If we pump the atoms out of the ground state at a faster rate than the metastable state decays, then we may achieve the desired population inversion. We will have more atoms in S_2 than in S_1. Transitions from S_2 to S_1 can now occur at a faster rate than transitions in the reverse direction. The stimulated transition is

$$S_2 + \gamma \rightarrow S_1 + \gamma + \gamma. \qquad (13.41)$$

The laser transition has an energy ($E_2 - E_1$). The photons are coherent and they have just the right energy to induce further transitions in other atoms.

Four-Level Laser

Figure 13-7 shows an energy level diagram for a four-level laser with energy levels for the ground state (S_1 with energy E_1), the laser final state (S_2 with energy E_2), the metastable state (S_3 with energy E_3), and the pump state (S_4 with energy E_4). The four-level laser differs from the three-level laser in that the stimulated transitions are from S_3 to S_2, and the ground state is not involved. This means that the population inversion must be maintained between the states S_2 and S_3. Since the S_2 depletes itself naturally through spontaneous transitions to the ground state S_1, it is much easier to maintain a population inversion in a four-level than in a three-level laser. Radiation quanta with energy ($E_4 - E_1$) must be supplied to the system in order to excite atoms from the ground state into the pump state. The pump state is very short-lived, with a decay by spontaneous emission into a long-lived metastable state.

13-6 EXAMPLES OF LASERS

The Ruby Laser

In 1960, Theodore H. Maiman built the first laser from a crystal of ruby (see color plate 3). The ruby crystal consists of aluminum oxide (Al_2O_3) in which some of the aluminum atoms ($Z = 13$), are replaced with chromium atoms ($Z = 24$). Aluminum has a single 3p electron. Chromium

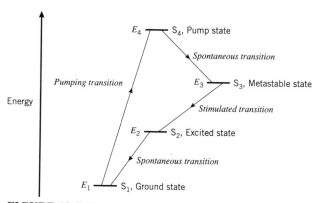

FIGURE 13-7 Energy-level diagram for a four-level laser.

is the fourth transition element, so triply ionized chromium (Cr^{3+}) has a single $3d$ electron and has properties similar to aluminum. Chromium atoms can replace aluminum atoms as an impurity in an occasional Al_2O_3 lattice position. In a ruby crystal there is about 0.05% chromium by weight. The chromium ion, Cr^{3+}, is the active atom in the ruby laser.

The relevant energy levels of Cr^{3+} are shown in Figure 13-8. There are two broad-bands of states with energies about 2.3 eV and 3.0 eV above the ground state. At an energy of 1.79 eV above the ground state is a narrow metastable state. (This narrow state actually has a fine structure splitting of about 3.6×10^{-3} eV). Since the high-energy states are so broad, they can be excited using a broadband light source such as a flash lamp. The electrons in the states E_3 or E_4 decay rapidly with a lifetime of about 10^{-7} s to the state E_2 with the excess energy going into vibrational modes of the crystal. The lifetime of the metastable state E_2 is about 5×10^{-3} seconds. A population inversion is achieved if the electrons in the chromium ions are pumped out of the ground state at a faster rate than they spontaneously decay back to the ground state. In this case there can be stimulated transitions with a photon energy of 1.79 eV ($\lambda = 694$ nm).

The geometry of the ruby crystal laser is indicated in Figure 13-9. A single crystal of ruby has the shape of a cylindrical rod. One end of the rod is made totally reflecting. The other end of the rod is made partially transmitting

FIGURE 13-9 The ruby laser.

to allow the laser beam to be emitted. The pumping radiation is supplied by a xenon flash lamp that is wound around the axis of the ruby crystal. When the chromium atoms are pumped, stimulated transitions are emitted in all directions. Photons that happen to go along the axis of the crystal are reflected back into the crystal by the mirrored ends and cause further stimulated transitions along the crystal axis. The duration of one flash from the xenon lamp lasts typically about 1 ms. Near the beginning of the flash cycle, the population of the metastable state increases rapidly until there is a population inversion sufficiently large to cause amplification. The amplification depletes the population inversion on the time scale of a few microseconds, much shorter than the time scale of the flash lamp. Therefore, the population inversion increases rapidly again until the point of amplification. Thus, one flash of the xenon lamp makes a laser output with several narrow spikes. The laser output is irregular because there are several oscillation modes in the ruby cavity.

EXAMPLE 13-3

Estimate the instantaneous laser power that can be delivered from a ruby crystal of length $L = 0.1$ m and diameter $d = 5$ mm, in a 5 μs pulse.

SOLUTION:

We need to determine the number of chromium atoms in the crystal. The density of aluminum oxide is

$$\rho = 3.7 \times 10^3 \text{ kg/m}^3.$$

The volume of the crystal is

$$V = \frac{\pi d^2 L}{4} = \frac{(3.14)(0.005 \text{ m})^2 (0.1 \text{ m})}{4} \approx 2 \times 10^{-6} \text{ m}^3.$$

The molecular mass number is

$$2 \times 27 + 3 \times 16 = 102.$$

The number of chromium atoms in the crystal is

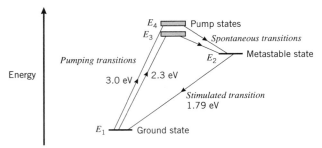

FIGURE 13-8 Energy level diagram for the ruby crystal laser.
Impurity chromium atoms present in a ruby crystal have energy levels as shown. This atom is responsible for the laser action. Photons with energies near 2.3 eV or 3.0 eV are supplied to excite or "pump" electrons from the ground state (E_1) to a multiplet of closely spaced states with energy E_3 or E_4. These states can decay spontaneously into a long-lived state with energy E_2. Photons with energy of 1.79 eV stimulate transitions to the ground state. Laser action can occur if there are more electrons in the metastable state than in the ground state.

$$N = \frac{N_A \rho V}{(102)(10^{-3} \text{ kg})}$$

$$= \frac{(6 \times 10^{23})(3.7 \times 10^3 \text{ kg/m}^3)(2 \times 10^{-6} \text{ m}^3)}{(102)(10^{-3} \text{ kg})}$$

$$\approx 4.3 \times 10^{23}.$$

The number of chromium atoms in the crystal is 5×10^{-4} N. If all of the chromium atoms are excited, then the stored energy is

$$E = (5 \times 10^{-4})(4.3 \times 10^{22})(1.8 \text{ eV}) \approx 3.9 \times 10^{19} \text{ eV}.$$

If we take the duration of the laser pulse to be 5 μs, the instantaneous laser power is

$$P = \frac{3.9 \times 10^{19} \text{ eV}}{5 \times 10^{-6} \text{ s}} = 8 \times 10^{24} \text{ eV/s} \approx 1.2 \times 10^6 \text{ W}. \blacksquare$$

The power output of a laser pulse may be significantly enhanced by a technique called *Q switching*. In a *Q*-switched system, amplification is prevented until the population inversion greatly exceeds the threshold value. Amplification can be prevented by allowing the stimulated radiation to escape until the pumping is complete and then suddenly containing the radiation. This may be accomplished in a variety of ways. One way is to make a shutter that shields the totally reflecting mirror until pumping is complete and then opening the shutter suddenly. Another way is to make the total reflector rotate so that it is aligned only momentarily, synchronized with the pumping flash. A third method is to place a "dye cell" at the end of the cavity. The dye is matched to the output of the laser and the concentration is adjusted so that stimulated radiation cannot reach the reflecting mirror because it is absorbed in the dye. The laser pumps the dye molecules, and when there is a population inversion in the dye, the dye is transparent to the laser and the radiation is contained for amplification. The technique of *Q* switching can provide very fast pulses (about 10–20 ns) and very large peak powers (about 100 MW).

The Nd:YAG Laser

The Nd:YAG laser medium consists of neodymium atoms, about 1%, incorporated into an yttrium-aluminum-garnet (YAG) crystal. The Nd:YAG laser is a four-level laser that is optically pumped. An energy-level diagram of

the Nd:YAG laser is shown in Figure 13-10. The output of the Nd:YAG laser has a wavelength

$$\lambda = 1.06 \text{ μm}. \qquad (13.42)$$

The size of the laser cavity is limited to about a centimeter because of the difficulties in growing the crystals. The YAG crystal has a high thermal conductivity, which allows for high-power laser operation. The Nd:YAG laser is capable of continuous operation at a power of 250 watts, or may be pulsed at a rapid rate with a peak power in the megawatt range. The Nd:YAG laser is rugged and is commonly used for cutting and drilling metals and other solids.

A similar laser may be made by implanting the neodymium atoms in a glass. These Nd:glass lasers can be very large (several meters) and are capable of very large energy outputs (100 kJ per pulse).

The Helium–Neon Gas Laser

The helium–neon laser was the first gas laser. A mixture of helium and neon gas have the proper combination of energy levels to make a practical laser. The relevant energy levels of the helium and neon atoms are shown in Figure 13-11. The long-lived state is the $2s$ state of helium. Recall that this notation means that the electron configuration of this state is $1s^1 2s^1$ (see Figure 9-11). This state is metastable because photon transitions to the ground state $(1s^2)$ are forbidden by the selection rule, $\Delta \ell = 1$. The $2s$ state of helium may be populated by running an electric current through the gas, which causes electrons to collide

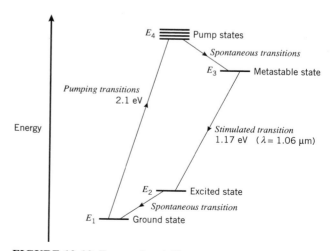

FIGURE 13-10 Energy-level diagram for the Nd:YAG laser.

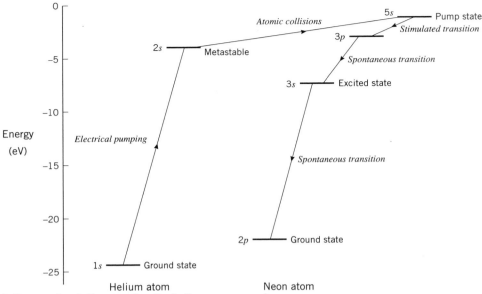

FIGURE 13-11 Energy-level diagram for the helium–neon gas laser.
Electrons in helium atoms are excited to the $2s$ state by means of an electrical current. This state is long-lived because its spontaneous decay is forbidden by the selection rule, $\Delta \ell = 1$. This energy is transferred to neon atoms through atomic collisions because the energy separation between the $1s$ and $2s$ states in helium is just about the same as the energy separation between the $2p$ and $5s$ states in neon. A population inversion is achieved and laser action can occur from the $5s$ to the $3p$ states.

with one another, leaving some of them with just enough energy to be elevated to the $2s$ state. Thus, the electrons are pumped electrically from the ground state to the metastable state. Since there is no p state for decay of the excited electrons, we cannot stimulate transitions from the $2s$ state in the helium atom. This is where the neon atom plays its role. There is an excited state of neon that just happens to have an energy nearly equal to the excitation energy of the $2s$ state of helium. The ground state of neon is $2p$ (an abbreviation for $1s^2 2s^2 2p^6$). The $5s$ excited state ($1s^2 2s^2 2p^5 5s^1$) is 20.66 eV above the ground state,

$$E_{5s} - E_{2p} = 20.66 \text{ eV}, \qquad (13.43)$$

which very nearly matches the separation between the $1s$ and $2s$ states of helium:

$$E_{2s} - E_{1s} = 20.61 \text{ eV}. \qquad (13.44)$$

Due to this accidental match in energy-level differences, an excited helium atom can transfer its energy to a neon atom in the ground state through an atomic collision:

$$\text{He}(2s) + \text{Ne}(2p) \rightarrow \text{He}(1s) + \text{Ne}(5s). \qquad (13.45)$$

Normally, there are essentially no electrons in either the

$3p$ or $5s$ states of neon. Through repeated helium–neon collisions, the $5s$ state is populated. The $3p$ state is then populated by spontaneous decays, but only briefly because the $3p$ states decay spontaneously into the $3s$ states. If the $5s$ state is populated at a faster rate than it decays, then a population inversion is established between the $5s$ and $3p$ states of neon:

$$n_{5s} > n_{3p}. \qquad (13.46)$$

Laser action can be established with $5s \rightarrow 3p$ transitions.

The energy difference between the $5s$ and $3p$ states of neon is

$$E_{5s} - E_{3p} = 1.96 \text{ eV}. \qquad (13.47)$$

Therefore, the wavelength of the laser light is

$$\lambda = \frac{hc}{E} = \frac{1240 \text{ eV} \cdot \text{nm}}{1.96 \text{ eV}} = 633 \text{ nm}, \qquad (13.48)$$

which is in the red region of the visible spectrum. There are other transitions in neon that are in the yellow, green, and infrared regions. The helium–neon gas laser is low power (milliwatt range). The helium–neon gas laser is widely used at supermarket checkout counters to read the product bar codes, and in many other applications. The geometry of the helium–neon laser is shown in Figure 13-12.

EXAMPLE 13-4

Calculate the rate at which neon atoms must be excited in order to produce a helium–neon gas laser beam with a power of 1 mW.

SOLUTION:

The rate (R) at which neon atoms must be excited is the laser power divided by the energy of a photon (E) in the laser light

$$R = \frac{P}{E},$$

or

$$R = \frac{10^{-3} \text{ J/s}}{1.96 \text{ eV}} \left(\frac{1 \text{ eV}}{1.6 \times 10^{-19} \text{ J}} \right) \approx 3 \times 10^{15} \text{ s}^{-1}.$$

Comparison with a Thermal Source

The photons from a laser are concentrated in a small wavelength interval ($\Delta\lambda$) called the laser *bandwidth*. The photon flux from the laser in this wavelength interval is immensely larger than can be achieved with a thermal source. The main reason for this is that $\Delta\lambda/\lambda$ is extremely small, a typical value being

$$\frac{\Delta\lambda}{\lambda} \approx 10^{-11}. \tag{13.49}$$

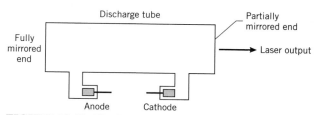

FIGURE 13-12 The helium–neon gas laser.
A gas discharge tube is filled with a mixture of helium ($\sim 10^{-3}$ atm) and neon ($\sim 10^{-4}$ atm). The gas is ionized by a direct current. Helium atoms are pumped to a metastable excited state ($2s$) by collisions with electrons. This excitation energy of the helium atoms is transferred to neon atoms through atomic collisions, leaving the neon atoms in the $5s$ state. Stimulated transitions occur between the $5s$ and $3p$ states of neon. One end of the discharge tube is made fully reflecting and the other end is made partially transmitting to allow the laser beam to exit.

Another reason for the large photon flux from a laser beam is the small angular divergence. Consider a laser beam emitted from a mirror of diameter d_m as indicated in Figure 13-13. The angular divergence (θ) of the laser beam is given by

$$\lambda = d_m \sin\theta \approx d_m\theta, \tag{13.50}$$

analogous to diffraction from a single slit (see Figure 1-2). The angular divergence is so small that a laser beam has been sent from the earth to a mirror of size 1/4 square meter set up on the moon by Apollo astronauts. The reflected beam was detected back on earth, and measurement of the transit time of the laser pulse gives a measurement of the distance to the moon to an accuracy of 10 centimeters!

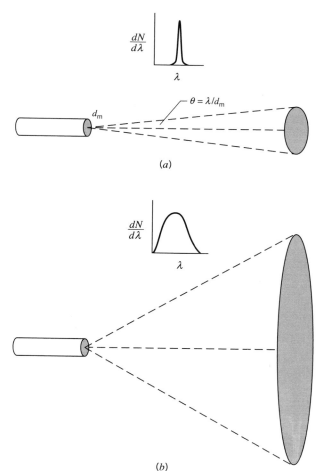

FIGURE 13-13 Comparison of a laser source with a thermal source.
(*a*) The laser source has a small bandwidth and a small angular divergence. (*b*) The thermal source has a large bandwidth and a large angular divergence.

EXAMPLE 13-5

Consider a typical helium–neon laser ($\lambda = 633$ nm) with an output power of 1 milliwatt and a fractional bandwidth of $\Delta\lambda/\lambda = 10^{-11}$. Estimate the temperature of a thermal source that would give the same photon flux as the laser in a small wavelength interval near 633 nanometers.

SOLUTION:

Let the diameter of the light source be d_m and let L be the distance from the light source to the point where the photon flux is measured. The diameter of the laser beam at the distance L is

$$d \approx L\theta = \frac{L\lambda}{d_m}.$$

The laser photon flux (Φ_{laser}) is

$$\Phi_{\text{laser}} = \frac{P_{\text{laser}}}{\left(\dfrac{hc}{\lambda}\right)\left(\dfrac{\pi d^2}{4}\right)} = \frac{4\,\lambda P_{\text{laser}}}{\pi d^2 hc},$$

where P_{laser} is the laser power. For a thermal source, the power is given by the universal formula. The power per area (ΔR) in a small wavelength interval ($\Delta\lambda$) is

$$\Delta R = \frac{dR}{d\lambda}\Delta\lambda = \frac{2\pi hc^2}{\lambda^5\left(e^{hc/\lambda kT}-1\right)}\Delta\lambda.$$

Taking the area of the thermal source to be the same as the area of the laser mirror, the power from the thermal source (P_{thermal}) is

$$P_{\text{thermal}} = \Delta R\left(\frac{\pi d_m^{\,2}}{4}\right) = \frac{2\pi hc^2}{\lambda^5\left(e^{hc/\lambda kT}-1\right)}\Delta\lambda\left(\frac{\pi d_m^{\,2}}{4}\right).$$

The thermal photons are isotropic, filling up the solid angle 2π. The thermal photon flux (Φ_{thermal}) is

$$
\begin{aligned}
\Phi_{\text{thermal}} &= \frac{P_{\text{thermal}}}{\left(\dfrac{hc}{\lambda}\right)\left(2\pi L^2\right)} \\[2ex]
&= \frac{\dfrac{2\pi hc^2}{\lambda^5\left(e^{hc/\lambda kT}-1\right)}\Delta\lambda\left(\dfrac{\pi d_m^{\,2}}{4}\right)}{\left(\dfrac{hc}{\lambda}\right)\left(2\pi L^2\right)} \\[2ex]
&= \frac{\pi c\Delta\lambda}{4\lambda^4\left(e^{hc/\lambda kT}-1\right)}\left(\frac{d_m}{L}\right)^2.
\end{aligned}
$$

Using $d_m/L = \lambda/d$, we have

$$\Phi_{\text{thermal}} = \frac{\pi c\Delta\lambda}{4\lambda^2 d^2\left(e^{hc/\lambda kT}-1\right)}.$$

In the limit of very large temperatures ($kT \gg hc/\lambda$), the thermal flux reduces to

$$\Phi_{\text{thermal}} = \frac{\pi\Delta\lambda kT}{4h\lambda d^2}.$$

Finally, we determine the temperature where the fluxes are equal:

$$\Phi_{\text{thermal}} = \Phi_{\text{laser}},$$

or

$$\frac{\pi\Delta\lambda kT}{4h\lambda d^2} = \frac{4\,\lambda P_{\text{laser}}}{\pi d^2 hc}.$$

Solving for kT we get

$$
\begin{aligned}
kT &= \frac{16\,\lambda P_{\text{laser}}}{\pi^2 c\left(\dfrac{\Delta\lambda}{\lambda}\right)} \\[2ex]
&= \left(\frac{16}{\pi^2}\right)\frac{\left(6.33\times10^{-7}\text{ m}\right)\left(10^{-3}\text{ J/s}\right)}{\left(3\times10^8\text{ m/s}\right)\left(10^{-11}\right)}\left(\frac{1\,\text{eV}}{1.6\times10^{-19}\text{ J}}\right) \\[2ex]
&\approx 2\times10^{12}\text{ eV},
\end{aligned}
$$

or

$$T \approx \frac{300\,\text{K}}{\left(\dfrac{1}{40}\,\text{eV}\right)}2\times10^{12}\text{ eV} \approx 2\times10^{16}\text{ K}.$$

The corresponding thermal temperature is very large. ∎

The foregoing example shows that even a low power laser has a photon flux far exceeding any laboratory thermal source. This property leads to many unique applications of the laser.

Holography

In ordinary photography, the intensity (amplitude squared) of the scattered light is recorded without any information about the phase, resulting in a two-dimensional image. If we record both the amplitude and the phase of scattered light, then a three-dimensional image may be made. Gas lasers have been used to create three-dimensional images or *holograms* of objects. The generation of holograms is made possible by the coherency of the laser light. The experimental setup for creation of a hologram is shown in Figure 13-14. Coherent light from a laser is incident on the object to be imaged. Laser light reflected from the object is collected on a photographic plate together with a part of the laser beam, which is used as a reference. The intensity on the photographic plate is the square of the sum of the amplitudes of the reference beam and the scattered light. The interference between the reference beam and the scattered light depends on their relative phases.

After development, the photographic plate is illuminated with a laser beam similar to that used as the reference beam. The reflected beam gives a three-dimensional image of the original object (see color plates 5 and 6).

The Length Standard

Laser technology has provided an accurate measurement of the speed of light, which has led to the current definition of the meter. A laser may be provided with a feedback mechanism that compares the laser wavelength with a reference wavelength from a spectral line. The laser feedback mechanism is used to tune the laser wavelength by mechanically adjusting the cavity length. Such a device is called a *stabilized laser*. The frequency of a stabilized laser may be accurately determined by comparison with the frequency of an atomic cesium clock (the time stan-

dard), and the wavelength may be accurately determined by comparison with the wavelength of a spectral line from ^{86}Kr (the former length standard). The product of the wavelength and frequency from a stabilized laser has provided a precise measurement of the speed of light,

$$c = 299,792,458 \pm 1 \text{ m/s}. \qquad (13.51)$$

The experimental uncertainty of this measurement of the speed of light is dominated by the natural width of the ^{86}Kr spectral line. In 1983, the meter was redefined such that the speed of light in vacuum is exactly

$$c = 299,792,458 \text{ m/s}. \qquad (13.52)$$

With this definition of the speed of light, stabilized lasers provide a length standard. The frequency (f) of the laser is determined by comparison with the standard cesium clock and the wavelength is calculated from $\lambda = c/f$, using the definition of c.

Other Lasers

There are many different types of lasers, having a wide variety of applications. The wavelengths of several lasers are summarized in Figure 13-15.

CHAPTER 13: PHYSICS SUMMARY

- In a system with two states S_1 and S_2, the probability per occupied state for resonant photon absorption ($S_1 + \gamma \rightarrow S_2$) is identical to the probability for stimulated emission ($S_2 + \gamma \rightarrow S_1 + \gamma + \gamma$). The photons emitted by stimulated emission are coherent.

- If the energy difference between atomic levels is much larger than kT, then spontaneous emission occurs with a much greater probability than stimulated emission. If the energy difference between atomic levels is much smaller than kT, then stimulated emission dominates over spontaneous emission.

- The necessary criteria for amplification by stimulated emission are: an intermediate state that has a long lifetime, an energy source to create a population inversion, and a method of containment of the photons from stimulated emission so that they can stimulate further transitions causing amplification of the radiation.

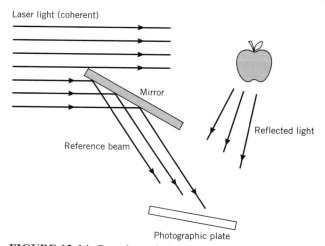

FIGURE 13-14 Creating a hologram.

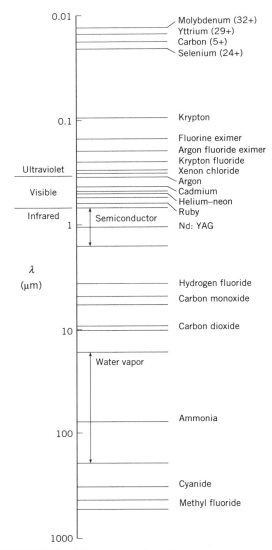

FIGURE 13-15 Laser wavelengths.

Encyclopedia of Lasers and Optical Technology, ed. Robert Meyers, Academic Press (1991).

H. P. Freund and R. K. Parker, "Free-Electron Lasers," *Sci. Am.* **260**, No. 4, 84 (1989).

J. P. Gordon, "The Maser," *Sci. Am.* **199**, No. 6, 42 (1958).

J. P. Gordon, H. J. Zeiger, and C. H. Townes, "Molecular Microwave Oscillator and New Hyperfine Structure in the Microwave Spectrum of NH_3," *Phys. Rev.* **95**, 282 (1954).

R. N. Hall et al., "Coherent Light Emission from GaAs Junctions," *Phys. Rev. Lett.* **9**, 366 (1962).

S. Haroche and D. Kleppner, "Cavity Quantum Electrodynamics," *Phys. Today* **42**, No. 1, 24 (1989).

A. Javan, W. R. Bennett, Jr., and D. R. Herriott, "Population Inversion and Continuous Optical Maser Oscillation in a Gas Discharge Containing a He-Ne Mixture," *Phys. Rev. Lett.* **6**, 106 (1961).

V. S. Letokhov, "Detecting Individual Atoms and Molecules with Lasers," *Sci. Am.* **259**, No. 3, 54 (1988).

T. H. Maiman, "Stimulated Optical Radiation in Ruby," *Nature* **187**, 493 (1960).

D. L. Matthews and M. D. Rosen, "Soft-X-Ray Lasers," *Sci. Am.* **259**, No. 6, 86 (1988).

M. I. Nathan et al., "Stimulated Emission of Radiation from GaAs *p-n* Junctions," *Appl. Phys. Lett.* **1**, 62 (1962).

W. D. Phillips and H. J. Metcalf, "Cooling and Trapping of Atoms," *Sci. Am.* **256**, No. 3, 50 (1987).

A. L. Schawlow, "Lasers and Coherent Light," *Phys. Today* **17**, No. 1, 28 (1964).

A. L. Schawlow, "Laser Light," *Sci. Am.* **218**, No. 3, 120 (1968).

A. L. Schawlow and C. H. Townes, "Infrared and Optical Masers," *Phys. Rev.* **112**, 1940 (1958).

Charles H. Townes, "Masers," *Proceedings of the International School of Physics Enrico Fermi*, Lake Como, Italy, Academic Press (1960).

W. T. Tsang, "The C^3 Laser," *Sci. Am.* **251**, No. 5, 148 (1984).

A. Yariv, *Optical Electronics,* Saunders (1991).

M. Young, *Optics and Lasers,* Springer-Verlag (1992).

REFERENCES AND SUGGESTIONS FOR FURTHER READING

M. W. Berns, "Laser Surgery," *Sci. Am.* **264**, No. 6, 84 (1991).

Steven Chu, "Laser Trapping of Neutral Particles," *Sci. Am.* **266**, No. 2, 84 (1992).

C. N. Cohen-Tannoudji and W. D. Phillips, "New Mechanisms for Laser Cooling," *Phys. Today* **43**, No. 10, 33 (1990).

R. S. Craxton, R. L. McCrory, and J. M. Soures, "Progress in Laser Fusion," *Sci. Am.* **255**, No. 2, 68 (1986).

T. F. Deutch, "Medical Applications of Lasers," *Phys. Today,* **41**, No. 10, 56 (October 1988).

QUESTIONS AND PROBLEMS

Summary of photon–atom interactions

1. The daytime sky is full of light from the sun, scattered by the atmosphere of the earth. Since the scattering varies as λ^{-4}, why doesn't the sky have a violet color?

2. Why are the processes (a), (b) and (c) of Figure 13-2 resonant processes while the others are not?

Stimulated emission of radiation

3. *Einstein's derivation of the Planck formula.* Show that the classical formula of Rayleigh-Jeans together

with the equality of the absorption and stimulated emission probabilities leads to the universal radiation formula first obtained by Planck.

4. Consider a system where transitions between two energy levels occur by absorption and emission of photons. At approximately what wavelength is the ratio of stimulated emission probability equal to the spontaneous emission probability at (a) room temperature, (b) liquid nitrogen temperature (77 kelvin), and (c) a temperature of 5800 kelvin?

5. (a) Why is a population inversion necessary for amplification by stimulated emission? (b) Is it easier to achieve a population inversion at higher or lower temperatures? Explain!

The ammonia maser

6. (a) Make an order-of-magnitude estimate for the size of the electric dipole moment of the ammonia molecule. (b) Make an order-of-magnitude estimate of the length of an electric quadrupole needed to deflect ammonia molecules by 1 centimeter. (*Hint:* You will need to make an estimate of the field gradient.)

*7. The maser of Gordon, Zeiger, and Townes used a cylindrical copper cavity of length $L = 12$ centimeters (see Figure 13-4). Ammonia molecules from the electrostatic separator with a speed v were directed into the cavity. (a) Calculate the most probable speed of the ammonia molecules at room temperature. (b) Standing electromagnetic waves were set up in the cavity such that the amplitude of the electric field varies with time (t) as

$$E(t) = \sin\left(\frac{\pi v t}{L}\right)\sin(\omega t),$$

where the range of t is from 0 to L/v and ω is the input frequency of the electromagnetic field inside the cavity. The transition probability is proportional to the square of the electric field. The output power is measured as a function of input frequency in the region of the ammonia inversion transition. Make an estimate of the width of the inversion line.

8. (a) Consider a wave guide of length L. Show that in thermal equilibrium the power flowing in each direction in a bandwidth Δf is given by

$$\Delta P = kT\Delta f .$$

(*Hint:* Calculate the number of modes per length and multiply by the average energy per mode.) (b) The power calculated in part (a) gives the noise of a microwave amplifier. The *noise temperature* of an amplifier may be defined by the input power needed to double the output power. The maser amplifier used by Penzias and Wilson to discover the cosmic microwave radiation at a wavelength of 7.35 centimeters had a noise temperature of 7 kelvin and a bandwidth of 25 MHz. Determine the input power needed to double the output power for this device.

Amplification at infrared and optical wavelengths

9. Why was the first maser built several years before the first laser?

10. How can the helium–neon laser be made to emit yellow or green light? (What energy levels are involved?)

11. In a certain four-level laser, the laser transition is to a state that is 0.042 eV above the ground state. Calculate the fraction of atoms in the excited state as a function of temperature in the range from 70 K to 300 K when no pumping radiation is present. Make a graph of your results. What does your result say about the efficiency of such a laser?

12. Laser action has been observed in the far ultraviolet region with quintuply ionized carbon (C^{+5}). The wavelength of the radiation is about 18 nanometers (see Figure 13-15). What transition produces the laser radiation?

13. Consider photon emission of a gas molecule moving with a nonrelativistic speed v. (a) Calculate the fractional shift in photon energy when the photon decay goes along the direction of motion of the molecule. (b) Show that the contribution to the fractional line width of a gas laser due to Doppler motion is

$$\frac{\Delta\lambda}{\lambda} = \sqrt{\frac{2kT}{Mc^2}}$$

where M is the molecular mass. (*Hint:* Take the typical molecular speed of a gas molecule to be the most probable speed.)

14. Estimate the contribution to the bandwidth ($\Delta\lambda$) from Doppler broadening for a helium–neon laser operating at $\lambda = 633$ nm at room temperature.

15. Why is the helium–neon laser not optically pumped?

16. *Optical Resonator.* Consider a laser designed for operation at $\lambda = 633$ nm. The cavity length is 0.01 m. (a) Estimate the photon lifetime in the cavity. (b) Use the uncertainty principle to estimate the laser bandwidth ($\Delta\lambda$). (c) Estimate the number of modes for a cavity volume of 10^{-6} m³. (d) Determine the Q factor

of the resonator, defined by $Q = \Delta\lambda/\lambda$. (e) The mirrors are designed so that only a small number of modes have a large Q factor. Find a relationship between the photon cavity lifetime and the Q factor.

17. Consider a helium–neon laser ($\lambda = 633$ nm) with a glass discharge tube of length 0.10 m. (a) Estimate the temperature change needed to destroy any given mode of vibration for laser operation. The linear coefficient of thermal expansion for glass is about 8×10^{-6} K^{-1}. (b) When the temperature changes by an amount more than that calculated in part (a), the power output of a helium–neon laser is observed to be nearly constant. Why?

18. A laser has mirrors with a reflection coefficient of 0.99. What is the maximum length of a laser cavity if the stimulated radiation is to be contained for 10^{-8} seconds?

19. A laser operating at $\lambda = 550$ nm has a bandwidth of 10^{-8} nm. The lifetime of radiation in the cavity and the spontaneous decay lifetime are both equal to 10^{-8} seconds. Estimate the pumping power.

Examples of lasers

20. Estimate the minimum volume of a ruby crystal that provides a peak laser power of 10 kilowatts in a 2 microsecond pulse.

21. A laser operates in the visible ($\lambda \approx 500$ μm) with a power output of 1 W. The area of the laser mirror is 10^{-4} m^2 and the laser bandwidth is $\Delta\lambda/\lambda = 10^{-11}$. (a) Estimate the photon flux from the laser at a distance of 1 m. (b) For a thermal source of the same size as the laser mirror and a temperature given by $kT = hc/\lambda$, estimate the total flux of visible photons at a distance of 1 m. (c) What is the photon flux from the thermal source within the laser bandwidth?

22. The CO_2 laser operates in the infrared at $\lambda = 10.6$ μm. What type of transitions are involved?

Additional problems

23. Consider a gas of hydrogen molecules at room temperature. (a) What fraction of the molecules are in the first rotational excited state? (b) What fraction of the excited molecules decay by spontaneous emission?

24. Consider a gas of H_2^+ molecules at a temperature of 2000 K. (a) What fraction of the molecules are in the first vibrational excited state? (The vibrational spectrum of H_2^+ is shown in Figure 10-6). (b) What fraction of the excited molecules decay by stimulated emission?

25. Consider a solid-state and a gas laser operating at nearly the same wavelength and with identical bandwidths. Make an estimate of the relative pumping powers of the two lasers.

26. A laser operating in the visible red region is not effective for surgery on blood-rich tissue. Why not?

27. Consider the measurement of the distance to the moon by detection of a reflected laser pulse. The laser wavelength is 600 nm with a 1-ns pulse, and the mirror size is 1 cm. (a) Why is a laser wavelength in the visible region chosen? (b) Why is a 1-nanosecond pulse chosen? (c) The reflector mirror on the moon has an area of 0.25 m^2. Estimate the fraction of the laser photons detected on the earth.

28. The Nd:YAG laser operates at $\lambda = 1.06$ μm. The metastable state has a lifetime of about 0.1 ms. The laser transition terminates at an energy about 1/4 eV above the ground state. (a) Does the laser need to be cooled for efficient operation? Explain. (b) If the crystal contains 10^{17} niodymium atoms, what is the maximum average power output of the laser?

29. *Electron spin resonance.* A sample of atoms is placed in a magnetic field which causes the ground state to be split by the Zeeman effect. Transitions are induced between the states by supplying electromagnetic radiation of the appropriate frequency. (a) When one mole of silver atoms is placed in a magnetic field of 3 tesla, how much energy can be absorbed by the sample at room temperature? (b) How much energy can be absorbed by the sample if it is cooled to liquid helium temperature (4 K)?

The manufacture of glass, along with the forming of metals, is an art that goes back to prehistoric times. It always seems to me remarkable that our first understanding of the ductility of metals in terms of atomic movements came *after* the discovery of the neutron.... The years that passed before anyone tried to get a theoretical understanding of electrons in glass surprises me even more. After all, the striking thing about glass is that it is transparent, and that one does not have to use particularly pure materials to make it so. But, in terms of modern solid-state physics, what does "transparent" mean? It means that, in the energy spectrum of electrons in the material, there is a gap of forbidden energies between the occupied states (the valence band) and the empty states (the conduction band); light quanta corresponding to a visible wavelength do not have the energy needed to make electrons jump across it. This gap is quite a sophisticated concept, entirely dependent on quantum mechanics....

Nevill F. Mott

In Chapter 10 we discussed the behavior of two atoms when they are close together in the formation of a molecule. In this chapter we shall be concerned with the behavior of a macroscopic quantity ($N_A \approx 10^{24}$) of atoms when they are *condensed* in the formation of a solid. We begin with a brief description of the possible forms that matter can take in the condensed state (see Figure 14-1). The simplest possibility is that the atoms form a regular array or crystalline structure. The atoms of a crystal form a periodic array, or lattice, that extends over distances thousands of times the size of an atom. In a crystal, the locations of all atoms are known with respect to the position of a given atom. This property is called *long-range order*. The motion of the atoms in a crystal is limited by neighboring atoms because the potential energy between a given atom and any one of its nearest neighbors is qualitatively the same as that of a simple molecule (see Figure 10-3). Each atom in a crystal is a quantum harmonic oscillator.

Another possibility is that the atoms are fixed but do not form a regular array. Such a material is called an *amorphous* solid. The motion of the atoms in an amorphous solid is limited by neighboring atoms just as in a crystal. The location of the neighboring atoms is known with respect to a given atom. This property is called *short-range order*. An amorphous solid differs from a crystal in that there is no long-range order. Glass and rubber are examples of amorphous solids.

A third possibility for condensed matter is that of a liquid. In a liquid the atoms are not held in a rigid pattern. A liquid has neither long-range nor short-range order.

We shall be concerned with crystalline solids in this chapter. Many materials fall into this category, including most metals. There are many possible geometries for the crystal structure. In a *simple cubic* lattice the Cartesian coordinates of the nuclei of the atoms are given by $(n_1 a, n_2 a, n_3 a)$, where a is the distance between any atom and its nearest neighbors and n_1, n_2, and n_3 are integers. The structure of a crystal is classified by a *unit cell*, which specifies the geometrical arrangement of neighboring atoms. The unit cell for the simple cubic structure is one in which the nuclei of neighboring atoms occupy the corners of a cube. Another possibility is that the atoms could be at the corners and center of a cube. This crystal lattice is called *body-centered cubic* (bcc). Since the space occupied by each atom in any crystal configuration is roughly constant (10^{-29} m^3), the size of the unit cell depends on the number of atoms it contains. In a *face-centered cubic* (fcc), atoms are located at the corners and faces of a cube. Other common crystal structures are *hexagonal close packed* (hcp), and *diamond*. The unit cells for these crystal structures are shown in Figure 14-2.

The structure of a crystal may be determined by x-ray scattering. Figure 14-3 shows the crystalline structure of the elements. The richness of the field of condensed matter physics lies in the fact that many interesting properties of compounds depend on the details of the crystalline lattice. Fortunately, however, many basic properties of crystalline solids do not depend on the exact structure of the crystal. In our discussion of solids, we shall focus on the energy distribution of the outer electrons in the atoms. The understanding of the energy distribution of the electrons leads to the explanation of many basic properties of solids. The electron energy distribution is specified by two parts: (1) the density of available states and (2) the occupation probability according to the Fermi-Dirac distribution function.

14-1 ELECTRONIC ENERGY BANDS

When two identical atoms are far apart, their electron probability densities do not significantly overlap. The electronic energy levels are degenerate because the atoms are identical. When the same two atoms are close together, the probability densities overlap and one cannot say with absolute certainty which electron belongs to which atom. The energy levels are slightly shifted because the electrons now feel the electrical force from the charges (electrons and nuclei) of both atoms. Two electrons that had the same energies in two separated atoms will have slightly different energies and orbitals when the atoms are together. When a large number of atoms are brought together in a solid (10^{29}/m^3), the allowed energy levels form *electronic bands*. An electronic energy band is a collection of a great number of energy levels that are so close together that the energy of the allowed levels may be usefully approximated as a continuous variable. The electrons that would be in 1s energy levels if the atoms were separated make up a band of levels called the electronic 1s

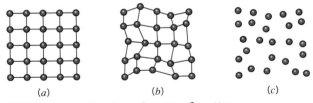

FIGURE 14-1 Condensed states of matter.
(*a*) Crystalline solid, (*b*) amorphous solid, and (*c*) liquid.

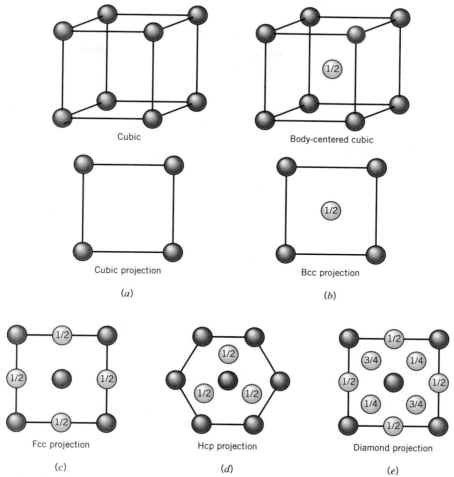

FIGURE 14-2 Crystal unit cells.
The cells are shown projected in a plane. Atoms labeled 1/4, 1/2, or 3/4 are that fraction of the distance to an adjacent plane. (*a*) Cubic, (*b*) body-centered cubic, (*c*) face-centered cubic, (*d*) hexagonal close packed, and (*e*) diamond.

band. Similarly, other electronic bands, $2s$ and $2p$, and so on, are formed. In many cases the bands overlap.

Consider the sodium atom ($Z = 11$), which has an electron configuration of $1s^2 2s^2 2p^6 3s^1$. In a sodium crystal, the outer electron in each atom is free to move. In the neighborhood of a nucleus, the potential energy of an electron is nearly spherically symmetric, and the Schrödinger equation can be solved by separation of variables and numerical integration. The main difference between a crystal and a single atom in the solution of the Schrödinger equation is the boundary condition. In a single atom, the wave function goes to zero at infinity, whereas in a crystal the wave function is periodic. The solution is similar to ordinary s, p, d wave functions near

the nuclei and plane waves between the nuclei. The Schrödinger equation may be solved by treating the spacing between atoms in the crystal as a free parameter. The actual crystal will correspond to a specific atomic separation. The result of such a calculation for sodium is shown in Figure 14-4. The horizontal axis (x) is the distance between adjacent atoms. For an atomic separation of 1 nm, the $3s$ and $3p$ levels are clearly separated. In a sodium crystal the actual separation of the atoms (a) corresponds to $x \approx 0.4$ nm. For an atomic separation of 0.4 nm, the $3s$ and $3p$ levels overlap to form a continuous band. Therefore, the electrons in this band are able to gain an infinitesimally small amount of energy (for example, due to collisions or by the application of an electric field) because of

1 **H** hcp																	2 **He** hcp
3 **Li** bcc	4 **Be** hcp											5 **B** rhom.	6 **C** diam.	7 **N** cub.	8 **O** comp.	9 **F**	10 **Ne** fcc
11 **Na** bcc	12 **Mg** hcp											13 **Al** fcc	14 **Si** diam.	15 **P** comp.	16 **S** comp.	17 **Cl** comp.	18 **Ar** fcc
19 **K** bcc	20 **Ca** fcc	21 **Sc** hcp	22 **Ti** hcp	23 **V** bcc	24 **Cr** bcc	25 **Mn** cub.	26 **Fe** bcc	27 **Co** bcc	28 **Ni** bcc	29 **Cu** fcc	30 **Zn** hcp	31 **Ga** comp.	32 **Ge** diam.	33 **As** rhom.	34 **Se** hex	35 **Br** comp.	36 **Kr** fcc
37 **Rb** bcc	38 **Sr** fcc	39 **Y** hcp	40 **Zr** hcp	41 **Nb** bcc	42 **Mo** bcc	43 **Tc** hcp	44 **Ru** hcp	45 **Rh** fcc	46 **Pd** fcc	47 **Ag** fcc	48 **Cd** hcp	49 **In** tet.	50 **Sn** diam.	51 **Sb** rhom.	52 **Te** hex	53 **I** comp.	54 **Xe** fcc
55 **Cs** bcc	56 **Ba** bcc	71 **Lu** hcp	72 **Hf** hcp	73 **Ta** bcc	74 **W** bcc	75 **Re** hcp	76 **Os** hcp	77 **Ir** fcc	78 **Pt** fcc	79 **Au** fcc	80 **Hg** rhom.	81 **Tl** hcp	82 **Pb** fcc	83 **Bi** rhom.	84 **Po** cub.	85 **At**	86 **Rn**
87 **Fr**	88 **Ra**	103 **Lr**	104 **Unq**	105 **Unp**	106 **Unh**	107 **Uns**	108 **Uno**	109 **Une**									

57 **La** hex.	58 **Ce** fcc	59 **Pr** hex.	60 **Nd** hex.	61 **Pm**	62 **Sm** comp.	63 **Eu** bcc	64 **Gd** hcp	65 **Tb** hcp	66 **Dy** hcp	67 **Ho** hcp	68 **Er** hcp	69 **Tm** hcp	70 **Yb** fcc
89 **Ac** fcc	90 **Th** fcc	91 **Pa** comp.	92 **U** comp.	93 **Np** comp.	94 **Pu** comp.	95 **Am** hex	96 **Cm**	97 **Bk**	98 **Cf**	99 **Es**	100 **Fm**	101 **Md**	102 **No**

FIGURE 14-3 Crystalline structure of the elements.
The abreviations are cub. = cubic, comp. = complex, hex. = hexagonal, diam. = diamond, and rhom. = rhombic. Adapted from C. Kittel, *Introduction to Solid State Physics,* Wiley (1986).

the availability of empty states. This property accounts for the electrical conductivity of metals.

The energy bands of sodium have been characterized in one dimension using the parameter x. The one-dimensional view is valid near the nucleus of a given atom because the potential is spherically symmetric. At locations corresponding to the midpoint between atoms, the potential is not spherically symmetric because the distance to the nearest atom depends on the direction that is chosen. Therefore, the details of the energy bands depend on the direction of x.

We now examine the consequences of a partially filled energy band in sodium by considering the propagation of an electron plane wave in a given direction, $\mathbf{i}_x$. Taking the nucleus of an arbitrary atom to be the origin, we choose the direction $\mathbf{i}_x$ to be in the direction of one of the closest neighbors, as indicated in Figure 14-5a. Since the crystal structure of sodium is bcc, this direction corresponds to propagation along the diagonal of the unit cell (see Figure 14-2). Figure 14-5b shows the electron energy versus its wave number (k). In this plot we define a negative wave number to be propagation in the $-\mathbf{i}_x$ direction. The energy at $k = 0$ corresponds to the minimum energy of an electron in the 3s band (see Figure 14-4). From this plot we see that the increase in energy with increasing k corresponds *very nearly to a free electron*, which is characterized by

$$E_k = \frac{p^2}{2m} = \frac{(\hbar k)^2}{2m}. \qquad (14.1)$$

This is especially true of electrons that reside near the bottom of the 3s band.

We also notice that there is a discontinuity in the allowed energy at certain values of the wave number. The energy gap occurs when an integer (n) number of electron half-wavelengths is equal to the spacing between atoms:

$$n\frac{\lambda}{2} = a. \qquad (14.2)$$

In terms of the electron wave number k, this corresponds to

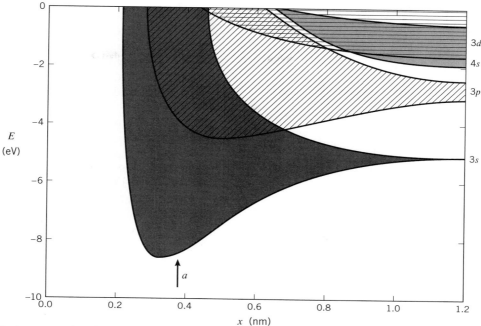

FIGURE 14-4 Outer energy bands in sodium.
The Schrödinger equation is solved for a sodium crystal treating the distance between atoms as a variable (x). At large distances the energy levels are separated, but at small distances they overlap. This explains the conductivity of metals. After J. C. Slater, "Electronic Energy Bands in Metals," *Phys. Rev.* **45**, 794 (1934).

$$k = \pm \frac{2\pi}{\lambda} = \pm \frac{n\pi}{a}. \qquad (14.3)$$

This is precisely the Bragg condition for electron scattering as first observed by Davison and Germer. Electrons that are incident on the crystal at a certain angle are totally reflected if they have an energy corresponding to the location of the band gap. These electrons cannot exist in the crystal because there are no states! The location of the energy gaps in k space separate the crystal into regions called *Brillouin zones*. The *first* Brillouin zone is bounded by

$$-\frac{\pi}{a} < k < \frac{\pi}{a}, \qquad (14.4)$$

the *second* Brillouin zone is bounded by

$$-\frac{2\pi}{a} < k < -\frac{\pi}{a} \quad \text{or} \quad \frac{\pi}{a} < k < \frac{2\pi}{a} \quad , \qquad (14.5)$$

and so on.

For energetic electrons corresponding to wave numbers in the second zone (see Figure 14-5b), the increase in energy with increasing k is smaller than that of a free

particle. Electrons in this zone may still be described as free particles provided we assign them an effective mass (m^*) that is larger than the electron mass. A particle treated in this fashion is referred to as a quasiparticle. (In some crystals m^* may be smaller than the electron mass.)

EXAMPLE 14-1
Use Figure 14-6b to make an estimate of the effective mass of an electron in zone 2.

SOLUTION:
For a free particle the energy difference (ΔE) between $k = 0$ and $k = 2\pi/a$ is

$$\Delta E \approx 14\,\text{eV} \approx \frac{(\hbar \Delta k)^2}{2m},$$

where $\Delta k = 2\pi/a$. For an electron in the crystal, the actual difference is about 6 eV. Therefore, the effective mass of the electron is given by

$$6\,\text{eV} \approx \frac{(\hbar \Delta k)^2}{2m^*},$$

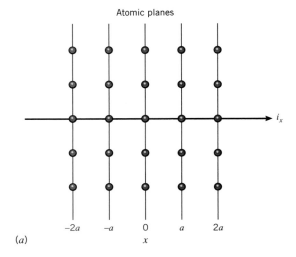

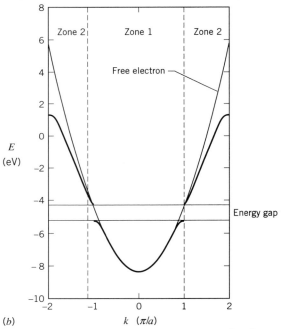

FIGURE 14-5 Electron plane waves propagating in a sodium crystal.
(a) The direction of propagation is chosen to be in the direction from one atom to its nearest neighbor, which is separated by a distance $x = a$. (b) Electron energy versus wave number k. The parabola corresponds to a free electron. After J. C. Slater, "Electronic Energy Bands in Metals," *Phys. Rev.* **45**, 794 (1934).

or

$$m^* \approx \left(\frac{14\,\text{eV}}{6\,\text{eV}}\right) m \approx \left(\frac{7}{3}\right)(0.5\,\text{MeV}) \approx 1.2\,\text{MeV}.\ \blacksquare$$

All metals have a similar electronic band structure. The outer electronic energy bands of copper ($Z = 29$), which has an electron configuration of $1s^2 2s^2 2p^6 3s^2 3p^6 3d^{10} 4s^1$, are shown in Figure 14-6. In copper the atomic separation is about 0.27 nanometer. At this separation distance, the $3d$, $4s$, and $4p$ bands overlap to form a single electronic band.

The properties of a solid depend on the details of its electronic band structure. There are three main classes of band structure in solids: (1) the outer band is partially filled, (2) the outer band is completely filled and there is a large energy gap between it and the next band, which is empty, and (3) the outer band is completely filled and there is a small energy gap between it and the next band. The generic band structure of solids is illustrated in Figure 14-7. The case of a partially filled outer band corresponds to a metal. The outer band is called the *conduction band*. The electrons that occupy states in the outer band are called conduction electrons. The electrons are free to move in the partially filled conduction band because there are empty energy levels with infinitesimally higher energy.

In an *insulator* the outermost band that contains electrons is completely full and is separated from the next available empty band by several electronvolts. The electrons cannot move without acquiring enough energy to cross the gap. In an *intrinsic semiconductor* the outer band is also completely full but is separated from the next band by a relatively small energy, roughly 1 electronvolt.

> In a metal, the outer electrons occupy a partially filled outer band and are free to move within the metal. In an insulator, the outer electrons completely fill the outer band and the next available band is separated by an energy gap.

EXAMPLE 14-2
Estimate the probability that an energetic electron in an insulator at room temperature is in the conduction band. Take the band gap to be 4 eV.

SOLUTION:
The occupation probability (P) for an electron to be in the next highest band is given by the Fermi-Dirac factor

$$P = \frac{1}{e^{(E-E_{\text{F}})/kT} + 1} \approx e^{-\Delta E/kT},$$

where ΔE is the energy difference between the top of the filled band and the Fermi energy. Since the Fermi energy is in the middle of the energy gap (E_{g}), we have

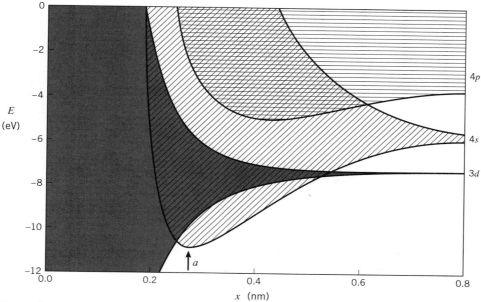

FIGURE 14-6 Energy bands of the outer electrons in copper.
After H. M. Kutter, "Energy Bands in Copper," *Phys. Rev.* **48**, 664 (1935).

$$\Delta E \approx \frac{E_g}{2} \approx 2 \text{ eV}.$$

At room temperature,

$$kT \approx \frac{1}{40} \text{eV},$$

and

$$P = e^{-\Delta E / kT} \approx e^{-80} \approx 10^{-35}. \qquad \blacksquare$$

14-2 FERMI ENERGY

Electrons are fermions and obey the Pauli exclusion principle: No two electrons can be found in the same state. The maximum occupation probability for any given state is 1. The Fermi-Dirac distribution function specifies the probability that a given state with energy E is occupied:

$$f_{FD} = \frac{1}{e^{(E-E_F)/kT} + 1}, \qquad (14.6)$$

where E_F is the Fermi energy of the metal. At $T = 0$ K, the factor $e^{(E-E_F)/kT}$ is either zero if $E < E_F$ or infinite if $E > E_F$. Thus, at $T = 0$ K, $f_{FD} = 1$ if $E < E_F$, $f_{FD} = 1/2$ if $E = E_F$, and $f_{FD} = 0$ if $E > E_F$. The physical interpretation of the Fermi energy at $T = 0$ K is the energy boundary between filled states and empty states. All states with an energy greater than E_F are empty, and all states with an energy less than E_F are filled.

The energy levels of electrons in a metal may be viewed in the context of a finite square-well, as shown in Figure 14-8. The electron energy levels are quantized and only one electron occupies each state, but the levels are very close together so that kinetic energy is treated as a continuous variable. The density of the levels is proportional to the square root of the kinetic energy, as in the case of an ideal gas. The levels are filled to the Fermi energy and empty above the Fermi energy. The work function of the metal (ϕ) is the difference between the height of the potential well (V_0) and the Fermi energy (E_F).

We now derive an expression for the Fermi energy of electrons in a metal. The Fermi energy depends on the number of conduction electrons per volume in the metal. Larger electron densities correspond to larger Fermi energies. In Chapter 12, we calculated the density of states (12.51) for electrons in a metal to be

$$\rho(E) = \frac{8\sqrt{2}\pi m^{3/2}}{h^3} \sqrt{E}, \qquad (14.7)$$

where m and E are the electron mass and energy. The number of conduction electrons per volume per unit en-

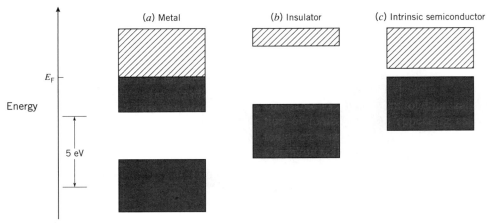

FIGURE 14-7 The generic band structure of solids.
(*a*) In a metal, the outer band is partially filled. (*b*) In an insulator, the outer band is completely full and is separated from the next band by several electronvolts. (*c*) In an intrinsic semiconductor, the outer band is completely full but is separated from the next band by a relatively small energy. The boundary between filled states and empty states is the Fermi energy (E_F).

ergy (dn/dE) is equal to the density of states times the Fermi-Dirac distribution function:

$$\frac{dn}{dE} = \rho(E) f_{FD}$$
$$= \frac{8\sqrt{2}\pi m^{3/2}}{h^3} \sqrt{E} \frac{1}{e^{(E-E_F)/kT}+1}. \quad (14.8)$$

An expression for the number of conduction electrons per volume in the metal is obtained by integrating dn/dE:

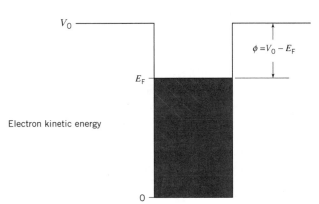

FIGURE 14-8 Electrons in a finite square-well.
The levels are filled up to the Fermi energy and empty above the Fermi energy. The work function of the metal (ϕ) is the difference between the height of the potential well (V_0) and the Fermi energy (E_F). For copper, $V_0 \approx 11$ eV, $E_F \approx 7$ eV, and $\phi \approx 4$ eV.

$$n = \int_0^\infty dE \, \rho(E) f_{FD}$$
$$= \frac{8\sqrt{2}\pi m^{3/2}}{h^3} \int_0^\infty dE \sqrt{E} \frac{1}{e^{(E-E_F)/kT}+1}. \quad (14.9)$$

At $T = 0$ K we may easily do the integration,

$$n = \frac{8\sqrt{2}\pi m^{3/2}}{h^3} \int_0^{E_F} dE \sqrt{E}$$
$$= \frac{8\sqrt{2}\pi m^{3/2}}{h^3} \left(\frac{2}{3} E_F^{3/2} \right). \quad (14.10)$$

Solving for the Fermi energy, we have

$$E_F = \left(\frac{h^2 c^2}{8mc^2} \right) \left(\frac{3}{\pi} \right)^{2/3} n^{2/3}. \quad (14.11)$$

The Fermi energy depends on the density of conduction electrons to the two-thirds power.

For atoms with one conduction electron and a diameter of 0.3 nm, the density of conduction electrons is

$$n \approx \left(\frac{1}{0.3 \text{ nm}} \right)^3, \quad (14.12)$$

and the Fermi energy (14.11) is

$$E_F \approx \left(\frac{3}{\pi} \right)^{2/3} \frac{(1240 \text{ eV}\cdot\text{nm})^2}{(0.3 \text{ nm})^2 (8)(5.11\times10^5 \text{ eV})}$$
$$\approx 4 \text{ eV}. \quad (14.13)$$

This is a typical value for the Fermi energy of a metal.

We have calculated the Fermi energy (14.11) at T = 0. At temperatures above zero kelvin but well below E_F/k, only those electrons that have energies near the Fermi energy may be thermally excited to higher energy levels because all states with lower energies are completely filled. Therefore, the Fermi energy does not depend on temperature, provided that $E_F \gg kT$, because the Fermi-Dirac distribution does not depend strongly on the temperature as shown in Figure 12-7 (and of course, the density of states is independent of temperature). This is the reason we have chosen to write the temperature dependence of the parameter A (12.35) as $e^{-E_F/kT}$.

EXAMPLE 14-3
The Fermi energy of potassium is determined to be 2.1 eV. The atomic mass of potassium is 39 and the density of potassium is 8.6×10^2 kg/m³. Calculate the number of conduction electrons per atom.

SOLUTION:
Since 0.039 kg of potassium is one mole of atoms, the number of atoms per volume (n_a) is

$$n_a = \frac{\rho N_A}{0.039\,\text{kg}} = \frac{(8.6\times10^2\ \text{kg/m}^3)(6.0\times10^{23})}{0.039\,\text{kg}}$$
$$\approx 1.3\times10^{28}\ \text{m}^{-3}.$$

The number of conduction electrons per volume is

$$n = \frac{8\pi}{3h^3c^3}(2E_F mc^2)^{3/2}$$
$$= \frac{8\pi[(2)(2.1\text{eV})(0.51\times10^6\ \text{eV})]^{3/2}}{(3)(1240\ \text{eV-nm})^3}$$
$$\approx 1.4\times10^{28}\ \text{m}^{-3}.$$

The number of conduction electrons per atom is

$$\frac{N}{\text{atom}} = \frac{n}{n_a} \approx \frac{1.4\times10^{28}\ \text{m}^{-3}}{1.3\times10^{28}\ \text{m}^{-3}} \approx 1.$$

The electron configuration of potassium ($Z = 19$) is $1s^22s^22p^63s^23p^64s$). The conduction electron is the 4s electron. ∎

EXAMPLE 14-4
Estimate the Fermi energy of copper, which has a density of 9×10^3 kg/m³.

SOLUTION:
Since copper has a single 4s electron (see Figure 9-7), there is one conduction electron per atom. The atomic mass of copper is $A = 63.5$. The density of conduction electrons is

$$n = \frac{N_A\rho}{A(10^{-3}\ \text{kg})} = \frac{(6.02\times10^{23})(9\times10^3\ \text{kg/m}^3)}{(63.5)(10^{-3}\ \text{kg})}$$
$$= 8.5\times10^{28}\ \text{m}^{-3}.$$

The Fermi energy (14.11) is

$$E_F = \left(\frac{h^2c^2}{8mc^2}\right)\left(\frac{3}{\pi}\right)^{2/3} n^{2/3}$$
$$= \frac{(1240\ \text{eV}\cdot\text{nm})^2}{(8)(5.11\times10^5\ \text{eV})}\left(\frac{3}{\pi}\right)^{2/3}(8.5\times10^{28}\ \text{m}^{-3})^{2/3}$$
$$\approx 7.0\ \text{eV}.$$ ∎

EXAMPLE 14-5
Estimate the average energy of a conduction electron in copper at low temperatures.

SOLUTION:
The average energy is

$$\langle E\rangle = \frac{\int_0^\infty dE\, E\rho(E)f_{FD}}{\int_0^\infty dE\, \rho(E)f_{FD}}.$$

At low temperatures the Fermi-Dirac distribution function is unity for

$$E < E_F,$$

and zero for

$$E > E_F.$$

The average conduction electron energy is

$$\langle E\rangle = \frac{\int_0^{E_F} dE\, E\rho(E)}{\int_0^{E_F} dE\, \rho(E)}.$$

Since the density of states is proportional to $E^{1/2}$, we have

$$\langle E \rangle = \frac{\int_0^{E_F} dE \, E^{3/2}}{\int_0^{E_F} dE \, \sqrt{E}} = \frac{\frac{2}{5} E_F^{5/2}}{\frac{2}{3} E_F^{3/2}} = \frac{3}{5} E_F.$$

The Fermi energy for copper was calculated in the previous example. The average energy is 3/5 of the Fermi energy:

$$\langle E \rangle \approx \left(\frac{3}{5} \right) (7.0 \text{ eV}) \approx 4 \text{ eV}. \qquad \blacksquare$$

14-3 HEAT CAPACITY

The Law of Dulong and Petit

The heat capacity is defined to be the amount of energy needed to raise the temperature of a specified amount of material (usually taken to be 1 mole) by 1 K. The heat capacity is usually expressed with either the volume (C_V) or the pressure (C_P) held constant. The heat capacity at constant volume is

$$C_V = \left(\frac{\partial E_{\text{tot}}}{\partial T} \right)_V, \qquad (14.14)$$

where E_{tot} is the total energy of the atoms. When energy is added to a solid, it takes the form of atomic vibrations. In the nineteenth century, motivated by the observation that the heat capacities of many solids at room temperature were identical, Dulong and Petit used the equipartition theorem (2.73) to calculate the energy associated with each atom. For six degrees of freedom corresponding to vibration and translation in three dimensions, atomic oscillators in a solid should have an average energy of $3kT$ per atom. This gives a heat capacity that is independent of temperature:

$$C_V = \frac{\partial E_{\text{tot}}}{\partial T} = \frac{\partial}{\partial T} \left(3kTN_A / \text{mole} \right)$$
$$= 3kN_A / \text{mole} \approx 25 \text{ J} / \text{K} \cdot \text{mole}. \quad (14.15)$$

For many solids *at room temperature*, the law of Dulong and Petit (14.15) gives the observed value of the heat capacity. When the technology became available to measure heat capacities of solids at low temperatures, an interesting puzzle was discovered. For insulators, the heat capacity is observed to vary as T^3 at low temperatures ($T \ll 300$ K). Figure 14-9a shows the heat capacity

of pure silicon plotted as a function of T^3. Pure silicon is an insulator at sufficiently low temperatures because it has a band gap of 1.1 eV (see Figure 14-7). All insulators have a similar variation of the heat capacity with temperature except that the coefficient of the T^3 term varies from material to material.

The law of Dulong and Petit ignores the contribution of electrons to the heat capacity. If each atom in a metal had

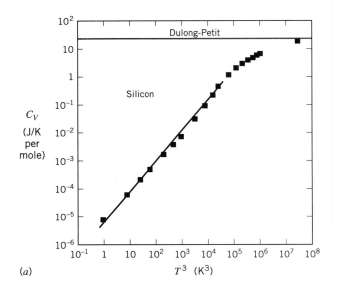

(a)

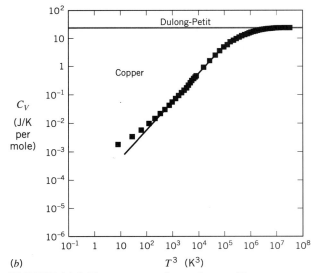

(b)

FIGURE 14-9 Temperature dependence of heat capacity.
The sloped solid line indicates a T^3 behavior. (a) Pure silicon, an insulator, and (b) copper, a metal.

one free electron, then the motion of these electrons would contribute $3N_A k/2$ per mole to the heat capacity. Free electrons should increase the heat capacity of a metal by 50% in this simple analysis! The lack of understanding of heat capacities was *the* central problem in physics at the turn of the century. *How can free electrons account for the electrical conductivity of metals without contributing significantly to the heat capacity?*

The measured heat capacity of copper as a function of temperature is shown in Figure 14-9*b*. The temperature dependence of the heat capacity of a metal is qualitatively the same as for an insulator, except that there is a deviation from the T^3 dependence at the very lowest temperatures ($T < 10$ K). This deviation is caused by the electrons. The temperature dependence of the heat capacity of insulators and metals is explained in the following sections.

The Result of Einstein

In 1907, Einstein was qualitatively able to explain the temperature variation in the heat capacity of solids by applying the Planck distribution to the energies of the atomic oscillators. The average energy of a one-dimensional atomic oscillator with a vibrational frequency f is

$$\langle E \rangle = \frac{hf}{e^{hf/kT} - 1}. \tag{14.16}$$

The total energy for a collection of oscillators vibrating in three dimensions is

$$E_{tot} = \langle E \rangle \left(\frac{N_A}{\text{mole}} \right) = \frac{3hfN_A}{e^{hf/kT} - 1} \text{ mole}^{-1}. \tag{14.17}$$

This gives a heat capacity of

$$C_V = \frac{\partial E_{tot}}{\partial T} = \frac{3N_A k \left(\frac{hf}{kT} \right)^2 e^{hf/kT}}{\left(e^{hf/kT} - 1 \right)^2} \text{ mole}^{-1}. \tag{14.18}$$

EXAMPLE 14-6

Show that the heat capacity (C_V) in the Einstein model (14.18) agrees with the law of Dulong and Petit (14.15) at high temperatures.

SOLUTION:

For large temperatures ($kT \gg hf$) we have

$$e^{hf/kT} \approx 1 + \frac{hf}{kT}.$$

The heat capacity (14.31) becomes

$$C_V \approx \frac{3N_A k \left(\frac{hf}{kT} \right)^2 \left(1 + \frac{hf}{kT} \right)}{\left(\frac{hf}{kT} \right)^2} \text{ mole}^{-1}$$

$$= 3N_A k \left(1 + \frac{hf}{kT} \right) \approx 3N_A k \text{ mole}^{-1}. \quad \blacksquare$$

At large temperatures, the Einstein heat capacity (14.18) approaches the result of Dulong and Petit ($3kN_A$/mole). The appropriate frequency (f) for each element was determined empirically. The quantity hf/k is called the *Einstein temperature* (T_E). The Einstein result for the temperature dependence gives reasonable but *not exact* agreement with experiments.

The Debye Model

In 1912, Peter Debye developed a model for solids in which the crystal vibrates as a whole. Debye considered the vibrations in the solid to be stationary sound waves analogous to electromagnetic standing waves in a cavity. The vibrational energy of the crystal is quantized in units of hf and one quantum is called a *phonon*. For phonons in a crystal, the calculation of the total energy is similar to that for electromagnetic waves in a cavity. The principle difference is that the number of modes is not infinite at small frequencies, as it is for electromagnetic waves, but is limited by the total number of degrees of freedom of all the atoms in the crystal

We first calculate the density of states, the number of photon states per volume per unit energy. The density of states is proportional to the phonon energy squared, the same as for a photon (i.e., the energy of a phonon is proportional to its momentum). We write this as

$$\rho(E) = AE^2, \tag{14.19}$$

where A is a constant. To account for the finite number of modes, Debye imposed a cutoff in the density of states corresponding to the maximum number of degrees of freedom of the atoms in the crystal. The maximum allowed phonon frequency is called the *Debye frequency* (f_D). The total number of degrees of freedom for a single atom is 3, corresponding to vibration in the x, y, and z directions. The total number of degrees of freedom for 1 mole ($3N_A$) is equal to the total number of states:

$$3N_A = V \int_0^{hf_D} dE\, \rho(E) = VA \int_0^{hf_D} dE\, E^2$$

$$= VA \frac{(hf_D)^3}{3}, \qquad (14.20)$$

where V is the volume. Solving for the constant A, we have

$$A = \frac{9N_A}{V(hf_D)^3}, \qquad (14.21)$$

and the density of states is

$$\rho(E) = AE^2 = \frac{9N_A E^2}{V(hf_D)^3}. \qquad (14.22)$$

The total energy per volume of 1 mole of atoms in the crystal is

$$\frac{E_m}{V} = \int_0^{hf_D} dE\, \frac{\rho(E)E}{e^{E/kT} - 1}$$

$$= \frac{9N_A}{V(hf_D)^3} \int_0^{hf_D} dE\, \frac{E^3}{e^{E/kT} - 1}. \qquad (14.23)$$

With the change of variables,

$$x \equiv \frac{E}{kT}, \qquad (14.24)$$

the total energy for 1 mole becomes

$$E_m = \frac{9N_A(kT)^4}{(hf_D)^3} \int_0^{hf_D/kT} dx\, \frac{x^3}{e^x - 1}. \qquad (14.25)$$

The integral may not be evaluated in closed form. At small temperatures, the total energy in phonons is proportional to the fourth power of the temperature. The same is true for the total energy of photons in a cavity; indeed, the two calculations are nearly identical.

The molar heat capacity (C_V) is obtained by differentiating E_m (14.25) with respect to T. At small temperatures, the integral is equal to $\pi^4/15$ and the molar heat capacity is

$$C_V = \frac{\partial E_m}{\partial T} \approx \frac{\partial}{\partial T}\left(\frac{9\pi^4 N_A(kT)^4}{15(hf_D)^3}\right)$$

$$= \frac{12\pi^4 N_A k(kT)^3}{5(hf_D)^3}. \qquad (14.26)$$

With the definition of *Debye temperature* (T_D),

$$T_D \equiv \frac{hf_D}{k}, \qquad (14.27)$$

the molar hear capacity becomes

$$C_V = \frac{12\pi^4}{5} N_A k \left(\frac{T}{T_D}\right)^3. \qquad (14.28)$$

The Debye model gives a quantitative account of the temperature dependence of the heat capacity of an insulator (see Figure 14-9a). The slope of C_V versus T^3 for small T in an insulator is proportional to T_D^{-3}. Measurement of the temperature dependence of the heat capacity gives a Debye temperature of 645 K for silicon and 2230 K for diamond.

The physical meaning of the Debye temperature is that kT_D is the maximum energy of a vibrational mode (phonon). There is a relationship between the temperature at which a solid melts (T_m) and the Debye temperature. The total energy that a solid can absorb at the melting point is proportional to kT_m. The potential energy of a single atom oscillating with frequency (ω) is

$$V = \frac{1}{2} M\omega^2 x^2, \qquad (14.29)$$

where M is the mass of the atom and x is its displacement. The total energy of all the atoms is proportional to $\omega^2 = (kT_D/\hbar)^2$. With the assumption that the crystal melts when the amplitude of the oscillations reaches a certain fraction of the spacing between atoms (d_{atom}), we have

$$kT_m \propto M(kT_D)^2 d_{atom}^2, \qquad (14.30)$$

or

$$T_D \propto \frac{1}{d_{atom}} \sqrt{\frac{T_m}{M}}. \qquad (14.31)$$

The relationship between Debye temperature and melting temperature of solids is illustrated in Figure 14-10 for two different crystal types, diamond and fcc.

> The vibrational energy in a solid is quantized; the quanta are called phonons. The quantization leads to a total vibrational energy which is proportional to T^4 and a heat capacity which is proportional to T^3 at low temperatures. The maximum phonon energy is a characteristic of the solid and is written as kT_D, where T_D is called the Debye temperature.

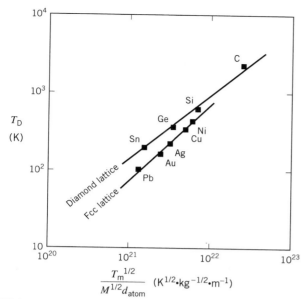

The plot axes are labeled T_D (K) on the vertical axis (from 10 to 10^4) and $\dfrac{T_m^{1/2}}{M^{1/2}d_{atom}}$ ($K^{1/2} \cdot kg^{-1/2} \cdot m^{-1}$) on the horizontal axis (from 10^{20} to 10^{23}). Data points are labeled C, Si, Ge, Sn, Ni, Cu, Ag, Au, Pb, with lines labeled "Diamond lattice" and "Fcc lattice".

FIGURE 14-10 Relationship between the Debye temperature and the melting point of solids.

Electronic Heat Capacity

The heat capacity of a metal deviates from the Debye model (14.28) at extremely small temperatures e.g., $T < 10$ K for copper as shown in Figure 14-9b. The Debye model works well for a metal except at the lowest temperatures. Evidently, the contribution of electrons to the heat capacity of a metal plays no role unless T is very small. We now examine this property in more detail.

The central problem is to understand how a conduction electron behaves as a free particle and does not contribute a relatively large amount to the heat capacity. A metal contains a large number of free electrons, one or more per atom. When thermal energy is supplied to the metal, the classical theory predicts that the free conduction electrons gain kinetic energy in proportion to the temperature increase of the metal. Thus, the electronic heat capacity is predicted to be large, that is, a large amount of energy would make a small increase in temperature because there are so many free electrons that can gain kinetic energy. Since the average kinetic energy of a free electron is $3kT/2$, we expect that the contribution to the electronic heat capacity (C_{el}) from conduction electrons to be

$$C_{el} = \frac{3kN_A}{2} \text{ mole}^{-1}. \qquad (14.32)$$

The observed electronic heat capacities of metals are smaller than this by more than an order of magnitude.

Furthermore, the electronic heat capacity is not the same for all metals and is temperature dependent.

The solution of the electronic heat capacity problem comes from the kinetic energy distribution of the conduction electrons in the metal. These conduction electrons do not obey Maxwell-Boltzmann statistics. It is the Maxwell-Boltzmann distribution that leads to an average particle kinetic energy of $3kT/2$ and therefore an electronic heat capacity of $3k/2$ per particle.

The electrons in the metal obey Fermi-Dirac statistics. When energy is supplied to the metal, the average conduction electron cannot be promoted to higher energy because the neighboring energy levels are all occupied. The Pauli exclusion principle prevents most of the electrons from gaining kinetic energy. Only those electrons that have energies near the Fermi energy can be excited to higher energy levels. The fraction of conduction electrons that can be excited is just the fraction that have an energy within a few times kT of the Fermi energy. We may write this fraction as

$$f_{el} \approx \frac{kT}{E_F}. \qquad (14.33)$$

The increase in kinetic energy for each conduction electron that is excited is approximately kT. Therefore, the electronic heat capacity of free electrons is roughly

$$C_{el} \approx N_A \frac{d}{dT}\left[(kT)\left(\frac{kT}{E_F}\right)\right] \text{mole}^{-1}$$
$$= \frac{2N_A k^2 T}{E_F} \text{ mole}^{-1}. \qquad (14.34)$$

This approximation of the electronic heat capacity gives the correct order of magnitude and the correct temperature dependence.

> The conduction electrons do not contribute significantly to the heat capacity of a metal at room temperature because only a small fraction of the electrons, those near the Fermi energy, may gain energy when the temperature is increased.

We now make a more accurate determination of C_{el}. Consider a metal at a temperature T_1. The average energy per conduction electron is

$$\langle E_k(T_1) \rangle = \frac{1}{n}\int_0^\infty dE\, E\, \rho(E) f_{FD}(E, T_1), \qquad (14.35)$$

where n is the density of electrons, $\rho(E)$ is the density of states, and f_{FD} is the probability that a given state is occupied. Recall that the density of states is proportional to $E^{1/2}$. At a slightly higher temperature, $T_2 = T_1 + \Delta T$, the average kinetic energy per conduction electron is

$$\langle E_k(T_2) \rangle = \frac{1}{n} \int_0^\infty dE\, E\, \rho(E) f_{FD}(E, T_2). \quad (14.36)$$

The electronic heat capacity is

$$C_{el} = \frac{N_A}{n} \frac{\langle E_k(T_2) \rangle - \langle E_k(T_1) \rangle}{T_2 - T_1} \text{mole}^{-1}. \quad (14.37)$$

The temperature dependence is due to the Fermi-Dirac distribution function. The change in electron energies occurs only near the Fermi energy. Thus, we may write the electronic heat capacity as

$$C_{el} = \frac{N_A}{n} \text{mole}^{-1} \int_0^\infty dE\, (E - E_F) \rho(E) \frac{df_{FD}}{dT}. \quad (14.38)$$

We now reduce this integral for the heat capacity to a simpler form. With the change of variables

$$x \equiv \frac{E - E_F}{kT}, \quad (14.39)$$

the Fermi-Dirac distribution function is

$$f_{FD} = \frac{1}{e^x + 1}. \quad (14.40)$$

The derivative is

$$\frac{df_{FD}}{dx} = -\frac{e^x}{(e^x + 1)^2}, \quad (14.41)$$

or

$$\frac{df_{FD}}{dT} = \frac{df_{FD}}{dx} \frac{dx}{dT}$$

$$= \frac{e^x}{(e^x + 1)^2} \frac{(E_F - E)}{kT^2}$$

$$= \frac{1}{T} \frac{xe^x}{(e^x + 1)^2}. \quad (14.42)$$

The electronic heat capacity is

$$C_{el} = \frac{N_A}{n} \text{mole}^{-1} \int_0^\infty dE\, (E - E_F) \rho(E) \frac{df_{FD}}{dT}$$

$$= \frac{N_A k^2 T}{n} \text{mole}^{-1} \int_{-E_F/kT}^\infty dx\, \rho(x) \frac{x^2 e^x}{(e^x + 1)^2}. \quad (14.43)$$

The integrand is almost zero except when x is near zero, which corresponds to E near E_F. This allows us to remove the density of states factor from the integrand and evaluate it at the Fermi energy. We can also change the limits of integration from minus infinity to infinity without significantly changing the result. Therefore,

$$C_{el} \approx \frac{N_A k^2 T}{n} \text{mole}^{-1} \rho(E_F) \int_{-\infty}^{+\infty} dx\, \frac{x^2 e^x}{(e^x + 1)^2}. \quad (14.44)$$

The integral is equal to $\pi^2/3$. The electronic heat capacity is

$$C_{el} \approx \frac{\pi^2 N_A k^2 T \rho(E_F)}{3n} \text{mole}^{-1}. \quad (14.45)$$

We now need to evaluate the density of states at the Fermi energy divided by the electron density. The number of states per volume (12.50) is

$$n_s = \frac{N}{V} = \frac{8\sqrt{2}\pi m^{3/2}}{h^3} \left(\frac{2}{3} E^{3/2} \right). \quad (14.46)$$

Taking the logarithm, we have

$$\ln(n_s) = \frac{3}{2} \ln E + B, \quad (14.47)$$

where B is a constant. Taking the derivative, we find

$$\frac{dn_s}{n} = \frac{3}{2} \frac{dE}{E}, \quad (14.48)$$

$$\rho(E) = \frac{dn_s}{dE} = \frac{3n_s}{2E}, \quad (14.49)$$

and

$$\frac{\rho(E_F)}{n} = \frac{3}{2E_F}. \quad (14.50)$$

Finally, we arrive at the result for the electronic heat capacity:

$$C_{el} = \frac{\pi^2 N_A k^2 T}{2 E_F} \text{ mole}^{-1}. \qquad (14.51)$$

The typical value of the Fermi energy is 4 eV. The contribution to the heat capacity per electron calculated using Fermi-Dirac statistics is smaller than the classical value $(3kN_A/2)$ by a factor of

$$\frac{\left(\dfrac{\pi^2 N_A k^2 T}{2 E_F} \right)}{\left(\dfrac{3 N_A k}{2} \right)} = \frac{\pi^2}{3} \frac{kT}{E_F}$$

$$\approx 3 \frac{\left(\dfrac{1}{40} \text{ eV} \right)}{4 \text{ eV}} \approx 0.02 . \quad (14.52)$$

Thus, the prediction of the electronic heat capacity at room temperature using quantum statistics is lower than the classical prediction by about a factor of 50.

Electronic heat capacities of metals at room temperature are shown in Figure 14-11. Our simple application of

Fermi-Dirac statistics gives the correct order of magnitude for the electronic heat capacity, while the classical prediction is too high by factors of 10 to 50. Our application of the Fermi-Dirac theory does not predict the electronic heat capacities exactly because we have ignored interactions of the electrons with the metallic lattice and with other electrons.

The total heat capacity of a metal is obtained by adding the electronic contribution (14.51) to the phonon contribution (14.28),

$$C_{tot} = C_{el} + C_V$$

$$= \left(\frac{\pi^2 k^2 N_A}{2 E_F} \text{ mole}^{-1} \right) T$$

$$+ \left(\frac{12 \pi^4 N_A k}{5 T_D^{\,3}} \text{ mole}^{-1} \right) T^3 . \quad (14.53)$$

Note that we have assumed that there is one conduction electron per atom. The electronic heat capacity is the dominant contribution to the total heat capacity of a metal at low temperatures because it varies with temperature as T, while the phonon contribution varies as T^3. If we plot C_{tot}/T versus T^2, we expect a linear relationship. This is shown for copper in Figure 14-12.

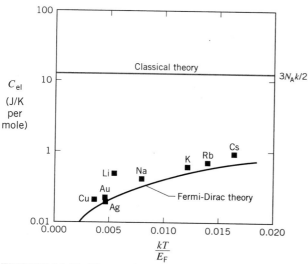

FIGURE 14-11 Electronic heat capacities of metals at room temperature.
The units are eV/K per conduction electron. The classical prediction is $(3/2)k$, independent of temperature. The prediction using Fermi-Dirac statistics is $(\pi^2/2)(T/T_F)k$. The measured electronic heat capacities are shown as solid squares. The Fermi-Dirac theory gives the correct order of magnitude for the electronic heat capacity, whereas the classical prediction is too high by factors of 10 to 50.

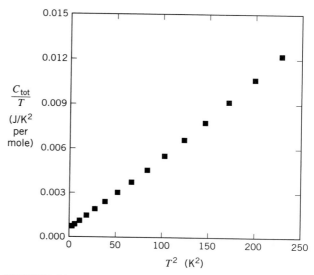

FIGURE 14-12 Heat capacity of copper at low temperature.

14-4 OHM'S LAW

We now discuss the electrical properties of metals and insulators. We shall concentrate on metals first. There is a remarkably simple relationship between the voltage and current in a conductor. Ohm's law states that the potential difference (V) along the length of an electrical conductor is directly proportional to the current (I) in the conductor. The constant of proportionality is called the resistance (R) of the conductor. The familiar form of Ohm's law is

$$V = IR. \tag{14.54}$$

Ohm's law is due to the properties of electrons in materials and hence does not have universal application. For example, Ohm's law does not hold for electrons in a vacuum! (Recall the example of the vacuum tubes discussed in Chapter 2.)

The resistance of a wire is proportional to the length (L) and inversely proportional to the cross-sectional area (A),

$$R = \frac{\rho L}{A}, \tag{14.55}$$

where the constant of proportionality (ρ) is called the *resistivity*. The resistivity is a property of the material. Ohm's law (14.54) for the wire becomes

$$V = \frac{I\rho L}{A}. \tag{14.56}$$

The electric field inside the wire is

$$E = \frac{V}{L}. \tag{14.57}$$

We define J to be the current per cross-sectional area in the wire,

$$J = \frac{I}{A}. \tag{14.58}$$

Combining these results, we get the following expression for Ohm's law:

$$J = \frac{E}{\rho}. \tag{14.59}$$

The *conductivity* (σ) is defined to be the reciprocal of the resistivity

$$\sigma \equiv \frac{1}{\rho}, \tag{14.60}$$

and Ohm's law (14.59) may be written

$$J = \sigma E. \tag{14.61}$$

The SI unit of resistance is the ohm (Ω). One ohm is equal to one volt per ampere:

$$1\Omega = 1 \text{ V/A} = 1 \text{ V} \cdot \text{s/C}. \tag{14.62}$$

The units of resistivity are ohm-meters. The units of conductivity are inverse ohm-meters. Metals have a large conductivity. The best electrical conductor is silver, which has a conductivity of $6.2 \times 10^7 \ \Omega^{-1} \cdot \text{m}^{-1}$ at room temperature. In contrast, the electrical conductivity of diamond is about $10^{-12} \ \Omega^{-1} \cdot \text{m}^{-1}$.

The conduction electrons in a metal may be thought of as a gas of particles in thermal equilibrium obeying Fermi-Dirac statistics. The electrons behave as free particles (see Figure 14-5b) except they interact with the lattice. We may view this electron–lattice interaction as electron–phonon scattering. The energy scale of the conduction electrons is set by E_F and the energy scale of the phonons is set by kT_D. Since $E_F \gg kT_D$, the energy of the electrons is much larger than the energy of the phonons. Therefore, in electron–phonon collisions the direction of the electron is easily changed but its energy does not change. The electron–phonon interaction is the cause of resistance in a metal because the time between collisions limits the length of time that an electron can be accelerated between collisions before its direction of travel is randomly changed. Recall that a typical value for the Fermi energy (E_F) in a metal is about 4 eV. The speed of an electron that has a kinetic energy equal to the Fermi energy is called the *Fermi speed* (v_F). The typical Fermi speed of an electron is

$$v_F = c\sqrt{\frac{2E_F}{mc^2}}$$

$$= \left(3 \times 10^8 \text{ m/s}\right)\sqrt{\frac{(2)(4 \text{ eV})}{5.1 \times 10^5 \text{ eV}}}$$

$$\approx 1.2 \times 10^6 \text{ m/s}. \tag{14.63}$$

The speed of the most energetic outer electrons in a metal is the same order of magnitude as the speed of an electron in the hydrogen atom.

EXAMPLE 14-7
Calculate the Fermi speed for copper.

SOLUTION:
The Fermi energy is 7.0 eV. The Fermi speed (v_F) in units of c is

$$\frac{v_F}{c} = \sqrt{\frac{2E_F}{mc^2}} = \sqrt{\frac{(2)(7 \text{ eV})}{5.1 \times 10^5 \text{ eV}}} \approx 5.2 \times 10^{-3}.$$

The electron is nonrelativistic. The Fermi speed is

$$v_F = (5.2 \times 10^{-3})(3.0 \times 10^8 \text{ m/s}) = 1.6 \times 10^6 \text{ m/s}. \blacksquare$$

When an electric field is applied, the electrons acquire an average *drift speed* (v_d). This drift speed is related to the current per area (J) and density of conduction electrons (n) by

$$J = nev_d. \tag{14.64}$$

A large current is made up of a very large number of electrons having a small drift speed. We now make an estimate of the drift speed in a copper wire that has a cross-sectional area of 1 millimeter squared and carries a current of 1 ampere. The number of conduction electrons per volume is approximately 10^{29} m^{-3}. The drift speed may be calculated from the current density (14.64) to be

$$v_d = \frac{J}{ne} \approx \frac{10^6 \text{ C} \cdot \text{s}^{-1} \cdot \text{m}^{-2}}{(10^{29} \text{ m}^{-3})(1.6 \times 10^{-19} \text{ C})}$$

$$\approx 10^{-4} \text{ m/s}. \tag{14.65}$$

The drift speed of the electrons is extremely small compared to the Fermi speed of the electrons. It is this property that explains Ohm's law.

The drift speed is equal to the acceleration times the average time between electron–phonon collisions, the *relaxation time*. The distance that the electron travels between collisions is the mean free path (d). The relaxation time (τ) depends on d and the Fermi speed. Writing the drift speed as the acceleration (force/mass) multiplied by the average time between collisions, we have

$$v_d = \frac{eE}{m}\tau = \frac{eE}{m}\frac{d}{v_F}. \tag{14.66}$$

The current per area (16.64) is

$$J = \frac{ne^2 d}{mv_F}E. \tag{14.67}$$

We see that Ohm's law (14.61) is satisfied with

$$\sigma = \frac{ne^2 d}{mv_F}. \tag{14.68}$$

There are two pieces of quantum mechanics in the description of the conductivity: (1) Fermi-Dirac statistics are needed to determine the Fermi speed (v_F), and (2) the mean free path between collisions (d) depends on the wave nature of the electron and the presence of impurities in the sample. The value of d is typically 100 times larger than the lattice spacing.

EXAMPLE 14-8

The conductivity of copper is measured to be 5.9×10^7 $\Omega^{-1} \cdot$m^{-1} at room temperature. Calculate the mean free path for a conduction electron in copper.

SOLUTION:
The mean free path is

$$d = \frac{\sigma m v_F}{ne^2}$$

$$= \frac{(5.9 \times 10^7 \ \Omega^{-1} \cdot \text{m}^{-1})(9.1 \times 10^{-31} \text{ kg})(1.6 \times 10^6 \text{ m/s})}{(8.5 \times 10^{28} \text{ m}^{-3})(1.6 \times 10^{-19} \text{ C})^2}$$

$$= 3.9 \times 10^{-8} \text{ m}.$$

The mean free path is two orders of magnitude larger than the distance between two copper atoms. $\blacksquare$

The electrical conductivity of a conductor is many orders of magnitude larger than that of an insulator. For example, at room temperature the electrical conductivity of diamond is smaller than copper by 20 orders of magnitude! It is rather remarkable then that Ohm's law is also found to hold for an insulator. To investigate this interesting property of insulators we introduce the *mobility* (μ) of an electron, defined to be the drift speed divided by the electric field. With this definition we have

$$v_d = \mu E. \tag{14.69}$$

In our derivation of Ohm's law in a metal, we have assumed that the mobility does not depend on the electric field. This is true for metals but not for insulators. The measured drift speed of electrons in diamond as a function of electric field is shown in Figure 14-13a. The slope of the curve is equal to the mobility. At small values of electric field the mobility is constant. In this region Ohm's law works. At large values of E the mobility is inversely proportional to E and Ohm's law does not hold. The mobility of electrons also depends on temperature. This is due to an interaction of the electrons with the vibrating lattice. The temperature dependence of the electron mobility in diamond is shown in Figure 14-13b.

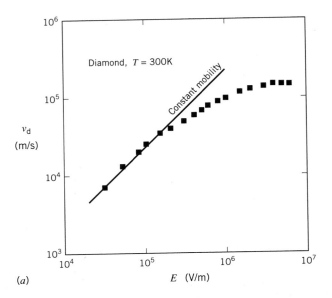

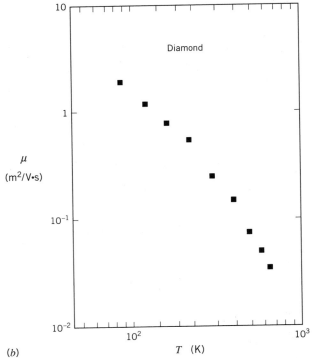

FIGURE 14-13 Electron drift speed in the insulator diamond.
(a) Electron drift speed (v_d) measured as a function of electric field strength (E). (b) The mobility ($\mu = v_d/E$) measured as a function of temperature. The measurements were taken at small values of electric field where the mobility does not depend on E. After F. Nava et al., *Solid State Com.* **33**, 475 (1980).

As a final remark we point out that in our derivation of Ohm's law we have also assumed that the number of conduction electrons remains fixed. This is true for a metal but not for an insulator at large E because additional electrons enter the conduction band. Therefore, in an insulator we may write the current density as

$$ J = n(E)e\, v_d(E). \qquad (14.70) $$

14-5 SEMICONDUCTORS

Intrinsic Semiconductors

An intrinsic semiconductor is an element in which the energy gap between the outermost filled band (called the *valence band*) and the next highest band is small compared to that of an insulator. The family of intrinsic semiconductors is made up of the first four elements in Group IV of the periodic table (see Figure 14-14). Silicon ($Z = 14$) is the most common intrinsic semiconductor. Its electron configuration is $1s^2 2s^2 2p^6 3s^2 3p^2$. In silicon, the $3s$ and $3p$ bands overlap to make a filled outer band (4 electrons per atom) with an energy gap of about 1.1 eV. The band structure of silicon is shown in Figure 14-15.

p-dopants	Intrinsic semiconductors	n-dopants
5 **B** $2s^2 2p^1$	6 **C** $2s^2 2p^2$ $E_g = 5.5$ eV	7 **N** $2s^2 2p^3$
13 **Al** $3s^2 3p^1$	14 **Si** $3s^2 3p^2$ $E_g = 1.1$ eV	15 **P** $3s^2 3p^3$
31 **Ga** $4s^2 4p^1$	32 **Ge** $4s^2 4p^2$ $E_g = 0.7$ eV	33 **As** $4s^2 4p^3$
49 **In** $5s^2 5p^1$	50 **Sn** $5s^2 5p^2$ $E_g = 0.1$ eV	51 **Sb** $5s^2 5p^3$

FIGURE 14-14 The intrinsic semiconductors diamond, silicon, germanium, and tin together with their neighboring elements in the periodic table.

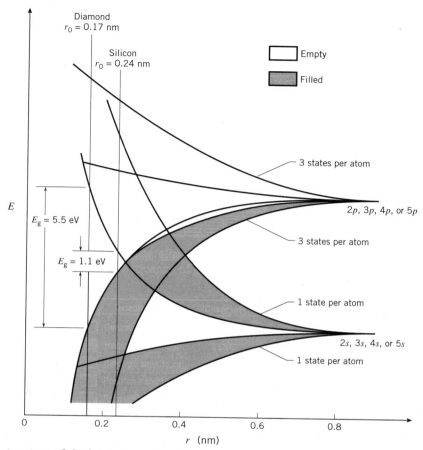

FIGURE 14-15 Band structure of the intrinsic semiconductor.
After R. L. Sproull and W. A. Phillips, *Modern Physics: The Quantum Physics of Atoms, Solids, and Nuclei,* Wiley (1980).

Carbon ($Z = 6$), germanium ($Z = 32$), and tin ($Z = 50$) have the same outer electron configurations as silicon, except the outer shells correspond to $n = 2$, $n = 4$, and $n = 5$. The crystal structure of silicon is a diamond lattice (see Figures 14-2 and 14-3). The band structures of diamond, germanium, and tin are similar to that of silicon except that the spacing between atoms is smaller in diamond and larger in germanium and tin. This leads to an energy gap of 5.5 eV in diamond, 0.7 eV in germanium, and 0.1 eV in tin.

We should point out one subtle feature of the energy gap. The maximum of the valence band need not coincide with the minimum of the conduction band. This is illustrated in Figure 14-16 which shows a plot of electron energy versus wave number (see Figure 14-5). If the maximum of the valence band coincides with the minimum of the conduction band as in Figure 14-16a, an

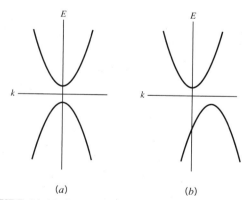

FIGURE 14-16 Semiconductor band gaps.
(*a*) Energy versus wave number for a semiconductor in which the minimum of the conduction band coincides with the maximum of the valence band. (*b*) The valence and conduction bands do not coincide.

electron can cross the gap with the absorption of a photon of energy E_g, a *direct transition*. If the valence and conduction bands do not coincide as in Figure 14-16b, an electron can still absorb a photon and cross the gap provided that the appropriate momentum is supplied by a phonon, an *indirect transition*. Since the phonon also has energy of order of magnitude kT_D, the indirect transitions are caused by photons with energies smaller than the energy gap. This effect can be especially important for those semiconductors with small band gaps.

In an intrinsic semiconductor, the band gap is small enough that some electrons are thermally excited to the conduction band, leaving unfilled states, or *holes*, in the valence band. Therefore, the conductivity of a semiconductor increases with temperature.

EXAMPLE 14-9

Make an estimate of the density of electrons in the conduction band in diamond, silicon, germanium, and tin at room temperature.

SOLUTION:

The occupation probability for an electron to be in the conduction band (see Example 14-2) is given by the Maxwell-Boltzmann factor

$$P = e^{-\Delta E / kT}.$$

where

$$\frac{\Delta E}{kT} = \frac{E_g}{2\,kT} \approx \frac{20\,E_g}{\text{eV}}.$$

The probability that an electron is excited to the conduction band is for diamond ($E_g = 5.5$ eV)

$$P \approx e^{-\Delta E / kT} \approx e^{-(20)(5.5)} \approx e^{-110} \approx 10^{-48}.$$

for silicon ($\Delta E = 1.1$ eV)

$$P \approx e^{-(20)(1.1)} \approx e^{-22} \approx 10^{-10},$$

for germanium ($\Delta E = 0.7$ eV)

$$P \approx e^{-(20)(0.7)} \approx e^{-14} \approx 10^{-6},$$

and for tin

$$P \approx e^{-(20)(0.1)} \approx e^{-2} \approx 10^{-1}.$$

We may estimate that about one electron per atom is near the top of the outer band. The density of these electrons is

$$n \approx \frac{1}{(0.2\,\text{nm})^3} \approx 10^{29}\ \text{m}^{-3}.$$

The density of conduction electrons (n_c) is for diamond

$$n_c \approx nP \approx (10^{-48})(10^{29}\ \text{m}^{-3}) = 10^{-19}\ \text{m}^{-3},$$

for silicon

$$n_c \approx nP \approx (10^{-10})(10^{29}\ \text{m}^{-3}) = 10^{19}\ \text{m}^{-3},$$

for germanium

$$n_c \approx nP \approx (10^{-6})(10^{29}\ \text{m}^{-3}) = 10^{23}\ \text{m}^{-3},$$

and for tin

$$n_c \approx nP \approx (10^{-1})(10^{29}\ \text{m}^{-3}) = 10^{28}\ \text{m}^{-3}. \quad \blacksquare$$

Doped Semiconductors

When impurity atoms are added to an intrinsic semiconductor, the resulting material is called a *doped* semiconductor. A modest doping fraction of 10^{-6} can greatly change the conduction properties of the material. There are two general classes of doped semiconductors as illustrated in Figure 14-17.

In an *n-type* (*n* for negative) doped semiconductor, the impurity atoms have one more outer electron than the intrinsic semiconductor. An example is germanium doped with arsenic (see Figure 14-14). The arsenic atoms occupy random positions in the crystal lattice. The energy levels of the arsenic atoms are slightly different from germanium. The outer electrons from arsenic reside in energy levels which are slightly below the conduction band. With a small amount of energy from thermal excitations, electrons can be boosted into the conduction band without leaving unfilled states (holes) in the valence band.

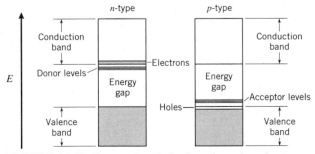

FIGURE 14-17 Energy bands in doped semiconductors.

In a *p-type* (*p* for positive) doped semiconductor, the impurity atoms have one less outer electron than the intrinsic semiconductor. An example is germanium doped with gallium (see Figure 14-14). The gallium atoms occupy random positions in the crystal lattice and leave holes in energy levels which are just above the valence band. Electrons can be thermally excited from the valence band. This creates holes in the valence band without putting electrons in the conduction band.

The *p–n* Junction

If we put a *p*-type and an *n*-type material in contact with each other, we create a *p–n junction*. An electrical device containing such a junction is called a diode. Some of the electrons near the junction boundary diffuse from the *n*-region into the *p*-region and some of the holes diffuse from the *p*-region into the *n*-region (see Figure 14-18). The movement of these mobile charges leaves behind an excess of fixed positive charge in the *n*-region near the boundary and an excess of fixed negative charge in the *p*-region near the boundary. These fixed charges create a strong electric field (10^6–10^8 V/m) that serves to keep the region free of mobile charge. This region is called the *depleted* region. The thickness of the depleted region is a few microns. A potential difference (V_0), called the *contact potential*, is created across the boundary.

The net current across a junction, after the initial movement that creates the contact potential, is zero. This current consists of four components, electrons and holes that can each travel in both directions. Let the current due to electrons moving from the *n*-region to the *p*-region be called I_n. The number of electrons traveling in the opposite direction, from the *p*-region to the *n*-region, creates a current $-I_n$. Let the current due to holes moving from the *p*-region to the *n*-region be called I_p. The number of holes traveling in the opposite direction, from the *n*-region to the *p*-region, creates a current $-I_p$. Equilibrium is established and no net current is flowing across the junction:

$$I = I_n - I_n + I_p - I_p = 0. \qquad (14.71)$$

Figure 14-19*a* shows an energy level diagram for a *p–n* junction. The Fermi energies for the *p*-type and the *n*-type coincide. The *p*-type region contains mobile holes at the top of the valence band that are kept from crossing into the *n*-type region by the contact potential. Similarly, the *n*-type material contains mobile electrons at the bottom of the conduction band that are kept from crossing into the *p*-type region. The contact potential and energy gap are related by $eV_0 = E_g$.

If we connect an external voltage source to the junction with the polarity selected such that an electric field is generated in the same direction as the field caused by the contact potential, this is called *reverse bias* (see Figure 14-19*b*). In reverse bias, the negative terminal of the voltage source is connected to the *p*-region and the positive terminal to the *n*-region. Since resistance to charge flow in the semiconductor is predominantly at the junction, the external applied voltage will appear at the junction. This destroys the equilibrium condition by making it essentially impossible for charges to cross the barrier in the direction opposing the potential, corresponding to electrons going from the *n*-region to the *p*-region or vice versa for holes. Mobile charges can still diffuse across in the other direction, creating a net current

$$I = -(I_n + I_p). \qquad (14.72)$$

This current is independent of the size of the reverse bias voltage and is relatively small because there are very few mobile electrons in the *p*-region or mobile holes in the *n*-region. If the reverse-bias voltage is made large enough, electrons that are not normally mobile because they are bound in the atoms may be freed, causing an avalanche breakdown.

If we connect an external battery such that the applied electric field is opposite that due to the contact potential, this serves to help charges cross the barrier (see Figure 14-19*c*). This is called *forward bias*. To achieve a forward bias, the positive voltage terminal is connected to the *p*-type semiconductor and the negative to the *n*-type. The potential of the external battery will appear as a lowering

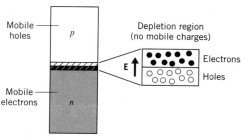

FIGURE 14-18 The *p–n* junction.
The depletion region is a few microns thick and contains the fixed charges left over after electrons diffuse from the *n*-region to the *p*-region and holes from the *p*-region to the *n*-region. These charges create an electric field that prevents further diffusion of charge.

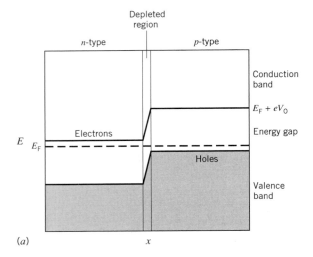

(a)

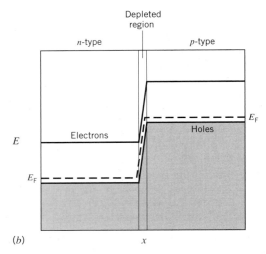

(b)

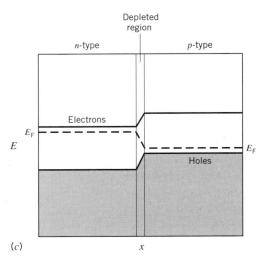

(c)

Figure 14-19 Energy levels of electrons and holes in an *p–n* junction.
(*a*) No external voltage applied, (*b*) reverse bias, and (*c*) forward bias.

of the contact potential of the junction. The effect of reducing the contact potential allows current to flow.

The mobile charge carriers, because they are the most energetic, are described by the Maxwell-Boltzmann distribution. A charge contributes to the current if it has enough energy to cross the barrier. The current due to the mobile electrons crossing from the *n*-region to the *p*-region plus that of mobile holes crossing from the *p*-region to the *n*-region is given by

$$I_1 = (I_n + I_p)e^{eV/kT}. \qquad (14.73)$$

The current due to mobile electrons that have diffused to the *p*-region and now cross back to the *n*-region does not depend on the junction voltage because it is energetically favorable for the charges to cross the junction. The same is true for the mobile holes that have diffused to the *n*-region and cross back to the *p*-region. This part of the current is just what it was before the external voltage was applied:

$$I_2 = -(I_n + I_p). \qquad (14.74)$$

The net current is

$$I = I_1 + I_2 = (I_n + I_p)(e^{eV/kT} - 1). \qquad (14.75)$$

The current–voltage characteristics of a diode are shown in Figure 14-20.

Transistors

The *bipolar junction transistor* (BJT) consists of three semiconductor layers of alternating types, either *n–p–n* or *p–n–p*. Transistors are capable of power amplification and as such have many uses. The transistor was invented in 1948 by John Bardeen, Walter H. Brattain, and William Shockley (see Figure 14-21).

The physics of the transistor may be understood by analyzing what happens to the electrons (or holes) at each junction. Figure 14-22 shows an energy band diagram for an *n–p–n* transistor. The three regions are called the emitter, base, and collector. When no external voltage is present (Figure 14-22*a*), there is an energy barrier between the *p*-type and *n*-type regions, as discussed earlier.

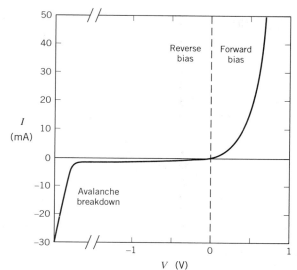

FIGURE 14-20 Current versus voltage in a *p–n* junction.

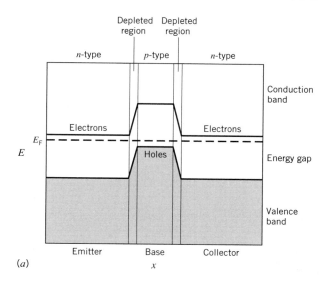

(a)

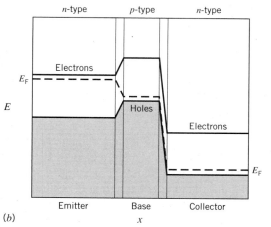

(b)

FIGURE 14-22 The *n–p–n* junction.
(a) No external voltage and (b) emitter–base forward biased and base–collector reverse biased.

This barrier prevents the flow of electrons between the two *n*-regions. In order to function as an amplifier, the transistor must be biased with external power supplies. These power supplies will provide the energy source for the power amplification. Figure 14-22*b* shows an energy band diagram for an *n–p–n* transistor that is connected to external voltage sources. The base–emitter junction is forward biased so that this energy barrier is reduced. The base–collector junction is reverse biased so that this energy barrier is raised. Electrons are ready to travel from the emitter to the collector, but they are hindered from doing so because of the potential barrier in the base. This barrier

FIGURE 14-21 The first transistor.
Courtesy A. T. & T. Bell Laboratories, reprinted by permission.

is made up of negative charge that has accumulated in the base. If electrons are allowed a path to escape from the base, then the emitter–base barrier is lowered and a small change in height of the barrier allows a large amount of charge to flow across it. A small base–emitter current causes a large emitter–collector current (see Figure 14-23). The *n–p–n* transistor is designed such that essentially all of the electrons from the emitter that enter the base pass through it into the collector. This is accomplished by making the base very thin and lightly *p*-doped while the emitter is heavily *n*-doped. The collector is usually lightly *n*-doped. The ratio (α) of collector (I_c) to emitter currents (I_e) is of order unity,

$$\alpha = \frac{I_c}{I_e} \approx 1. \qquad (14.76)$$

The ratio of collector to base (I_b) currents is called the *gain* (β) of the transistor,

$$\beta = \frac{I_c}{I_b}. \qquad (14.77)$$

The numerical value of β is in the range 10–1000 with 100 being a typical value. The gains of transistors, even of the same model number, can easily vary by factors of 2 from component to component. Therefore, circuit design must not depend on the exact value of the transistor gain. In addition, every transistor has current and voltage limits beyond which it will "burn out."

A *p–n–p* transistor consists of a heavily *p*-doped emitter and a lightly *p*-doped collector separated by a lightly *n*-doped base. The operation of a *p–n–p* transistor is similar to that of an *n–p–n* transistor except that the bias voltages are reversed and the current is dominated by the motion of holes.

The normal operation of the transistor may be summarized by noting that a small current in the base controls a large current through the collector. The base–emitter junction is forward-biased while the base–collector junction is reverse biased. Therefore, there is a small resistance between emitter and base and a large resistance between collector and base. Since almost every charge that enters the base from the emitter continues into the collector, the current across each junction is the same. Since the collector junction has higher resistance, it has a higher power ($P = I^2R$), and the transistor serves as a power amplifier. The current–voltage characteristics of a common transistor are shown in Figure 14-23*b*. To operate the transistor as an amplifier, a voltage of about 0.5 V is maintained between the collector and emitter (V_{ce}). The collector current (I_c) is then about 100 times the base current (I_b).

Field-Effect Transistors

The collector current in a bipolar transistor is controlled by the base current. In another type of transistor, the *field-effect transistor* (FET), current is controlled by an electric field. There are two main categories of FETs, the junction field-effect transistor (JFET) and the metal-oxide-semiconductor field-effect transistor (MOSFET). Both the JFET and the MOSFET come in two varieties, *p*-channel

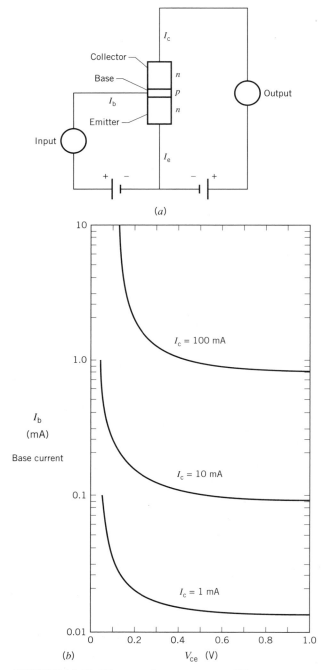

FIGURE 14-23 The transistor as an amplifier.
(*a*) Circuit diagram for an *n-p-n* transistor in which the base-emitter junction is forward biased and the base–collector junction is reverse biased. (*b*) Current-voltage characteristics of a typical transistor amplifier. For a given collector–emitter voltage (e.g., $V_{ce} = 0.5$ V), the collector current is approximately 100 times the base current.

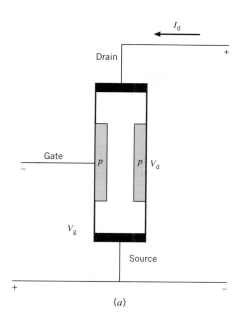

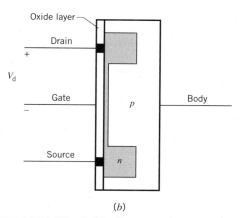

FIGURE 14-24 The field-effect transistor.
(*a*) JFET, and (*b*) MOSFET.

and *n*-channel, analogous to *p–n–p* and *n–p–n* bipolar transistors. The *n*-channel JFET is shown in Figure 14-24*a*. The JFET consists of a semiconductor bar, most commonly silicon, with electrical contacts, called the source and the drain, on each end. The middle of the bar contains a second semiconductor with the opposite doping, called the *gate*. The gate is designed to draw essentially no current. The gate–source is reversed biased, creating an electric field in the semiconductor bar. The term *gate* arises because the amount of voltage bias determines the size of the energy gap that the mobile charges pass through. The three terminals gate, source,

and drain are analogous to the base, emitter, and collector of the bipolar transistor.

The MOSFET differs from the JFET in that the gate consists of a metal electrode that is insulated from the semiconductor by a layer of glass (SiO_2). These devices are sometimes referred to as insulated-gate FETs (IGFETs). The *n*-channel MOSFET is shown in Figure 14-24*b*.

14-6 THE HALL EFFECT

Consider a conductive sheet placed in a magnetic field as shown in Figure 14-25. A voltage is applied to the sheet causing a current in the *x* direction. The current may be due to either mobile electrons (Figure 14-25*a*) or mobile holes (Figure 25*b*). For the case of mobile electrons drifting in the negative *x* direction, there is a magnetic force ($\mathbf{F}_m$),

$$\mathbf{F}_m = -e\mathbf{v}_d \times \mathbf{B}, \qquad (14.78)$$

on electrons in the negative *y* direction. This magnetic force causes an accumulation of negative charge on the lower edge of the sheet (Figure 14-25*a*). This charge generates an electric force ($\mathbf{F}_e$) on the drifting electrons that just cancels the magnetic force. This is called the *Hall effect* after Edwin H. Hall, who first observed it in 1879. If the current is due to mobile holes (Figure 14-25*b*), then the magnetic force on the holes is in the negative *y* direction and positive charge accumulates on the lower edge of the sheet. The magnitudes of the forces are

$$F_e = eE = F_m = ev_dB. \qquad (14.79)$$

The Hall voltage (V_H) is defined to be the potential difference across the width of the sheet from the accumulated charges:

$$V_H = Ey = v_dBy, \qquad (14.80)$$

where *y* is the width of the sheet. The Hall voltage is very small ($V_H \ll 1$ V) even for large magnetic fields because the electron drift speed is small. The polarity of the Hall voltage is determined by the sign of the mobile charges (electrons or holes). Measurement of the Hall voltage provides a determination of the drift velocity. If the current density ($J = nev_d$) is also measured, then measurement of the Hall voltage provides a determination of the density of mobile charges:

$$n = \frac{J}{ev_d} = \frac{JBy}{eV_H}. \qquad (14.81)$$

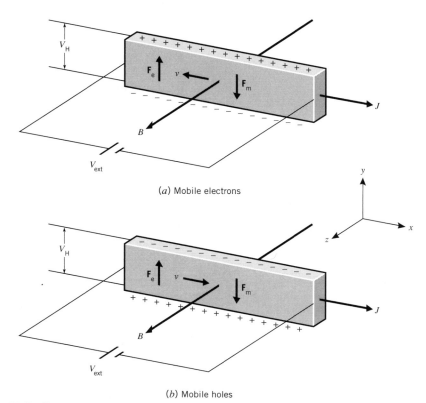

(a) Mobile electrons

(b) Mobile holes

FIGURE 14-25 The Hall effect.
A thin conducting sheet is placed in a uniform magnetic field (B) which is in the z direction. A voltage is applied so that a current (J) flows in the x direction. (a) The current is due to electrons moving in the negative x direction. There is a magnetic force ($\mathbf{F}_m$) on the electrons in the negative y direction. Negative charges accumulate on the lower edge of the conducting sheet until the resulting electrical force ($\mathbf{F}_e$) on an electron just cancels the magnetic force. (b) The current is due to holes moving in the x direction. There is a magnetic force ($\mathbf{F}_m$) on the holes in the negative y direction. Positive charges accumulate on the lower edge of the conducting sheet until the resulting electrical force ($\mathbf{F}_e$) on a hole just cancels the magnetic force.

If the sheet has a thickness d, we may write the current density as

$$J = \frac{I}{yd}, \tag{14.82}$$

to get

$$V_H = \frac{IB}{ned}. \tag{14.83}$$

The Hall effect provides a practical method for magnetic field measurement. If a device in which the density of mobile charges is known is placed in an unknown external magnetic field, then measurement of the current and the Hall voltage determines B. Such a device is called a *Hall probe*.

EXAMPLE 14-10

A sheet of copper with a thickness of 100 μm carries a current of 1 A. What is the Hall voltage when the copper sheet is placed in an external magnetic field of 1 T? The density of conduction electrons in copper is 8.45×10^{29} m⁻³.

SOLUTION:

The Hall voltage times the electron charge is

$$eV_H = \frac{IB}{nd} = \frac{(1\,\text{A})(1\,\text{T})}{\left(8.45 \times 10^{28}\ \text{m}^{-3}\right)\left(10^{-4}\ \text{m}\right)}$$

$$= 1.18 \times 10^{-25}\ \text{J}.$$

The Hall voltage is

$$V_H = \left(\frac{1.18 \times 10^{-25} \text{ J}}{e} \right) \left(\frac{1 \text{ eV}}{1.6 \times 10^{-19} \text{ J}} \right)$$

$$= 7.40 \times 10^{-7} \text{ V.} \qquad \blacksquare$$

The *Hall resistance* (R_H) is defined to be the Hall voltage divided by the current.

$$R_H \equiv \frac{V_H}{I} = \frac{B}{ned}. \qquad (14.84)$$

Quantum Hall Effect

In 1980, a surprising discovery was made concerning the Hall effect by Klaus von Klitzing and his collaborators, Gerhardt Dorda and Michael Pepper. Measurements were made of the Hall voltage in a MOSFET at low temperature (about 2 K) in a strong magnetic field (about 15 T). The geometry of their MOSFET is shown in Figure 14-26a (insert). The current in the MOSFET was kept constant at 1 μA and the gate voltage (V_g) was varied from 0 to 25 V. Two voltages were measured, the Hall voltage (V_H) and the voltage along the strip (V_s). Figure 14-26a shows a plot of V_H and V_s versus V_g. For certain values of V_g, V_s vanishes and therefore the resistance of the strip vanishes! When $V_s = 0$, the Hall voltage is observed to have a plateau. The corresponding values of the Hall resistance are

$$R_H = \frac{h}{ve^2}, \qquad (14.85)$$

where $v = 1,2,3 \dots$. This result is known as the *quantum Hall effect* (QHE). The Hall resistance (14.85) does not depend on the detailed properties of the material!

Figure 14-26b shows a detail of one of the voltage plateaus. These data show a plateau in the Hall resistance (14.85) at a value of 6453.3 ± 0.1 Ω corresponding to $v = 4$. It was immediately appreciated that the QHE allowed a very accurate determination of the electromagnetic coupling constant (α), through the relationship

$$\alpha = \frac{ke^2}{\hbar c} = \frac{2\pi k}{vcR_H}. \qquad (14.86)$$

The value of the quantized Hall resistance for $v = 1$ is known as the *von Klitzing resistance* (R_K),

$$R_K = \frac{h}{e^2} = \frac{6.6262 \times 10^{-34} \text{ J} \cdot \text{s}}{\left(1.6022 \times 10^{-19} \text{ C} \right)^2}$$

$$= 25813 \ \Omega. \qquad (14.87)$$

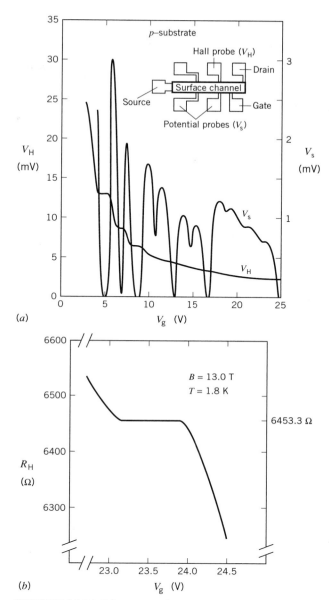

FIGURE 14-26 Discovery of the quantum Hall effect. A MOSFET (insert) is cooled with liquid helium and placed in a strong magnetic field. (*a*) The voltage along the strip V_s (right-hand scale) and the Hall voltage V_H (left-hand scale) are measured as a function of the gate voltage V_g. At certain values of V_g, V_s is observed to vanish and V_H is observed to have a plateau. ($T = 1.5$ K and $B = 18$ T.) (*b*) The Hall resistance R_H is plotted as a function of V_g in the region of one of the plateaus. The data show a plateau at $R_H = h/4e^2 = 6453$ Ω. ($T = 1.8$ K and $B = 13$ T.) From K. v. Klitzing, G. Dorda and M. Pepper, "New Method for High-Accuracy Determination of the Fine-Structure Constant Based on Quantized Hall Resistance," *Phys. Rev. Lett.* **45**, 494 (1980).

The experimental technique of determining the Hall resistance is so accurate (about 1 part per billion), that in 1990 a new definition of the ohm was adopted,

$$R_K \equiv 25812.807 \ \Omega . \qquad (14.88)$$

The important feature of the MOSFET relevant for the QHE is that the current is confined to a thin layer of silicon; the electrons move in two dimensions. The number of states per area (n_2, the two-dimensional analogy of n_s) is directly proportional to the magnetic field. The magnetic field dependence of n_2 is a consequence of the fact that the electrons are constrained to move in two dimensions. The Bohr model may be used to deduce the magnetic field dependence of n_2. The Bohr-Sommerfeld quantization condition on an electron constrained to move in a circle in a magnetic field may be written

$$\oint d\mathbf{r} \cdot \mathbf{p} = \nu h = e\Phi , \qquad (14.89)$$

where ν is a positive integer and Φ is the magnetic flux through a loop of radius r. (The quantization of magnetic flux is discussed in more detail in Chapter 15.) The magnetic flux may be written

$$\Phi = \pi r^2 B = \pi \left(\frac{p}{eB} \right)^2 B = \frac{\pi p^2}{e^2 B}, \qquad (14.90)$$

where we have used the relationship $p=erB$. The number of states per area is

$$n_2 = \frac{\pi p^2}{h^2}, \qquad (14.91)$$

which gives

$$n_2 = \frac{\nu e B}{h}. \qquad (14.92)$$

The number of states per area is an integer multiple of eB/h. The integer ν is called the *filling factor*.

For fixed magnetic field, the Hall resistance (14.84) for the two dimensional strip is

$$R_H = \frac{B}{n_2 e}. \qquad (14.93)$$

The quantization condition on n_2 (14.92) gives the quantized Hall resistance (14.85) discovered by von Klitzing et al.

The energy levels of electrons whose motion is confined to two dimensions are called *Landau* levels. The electrons that are free to move in the silicon strip occupy energy levels called *extended* levels. Above and below each extended energy level are *localized* levels caused by impurities. Electrons that occupy the localized levels are not free to move. The Landau levels in a MOSFET depend on both gate voltage (V_g) and the magnetic field. Figure 14-27 shows Landau levels in a fixed magnetic field as a function of V_g. An increase in V_g causes a decrease in the Fermi energy. When the Fermi energy resides in the region of localized energy levels, electrical conduction occurs normally; electrons with energies near the Fermi energy can move into nearby empty states and the motion of electrons causes a transfer of energy to phonons resulting in resistance. When the Fermi energy resides in the region of localized energy levels, however, the electrons cannot move and the resistance along the strip vanishes. It is important to note that the resistance vanishes because the extended Landau level is completely full. This is possible because the localized levels constitute a reservoir of electrons. The charge reservoir of the localized states is large enough to keep the extended Landau level full for small changes in the Fermi energy, hence a vanishing strip resistance coincides with a plateau in the Hall resistance. The ratio of Hall resistances (14.93) at any two plateaus is the ratio of integers because n_2 (14.92) depends on the number of filled extended Landau levels. The plateaus occur when $n_2 h/eB$ is an integer.

The QHE can also be observed by keeping the Hall voltage fixed and varying the magnetic field. (The Hall voltage can be controlled with V_g, see Figure 14-26a.) Since the density of states increases with increasing B, the Fermi level decreases with increasing B. The QHE occurs when the Fermi energy resides in the region of localized energy levels.

Fractional Quantum Hall Effect

In 1982, Daniel T. Tsui, Horst L. Störmer, and Arthur C. Gossard discovered a second remarkable phenomenon in the Hall effect. Tsui et al. were studying the quantum Hall effect in a *heterojunction*, an interface of two semiconducting crystals. The junction studied was GaAs–AlGaAs. The GaAs semiconductor was doped with silicon atoms resulting in mobile electrons with a density of 1.23×10^{15} m^{-2}, and an electron mobility of 9 m^2/V·s. The device current was kept constant at 1 μA. The data of Tsui, Störmer, and Gossard are shown in Figure 14-28, which shows the Hall resistance and the strip resistivity as a function of the magnetic field strength. A plateau in the Hall resistance is expected from the quantized Hall effect when

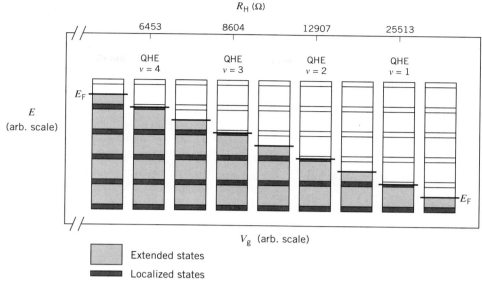

FIGURE 14-27 Band diagram of Landau levels in a MOSFET.
Electrons that are free to move occupy extended states. Localized states, which are caused by impurities, are occupied by electrons that cannot move. For fixed magnetic field, the Fermi energy decreases with increasing gate voltage V_g. When the Fermi energy lies in a band of extended states, normal conduction occurs. When the Fermi energy coincides with a band of localized states, the resistance along the strip vanishes and the QHE occurs. The quantized Hall resistance is inversely proportional to the number of filled extended bands.

$$B = \frac{n_2 h}{e v}, \qquad (14.94)$$

where $v = 1, 2, 3 \ldots$. These plateaus are observed. An additional plateau is observed at

$$B = \frac{3 n_2 h}{e}, \qquad (14.95)$$

and

$$R_H = \frac{3 h}{e^2}, \qquad (14.96)$$

FIGURE 14-28 Discovery of the fractional quantum Hall effect.
A GaAs-AlGaAs heterojunction is cooled with liquid helium and placed in a strong magnetic field. The Hall resistance R_H and strip resistivity ρ are measured as a function of the magnetic field at temperatures from 0.48 to 4.15 K. The QHE is observed for integer values of the filling factor v. An additional plateau in R_H and a corresponding drop in ρ is observed when $v = 1/3$. From D. C. Tsui, H. L. Stormer and A. C. Gossard, "Two-Dimensional Magnetotransport in the Extreme Quantum Limit," *Phys. Rev. Lett.* **48**, 1559 (1982).

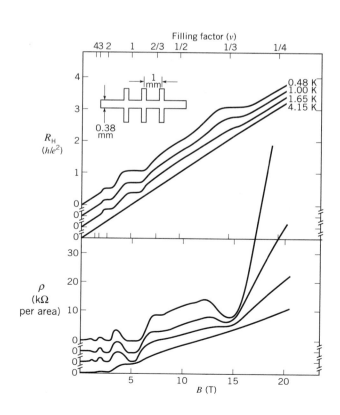

corresponding to a filling factor of

$$v = \frac{1}{3}. \tag{14.97}$$

This unpredicted phenomenon is called the *fractional quantized Hall effect* (FQHE). The FQHE has also been observed at other values of the filling factor (for example, $v = 2/3, 2/5, 3/5, 4/5,$ and $2/7$).

CHAPTER 14: PHYSICS SUMMARY

- Electron energy levels in a solid form bands. In a conductor, the highest energy electronic band is partially filled. In an insulator, there is an energy gap of several electronvolts between the highest energy band, which is filled, and the next available band, which is empty. The band structure of an intrinsic semiconductor is similar to that of an insulator except that the energy gap is typically about 1 electronvolt.

- Metals are good electrical conductors because they contain a large number of electrons, roughly 10^{29} m^{-3}, that are easily displaced when an electric field is applied.

- In a metal, the current density (J) and electric field are related by Ohm's law,

$$J = \sigma E.$$

Ohm's law holds because the electric field accelerates electrons for a characteristic time,

$$\tau = \frac{d}{v_F},$$

where d is the mean free path between collisions and v_F is the Fermi speed.

- Metals are good thermal conductors and have low electronic heat capacities. When energy is supplied to the metal in the form of heat, only those electrons with energies near the Fermi energy can gain kinetic energy.

- The properties of semiconductors are greatly enhanced by the addition of impurity atoms. If the dopants have an excess of electrons, then an *n*-type semiconductor is created. If the dopants have fewer electrons, then a *p*-type semiconductor is created. A

semiconductor diode is made by joining *p*-type and *n*-type materials. At the semiconductor junction, diffusion of charges creates a contact potential difference across a thin region that is depleted of mobile charges. If an external voltage is applied with a polarity that increases the contact potential (reverse bias), then essentially no current flows. If an external voltage is applied with a polarity that decreases the contact potential (forward bias), then the current increases exponentially with the voltage.

- A transistor is made with a *p–n–p* or an *n–p–n* junction. The three terminals are called the collector, base, and emitter. The base–emitter junction is forward biased and the base–collector junction is reverse biased. A small current through the base (I_b) controls a large current in the collector (I_c). Typically, $I_c/I_b \approx 100$.

- When a voltage is applied to a conducting sheet, the magnetic force on the moving charges causes an accumulation of charge on the edge of the sheet that creates a potential difference called the Hall voltage. The resulting electric force on the moving charges balances the magnetic force. The Hall resistance is defined to be the Hall voltage divided by the current.

- In semiconductors at low temperatures, it was discovered that the Hall resistance has a plateau at

$$R_H = \frac{h}{ve^2},$$

where

$$v = 1, 2, 3, \dots$$

This result is known as the quantum Hall effect. Measurement of the Hall resistance allows a very accurate determination of α through the relationship,

$$\alpha = \frac{ke^2}{\hbar c} = \frac{2\pi k}{vcR_H}.$$

REFERENCES AND SUGGESTIONS FOR FURTHER READING

N. W. Ashcroft and N. D. Mermin, *Solid State Physics,* Saunders (1976).

L. Bragg, "X-Ray Crystallography," *Sci. Am.* 219, No. 1, 58 (1968).

P. W. Bridgeman, "Synthetic Diamonds," *Sci. Am.* 193, No. 5, 42 (1955).

M. H. Brodsky, "Progress in Gallium Arsenide Semiconductors," *Sci. Am.* 262, No. 2, 68 (1990).

J. Demuth and P. Avouris, "Surface Spectroscopy," *Phys. Today* 36, No. 11, 62 (1983).

W. R. Frensley, "Gallium Arsenide Transistors," *Sci. Am.* 257, No. 2, 80 (1987).

M. W. Geis and J. C. Angus, "Diamond Film Semiconductors," *Sci. Am.* 267, No. 4, 84 (1992).

D. L. Goodstein, *States of Matter,* Dover (1985).

B. I. Halperin, "The Quantized Hall Effect," *Sci. Am.* 254, No. 4, 52 (1986).

H. A. Hauptman, "The Phase Problem of X-Ray Crystallography," *Phys. Today* 42, No. 10, 24 (1990).

M. Heiblum and L. F. Eastman, "Ballistic Electrons in Semiconductors," *Sci. Am.* 256, No. 2, 102 (1987).

L. Hoddeson, G. Baym, and M. Eckert, "The Development of the Quantum-Mechanical Electron Theory of Metals: 1928-1933," *Rev. Mod. Phys.* 59, 287 (1987).

P. Horowitz and W. Hill, *The Art of Electronics,* Cambridge Univ. Press (1980).

C. Kittel, *Introduction to Solid State Physics,* Wiley (1986).

K. v. Klitzing, G. Dorda, and M. Pepper, "New Method for High-Accuracy Determination of the Fine-Structure Constant Based on Quantized Hall Resistance," *Phys. Rev. Lett.* 45, 494 (1980).

A. R. MacKintosh, "The Fermi Surface of Metals," *Sci. Am.* 209, No. 1, 110 (1963).

N. Mott, "Electrons in Glass," *Rev. Mod. Phys.* 50, 203 (1978).

N. Mott, "The Solid State," *Sci. Am.* 217, No. 3, 80 (1960).

D. R. Nelson, "Quasicrystals," *Sci. Am.* 255, No. 2, 42 (1986).

F. H. Rockett, "The Transistor," *Sci. Am.* 179, No. 3, 52 (1948).

J. C. Slater, "Electronic Energy Bands in Metals," *Phys. Rev.* 45, 794 (1934).

R. L. Sproull, "The Conduction of Heat in Solids," *Sci. Am.* 207, No. 6, 92 (1962).

R. L. Sproull and W. A. Phillips, *Modern Physics: The Quantum Physics of Atoms, Solids and Nuclei,* Wiley (1980).

D. L. Stein, "Spin Glasses," *Sci. Am.* 261, No. 1, 52 (1989).

S. M. Sze, *Physics of Semiconductor Devices,* Wiley (1969).

D. M. Trotter, Jr., "Photochromic and Photosensitive Glass," *Sci. Am.* 264, No. 4, 124 (1991).

D. C. Tsui, H. L. Störmer, and A. C. Gossard, "Two-Dimensional Magnetotransport in the Extreme Quantum Limit," *Phy. Rev. Lett.* 48, 1559 (1982).

D. R. Yennie, "Integral Quantum Hall Effect for Nonspecialists," *Rev. Mod. Phys.* 58, 781 (1987).

QUESTIONS AND PROBLEMS

Electronic energy bands

1. Why are most of the elements (about three-fourths) metals?

Fermi energy

2. What is the relationship between the work function and the Fermi energy of a metal?

3. In 1 gram of copper, how many conduction electrons have kinetic energies in the interval 1–2 eV?

4. Determine the energy E_m where one-half of the conduction electrons in copper have an energy greater than E_m.

5. The Fermi energy for zinc is 9.39 eV. Calculate the density of conduction electrons. How many conduction electrons are there per atom?

6. The density of calcium ($Z = 20$) is 1.8 times the density of potassium ($Z = 19$). The Fermi energy in potassium is 2.1 eV. Estimate the Fermi energy in calcium.

7. Copper is an excellent conductor with one valence electron per atom and a Fermi energy of about 7 eV. Make an estimate of the fraction of conduction electrons that have energies larger than E_F at room temperature.

8. One gram of gold is cooled to liquid helium temperature. Calculate the total kinetic energy of the conduction electrons.

Heat capacity

9. The Fermi energy of aluminum is 11.6 eV. Calculate (a) the Fermi speed and (b) the electronic heat capacity of 1 gram of aluminum at room temperature.

10. Determine the temperature where the contribution to the heat capacity from electrons is equal to that from phonons in copper. The Fermi energy of copper is 7.0 eV and the Debye temperature is 343 K.

11. An accelerator is capable of delivering 10^{12} protons each with a kinetic energy of 1 GeV in a pulse of short duration. The protons are directed into a copper target (initially at room temperature) having a volume of 10^{-3} m^3 where they deposit all their kinetic

energy. Estimate the number of pulses needed to raise the temperature of the copper target by 10 kelvin. You may neglect the cooling of the target between pulses.

12. What is the typical energy of a phonon in a copper crystal at room temperature? What is the typical energy of a conduction electron?

13. Aluminum has a melting point of 934 kelvin and a density of 2.7×10^3 kg/m^3. Use the data of Figure 14-10 to estimate the Debye temperature of aluminum.

Ohm's law

14. The mean free path of electrons in a very pure sample of copper at 4 kelvin is about 3 millimeters. Calculate the resistivity of the sample.

15. Calculate the mobility of electrons in copper at room temperature. Take the conductivity to be 5.9×10^7 $\Omega^{-1} \cdot$m^{-1}. Compare the mobility of copper with that of diamond at room temperature (see Figure 14-13).

16. Make a rough sketch of the current density (14.70) as a function of electric field in an insulator. Indicate the region where Ohm's law holds and the effect of a saturating drift speed. Assume that at some very large value of electric field that the number of conduction electrons grows exponentially.

17. Use the measured mobility in diamond (Figure 14-13b) to calculate the relaxation time at room temperature.

18. The mobility of silicon at room temperature is measured to be 0.19 m^2/V$\cdot$s. (a) Calculate the relaxation time. (b) Determine the drift speed when an electric field of 10^3 V/m is applied.

Semiconductors

19. The next element after tin in Group IV of the periodic table is lead. Why is lead not an intrinsic semiconductor?

20. Estimate the temperature where the density of conduction electrons in undoped germanium is 10^{15} per cubic meter.

21. Why doesn't Ohm's law hold for conduction in a diode?

22. Why is diamond transparent to visible light whereas a silicon crystal is not?

23. When a germanium crystal is doped with indium, what type of semiconductor results?

24. Calculate the ratio of forward to reverse current in a diode at room temperature for an applied voltage of plus or minus 1 volt.

25. Draw an energy level diagram similar to Figure 14-23 for a *p–n–p* transistor that is (a) unbiased and (b) biased for operation as an amplifier.

26. The conductivity of a sample of germanium is measured to be 2 $\Omega^{-1} \cdot$m^{-1} at room temperature. Give an expression for the conductivity as a function of temperature (T) for $T > 300$ K. What is the conductivity at $T = 1000$ K?

The Hall effect

27. Why do you think it took 101 years from the time the Hall effect was discovered until the quantized Hall effect was discovered?

28. Why was the discovery of the quantized Hall effect made at low temperature?

29. Silver has a density of 1.05×10^4 kg/m^3 and an atomic number of 108. (a) Calculate the density of conduction electrons. (b) A current density of 10^6 A/m^2 flows in a thin sheet of width 10^{-2} m. Determine the strength of magnetic field needed to generate a Hall voltage of 1 µV.

30. Von Klitzing, Dorda, and Pepper (*Phys. Rev. Lett.* **45**, 494, 1980) measured the $n = 4$ quantized Hall resistance with high accuracy and obtained a value of R_H = 6453.17 ± 0.02 Ω. With what accuracy did they determine the electromagnetic coupling constant, α?

31. Tsui, Störmer, and Gossard (*Phys. Rev. Lett.* **48**, 1559, 1982) discovered the fractional quantum Hall effect for filling factors of $v = 1/3$ and 2/3 in a device that had a density of charge carriers equal to 1.23×10^7 m^{-2}. Calculate the values of magnetic field where the FQHE was observed.

32. (a) Prove that the filling factor is dimensionless. (b) A quantum Hall device is designed to have a filling factor of one at $B = 10$ tesla. What is the charge carrier concentration?

Additional problems

33. Calculate the average energy of a conduction electron in gold.

34. What type of semiconductor (p or n) is obtained when silicon is doped with (a) antimony, (b) phosphorous, (c) aluminum, and (d) indium?

35. The longest wavelength that a certain semiconductor can absorb is 1.85 µm. Calculate the energy gap of the semiconductor.

36. The Fermi temperature (T_F) is defined by

$$E_F \equiv kT_F.$$

Calculate the Fermi temperature for silver ($E_F = 5.48$ eV). What is the physical significance of the Fermi temperature?

*37. (a) For what approximate value of the electron density does the nonrelativistic expression for the Fermi energy break down? (b) Derive an expression for the Fermi energy of a gas of relativistic electrons, in terms of the electron density. (c) A star of mass 2×10^{30} kg collapses to a radius of 10^4 m. Estimate the Fermi energy of the electrons in the collapsed star.

*38. *Thermal conductivity and the Wiedemann-Franz law.* (a) Consider the flow of charge q in a metal rod of gradient cross-sectional area A when a voltage V is applied. Show that Ohm's law may be written

$$\frac{dq}{dt} = \sigma A \frac{dV}{dx}.$$

(b) The heat flow in the rod is described with a similar equation. If energy in the form of heat (Q) is supplied at one end of the rod, there will be a temperature gradient (dT/dx) along the length of the rod. The rate at which energy is transferred to the other end of the rod (dQ/dt) may be written

$$\frac{dQ}{dt} = \kappa A \frac{dT}{dx},$$

where the constant of proportionality (κ) is called the *thermal conductivity*. Show that for an ideal gas the thermal conductivity may be written

$$\kappa = \frac{C_V \langle v \rangle d}{3V},$$

where $\langle v \rangle$ is the average speed of a gas molecule, d is the mean free path and V is the volume. (c) Calculate κ for a metal by treating the electrons as a gas and show that

$$\frac{\kappa}{\sigma T} = \frac{\pi^2 k^2}{3 e^2} = 2.45 \times 10^{-8} \text{ W} \cdot \Omega / \text{K}^2.$$

39. *Light-emitting diode* (LED). Consider a heavily doped p–n junction that is forward biased so that a large number of electrons and holes diffuse into the junction region. Electrons can combine with the holes. (a)

How much energy is released when an electron falls into a hole? (b) In some semiconductors, such as cadmium selenium, the electron–hole combining process occurs with the emission of a photon. Calculate the wavelength of the photons that are emitted if the band gap in CdSe is 1.8 eV. (c) In other semiconductors, such as silicon and germanium, the energy from the electron–hole combining process does not produce photons. What happens to the energy?

*40. *The photovoltaic effect.* When light is incident on a p–n junction, photons can create electron–hole pairs. The electrons are able to a move a distance d_e and holes are able to move a distance d_p before recombining. (a) Photons incident on the junction create electron-hole pairs at a rate R. Calculate the current density in a forward-biased junction. (b) Calculate the current density in a reverse-biased junction as a function of R. (c) For illumination of an open-circuit diode, a voltage will build up between the terminals. Calculate the voltage as a function of R. (d) What is the maximum possible voltage? (e) Gallium arsenide, which has a band gap of 1.4 eV, is well suited for conversion of solar energy to electrical energy. Why is GaAs better than germanium for solar power conversion?

*41. *The Richardson-Dushman equation.* Consider the escape of electrons from the surface of a metal. Let the x direction be normal to the surface. (a) Show that the condition for electrons to escape is that the x component of momentum be greater than some critical value (p_c) given by

$$p_c = \sqrt{2m(E_F + \phi)},$$

where ϕ is the work function of the metal. (b) Writing the current density (J) in terms of the number of electrons per volume (n) and making use of the density of states for electrons in a metal, show that

$$J = \frac{2e}{h^3 m} \int_{p_c}^{+\infty} dp_x \, p_x \int_{-\infty}^{+\infty} dp_y \int_{-\infty}^{+\infty} dp_z$$

$$\times \frac{1}{e^{(p_x^2 + p_x^2 + p_x^2)/2mkT - E_F/2mkT} + 1}.$$

(c) Use the fact that $\phi \gg kT$ to show that

$$J \approx \left(\frac{4\pi me}{h^3} \right)(kT)^2 \, e^{-\phi/kT}.$$

42. (a) Show that the pressure (P) of the conduction electrons in a metal is

$$P = \frac{2}{5}nE_F,$$

where n is the density of conduction electrons and E_F is the Fermi energy. (b) Evaluate the electron pressure for copper and compare your answer to atmospheric pressure.

43. The heat capacity per volume of NaCl is 1.9×10^6 $J \cdot m^{-3} \cdot K^{-1}$. Treating the phonons as an ideal gas (see problem 38) with an average speed of 500 m/s and a mean free path of 2.3 nm, calculate the thermal conductivity.

SUPERCONDUCTIVITY

… The ground-state wave function of a superconductor can be considered to be a sum of low-lying normal configurations in which the quasiparticle states are paired such that if the state $k\uparrow$ is occupied, then $-k\downarrow$ is also occupied. The states of opposite spin and momentum are either both occupied or both empty.

When there is current flow, the paired states $(k_1\uparrow, k_2\downarrow)$ all have exactly the same net momentum, $m^*v_s = 2mv_s = \hbar(k_1+k_2)\dots$. Scattering of individual particles does not change the common momentum of the paired states, so the current persists in time.

John Bardeen

The phenomenon of *superconductivity*, in which the resistance of certain materials completely vanishes at low temperatures, is one of the most interesting and sophisticated in condensed matter physics. In the superconducting state, pairs of electrons with spins and momenta antialigned develop an extremely weak electrical attraction for each other by interaction with the lattice. The attraction of the two electrons may be thought of as a two-step process in which the passage of one electron distorts the lattice to which the second electron is attracted. The bound state of two electrons in a superconductor is called a *Cooper pair*. The size of a Cooper pair is a few hundred nanometers, far larger than the size of an atom. A Cooper pair behaves like a boson, that is, the total intrinsic angular momentum is zero. The Cooper pairs in a superconductor all have the same drift speed. The formation of the Cooper pairs causes the appearance of a gap in the electron energy distribution. The energy gap prevents Cooper pairs from moving to higher energy levels. A consequence of this is that the current in a superconductor is not limited by scattering, as it is in a normal conductor. We shall first discuss some of the basic properties of superconductors before returning to a discussion of the theory.

15-1 BASIC EXPERIMENTAL PROPERTIES OF SUPERCONDUCTORS

The temperature dependence of the resistivity of a typical metal is shown in Figure 15-1. The resistivity is caused by collisions of the electrons when they are accelerated by an electric field applied to the metal. The resistivity ρ is defined to be the inverse of the conductivity (14.60). Since the conductivity (14.68) is directly proportional to the mean free path between collisions (d), ρ is inversely proportional d. At temperatures greater than about 15 K, d is limited by thermal vibrations of the atoms. The electron scattering cross section, which is inversely proportional to d (2.54), is proportional to the square of the amplitude of the atomic vibrations, which is proportional to the temperature T. Therefore, for $T > 15$ K,

$$\rho \propto \frac{1}{d} \propto T. \tag{15.1}$$

At low temperatures ($T < 15$ K), the mean free path does not become infinite, because there is a contribution to d from impurities or other defects in the metal. If the metal were a perfect crystal, this contribution would theoreti-

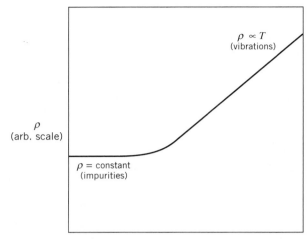

FIGURE 15-1 Resistivity in a metal.
The resistivity (ρ) is inversely proportional to the electron mean free path. At low temperatures (below about 15 kelvin) the resistivity is a constant due to the presence of impurities that limit the mean free path. At large temperatures, the resistivity is proportional to the temperature because the mean free path is limited by atomic vibrations.

cally not be present. In practice, however, even the purest samples contain some defects, and scattering will occur at its boundaries, leading to

$$\rho \propto \frac{1}{d} = \text{constant} \tag{15.2}$$

at low temperatures.

What happens to the resistivity of a very pure element near zero kelvin? This interesting question had no known answer at the beginning of the twentieth century. The cooling of metals to very low temperatures was first achieved by Heike Kamerlingh Onnes in 1908 when he liquefied helium. Helium boils at a temperature of about 4.2 kelvin (at standard pressure). Onnes chose to investigate the low-temperature resistivity of mercury, because mercury could be made very pure by distillation. In 1911, Onnes discovered that the resistivity of mercury suddenly dropped to *zero* when cooled below 4.2 kelvin. The data of Onnes are shown in Figure 15-2.

A new physical phenomenon, which Onnes called superconductivity, had appeared at low temperatures. Superconductivity is a state of zero resistivity (or infinite conductivity). The abrupt change in resistance at about 4.2 kelvin is a *phase transition* from the normal to the superconducting state. The temperature at which the phase

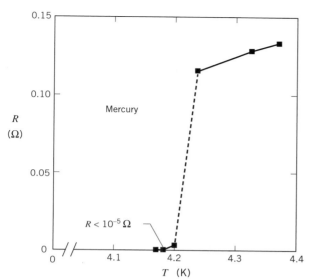

FIGURE 15-2 The discovery of superconductivity by H. K. Onnes.

The electrical resistance R of a sample of mercury is plotted as a function of the temperature T. At a temperature of about 4.2 K, the resistance suddenly vanishes. The temperature at which the resistance falls to zero is called the critical temperature. From H. K. Onnes, *Commun. Phys. Lab.* **12**, 120 (1911).

transition occurs is called the *critical temperature*. When the phase transition occurs, the density of states available to the electrons undergoes a significant change. Not all metals become superconductors, even near zero kelvin. For those pure elements that do become superconducting at low temperatures, the critical temperature below which the resistance remains zero varies from near 0 K to about 10 K.

Onnes also observed superconductivity in lead and tin. He was able to demonstrate that the resistance was really zero by measuring the decay time of a current in a superconducting ring. Such a current was not observed to decay at all, even after many hours! Onnes also investigated the possibility of making superconducting magnet coils. The attempt to produce large magnetic fields with these superconducting coils always failed at magnetic fields above 0.1 tesla. We shall investigate the reason for this in more detail.

Type-I Superconductors

The superconductors discovered by Onnes are not merely perfect conductors; they have the additional property that there is zero magnetic field inside the superconductor

when in the superconducting state. Such a material is called a *perfect diamagnet*. The vanishing of the magnetic field inside a superconductor is an independent property that is consistent with zero resistance. (In other words, zero resistance does not imply zero magnetic field.) This class of superconductors is called *type-I*. Thirty pure elements are type-I superconductors at low temperatures. The superconducting elements and their critical temperatures are indicated in Figure 15-3. The best conductors like copper, silver and gold are not superconductors. The reason is that the mechanism for superconductivity is an interaction of a pair of electrons with the lattice of atoms in the metal. This interaction is very weak and depends on lattice vibrations at low temperature. The best conductors at normal temperatures have the smallest lattice vibrations and do not superconduct! We will return to a discussion of this important physics later in the chapter. The properties of superconductors are very sensitive to impurities and how the material has been annealed.

> A type-I superconductor has zero resistance to the flow of current and zero magnetic field inside the superconductor.

Critical Temperature

In an ordinary conductor, the current (I) is accompanied by power loss (P) that is proportional to the resistance (R):

$$P = I^2 R. \qquad (15.3)$$

The energy dissipated heats the conductor. If energy from an external source is not supplied to the conductor, then the current rapidly decays to zero. In the superconducting state, the resistance is zero and current can exist without power loss. If a current is started in a superconducting loop, the current will flow forever without attenuation as long as the material remains in the superconducting state. The state of superconductivity exists only below the critical temperature (T_c). Each superconductor has its own critical temperature. The critical temperatures of the elements vary from just a small fraction of a kelvin in rhodium to 9.3 kelvin in niobium. The typical order of magnitude of the critical temperatures for the superconducting elements is the boiling point of liquid helium.

The Meissner Effect

Consider what happens to a superconducting material when we suddenly turn on an external magnetic field. The changing magnetic field induces currents on the surface of the superconductor. These surface currents generate a

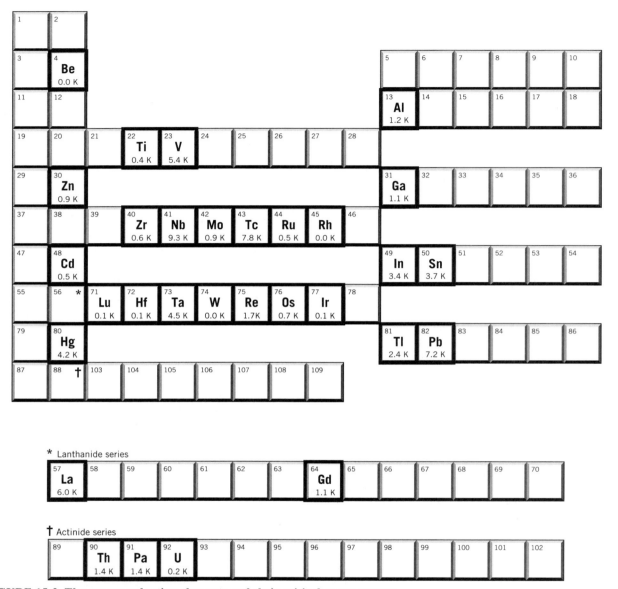

FIGURE 15-3 The superconducting elements and their critical temperatures.

component of magnetic field inside the superconductor that exactly cancels the contribution of the external field. The net magnetic field inside the superconductor is zero. This is possible in a superconductor because currents are able to flow without resistance. The phenomenon of the exclusion of a magnetic field from the interior of a super-conductor is analogous to the exclusion of an electric field from the interior of an ordinary conductor. In the presence of an electric field, charges in the conductor are free to

move and arrange themselves to create a component of electric field inside the conductor that exactly cancels the contribution of the external electric field (see Figure 15-4).

If a conductor in the normal phase (nonsuperconducting) is subjected to an external magnetic field, the field will uniformly penetrate the material. We now examine what happens as the temperature is lowered below critical temperature. If the material were to suddenly become *merely* a perfect conductor, no current would flow until

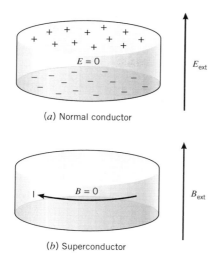

(a) Normal conductor

(b) Superconductor

FIGURE 15-4 Conductors in an external field.
(a) An external electric field does not penetrate an ordinary conductor because the charges arrange themselves on the surface in such a manner that the net electric field inside the conductor is zero. (b) An external magnetic field does not penetrate a superconductor because currents flow on the surface in such a manner that the net magnetic field inside the superconductor is zero.

the external field was removed. The changing magnetic field would induce currents that would trap the magnetic flux inside the conductor. This is not what happens for a superconductor. When the material is cooled into the superconducting phase in the presence of an external magnetic field, something rather remarkable happens. The magnetic flux is *completely expelled* from the interior of the superconductor, provided that the magnetic field is not too strong. This phenomenon of magnetic field expulsion inside a superconductor is called the *Meissner effect* after Walter Meissner, who together with R. Ochsenfeld discovered this fundamental property in 1933. A type-I material that is superconducting in the presence of an external magnetic field is said to be in the *Meissner state*. A type-I superconductor is a perfect diamagnet and it will repel a permanent magnet (see color plate 9). We shall investigate this interesting property of superconductors further.

Critical Magnetic Field

Type-I materials are not able to remain superconducting in the presence of a sufficiently large magnetic field. Each superconductor has a *critical magnetic field*, $B_c(T)$, which depends on the temperature. When the external magnetic

field is made larger than the critical field, the state of superconductivity is destroyed and the material returns to its normal conducting state. The temperature dependence of the critical field is approximately of the form

$$B_c(T) \approx B_c(0)\left[1 - \left(\frac{T}{T_c}\right)^2\right]. \qquad (15.4)$$

The critical magnetic field at zero temperature, $B_c(0)$, is different for each superconducting material. The values of $B_c(0)$ for the elements range from about 5×10^{-6} tesla in ruthenium to 0.2 tesla in niobium. The critical magnetic field is strongly correlated with the critical temperature. Figure 15-5 shows a plot of $B_c(0)$ versus T_c for the superconducting elements. The typical value of critical magnetic field at zero kelvin is

$$B_c(0) \approx (0.01\,\text{tesla/K})T_c. \qquad (15.5)$$

Figure 15-6 shows a plot of the critical magnetic field (15.4) as a function of temperature for a type-I superconductor. This type of plot is called *phase diagram*. The function $B_c(T)$ (15.4) defines a boundary between the normal state and the superconducting state. The superconducting state is a state of lower total energy than the normal state. The energy difference between the normal and superconducting states at zero magnetic field is called the *condensation energy*.

The critical magnetic field is directly related to the condensation energy of the superconductor. For a simple geometry such as a cylinder, the magnetic energy density (u_B) is a direct measure of the condensation energy per unit volume (u_c) of the superconductor:

$$u_c = u_B = \frac{B_c^2(T)}{2\mu_0}. \qquad (15.6)$$

For this geometry, the critical field, $B_c(T)$, is referred to as the *thermodynamic critical field*. (For other geometries, the amount of energy required to expel the magnetic field can be calculated and a geometrical factor called the *demagnetization factor* is introduced to relate u_c and B_c.) If the superconducting transition occurs in the presence of a magnetic field, an amount of energy equal to the condensation energy is released. The energy released in a phase transition is referred to as *latent heat*, and such a transition is called a *first-order* phase transition. The transition is characterized by the coexistence of normal and superconducting domains (analogous to ice in water).

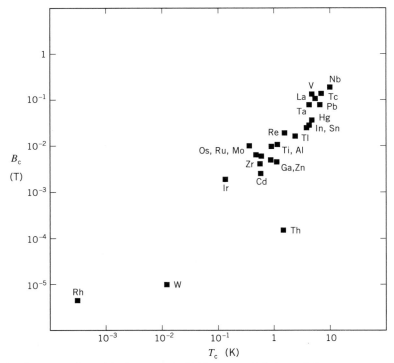

FIGURE 15-5 Correlation between the critical magnetic field at zero kelvin and the critical temperature for the superconducting elements.

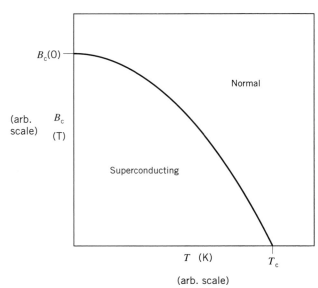

FIGURE 15-6 Phase diagram for a type-I superconductor. The external magnetic field defines a temperature-dependent boundary called the critical magnetic field, $B_c(T)$. For magnetic fields below the critical value, the material is in the superconducting state.

At the critical temperature, the condensation energy goes to zero, and above the critical temperature there is no superconducting state. At zero magnetic field, the transition is thermodynamically reversible. There is, however, a discontinuity in the heat capacity, as will be discussed in more detail. Such a transition is called a *second-order* phase transition.

EXAMPLE 15-1

The critical temperature for niobium is 9.3 K and the critical field at zero kelvin is 0.20 tesla. Calculate the critical magnetic field and condensation energy per volume for a cylinder of niobium at liquid helium temperature.

SOLUTION:

At liquid helium temperature (4.2 K), the critical field (15.4) is

$$B_c \approx B_c(0)\left[1 - \left(\frac{T}{T_c}\right)^2\right]$$

$$= (0.20 \text{ tesla})\left[1 - \left(\frac{4.2 \text{ K}}{9.3 \text{ K}}\right)^2\right] \approx 0.16 \text{ tesla}.$$

The condensation energy per volume (15.6) is

$$u_c = \frac{B_c{}^2}{2\mu_0} = \frac{(0.16 \text{ tesla})^2}{(2)(4\pi \times 10^{-7} \text{ N/A}^2)}$$

$$= 1.0 \times 10^4 \text{ J/m}^3 .\qquad \blacksquare$$

Critical Current

The critical field also sets a limit on the current-carrying capacity of a type-I superconducting wire. By Ampère's law, a current (I) in a long wire generates a magnetic field (B) outside the wire of magnitude

$$B = \frac{\mu_0 I}{2\pi r},\qquad (15.7)$$

where r is the distance from the center of the wire. This result does not depend on how the current is distributed in the cross section of the wire as long as it is symmetrical. Type-I superconductors cannot carry an indefinitely large current and remain superconducting because the current will create a B-field at the surface (r_0), which will destroy the superconductivity. Nevertheless, the currents can be surprisingly large.

EXAMPLE 15-2

Consider a wire made of tin with a radius about the size of a hair (40 μm). How much current can the wire carry at zero kelvin and remain superconducting?

SOLUTION:

The critical magnetic field of tin is 0.03 tesla. The superconductivity of tin is destroyed when the magnetic field created by the current at the surface of the wire exceeds this value. The maximum superconducting current is

$$I = \frac{2\pi r_0 B_c}{\mu_0}$$

$$= \frac{(2)(\pi)(4 \times 10^{-5} \text{ m})(0.03\,\text{T})}{4\pi \times 10^{-7} \text{ N/A}^2}$$

$$= 6\,\text{A}.\qquad \blacksquare$$

The current calculated in Example 15-2 is quite large. It is even more surprising considering that the current must reside on the surface of the hair-thick wire! If there was current inside the wire, Ampère's law would require that the magnetic field would be nonzero inside the wire. We know that this cannot be the case because there is zero field inside a type-I superconductor.

Type-II Superconductors

In 1930, superconductivity was discovered in lead-bismuth alloys by W. J. deHaas and J. Voogd. Many alloys are now known to superconduct. This class of superconductors is called *type-II*. Because of their mechanical properties, the alloys became known historically as *hard* superconductors and the pure elements were referred to as *soft* superconductors. The important distinction between type-I and type-II superconductors, however, is not in their mechanical properties but rather in their magnetic properties.

In 1953, the critical temperature of niobium-tin (Nb_3Sn) was measured by B. T. Matthias and collaborators. The experimenters wound a copper coil around the sample of niobium-tin as indicated in Figure 15-7 and measured the inductance (L) of the device. The inductance is measured by switching on an external magnetic field and measuring the induced current (I), which is given by the expression

$$N\frac{d\Phi_m}{dt} = L\frac{dI}{dt},\qquad (15.8)$$

where N is the number of turns in the coil and Φ_m is the magnetic flux through the coil. At normal temperatures, the inductance of the device depends only on its geometry. For a solenoid of length d and cross-sectional area A, the inductance is

$$L = \frac{A\mu_0 N^2}{d}.\qquad (15.9)$$

Now consider what happens when the temperature of the device is lowered. When the niobium-tin becomes superconducting, magnetic flux is expelled from the solenoid and

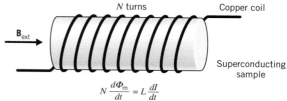

$$N\frac{d\Phi_m}{dt} = L\frac{dI}{dt}$$

FIGURE 15-7 Arrangement for determination of the critical temperature by magnetic measurements. A copper coil is wound around a superconductor. In the normal state the inductance of the device is proportional to the cross-sectional area of the superconductor. The inductance of the device is reduced when the core becomes superconducting because the cross-sectional area through which the magnetic flux can penetrate is restricted by the superconductor.

the inductance drops. The area that the magnetic field can penetrate has been reduced. The data of Matthias et al., are shown in Figure 15-8. The critical temperature of Nb$_3$Sn is observed to be 18 kelvin. At the time of this measurement, this was the highest known critical temperature.

The existence of superconductivity in the presence of a very large magnetic field and at large current densities was discovered by J. Eugene Kunzler and collaborators in 1958. Kunzler et al., discovered that the alloy Nb$_3$Sn could superconduct in very large magnetic fields and large current densities. Kunzler also discovered that wires made of niobium-clad niobium-tin could carry a larger current density than bars of niobium-tin. The data of Kunzler et al. are shown in Figure 15-9. The niobium-clad wire is observed to superconduct at a magnetic field of 8.8 tesla and a current density (J) greater than 10^9 A/m^2. The observed critical field of the alloy is much larger than that of either niobium (0.2 T) or tin (0.03 T). The important experimental distinction between the magnetic properties of type-I and type-II superconductors had been discovered.

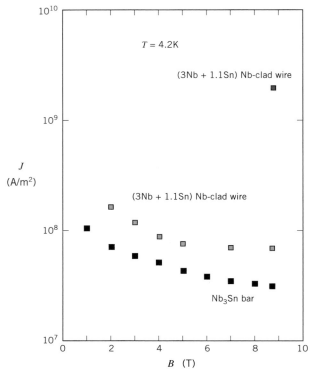

FIGURE 15-9 Discovery of superconductivity at high current densities ($J > 10^9$ A/m^2) and high magnetic fields ($B = 8.8$ T).

From J. E. Kunzler et al., *Phys. Rev. Lett.* **6**, 89 (1961).

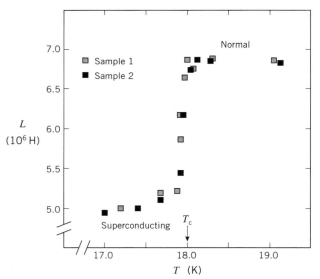

FIGURE 15-8 Magnetic measurement of the critical temperature of Nb$_3$Sn.

A copper coil is wound around a sample of niobium-tin (see Figure 15-7). The device is emersed in liquid hydrogen and the inductance (L) is measured as a function of temperature. The inductance drops when the niobium-tin becomes superconducting. At the time of this measurement, this was the highest observed critical temperature of any material. From B. T. Matthias et al., "Superconductivity of Nb$_3$Sb," *Phys. Rev.* **95**, 1435 (1954).

Critical Magnetic Fields

All type-II superconductors are compounds and alloys. Type-II superconductors have two critical magnetic fields, B_{c1} and B_{c2}. The relationship between the type-II critical magnetic fields and the thermodynamic critical field (B_c) of a type-I superconductor is

$$B_{c1} < B_c < B_{c2}. \tag{15.10}$$

Both B_{c1} and B_{c2} are a function of temperature. The value of B_{c2} for several compounds exceeds 10 tesla at liquid helium temperature. Below the lower critical field (B_{c1}), the material is superconducting as for type-I. Above the higher critical field (B_{c2}), the material is normal as for type-I. For values of magnetic field between the two critical fields, the material is in a special finely divided intermediate state called a *mixed state*. The mixed state, also called the *vortex state*, consists of normal cores surrounded by circulating superconducting currents, as indicated in Figure 15-10. The size of the normal core is

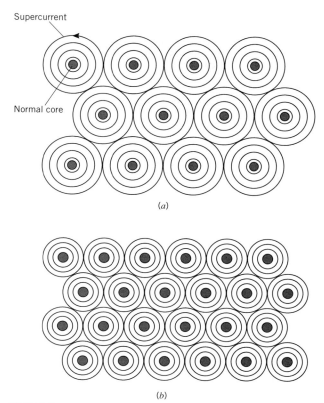

(a)

(b)

FIGURE 15-10 The mixed state.
(a) For a range of external magnetic fields between B_{c1} and B_{c2}, the cross section of the interior of a type-II superconductor consists of an array of supercurrents circulating around a normal core. The size of the normal core is typically 300 nm. *(b)* The spacing between the vortices depends on the external field. For larger values of external magnetic field, the vortices are closer together.

$$\mathbf{B}_m = \mu_0 \mathbf{M}. \qquad (15.11)$$

The net field is given by

$$\mathbf{B} = \mathbf{B}_{ext} + \mathbf{B}_m = \mathbf{B}_{ext} + \mu_0 \mathbf{M}. \qquad (15.12)$$

For a type-I superconductor, the magnetization field is exactly equal to $(-\mathbf{B}_{ext})$, so that B is identically equal to zero inside the superconductor. Figure 15-11a shows the magnetization in a type-I superconductor. In a type-II superconductor, the field can penetrate the material for values of magnetic field above B_{c1}. This means that the magnetization does not completely cancel the external field, as shown in Figure 15-11b. When the field is equal to (B_{c2}), the magnetization is zero because the magnetic flux density is so large that the normal cores of the vortices completely overlap and the normal state is recovered. The phase diagram for a type-II superconductor is shown in Figure 15-12.

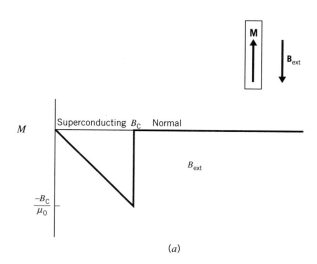

(a)

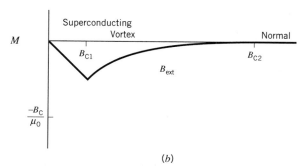

(b)

FIGURE 15-11 Magnetization versus external field in a thin superconducting rod.
(a) Type-I superconductor. *(b)* Type-II superconductor.

typically 300 nm. A type-II superconductor can develop a large array of internal surfaces so that the magnetic field can be distributed throughout the material. As long as the vortices do not move (a phenomenon called *pinning*), the mixed state can still transport current with zero resistance. If the external field is increased, then the normal cores are closer together and stronger superconducting currents are needed to keep the field expelled from the superconductor. For sufficiently large magnetic fields, the vortices become so close together that the normal cores completely overlap and the material does not superconduct.

In the presence of an external magnetic field ($\mathbf{B}_{ext}$), the field inside a material is characterized by the magnetization vector ($\mathbf{M}$). The portion of the field ($\mathbf{B}_m$) due to this magnetization is

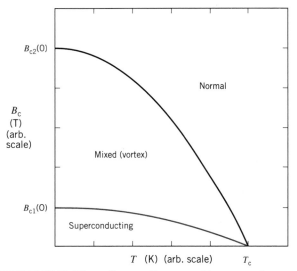

FIGURE 15-12 Phase diagram for a type-II superconductor.
There are two critical magnetic fields, $B_{c1}(T)$ and $B_{c2}(T)$. When
the external magnetic field is below the lower critical field, the
material is in the superconducting state and there is no magnetic
field inside the superconductor, just like a type-I superconductor. When the external magnetic field is above the upper critical
field, the material is in the normal state. For values of the
external field between $B_{c1}(T)$ and $B_{c2}(T)$, the material is in a
mixed state that is partially normal and partially superconducting. The magnetic field penetrates the material in the normal
region.

Until 1986, the highest known critical temperature of
any superconductor was 23.2 K, in Nb_3Ge. Various theories predicted that it was *impossible* to have much higher
critical temperatures. These theories were wrong! In 1986,
a new category of type-II superconductors, called *high-T_c*
superconductors, were discovered with critical temperatures significantly above that of liquid nitrogen (77 K)!
This was a major breakthrough in the field of superconductivity. High-T_c superconductors are discussed in Section 15-4.

Critical Currents
Type-II materials can carry a superconducting current
only if the vortices are fixed. The vortices repel each other
and if there is no net current, the vortices will arrange
themselves in a regular array. When a current is present,
the vortices feel a force, causing them to move. The
motion of the vortices is called *flux flow*. Only if the
material has numerous microscopic defects will enough of
the vortices bind to energetically favorable locations to

prevent motion of the flux lattice. The process of squeezing and drawing billets of type-II superconducting material into long wires improves the ability of those wires to
carry a substantial superconducting current. Critical currents in a given material are observed to vary by several
orders of magnitude depending on the details of the
microscopic structure.

We may summarize the properties of type-II superconductors by plotting a three-dimensional surface in T_c, B_{c2},
J_c space. This is shown in Figure 15-13 for niobium-
titanium (NbTi). Below the surface in T_c, B_{c2}, J_c space,
NbTi is in the superconducting state and above the surface, NbTi is in the normal state. If a conductor originally
in the superconducting state crosses the T_c, B_{c2}, J_c boundary due to external conditions, the conductor abruptly
becomes resistive. This is called a *quench*.

15-2 DEVELOPMENT OF THE THEORY OF SUPERCONDUCTIVITY

The development of the theory of superconductivity has a
history that is almost as rich and varied as that of the
experimental discoveries. The reason for this is that some
aspects of the theory address *microscopic* phenomena,

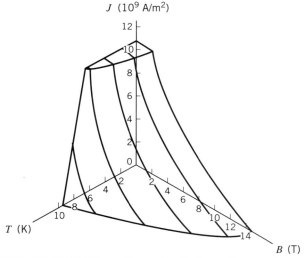

FIGURE 15-13 Three-dimensional phase diagram for niobium-titanium (NbTi).
Below the surface in T_c, B_{c2}, J_c space, NbTi is in the supercon-
ducting state and above the surface it is in the normal state. From
D. Larbalestier et al., "High-Field Superconductivity," *Phys.
Today* **39**, 24 (1986).

whereas others address *macroscopic* phenomena. Thus, over the decades the theoretical progress in the understanding of superconductivity has not always had an obviously logical progression; however, there has been one common thread in the understanding of superconductivity, the interplay between experiment and theory. The major theoretical breakthroughs have been:

1. Phenomenological description of perfect conductivity and perfect diamagnetism with the introduction of the magnetic field penetration depth (λ).

2. Description of the density of superconducting electrons with a Schrödinger-like wave equation and the introduction of the coherence length (ξ), the characteristic distance scale of superconductivity.

3. Development of the theory of superconductivity as an electron–lattice–electron interaction where electrons form pairs with opposite spins and momenta, allowing a quantitative prediction of the appearance of an energy gap in the allowed electron energy states.

4. Prediction of quantum tunneling of a superconducting electron pair through an insulating junction between two superconductors.

* *Challenging*

The London Equation

In 1935, Fritz and Heinz London found a phenomenological expression that contained both infinite conductivity and the Meissner effect in superconductors. The *London equation* states that the time derivative of the superconducting current density (**J**) is proportional to the electric field:

$$\frac{\partial \mathbf{J}}{\partial t} = a\mathbf{E}, \qquad (15.13)$$

where a is constant. The London equation may be compared to the relationship between the current density and the electric field in a normal conductor (Ohm's law):

$$\mathbf{J} = \sigma \mathbf{E}. \qquad (15.14)$$

The current is directly proportional to the drift speed of the electrons. Recall how the electrons acquire their drift speed. The force on the electrons due to an electric field causes an acceleration (eE/m) that acts for a time (τ) that is proportional to the electron mean free path (d). This

results in a drift speed proportional to the electric field ($v_d = eE\tau/m$). The conductivity (σ) is directly proportional to the mean free path of the electrons, which is of finite size in a normal conductor ($\sigma \propto \tau \propto d$). In a normal conductor the electric field provides the force needed to keep the electrons moving. According to the London equation (15.13), the current in a superconductor can be nonzero even if the electric field is zero. The superconducting state is a state of infinite conductivity. An electric field in a superconductor causes an acceleration of the electrons. There is no limit to the current from the mean free path because all the Cooper pairs have the same drift speed. If some electrons get scattered, the rest keep going.

We write the superconducting current in terms of the density of superconducting electrons (n) and their drift velocity ($\mathbf{v}_d$)

$$\mathbf{J} = ne\,\mathbf{v}_d. \qquad (15.15)$$

Taking the time-derivative, we have

$$\frac{\partial \mathbf{J}}{\partial t} = ne\frac{d\mathbf{v}_d}{dt}. \qquad (15.16)$$

Writing the electron acceleration as the electric force divided by the electron mass,

$$\frac{d\mathbf{v}_d}{dt} = \frac{e\mathbf{E}}{m}, \qquad (15.17)$$

we have

$$\frac{\partial \mathbf{J}}{\partial t} = \frac{ne^2\mathbf{E}}{m}. \qquad (15.18)$$

Therefore, the constant a in the London equation (15.13) is

$$a = \frac{ne^2}{m}. \qquad (15.19)$$

The London equation predicts the Meissner effect. Taking the curl of both sides of the London equation (15.13), we have

$$\frac{\partial(\nabla \times \mathbf{J})}{\partial t} = a\nabla \times \mathbf{E}. \qquad (15.20)$$

Using Faraday's law (B.15), we have

$$\frac{\partial(\nabla \times \mathbf{J})}{\partial t} = -a\frac{\partial \mathbf{B}}{\partial t}. \qquad (15.21)$$

or

$$\nabla \times \mathbf{J} = -a\mathbf{B}. \qquad (15.22)$$

Ampère's law (B.17) with $\partial \mathbf{E}/\partial t = 0$ is

$$\nabla \times \mathbf{B} = \mu_0 \mathbf{J}. \qquad (15.23)$$

Taking the curl of the curl of $\mathbf{B}$ (15.23) and using the curl of $\mathbf{J}$ (15.22), we have

$$\nabla \times \nabla \times \mathbf{B} = -\mu_0 a\mathbf{B}, \qquad (15.24)$$

which we may write as

$$\nabla^2 \mathbf{B} = \mu_0 a\mathbf{B}. \qquad (15.25)$$

The only uniform magnetic field that satisfies this expression is

$$B = 0. \qquad (15.26)$$

The magnetic field is zero inside a superconductor.

Penetration Depth

The equation for the magnetic field inside a superconductor (15.25) has an exponential solution. This means that an external magnetic field will actually penetrate the superconductor with an exponential attenuation. The *penetration depth* is a measure of the distance that a magnetic field can penetrate a type-I superconductor. We may define the *London* penetration depth (λ_L) by writing $\nabla^2\mathbf{B}$ (15.25) as

$$\nabla^2 \mathbf{B} \equiv \frac{\mathbf{B}}{\lambda_L^2}. \qquad (15.27)$$

With this definition and our expression for the constant a (15.19) appearing in the London equation, the London penetration depth may be written

$$\lambda_L = \sqrt{\frac{m}{\mu_0 n e^2}}. \qquad (15.28)$$

The actual penetration depth (λ) can be measured and reasonable agreement with the London formula is obtained ($\lambda \approx \lambda_L$).

For a type-I superconductor, the magnetic field is attenuated exponentially in the superconductor:

$$B(x) = B_0 e^{-x/\lambda}. \qquad (15.29)$$

The penetration depth varies with temperature. The penetration depth is a minimum at $T = 0$ and becomes infinite at $T = T_c$, that is, the magnetic field completely penetrates the superconductor at the critical temperature. Type-I superconductors have penetration depths that are typically about 50 nm.

> The penetration depth (λ) is the characteristic distance that a magnetic field penetrates a superconductor.

EXAMPLE 15-3

Estimate the magnetic field penetration depth in lead at zero kelvin, assuming that the density of superconducting electrons is about $2 \times 10^{28}/m^3$.

SOLUTION:

The penetration depth is

$$\lambda_L = \sqrt{\frac{m}{\mu_0 n e^2}}$$

$$= \sqrt{\frac{mc^2}{4\pi nke^2}}$$

$$= \sqrt{\frac{0.511 \times 10^6 \text{ eV}}{(4)(\pi)(2 \times 10^{28} \text{ m}^{-3})(1.44 \text{ eV} \cdot \text{nm})}}$$

$$\approx 40 \text{ nm}. \qquad \blacksquare$$

*

Coherence Length

In 1950, Vitaly L. Ginzburg and Lev D. Landau introduced a wave function (ψ) whose square represents the Cooper pair density (n),

$$|\psi|^2 = n. \qquad (15.30)$$

The wave function ψ satisfies a Schrödinger-like equation called the *Ginzburg-Landau equation,* which may be written in one dimension as

$$\frac{d^2\psi}{dx^2} = \frac{2m^*a}{\hbar^2}\psi, \qquad (15.31)$$

where m^* is the mass of a Cooper pair and a is a temperature-dependent constant. The solution is of the form

$$\psi = Ce^{-x/\xi}, \qquad (15.32)$$

where C is a constant and

$$\xi = \frac{\hbar}{\sqrt{2ma}}. \qquad (15.33)$$

The parameter ξ is called the *coherence length*. The coherence length is the characteristic distance scale over which Cooper pairs are correlated. The Ginzburg-Landau equation puts the phenomenology of superconductivity in the language of quantum mechanics.

The concept of an *intrinsic* coherence length was introduced by A. Brian Pippard in 1953. The motivation for an intrinsic coherence length is that only electrons that have kinetic energies within a few times kT_c of the Fermi energy can participate in the onset of superconductivity at the critical temperature. The momentum distribution of the electrons that form Cooper pairs will have a certain width (Δp). Since the Cooper pairs have correlated momenta, the uncertainty principle gives a minimum distance between electrons of Cooper pairs of $\Delta x = \hbar/2\Delta p$. This distance scale in a pure superconductor, called the intrinsic coherence length (ξ_0), is defined to be

$$\xi_0 \equiv \frac{\hbar v_F}{7\,kT_c}. \qquad (15.34)$$

The order of magnitude of ξ_0 is given by the uncertainty principle. The distribution of the distance between the electrons in Cooper pairs has a width characterized by $\Delta x = \xi_0$. The Cooper pairs are formed from electrons near the Fermi energy that have a spread in energies (ΔE). The momentum distribution of the Cooper pairs has a width (Δp) given by

$$\Delta E = \Delta\left(\frac{p^2}{2m}\right) = \frac{2\,p\Delta p}{2\,m} = v_F\,\Delta p. \qquad (15.35)$$

From the uncertainty principle we have

$$\Delta p\Delta x \approx \Delta p\xi_0 \approx \frac{\hbar}{2}. \qquad (15.36)$$

The uncertainty principle gives

$$\Delta p\xi_0 \approx \left(\frac{\Delta E}{v_F}\right)\xi_0 \approx \left(\frac{\Delta E}{v_F}\right)\left(\frac{\hbar v_F}{7kT_c}\right) \approx \frac{\hbar}{2}. \qquad (15.37)$$

We have agreement with the uncertainty principle if

$$\Delta E \approx \frac{7}{2}kT_c. \qquad (15.38)$$

that is, the electrons that form Cooper pairs have kinetic energies within a few times kT_c of the Fermi energy at the onset of superconductivity.

EXAMPLE 15-4

Estimate the intrinsic coherence length for lead, if the electron speed at the Fermi surface is 2×10^6 m/s.

SOLUTION:

The Fermi speed in lead is

$$v_F \approx 2\times10^6 \text{ m/s},$$

or

$$\frac{v_F}{c} = \frac{2\times10^6 \text{ m/s}}{3\times10^8 \text{ m/s}} = 0.67\times10^{-2}.$$

The transition temperature is 7.2 K, corresponding to

$$kT_c = \left(\frac{1}{40}\text{eV}\right)\left(\frac{7.2\text{ K}}{300\text{ K}}\right) \approx 6\times10^{-4}\text{ eV}.$$

The intrinsic coherence length (15.35) is

$$\xi_0 = \frac{\hbar v_F}{7\,kT_c} = \frac{\hbar c\left(\dfrac{v_F}{c}\right)}{7\,kT_c}$$

$$= \frac{(200\text{ eV}\cdot\text{nm})(0.67\times10^{-2})}{(7)(6\times10^{-4}\text{ eV})}$$

$$\approx 300\,\text{nm}. \qquad \blacksquare$$

EXAMPLE 15-5

Estimate the order of magnitude of the binding energy of a Cooper pair.

SOLUTION:

The attraction of the Cooper pair is electromagnetic, even though it is a complicated interaction involving the positive charges in the lattice. Since the electrons in a Cooper pair have a separation that is characterized by ξ_0, the order of magnitude of the binding energy is

$$E_b \approx \frac{ke^2}{\xi_0} \approx \frac{1.44\,\text{eV}\cdot\text{nm}}{300\text{ nm}}$$

$$\approx 5\times10^{-3}\text{ eV} = 5\text{ meV}. \qquad \blacksquare$$

EXAMPLE 15-6

Calculate the size of a hydrogen atom that has the same binding energy as a Cooper pair.

SOLUTION:

From the Bohr model, the principle quantum number n is given by

$$E_b = \frac{13.6\,\text{eV}}{n^2} \approx 5\,\text{meV}.$$

or

$$n = \sqrt{\frac{13.6\,\text{eV}}{5\,\text{meV}}} \approx 50.$$

The diameter (d) of the nth Bohr orbit is

$$d = 2r_n = 2n^2 a_0 \approx (2)(50)^2 (0.05\,\text{nm}) \approx 300\,\text{nm}.$$

The strength of the electromagnetic force is a few meV per 300 nm. ■

If the superconductor is not pure, the coherence length will be degraded by the scattering of the electrons from the impurities. The characteristic length for this scattering is the mean free path (d) of the conduction electrons in the normal state. We may combine this scattering rate, which is inversely proportional to d, with the intrinsic coherence length (ξ_0) to produce a net coherence length (ξ):

$$\frac{1}{\xi} = \frac{1}{\xi_0} + \frac{1}{d}. \qquad (15.39)$$

The coherence length (ξ) is a measure of the distance over which a pair of electrons have correlated spins and momenta. The order of magnitude of the coherence length is 10^{-6} m, so that superconducting electron pairs are separated by many atoms. The relationship between ξ and d (15.39) shows that the coherence length is larger for larger values of the mean free path.

> The coherence length (ξ) is a measure of the distance over which a pair of electrons have correlated spins and momenta in a superconductor.

In a type-I superconductor, the coherence length is longer than the penetration depth. When the mean free path is shortened, which can happen when an alloy is formed, the coherence length is shortened and penetration depth is lengthened. The ratio of the penetration depth to the coherence length determines whether a superconductor is type-I or type-II. For type-I superconductors

$$\xi > \sqrt{2}\lambda, \qquad (15.40)$$

and for type-II superconductors

$$\xi < \sqrt{2}\lambda. \qquad (15.41)$$

The penetration depth is illustrated in Figure 15-14.

The Isotope Effect

A major conceptual advance was made during the 1950s, with the discovery that the critical temperatures of different isotopes of mercury varied with atomic mass (A). The observed dependence of T_c for isotopes of mercury was of the form

$$T_c \propto \frac{1}{\sqrt{A}}. \qquad (15.42)$$

Figure 15-15 shows the superconducting transition temperature for different isotopes of mercury, plotted as a function of $A^{-1/2}$. These data demonstrated that the physical phenomenon of superconductivity had something to do with the lattice and was not an entirely electronic effect. Other superconducting materials are found to have a different A dependence.

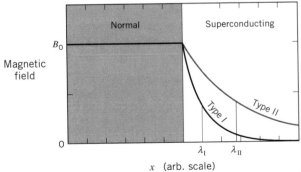

FIGURE 15-14 Magnetic field penetration in a superconductor.
The values of the penetration depth and coherence length of the superconducting state depend on the mean free path of conduction electrons in the normal state. Type-I superconductors have longer mean free paths resulting in $\xi > \sqrt{2}\lambda$. Type-II superconductors have shorter mean free paths resulting in $\xi < \sqrt{2}\lambda$.

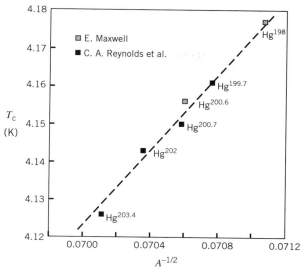

FIGURE 15-15 The isotope effect.
Isotopes of mercury are observed to have different critical temperatures. The transition temperature is plotted as a function of $A^{-1/2}$. The solid line is a linear fit. These data were the first to prove that superconductivity is not a purely electronic effect, but also involves lattice vibrations. Data from E. Maxwell, *Phys. Rev.* **78**, 477 (1950) and C.A. Reynolds et al., *Phys. Rev.* **78**, 487 (1950).

BCS Theory

A microscopic theory of superconductivity was developed by John Bardeen, Leon Cooper, and J. Robert Schrieffer in 1957. This theory (BCS) provided a major conceptual breakthrough in the quantitative understanding of the mechanism of superconductivity. The central ingredient of the BCS theory is that electrons form Cooper pairs. By interaction with the lattice, the electrons have a weak attraction for each other. This may be thought of in the following way. One electron passes through the lattice and the positive ions are attracted to it, causing a distortion in their nominal positions (see Figure 15-16). The second electron (the Cooper pair partner) comes along and is attracted by the displaced ions. The electrons that make up a Cooper pair have their momentum and spin vectors antialigned. The Cooper pairs behave like bosons. We must keep in mind, however, that the typical distance over which the electron pair is correlated is several hundred nanometers. The attraction between the electrons is extremely weak. In spite of the weakness of the attraction, the superconducting phase transition occurs because the electrons close to the Fermi energy have a net attraction.

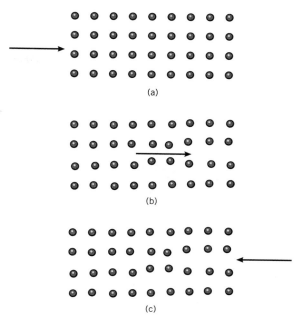

FIGURE 15-16 The electron-lattice-electron interaction.
(*a*) Undistorted lattice. (*b*) Distortion of a lattice by passage of an electron. (*c*) The distorted lattice provides a weak attraction for a second electron.

The attraction may be pictured in a Feynman diagram (Figure 15-17), where two electrons exchange a phonon.

We now examine the energy distribution of the electrons in a superconductor. Figure 15-18 shows the electron energy distribution near the Fermi energy in a superconductor at three temperatures. At a temperature just above the critical temperature the material is in the normal phase. At a temperature just below the transition temperature (T_c), a small energy gap develops where there are no states. The energy gap is caused by the binding energy due

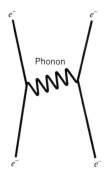

FIGURE 15-17 Feynman diagram for the attraction of two electrons by phonon exchange.

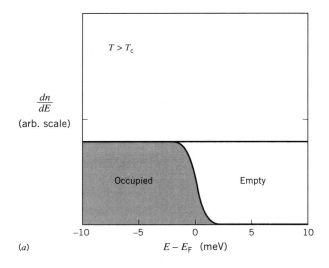

(a)

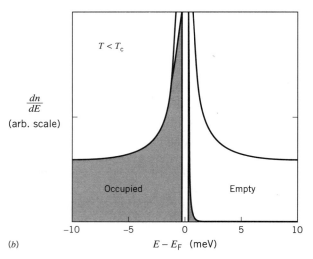

(b)

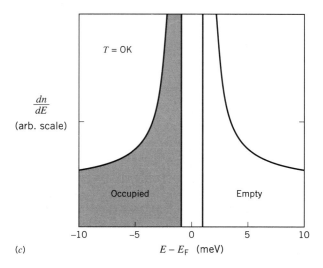

(c)

FIGURE 15-18 Electron energy distribution in a superconductor.
The density of states and occupation probability are shown for energies near the Fermi energy at three temperatures: (a) just above critical temperature (normal phase), (b) just below the critical temperature, and (c) at zero kelvin.

to the formation of Cooper pairs. The formation of Cooper pairs is hindered by the thermal excitation of electrons. As the temperature is lowered, the number of electrons that can cross the gap is significantly reduced, a greater number of Cooper pairs are formed, and the energy gap becomes larger. Thus, the appearance of the energy gap as the temperature is lowered below T_c has a threshold effect. The order of magnitude of the size of the energy gap at zero kelvin is

$$E_g(0) \approx 10^{-4} E_F. \tag{15.43}$$

Those states that had occupied the energies where the gap has appeared have been shifted away from the Fermi energy to make a sharp increase in the density of states just above and just below the gap. The total number of states has not changed but their energy distribution has changed. At a temperature of zero kelvin, all the states below the gap are filled, and the total energy is lower than in the normal state. The Cooper pairs are unable to scatter to higher energy states because of the presence of the gap. The energy gap is indicated schematically in Figure 15-19.

The BCS theory makes a quantitative prediction for the size of the energy gap as a function of the temperature. The BCS theory predicts that the gap size at zero kelvin is proportional to the critical temperature,

$$E_g \approx 3.5 kT_c. \tag{15.44}$$

Figure 15-20 shows the comparison of measured energy gaps with the BCS theory.

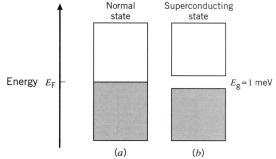

Figure 15-19 Band diagrams for (a) normal conductor and (b) superconductor.

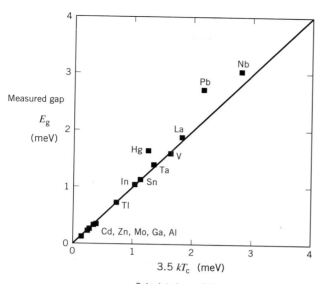

FIGURE 15-20 Energy gaps of superconducting elements.
The vertical axis indicates the measured energy gap and the horizontal axis is the calculation from BCS theory.

EXAMPLE 15-7

Estimate the energy gap at zero kelvin for tin.

SOLUTION:

The critical temperature for tin is

$$T_c = 3.7 \text{ K}.$$

The energy gap from BCS theory (15.44) is

$$E_g \approx 3.5 \, kT_c$$

$$\approx (3.5)(0.026 \text{ eV}) \left(\frac{3.7 \text{ K}}{300 \text{ K}} \right)$$

$$\approx 1.1 \text{ meV}. \qquad \blacksquare$$

The size of the energy gap in a superconductor is related to the critical magnetic field. The change in energy per volume of the electrons in a superconductor due to the appearance of the energy gap is equal to the energy per volume of the magnetic field B_c that destroys the superconductivity.

EXAMPLE 15-8

The critical magnetic field in lead at zero kelvin is 0.080 tesla. Make an order-of-magnitude estimate of the energy gap for lead at zero kelvin. The density of conduction electrons in lead is 1.3×10^{29} m^{-3} and the Fermi energy is 9.4 eV.

SOLUTION:

The condensation energy (15.6) is

$$u_c = \frac{B_c^2}{2\mu_0} = \frac{(0.080 \text{ T})^2}{(2)(4\pi \times 10^{-7} \text{ N/A}^2)} = 2.5 \times 10^3 \text{ J/m}^3.$$

The fraction of electrons (f) that are moved out of the region of the gap is

$$f \approx \frac{\left(\dfrac{E_g}{2} \right)}{E_F} = \frac{E_g}{2E_F}.$$

The approximate change in energy per electron is

$$\Delta E \approx -\frac{E_g}{2}.$$

By conservation of energy,

$$u_c + nf\Delta E = 0,$$

where n is the number of conduction electrons per volume. Therefore,

$$E_g \approx \sqrt{\frac{2u_c E_F}{n}}$$

$$\approx \sqrt{\frac{(2)(2.5 \times 10^3 \text{ J/m}^3)(9.4 \text{ eV})}{1.3 \times 10^{29} \text{ m}^{-3}} \frac{\text{eV}}{1.6 \times 10^{-19} \text{ J}}}$$

$$\approx 2 \text{ meV}. \qquad \blacksquare$$

When electrons cross the energy gap because of thermal excitations, the size of the gap is reduced because the number of Cooper pairs is reduced. At temperatures well below the critical temperature, the size of the energy gap is nearly constant, $E_g(0)$, because an insignificant number of electrons are in states above the gap. At temperatures close to the critical temperature, enough electrons cross the gap to significantly reduce the number of Cooper pairs, in turn reducing the size of the gap. Therefore, there is a threshold effect. To a reasonable approximation the size of the gap is zero above the critical temperature and is constant below the critical temperature. The temperature dependence of the energy gap is predicted from BCS theory. For temperatures near the critical temperature, the size of the energy gap varies as

$$E_g(T) \approx E_g(0)\sqrt{1 - \frac{T}{T_c}} \quad . \qquad (15.45)$$

When a material undergoes a superconducting transition, only a relatively small fraction of the conduction electrons have a change in energy. At sufficiently low temperature, however, all the conduction electrons form Copper pairs, not just those near the energy gap.

Heat Capacity

The heat capacity of vanadium as a function of temperature is shown in Figure 15-21. In the normal state near the critical temperature, the heat capacity (14.53) contains a term proportional to T^3. When vanadium is cooled into the superconducting state, the heat capacity suddenly increases at the critical temperature of 5.03 K. This increase occurs because both the occupation probability and the density of states change with temperature. At low temperatures, the energy gap becomes independent of temperature and it also becomes exponentially more difficult for electrons to be excited across the gap. Figure 15-22

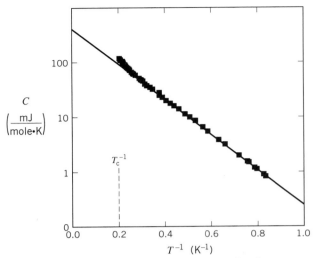

FIGURE 15-22 Heat capacity of superconducting vanadium.
When plotted versus $(1/T)$ the heat capacity of superconducting vanadium has as exponential behavior (solid line). These data indicate the presence of an energy gap. From W.S. Corak et al., *Phys. Rev.* **102**, 656 (1956).

shows the heat capacity of superconducting vanadium as a function of $(1/T)$. At low temperatures, the heat capacity is of the form

$$C = Ae^{-b/kT}, \qquad (15.46)$$

where A and b are constants. The exponential behavior indicates the presence of an energy gap.

15-3 FURTHER PROPERTIES OF SUPERCONDUCTORS

Persistent Currents and Trapped Magnetic Flux

In a ring-shaped conductor in the normal state in the presence of an external magnetic field, the magnetic field lines pass through the ring and through the conducting material (see Figure 15-23). When the conductor is cooled into the superconducting state, currents are generated on the surface of the ring. The additional magnetic field due to this current cancels the external field inside the boundary of the superconductor so that the magnetic field does not penetrate the superconductor. The magnetic field lines do, however, pass though the opening in the ring. Suppose now that the external magnetic field is removed. Addi-

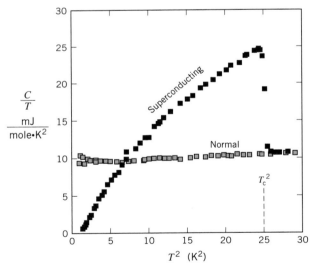

FIGURE 15-21 Heat capacity of a vanadium.
The measured heat capacity of vanadium is measured at low temperatures in the presence of zero field (where vanadium superconducts) and fields above the critical field (where vanadium remains normal). The heat capacity in the normal state is found to fit the form, $C = \gamma T + aT^3$. In the superconducting state the heat capacity is found to have a discontinuity at the critical temperature. From W.S. Corak et al., *Phys. Rev.* **102**, 656 (1956).

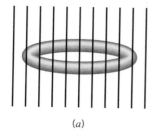

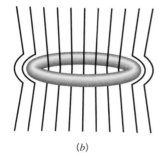

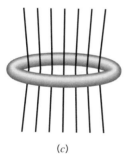

FIGURE 15-23 Magnetic flux in the region of a thin ring.
(*a*) The ring is normal in the presence of an external field, (*b*) the ring is superconducting in the presence of an external field, and (*c*) the ring is superconducting and the external field is removed.

tional currents will be generated in the superconductor to insure that B remains zero inside the superconductor. These currents generate a magnetic flux that is equal to the flux through the ring before the field was turned off. Thus, the magnetic flux through the ring is *trapped*. The magnetic flux is trapped when the ring becomes superconducting because a changing magnetic flux would create an electromotive force in the ring according to Faraday's law. Such a voltage cannot be generated in the superconducting ring because the superconductor has zero resistance. The energy stored in the trapped magnetic field has come from the decrease in energy of the electrons in the superconductor.

A quantitative measurement of the persistent current was made by J. File and R. G. Mills. They constructed a solenoidal coil from superconducting wire made of niobium-zirconium (25%) alloy. The diameter of the superconducting wire was 5.4×10^{-4} m. The coil consisted of 984 turns in a double layer about 0.1 m diameter and 0.25 m in length. The two ends of the coil were welded together. A persistent current was induced in the solenoid and the trapped magnetic field was accurately measured as a function of time with a nuclear magnetic resonance probe. The value of trapped magnetic field was

$$B_0 = 0.210488 \text{ T}. \tag{15.47}$$

The change in this trapped field was measured over a period of more than one month (3×10^6 s). The results of this measurement are shown in Figure 15-24. The measured change in the magnetic field (ΔB) during this time was only

$$\Delta B \approx -1.7 \times 10^{-7} \text{ T}. \tag{15.48}$$

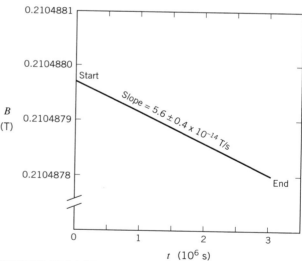

FIGURE 15-24 Measurement of the lifetime of a persistent current.
A solenoid has its terminals welded together and is placed in an external magnetic field and then cooled into the superconducting state. The external field is then removed, causing a trapped magnetic flux through the solenoid that is sustained by a persistent current. The decay of this field as a function of time is accurately measured with an NMR probe. The lifetime of the magnetic field and the persistent current that generates it is measured to be 1.2×10^5 years. From J. File and R.G. Mills, *Phys. Rev. Lett.* **10**, 93 (1963).

The lifetime of the trapped field, the time that it would take for it to decay to $1/e$ of its starting value, is the same as the lifetime of the superconducting current since the magnetic field is proportional to the current. The lifetime (τ) is given by

$$\frac{\Delta B}{\Delta t} = -\frac{B_0}{\tau}, \qquad (15.49)$$

which gives

$$\tau = -\frac{B_0 \Delta t}{\Delta B} = \frac{(0.21\,\text{T})(3\times 10^6\,\text{s})}{1.7\times 10^{-7}\,\text{T}}$$
$$= 3.7\times 10^{12}\,\text{s} = 1.2\times 10^5\,\text{y}. \qquad (15.50)$$

The current *persists* for 120,000 years!

The inductance (L) of the solenoid is defined by the relationship between the rate of change of the magnetic flux (Φ_m) and the rate of change of the current (I):

$$N\frac{d\Phi_m}{dt} = L\frac{dI}{dt}, \qquad (15.51)$$

where N is the number of turns in the coil. At any given time, we have

$$L = \frac{N\Phi_m}{I}. \qquad (15.52)$$

The SI unit of inductance is the henry (H),

$$1\,\text{H} = 1\,\text{T}\cdot\text{m}^2/\text{A} = 1\,\text{V}\cdot\text{s/A}. \qquad (15.53)$$

The superconducting solenoid of File and Mills is an LR circuit. From Faraday's law of induction and Ohm's law we have

$$RI = -L\frac{dI}{dt}. \qquad (15.54)$$

The current as a function of time may be written as

$$I(t) = I_0 e^{-\frac{R}{L}t}. \qquad (15.55)$$

Clearly the resistance of the solenoid is very small because the lifetime of the current is very large. File and Mills determined that this persistent current provided an upper limit of the resistivity of the superconducting wire equal to

$$\rho < 4.3\times 10^{-24}\,\Omega\cdot\text{m}. \qquad (15.56)$$

EXAMPLE 15-9

Calculate the value of the persistent current, the resistance of the superconducting coil circuit, and the maximum power dissipated in the superconducting coil circuit in the File-Mills experiment.

SOLUTION:

The current in a solenoid and the resulting magnetic field are related by

$$B = n\mu_0 I,$$

where n is the number of turns per length of the coil. The value of the persistent current is

$$I = \frac{B}{n\mu_0} = \frac{0.21\,\text{T}}{\left(\dfrac{984}{0.25\,\text{m}}\right)(4\pi\times 10^{-7}\,\text{N/A}^2)} = 42.5\,\text{A}.$$

The resistance (R) of the superconducting coil is

$$R = \frac{L}{\tau} = \frac{N\Phi_m}{I_0\tau}.$$

The flux is

$$\Phi_m = \pi r^2 B,$$

where r is the radius of the solenoid (0.05 m). Therefore,

$$R = \frac{N\pi r^2 B}{I_0\tau} = \frac{(984)(\pi)(0.05\,\text{m})^2(0.21\,\text{T})}{(42.5\,\text{A})(1.2\times 10^5\,\text{y})}.$$

The number of seconds per year is

$$1\,\text{y} = (365\,\text{d})(24\,\text{h/d})(60\,\text{m/h})(60\,\text{s/m})$$
$$= 3.15\times 10^7\,\text{s} \approx \pi\times 10^7\,\text{s},$$

so we have

$$R = \frac{(984)(\pi)(0.05\,\text{m})^2(0.21\,\text{T})}{(42.5\,\text{A})(1.2\times 10^5\,\text{y})(\pi\times 10^7\,\text{s/y})} \approx 10^{-14}\,\Omega,$$

where we have used

$$\frac{\text{T}\cdot\text{m}^2}{\text{A}\cdot\text{s}} = \frac{\left(\dfrac{\text{s}\cdot\text{V}}{\text{m}^2}\right)\text{m}^2}{\text{A}\cdot\text{s}} = \frac{\text{V}}{\text{A}} = \Omega.$$

This resistance is most likely dominated by the resistance of the weld joint of the solenoid terminals. The power dissipated in the circuit is

$$P = I^2 R = (42.5\,\text{A})^2 \left(10^{-14}\,\Omega\right) \approx 2 \times 10^{-11}\,\text{W}.\ \blacksquare$$

It is possible to estimate the lifetime of a superconducting current by estimating the probability that some minimum portion of the superconducting ring goes normal due to a thermal fluctuation. For type-I superconductors, the estimate of the lifetime (see problem 26) is far, far in excess of the age of the sun except when the temperature or magnetic field is close to the critical value. In the experiment of File and Mills, the small decrease observed in the persistent current was most likely due to a finite resistance of the welded joint of the solenoid terminals (about $10^{-14}\,\Omega$).

Discovery of Magnetic Flux Quantization

The magnetic flux (Φ_m) passing through the area enclosed by a superconducting ring is observed to be quantized:

$$\Phi_m = \oint_S d\mathbf{a} \cdot \mathbf{B} = n\Phi_0, \qquad (15.57)$$

where n is an integer and Φ_0 is the elementary *flux quantum*.

Magnetic flux quantization was discovered in 1961 by Bascom S. Deaver, Jr., and William M. Fairbank and independently by R. Doll and M. Näbauer. Deaver and Fairbank made a tiny hollow superconducting cylinder by electroplating tin on a copper wire. They placed the cylinder in a known magnetic field and then cooled it below the critical temperature so that the tin became superconducting. The magnetic field was then turned off. The trapped magnetic flux through the cylinder was then measured by mechanically oscillating the superconducting cylinder (1 mm amplitude and 100 Hz) and measuring the electrical pickup in two coils placed at the ends of the cylinder. Doll and Näbauer made a 10 μm-diameter superconducting lead cylinder by evaporating lead onto a quartz fiber. Doll and Näbauer measured the trapped magnetic flux by measuring the torque on the superconducting ring when placed in an external magnetic field.

Some of the results of Deaver and Fairbank are shown in Figure 15-25. The inner area of the superconducting cylindrical shell was $1.7 \times 10^{-10}\,\text{m}^2$. For external magnetic fields less then about $7 \times 10^{-6}\,\text{T}$, there is zero magnetic flux through the superconducting cylinder. Superconducting currents in the cylinder generate a magnetic field that opposes the external field such that the field is zero inside the superconductor and flux through the cylinder is zero. At an external magnetic field just greater than $7 \times 10^{-6}\,\text{T}$, the flux takes a quantum step from zero to a value of

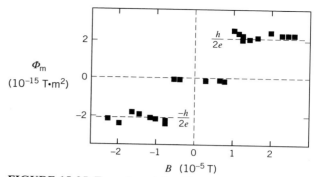

FIGURE 15-25 Experimental evidence of magnetic flux quantization through a superconducting ring.
From B.S. Deaver and W.M. Fairbank, *Phys. Rev. Lett.* **7**, 43 (1961).

$$\Phi_0 \approx 2 \times 10^{-15}\,\text{T} \cdot \text{m}^2. \qquad (15.58)$$

For larger external magnetic fields $(3 \times 10^{-5}\,\text{T})$, the flux jumps to twice the above value, and for still larger fields $(4 \times 10^{-5}\,\text{T})$ the flux jumps to three times the above value. The magnetic flux through the cylinder is observed to be quantized

$$\Phi_m = n\Phi_0, \qquad (15.59)$$

where n is an integer (positive, negative, or zero). Within measured errors, the experimenters determined that the elementary flux quantum was

$$\Phi_0 = \frac{h}{2e} = \frac{4.14 \times 10^{-15}\,\text{eV} \cdot \text{s}}{(2)(e)}$$

$$= 2.07 \times 10^{-15}\,\text{V} \cdot \text{s}$$

$$= 2.07 \times 10^{-15}\,\text{T} \cdot \text{m}^2. \qquad (15.60)$$

EXAMPLE 15-10

Compare the magnetic flux trapped by the superconducting state to that for the normal state at the lowest value of magnetic field where $n = 1$.

SOLUTION:

The change in magnetic flux occurs at

$$B = 7 \times 10^{-6}\,\text{T},$$

for a cylinder with inner area of

$$A = 1.7 \times 10^{-10}\,\text{m}^2.$$

The flux in the normal state would be

$$\Phi_m = BA = \left(7 \times 10^{-6} \text{ T}\right)\left(1.7 \times 10^{-10} \text{ m}^2\right)$$
$$= 1.2 \times 10^{-15} \text{ T} \cdot \text{m}^2 .$$

The flux in the superconducting state is

$$\Phi_m = \frac{nh}{2e} = (1)\left(\frac{h}{2e}\right) = \Phi_0 = 2.07 \times 10^{-15} \text{ T} \cdot \text{m}^2 .$$

This flux is larger than the flux of the external field because of the superconducting currents that are generated in the cylinder. ∎

EXAMPLE 15-11

Calculate the minimum value of the magnetic charge of a monopole consistent with the elementary flux quantum.

SOLUTION:

According to Gauss's law (see Appendix B), the magnetic field at a distance r from a magnetic monopole of charge g is

$$B = \frac{g}{4\pi r^2} .$$

The magnetic flux at a distance r from the monopole is

$$\Phi_m = \left(4\pi r^2\right)\left(\frac{g}{4\pi r^2}\right) = g .$$

Since the size of the flux quantum is $h/2e$, we have

$$g = \frac{h}{2e} . \qquad ∎$$

* Challenging

Reason for Flux Quantization

Magnetic flux quantization (15.59) arises from the boundary condition on the wave function of the electrons in the superconductor. Consider a thin superconducting ring. We may write the wave function as

$$\psi(\phi) = \sqrt{n_C} \, e^{i\phi} , \qquad (15.61)$$

where n_C is the density of superconducting electron *pairs* (Cooper pairs) inside the material of the ring. (Any complex number can always be written in this fashion.) The wave function (15.61) squared is equal to the density of the Copper pairs. The Cooper pairs have a com-

mon momentum $\mathbf{p}_C$. The boundary condition on the wave function may be written using the Bohr-Sommerfeld quantization condition (5.4), which we write in the form of a line integral of $\mathbf{p}_C$ around the ring:

$$\oint d\mathbf{l} \cdot \mathbf{p}_C = nh , \qquad (15.62)$$

where n is an integer. To make use of the quantization condition (15.62), we need to relate the common momentum of the Cooper pairs to the current density $\mathbf{J}$ in the ring. Since the current density is

$$\mathbf{J} = qn_C \mathbf{v}_C , \qquad (15.63)$$

where q is the charge of a Cooper pair (twice the electron charge), the common momentum is

$$\mathbf{p}_C = m^* \mathbf{v}_C = \frac{m^*}{qn_C} \mathbf{J} , \qquad (15.64)$$

where m^* is the mass of a Cooper pair (twice the electron mass) and $\mathbf{v}_C$ is the speed of the Cooper pair. Using Stokes's theorem (C.10) and the relationship between $\mathbf{p}_C$ and $\mathbf{J}$ (15.64), the quantization condition (15.62) may be written

$$nh = \oint d\mathbf{l} \cdot \mathbf{p}_C = \oiint d\mathbf{a} \cdot \left(\nabla \times \mathbf{p}_C\right)$$
$$= \frac{m^*}{qn_C} \oiint d\mathbf{a} \cdot \left(\nabla \times \mathbf{J}_C\right) . \qquad (15.65)$$

The London equation (15.22) relates the curl of $\mathbf{J}$ with $\mathbf{B}$. For Cooper pairs this relationship is

$$\nabla \times \mathbf{J} = -\frac{n_C q^2}{m^*} \mathbf{B} . \qquad (15.66)$$

Therefore, the quantization condition (15.65) becomes

$$nh = -q \oiint d\mathbf{a} \cdot \mathbf{B} = -q\Phi_m , \qquad (15.67)$$

or

$$\Phi_m = -\frac{nh}{q} = \frac{nh}{2e} . \qquad (15.68)$$

The elementary flux quantum Φ_0 is also associated with each vortex inside a type-II superconductor (see Figure 15-10). At the midpoint between two adjacent vortices, the current density must be zero by symmetry. The flux inside one vortex cell bounded by such paths must also be an integer multiple of Φ_0. The experimental determination of the elementary flux quantum, Φ_0, was a dramatic demonstration of the formation of Cooper pairs ($q = -2e$) in the superconductor. Figure 15-26 shows the expected flux quantization for a thin superconducting ring in a weak magnetic field.

Junctions and Tunneling

Normal-Superconducting Junction

Consider a normal metal separated from a superconducting material by a thin insulator as pictured in Figure 15-27a. Electrons in the superconductor have an energy smaller than the Fermi energy by an amount at least equal to $E_g/2$ because of the presence of the energy gap. When a very small voltage is applied across the insulator, no current flows. When the applied voltage is equal to

$$V = \frac{E_g}{2e}, \tag{15.69}$$

then a current is caused by electrons entering the superconductor. The current as a function of the applied voltage is shown in Figure 15-27b. Since there is a sharp peak in the density of states (see Figure 15-18), the current rises rapidly at a voltage of $E_g/2e$. The transition is not perfectly sharp because of the tunneling of normal electrons when the voltage is slightly below threshold. The current–voltage characteristic of such a junction

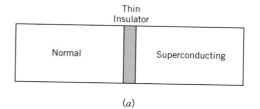

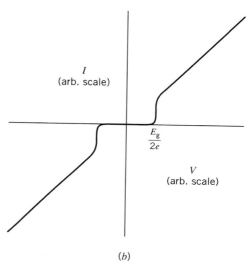

(b)

FIGURE 15-27 Normal–superconducting junction. (a) Normal metal and superconducting metal separated by a thin insulator. (b) Current versus voltage in the junction. No current flows for very small voltages because of the superconducting energy gap. The threshold voltage is $E_g/2e$.

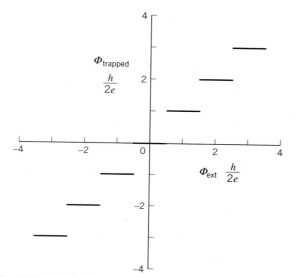

FIGURE 15-26 Trapped flux in a thin superconducting ring as a function of the external magnetic flux.

provides a convenient method for determination of the energy gaps and density of states. The first measurements of energy gaps were made in 1960 by Ivar Giaever.

Josephson Junction

Consider two superconductors separated by a thin insulator as indicated in Figure 15-28a. Electrons may tunnel from one side of the insulator to the other side. In 1962, Brian Josephson predicted that a *Cooper pair* could tunnel between the superconductors with the *same probability* as ordinary electrons rather than the square of the single electron probability. To see how the tunneling of Cooper pairs occurs, we must examine the time dependence of the superconducting electron wave functions, Ψ_1 and Ψ_2, in the two superconductors. We shall write the time-dependent wave functions in terms of the densities (n_1 and n_2) of the superconducting electron pairs:

$$\Psi_1 = \sqrt{n_1}\, e^{i\phi_1} \tag{15.70}$$

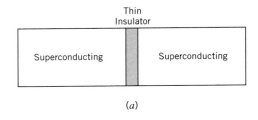

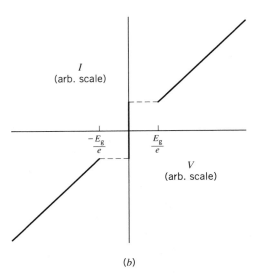

(b)

FIGURE 15-28 Junction of two superconductors.
(a) Two superconductors separated by a thin insulator. (b)
Current versus voltage in the junction.

and

$$\Psi_2 = \sqrt{n_2}\, e^{i\phi_2}, \qquad (15.71)$$

where ϕ_1 and ϕ_2 are the phases of the wave functions. The time dependence of the wave functions is given by the time-dependent Schrödinger equation (7.157). Let T represent the tunneling rate for Cooper pairs. The time rate of change of Ψ_1 depends on Ψ_2 and vice versa:

$$i\hbar \frac{\partial \Psi_1}{\partial t} = \hbar T \Psi_2, \qquad (15.72)$$

and

$$i\hbar \frac{\partial \Psi_2}{\partial t} = \hbar T \Psi_1. \qquad (15.73)$$

Substituting the wave functions (15.70) and (15.71) into the Schrödinger equation (15.72), we get

$$\frac{i\hbar}{2\sqrt{n_1}} e^{i\phi_1} \frac{\partial n_1}{\partial t} - \hbar \sqrt{n_1}\, e^{i\phi_1} \frac{\partial \phi_1}{\partial t}$$
$$= \hbar T \sqrt{n_2}\, e^{i\phi_2}. \qquad (15.74)$$

Solving for the time derivative of n_1, we have

$$\frac{\partial n_1}{\partial t} = -2\,i n_1 \frac{\partial \phi_1}{\partial t} - 2\,iT\sqrt{n_1 n_2}\, e^{i(\phi_2 - \phi_1)}. \qquad (15.75)$$

The real part of $\partial n_1/\partial t$ (15.75) gives

$$\frac{\partial n_1}{\partial t} = 2\,T\sqrt{n_1 n_2}\, \sin(\phi_2 - \phi_1). \qquad (15.76)$$

(The imaginary part of $\partial n_1/\partial t$ gives an expression for $\partial \phi_1/\partial t$.) Since the current flowing in the junction is proportional to the time derivative of n_1, Josephson predicted that the junction current (I) is proportional to the sine of the phase difference:

$$I = I_0 \sin(\phi_2 - \phi_1). \qquad (15.77)$$

This phenomenon is called the *dc Josephson effect*.

If a voltage (V) is applied across the junction, the Schrödinger equations become

$$i\hbar \frac{\partial \Psi_1}{\partial t} = \hbar T \Psi_2 - eV\Psi_1 \qquad (15.78)$$

and

$$i\hbar \frac{\partial \Psi_2}{\partial t} = \hbar T \Psi_1 + eV\Psi_2. \qquad (15.79)$$

Substituting the wave functions (15.70) and (15.71) into the Schrödinger equation (15.78), we have

$$\frac{i\hbar}{2\sqrt{n_1}} e^{i\phi_1} \frac{\partial n_1}{\partial t} - \hbar \sqrt{n_1}\, e^{i\phi_1} \frac{\partial \phi_1}{\partial t}$$
$$= \hbar T \sqrt{n_2}\, e^{i\phi_2} - eV\sqrt{n_1}\, e^{i\phi_1}. \qquad (15.80)$$

Solving for $\partial \phi_1/\partial t$ and taking the imaginary part, we get

$$\frac{\partial \phi_1}{\partial t} = \frac{eV}{\hbar} - T\sqrt{\frac{n_2}{n_1}} \cos(\phi_2 - \phi_1). \qquad (15.81)$$

Similarly the expression for $\partial \phi_2/\partial t$ is

$$\frac{\partial \phi_2}{\partial t} = -\frac{eV}{\hbar} - T\sqrt{\frac{n_1}{n_2}} \cos(\phi_2 - \phi_1). \qquad (15.82)$$

Since the quantum tunneling of the electron pairs does not significantly alter the densities n_1 and n_2, we have $n_1 \approx n_2$. Taking the difference of the time derivatives of the phases, we have

$$\frac{\partial\left(\phi_2 - \phi_1\right)}{\partial t} = -\frac{2\,eV}{\hbar}. \tag{15.83}$$

Integrating we get the time dependence of the phase difference to be

$$\phi_2 - \phi_1 = \delta_0 - \frac{2\,eV}{\hbar}\,t, \tag{15.84}$$

where we have written the initial phase as δ_0. Therefore, the current in the junction oscillates with a frequency of $2eV/\hbar$,

$$I = I_0 \sin\left(\delta_0 - \frac{2\,eV}{\hbar}\,t\right). \tag{15.85}$$

This phenomenon is called the *ac Josephson effect*. The current–voltage characteristics of a Josephson junction are shown in Figure 15-28b.

The first observation of the tunneling of electron pairs through a thin insulator separating two superconductors was made in 1963 by Phillip W. Anderson and John M. Rowell. Their Josephson junction was made of 200-nm-thick superconducting tin and lead separated by a thin layer of tin oxide. The size of the junction was 250 μm × 650 μm. The resistance of the junction with both metals in the normal state was 0.4 Ω. When the junction was cooled to 1.5 kelvin, a current of 0.65 mA was observed to flow through the junction with zero voltage across the junction. The tunneling current was also observed to be very sensitive to the presence of a small magnetic field. The latter observation was important in identifying the Josephson effect. Current at zero voltage had undoubtedly been seen before in low-resistance junctions, but it was attributed to superconducting shorts through small pinholes in the insulator, and the devices were discarded!

If a constant voltage is applied, then an alternating current is created with frequency

$$f = \frac{2\,eV}{h}. \tag{15.86}$$

This technique has been used to make a very precise measurement of e/h. The Josephson junction is now used

to define the standard volt, and series arrays of Josephson junctions are used to make precision voltage comparitors.

EXAMPLE 15-12

Calculate the frequency of current oscillation in a Josephson junction with a voltage of 1.000 μV.

SOLUTION:

The oscillation frequency (15.86) is

$$f = \frac{2\,eV}{h} = \frac{(2)(e)\left(10^{-6}\ \text{V}\right)}{4.136 \times 10^{-15}\ \text{eV} \cdot \text{s}} = 483.6\ \text{MHz}. \ \blacksquare$$

The Josephson tunnel current is dramatically effected by an applied magnetic field. The magnetic field dependence was first measured by Rowell in a Pb–I–Pb junction. The tunneling current as a function of the applied magnetic field in shown is Figure 15-29. The data show pronounced maxima and minima in the measured current. The minima occur when the Josephson junction contains an integral number of flux units (Φ_0).

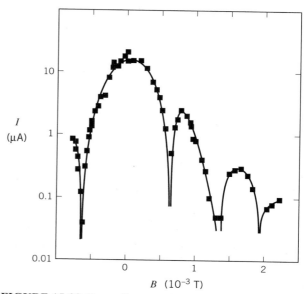

FIGURE 15-29 Tunneling current in a Pb–I–Pb Josephson junction in the presence of a weak applied magnetic field.
The tunneling current is a minimum whenever the junction contains an integral number of elementary flux units, $h/2e$. From J.M. Rowell, *Phys. Rev.* **11**, 200 (1963).

Josephson Junctions and Radiation

Sidney Shapiro was the first to study the Josephson tunnel current in the presence of microwave radiation. Shapiro made his junctions out of superconducting aluminum and tin separated by a thin layer of aluminum oxide (Al_2O_3). The energy gap in such a junction was measured to be about 0.2 meV. The frequency of the microwave radiation used was 9.3 GHz, corresponding to photon energies of

$$E = hf$$
$$= \left(9.3 \times 10^9 \text{ s}^{-1}\right)\left(4.14 \times 10^{-15} \text{ eV} \cdot \text{s}\right)$$
$$\approx 0.04 \text{ meV}. \qquad (15.87)$$

Shapiro discovered a significant change in the current–voltage characteristics of the Josephson junction in the presence of the microwave radiation. It was discovered that a tunnel current not only existed at zero voltage as in the case when no radiation was present, but that it also existed at voltages of

$$V = \pm \frac{hf}{2e}, \qquad (15.88)$$

where hf is the energy of the microwave photons. This effect is due to a frequency locking of the Josephson current to the microwave radiation. This interesting effect had been predicted by Josephson.

When higher amplitudes of radiation are used, many quantized voltage steps are observed to appear. The quantized voltage levels occur at

$$V = \frac{nhf}{2e}, \qquad (15.89)$$

where n is an integer. Occasionally a step is observed to be missing. The results of Shapiro are shown in Figure 15-30.

Superconducting Quantum Interference

Superconducting Quantum Interference was first observed by Robert Jaklevic, John Lambe, Arnold Silver, and James Mercereau in 1964. The experimenters built the first superconducting quantum interference device (SQUID). On a quartz substrate, they vacuum-deposited two 100-nm tin films separated by a thin insulating tin-oxide layer as shown in Figure 15-31a to form two parallel Josephson junctions. The electrical circuit of the SQUID is shown in Figure 15-31b. The two junctions are represented by an internal capacitance (C) and a small

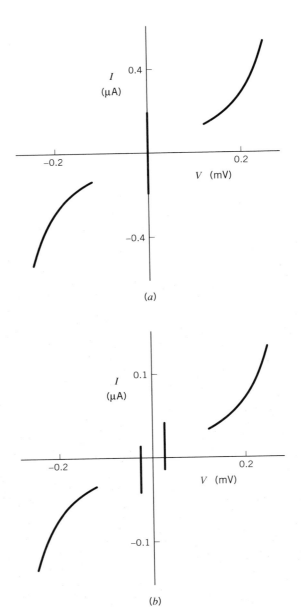

FIGURE 15-30 Josephson junction in the presence of microwave radiation.

(a) Current–voltage characteristic in a Al–Al_2O_3–Sn junction without radiation present. (b) Current–voltage characteristic in a 9.3 GHz radiation field. The tunnel current no longer occurs at $V = 0$, but at $V = \pm hf/2e$, where f is the frequency of the radiation. After S. Shapiro, *Phys. Rev. Lett.* **11**, 80 (1963).

external resistance (R). The inductance (L) is that of the loop containing the two junctions.

We have already seen that the Josephson current (I) through a junction will depend on the magnetic flux

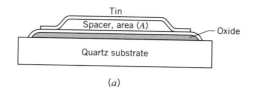

(a)

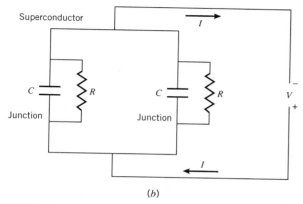

(b)

FIGURE 15-31 Superconducting quantum interference device (SQUID).
(a) The SQUID consists of two superconductors separated by thin insulators in two places to form two Josephson junctions. After R. C. Jaklevic, et al., *Phys. Rev. Lett.* **12**, 159 (1964). (b) The electrical circuit diagram of the SQUID consists of two Josephson junctions connected in parallel. Each junction has a resistance R and capacitance C and the device has a total inductance L.

through the junction. In the present case of a superconducting loop containing two junctions, something even more remarkable happens. The current through the two junctions will depend on the magnetic flux through the loop enclosed by the two junctions because of a quantum mechanical interference of the two currents through separate junctions. The phase difference around a loop that contains a magnetic flux Φ_m is

$$\phi_2 - \phi_1 = \frac{e\Phi_m}{\hbar}. \tag{15.90}$$

The total current in the loop that contains two junctions is

$$I = I_0 \sin\left(\delta_0 + \frac{e\Phi_m}{\hbar}\right) + I_0 \sin\left(\delta_0 - \frac{e\Phi_m}{\hbar}\right)$$

$$= 2 I_0 \sin\delta_0 \cos\left(\frac{e\Phi_m}{\hbar}\right). \tag{15.91}$$

The Josephson current as a function of magnetic field is shown in Figure 15-32. The minima in the current occur when the magnetic flux through the SQUID is equal to an integer multiple of h/e. Location of the maxima and minima in a SQUID circuit can be used to measure very small values of magnetic field (10^{-14} T).   ✳

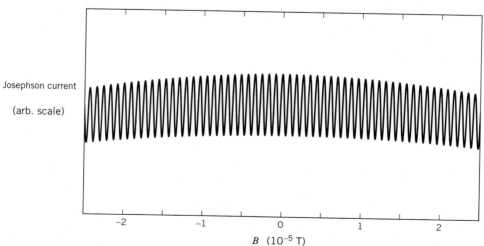

FIGURE 15-32 First observation of superconducting quantum interference.
From R. C. Jaklevic, et al., *Phys. Rev. Lett.* **12**, 159 (1964).

15-4 HIGH-T_c SUPERCONDUCTORS

In 1986, J. Georg Bednorz and K. Alex Müller made a startling discovery while studying the conductivity of the ceramic, lanthanum-barium-copper-oxide ($La_{2-x}Ba_xCuO_4$). The compound $La_{2-x}Ba_xCuO_4$ is La_2CuO_4 with some of the lanthanum atoms replaced at random by barium atoms. Bednorz and Müller discovered superconductivity in the ceramic at the relatively high critical temperature of 30 kelvin. Figure 15-33 shows the data of Bednorz and Müller. This was a very surprising result for two reasons: (1) Ceramics are usually insulating rather than conducting, so that superconductivity was not expected at *any* temperature, and (2) such a high value of T_c had never

been observed in any material in 75 years of active research on superconductivity.

By directing attention to a new class of materials to be investigated, the discovery of Bednorz and Müller triggered worldwide interest, and soon even more remarkable results were obtained. Within a few months, a related material, $YBa_2Cu_3O_7$, was discovered with a critical temperature of 92 K! This was an important milestone in superconductivity because this material can be cooled into the superconducting state with liquid nitrogen, which boils at 77 K. These data are shown in Figure 15-34. This material ($YBa_2Cu_3O_7$) has become known as the *1-2-3* superconductor because of the ratio of yttrium to barium to copper atoms.

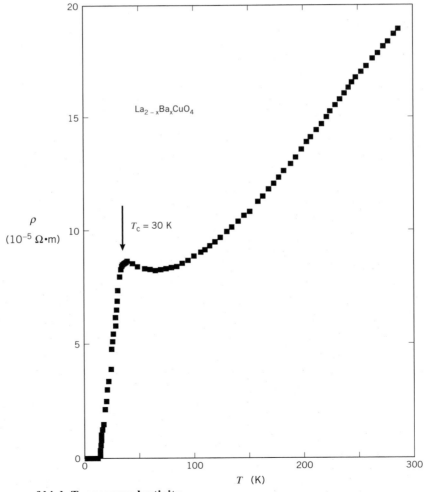

FIGURE 15-33 Discovery of high-T$_c$ superconductivity.
From J. G. Bednorz and K. A. Müller, *Z. Phys.* **B64**, 189 (1986).

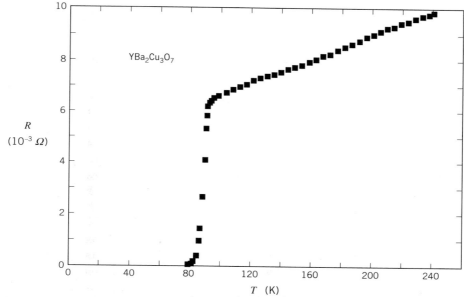

FIGURE 15-34 First observation of superconductivity at temperatures above that of liquid nitrogen.
From M.K. Wu et al., *Phys. Rev.* **58**, 908 (1987).

Many additional high-T_c superconductors have been discovered with critical temperatures above 30 K. Some of these materials are listed in Table 15-1. The material with the highest critical temperature (125 K) is $Tl_2Ba_2Cu_3O_{10}$. Most high-T_c superconductors are composed of layers of copper-oxides, in various configurations, separated by layers of other atoms that serve as spacers and also charge reservoirs. At superconducting temperatures, the copper-oxide layers provide the path for the current. The critical temperature is sensitive to the layers of the spacer atoms.

The crystal structure of $La_{2-x}Ba_xCuO_4$ has layers of copper-oxide spaced by lanthanum-barium layers. The copper-oxide layers are in the form of elongated octahedrons. Because of the shape of the octahedron, the interaction of electrons in the crystal is strongly sensitive to the position of the copper atom. Since the basic mechanism of (low-T_c) superconductivity was known be to an electron–lattice–electron interaction, this was the primary motivation that led Bednorz and Müller to investigate the copper-oxides.

The material La_2CuO_4 is by itself not a superconductor. The outer electron configuration of copper is $3d^{10}4s^1$. In the molecular bonding, the copper atom donates two valence electrons to the oxygen atoms. This leaves an odd number of electrons in the outer shell. These electrons all have their spins antialigned in pairs, except the "odd" electron. The odd electrons in the copper atoms interact with each other and tend to have their spins alternating up and down, a phenomenon called *antiferromagnetism*. Antiferromagnetism destroys any possibility of superconductivity. The addition of barium atoms removes the antiferromagnetism. Barium contains one less electron than lanthanum. The outer electron configuration of barium is $6s^2$ and lanthanum is $5d^16s^2$ (see Figure 9-7). Since the

TABLE 15-1
CRITICAL TEMPERATURES FOR SOME "HIGH-T_c" SUPERCONDUCTING OXIDE-COMPOUNDS.

Material	T_c (K)
$La_{2-x}Ba_xCuO_4$	30
$La_{2-x}Sr_xCuO_4$	38
$La_{2-x}Sr_xCaCuO_6$	60
$YBa_2Cu_3O_7$	92
$Bi_2Sr_2Ca_2Cu_3O_{10}$	110
$Tl_2Ba_2Ca_2Cu_3O_{10}$	125

lanthanum and barium atoms have the same size, the barium atoms occupy the same crystal locations as the lanthanum. This is known as a *solid solution*. The copper atoms can donate electrons to the barium atoms, leaving conducting holes in the conduction band. If just the right amount of lanthanum is replaced by barium, roughly 10%, the compound can superconduct. The phase diagram of such a superconductor is shown in Figure 15-35. Barium is not the only dopant for making La_2CuO_4 superconducting. Other possibilities are strontium (outer electron configuration of $5s^2$) and calcium (outer electron configuration of $4s^2$).

The 1-2-3 superconductor ($YBa_2Cu_3O_7$) is not a solid solution, but rather has a regular crystal structure as shown in Figure 15-36. Like the compound $La_{2-x}Ba_xCuO_4$, one

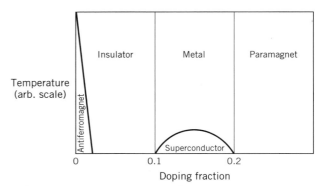

FIGURE 15-35 Phase diagram for the high-T_c superconductors.
After P. W. Anderson and R. Schrieffer, *Phys. Today* **44**, 54 (1991).

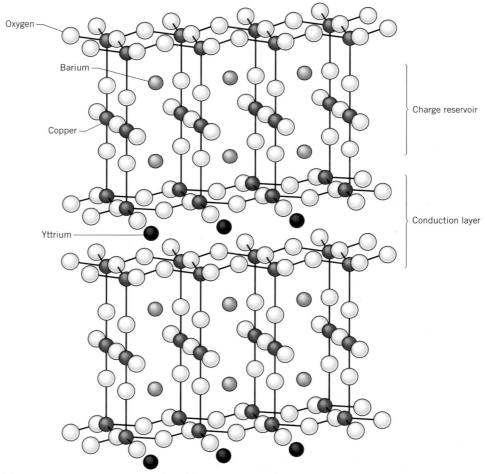

FIGURE 15-36 Molecular structure of the high-T_c 1-2-3 superconductor ($YBa_2Cu_3O_7$).
This was the first superconductor discovered with a critical temperature above liquid nitrogen temperature. After J. D. Jorgensen, *Phys. Today* **44**, 34 (1991).

property crucial for the superconductivity is the copper-oxide planes of atoms. The copper-oxide chains are a source of conducting holes. If oxygen is removed from the material by heating in a vacuum, the oxygen is first removed from the chains and the critical temperature drops. By the time 60% of the oxygen atoms are removed from the chains, the material becomes a semiconductor that is insulating at low temperatures.

15-5 APPLICATIONS OF SUPERCONDUCTIVITY

The applications of superconductivity are rich and varied. In this section we will mention some of the ingenious devices that have been made possible by superconductor technology.

Superconducting Magnets

The quest to make a superconducting magnet began immediately with the efforts of Onnes in 1911. The motivation was clear: High-field normal magnets have large heat dissipations and huge power requirements. For example, consider the magnet that Kunzler et al. used to measure the magnetic properties of niobium–tin. The magnet was a solenoid with copper coils. The maximum field was 8.8 T in a working volume 0.1 m long and 0.05 m in diameter. The power consumption was 1.5 MW and the magnet required 60 liters of water per second for cooling! The magnet and cooling apparatus occupied several large rooms of laboratory space. A superconducting version of the same magnet is about 1 m in size, has essentially zero power consumption, and requires just a couple liters of liquid helium per day for refrigeration!

One extremely important technical aspect of superconducting magnet design is the cables for the superconducting coils. Superconducting magnet-coil cable is made up of a large number of strands of superconducting material of 20 μm size embedded in a copper matrix. The reason for this design is twofold. First, the conductor must be mechanically stable under the electromagnetic forces it experiences. To maintain magnetic field stability to a precision of

$$\frac{\Delta B}{B} = 10^{-4}, \qquad (15.92)$$

requires stability of the conductors to about 25 μm. Second, the cable must be able to carry the normal current in the case of a quench.

The superconductor most commonly used to make magnet coils is niobium–titanium (NbTi), which is a type-II superconductor. In order to maintain temperatures below the critical temperature of 9 K, a magnet using NbTi coils must be cooled with liquid helium. The reasons that NbTi is suitable for magnet coils are as follows:

1. NbTi can support large currents in high magnetic fields at 4 K (for example, $J_c = 10^9$ A/m² for $B = 8.5$ T).
2. NbTi can be fabricated into multistrand cables, which are mechanically stable under large electromagnetic forces.
3. NbTi is not granular, which makes it possible to transport currents over reasonable distances (tens of meters).

High-field magnets are also made using the type-II superconductor, Nb_3Sn.

Another important technical aspect of superconducting magnet design is the cryogenic system needed to maintain the magnet below T_c. The efficiency for extracting heat at 4 K is about 1%, and the typical mechanical efficiency of refrigeration is 20%. This means that it requires 500 W at room temperature to extract 1 W at liquid helium temperature. Great care must be used in the design of supporting structures and cable leads connecting the superconducting magnet to the normal environment. A buffer zone at liquid nitrogen temperature (77 K) must be used to limit liquid helium costs to a reasonable level.

Particle Physics

Superconducting magnets are used in high-energy physics for accelerators and particle spectrometers. There are two reasons for designing an accelerator using superconducting magnets: (1) to obtain high magnetic fields and therefore correspondingly higher particle energies and (2) to have low power costs for operation. A good example of superconducting magnet technology is the "energy doubler/saver" proton accelerator at the Fermi National Accelerator Laboratory. The name "energy doubler/saver" comes from the fact that the original machine accelerated protons to 500 GeV with conventional magnets whereas the second-generation superconducting version accelerated protons to 1000 GeV with substantial power savings. The accelerator uses 990 superconducting magnets in a ring with a circumference of 6 kilometers.

Magnetic-Resonance Imaging

The first large-scale commercial application of superconductivity is in magnetic resonance imaging (MRI). A magnetic resonance image is a two-dimensional picture of the amplitude and phase of the nuclear magnetic resonance signal from an object. Images of the human body (see Figure 15-37) are used for medical purposes to detect tumors and other abnormalities. The magnetic field requirements for MRI include (1) uniform magnetic field (1 ppm) over a region large enough to contain a person, (2) field strengths of about 1 T, and (3) stability of the magnetic field with time (1 ppm/h). Magnetic resonance imaging can be performed with conventional magnets, but superconducting magnets are better suited to meeting the foregoing requirements. Hundreds of superconducting magnets are currently in use for MRI.

Other Applications of Superconducting Magnets

Superconducting magnets are useful whenever very high fields are needed and the benefits of low power consumption outweigh the inconveniences of the necessary cryogenics. Examples include trains that have been built with superconducting magnets in their bases, and large magnets used to contain plasmas for fusion research.

Superconducting Quantum Interference Devices

There are very many practical and ingenious applications of superconducting quantum interference devices (see

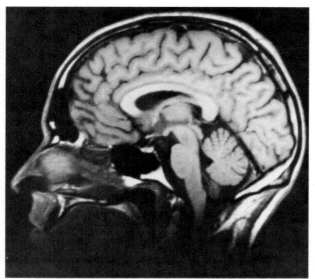

FIGURE 15-37 Magnetic resonance image of human brain.
Courtesy Siemans Medical Systems, Inc.

color plate 10). An incredibly sensitive magnetic flux detector may be built using SQUIDs. Superconducting quantum interference devices may be sensitive enough to detect magnetic fields from the heart (10^{-10} T) and even the brain (10^{-13} T). The SQUID may also be used to measure precisely a voltage or a position. For example, in the search for gravitational radiation, a large aluminum mass (4800 kg) *Weber bar* is cooled to liquid helium temperatures in order to search for longitudinal oscillations induced by gravity waves. Oscillations are detected with a tranducer made from a SQUID.

Electronics

Josephson junctions are useful in electronic applications because they can be designed to switch very quickly from superconducting to resistive. The intrinsic switching time of a Josephson junction is given by

$$\Delta t \approx \frac{h}{E_g} . \tag{15.93}$$

For an energy gap of 1 meV, the intrinsic switching speed is

$$\Delta t \approx \frac{4 \times 10^{-15} \text{ eV} \cdot \text{s}}{10^{-3} \text{ eV}} \approx 4 \text{ ps.} \tag{15.94}$$

In a practical device, the switching time of the junction is limited by its resistive-capacitive (*RC*) time constant, which can be a few picoseconds. In order for semiconductor junctions to compete with the speed of Josephson junctions, they need to be made smaller than current fabrication techniques allow.

Another advantage of Josephson junctions over semiconductor devices is their very low power dissipation. Josephson junctions can be made with power dissipations of a few microwatts. It is important to have low power dissipation in a high-speed computer, because the power dissipation limits the density with which components can be assembled. High density is needed to reduce the transit time for the electronic signals between various parts of the circuit.

CHAPTER 15: PHYSICS SUMMARY

- Certain metals undergo a phase transformation to the superconducting state when cooled below a critical temperature (T_c). In the superconducting state, the resistance of the metal is zero and the conductivity is infinite.

- All superconducting elements are of the type-I variety. Critical temperatures are typically a few kelvin. The magnetic field is zero inside a type-I superconductor. An external magnetic field above a critical value, $B_c(T)$, will cause a transformation back to the normal resistive state. A typical value of the critical field at zero kelvin is 0.1 T. If a type-I superconductor is cooled into the superconducting state in the presence of an external magnetic field, the field is expelled from the interior of the superconductor. This phenomenon is called the Meissner effect.

- Type-II superconductors are compound materials having two critical magnetic fields, B_{c1} and B_{c2}. The typical value of B_{c2} is tens of tesla. Below B_{c1} and above B_{c2}, the material is superconducting and normal, respectively. For fields between B_{c1} and B_{c2} the compound is in an intermediate state that has zero resistance, with the magnetic field penetrating the material in quantized magnetic flux tubes.

- Superconductors are characterized by two length parameters, the penetration depth (λ), which specifies the distance that a magnetic field penetrates a superconductor, and the coherence length (ξ), which specifies the correlation distance of two electrons that form a superconducting pair.

- The physical mechanism of superconductivity is an electron–lattice–electron interaction. Cooper pairs of electrons with opposite values of spin and momentum behave like a system of bosons separated from the normal electron states by an energy gap of about $3.5kT_c$.

- The magnetic flux through any superconducting ring is quantized,

$$\Phi_m = \frac{nh}{2e} = n\left(2.07 \times 10^{-15}\ \mathrm{T \cdot m^2}\right),$$

where n is an integer.

- High T_c type-II superconductor oxide-compounds exist with values of T_c up to 125 K.

REFERENCES AND SUGGESTIONS FOR FURTHER READING

P. W. Anderson and J. M. Rowell, "Probable Observation of the Josephson Tunneling Effect," *Phys. Rev. Lett.* **10**, 230 (1963).

P. W. Anderson and R. Schrieffer, "A Dialogue on the Theory of High T_c," *Phys. Today* **44**, 54 (1991).

J. Bardeen, "Superconductivity and Other Macroscopic Quantum Phenomena," *Physics Today* **43**, 25 (1990).

J. Bardeen, L. N. Cooper, and J. R. Schrieffer, *Phys. Rev.* **106**, 162 (1957); *Phys. Rev.* **108**, 1175 (1957).

B. Batlogg, "Physical Properties of High-T_c Superconductors," *Phys. Today* **44**, 44 (1991).

J. G. Bednorz and K. A. Müller, "Perovskite-type Oxides–The New Approach to High-T_c Superconductivity," *Rev. Mod. Phys.* **60**, 585 (1988).

J. G. Bednorz and K. A. Müller, "Possible High T_c Superconductivity in the Ba-La-Cu-O System," *Z. Phys* **B64**, 189 (1986).

D. J. Bishop, P. L. Gammel, and D. A. Huse, "Resistance in High-Temperature Superconductors," *Sci. Am.* **268**, No. 2, 48 (1993).

N. Byers and C. N. Yang, "Theoretical Considerations Concerning Quantized Magnetic Flux in Superconducting Cylinders," *Phys. Rev. Lett.* **7**, 46 (1961).

R. J. Cava, "Superconductors Beyond 1-2-3," *Sci. Am.* **263**, No. 2, 42 (1990).

J. Clarke, "SQUIDs, Brains and Gravity Waves," *Phys. Today* **39**, No. 3, 36 (1986).

W. S. Corak et al., "Atomic Heats of Normal and Superconducting Vanadium," *Phys. Rev.* **102**, 656 (1956).

Bascom S. Deaver, Jr. and William M. Fairbank, "Experimental Evidence for Quantized Flux in Superconducting Cylinders," *Phys. Rev. Lett.* **7**, 43 (1961).

P. G. DeGennes, *Superconductivity of Metals and Alloys,* Benjamin (1966).

R. Doll and M. Näbauer, "Experimental Proof of Magnetic Flux Quantization in a Superconducting Ring," *Phys. Rev. Lett.* **7**, 51 (1961).

R. P. Feynman, R. B. Leighton, and M. Sands, *The Feynman Lectures on Physics,* Vol. III, Chap. 21, Addison-Wesley (1965).

J. File and R. G. Mills, "Observation of Persistent Current in a Superconducting Solenoid," *Phys. Rev. Lett.* **10**, 93 (1963).

I. Giaever, "Electron Tunneling Between Two Superconductors," *Phys. Rev. Lett.* **5**, 464 (1960).

I. Giaever, "Energy Gap in Superconductors Measured by Electron Tunneling," *Phys. Rev. Lett.* **5**, 147 (1960).

G. F. Hardy and J. K. Hulm, "The Superconductivity of Some Transition Metal Compounds," *Phys. Rev.* **93**, 1004 (1954).

H. Hayakawa, "Josephson Computer Technology," *Phys. Today* **39**, No. 3, 46 (1986).

A. F. Hebard, "Superconductivity in Doped Fullerenes," *Phys. Today* **45**, No. 11, 26 (1992).

R. C. Jaklevic, J. Lambe, A. H. Silver, and J. E. Mercereau, "Quantum Interference Effects in Josephson Tunneling," *Phys. Rev. Lett.* **12**, 159 (1964).

J. D. Jorgensen, "Defects and Superconductivity in the Copper Oxides," *Phys. Today* **44**, 34 (1991).

B. D. Josephson, "Possible New Effects in Superconducting Tunneling," *Phys. Lett.* **1**, 251 (1962).

C. Kittel, *Introduction to Solid State Physics,* Wiley (1986).

J. E. Kunzler, "Superconductivity in High Magnetic Fields at High Current Densities," *Rev. Mod. Phys.* **33**, 501 (1961).

J. E. Kunzler, "Superconducting Materials and High Magnetic Fields," *J. App. Phys.* **33**, 1042 (1962).

J. E. Kunzler et al., "Superconductivity in Nb_3Sn at High Current Density in a Magnetic Field of 88 kgauss," *Phys. Rev. Lett.* **6**, 89 (1961).

J. E. Kunzler and M. Tanenbaum, "Superconducting Magnets," *Sci. Am.* **206**, No. 6, 60 (1962).

D. Larbalestier, "Critical Currents and Magnet Applications of High-T_c Superconductors," *Phys. Today* **44**, 74 (1991).

D. Larbalestier et al., "High-Field Superconductivity," *Phys. Today* **39**, No. 3, 24 (1986).

O. V. Lounasmaa, "New Methods for Approaching Absolute Zero," *Sci. Am.* **221**, No. 6, 26 (1969).

B. T. Matthias et al., "Superconductivity of Nb_3Sb," *Phys. Rev.* **95**, 1435 (1954).

E. Maxwell, "Isotope Effect in the Superconductivity of Mercury," *Phys. Rev.* **78**, 477 (1950).

Lars Onsager, "Magnetic Flux Through a Superconducting Ring," *Phys. Rev. Lett.* **7**, 50 (1961).

R. Palmer and A. Tollestrup, "Superconducting Magnet Technology for Accelerators," *Ann. Rev. Nucl. Part. Sci.* **34**, 247 (1984).

N. E. Phillips, "Low-Temperature Heat Capacities of Gallium, Cadmium, and Copper," *Phys. Rev.* **134**, 385 (1964).

T. G. Phillips and D. B. Rutledge, "Superconducting Tunnel Detectors in Radio Astronomy," *Sci. Am.* **254**, No. 5, 96 (1986).

F. Reif, "Superfluidity and 'Quasi-Particles'," *Sci. Am.* **203**, No. 5, 138 (1960).

C. A. Reynolds et al., "Superconductivity of Isotopes of Mercury," *Phys. Rev.* **78**, 487 (1950).

P. L. Richards, "Analog Superconducting Electronics," *Phys. Today* **39**, No. 3, 54 (1986).

J. M. Rowell, "Magnetic Field Dependence of the Josephson Tunnel Current," *Phys. Rev. Lett.* **11**, 200 (1963).

J. Rowell, "High-Temperature Superconductivity," *Phys. Today* **44**, 22 (1991).

R. Schrieffer, "John Bardeen and the Theory of Superconductivity," *Phys. Today* **45**, No. 4, 46 (1992).

Sidney Shapiro, "Josephson Currents in Superconducting Tunneling: The Effect of Microwaves and Other Observations," *Phys. Rev. Lett.* **11**, 80 (1963).

M. Tinkham, *Introduction to Superconductivity,* McGraw-Hill (1975).

A. M. Wolsky, R. F. Glese, and E. J. Daniels, "The New Superconductors: Prospects for Applications," *Sci. Am.* **260**, No. 2, 61, (1989).

M. K. Wu et al., *Phys. Rev. Lett.* **58**, 908 (1987).

QUESTIONS AND PROBLEMS

Basic experimental properties of superconductors

1. Why is the heat capacity of a metal greater in the superconducting state than in the normal state at the critical temperature?

2. How can one measure the normal state conductivity of a superconductor at temperatures below the critical temperature?

3. What happens to a thin type-II superconducting wire that is carrying a large current if the wire suddenly becomes normal? Why are the superconducting strands of wire embedded in a copper jacket in the construction of a superconducting magnet?

4. Why do the elements with higher critical temperatures also have higher critical magnetic fields at zero kelvin?

5. Vanadium has a critical temperature of 5.4 K and a critical magnetic field at zero kelvin of 0.14 T. Estimate the magnetic field necessary to destroy superconductivity in a sample of vanadium at 4.2 K.

6. The critical temperature of lead is 7.2 K and the critical magnetic field (at zero kelvin) is 0.08 T. (a) A lead cylinder with a volume of $10^{-9} m^3$ is cooled to 4.2 K. Estimate the condensation energy. (b) What magnetic field is necessary to reduce the condensation energy by 10%?

7. Estimate the minimum radius of a niobium wire that can carry a superconducting current of 10 A at a temperature of 4.2 K.

8. Which element can superconduct in the largest magnetic field at $T = 4$ K? Estimate the value of the critical magnetic field at $T = 4$ K.

9. A lead wire of 100 μm diameter is cooled into the superconducting state at a temperature of 4.2 K. The critical temperature of lead is 7.2 K and the critical magnetic field of lead is $B_c(0) = 0.08$ T. Calculate the maximum current the wire can carry and remain superconducting.

10. The superconducting transition temperature for niobium is 9.3 K, and the critical magnetic field at zero kelvin is 0.20 tesla. Calculate the temperature at which the critical magnetic field is lower by a factor of 2.

Development of the theory of superconductivity

11. Which superconducting element has the largest energy gap at zero kelvin? Calculate the value of the energy gap.

12. The magnetic field penetration depth in tin at zero kelvin is about 30 nm. Calculate the density of superconducting electrons in tin at zero kelvin.

13. Does the London equation hold for Cooper pairs? Why or why not?

*14. (a) Use Gauss's law for a magnetic field, $\nabla \cdot \mathbf{B} = 0$, to show that $\mathbf{B}$ may be written as the curl of a vector $\mathbf{A}$, $\mathbf{B} = \nabla \times \mathbf{A}$. (b) Show that the London equation may be written as

$$\mathbf{J} = -\frac{n_C e^2}{m} \mathbf{A}.$$

15. When a certain type-I superconductor is placed in a magnetic field, the field penetrates a distance of 50 μm into the superconducting material. Estimate the density of superconducting electrons.

16. The Fermi energy of gallium ($T_c = 1.1$ K) is about 10 eV. Estimate the intrinsic coherence length.

17. Determine the energy gap for vanadium from the heat capacity data of Figure 15-22.

Further properties of superconductors

18. (a) Calculate the inductance of a solenoid of 10^3 turns, 1 m in diameter and 2 m long. (b) The solenoid is superconducting with the persistent current having a lifetime of 10^5 years. Calculate the resistance of the solenoid.

19. Consider a thin superconducting lead cylindrical shell with a diameter of 10 μm and length of 1 mm, such as the one used by Doll and Näbauer to discover magnetic flux quantization. Calculate the minimum magnetic field needed to trap flux of magnitude h/e.

20. The size of the elementary flux quantum is roughly 2 fT·m². If we want to detect a magnetic field of 1 fT by superconducting quantum interference, why can the area of the SQUID be many orders of magnitude smaller than 2 m²?

21. A Pb–I–Pb Josephson junction is operated at liquid helium temperatures with a superconducting energy gap measured to be 2.7 meV. Calculate the frequency of the radiation needed to cause the tunneling current to occur at voltages of ±1.35 mV.

22. A SQUID is designed for use as a magnetometer to measure small variations in the earth's magnetic field. The experimenters design the SQUID so that a fractional change in the earth's magnetic field of 10^{-4} corresponds to one flux quantum. What is the area of the SQUID? (Take the earth's magnetic field to be 5×10^{-5} T.)

23. Consider the data of J. M. Rowell in measuring the Josephson current as a function of external magnetic field. Calculate the area of the Josephson junction perpendicular to the applied magnetic field.

24. A constant voltage is applied across a Josephson junction and the current is observed to oscillate with a frequency of 0.5 GHz. What is the value of the voltage?

Additional problems

25. An experimenter wishes to study the properties of a metal at low temperatures. The experimenter solders the sample to a copper rod and places the assembly in a cryostat. The copper rod is then cooled first with liquid nitrogen and then with liquid helium. Since copper is an excellent thermal conductor, as the copper rod cools, the sample also cools very efficiently. At very low temperatures, however, the sample never gets as cold as the copper rod. Why?

26. *Lifetime of a persistent current* (after Kittel, p. 343.). Consider a persistent current in a superconducting ring. We may write the probability per unit time (P) that a flux quantum will leak out of the ring (thereby causing a reduction in the current as $P = P_0 e^{-\Delta E/kT}$. (*a*) Estimate the size of the exponential factor by assuming that the minimum volume of the superconductor that must fluctuate to the normal state is equal to the thickness (x) of the superconductor times the square of the coherence length (ξ). The change in energy when this volume becomes normal is the volume times energy density of the critical field. Make estimates of x, ξ, B_c, and T to get a numerical answer for the exponential term. (*b*) Estimate the value of P_0 by assuming a characteristic decay time equal to Planck's constant divided by the size of the superconductor energy gap (E_g). (*c*) How does your answer for the lifetime of the current ($1/P$) compare with the age of the sun? (*d*) What happens if you make a different set of assumptions for the parameters of the superconductor?

27. A SQUID is constructed with an area of 10^{-8} m² between the Josephson junctions. When placed in a weak external magnetic field, the tunneling current is observed to oscillate as a function of the applied field. Calculate the spacing between maxima.

28. An engineer designs a long, large-radius supercon-ducting solenoid to operate with a magnetic field strength of 10 T at a temperature of 4 K. The type-II superconductor, niobium-titanium (NbTi), is chosen for the coil, which is wound with a density of 10^4 turns per meter. The critical temperature for NbTi is 9.3 K and the upper critical field is 15 T at zero kelvin. (a) Use Ampère's law to calculate the current in the coil for the design field of 10 tesla. (b) A quench will occur if the upper critical field is exceeded. Calculate the maximum current that the coil can carry at 4 K without quenching. (c) It may be that the magnet will quench at a lower current than calculated in part (b). Why would this happen?

... a dramatic race was under way to see who would be the first to accelerate protons to an energy high enough to disintegrate the atomic nucleus. This contest could be considered as the beginning of what was to become a Golden Age of high-energy physics. The race might also be taken to mark the end of an Age of Innocence of nuclear physicists. Heretofore during an era to which all physicists look back with nostalgia, much of the fundamental knowledge about the nucleus had been obtained by the use of rather primitive experimental devices, followed by sophisticated analysis. Rutherford's famous α-particle scattering experiment is a case-in-point, a little string and sealing wax and not much else. Not much, that is, except great leaps of reason and imagination... .

The race had already taken on many of the characteristics, and a few of the idiosyncrasies, of present high-energy physics: It was exciting, competitive, international... and romantic... . By all rights Ernest Lawrence and M. Stanley Livingston were the first to accelerate protons to 1 MeV in 1932 with their cyclotron, but they lost out nevertheless. Cockcroft and Walton had built a voltage multiplier at the Cavendish Laboratory in England. To their despair the accelerating column could hold off less than half a million volts. Sparking was their nemesis. Rutherford suggested they give it a try, notwithstanding the low energy. They did so. Their observation of the disintegration of lithium into α-particles, published in 1932, reminded everyone that the race was about physics, not gadgets. They had won the race.

Robert Rathbun Wilson

16-1 PARTICLE ACCELERATORS
16-2 PARTICLE DETECTORS

High-energy physics or particle physics is the study of fundamental particles and their forces. The origin of the name is clear from the relationship between momentum and wavelength of a particle:

$$p = \frac{h}{\lambda}. \qquad (16.1)$$

Investigation of particles and forces involves probing short distances, this is accomplished with probes of short wavelength, i.e., high momentum and energy. "High energy" is defined by the technology of the period, varying from a few keV in the spectrometer of J. J. Thomson, in which he discovered the electron, to 100 GeV in the spectrometer of UA1, where the Z^0 was discovered. In this sense, high-energy physics is the frontier of scientific exploration.

The "gadgets" of high-energy physics are of two types: (1) particle accelerators and (2) particle detectors.

16-1 PARTICLE ACCELERATORS

Particles are accelerated through their electromagnetic interactions. Only electrically charged particles that are stable against spontaneous decay may be readily accelerated. The only particles that fit this description are the electron, the proton, and stable nuclei (heavy ions) plus their antiparticles. All accelerators must have a source of charges, a means for accelerating the charges, and an evacuated path for the charges to travel until they reach their target. The vacuum is needed so that the particles are not scattered by collisions with air molecules.

Besides the particle species, the two parameters that dictate the physics that is accessible with a given accelerator are the energy and the particle flux. The particle flux in an experiment is often referred to as the *luminosity*. The units of luminosity are $m^{-2}s^{-1}$. From the definition of cross section (6.9), the rate (R) of occurrence for a particular process is the luminosity (L) times the cross section (σ) for that process:

$$R = L\sigma. \qquad (16.2)$$

Energy is important for the study of short-distance phenomena and the production of massive particles, while luminosity is important for the study of rare processes. To make a sensitive measurement, one needs both adequate energy and luminosity. Due to many innovative technological advancements in the field of accelerators, the maximum achievable particle energy has grown exponentially over many decades since the early accelerators of

1930 (see Figure 16-1). Each new accelerator technique has reached a limit in the maximum energy within a few years, only to be overtaken by a new invention. Accelerators have found numerous applications in medicine and condensed matter physics, but the driving force for accelerator development has always been the quest for higher energies to probe the structure of matter in yet finer detail. Most of what we have learned about particles and forces is directly connected with the construction of high-energy accelerators.

Cockroft-Walton

In 1930, an international race was underway to produce the first artificial disintegration of the nucleus. This goal was achieved at the Cavendish Laboratory in England by John D. Cockroft and Ernest T. S. Walton, who built a huge voltage divider and used it to linearly accelerate protons to kinetic energies of several hundred keV. With

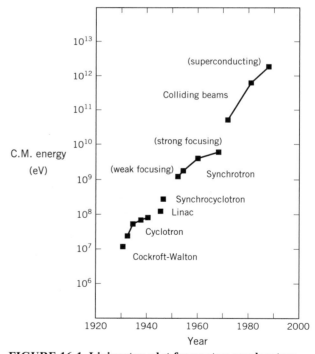

FIGURE 16-1 Livingston plot for proton accelerators. The plot shows the center-of-mass energy achieved at the time when the indicated technology gave the world's highest energy. The data show a remarkable exponential growth with time over several decades. This was made possible by the development of the technologies indicated. After M. Stanley Livingston.

this first proton accelerator, Cockroft and Walton observed the reaction

$$p + {}^7\text{Li} \rightarrow \alpha + \alpha. \qquad (16.3)$$

The maximum energy obtainable with the Cockroft-Walton technique, roughly 1 MeV, is limited by electrical discharge. The acceleration technique is schematically illustrated in Figure 16-2. Acceleration of a proton to a kinetic energy of 1 MeV requires a potential difference of one million volts. The insulation for this high voltage makes the Cockroft-Walton accelerator large and bulky. In spite of this, the Cockroft-Walton is still widely used in the first stage of high-energy accelerators. The Cockroft-Walton accelerator at Fermilab is shown in color plate 11.

EXAMPLE 16-1

Make an order-of-magnitude estimate of the proton energy required to initiate the reaction $p + {}^7\text{Li} \rightarrow \alpha + \alpha$.

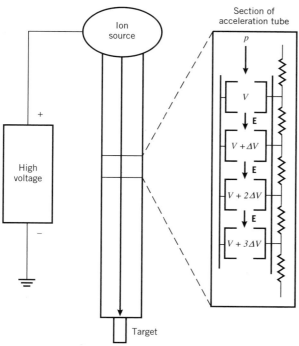

FIGURE 16-2 Schematic of the Cockroft-Walton accelerator.
The protons pass through a large number of electrodes each of which is at a successively higher voltage. The kinetic energy gained by the protons in passing from one electrode to the next is $e\Delta V$.

SOLUTION:

First consider conservation of energy. If the proton has zero kinetic energy, then the Q value of the decay is

$$Q = m_p c^2 + m_{\text{Li}} c^2 - m_\alpha c^2 - m_\alpha c^2,$$

or

$$Q \approx 938.3 \text{ MeV} + 6533.9 \text{ MeV}$$
$$-3727.4 \text{ MeV} - 3727.4 \text{ MeV} \approx 17 \text{ MeV}.$$

The Q value is positive, so conservation of energy does not place a limit on the proton energy threshold. The energy threshold limit comes from the Coulomb barrier presented by the ${}^7\text{Li}$ nucleus. If R is the nuclear radius, the energy threshold for the proton to surmount this barrier is

$$E_k = \frac{Zke^2}{R} = \frac{\alpha Z\hbar c}{R}.$$

If we estimate R to be about 2 fm and use $Z = 3$ for lithium, then

$$E_k = \left(\frac{1}{137}\right)(3)\left(\frac{200 \text{ MeV} \cdot \text{fm}}{2 \text{ fm}}\right) \approx 2 \text{ MeV}.$$

Thus, the order of magnitude of the proton kinetic energy needed to cause the nuclear reaction is 2 MeV. This is only an order-of-magnitude estimate because of the wave nature of the particles. The reaction rate will be energy dependent. A proton with an energy less than 2 MeV can still produce the reaction (16.3), although with reduced probability. ■

Example 16-1 illustrates the trade-off between energy and luminosity in a high energy physics experiment. In many experiments the event rate is maximum with the largest possible luminosity and the largest possible energy. In practice, a larger luminosity may often be obtained at a lower energy. For a given experiment, the energy and luminosity must be optimized within the constraints of the available accelerators. Cockroft and Walton had a lower energy than they desired, but they had sufficient luminosity to observe the reaction of interest (16.3). Naturally, there are threshold effects: if the energy is too low, then no luminosity may be sufficient to observe the reaction.

The brute force acceleration by direct current (DC) high voltage was continued with the development of the Van de Graaff accelerator. The rectifier circuit of Cockroft-Walton was replaced with an electrostatic charging belt.

The effects of sparking were reduced by insulating the accelerating tube with compressed gases. Van de Graaff accelerators can produce particle kinetic energies of about 10 MeV. Recall that the kinetic energies of particles from nuclear decays are typically a few MeV. The Van de Graaff can produce kinetic energies that are larger than those present in the natural radioactive decays.

Cyclotron

A new concept in particle acceleration was introduced by R. Wideröe in 1928. In his pioneering apparatus, Wideröe passed electrons across two gaps, as indicated in Figure 16-3. The voltage across both gaps is made to oscillate such that the electrons arrive at each gap when the voltage is a maximum. The electrons are accelerated in each of the two gaps. This condition requires the voltage to oscillate at a radio frequency (several megahertz). The innovation is that the gain in kinetic energy of the electron (ΔE) is

proportional to twice the maximum voltage (V_{max}) in the system,

$$\Delta E = 2\,eV_{max}.\qquad(16.4)$$

There is no limit in principle to the number of acceleration gaps that can be added. The electron would gain an energy (eV_{max}) in each gap. The radio frequency (RF) technique allows acceleration of a charged particle to a large energy using an oscillating voltage of small amplitude (V_{max}).

EXAMPLE 16-2

An electron with a very small initial kinetic energy is accelerated in two narrow gaps separated by a distance of 1 m (see Figure 16-3). If the maximum voltage is equal to 100 kV, calculate the oscillation frequency of the voltage.

SOLUTION:

The acceleration condition is that the electron must arrive at the second gap when the voltage is maximum. The kinetic energy of the electron after passing the first gap is

$$E_k = e(100\ \text{kV}) = 100\ \text{keV} = 0.10\ \text{MeV}.$$

The total energy of the electron is

$$E = mc^2 + E_k = 0.51\ \text{MeV} + 0.10\ \text{MeV} = 0.61\ \text{MeV}.$$

The momentum of the electron is given by

$$\begin{aligned}
pc &= \sqrt{E^2 - \left(mc^2\right)^2}\\
&= \sqrt{(0.61\ \text{MeV})^2 - (0.51\ \text{MeV})^2}\\
&= 0.33\ \text{MeV}.
\end{aligned}$$

The speed of the electron is given by

$$\frac{v}{c} = \frac{pc}{E} = \frac{0.33\ \text{MeV}}{0.61\ \text{MeV}} = 0.54.$$

To be accelerated in both gaps, the voltage oscillation frequency must be

$$\begin{aligned}
f &= \frac{v}{d} = \frac{(0.54)\left(3\times10^8\ \text{m/s}\right)}{1\,\text{m}}\\
&= 1.6\times10^8\ \text{s}^{-1} = 160\,\text{MHz}.
\end{aligned}$$

A radio frequency is required.   ∎

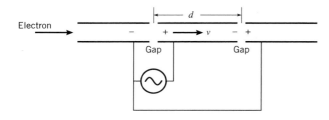

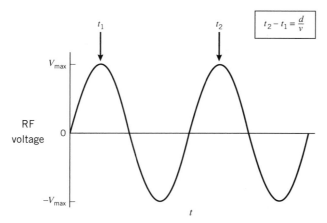

FIGURE 16-3 Schematic of the Wideröe accelerator. An oscillating voltage is applied to three sections of a conducting evacuated tube. The voltage oscillates at a fixed radio frequency (several megahertz). The frequency and length of the tube are chosen such that electrons are accelerated in each of two gaps. The gain in energy from each gap is eV_{max}.

The RF acceleration technique is useful in practice largely because of a phenomenon called longitudinal *phase stability*. Consider a group or *bunch* of charged particles that are to be collectively accelerated by the RF technique. The particle bunch will have a finite length. It is highly desirable to choose the phase (ϕ_0) of the voltage so that the acceleration occurs at a voltage (V_0) smaller than V_{max}. This is illustrated in Figure 16-4. This means that particles in the *center* of the bunch are accelerated by the voltage V_0. Particles that arrive *early* are accelerated by a *smaller voltage*, thus keeping them synchronized with the center of the bunch. Particles that arrive *late* are accelerated by a *larger voltage*, which also keeps them synchronized with the center of the bunch. This phenomenon results in a range of phases (ϕ_{min} to ϕ_{max}) that produce stable acceleration. This is especially important for those applications where the particles spend a long time in the accelerator.

In 1930, Ernest O. Lawrence used the radio frequency technique to accelerate protons in an ingenious scheme where the protons were made to repeatedly pass through the acceleration gap by use of a magnetic field. The device of Lawrence, called the *cyclotron*, was the first circular accelerator. The cyclotron is shown in Figure 16-5. The shape of a cyclotron is that of a pillbox cut into two equal-sized pieces across its diameter. Because of their shape, the two pieces are called "Ds." The gap between the Ds has an oscillating potential difference of about 150 kV. The frequency is chosen so that the protons are accelerated each time they cross the gap. The protons are made to

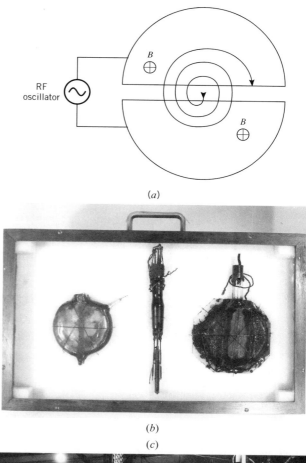

(a)

(b)

(c)

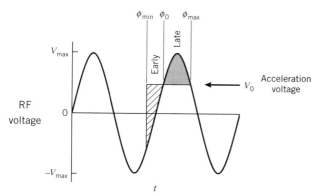

FIGURE 16-4 Longitudinal phase stability in RF acceleration.
All particles do not arrive at the acceleration gap at the same time. For this reason the acceleration voltage (V_0) is chosen to be smaller than the maximum voltage. Particles that arrive early are accelerated by a smaller voltage. Particles that arrive late are accelerated by a larger voltage. This results in a stable acceleration for particles, with phases in the range ϕ_{min} to ϕ_{max}.

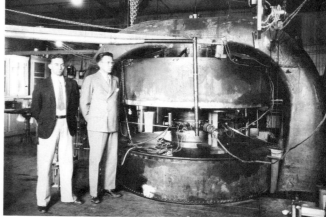

FIGURE 16-5 The cyclotron.
Protons are accelerated by a voltage which oscillates between two D-shaped electrodes at a radio frequency. A magnetic field gives the protons a spiral trajectory. (*a*) Schematic. (*b*) The first "4-inch" cyclotron of E. O. Lawrence and M. S. Livingston. (*c*) Livingston (left) and Lawrence with the "27-inch" cyclotron. Photographs courtesy Lawrence Berkeley Laboratory.

repeatedly cross the accelerating gap by application of a magnetic field. As the protons gain energy, they have a larger radius of curvature in the fixed magnetic field. The resulting motion is that of a spiral. The first cyclotron built by Lawrence was a mere one-tenth of a meter in size. As part of the race to split the nucleus, Lawrence and M. Stanley Livingston constructed a larger cyclotron (about one-fourth of a meter in size) that accelerated protons to 1 MeV.

In a cyclotron, the time (T) between accelerations is the time for the particle to make one-half of a revolution. For a particle traveling at a speed v in a circle of radius r,

$$T = \frac{\pi r}{v}. \tag{16.5}$$

Therefore, the frequency (ω_{cyc}) of the oscillating voltage is chosen to be

$$\omega_{cyc} = \frac{\pi}{T} = \frac{v}{r}, \tag{16.6}$$

so that the voltage changes sign for each half-revolution. This frequency is called the *cyclotron frequency*. The magnetic field (B) is fixed, and the relationship between the particle momentum (p) and the radius of curvature (r) is

$$p = eBr. \tag{16.7}$$

For nonrelativistic particles, the orbital frequency (ω_{orb}) is a constant:

$$\omega_{orb} = \frac{v}{r} = \frac{veB}{p} = \frac{eB}{m}. \tag{16.8}$$

The particles may be accelerated each time they cross the gap by selecting the cyclotron frequency to be equal to the orbital frequency ($\omega_{cyc} = \omega_{orb} = eB/m$).

For relativistic particles the orbital frequency depends on the particle speed because the momentum depends on the Lorentz γ factor ($p = \gamma m v$):

$$\omega_{orb} = \frac{v}{r} = \frac{veB}{p} = \frac{eB}{\gamma m}. \tag{16.9}$$

Thus, if the frequency (ω_{cyc}) is kept constant, particles may not be accelerated to extreme relativistic energies because the γ factor becomes important.

We now make an estimate of the maximum particle energy in a cyclotron. For a D-voltage V, the energy gained per revolution is $e \times 2V$. The increase in γ ($\Delta\gamma$) after the first revolution is

$$\Delta\gamma = \frac{(e)(2V)}{mc^2}. \tag{16.10}$$

This increase in γ causes orbital frequency to be out of phase with the cyclotron frequency by $\Delta\gamma$ radians. In the next revolution, γ increases further by $\Delta\gamma$ and the total phase difference of the orbital and cyclotron frequencies is $\Delta\gamma + 2\Delta\gamma = 3\Delta\gamma$ radians. After N revolutions the phase difference is

$$\Delta\gamma + 2\,\Delta\gamma + 3\,\Delta\gamma + \ldots N\Delta\gamma$$
$$= \frac{N(N+1)\Delta\gamma}{2}. \tag{16.11}$$

When the phase difference is π, the particles are no longer accelerated because they arrive at the gap when the voltage has the wrong polarity. The maximum number of revolutions that the particles make while still being accelerated is

$$N \approx \sqrt{\frac{2\pi}{\Delta\gamma}} = \sqrt{\frac{\pi mc^2}{eV}}. \tag{16.12}$$

The kinetic energy of the particle after N revolutions is

$$E_k \approx (e)(2V)\sqrt{\frac{\pi mc^2}{eV}}$$
$$= 2\sqrt{\pi Vemc^2}. \tag{16.13}$$

For a D-voltage of 150 kV, the maximum kinetic energy for protons is

$$E_k \approx 2\sqrt{\pi(0.15\,\text{MeV})(940\,\text{MeV})}$$
$$\approx 40\,\text{MeV}. \tag{16.14}$$

For electrons, the maximum kinetic energy is much smaller, so that the conventional cyclotron is not useful for electron acceleration.

EXAMPLE 16-3

(a) Estimate the size of the magnetic field in the 1-MeV cyclotron of Lawrence and Livingston. (b) Estimate the cyclotron frequency.

SOLUTION:

(a) The magnetic field is

$$B = \frac{p}{er} = \frac{\sqrt{2mc^2 E_k}}{cer},$$

where p is the maximum momentum and E_k is the maximum kinetic energy of the proton corresponding to a maximum radius r. Numerically, we have

$$B = \frac{\sqrt{2mc^2 E_k}}{cer}$$

$$\approx \frac{\sqrt{(2)(9.4 \times 10^8 \text{ eV})(10^6 \text{ eV})}}{(3 \times 10^8 \text{ m/s})(e)(0.25 \text{ m})} \approx 0.58 \text{ T}.$$

(b) The particle speed is

$$v = \sqrt{\frac{2E_k}{m}}.$$

The cyclotron frequency is

$$\omega_{cyc} = \frac{v}{r} = \frac{1}{r}\sqrt{\frac{2E_k}{m}} = \frac{c}{r}\sqrt{\frac{2E_k}{mc^2}},$$

or

$$\omega_{cyc} \approx \frac{3 \times 10^8 \text{ m/s}}{0.25 \text{ m}}\sqrt{\frac{(2)(1 \text{ MeV})}{940 \text{ MeV}}} \approx 55 \text{ MHz.} \quad \blacksquare$$

Synchrocyclotron

In 1945, a way was invented to accelerate particles to relativistic speeds in a cyclotron by varying the frequency of the field synchronously with the inverse of the particle energy (γmc^2). Such a device is called a *synchrocyclotron*. In a synchrocyclotron there is no longer an advantage to having a very large voltage per turn as in a conventional cyclotron. The acceleration per revolution is usually chosen to be about 10 keV per turn. Since there is a limit of how large a magnetic field can be produced, the maximum energy achievable is limited by how big one can make the device. The practical energy limit of the synchrocyclotron is about 700 MeV.

Betatron

The first circular accelerator for electrons, the *betatron*, was invented by D. W. Kerst in 1940. In the betatron, electrons are in an orbit of fixed radius (r). The acceleration comes from the application of a changing magnetic flux (Φ) through the loop of the electron orbit. By Faraday's law of induction (B.11)

$$\oint d\mathbf{l} \cdot \mathbf{E} = -\frac{\partial \Phi}{\partial t}, \tag{16.15}$$

or

$$2\pi r E = -\frac{\partial \Phi}{\partial t}. \tag{16.16}$$

The induced electric field is

$$E = -\frac{1}{2\pi r}\frac{\partial \Phi}{\partial t}. \tag{16.17}$$

The change in momentum is

$$\frac{dp}{dt} = -eE = \frac{e}{2\pi r}\frac{\partial \Phi}{\partial t}. \tag{16.18}$$

In a betatron, the electrons are accelerated by the magnetic field that keeps them in a circular orbit. This acceleration causes the electrons to radiate. Radiation from a charged particle in uniform circular motion is called *synchrotron radiation*. The power radiated by an orbiting charge may be obtained from the classical formula (B.30),

$$P = \frac{2ke^2}{3c^3}a^2, \tag{16.19}$$

provided that we account for relativistic effects. For a nonrelativistic particle, the acceleration is v^2/r, where v is the speed and r is the radius of the orbit. For a relativistic particle, the acceleration is

$$a = \frac{1}{m}\frac{dp}{d\tau} = \left(\frac{1}{m}\right)\gamma\left(\frac{v}{r}\right)p = \gamma^2\frac{v^2}{r}, \tag{16.20}$$

where τ is the proper time and γ is the Lorentz gamma factor of the particle. The acceleration of a relativistic particle is larger than for the nonrelativistic case by a factor of γ^2; one power of γ comes from time dilation and another comes from relativistic momentum (because the change in momentum is proportional to the momentum for a circular orbit). The radiated power (16.19) is

$$P = \frac{2ke^2}{3c^3}\left(\frac{\gamma^2 v^2}{r}\right)^2 = \frac{2ke^2\gamma^4 v^4}{3c^3 r^2}. \tag{16.21}$$

The time for one revolution is $v/2\pi r$, so the energy lost per revolution is

$$\Delta E = P\left(\frac{2\pi r}{v}\right)$$

$$= \left(\frac{2ke^2\gamma^4 v^4}{3c^3 r^2}\right)\left(\frac{2\pi r}{v}\right)$$

$$= \frac{4\pi}{3}\alpha\frac{\hbar c}{r}\gamma^4\beta^3, \tag{16.22}$$

where $\beta = v/c$.

For a given particle energy, the Lorentz γ factor is inversely proportional to the particle mass. Therefore, sychrotron radiation is much more important for electrons than for protons. Synchrotron radiation places a practical limit on the maximum energy of a betatron of a few hundred MeV.

EXAMPLE 16-4

Calculate the energy lost per revolution by electrons and protons of kinetic energy 300 MeV in a circular orbit of 1 m radius.

SOLUTION:

The electrons have

$$\beta \approx 1.$$

The gamma factor of the electron is

$$\gamma = \frac{E}{mc^2} = \frac{E_k + mc^2}{mc^2} = \frac{300\,\text{MeV} + 0.511\,\text{MeV}}{0.511\,\text{MeV}} = 588.$$

The energy lost per revolution is (16.22)

$$\Delta E = \frac{4\pi}{3}\alpha\frac{\hbar c}{r}\gamma^4\beta^3$$
$$= \left(\frac{4\pi}{3}\right)\left(\frac{1}{137}\right)\left(\frac{200\,\text{MeV}\cdot\text{fm}}{10^{15}\,\text{fm}}\right)(588)^4$$
$$\approx 730\,\text{eV}.$$

The proton has

$$\gamma = \frac{E}{mc^2} = \frac{940\,\text{MeV} + 300\,\text{MeV}}{940\,\text{MeV}} = 1.3,$$

and

$$\beta = \frac{\sqrt{E^2 - mc^2}}{E}$$
$$= \frac{\sqrt{E_k^2 + 2E_k mc^2}}{E}$$
$$= \frac{\sqrt{(300\,\text{MeV})^2 + 2(300\,\text{MeV})(940\,\text{MeV})}}{1240\,\text{MeV}}$$
$$= 0.65.$$

The energy lost per revolution is

$$\Delta E = \frac{4\pi}{3}\alpha\frac{\hbar c}{r}\gamma^4\beta^3$$
$$= \left(\frac{4\pi}{3}\right)\left(\frac{1}{137}\right)\left(\frac{200\,\text{MeV}\cdot\text{fm}}{10^{15}\,\text{fm}}\right)(1.3)^4(0.65)^3$$
$$\approx 5\times10^{-9}\,\text{eV}. \qquad\blacksquare$$

In the betatron accelerator, there are small displacements of the particles due to collisions with the gas inside the beam pipe. The accelerator must be designed to have electromagnetic restoring forces so that a particle slightly out of orbit gets pulled back into orbit. This results in the particles undergoing oscillations as they orbit the machine. Oscillations perpendicular to the trajectory of the orbiting particle (radial and vertical oscillations) are called *betatron oscillations*. Oscillations along the particle trajectory are referred to as *phase oscillations*. The understanding and design of magnetic fields to provide the restoring forces was one of the most important results from betatron research.

Synchrotron

In a synchrotron, charged particles travel in a vacuum pipe passing through a ring of magnets at fixed radius. At one or more places in the circular orbit, the particles pass through a cavity where a radio frequency (RF) is applied for acceleration. Each particle gets a small increase in energy with each passage through the RF cavity (see color plate 14). Since the radius of the particle trajectory is fixed, the magnetic field must be increased "in synchronization" with the increase in energy of the particle. The synchrotron may be used to accelerate electrons, protons, and heavy ions.

The first technological advancement that made the synchrotron feasible was the invention of phase stability by V. Veksler in 1945 and E. M. McMillan in 1946. Particles that are just slightly out of phase, because they have an orbit that is too large or too small, can oscillate around the synchronous phase without losing their orbits. Phase stability is achieved by arranging the fringe magnetic fields so that the particles automatically arrive in phase with the acceleration mechanism. This arrangement is called *weak focusing* and is illustrated in Figure 16-6. Consider a proton that has a nominal orbit with $r = r_0$ and $z = 0$. The magnetic field is oriented in the z direction so that a proton is kept at a radius r_0. Now consider a charged particle with an orbit that varies from the nominal orbit, so that $r = r_0 + \Delta r$ and $z = \Delta z$. The field at large radius is shaped

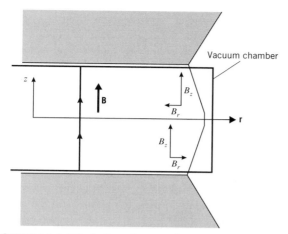

FIGURE 16-6 Weak-focusing synchrotron.
A synchrotron contains a ring of magnets at a fixed radius. At one or more places along the circumference, an RF accelerating cavity is introduced. The particles are accelerated when they pass through the cavity and the magnetic field is increased at the same time in order to keep the radius of the orbit fixed. In a weak-focusing machine, vertical and radial restoring forces on a particle slightly out of orbit are provided by shaping the magnetic field. All bending magnets are shaped in the same manner, requiring a large vacuum chamber to contain the particle oscillations.

to have a larger z component plus a radial component. The larger z component gives a force on the particle that pushes it to a smaller radius, and the radial component gives a force on the particle that pushes it to a smaller z coordinate. The weak focusing technique requires a very large cross-sectional area of the vacuum chamber (typically 1.3 by 0.3 meters) because the oscillations in the particle orbits are large. The first electron synchrotron was constructed from a converted betatron in 1946.

The first proton synchrotron, the *cosmotron*, was built at Brookhaven National Laboratory in 1952 and achieved an energy of 3 GeV. It did not go unnoticed that the size of the accelerators was increasing with increasing energy. In a 1952 review paper on synchrotrons I. S. Blumenthal remarks:

> *The Brookhaven cosmotron is of the Racetrack variety. The name fits this colossus with no stretch of the imagination, since it is to be a racetrack not only in shape but in size as well! The orbital diameter is of the order of 50 feet. The solution of high vacuum in this machine represents a tremendous challenge.*

Proton synchrotrons would reach a diameter of 1 kilometer within two decades!

The second technological advancement that allowed higher energies in synchrotrons was the invention of *strong focusing* by N. Christophilos in 1950 and by H. Courant, Livingston, and Snyder in 1952. In the strong-focusing technique, oscillations are greatly reduced by the introduction of magnets that alternately focus and defocus in both the horizontal and vertical planes. The magnets can be combined to yield a net focusing effect; the first strong-focusing synchrotrons were of this type. In the designs of higher energy machines, the focusing is provided by quadrupole magnets. The concept of strong focusing is illustrated in Figure 16-7. Strong focusing is especially important in a large-radius (high-energy) machine where

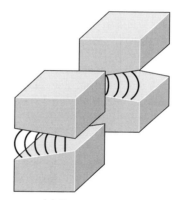

(a) Alternating gradient

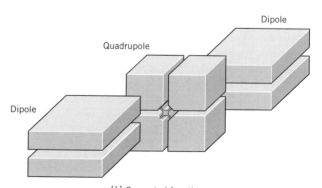

(b) Separated functions

FIGURE 16-7 Strong-focusing synchrotron.
(a) In a strong-focusing synchrotron, the magnetic field gradients alternate between focusing and defocusing. This results in smaller orbit oscillations and allows the use of a much smaller vacuum chamber than weak-focusing. This in turn allows the construction of much larger and higher energy synchrotrons. *(b)* Strong focusing can also be achieved using quadruple magnets.

the particle path length is very long and where one needs to keep the magnet aperture as small as possible to reduce the cost of the magnets. The typical cross-sectional area of the vacuum chamber needed to contain the beam in a strong-focusing machine is significantly smaller than that needed for a weak-focusing machine. The Brookhaven alternating gradient synchrotron (AGS) and the proton synchrotron (PS) at CERN were among the first successful strong-focusing proton machines. The first strong-focusing machines had energies of about 30 GeV.

In a proton synchrotron, the energy is limited by the field strength of the bending magnets for a fixed radius. Thus, a very high energy machine is made with high-field magnets at a large radius. The largest proton synchrotrons are the *Main Ring* and *Tevatron* at Fermilab and the *Super Proton Synchrotron* (SPS) at CERN (see color plate 12) which have radii of 1 km. The Fermilab main ring has an energy of 500 GeV and the CERN SPS has an energy of 450 GeV. The Fermilab Tevatron, built in the same tunnel just under the main ring, is the world's first superconducting synchrotron and has reached an energy of 1 TeV.

In an electron synchrotron, the energy is limited by synchrotron radiation. It has already been mentioned that the amount of synchrotron radiation varies as the fourth power of the particle energy and is inversely proportional to the radius. Therefore, a high-energy electron machine is made large in order to reduce the synchrotron radiation. Although the synchrotron radiation is the high-energy physicist's nemesis, it does have useful applications in the field of condensed matter physics. Old "high-energy" electron synchrotron machines are often outfitted with extra "wiggler" magnets for the purpose of enhancing the intensity of the synchrotron radiation. The world's largest electron synchrotron is the *Large Electron Project* (LEP) at CERN, which has a radius of about 4 km. It seems rather unlikely that a larger-radius conventional electron synchrotron will ever be built. To achieve electron energies beyond the capabilities of LEP, a new technology must be developed.

Storage Rings

A *storage ring* consists of a vacuum pipe passing through a ring of magnets that maintain a constant field so that particles may circulate continuously. The storage ring usually doubles as a synchrotron so that particles are both accelerated and stored in the same machine. Two storage rings that intersect at one or more places can be used to study the collisions of two stored beams. Particle–antiparticle collisions may be studied with one storage ring, with the particles and antiparticles circulating in opposite di-

rections. Antiparticles have the same mass but opposite electric charge as the particles. Examples of antiparticles are the positron (see Chapter 4) and the antiproton. The discovery of the antiproton is discussed in Chapter 17. The particles and antiparticles circulate in the storage ring in multiple bunches and must be separated at those places where they cross one another, except at special locations, where the bunches are focused into collision with one another.

For a colliding beam machine to work, the particles must be stored in stable orbits on the time scale of hours compared to the few seconds they spend in a synchrotron in the acceleration process (hence, the name "storage rings"). This requires an extremely high vacuum compared to that needed in a synchrotron. In a storage ring, the magnets are continuously operating, whereas in the normal operation of a synchrotron they are pulsed briefly every few seconds. Also, the beams in a storage ring must be focused to a small cross-sectional area and contain a large number of particles in order to have a useful luminosity.

The first colliding beam machines were designed for electron–positron collisions with an energy of 250 MeV per beam at Frascati and 500 MeV per beam at Princeton. Both were operational in 1961. Since then many electron–positron colliders have been built. By far the most successful machine was the 4 GeV per beam *Stanford Positron Electron Accelerator Rings* (SPEAR), where the charm quark and the tau lepton were discovered. The largest electron–positron collider is the LEP machine at CERN. The LEP machine was designed to study electron–positron collisions at a center of mass energy equal to the Z^0 mass energy (about 100 GeV).

The first proton–proton collider was the CERN Intersecting Storage Rings (ISR), which came into operation in 1971. The CERN ISR had a maximum proton energy of 31 GeV per beam. The operation of the ISR was made possible by the vacuum technology available at CERN and the technique of momentum stacking, which allowed the accumulation of an intense proton beam.

A major increase in center-of-mass energy was achieved with the invention of colliding beam machines (see Figure 16-1). Two colliding protons, each with an energy E^*, have the same center-of-mass energy as a proton of energy

$$E \approx \frac{2E^{*2}}{mc^2}, \qquad (16.23)$$

colliding with a stationary proton, where m is the proton mass. Therefore, colliding beams are used to achieve the highest center-of-mass collision energies.

EXAMPLE 16-5

The CERN ISR had a proton–proton center-of-mass energy of 62 GeV. Calculate the proton energy needed to achieve the same center-of-mass energy by collision with a stationary proton.

SOLUTION:

The equivalent proton laboratory energy is

$$E \approx \frac{2E^{*2}}{mc^2} \approx \frac{(2)(62\,\text{GeV})^2}{0.94\,\text{GeV}} \approx 8.2\,\text{TeV} \,. \quad \blacksquare$$

In 1981, the CERN SPS was converted to the world's first proton–antiproton collider. The energy was 270 GeV per beam. The proton–antiproton collider was made possible by the invention of *stochastic cooling* by Simon van der Meer in 1968. Stochastic cooling is the technique which allows accumulation of an intense beam of antiprotons suitable for injection into the collider ring.

Linear Accelerators

The radio frequency technique pioneered by Wideröe has been developed and refined to make high-energy linear accelerators (*linacs*). The key ingredient of the RF acceleration technique is the high-frequency power source. The klystron amplifier invented in 1937 and developed in the 1940s can provide several megawatts at a frequency of a few gigahertz.

One refinement of the RF technique is to accelerate relativistic particles using electromagnetic waves in a wave-guide cavity. Acceleration occurs in such a cavity when the phase velocity of the wave is equal to the particle velocity. The phase velocity of the electromagnetic waves is adjusted by appropriate geometric shaping of the wave-guide cavity. The RF acceleration technique is indicated in Figure 16-8.

The largest linear accelerator is located at the Stanford Linear Accelerator Center (SLAC). It is 3.2 kilometers long and accelerates electrons to 25 GeV. In an upgraded and modified mode of operation, it has been converted to a "single-pass" electron–positron linear collider (SLC) with an energy of 50 GeV per beam.

EXAMPLE 16-6

One figure of merit for a linear accelerator is the amount of energy that can be transferred to a beam of particles per length. Calculate this quantity for SLAC.

SOLUTION:

The electron energy per length in the original design of SLAC is

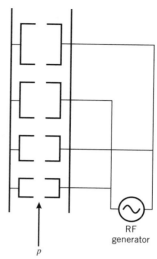

FIGURE 16-8 Schematic of RF acceleration.
The particle sees no voltage when it is inside the cell. The particle is accelerated when it is between two cells if it has a speed such that it arrives in phase with the oscillating voltage. Since the speed of the particle increases after acceleration, the cell length must increase. The RF generator in the early accelerators operated at a frequency of 7 MHz.

$$\frac{25\,\text{GeV}}{3200\,\text{m}} \approx 8\,\text{MeV/m} \,.$$

The upgraded beam line provides

$$\frac{50\,\text{GeV}}{3200\,\text{m}} \approx 16\,\text{MeV/m} \,. \quad \blacksquare$$

Accelerator Complexes

Fermilab

In the acceleration of particles to high energies, many accelerators are used in sequence. Figure 16-9 shows the layout of the accelerators at Fermilab. Protons are accelerated first by the old-fashioned DC method in a Cockroft-Walton device. They then are accelerated in a proton linac to an energy of 200 MeV. After the linac, they are injected into a booster synchrotron where they are accelerated to 8 GeV. After the booster, they are injected into a large synchrotron, and accelerated up to 500 GeV with the original main ring and, since 1986, to 800 GeV with the superconducting Tevatron. In addition, the Tevatron is used as a proton–antiproton colliding beam machine. For fixed-target physics, the protons are extracted into three experimental areas where multiple secondary beams of pions, kaons, neutrons, muons, neutrinos, electrons, and photons are generated.

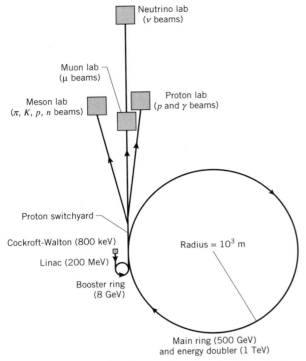

FIGURE 16-9 The fixed-target accelerator layout at Fermilab.

CERN

Figure 16-10 shows the layout of the CERN proton–antiproton accelerator complex. Protons are accelerated in a linear accelerator and a booster synchrotron up to 3.5 GeV. The protons are injected into the proton synchrotron (PS), where they are accelerated to 26 GeV. From the proton synchrotron they are injected into the Super Proton Synchrotron (SPS), where they can be accelerated to 450 GeV. For proton–proton colliding beam physics, the protons are injected into the ISR. For fixed-target physics, the proton beam is extracted into areas where various secondary beams are generated. When it is used for studying proton–antiproton collisions, protons from the PS are sent to a special target to produce antiprotons. The antiprotons are produced with a spread in momentum around a typical value of 3.5 GeV/c. These antiprotons are collected in a special storage ring called the *antiproton accumulator* (see color plate 13). In the antiproton accumulator, the spread in momentum of the particles is reduced by stochastic cooling. In the stochastic cooling process, the position of antiprotons is sensed electronically in the ring. An electrical signal is sent across the diameter of the ring to the opposite side, arriving ahead of the antiprotons. This signal tells how to accelerate the antiprotons, to make

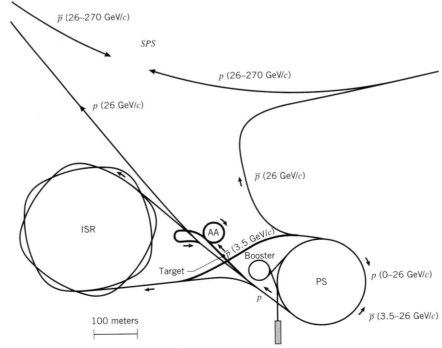

FIGURE 16-10 The proton–antiproton facility at CERN.

the average momentum closer to the central value of 3.5 GeV/c. When a dense enough antiproton beam is accumulated, it is sent to the PS and then into the SPS. Protons and antiprotons are injected into the SPS traveling in opposite directions. Multiple bunches of the two particle types are then accelerated simultaneously to high energy. In addition, the CERN complex has the gigantic 25-km circumference LEP electron–positron colliding beam machine.

Tables 16-1 and 16-2 list some accelerators together with the famous physics discoveries that were made at each facility.

16-2 PARTICLE DETECTORS

The Principles of Particle Detection

Ionization Loss

When a relativistic charged particle passes through matter, it loses energy by its electromagnetic interaction with the atomic electrons in the material through which it is moving. If the incoming particle is not an electron, then collisions with the nuclei are not important because energy cannot be efficiently transferred to the nuclei due to their large masses. For an incoming electron, collisions with nuclei are important because the electron can radiate when it is accelerated. We shall treat the case of an incoming electron separately. Atoms are ionized by the

TABLE 16-1
PARTICLE ACCELERATORS WITH FIXED-TARGET EXPERIMENTS THAT HAVE MADE MAJOR DISCOVERIES IN PARTICLE PHYSICS.
All the machines are synchrotrons, except SLAC which is a linear accelerator.

Name	Location	Type	Energy (GeV)	Major Discovery(s)
Bevatron	Berkeley, CA	Proton	7	Antiproton
SLAC	Stanford, CA	Electron	20	Proton structure
PS	Geneva, Switz.	Proton	26	Neutral currents
AGS	Long Island, NY	Proton	40	Ω, 2 neutrinos, CP violation, J (c quark)
FNAL	Batavia, IL	Proton	500–1000	Υ(b quark)

TABLE 16-2
PARTICLE ACCELERATORS WITH COLLIDING BEAM EXPERIMENTS THAT HAVE MADE MAJOR DISCOVERIES IN PARTICLE PHYSICS.

Name	Location	Type	CM Energy (GeV)	Major Discovery
SPEAR	Stanford, CA	e^+e^-	3–8	ψ (c quark), τ lepton
ISR	Geneva, Switz.	pp	62	high-p_T hadrons
PETRA	Hamburg, Ger.	e^+e^-	15–46	gluon
PEP	Stanford, CA	e^+e^-	15–30	b quark lifetime
CESR	Ithaca, NY	e^+e^-	8–16	B mesons
SPS	Geneva, Switz.	$\bar{p}p$	540–900	W and Z^0
LEP	Geneva, Switz.	e^+e^-	100	3 neutrinos

passage of a charged particle and the energy loss per distance traversed is called ionization loss, $-dE/dx$. In Chapter 11 we used the equation for ionization loss to calculate how much energy charged particles deposit in matter. We now examine in more detail the form of the expression for ionization loss. We shall obtain the essential physics of the expression for ionization loss by considering how much energy is lost by a collision with a single electron at a distance b, and then summing the contributions to all electrons along the path of the incoming particle.

We define the number of electrons encountered per unit distance to be

$$N_x \equiv \frac{dN}{dx}. \qquad (16.24)$$

The number of collisions with electrons is proportional to the density of electrons in the material (n). In going an infinitesimal distance dx, the number of electrons (N_x) with an impact parameter between b and $b + db$ that the particle encounters is given by

$$dN_x = n2\pi b\,db. \qquad (16.25)$$

The electrons are not relativistic, except in very rare collisions. Each electron that is encountered gets a momentum kick (Δp) that is inversely proportional to its impact parameter and inversely proportional to the particle speed (v). We have obtained the expression for the momentum transfer (6.38) in our discussion of Rutherford scattering. The case of a heavy particle of charge e scattering with an electron is Rutherford scattering in the small angle limit. Therefore, the momentum transfer to the electron is

$$\Delta p = \frac{2ke^2}{bv}. \tag{16.26}$$

We define the change in energy of the incoming particle per distance to be

$$E_x \equiv \frac{dE}{dx}. \tag{16.27}$$

The change in energy per collision is

$$\frac{dE_x}{dN_x} = -\frac{(\Delta p)^2}{2m} = -\frac{2k^2e^4}{mb^2v^2}. \tag{16.28}$$

In terms of the impact parameter b, differential distribution for the change in energy is

$$\begin{aligned} \frac{dE_x}{db} &= \frac{dE_x}{dN_x}\frac{dN_x}{db} \\ &= \left(-\frac{2k^2e^4}{mb^2v^2}\right)(n2\pi b). \end{aligned} \tag{16.29}$$

The change in energy of the incoming particle per distance is obtained by integrating over all impact parameters. The maximum impact parameter (b_{max}) corresponds to the minimum energy that can be transferred to an electron. The minimum impact parameter (b_{min}) corresponds to the maximum energy that can be transferred to an electron. The change in energy is

$$\begin{aligned} \frac{dE}{dx} = E_x &= -\frac{4\pi nk^2e^4}{mv^2}\int_{b_{min}}^{b_{max}} db\frac{1}{b} \\ &= -\frac{4\pi nk^2e^4}{mv^2}\ln\left(\frac{b_{max}}{b_{min}}\right). \end{aligned} \tag{16.30}$$

The quantity dE/dx is negative because the particle loses energy when it passes through matter. The quantity $-dE/dx$ is called the *ionization loss*. The important physics of the ionization formula (16.30) is that at small speeds the energy loss per distance is inversely proportional to the square of the speed of the incoming particle.

Slower moving particles ionize more heavily because they spend more time near each electron that they encounter. The ionization loss does not depend on the mass of the particle, only on its speed. The density of electrons in the medium is given by

$$n = \frac{N_A Z\rho}{A\left(10^{-3}\ \text{kg}\right)}. \tag{16.31}$$

Using $\alpha = ke^2/\hbar c$ and $\beta = v/c$, we may write the ionization loss as

$$-\frac{dE}{dx} = \frac{4\pi\alpha^2\left(\hbar c\right)^2}{mc^2\beta^2}\frac{N_A Z\rho}{A\left(10^{-3}\ \text{kg}\right)}\ln\left(\frac{b_{max}}{b_{min}}\right). \tag{16.32}$$

We now examine the limits b_{max} and b_{min}. There is a physical limit on the minimum energy transfer due to quantization of energy levels in the atom. If we define I as the minimum energy that the electron can gain, then we may approximate I to be the potential energy at the distance b_{max}:

$$I \approx \frac{\gamma ke^2}{b_{max}}, \tag{16.33}$$

where γ is the Lorentz gamma factor of the incoming particle. The γ factor appears because the component of electric field due to a moving charge transverse to the direction of motion is proportional to γ. Solving for b_{max},

$$b_{max} \approx \frac{\gamma ke^2}{I}. \tag{16.34}$$

The maximum momentum that can be transferred to the electron is $\Delta p = 2\gamma mv$, corresponding to an energy transfer of $\Delta E = 2\gamma^2 mv^2$. The relationship between energy transferred to the electron and impact parameter is

$$\Delta E = \frac{2k^2e^4}{mv^2b^2}. \tag{16.35}$$

Therefore,

$$2\gamma^2 mv^2 = \frac{2k^2e^4}{mv^2b_{min}^2}. \tag{16.36}$$

Solving for b_{min}, we get

$$b_{min} = \frac{ke^2}{\gamma mv^2}. \tag{16.37}$$

Dividing b_{max} (16.34) by b_{min} (16.37) and taking the logarithm, we have

$$\ln\left(\frac{b_{max}}{b_{min}}\right) \approx \ln\left(\frac{\gamma ke^2}{I}\frac{\gamma mv^2}{ke^2}\right)$$

$$= \ln\left(\frac{mc^2\gamma^2\beta^2}{I}\right). \quad (16.38)$$

The ionization loss (16.32) becomes

$$-\frac{dE}{dx} = \frac{4\pi\alpha^2(\hbar c)^2 N_A Z\rho}{mc^2\beta^2 A(10^{-3}\text{ kg})}\ln\left(\frac{mc^2\gamma^2\beta^2}{I}\right). \quad (16.39)$$

We emphasize that in this formula β corresponds to the speed of the ionizing particle and m is the electron mass. The effective ionization potential per electron varies slightly with the medium type, but is approximately equal to

$$I \approx (16\,\text{eV})Z^{0.9}. \quad (16.40)$$

The formula for the ionization loss (16.39) was obtained by Bohr in 1915 including a correction term of order β^2. The expression (16.39) is not exact because we have not taken into account the quantum mechanical wave nature of the electrons. A quantum mechanical calculation was performed in 1930 by Bethe with the result

$$-\frac{dE}{dx} = \frac{4\pi\alpha^2(\hbar c)^2 N_A Z\rho}{mc^2\beta^2 A(10^{-3}\text{ kg})}\left[\ln\left(\frac{2mc^2\gamma^2\beta^2}{I}\right)-\beta^2\right].$$

$$(16.41)$$

The more precise formula of Bethe is only a small correction to the result of Bohr (16.39).

The value of $-dE/dx$ reaches a minimum when $\beta\gamma$ is approximately equal to 3.5, and then rises slowly. The rise is due to the relativistic effect of the ionizing particle experiencing a larger electric field transverse to its direction of motion. (An increasing portion of the energy loss also goes into Cerenkov radiation. Cerenkov radiation is discussed later in this section.) There is a further correction to be made to the ionization loss (16.41), which is the inclusion of a phenomenological term that opposes the relativistic rise. The origin of this additional term is the polarization of the medium, whereby the fields become so strong that electric charges arrange themselves to cancel the effect of the transverse electric field. This squelching of the relativistic rise is called the *density effect* because in denser materials the polarization effect is much stronger.

The amount of relativistic rise above minimum ionization varies from about 10% in solids to about 50% in gases.

The constant appearing in the ionization formula (16.41) is

$$\frac{4\pi\alpha^2(\hbar c)^2 N_A}{mc^2(10^{-3}\text{ kg})} = 30.7\text{ keV}\cdot\text{m}^2/\text{kg}. \quad (16.42)$$

Thus, we may write the ionization formula as

$$-\frac{dE}{dx} = (30.7\text{ keV}\cdot\text{m}^2/\text{kg})$$

$$\frac{Z\rho}{A\beta^2}\left[\ln\left(\frac{2mc^2\gamma^2\beta^2}{I}\right)-\beta^2\right]. \quad (16.43)$$

Figure 16-11 shows a plot of $-dE/dx$ versus $\beta\gamma$ in liquid argon. The quantity $\beta\gamma$ is proportional to the particle momentum. The density of liquid argon is 1.4×10^3 kg/m^3. The minimum value of $-dE/dx$ is about 210 MeV/m. The shape of this ionization loss curve is universal for all materials. The value of minimum ionization is in the range

$$\left(-\frac{dE}{dx}\right)_{min} \approx (0.1-0.2\text{ MeV}\cdot\text{m}^2/\text{kg})\rho. \quad (16.44)$$

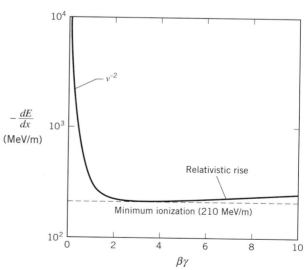

FIGURE 16-11 Formula for ionization loss (−dE/dx).
The energy loss by ionization of a charged particle passing through matter is a universal function depending on the velocity of the particle and the density of electrons in the medium. The universal function $-dE/dx$ is plotted versus $\beta\gamma$ of the ionizing particle and is normalized for liquid argon, which has a density of 1.4×10^3 kg/m^3.

Table 16-3 gives $(-dE/dx)_{min}$ for several materials. Figure 16-12 shows the ionization loss as a function of electron density for several different materials.

In the measurement of dE/dx, there are fluctuations about the average value that are caused by a small number of collisions that cause large energy transfers. In a thin layer of material where the energy loss is less than 1 MeV, the distribution of dE/dx from several measurements will be asymmetric, with some measurements giving large energy losses, referred to as the *Landau tail*. In a thick layer where the energy loss is much greater than 1 MeV, the fluctuations in the energy loss will be nearly Gaussian.

The momentum of a particle of known charge may be determined by its curvature in a magnetic field. If the average value of dE/dx is also measured, the particle speed is determined and therefore its mass is known. Figure 16-13 shows the measurement of ionization loss versus momentum for a sample of particles that were produced in electron–positron collisions at a center-of-mass energy of 29 GeV. The average ionization loss was measured in a gas mixture of 80% Ar and 20% CH_4 at 8.5 atmospheres. Superimposed on the data are the universal curves for ionization loss expected for electrons, muons, pions, kaons, and protons.

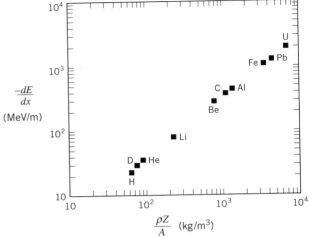

FIGURE 16-12 Ionization loss as a function of electron density.
When a charged particle passes through matter, the average energy loss by electromagnetic interaction with electrons is proportional to the density of electrons. The measured average ionization loss is plotted versus $\rho Z/A$ for various materials from liquid hydrogen to uranium.

TABLE 16-3
PROPERTIES OF SELECTED MATERIALS.

Material	Density (kg/m³)	Int. length (m)	Rad. length (m)	$(dE/dx)_{min}$ (MeV/m)
Air	1.205	747.	304.	0.2
Liq. H_2	70.8	7.18	8.65	29.
Liq. He	125.	5.21	7.55	24.
Liq. D_2	162.	3.38	7.57	34.
H_2O	1000	0.849	0.361	203.
Be	1848	0.407	0.353	298.
C	2265	0.381	0.188	403.
Concrete	2500	0.400	0.107	425.
Al	2700	0.394	0.24	437.
Fe	7870	0.168	0.0176	1160
Pb	11350	0.171	0.0056	1280
U	18950	0.105	0.0032	2070

Data from the "Review of Particle Properties," *Phys. Lett.* **B204** (1988).

EXAMPLE 16-7

Estimate the order of magnitude of the range of a 5-MeV alpha particle in air.

SOLUTION:

The speed of a 5-MeV alpha particle is given by

$$E_k = \frac{mv^2}{2},$$

or

$$\frac{v}{c} = \sqrt{\frac{2E_k}{mc^2}} \approx \sqrt{\frac{10\,\text{MeV}}{3700\,\text{MeV}}} \approx 0.05.$$

The alpha particle is heavily ionizing because of its small speed. The rate of energy loss by ionization is

$$-\frac{dE}{dx} = \left(30.7\ \text{keV} \cdot \text{m}^2/\text{kg}\right)$$

$$\frac{Z\rho}{A\beta^2}\left[\ln\left(\frac{2mc^2\gamma^2\beta^2}{I}\right)-\beta^2\right].$$

The logarithmic factor is

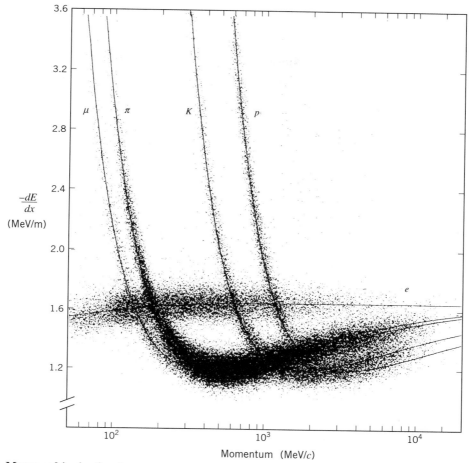

FIGURE 16-13 Measured ionization loss versus momentum.
The particles were produced in electron–positron collisions at a center-of-mass energy of 29 GeV. The ionization was measured in a gas mixture of 80% Ar and 20% CH_4 at 8.5 atm. For each particle type, the ionization is expected to follow the universal dE/dx curve. The solid lines indicate expectations for electrons, muons, pions, kaons, and protons. From H. Aihara et al., *Phys. Rev. Lett.* **61**, 1263 (1988).

$$\ln\left(\frac{2mc^2\gamma^2\beta^2}{I}\right)$$

$$= \ln\left[\frac{(2)(0.5\,\text{MeV})(1)^2(0.05)^2}{(16\,\text{eV})(7)^{0.9}}\right] \approx 3.$$

The density of air is about 1 kg/m³. For air,

$$\frac{Z}{A} = 0.5.$$

The initial rate of ionization loss is

$$-\frac{dE}{dx} = \left(30.7\ \text{keV} \cdot \text{m}^2/\text{kg}\right)(0.5)$$

$$\left[\frac{1\,\text{kg/m}^3}{(0.05)^2}\right](3) \approx 20\,\text{MeV/m}.$$

The rate of ionization loss increases rapidly as the alpha particle loses energy. Therefore the range of the alpha particle in air is only a few millimeters. ∎

Cerenkov Radiation
The speed of light in a medium with index of refraction n is c/n. A charged particle may have a speed up to the

limiting value of the speed of light in vacuum (c) and can therefore go faster than the speed of light in a medium. When a charged particle goes faster than the speed of light in a medium, a portion of the electromagnetic radiation emitted by excited atoms along the path of the particle is coherent. The coherent radiation is emitted at a fixed angle with respect to the particle trajectory. This radiation was first observed by Pavel Cerenkov in 1935. The condition for the emission of Cerenkov radiation is

$$v > \frac{c}{n}. \qquad (16.45)$$

Using $\beta = v/c$ we have

$$\frac{1}{\beta n} < 1. \qquad (16.46)$$

The wave front of Cerenkov radiation is shown in Figure 16-14. The Cerenkov light is emitted at a characteristic angle (θ) given by

$$\cos \theta = \frac{1}{\beta n}. \qquad (16.47)$$

The energy spectrum of the Cerenkov radiation is continuous, and a significant fraction appears in the visible region. The total amount of energy appearing in Cerenkov radiation is small compared to the total ionization loss. For a relativistic particle ($\beta \approx 1$) in water, the portion of

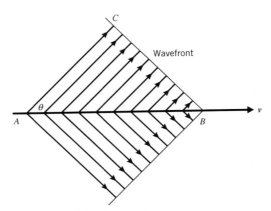

FIGURE 16-14 Cerenkov radiation.
Cerenkov radiation occurs when a charged particle travels faster than the speed of light in a medium characterized by the index of refraction (n). The charged particle travels from point A to point B with the speed v and light travels from point A to point C with a speed c/n, where $v > c/n$. Coherent radiation is emitted at an angle given by $\cos\theta = 1/\beta n$.

ionization loss that appears in visible Cerenkov radiation $(-dE/dx)_C$ is

$$\left(\frac{-dE}{dx} \right)_C \approx 40 \, \text{keV/m}, \qquad (16.48)$$

compared to 200 MeV/m total ionization loss. The fraction of ionization loss appearing as visible Cerenkov radiation is only about 1/5000; however, the presence of Cerenkov radiation provides a signature of the speed of a particle.

EXAMPLE 16-8
The index of refraction of water is $n = 1.33$. (a) Determine the energy threshold for an electron to cause Cerenkov radiation in water. (b) Calculate the angle of emission of Cerenkov radiation for a 500-MeV electron in water.

SOLUTION:
(a) The speed threshold for Cerenkov radiation is

$$v = \frac{c}{n},$$

or

$$\beta = \frac{v}{c} = \frac{1}{n}.$$

The energy threshold given by

$$E = \gamma mc^2 = \frac{mc^2}{\sqrt{1 - \beta^2}} = \frac{mc^2 n}{\sqrt{n^2 - 1}}.$$

Numerically, we have

$$E = \frac{(0.511 \, MeV)(1.33)}{\sqrt{(1.33)^2 - 1}} = 0.775 \, \text{MeV}.$$

Cerenkov radiation occurs for electrons in water when $E > 0.775$ MeV.
(b) The Cerenkov angle is given by

$$\cos \theta = \frac{1}{\beta n}.$$

A 500-MeV electron has β near unity. Therefore,

$$\theta = \arccos\left(\frac{1}{n}\right) = 0.720,$$

or about 41.2 degrees.   ∎

Figure 16-15 shows a photomultiplier tube specially developed to detect Cerenkov radiation.

Transition Radiation

When a relativistic charged particle traverses the boundary between two media, a small amount of electromagnetic radiation may also be emitted. This radiation is called transition radiation. The radiation is caused by the time-dependent electric field of the moving particle, which creates a time-dependent polarization in the region of the particle's trajectory. The transition radiation may be thought of as arising from the apparent acceleration of the particle due to the change in index of refraction at the media boundary. The transition radiation is coherent and is concentrated at an angle equal to $1/\gamma$ with respect to the

FIGURE 16-15 Photomultiplier tube with large-area photocathode.

This photomultiplier was developed for use in the Kamiokande experiment to search for proton decay. Several hundred of these tubes were used to monitor a tank containing several kilotons of water. The tubes were designed to detect the Cerenkov radiation from the decay products of an unstable proton. Courtesy Hamamatsu Corp.

particle direction. The probability of photon emission per transition is roughly equal to α (1/137). The energy spectrum of the transition photons is characterized by a quantity called the *plasma frequency*. The plasma frequency is proportional to the square root of the electron density:

$$\omega_p = \sqrt{\frac{4\pi\alpha^2 c^2 (\hbar c)^2 N_A Z\rho}{mc^2 A(10^{-3} \text{ kg})}}. \qquad (16.49)$$

The maximum energy (E_γ) of transition photons is approximately

$$E_\gamma \approx \gamma\hbar\omega_p. \qquad (16.50)$$

For materials of density equal to 10^3 kg/m³

$$\hbar\omega_p \approx 20\,\text{eV}. \qquad (16.51)$$

For values of γ greater than about 100, x rays are produced. The total energy (E_{tot}) emitted per transition is approximately

$$E_{tot} \approx \frac{\gamma\hbar\omega_p\alpha}{3}, \qquad (16.52)$$

which is a very small amount of energy compared to the total ionization energy loss.

Bremsstrahlung

When electrons or positrons pass through matter, they lose energy by ionization like all other charged particles. In addition, the electron and positron lose energy by bremsstrahlung ("braking radiation") as a consequence of their small mass. The bremsstrahlung process is caused by an electromagnetic interaction with an atomic nucleus in which a photon is radiated from the accelerated electron or positron,

$$e^- + N \rightarrow e^- + N + \gamma, \qquad (16.53)$$

or

$$e^+ + N \rightarrow e^+ + N + \gamma. \qquad (16.54)$$

Collisions with nuclei are much more effective in producing bremsstrahlung than collisions with electrons because the interaction probability is proportional to the electric charge squared. Since the power radiated (B.30) is proportional to the square of the acceleration, energy loss by bremsstrahlung is inversely proportional to the particle mass squared. For electrons, the bremsstrahlung process completely dominates dE/dx at energies larger than a few MeV. For the next heaviest particle, the muon, the bremsstrahlung probability is lower than that for electrons by a factor of more than 10^4 and is not important except for muons with energies in the TeV range or greater.

Materials are characterized by their radiation lengths (χ_0), defined to be the distance in which the energy of an energetic electron is reduced to a factor of $1/e$ of the incident energy due to bremsstrahlung:

$$\frac{dE}{E} = -\frac{dx}{\chi_0}. \qquad (16.55)$$

The radiation length of a material is very nearly inversely proportional to both Z and ρ. The relative importance of ionization and bremsstrahlung for electrons in lead is shown in Figure 16-16. The critical energy (E_c), the energy above which the radiation process dominates over ionization and Compton effect, is given by

$$E_c = \frac{600 \text{ MeV}}{Z}. \qquad (16.56)$$

For lead $(Z = 82)$, the critical energy is about 7 MeV. Radiation lengths for several materials are given in Table 16-3.

Pair Production
When photons pass through matter they interact with electrons by the photoelectric effect and Compton scatter-ing. When the photon energy is above a few MeV, a third process dominates the photon interaction. The photon may convert to an electron–positron pair:

$$\gamma + N \rightarrow e^+ + e^- + N. \qquad (16.57)$$

The pair production process is closely related to that of bremsstrahlung. When a photon converts to an e^+e^- pair, the photon has disappeared from the universe so that a beam of photons will be attenuated when passing through matter. The attenuation factor is observed to be $1/e$ when traversing a distance of $(9/7)\chi_0$. Thus, the characteristic length-scale for the interaction of electrons, positrons, and photons in matter is the radiation length of the material, χ_0. The cross sections for photoelectric effect, Compton scattering, and pair production are shown in Figure 16-17.

Nuclear Interactions
When an energetic hadron (i.e., a pion, kaon, neutron, or proton) passes through matter, it will ultimately have a strong interaction with a nucleus. At high energies, all hadron–hadron cross sections are approximately equal. The hadron–hadron total cross sections at high energy are given in Table 16-4. A material may be characterized by its interaction length (λ_0), the distance through which a 100–GeV neutron has a probability $1/e$ of not having an inelastic interaction with a nucleus. In a given material, the value of λ_0 is much larger than the value of χ_0, except for the lightest elements. Interaction lengths of materials are given in Table 16-3.

Weak Interactions
Neutrinos are the only known stable particles that produce none of the following effects when passing through matter: ionization loss, Cerenkov radiation, transition radiation, bremsstrahlung, pair production, or nuclear interactions. Matter is *almost* transparent to the neutrino. We say "almost" because the neutrino does have a weak interaction. The weak interaction can be detected with only a tiny efficiency even using a gigantic detector, unlike the other particles, which can be detected with 100% probability in a relatively small chunk of matter. In practice this means that a neutrino in the final state of an interaction cannot be directly detected. The interaction cross section for a 100-GeV neutrino is about 9 orders of magnitude smaller than that of a neutron. Thus, it would take about 10^9 m of concrete to absorb such a neutrino! In some cases the neutrino can be inferred to exist by invoking conservation of momentum. Neutrinos can be directly detected only when intense beams are incident on large detectors.

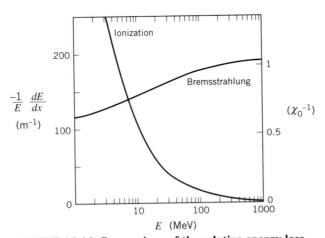

FIGURE 16-16 Comparison of the relative energy loss by ionization and bremsstrahlung for electrons in lead. Below about 7 MeV, ionization dominates and above 7 MeV bremsstrahlung dominates. Adapted from the Particle Data Group, "Review of Particle Properties," *Phys. Rev.* **D45**, 1 (1992).

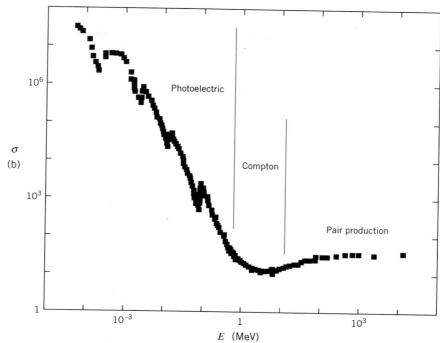

FIGURE 16-17 The absorption of photons in lead.
At low energies the photoelectric effect dominates and the cross section varies as approximately
E^{-3}. Near 1 MeV the Compton scattering process dominates and the cross section varies as $1/E$. At
large energies, pair production dominates and the cross section is roughly constant. Adapted from the
Particle Data Group, "Review of Particle Properties," *Phys. Rev.* **D45**, 1 (1992).

Components of Particle Detectors

An ionization detector consists of an anode and a cathode
with a large voltage difference between them. The space
between the anode and the cathode is filled with a gas or
liquid medium. When a charged particle passes through
the detector, atoms from the medium are ionized; the freed
electrons and ions drift in the applied electric field, resulting in an electric current. If the voltage is low, the current
is proportional to the voltage (see Figure 16-18). At larger
voltages a plateau is reached. At still higher voltages,
secondary ionization is produced, resulting in a multiplication factor and corresponding current that is again
proportional to the voltage.

Geiger Counters

The simplest ionization counter is the Geiger-Müller tube
or *Geiger counter,* invented in 1908 by Hans Geiger and
subsequently modified by Wilhelm Müller. A Geiger
counter consists of an electrically isolated wire inside a
tube filled with gas, usually an argon–alcohol mixture (see
Figure 16-19). The wire is placed at positive high voltage,

typically 1000 V, relative to the tube wall. When a charged
particle enters the counter, ionization is produced in the
gas and a discharge occurs. The alcohol serves to quench
this discharge resulting in a current that is independent of
the amount of primary ionization. The detection efficiency for a charged particle is 100%, although the counter
has a dead time of about 100 μs after each count while the
ionization is cleared from the counter by the electric field.
The signal is amplified and usually fed into a circuit that
produces an audible "click." Geiger counters may be used
for neutron detection by filling them with boron trifluoride.
The boron nucleus has a large cross section for producing

TABLE 16-4
HADRONIC CROSS SECTIONS AT 100 GEV.

Process	Cross section (mb)
$\pi^+ p$ or $\pi^- p$	25
$K^+ p$ or $K^- p$	20
pp, np, or $\bar{n}p$	40

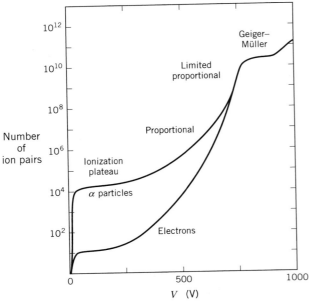

FIGURE 16-18 Ionization in a gas counter.
The number of ion pairs created by α particles and electrons upon passing through a gas with an applied voltage V. Fashioned after K. Kleinknecht, *Detectors for Particle Radiation*, Cambridge University Press (1986).

an alpha particle when struck by a slow neutron. Geiger counters are made sensitive to γ-ray photons by making the cathode tube out of a large-Z metal such as lead so the photons may convert to e^+e^-, typically with 1% efficiency, upon entering the counter.

Cloud Chamber

The *cloud chamber* invented by C. T. R. Wilson (see Chapter 1) was the first in a series of charged-particle detectors operating on a similar principal: The trajectory of a charged particle is precisely measured by detection of

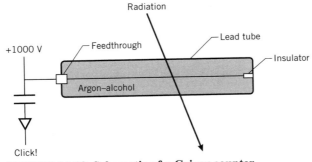

FIGURE 16-19 Schematic of a Geiger counter.

the ions it produces and its curvature in a magnetic field is used to determine both the sign of the electric charge and the component of particle momentum transverse to the field. The result is a detailed picture of the trajectory of all charged particles in the chamber in a given interaction. The cloud chamber was used to discover antimatter (the positron), the second lepton family (the muon), and the strange quark (the kaon and the lambda).

Emulsions

As early as 1910, it was discovered that a *photographic emulsion* containing silver halide grains was sensitive to ionizing radiation. After development, a photographic plate will have a trail of silver grains along the charged particle trajectory that may be observed under a microscope. A handy method for confirming the cross-sectional area and location of an incoming beam in a fixed target experiment is to place a piece of Polaroid film in the beam for a few pulses. Ordinary photographic film is not suitable for charged-particle tracking because the efficiency is low and the emulsions are only a few microns thick. In the 1930s, emulsions were specifically prepared for charged-particle detection that were thicker and contained a greater density of silver halides. C. F. Powell and G. P. S. Occhialini were pioneers in this effort; in collaboration with Ilford Laboratories they developed the Nuclear Research emulsion, which they used to discover the charged pion (together with C. Lattes and H. Muirhead) and observe its weak interaction decay. In 1948, the first emulsion sensitive to minimally ionizing particles was developed at Kodak Laboratories.

Bubble Chambers

The *bubble chamber* was invented in 1952 by D. Glaser. Its principle of operation is similar to that of the cloud chamber. A pressurized liquid such as hydrogen is brought to a superheated state by an abrupt lowering of the pressure. Bubbles are formed preferentially around the ions created by the passage of charged particles. The bubbles are photographed and measurement of their position gives a measurement of the charged particle trajectories to an accuracy of about 30 µm.

The liquid in the bubble chamber also serves as a target for the interaction to be studied. Bubble chamber liquids that have been used include hydrogen, deuterium, helium, propane, freon, neon-hydrogen, and xenon. Large bubble chambers have been built to measure neutrino interactions such as the 38 m³ Big European Bubble Chamber (BEBC) and the 32 m³ 15 Foot Bubble Chamber at Fermilab. The bubble chamber was the single most powerful instrument in the era 1950–70 for the discovery and classification of the hadrons. In 1973, the first examples of *neutral currents*, neutrino interactions by Z^0 exchange, were discovered in the Gargemelle bubble chamber at CERN. The main disadvantage of bubble chambers are their low repetition rate, typically 1 Hz. Furthermore, the analysis of bubble chamber data is a very lengthy process, making it impractical to analyze more than 10^5 pictures from a given experiment.

Spark and Streamer Chambers

In a spark chamber, a large electric field (~1 MV/m) is suddenly applied over a small gap resulting in a discharge or "spark" at the location of the ionization trail. The sparks may be photographed or recorded electronically. The first spark chambers were built by M. Conversi and A. Gozzini in 1955. Large arrays of spark chambers may be used to track particles over distances of several meters. In the streamer chamber, a large electric field (~1.5 MV/m) is rapidly applied over a large volume. The ionization trails, or *streamers*, left by charged tracks may be photographed.

Proportional and Drift Chambers

In 1968, Georges Charpak invented the *multiwire proportional chamber* (MWPC). The MWPC contains a plane of anode wires with regular spacing (typically, one to a few millimeters). Ionization is collected on the wire nearest the particle path. Each wire has separate electronics and the spatial resolution is determined by the wire spacing. The wires operate with high gain so relatively inexpensive electronics may be used. This is necessary because of the large number of wires in such a chamber. The amplification within the chamber comes from the very large electric field within 100 μm of a wire. The drifting electrons are accelerated and produce ionization cascades. This multiplication is called *wire gain* and is typically a factor of 10^5.

In a *drift chamber*, also developed by Charpak, ionization is collected over distances ranging from a fraction of a few millimeters to several centimeters. Spatial resolution is achieved by measurement of the time of arrival of the charge. Argon-based gas mixtures have been widely used in drift chambers because the electron drift velocity, about 50 μm/ns, is almost independent of the electric field strength.

Silicon Detectors

An ionization detector may also be made out of a semiconductor such as silicon. One form of the *silicon detector* consists of a large number of *n–p* diodes in a geometry of parallel narrow strips. A strip in a silicon detector plays the role of a wire in a proportional chamber. The diodes are reverse biased and are made thin so that they are fully depleted. When a charged particle passes through the silicon, the ionization is collected. The resolution can be very good (< 10 μm) because the strips can be made close together.

Table 16-5 summarizes the resolution of tracking chambers. Table 16-6 summarizes the major discoveries made by tracking particles in a magnetic field.

Scintillation Counters

Scintillation counters are detectors for charged particles. They consist of scintillating material, usually a doped plastic, that emits light in response to molecular excitations by the passage of a charged particle. The scintillation light may be detected with photomultipliers or photodiodes. The typical light yield in a plastic scintillator in the range sensitive to photon detectors is 10^4 photons per MeV of deposited energy. Because the light yield is large, scintillation counters have a single particle detection efficiency of nearly 100%. The high detection efficiency makes them useful as a veto counter, an application where one needs to confirm the absence of a charged particle.

Scintillation counters range in size from very small (a few square millimeters) to very large (a few square meters). One important feature of scintillation counters is their speed, which is in the nanosecond range. Scintillation counters are often designed to be rugged. In colliding beam experiments, scintillation counters are used to detect interactions of the primary particles. In some experiments they are used to accurately measure the time of arrival and therefore the speed of a particle. In other applications they are used to measure energy.

**TABLE 16-5
CHARACTERISTIC SPATIAL RESOLUTION OF
TRACKING CHAMBERS.**

Technique	Resolution (μm)
Emulsion	1
Silicon	10
Bubble	10–150
Drift	50–300
Streamer	300
Proportional	300
Spark	500
Cloud	500

TABLE 16-6
MAJOR DISCOVERIES MADE BY TRACKING
PARTICLES IN A MAGNETIC FIELD.

Technique	Year	Discovery
Phosphorescence	1897	Electron
Cloud chamber	1932	Positron
	1937	Muon
	1947	Strange particles
Bubble chamber	1950–70	Classification of the hadrons
	1973	Neutral currents
Spark chamber	1974	Charm quark
	1977	Tau lepton
Proportional chamber	1977	Bottom quark
Drift chamber	1978	Gluon
	1983	W^+, W^-, and Z^0 bosons

Cerenkov Counters

Cerenkov counters are used to gather information on the speed of a particle. When combined with a measurement of momentum, this reveals the mass or identity of a particle. Cerenkov counters are primarily used to separate pions, kaons, and protons. A Cerenkov counter is made by filling a large container with an appropriate gas and mirrors to image the light onto a photomultiplier cathode. There are two principal types of Cerenkov counters: threshold counters designed to detect light above some speed, and ring-imaging Cerenkov counters designed to measure the angle of emission of the light.

Electromagnetic Calorimeters

Calorimeters are devices for measurement of the total energy of a particle. There are two types of calorimeters, electromagnetic and hadronic. *Electromagnetic calorimeters* are used to measure electron, positron, and photon energies by means of their electromagnetic interactions with nuclei. The interaction processes are bremsstrahlung,

$$e^+ + N \rightarrow e^+ + N + \gamma, \quad (16.58)$$

or

$$e^- + N \rightarrow e^- + N + \gamma, \quad (16.59)$$

and pair production

$$\gamma + N \rightarrow e^+ + e^- + N. \quad (16.60)$$

Once initiated, there is an electromagnetic cascade or *shower* of bremsstrahlung alternating with pair production. These are the dominant mechanisms by which high-energy electrons, positrons, and (gamma ray) photons lose energy when passing through matter (see Figures 16-16 and 16-17). The net result of the shower process is the conversion of the kinetic energy of the incident particle into a large number of electrons, positrons, and photons. The number of electrons, positrons, and photons produced is directly proportional to the energy of the incident particle. In an electromagnetic calorimeter, the energy is determined by measurement of the amount of ionization (dE/dx) produced by the charged particles in the shower. The number of electrons produced at the peak of a 10-GeV shower in lead is about 10^3.

Figure 16-20 shows an electromagnetic shower measured in a cloud chamber that contains lead plates of thickness 0.0127 meter. The shower is initiated by gamma ray passing through the top lead plate. The energy of the incoming particle is a few GeV. The trajectories of the electrons and positrons that are produced in the shower are measured in the cloud chamber. (The gamma rays leave no tracks in the cloud chamber.) The number of electrons and positrons is multiplied in the second and third lead plates by the bremsstrahlung and shower processes. In the multiplication process the average energy of each electron and positron decreases until finally there is insufficient energy to penetrate the lead plates.

The electron radiation probability varies as the square of the acceleration, and the force that causes the acceleration is proportional to the atomic number of the medium (Z). Therefore, heavy nuclei such as lead ($Z = 82$) are frequently used as absorbers in electromagnetic calorimeters. The radiation length of lead is about 5.6 mm. Electromagnetic calorimeters must have sufficient material, typically 15–25 radiation lengths, to completely absorb electron and photon showers.

An electromagnetic calorimeter consists of an absorber as well as a mechanism for detection of the resulting ionization. There are two main techniques in use for detection of this ionization: (1) direct collection by drifting electrons in an electric field and (2) the scintillation process. In some devices a single material plays the role of both shower initiator and ionization detector. Such devices are called *total absorption* calorimeters. An example of a material that is used in a total absorption electromagnetic calorimeter is NaI. Sodium iodide has a radiation length of 0.026 m and produces scintillation light in

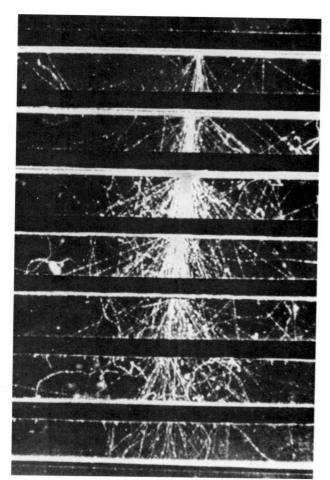

FIGURE 16-20 Electromagnetic shower observed in a cloud chamber.
An energetic photon enters the chamber at the top and interacts in a lead plate of thickness 0.0127 m. An electromagnetic shower of e^+e^- pair production and photon brehmsstrahlung is initiated and several electrons and positrons are observed to emerge from the first plate. These electrons and positrons interact in the second plate and the shower is multiplied. The shower reaches a maximum (between the third and fourth plates) when the brehmsstrahlung photons no longer have enough energy to produce additional e^+e^- pairs. Photograph by C. Y. Chao. From B. Rossi, *High-Energy Particles*, Prentice-Hall (1952).

response to the ionization. The total light yield is about 4×10^4 photons per MeV deposited. The electromagnetic energy resolution of a NaI calorimeter can be exceptionally good:

$$\frac{\Delta E}{E} \approx \frac{0.02 \text{ GeV}^{1/4}}{E^{1/4}}. \tag{16.61}$$

It is not necessary to detect all of the ionization in order to measure the energy of a showering particle. One can sample the ionization at regular intervals, for example, every 1/4 radiation length. Such a device is called a sampling calorimeter. A sampling calorimeter consisting of a repeated sandwich of plastic scintillator (a few millimeters thick) and lead (1 millimeter thick) has an energy resolution of approximately

$$\frac{\Delta E}{E} \approx \frac{0.17 \text{ GeV}^{1/2}}{\sqrt{E}}. \tag{16.62}$$

There are many other considerations in the practical design of an electromagnetic calorimeter such as dynamic range, linearity, radiation hardness, and segmentation (just to mention a few!). Particle physicists have ingeniously constructed electromagnetic calorimeters out of many different materials but the principle is always the same: absorption by bremsstrahlung and pair production and measurement of the ionization produced by the resulting electrons.

Hadronic Calorimeters

Hadronic calorimeters, as their name implies, are used to measure hadron energies. The hadrons that live long enough to enter a calorimeter are the proton, charged pion, charged kaon, neutron, and long-lived neutral kaon. The process is similar to that which occurs in an electromagnetic calorimeter except that the shower of a hadron is initiated by its strong interaction. The absorbing material in a hadron calorimeter is specified by its interaction length (λ_0). In a given material, the value of λ_0 is much larger than the value of χ_0 except for the lightest elements (see Table 16-3). The nuclear interaction produces additional hadrons, including both charged and neutral pions, protons, neutrons, and nuclear fragments that all generate further interactions. The net result is the creation of a large number of ionizing particles that may be detected. Since the neutral pions decay promptly into two photons and these photons generate showers by pair production and bremsstrahlung, the hadronic shower has an electromagnetic component.

Iron, which has an interaction length of about 0.17 meter is commonly used as an absorber in hadronic calorimeters. Hadronic calorimeters must have several interaction lengths, typically about 10, to absorb strong interactions. Because of the relatively large values of λ_0, hadron calorimeters are almost always sampling devices. The most common hadron calorimeter has been a sandwich of iron and plastic scintillator. Such a device typically has a resolution of

$$\frac{\Delta E}{E} \approx \frac{1}{\sqrt{E}} \, \text{GeV}^{1/2}. \qquad (16.63)$$

In order to achieve the best resolution and linearity at high energies, the hadron calorimeter must be designed to have a uniform response to both hadronic and electromagnetic shower components. This is important because the number of π^0s fluctuates from shower to shower. A device with such a uniform response to electrons and hadrons is called a *compensated* calorimeter. Compensated calorimeters have been built that have an energy resolution that is a factor of 3 better than noncompensated calorimeters.

Particle Spectrometers

The only known absolutely stable particles are the photon, the electron and positron, the neutrinos and antineutrinos, and the proton and antiproton. All other particles are observed to have finite lifetimes, and eventually decay, often in multiple stages, into two or more of these particles. For example, the decay chain of a charged pion is

$$\pi^+ \rightarrow \mu^+ + \nu_\mu \qquad (16.64)$$

followed by

$$\mu^+ \rightarrow e^+ + \nu_e + \overline{\nu}_\mu. \qquad (16.65)$$

Table 16-7 lists those particles that live long enough to be directly detected before decaying. A useful parameter is the particle lifetime times the speed of light ($c\tau$). A relativistic particle travels a typical distance $\gamma c\tau$ before decaying. The photon, electron, and muon are detected only by their electromagnetic interactions. The charged pion, kaon, and proton are detected through both their electromagnetic and strong interactions. The neutron and K^0_L are detected only by the strong interaction. These interactions are strong enough so that every particle listed interacts provided the detector is large enough. Detection efficiencies can be nearly 100%. The neutrino interacts so weakly with matter that a detector larger than the size of the earth would be required to have a reasonably large detection efficiency at an energy of 1 GeV. Not included in the list are very short-lived particles such as charm and bottom mesons, which in certain cases may be directly detected in emulsions.

The particles that live long enough to traverse several meters across a particle detector are the electron, photon, proton, charged pion, neutron, long-lived neutral kaon, muon, and neutrino. All other known particles decay into these particles. A modern high-energy physics detector is

TABLE 16-7
DETECTING THE VARIOUS PARTICLE TYPES.
The particles listed have lifetimes long enough to traverse a particle detector (typical distance scale of meters). All the hundreds of other known particles decay into these particles. Neutrino interactions are not directly detected as final state particles because their interaction is too weak.

Particle	$c\tau$	Detection method
γ	Stable	Electromagnetic calorimeter
e^+, e^-	Stable	Tracking and electromagnetic calorimeter
μ^+, μ^-	6.59×10^2 m	Tracking with penetration of many interaction lengths
π^+, π^-	7.80 m	Tracking and hadronic calorimeter
K^+, K^-	3.71 m	Tracking and hadronic calorimeter
K^0_L	15.5 m	Hadronic calorimeter
p	Stable	Tracking and hadronic calorimeter
n	2.69×10^{11} m	Hadronic calorimeter
ν	Stable	Momentum conservation

designed to distinguish these particles by their interactions in matter. A "picture book" of the interactions of these eight particles is given in Figure 16-21. The components of our ideal detector are (1) charged particle tracking in a magnetic field to determine momentum, (2) electromagnetic calorimeter to measure photon and electron energy, (3) hadronic calorimeter to measure hadron energy, and (4) muon chambers.

Most experiments require the identification of relatively rare primary particles (electrons, muon, and photons) in the presence of a very large background of hadrons (mainly pions). The electron is tracked in a magnetic field where its charge-sign and momentum are measured by magnetic curvature. The electron creates a shower in the electromagnetic calorimeter, where its energy is measured. The shower does not penetrate into the hadronic calorimeter. The energy from the calorimeter is equal to pc from the magnetic measurement. The photon is identified by the presence of an electromagnetic shower together with the absence of ionization in the tracking chamber. The shower does not penetrate into the hadronic

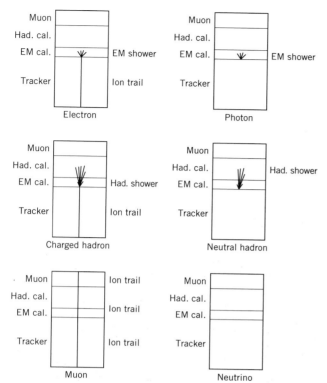

FIGURE 16-21 Picture book of high-energy particle interactions.

A particle detector for high energy physics typically contains the following components: particle tracking in a magnetic field, electromagnetic calorimeter, hadronic calorimeter, and muon chambers. In a set of six pictures we indicate the interaction of the various particle types in an idealized detector. In each case, the particle enters the detector from the bottom of the picture.

calorimeter and the energy is measured in the electromagnetic calorimeter. The muon is identified by its lack of electromagnetic or hadronic shower. An ionization trail is observed throughout the detector. The charge and momentum of the muon are measured by magnetic curvature. The charged hadron is identified by its hadronic shower. The charge and momentum of the hadron are measured by magnetic curvature. The neutral hadron is identified by its hadronic shower together with the absence of a track. The neutrino has no interaction in the detector. In a detector that covers nearly all of the solid angle, the presence of a neutrino may be inferred by momentum conservation.

The Fermilab Multi-Particle Spectrometer

Figure 16-22 shows an example of a fixed target detector, the Multi-Particle Spectrometer (MPS) at Fermilab. This spectrometer is used to study the structure of hadrons. The

incoming beam is either positive (p, K^+, π^+) or negative ($\bar{p}$, K^-, π^-) with an energy of 200 GeV, and the target is liquid hydrogen. Following the hydrogen target are multiwire proportional chambers for charged particle tracking, a large acceptance superconducting magnet, and additional proportional chambers and spark chambers. This is followed by a large multicell Cerenkov counter, more spark chambers, and then lead-scintillator electromagnetic and iron-scintillator hadronic calorimeters. The calorimeters are placed at ±90° in the center of mass, which for a 200-GeV beam corresponds to laboratory angles of ±100 mr. The calorimeters are also used as an *event trigger*, the process by which certain rare events are selected for electronic recording. The Fermilab Multi-Particle Spectrometer was used to make the first measurements of the quark–quark scattering cross section.

The UA1 Spectrometer

Figure 16-23 shows the layout of the UA1 spectrometer at CERN (see also color plate 16). This detector was used to study high-energy proton–antiproton collisions. The UA1 detector is a 2.4 kiloton spectrometer covering the full solid angle around the interaction point. The spectrometer magnet is a dipole with a field of 0.7 tesla. Surrounding the beam line is a 6-meter-long drift chamber (see Figure 16-23b and color plate 17). The drift chamber is followed by a lead-scintillator electromagnetic calorimeter (Figure 16-23c), an iron-scintillator hadronic calorimeter (Figure 16-23c) and finally muon chambers (Figure 16-23d). The principal event triggers required the presence of an electron, muon, or jet. The UA1 Spectrometer was used to discover the W and Z^0 particles, as well as to measure many properties of the strong interaction such as the search for quark substructure described in Chapter 6. (The discovery of the W and Z^0 particles is described in Chapter 18.)

CHAPTER 16: PHYSICS SUMMARY

- The physics accessible with a given accelerator depends on the particle type, energy, and luminosity (flux). Invention of new accelerator technologies has provided an exponential growth in energy over several decades.

- The largest center-of-mass energies are obtained with colliding beam machines. The performance of high-energy electron–positron colliders is limited by synchrotron radiation. The performance of high-energy hadron colliders is limited by the strength of the bending magnets.

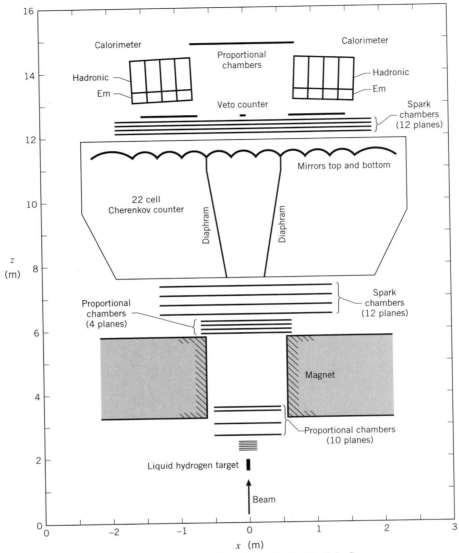

FIGURE 16-22 Example of a fixed-target detector, the Fermilab Multi-Particle Spectrometer (MPS).
The MPS was configured to study high-transverse-momentum hadrons. The main features are (1) liquid hydrogen target, (2) proportional chambers for tracking, (3) superconducting magnet with large "picture window" acceptance, (4) spark chambers with magnetostrictive readout for tracking, (5) Cerenkov counters for particle identification, and (6) electromagnetic and hadronic calorimeters for energy measurement. From the author's Ph.D. Thesis, California Institute of Technology (1980).

- Charged particles lose energy when they pass through matter by their interaction with the atomic electrons in the matter. Most of this energy loss goes into ionization of the atoms along the trajectory of the charged particles. The energy loss of a charged particle passing through matter due to ionization is proportional to the density of electrons in the matter and is roughly equal to $(0.1-0.2$ MeV·m^2/kg$)$ ρ.

- Particles are detected by their interactions with matter. The interactions of the stable particles are

electron   ionization loss, electromagnetic shower

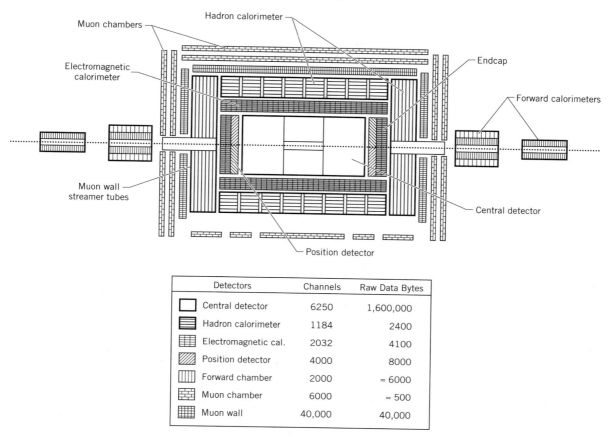

UA 1 Experiment layout

Detectors		Channels	Raw Data Bytes
	Central detector	6250	1,600,000
	Hadron calorimeter	1184	2400
	Electromagnetic cal.	2032	4100
	Position detector	4000	8000
	Forward chamber	2000	≈ 6000
	Muon chamber	6000	≈ 500
	Muon wall	40,000	40,000

(a)

FIGURE 16-23 The colliding-beam detector, UA1.
The detector completely surrounds the proton–antiproton collision point. (a) Plan view, (b) detail of the central ionization detector, (c) detail of the calorimeter, and (d) detail of the muon chambers.

photon	electromagnetic shower
proton	ionization loss, hadronic shower
charged pion	ionization loss, hadronic shower
charged kaon	ionization loss, hadronic shower
neutron	hadronic shower
K^0_L	hadronic shower
muon	ionization loss (no shower)

REFERENCES AND SUGGESTIONS FOR FURTHER READING

I. S. Blumenthal, "The Operating Principles of Synchrotron Accelerators," *Amer. Jour. Phys.* **21**, 164 (1953).

H. Breuker et al., "Tracking and Imaging Elementary Particles," *Sci. Am.* **265**, No. 2, 58 (1991).

G. Brianti and W. Scandale, "CERN's Large Hadron Collider: A New Tool for Investigating the Microcosm," *Part. World* **3**, 101 (1992).

G. Charpak, "Multiwire and Drift Proportional Chambers," *Phys. Today* **31**, No. 10, 23 (1978).

G. Charpak and F. Sauli, "Multiwire Proportional Chambers and Drift Chambers," *Nucl. Inst. Meth.* **162**, 405 (1979).

J. D. Cockcroft and E. T. S. Walton, "Experiments with High Velocity Positive Ions. II The Disintegration of Elements by High Velocity Protons," *Proc. of the Royal Society, Series A,* **137**, 229 (1932).

R. C. Fernow, *Introduction to Particle Physics,* Cambridge Univ. Press (1986).

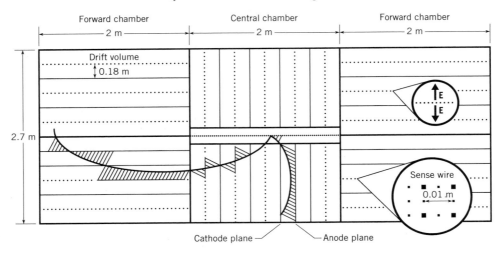

Central Detector
Cylindrical Drift Chamber with "Image" Readout

Central Detector Parameters	
Electric field	1.5 KV/cm
Magnetic field	0.7 Tesla
Drift velocity	53 μm/ns
Drift angle (B = 0.7 T)	23°
Drift gap	0.18 m
Sense wire spacing	0.01 m
Sense wires in drift volume	175 (Forward) 80 (Central)
Drift volumes	46
Sense wires	6250

(b)

E. L. Ginzton, "The Klystron," *Sci. Am.* **190**, No. 3, 84 (1954).

D. A. Glazer, "The Bubble Chamber," *Sci. Am.* **192**, No. 2, 46 (1955).

K. Kleinknect, *Detectors for Particle Radiation,* Cambridge Univ. Press (1986).

E. O. Lawrence and M. S. Livingston, "Production of High Speed Ions," *Phys. Rev.* **42**, 20 (1932).

L. M. Lederman, "The Tevatron," *Sci. Am.* **264**, No, 3, 48 (1991).

W. R. Leo, *Techniques for Nuclear and Particle Physics Experiments,* Springer-Verlag (1987).

J. N. Marx and D. R. Nygren, "The Time Projection Chamber," *Phys. Today* **31**, No. 10, 46 (1978).

S. van der Meer, "Stochastic Cooling and the Accumulation of Antiprotons," *Rev. Mod. Phys.* **57**, 689 (1985).

S. Myers and E. Picasso, "The LEP Collider," *Sci. Am.* **263**, No. 1, 54 (1990).

Proceedings of the CERN Accelerator School, CERN Yellow Report 85-19, ed. P. Bryant and S. Turner (1985) and CERN Yellow Report 87-10, ed. S. Turner (1987).

J. R. Rees, "The Stanford Linear Collider," *Sci. Am.* **261**, No. 4, 58 (1989).

B. Rossi, *High Energy Particles,* Prentice-Hall (1952).

M. Sands, "The Physics of Electron Storage Rings: an Introduction," SLAC-121 (1970).

E. Segrè, *Nuclei and Particles,* Benjamin (1977).

A. M. Sessler. "New Particle Acceleration Techniques," *Phys. Today,* January 1988, p. 26.

R. J. Van de Graaff, K. T. Compton and L. C. Van Atta, "The Electrostatic Production of High Voltage for Nuclear Investigations," *Phys. Rev.* **43**, 149 (1933).

W. J. Willis, "The Large Spectrometers," *Phys. Today* **31**, No. 10, 32 (1978).

R. R. Wilson, "U.S. Particle Accelerators: an Historical Perspective," AIP Conference Proceedings, No. 92, 297 (1982).

H. Winick, "Synchrotron Radiation," *Sci. Am.* **257**, No. 5, 88 (1987).

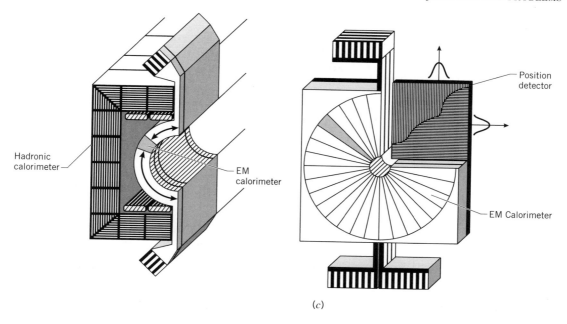

Hadronic calorimeter

EM calorimeter

Position detector

EM Calorimeter

(c)

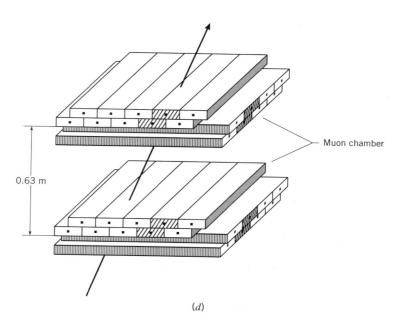

Muon chamber

0.63 m

(d)

QUESTIONS AND PROBLEMS

Particle accelerators

1. Why is the strong force not used to accelerate protons?

2. (a) The proton synchrotron at Fermilab was originally designed for an energy of 400 GeV. The radius of the machine is 1 km. Estimate the magnetic field strength of the dipole magnets. (b) The LEP electron–positron colliding beam machine was designed to study the properties of the Z^0 boson. The mass energy of the Z^0 boson is approximately 100 GeV. The radius of the LEP machine is about 4 km. Estimate the magnetic field strength of the dipole magnets. (c) For the two accelerators, compare the ratio E/B, where E is the beam energy and B is the

magnetic field. Why are the ratios so different? What determines the size of each accelerator?

3. Calculate the energy loss per revolution for 50-GeV electrons in the LEP accelerator, which has a radius of 4 km.

4. (a) Calculate the energy loss per revolution for 1-TeV protons in a synchrotron with a radius of 1 km. (b) Calculate the energy loss per revolution for 10-TeV protons in a synchrotron with a radius of 4 km.

5. (a) Show that in a linear accelerator, the particle acceleration is proportional to the increase in energy per unit distance ($\Delta E/\Delta x$). (b) How large must $\Delta E/\Delta x$ be for the radiation losses to be important? Give a numerical estimate in MeV per meter.

6. Consider the design of a proton–proton "supercollider" using superconducting bending magnets with a field strength of 5 tesla. Calculate the radius of the ring as a function of the total center-of-mass energy. What size ring is necessary to get 10 TeV, 20 TeV, and 40 TeV?

7. In the operation of the CERN proton–antiproton collider the protons and antiprotons are injected into the SPS in six bunches each, equally spaced around the machine, which has a radius of 1 km. The particles are accelerated to 270 GeV and the bunches are focused to a cross-sectional area of πr^2, where the radius r of each bunch is about 100 μm. The particle bunches collide head-on. The luminosity depends on the product of the number of stored protons (N_1) and the number of stored antiprotons (N_2). (a) Determine the product $N_1 N_2$ necessary to produce a luminosity of 10^{34} m^{-2}s^{-1}. (b) What is the average number of events per day from a reaction that has a cross section of 10 nb?

Particle detectors

8. (a) Why does it take a larger accelerator to probe shorter distances? (b) Why does it also take a larger spectrometer to probe shorter distances? Compare the size of the UA1 spectrometer with the spectrometer of J. J. Thomson.

9. Why are Cerenkov counters often so large?

10. Why are energy loss processes for electrons in matter compared as a relative energy loss ($-1/E)dE/dx$ whereas photon interactions in matter are compared as cross sections?

11. Estimate the length of iron needed to absorb a beam of 100-GeV muons.

12. The proton–proton cross section is about 40 mb for 100-GeV protons incident on a stationary target. From this, estimate the interaction length of iron.

13. Consider the interaction of a heavy charged particle with a single electron as the particle passes through matter. (a) Show that the momentum transferred to the electron may be written as

$$\Delta p = \frac{e}{v} \int_{-\infty}^{+\infty} dx\, E_t \,,$$

where v is the speed of the particle and E_t is the component of electric field due to the incoming particle in the direction of the momentum transfer. (b) Use Gauss's law to show that

$$\Delta p = \frac{2\,ke^2}{vb} \,.$$

14. A relativistic muon travels a distance of 1 m in water. Estimate the number of visible photons that are emitted as Cerenkov radiation.

15. A threshold Cerenkov counter is designed to separate kaons from protons at an energy of 200 GeV. What value of index of refraction should be chosen for the Cerenkov medium?

16. The index of refraction of air is $n = 1.00029$. Determine the energy threshold for a proton to cause Cerenkov radiation in air.

17. Make a comparison of the proton, photon, and neutrino cross sections in matter. What do these data tell us about the relative strengths of the interactions?

18. The cross section for a certain physical process is 1 nb. What average luminosity is needed in order to produce 10^6 events in one year?

19. (a) Estimate the minimum energy that an electron must have to penetrate a lead plate in the cloud chamber of Figure 16-20. (b) What fraction of its energy does an energetic electrons lose when passing through the lead plate?

Additional problems

*20. The b and anti-b quarks make a bound state called the Υ particle which has a mass of about 10 GeV/c^2. The Υ particle was discovered at Fermilab in a fixed target experiment using an intense proton beam. (a) Design a spectrometer to search for the Υ particle in the presence of a large hadronic background, assuming that it decays into $e^+ + e^-$. (b) Design the spectrometer assuming the Υ particle decays into $\mu^+ + \mu^-$.

*21. Consider an ionization (dE/dx) detector designed to distinguish pions ($m_\pi \approx 140$ MeV/c^2), kaons ($m_K \approx 500$ MeV/c^2), and protons ($m_p \approx 940$ MeV/c^2) at relativistic momenta. (a) What is the relative rate of ionization for pions, kaons, and protons at a momentum of 3 GeV/c? (b) In a certain experiment the pion to kaon ratio is 20 and the ionization detector has a resolution of 4%. Design a set of criteria (dE/dx "cuts") to identify kaons. Estimate the kaon identification efficiency and the pion background as the cuts are varied.

22. Alpha particles are accelerated in a cyclotron using a radio frequency of 7.0 MHz and extracted at a radius of 1 m. Calculate the kinetic energy of the α particles.

HIGH-ENERGY PHYSICS: CLASSIFICATION OF THE PARTICLES

It is fun to speculate about the way quarks would behave if they were physical particles (instead of purely mathematical entities as they would be in the limit of infinite mass). Since charge and baryon number are exactly conserved, one of the quarks (presumably $u^{2/3}$ or $d^{-1/3}$) would be absolutely stable, while the other member of the doublet would go to the first member very slowly by β-decay or K-capture. The isotopic singlet quark would presumably decay into the doublet by weak interactions, much as Λ goes into N. Ordinary matter near the earth's surface would be contaminated by stable quarks as a result of high energy cosmic ray events throughout the earth's history, but the contamination is estimated to be so small that it would have never have been detected. A search for stable quarks of charge $-1/3$ or $+2/3$ and/or stable di-quarks of charge $-2/3$ or $+1/3$ or $+4/3$ at the highest energy accelerators would help to reassure us of the non-existence of real quarks.

Murray Gell-Mann

In this chapter and the next, we examine the major discoveries that have been made in high-energy physics using the "gadgets" that have been discussed in the last chapter. The great discoveries in the field of high-energy physics have involved a close interplay between theory and experiment. The discoveries fall into two general categories: (1) new particles and (2) properties of the particles and forces. The two categories are closely related; very often the discovery of a new particle teaches us something fundamental about the forces. Figure 17-1 shows the energy at which the fundamental particles were discovered plotted versus the year of their discovery (compare to the Livingston plot of Figure 16-1). We have gone a long way from the alpha, beta, and gamma particles at the turn of the century to the discovery of the Z^0 in 1983. From this plot it is evident that part of the experimental program for the advancement of our fundamental knowledge of particles and forces must include accelerators of higher energy than are currently available.

Many names are used in the description of particles. Several of these have already been encountered in this text; others will be defined in this chapter. Table 17-1 gives a short glossary of some of the commonly used names.

17-1 DISCOVERY OF THE MESONS

The Pion

In 1935, Hideki Yukawa formulated a theory of the strong interaction between two nucleons. The Yukawa theory was motivated by quantum electrodynamics (QED). In QED charged particles interact electromagnetically by the exchange of photons. The range of the electromagnetic force is infinite because the photon is massless, that is, a photon may have arbitrarily small energy. In Yukawa's theory, nucleons interact strongly through the exchange of massive particles called *mesons*. The exchange of a massive particle results in a force that has a finite range.

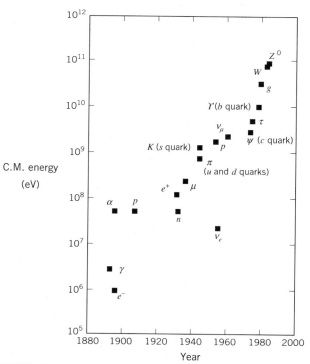

FIGURE 17-1 The energy at which the fundamental particles were discovered plotted versus the year of discovery.

TABLE 17-1
GLOSSARY OF PARTICLE NAMES.

Term	Meaning
Boson	Particle with integer spin
Fermion	Particle with half-integer spin
Hadron	Strongly interacting particle
Quark	Fractionally charged constituent of strongly interacting particles
Gluon	Strong interaction fundamental boson
Meson	Bound state (boson) of quark–antiquark
Baryon	Bound state (fermion) of three quarks
Leptons	Electron, muon, tau, neutrinos, and their antiparticles
Pion	A type of meson made out of u and d quarks
Kaon	A type of meson containing an s quark

Yukawa used the uncertainty principle to predict the mass of his hypothetical meson. The meson exists by "borrowing" an amount of energy mc^2 for a time Δt. The order-of-magnitude distance that the meson can travel in the time Δt is

$$R \approx c\Delta t \approx \frac{\hbar c}{2\,\Delta E} \approx \frac{\hbar c}{2\,mc^2}. \quad (17.1)$$

Taking the range of the strong force to be 1 fm, Yukawa predicted the mass energy of the meson to be

$$mc^2 \approx \frac{\hbar c}{2R} \approx \frac{200\,\text{MeV} \cdot \text{fm}}{2\,\text{fm}} \approx 100\,\text{MeV}. \quad (17.2)$$

The prediction of Yukawa went largely unnoticed until a new particle with mass of about 100 MeV/c^2 was discovered in 1937 in cosmic rays by Anderson and Neddermeyer and independently by Street and Stevenson. The particle was observed by its track in a cloud chamber (see Figure 17-2). By coincidence the new particle had the mass predicted by Yukawa. It was soon learned that this particle could not be the Yukawa meson because the new particle was observed *not* to have a strong interaction with nuclei.

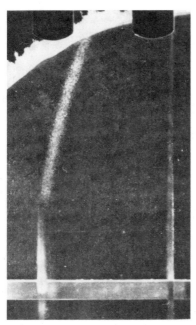

FIGURE 17-2 Discovery of the muon.
This cosmic ray muon was observed in a cloud chamber. From J. C. Street and E. C. Stevenson, "New Evidence for the Existence of a Particle of Mass Intermediate Between the Proton and Electron," *Phys. Rev.* **52**, 1003 (1937).

The new particle eventually became known as the *muon* (μ). Two types of muons were observed, one with positive charge (μ^+) and one with negative charge (μ^-). The μ^+ was observed to decay into a positron and the μ^- into an electron with a lifetime of about 2 µs.

Hans A. Bethe and Robert Marshak predicted that the muon could be the decay product of a Yukawa meson. In 1947, C. Lattes, H. Muirhead, G. Occhialini, and C. F. Powell discovered a particle called the *pion* (π) which was observed to decay into a muon and a neutrino:

$$\pi^+ \rightarrow \mu^+ + \nu. \quad (17.3)$$

and

$$\pi^- \rightarrow \mu^- + \bar{\nu}. \quad (17.4)$$

The decay was observed in emulsions that were exposed to cosmic rays at an altitude of 3000 m. Other experiments showed that the pion had a mass of about 140 MeV/c^2, a lifetime of 2.6×10^{-8} s, and a strong interaction with nuclei. The pions were observed to interact with nuclei and transform a neutron into a proton and vice versa:

$$\pi^+ + {}^A_Z N \rightarrow p + {}^{A-1}_Z N. \quad (17.5)$$

and

$$\pi^- + {}^A_Z N \rightarrow n + {}^{A-1}_{Z-1} N. \quad (17.6)$$

The Yukawa meson had been found! The decay of pions in emulsions are shown in Figure 17-3.

It was once believed (perhaps hoped?) that the pions would provide a fundamental explanation for the quantum nature of the strong force, analogous to the role that the photon provides for the electromagnetic force. This was not to be the case; it would take another four decades for physicists to uncover the secrets of the strong interaction. We now know that the Yukawa theory is not correct. The nucleons and the pion are composite objects and the fundamental interaction is between quarks and gluons. Nevertheless, the Yukawa theory provided a major advance in the understanding of the strong interaction.

More Mesons

In 1946, a new type of particle was discovered by G. D. Rochester and C. C. Butler in cloud chamber pictures of cosmic ray tracks. The new particles were electrically neutral and decayed into two charged particles producing a "*V*" pattern in the cloud chamber (see Figure 17-4). One of the new particles was called the *kaon* (K). With the

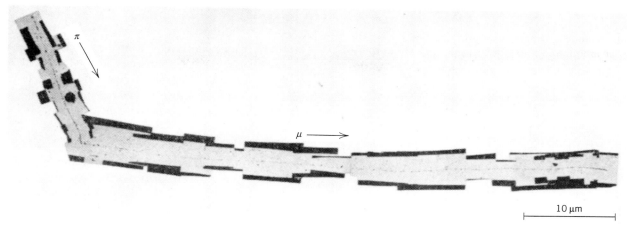

(a)

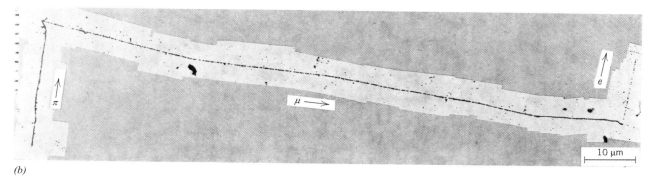

(b)

FIGURE 17-3 Discovery of the pion.
(*a*) The decay $\pi \rightarrow \mu$ observed in emulsions. From C. M. G. Lattes, H. Muirhead, G. P. S. Occhialini
and C. F. Powell, *Nature* **159**, 694 (1947). (*b*) Observation of the $\pi \rightarrow \mu \rightarrow e$ decay chain by the same
technique in emulsion with greater sensitivity. From R. H. Brown et al., *Nature* **163**, 47 (1949).

discovery of the pion, the kaon was interpreted as decaying by

$$K^0 \rightarrow \pi^+ + \pi^-. \qquad (17.7)$$

The kaon is a strongly interacting particle, another type of
meson.

With the commissioning of the Cosmotron in 1952 and
the invention of the bubble chamber in the same year, a

FIGURE 17-4 Discovery of "*V*" particles.
The *V* particles are neutral particles that decay into two charged
particles. The *V* particle observed in this cloud chamber is
produced by the interaction of a cosmic ray in a lead plate. The
lifetime of the *V* particle is long enough so that it travels out of
the lead plate before it decays. From G. D. Rochester and C. C.
Butler, *Nature* **160**, 855 (1947).

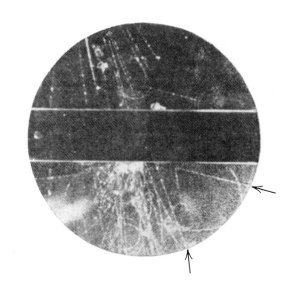

horde of new particles was discovered in the following decade. Sixteen strongly interacting particles were discovered with masses smaller than the proton and more than a hundred were discovered with masses larger than the proton!

17-2 QUANTUM NUMBERS OF THE PION

The discovery of the mesons raised many questions. At the top of the list was, *what are the properties of the lightest meson, the pion?*

Mass

The first accelerator to produce pions was the 184-inch cyclotron at Berkeley. The mass energy of the charged pions (π^+ and π^-) were determined by measurement of their momentum (p) and kinetic energy (E_k) and using the relationship

$$E_k = \sqrt{\left(mc^2\right)^2 + \left(pc\right)^2} - mc^2, \qquad (17.8)$$

or

$$mc^2 = \frac{\left(pc\right)^2 - E_k^2}{2E_k}. \qquad (17.9)$$

Pions of fixed momentum were selected by their curvature in a magnetic field. The pion energy was determined by measuring how far it traveled in nuclear emulsions. The distance that the particle travels in a material is called the particle *range* (R). Recall that the rate of energy loss of a charged particle in matter, $-dE/dx$ (16.41), does not depend on the mass of the particle, only on its speed. Therefore, the relationship between $-dE/dx$ and the particle speed can be empirically determined with protons. Pions and protons of the same speed have the same $-dE/dx$. The kinetic energy of the particle is related to its range by

$$E_k = \int_0^R dx\left(-\frac{dE}{dx}\right). \qquad (17.10)$$

Measurement of R determines E_k because $-dE/dx$ is known. The π^+ and π^- were found to have identical masses of 140 MeV/c^2.

Lifetime

The first accurate measurement of the π^+ lifetime was made in 1950 by Owen Chamberlain, R. F. Mozley, Jack Steinberger, and C. Wiegand. A beam of photons produced at a 340-MeV synchrotron were directed into a paraffin target. Pions are created by the process

$$\gamma + p \rightarrow \pi^+ + n. \qquad (17.11)$$

The pions are directed into a scintillating crystal, where they lose all their kinetic energy by ionization loss ($-dE/dx$). The π^+ mesons decay at rest producing muons ($\pi^+ \rightarrow \mu^+ + v$). The muons lose all their kinetic energy by ionization loss and decay at rest by the process

$$\mu^+ \rightarrow e^+ + v + \bar{v}. \qquad (17.12)$$

(see Figure 17-3b). The time taken for the pion and muon to stop is small compared to the lifetimes of the pion and muon. The π–μ–e decay chain produces three flashes of scintillation light caused by (1) the ionization from the π^+, (2) the ionization from the μ^+, and (3) the ionization from the e^+. The scintillation light was detected with photomultipliers.

The pion lifetime is short compared to the muon lifetime. Therefore, the pion decay signature is three flashes of light with the time between the last two flashes equal to the muon lifetime (about 2 μs). The pion lifetime was determined by measuring the time interval between the first two flashes. The experimenters could distinguish the first two flashes if they were separated by at least 22 ns. The distribution of the observed number of decays per time interval between the flashes (dN/dt) for 554 pion decays is shown in Figure 17-5. The slope gives the π^+ lifetime to be 26 ns. The π^- is found to have the same lifetime as the π^+.

Intrinsic Angular Momentum

The intrinsic angular momentum ("spin") of a particle gives the number of internal degrees of freedom. A particle with intrinsic angular momentum quantum number s has $2s + 1$ possible states. The factor $2s + 1$ appears in the density of states for a reaction in which the particle is produced; if there are more possible states for the particle, there is a greater probability to create it. The intrinsic angular momentum of the π^+ was determined by measuring the cross sections for the two processes:

$$p + p \rightarrow \pi^+ + d \qquad (17.13)$$

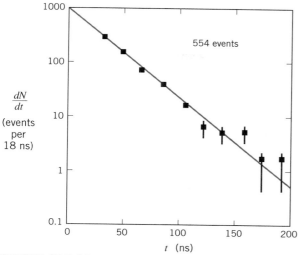

FIGURE 17-5 Measurement of the π^+ lifetime.
Pions are stopped in a scintillating crystal giving three flashes of
light from the $\pi-\mu-e$ decay chain. The distribution of the time
interval between the first two flashes gives the π^+ lifetime. From
O. Chamberlain et al., *Phys. Rev.* **81**, 284 (1950).

and

$$\pi^+ + d \rightarrow p + p. \qquad (17.14)$$

The intrinsic angular momentum quantum number of the
proton (s_p) is 1/2. Two protons must have an integer total
angular momentum quantum number (j). The intrinsic
angular momentum quantum number of the deuteron is
(s_d) is 1. Therefore, the existence of the reaction (17.13)
shows that the pion must have an integer intrinsic angular
momentum quantum number (s_π). The simplest possibilities are $s_\pi=0$ or $s_\pi=1$. The spin-dependence of the cross
section for the first reaction (17.13) is

$$\sigma_{p+p\rightarrow\pi^++d} \propto (2s_\pi+1)(2s_d+1)$$
$$= 3(2s_\pi+1). \qquad (17.15)$$

In writing the spin-dependence of the cross section in this
fashion we are assuming that the interacting protons are in
random spin states, that is, that the protons are not polarized. The spin-dependence of the cross section for the
second reaction (17.14) is

$$\sigma_{\pi^++d\rightarrow p+p} \propto \left(\frac{1}{2}\right)(2s_p+1)(2s_p+1) = 2, \qquad (17.16)$$

where the factor of 1/2 accounts for the fact that we are
creating two identical fermions, which cannot have iden-

tical quantum numbers. The spin-dependence of the ratio
of cross sections is

$$\frac{\sigma_{p+p\rightarrow\pi^++d}}{\sigma_{\pi^++d\rightarrow p+p}} \propto \frac{3(2s_\pi+1)}{2}. \qquad (17.17)$$

This ratio is three times smaller if $s_\pi=0$ compared to the
case $s_\pi=1$. The measurement of the cross-section ratio
determines the value of s_π. In comparing the cross sections
we must keep in mind that there also is a momentum
dependence to the density of states (see problem 7). The
cross-section measurements show clearly that $s_\pi=0$. The
π^- is found to have the same spin as the π^+.

Intrinsic Parity

We have seen that parity is a property of the wave function
that describes a particle. The parity of a *pair* of particles
depends on their relative orbital angular momentum. Each
particle also may be assigned an *intrinsic parity* (P). The
parity of a particle pair is written

$$P = P_a P_b (-1)^\ell. \qquad (17.18)$$

where P_a and P_b are the intrinsic parities of the two
particles and ℓ is the orbital angular momentum quantum
number. The value of P is +1 when the total wave function
is even under particle exchange and −1 when the total
wave function is odd under particle exchange.

The intrinsic parity of the nucleons is *defined* to be +1
(analogous to the electric charge of a proton defined to be
positive). The relative intrinsic parity of particles may be
determined from experiment with the assumption that
parity is conserved. This assumption is demonstrated by
experiment to be valid for the strong interaction. The
parity of the π is determined by stopping pions in a
deuteron target and observing the reaction

$$\pi^- + d \rightarrow n + n. \qquad (17.19)$$

When the pion stops it goes into an atomic s state by the
emission of photons. The "capture" of the π by the deuteron is from a state of zero orbital angular momentum.
The total angular momentum quantum number of the
initial state is $j=1$ ($s_\pi=0, s_d=1$ and $\ell=0$). By conservation
of angular momentum, $j=1$ in the final state. The possibilities for $j=1$ in terms of the total spin quantum number
and orbital angular momentum quantum number of the
two neutrons are ($\ell=0, s=1$), ($\ell=1, s=0$), ($\ell=1, s=1$),
or ($\ell=2, s=1$). Since the neutrons are fermions, they must
have an antisymmetric total wave function. The symmetry

of the spin part is even if $s = 1$ (spins aligned) and odd if $s = 0$ (spins antialigned). The symmetry of the space part is even if ℓ is even and odd if ℓ is odd. Therefore, even–even and odd–odd combinations of the spin-space symmetry are excluded. There is only one possibility for an antisymmetric total wave function ($\ell = 1, s = 1$). The parity (17.18) of the final state is -1. By parity conservation, the parity of the initial state is also -1. The intrinsic parity of the deuteron is $+1$ because the neutron and proton are in a bound state with $\ell = 0$. Since the π^- orbital angular momentum in the initial state is zero, the parity (7.18) of the pion is -1. The π^+ is found to have the same intrinsic parity as the π^-.

17-3 THE ANTIPROTON

In 1927, Dirac formulated a relativistic version of the Schrödinger equation that incorporated electron intrinsic angular momentum. The *Dirac equation* is a statement of energy conservation in matrix form that contains the four components of energy and momentum and a four-component wave function. Two of the components of the wave function correspond to the two projections of the electron intrinsic angular momentum ($m_s = \pm 1/2$). The Dirac equation predicts the correct size of the electron magnetic moment ($g = 2$). The other two components of the wave function correspond mathematically to a negative energy electron state. The beauty and consistency of the formulation led Dirac to postulate the existence of a state of the electron with positive electric charge (e^+). The positron (see Chapter 4) is the antiparticle of the electron. The positron has the same mass, intrinsic angular momentum, and wave properties as the electron.

The proton has an intrinsic angular momentum quantum number $s = 1/2$, the same as the electron. After the discovery of the Dirac antiparticle of the electron (e^+), it was natural to postulate that the proton might also have a Dirac antiparticle. It was soon learned from experiments, however, that the magnetic moment of the proton was not given by the Dirac equation, that is, the g factor (11.100) for a proton is *not* equal to 2. The magnetic moment of the proton is larger than that of a pointlike particle by a factor of 1.4. For this reason, the proton was known not to be a pointlike particle even before the structure of the proton was directly observed. Even though the proton is not pointlike, the antiparticle of the proton does exist! The *antiproton* ($\bar{p}$) has the same mass and intrinsic angular momentum as the proton, but has a negative charge. In the next section we shall discuss other quantum numbers of the proton and antiproton in terms of their constituent quarks. The conservation laws associated with the quantum numbers allow a proton–antiproton *pair* to be created in particle collisions, provided that energy and momentum are conserved.

The antiproton was first observed in 1955 by Owen Chamberlain, Emilio Segrè, Clyde Wiegand, and Thomas Ypsilantis by colliding protons with copper nuclei. The antiprotons are produced by the reactions

$$p + p \rightarrow p + p + p + \bar{p}, \qquad (17.20)$$

and

$$p + n \rightarrow p + n + p + \bar{p}. \qquad (17.21)$$

The signature of the antiproton is a particle with negative charge and the same mass as the proton. The experimental layout of Chamberlain et al., is shown in Figure 17-6. Protons with kinetic energy of 6.2 GeV are directed into a copper target. Energy is converted into mass and several particles are created in the collisions. Particles with negative electric charge and momenta of $p = 1.19$ GeV/c are selected with two dipole magnets. The trigger for the experiment is a signal in each of three scintillation counters (S1, S2, and S3) signifying the passage of a relativistic negatively charged particle with $p = 1.19$ GeV/c. The speed of particles that enter the spectrometer is measured. The measurement of both momentum and speed determines the particle mass.

The main difficulty of the experiment is identification of the small number of antiprotons in the presence of a large background of pions (π^-). Pions are copiously produced through reactions such as

$$p + p \rightarrow p + p + \pi^+ + \pi^- \qquad (17.22)$$

or

$$p + n \rightarrow p + n + \pi^+ + \pi^-. \qquad (17.23)$$

For every antiproton produced, about 62,000 negatively charged pions are produced! The antiproton is distinguished from the pion because the pion mass (140 MeV/c^2) is significantly smaller than the antiproton mass (940 MeV/c^2).

To ensure that the small number of antiprotons could be disentangled from a large number of pions, the experimenters measured the particle speed with two independent methods. The first method for measuring speed was by time of flight. The distance between scintillation counters S1 and S2 was about 12 m. The speed of a pion that has a momentum of $p = 1.19$ GeV/c is

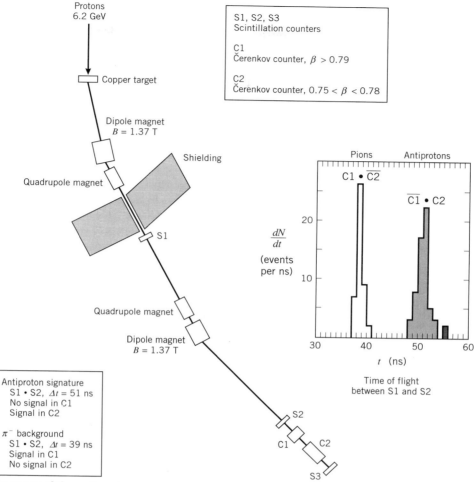

FIGURE 17-6 Discovery of the antiproton.
After O. Chamberlain, E. Segrè, C. Wiegand, and T. Ypsilantis, *Phys. Rev.* **100**, 947 (1955).

$$v = \frac{pc}{E}c$$

$$= \frac{1.19\,\text{GeV}}{\sqrt{(1.19\,\text{GeV})^2 + (0.14\,\text{GeV})^2}}\,c$$

$$\approx 0.99\ c. \tag{17.24}$$

The speed of an antiproton that has the same momentum is

$$v = \frac{pc}{E}c$$

$$= \frac{1.19\,\text{GeV}}{\sqrt{(1.19\,\text{GeV})^2 + (0.94\,\text{GeV})^2}}\,c$$

$$\approx 0.78\ c. \tag{17.25}$$

Since the pion and antiproton speeds are different, the travel times between scintillation counters S1 and S2 (see Figure 17-6) are also different. Pions traverse this distance in about 39 ns, whereas antiprotons require 51 ns. The time resolution of the scintillation counters is about 1 ns. The second method for measuring the particle speed is with two Cerenkov counters. The first Cerenkov counter (C1, a threshold type) is designed to count particles with $\beta > 0.79$. The second Cerenkov counter (C2, a differential type) is designed to count particles with $0.75 < \beta < 0.78$. Taking into account the energy loss by ionization in the spectrometer, the speed divided by c of an antiproton that reaches C2 is $\beta = 0.765$. The speed divided by c of a pion that reaches C2 is $\beta = 0.99$. Thus, C1 gives a signal for pions but not antiprotons and C2 gives a signal for antiprotons but not pions.

A total of 60 antiprotons were observed in the original experiment. The time of flight of these antiprotons (defined by a signal in C2 and no signal in C1) is shown in Figure 17-6. From the momentum and speed measurements of these 60 particles, the antiproton mass was determined to a precision of about 1.5% and found to be equal to the proton mass. Further experiments have very accurately determined the antiproton mass. The antiproton mass is *identical* to the proton mass:

$$m_{\text{antiproton}} = m_{\text{proton}}. \qquad (17.26)$$

All particles that have half-integer intrinsic angular momentum (fermions) have unique antiparticles. A particle does not have to have nonzero electric charge to have a unique antiparticle. For example, the neutron has a unique antiparticle, the antineutron ($\bar{n}$). The neutrino also has a unique antiparticle ($\bar{v}$). All the fundamental quarks and leptons have unique antiparticles. Some examples of antiparticles are given in Table 17-2.

17-4 CLASSIFICATION OF THE HADRONS: THE QUARK MODEL

Free particles that participate in the strong interaction are called *hadrons*. The first hadrons to be discovered were the proton and the neutron, followed by the pion. There are three different pions. The pions all have masses of about 140 MeV/c^2. The three pions have electric charge quantum numbers of +1, 0 and −1 (π^+, π^0, π^-). The strong interaction properties of the three pions are identical, just as the strong interaction properties of the proton and neutron are identical (see Chapter 11). The pions are the least massive of all hadrons. The large number of hadrons

and their similar properties led physicists to believe that the hadrons were not fundamental particles. The situation with the understanding of the hadrons and the strong force in 1960 was similar to the situation with the understanding of atoms and the electromagnetic force at the beginning of the twentieth century. The strong interaction between two hadrons is somewhat analogous to the electromagnetic interaction between neutral atoms to form molecules.

The hadrons are divided into two general classes according to their intrinsic angular momentum: (1) fermions (like the proton and neutron) and (2) bosons (like the pions). The fermions are all more massive than the lightest bosons. Hadrons that are fermions are called *baryons* (from the Greek meaning "heavy"). Examples of the baryons are the proton (p), neutron (n), lambda (Λ), sigmas (Σ^+, Σ^0, Σ^-), cascades (Ξ^0, Ξ^-) and the omega (Ω^-). Hadrons that are bosons are called *mesons* (from the Greek meaning "middle" because the first to be discovered, the pions, have masses between that of the proton and the electron). Examples of the mesons are the pions (π^+, π^0, π^-), the kaons (K^+, K^0, $\bar{K}^0$, K^-), and the eta (η^0).

Baryon Number Conservation

An important empirical law has been established governing the interactions of the hadrons. The rule is that a fermion cannot be created except in association with an antifermion or by the decay of another fermion. Similarly, a fermion cannot be destroyed except in association with an antifermion. This experimentally established rule may be summarized by assigning each particle a quantum number called *baryon number*. The baryon number (B) for the fermions (p, n, Λ, Δ, Σ, Ξ, Ω, …) is defined to be

$$B \equiv +1. \qquad (17.27)$$

The baryon number for the antifermions ($\bar{p}$, $\bar{n}$, $\bar{\Lambda}$, $\bar{\Delta}$, $\bar{\Sigma}$, $\bar{\Xi}$, $\bar{\Omega}$ …) is defined to be −1. The particles with $B = 1$ are called baryons and the particles with $B = -1$ are called antibaryons. The baryon number for the mesons is defined to be zero. The definition of particles (p, n, etc.) and antiparticles ($\bar{p}$, $\bar{n}$, etc.) is made so that the quantity B is conserved; however, it is still a very significant empirical result that such a classification *can* be made that will conserve B!

Baryon number is an additive quantum number that is conserved in the strong interactions. The baryon conservation rule states that the number of baryons minus the number of antibaryons cannot change in any interaction:

$$\sum B = \text{constant}. \qquad (17.28)$$

TABLE 17-2
PARTICLES AND THEIR ANTIPARTICLES.

Particle	Antiparticle
Electron (e^-)	Positron (e^+)
Proton (p)	Antiproton ($\bar{p}$)
Neutron (n)	Antineutron ( )
Muon (μ^-)	Mu-plus (μ^+)
Neutrino (v)	Antineutrino ($\bar{v}$)

If particles are created or annihilated in an interaction, there must be no net change of B. The law of baryon number conservation (17.28) is observed to apply for all the forces, with the simple extension that all particles (electron, photon, muon, etc.) except the baryons and antibaryons have $B=0$. A violation of baryon number has never been observed.

EXAMPLE 17-1

Which of the following interactions are allowed by baryon number conservation: (a) $p \rightarrow \pi^+ + \pi^0$, (b) $\pi^- + p \rightarrow \pi^0 + n$, (c) $\Lambda + p \rightarrow \pi^+ + K^- + p + n$, (d) $p + p \rightarrow n + p + p + \bar{p} + \pi^+$?

SOLUTION:

Baryon number conservation gives

(a) $\qquad 1 \rightarrow 0 + 0 \qquad$ (not allowed).

(b) $\qquad 0 + 1 \rightarrow 0 + 1 \qquad$ (allowed).

(c) $\qquad 1 + 1 \rightarrow 0 + 0 + 1 + 1 \qquad$ (allowed) .

(d) $\qquad 1 + 1 \rightarrow 1 + 1 + 1 - 1 + 0 \qquad$ (allowed) . ∎

Baryon number conservation together with energy conservation implies that the least massive baryon, the proton, cannot decay. If baryon number is *absolutely* conserved, then the proton has an *infinite* lifetime. The proton lifetime is observed to be longer than about 10^{40} seconds. Our observation that the proton is stable provides the empirical foundation of baryon number conservation. No other free baryon is stable, and so all other free baryons ultimately end up as protons. Some examples of baryon decay are

$$\Delta^{++} \rightarrow \pi^+ + p, \qquad (17.29)$$

$$n \rightarrow p + e^- + \bar{v}_e, \qquad (17.30)$$

and

$$\Lambda \rightarrow \pi^- + p. \qquad (17.31)$$

The first of these decays is a strong interaction; the last two are weak interactions.

Baryon number is conserved in all interactions.

Strangeness

Certain particles, such as the kaon, are produced only in *association* with certain other particles. To classify these interactions, an empirical law was established by assigning each hadron a new quantum number called *strangeness* (S). The K^+, and K^0 are assigned a strangeness of

$$S = +1. \qquad (17.32)$$

The K^-, $\bar{K}^0$, Λ, Σ^+, Σ^-, and Σ^0 are assigned a strangeness of -1, the Ξ^- and Ξ^0 are assigned a strangeness of -2, and the Ω^- is assigned a strangeness of -3. Note that both baryons and mesons can have nonzero strangeness. The p, n, π, and η have a strangeness of zero. The strangeness of an antiparticle is (-1) times the strangeness of the corresponding particle. Strangeness is an additive quantum number:

$$\sum S = \text{constant}. \qquad (17.33)$$

The law of strangeness conservation (17.33) holds for strong and electromagnetic interactions but *not for weak interactions*. Strangeness is not absolutely conserved.

EXAMPLE 17-2

Which of the following strong interactions are allowed by strangeness number conservation: (a) $\pi^- + p \rightarrow K^- + p$, (b) $\pi^- + p \rightarrow K^+ + \Sigma^-$, (c) $K^- + p \rightarrow K^+ + \Xi^- + \pi^-$, (d) $K^+ + p \rightarrow K^- + \Xi^0 + \pi^+$?

SOLUTION:

Strangeness number conservation gives

(a) $\qquad 0 + 0 \rightarrow -1 + 0 \qquad$ (not allowed).

(b) $\qquad 0 + 0 \rightarrow 1 - 1 \qquad$ (allowed).

(c) $\qquad -1 + 0 \rightarrow 1 - 2 + 0 \qquad$ (allowed).

(d) $\qquad 1 + 0 \rightarrow -1 - 2 + 0 \qquad$ (not allowed). ∎

Isospin

The motivation for the quark model was rooted in the internal symmetry displayed by the hadrons. This symmetry is suggested in Figure 17-7, which shows the mass of the lightest hadrons versus their intrinsic angular momentum quantum number and intrinsic parity (s^P). Groups of hadrons with the same quantum numbers s^P tend to have roughly the same mass. Within the groups, there is even a

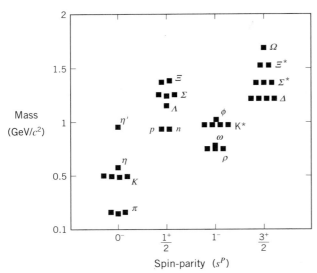

FIGURE 17-7 Classifying the hadrons by their mass, intrinsic angular momentum, and intrinsic parity.

more striking relationship: several particles tend to have nearly identical masses. For example, the proton and neutron have very nearly the same mass. As far as the strong interaction is concerned, the proton and neutron may be considered as two different states of the same particle, the nucleon. This is analogous to the electron having two spin states. The nucleon is assigned a quantum number, analogous to spin, called *isospin*. The isospin (*I*) quantum number of the nucleon is

$$I = \frac{1}{2}. \qquad (17.34)$$

There are two possibilities for the projection of the isopin along an arbitrary axis. These projections correspond to the quantum numbers (m_I)

$$m_I = +\frac{1}{2} \qquad (17.35)$$

for the proton, and

$$m_I = -\frac{1}{2} \qquad (17.36)$$

for the neutron. Isospin is conserved in strong interactions. Notice that for the nucleons, the relationship between electric charge, m_I, and baryon number is

$$\frac{q}{e} = m_I + \frac{B}{2}. \qquad (17.37)$$

We shall need to modify this expression only slightly to accommodate all the hadrons.

The other hadrons have a symmetry similar to that displayed by the proton and neutron. There are three pions, all having nearly the same mass. The π^+ and π^- masses are identical; the π^0 mass is slightly smaller. The slight difference in the pion masses is caused by the electromagnetic interaction. (The same is true for the proton and neutron.) The three pions have an identical strong interaction behavior. For the strong interaction, the pions (π^+, π^0, π^-) behave like three different states of the same particle (π). The pion has an isospin of one (*I*=1). There are three possibilities for the quantum number m_I: +1 for the π^+, 0 for the π^0, and −1 for the π^-. With these assignments of m_I, we see that the relationship between charge, isospin, and baryon number (17.37) also holds for the pions.

Quarks

The symmetry displayed in Figure 17-7 provided the motivation for Gell-Mann and Zweig (independently) to invent a composite model of the hadrons in 1964. In this model, all the known hadrons were constructed from three basic constituents. Gell-Mann coined the term *quark* for the constituents. The quarks each have an intrinsic angular momentum quantum number,

$$s = \frac{1}{2}. \qquad (17.38)$$

Each quark is assigned a fractional electric charge. One of these quarks, called the *up* (*u*) quark has an electric charge,

$$q = +\frac{2}{3} e, \qquad (17.39)$$

while another of the quarks, called the *down* (*d*) quark has an electric charge,

$$q = -\frac{1}{3} e. \qquad (17.40)$$

Each quark has a corresponding antiquark. The electric charge of the antiup quark ($\bar{u}$) is −2e/3 while the electric charge of the antidown quark ($\bar{d}$) is +e/3.

The lightest mesons and baryons, the nucleons and the pions, may be constructed from the u and d quarks and their antiquarks. The total charge of a particle is the sum of the charges of the quarks. The intrinsic angular momentum of the particle (**J**) is the vector sum of the orbital plus intrinsic angular momenta (**L** + **S**) of the quarks. The proton has a charge of $+e$ and is made up of two u quarks plus one d quark. The intrinsic angular momentum quantum number of the proton is 1/2. The total angular momentum quantum number of the three quarks is $j = 1/2$. For this reason the intrinsic angular momentum quantum number of a composite particle is often labeled j, that is, we use j and s interchangeably. The neutron is a $j = 1/2$ combination of one up quark and two down quarks (udd). The pions (π^+, π^-, and π^0) are $j = 0$ combinations of one up quark and one antidown quark ($u\bar{d}$), one down quark and one antiup quark ($d\bar{u}$), and a mixture of one up quark and one antiup quark plus one down quark and one antidown quark ($u\bar{u}$ and $d\bar{d}$).

In the quark model, all the mesons are constructed from a quark–antiquark pair and all the baryons are constructed from three quarks. Antibaryons are constructed from three antiquarks. There are more hadrons than can be constructed from the up and down quarks plus their antiquarks. Examination of Figure 17-7 shows that there are nine mesons with zero spin (and negative parity). This suggests that *three* quarks are needed to construct the hadrons because there are $3 \times 3 = 9$ quark–antiquark combinations of three quarks. All the other hadrons of Figure 17-7 may be constructed out of three types of quarks.

Quantum Numbers of the Quarks

The masses of the up and down quarks are nearly identical. The hadrons that contain them in the same total angular momentum combinations have nearly identical masses. For strong interaction processes, the up and down quarks behave like two different states of the same quark. The up and down quarks form an isospin doublet ($I = 1/2$). The values of m_I are $+1/2$ for the up quark and $-1/2$ for the down quark.

The third quark, the *strange* quark, is in a class all by itself. The strange quark has a much greater mass than the up and down quarks. The strange quark does not have a partner of almost equal mass; it is an isospin singlet ($I = 0$), $m_I = 0$. The strange quark has the quantum of strangeness that the other two quarks do not have. The strangeness assignments of the quarks are zero for the up quark, zero for the down quark, and -1 for the strange quark. The strangeness quantum number of the antistrange quark is $+1$.

All of the quarks have an intrinsic angular momentum quantum number of 1/2 and baryon number of 1/3. The electric charge of the strange quark is $-1/3$. The quantum numbers of the quarks are summarized in Table 17-3. The

TABLE 17-3
QUANTUM NUMBERS OF THE FIRST THREE QUARKS.

	Up	Down	Strange
Spin (s)	1/2	1/2	1/2
Isospin (I)	1/2	1/2	0
m_I	1/2	$-1/2$	0
Electric charge (q/e)	2/3	$-1/3$	$-1/3$
Baryon number (B)	1/3	1/3	1/3
Strangeness (S)	0	0	-1
Strong charge (color)	R, G, or B	R, G, or B	R, G, or B

relationship between the quantum numbers of electric charge, m_I, strangeness, and baryon number is

$$\frac{q}{e} = m_I + \frac{S + B}{2}. \tag{17.41}$$

This expression holds for the quarks as well as the hadrons.

Constructing the Hadrons from the Quarks

The quantum numbers of the hadrons (baryon number, electric charge, spin, isospin, and strangeness) are equal to the sums of the quantum numbers of their constituent quarks.

Mesons

Mesons are constructed from quark–antiquark pairs. Consider the mesons with quark–antiquark spins antialigned ($s = 0$) and zero orbital angular momenta ($\ell = 0$). The total angular momentum quantum number (j) of these mesons is zero. These particles are the lowest energy states of the quark–antiquark pairs. The intrinsic parity of these states is defined to be the product of three terms: the intrinsic parities of the quark and antiquark and a term arising from the space symmetry of the meson wave function:

$$P = P_q P_{\bar{q}} (-1)^\ell. \tag{17.42}$$

The intrinsic parity of each quark is defined to be $+1$, and the intrinsic parity of each antiquark is -1. Therefore, the intrinsic parity of a meson is

$$P = -(-1)^\ell = (-1)^{\ell+1}. \tag{17.43}$$

For $\ell = 0$ the intrinsic parity is negative. The lowest energy states of quark–antiquark pairs have spin zero and negative (intrinsic) parity,

$$j^P = 0^-. \tag{17.44}$$

There are nine quark–antiquark combinations with spin zero and negative parity (see Figure 17-7). These particles are called the *pseudoscalar* mesons. Consider the states with zero strangeness. Zero strangeness includes the possibility of the strange–antistrange combination. There are five such states. The quark combinations of these five states are $u\bar{d}, d\bar{u}, u\bar{u}, d\bar{d}$, and $s\bar{s}$. There is a triplet of states corresponding to an isospin of one ($m_I = 1, 0,$ or -1) that does not involve the $s\bar{s}$ combination. These particles are the pions. There are two singlet states with an isospin of zero. These particles are called the eta and the eta-prime. The eta and the eta-prime are a mixture of the states $u\bar{u}, d\bar{d}$, and $s\bar{s}$. There are two states with an isospin of 1/2 and a strangeness of +1 ($u\bar{s}$ and $d\bar{s}$). These states are called the K^+ and K^0. Similarly, there are two states with an isospin of 1/2 and a strangeness of -1 ($s\bar{u}$ and $s\bar{d}$). These states are called the K^- and $\bar{K}^0$. This completes the *nonet* of pseudoscalar mesons. Figure 17-8a shows a plot of S versus m_I for these mesons.

There are also quark–antiquark excited states. There is a countless number of them just as there is an infinite number of energy levels in the hydrogen atom! Consider the quark–antiquark pairs that have spins aligned (the total intrinsic angular momentum quantum number is $s = 1$) and zero orbital angular momenta ($\ell = 0$). The total angular momentum quantum number is $j = 1$, that is, these mesons have a spin of one. The parity (17.42) of these mesons is negative, so that

$$j^P = 1^-. \tag{17.45}$$

These particles are called the *vector* mesons. The vector mesons have the same spin and parity as the photon. There is a one-to-one correspondence between the vector meson states and the pseudoscalar states. Figure 17-8b shows a plot of S versus m_I for the vector mesons.

Baryons

In the quark model, the baryons are constructed out of three quarks. Consider those states with all the quark spins aligned and zero orbital angular momentum. The total intrinsic angular momentum quantum number of these states is

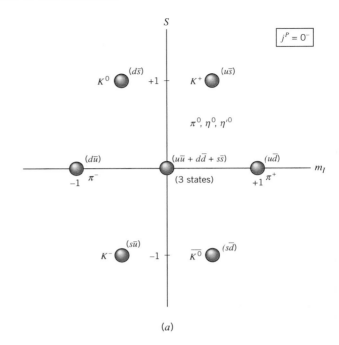

(a)

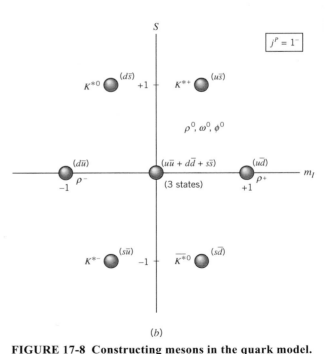

(b)

FIGURE 17-8 Constructing mesons in the quark model. (a) The pseudoscalar mesons have quark and antiquark spins antialigned and zero orbital angular momentum resulting in $j = 0$. (b) The vector mesons have quark and antiquark spins aligned and zero orbital angular momentum resulting in $j = 1$.

$$j = \frac{3}{2}. \qquad (17.46)$$

There are ten such baryon states (a *decimet* of baryons, see Figure 17-9). All of these states have comparable masses (1.2 GeV/c^2 to 1.7 GeV/c^2). We now classify the states in the context of the quark model. There are four states with no strange quarks; these quark combinations are *uuu*, *uud*, *udd*, and *ddd*. This *quartet* of states has $I = 3/2$ and the possible values of m_I are 3/2, 1/2, −1/2, and −3/2. These particles are called Δ^{++}, Δ^{+}, and Δ^{0}, Δ^{-}. Figure 17-9 shows the location of these states in a plot of S verus m_I. There are three states that have one strange quark; these quark combinations are *uus*, *uds*, and *dds*. This triplet of states has $I = 1$ and the possible values of m_I are 1, 0, and −1. These particles are called Σ^{*+}, Σ^{*0}, and Σ^{*-}. They make up the second row of particles in Figure 17-9. There are two states that have two strange quarks; the quark combinations are *uss* and *dss*. This doublet of states has $I = 1/2$ and the possible values of m_I are 1/2 and −1/2. These particles are called Ξ^{*0} and Ξ^{*-}. They make up the third row of

particles in Figure 17-9. Finally, there is only one state with three strange quarks (*sss*). This state is an isosinglet ($I = 0$) and the value of m_I must be zero. This particle is called the omega (Ω^-).

The existence of the omega particle provided the crowning achievement of the quark model because it was searched for and discovered only after the quark model was used to make a prediction of its mass and decay modes. The omega particle (Ω^-) was discovered at Brookhaven National Laboratory in 1964 (see Figure 17-10). The Ω^- was produced by the strong interaction

$$K^- + p \rightarrow \Omega^- + K^+ + K^0. \qquad (17.47)$$

Strangeness is conserved in this interaction:

$$(-1) + (0) = (-3) + (+1) + (+1). \qquad (17.48)$$

The Ω^- is stable against strong decay because strangeness and baryon number are conserved in the strong interactions and the Ω^- mass is smaller than the mass of any combination of hadrons that have $B = 1$, $q = -e$ and $S = -3$ (e.g., $p + K^- + K^- + \overline{K}{}^0$). The Ω^- is long-lived and its ionization track is observed in the bubble chamber up to the point where it decays by the weak interaction

$$\Omega^- \rightarrow \Xi^0 + \pi^-, \qquad (17.49)$$

followed by the additional weak decays

$$\Xi^0 \rightarrow \Lambda^0 + \pi^0, \qquad (17.50)$$

and

$$\Lambda^0 \rightarrow p + \pi^-. \qquad (17.51)$$

The net result is that through three sequential weak decays, one for each of the three strange quarks, the Ω^- ends up as a proton and three pions.

There is also a family of baryons with spin = 1/2 and wave functions that are space symmetric:

$$j^P = \frac{1}{2}^+. \qquad (17.52)$$

These states have two quarks with parallel spins and one antiparallel and zero orbital angular momentum. There are eight such states. All of these particles have similar masses, from about 0.9 GeV/c^2 to about 1.3 GeV/c^2. We now construct these states from the quarks. Consider first the states with zero strangeness. Apart from symmetry considerations, there are four states with zero strangeness (*uuu*, *uud*, *udd*, *ddd*). Two of these states, the *uuu* and *ddd*

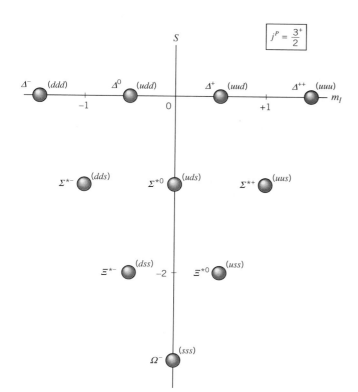

FIGURE 17-9 Constructing the $j^P = 3/2^+$ baryons in the quark model.

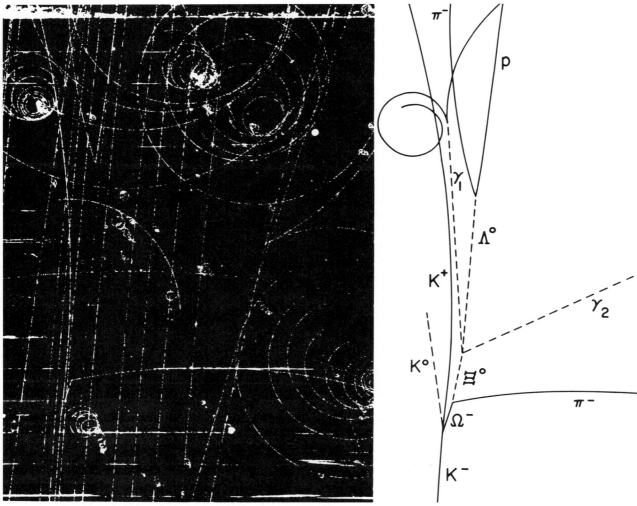

FIGURE 17-10 Discovery of the Ω^- particle.
The Ω^- observed in this bubble chamber photograph is produced together with two kaons in a K^-p collision. This is a strangeness conserving strong interaction. The Ω^- decays by weak interaction which does not conserve strangeness. The Ξ^0 and Λ^0 also have weak-interaction decays. The observation of the Ω^-, which is composed of three strange quarks, was a great triumph for the quark model of hadrons. From V. E. Barnes et al., *Phys. Rev. Lett.* **12**, 204 (1964). Courtesy Brookhaven National Laboratory.

states, are not allowed. The reason that the states with three identical quarks are forbidden is that the overall symmetry of the wave function must be antisymmetric for a fermion. This overall symmetry is the product of four symmetries:

$$\text{Symmetry} = (\text{spin}) \times (\text{space}) \times (\text{flavor}) \times (\text{color}). \quad (17.53)$$

For all the hadrons, the color part is antisymmetric. Since the net spin is 1/2, two of the quarks have spin aligned and

one is antialigned so that the spin part is antisymmetric. By definition of the parity of this group of particles, the space part is symmetric. Therefore, the overall symmetry requires that the flavor part of the wave function be antisymmetric. If the three quarks are all the same, then the flavor part is symmetric. Thus, quantum mechanics does not allow a state of three identical quarks with $j^P = 1/2^+$. (For $j^P = 3/2^+$, the spin part is symmetric and the state exists, the Δ^{++} particle.) There are only two states with zero strangeness. These states are the proton (*uud*) and the neutron (*udd*). The proton and neutron have $I = 1/2$ ($m_I = 1/2$ for the proton and $m_I = -1/2$ for the neutron). Figure 17-11 shows the plot of S versus m_I for the $j^P = 1/2^+$ states. There is a triplet of states with one strange quark (*uds*, *uds*, and *dds*). For all these states the space part is symmetric, the spin part is antisymmetric, and the flavor part is antisymmetric, and so they are all allowed. These states have $I = 1$ and $m_I = 1, 0$ or -1. ($I = 1$ means that the isospin of the two light quarks are aligned.) These particles are called the Σ^+, Σ^0, and Σ^-. In addition there is a state with $I = 0$. ($I = 0$ means that the isospin of the two light quarks are antialigned.) This particle is called the Λ^0. There are two states with two strange quarks (*uss* and *dss*). These particles are called Ξ^0 and Ξ^-. This completes the set of baryons with $j^P = 1/2^+$.

Decays of the Hadrons

Stable Hadrons
The only hadron that is observed to be absolutely stable against strong, electromagnetic, and weak decay is the proton. Several hadrons, however, are stable against strong decay. They have only weak decays or in a few cases electromagnetic decays. Hadrons that do not have a strong

interaction decay are often referred to as "stable" hadrons, and they have lifetimes that are many orders of magnitude larger than unstable hadrons. The stable hadrons include both baryons and mesons.

A good way to think about the decays of the hadrons is that all three forces are working to make the hadron decay. The strongest "worker" is the strong force, followed by the electromagnetic and then the weak force. The decay rate is the sum of the decay rates from the three forces. Often, however, one or more of the decay rates is identically zero because of a conservation law. If the strong decay is allowed, the weak and electromagnetic decays may still occur, but we usually can't detect them because their rate is so slow compared to the strong rate. If the strong decay is forbidden and the electromagnetic decay is allowed, then the weak decay may also occur but is usually not detected because the electromagnetic decay rate is so fast. If the strong and electromagnetic decays are both forbidden, then the weak decay is readily observable unless it too is forbidden.

Strong Decays of Hadrons
The $j = 1$ mesons have larger masses than the $j = 0$ mesons (see Figure 17-7). The lightest of the $j = 1$ mesons are the rho mesons (ρ). The rho mesons decay by strong interaction into pions

$$\rho^0 \rightarrow \pi^+ + \pi^-, \tag{17.54}$$

$$\rho^+ \rightarrow \pi^+ + \pi^0, \tag{17.55}$$

and

$$\rho^- \rightarrow \pi^- + \pi^0. \tag{17.56}$$

The lifetime of the rho meson is about 4×10^{-24} s.

EXAMPLE 17-3
Show that the decay $\rho^0 \rightarrow \pi^0 + \pi^0$ cannot occur.

SOLUTION:
The ρ^0 has an intrinsic angular momentum quantum number of $j = 1$. The decay products have zero intrinsic angular momentum, so by conservation of angular momentum the two pions are in an $\ell = 1$ state. The two pions have a total wave function (space × spin) that is antisymmetric under particle exchange. This is forbidden because two identical bosons always have a symmetric total wave function! The decay is not observed. (Note that the restriction does not apply to the decay, $\rho^0 \rightarrow \pi^+ + \pi^-$, because the π^+ and π^- are not identical particles.) ∎

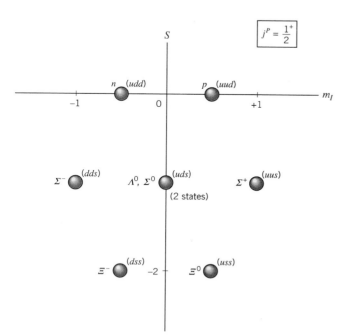

FIGURE 17-11 Constructing the $j^P = 1/2^+$ baryons in the quark model.

The $j = 3/2$ baryons all have masses larger than that of a light nucleon plus a pion. These baryons (except the Ω^-) decay by the strong interaction. For example, the Δ^- baryon decay processes are

$$\Delta^{++} \rightarrow p + \pi^+, \qquad (17.57)$$

$$\Delta^+ \rightarrow n + \pi^+ \quad \text{or} \quad \Delta^+ \rightarrow p + \pi^0, \quad (17.58)$$

$$\Delta^0 \rightarrow p + \pi^- \quad \text{or} \quad \Delta^0 \rightarrow n + \pi^0, \quad (17.59)$$

and

$$\Delta^- \rightarrow n + \pi^-. \qquad (17.60)$$

EXAMPLE 17-4

Estimate the lifetime of the Δ baryon as the time that it takes a relativistic pion to travel a distance equal to the particle size. Calculate the lifetime of the Δ baryon using the uncertainty principle and the measured width (Γ) of 115 MeV.

SOLUTION:

The size (R) of the Δ baryon is about

$$R = 1\,\text{fm}.$$

The speed of a relativistic pion is the speed of light (c). The lifetime is estimated as

$$\tau = \frac{R}{c} = \frac{10^{-15}\,\text{m}}{3 \times 10^8\,\text{m/s}} \approx 3 \times 10^{-24}\,\text{s}.$$

From the uncertainty principle, we have

$$\tau = \frac{\hbar}{\Gamma} = \frac{\hbar c}{\Gamma c}$$

$$= \frac{200\,\text{MeV} \cdot \text{fm}}{(115\,\text{MeV})(3 \times 10^{23}\,\text{fm/s})}$$

$$\approx 6 \times 10^{-24}\,\text{s}.$$

This is the typical time scale for the strong interaction. ∎

Electromagnetic Decays of Hadrons

The typical strong interaction lifetime is 10^{-23} s. The typical order of magnitude of the lifetime of a particle that is stable against strong decay but not electromagnetic decay is expected to be longer by the square of the ratio of the strong and electromagnetic coupling constants:

$$\tau_{\text{EM}} \sim \left(\frac{\alpha_s}{\alpha}\right)^2 \left(10^{-23}\,\text{s}\right) \sim 10^{-17}\,\text{s}. \quad (17.61)$$

The neutral pion decays by

$$\pi^0 \rightarrow \gamma + \gamma \qquad (17.62)$$

with a lifetime of the only 8.7×10^{-17} seconds. Other examples of electromagnetic decays of hadrons are

$$\eta^0 \rightarrow \gamma + \gamma \quad \text{or} \quad \eta^0 \rightarrow \pi^0 + \pi^0 \quad (17.63)$$

which has a mean lifetime of 6.6×10^{-19} s, and

$$\Sigma^0 \rightarrow \Lambda^0 + \gamma \qquad (17.64)$$

which has a mean lifetime of 5.8×10^{-20} s.

Weak Decays of the Hadrons

All hadrons (except the proton) that do not decay by strong or electromagnetic interactions decay by the weak interaction. One signature of a weak decay is the presence of a neutrino in the final state. Another signature is non-conservation of strangeness. The typical order of magnitude of the lifetime of a particle that decays by the weak interaction is

$$\tau_{\text{weak}} \sim \left(\frac{\alpha_s}{\alpha_w}\right)^2 \left(10^{-23}\,\text{s}\right) \sim 10^{-11}\,\text{s}. \quad (17.65)$$

Examples of weak decays of hadrons are

$$\pi^+ \rightarrow \mu^+ + \nu_\mu, \qquad (17.66)$$

which has a mean lifetime of 2.6×10^{-8} s,

$$K^+ \rightarrow \mu^+ + \nu_\mu \qquad (17.67)$$

which has a mean lifetime of 1.2×10^{-8} s,

$$K^0 \rightarrow \pi^+ + \pi^- \qquad (17.68)$$

which has a mean lifetime of 8.9×10^{-11} s, and

$$\Lambda^0 \rightarrow p + \pi^- \qquad (17.69)$$

which has a mean lifetime of 2.6×10^{-10} s.

Weak decays are discussed in more detail in Chapter 18.

17-5 LEPTONS

The leptons are fermions with spin = 1/2 that have no strong interaction. The leptons also have no detected structure. The electron was the first lepton to be discovered. With the understanding of beta decay came the discovery of the neutrino. The neutrino is similar to the electron except that it has zero mass and zero electric charge. The electron and its neutrino make a *doublet* of particles, according to a conservation law that is observed to hold in all interactions. Each lepton is assigned a quantum number called *lepton number*. The lepton numbers (L_e) for the electron and the neutrino are defined to be

$$L_e = +1. \tag{17.70}$$

The positron and its antineutrino make a doublet of antileptons; the lepton numbers for the positron and the antineutrino are defined to be −1. The conservation law is called lepton number conservation:

$$\sum L_e = \text{constant}. \tag{17.71}$$

The net lepton number does not change in any interaction. A violation of lepton number conservation has never been observed.

EXAMPLE 17-5

Which of the following interactions are allowed by lepton number conservation (a) $p \rightarrow n + e^+$, (b) $e^+ + e^- \rightarrow v + \bar{v}$, (c) $e^+ + n \rightarrow p + v$, (d) $e^- + p \rightarrow n + v$?

SOLUTION:

Lepton number conservation gives

(a) $\qquad 0 \rightarrow 0 - 1 \qquad$ (not allowed).

(b) $\qquad -1 + 1 \rightarrow 1 - 1 \qquad$ (allowed).

(c) $\qquad -1 + 0 \rightarrow 0 + 1 \qquad$ (not allowed).

(d) $\qquad 1 + 0 \rightarrow 0 + 1 \qquad$ (allowed). ∎

The Muon

A second type of charged lepton, the muon, was discovered in cosmic ray experiments (see Figure 17-2). The muon is a fermion with spin = 1/2. The muon comes in two electric charges, μ^+ and μ^-. The μ^+ and μ^- are particle–antiparticle. The muon is much more massive than the electron. The mass energy of the muon is

$$mc^2 = 105.7 \text{ MeV}. \tag{17.72}$$

The muon is not stable. It is observed to decay by the weak interaction,

$$\mu^- \rightarrow e^- + v + \bar{v}, \tag{17.73}$$

with a lifetime of about 2.2 μs. Notice that this decay violates lepton number conservation unless the muon has the same lepton number as the electron, or the neutrino and antineutrino do not have a total lepton number of zero! It was discovered that lepton number is conserved in muon decay (17.73) and the v and $\bar{v}$ are different species; the $\bar{v}$ is associated with the electron and the v is associated with the muon.

Discovery of Two Neutrinos

An experiment critical for our understanding of leptons was performed in 1961 by a group of physicists led by Leon Lederman, Melvin Schwartz, and Jack Steinberger. The goal of the experiment was to see if the neutrino that was present with the muon in pion decay ($\pi^+ \rightarrow \mu^+ + v$) was the same as the neutrino that was present with the positron in beta-plus decay. The concept of the experiment was to generate a beam of "muon neutrinos" from pion decay, observe their interactions in matter, and see if the "muon neutrinos" would produce both muons and positrons. This experiment was the first to observe the interaction of high-energy neutrinos. The feasibility of the experiment was first recognized by Schwartz and Bruno Pontecorvo. The great challenge of the experiment lies in the fact that the neutrino cross section is extremely small.

EXAMPLE 17-6

The weak interaction coupling strength at an energy of 1 GeV is roughly $\alpha_w = 10^{-6}$. Estimate the order of magnitude of the cross section for a 1 GeV neutrino interacting with a nucleon.

SOLUTION:

The cross section is given by the size (R) of the neutrino times the square of the coupling constant (α_w) because the weak interaction has a short range.

$$\sigma \approx \alpha_w^2 \pi R^2.$$

The size of the neutrino is given by its deBroglie wavelength,

$$R = \frac{h}{p} = \frac{hc}{pc} = \frac{hc}{E}.$$

Therefore, our estimate of the cross section is

$$\sigma \approx \alpha_w^{\,2} \, \pi \left(\frac{hc}{E} \right)^2$$

$$\approx \left(10^{-6} \right)^2 \pi \left[\frac{1.2 \ \text{GeV} \cdot \text{fm}}{1 \ \text{GeV}} \right]^2$$

$$\approx 10^{-42} \ \text{m}^2 . \qquad \blacksquare$$

The neutrino cross section (per nucleon) at neutrino energies of 1 GeV is about $10^{-42} \ \text{m}^2$.

A large flux of neutrinos and a massive detector are needed to observe neutrino interactions. We should point out that the neutrino cross section *increases* with energy even though the neutrino size is decreasing as the energy squared. This is because the weak coupling (α_w) is also energy dependent. The weak coupling increases with energy. This important concept is discussed in Chapter 18.

EXAMPLE 17-7

Estimate the neutrino flux necessary (at 1 GeV) to have an interaction rate of one per hour in a 10-ton detector.

SOLUTION:

The required neutrino flux (Φ) is the scattering rate (R) divided by the total cross section (σ_T):

$$\Phi = \frac{R}{\sigma_T}.$$

The total cross section is the cross section per nucleon times the number of nucleons in the target (N_n):

$$\sigma_T = \sigma N_n .$$

The number of nucleons in a target of mass $M = 10^4$ kg is

$$N_n \approx \frac{MN_A}{10^{-3} \ \text{kg}} = \frac{\left(10^4 \ \text{kg} \right) \left(6 \times 10^{23} \right)}{10^{-3} \ \text{kg}} = 6 \times 10^{30} .$$

The flux corresponding to a scattering rate of one per hour is

$$\Phi = \frac{R}{\sigma N_n} = \frac{1}{\left(3600 \, \text{s} \right) \left(10^{-42} \ \text{m}^2 \right) \left(6 \times 10^{30} \right)}$$

$$\approx 5 \times 10^7 \ \text{m}^{-2} \cdot \text{s}^{-1} .$$

A large number of neutrinos and a large target are necessary to cause a measurable interaction rate! $\blacksquare$

The neutrino experiment proposed by Schwartz was carried out at the Brookhaven AGS . The experimental arrangement is indicated in Figure 17-12. An intense beam of 15-GeV protons is sent into a beryllium target, where charged pions (π^+ and π^-) are copiously produced by the strong interaction. The pions decay by the weak interaction,

$$\pi^+ \rightarrow \mu^+ + \nu \qquad (17.74)$$

and

$$\pi^- \rightarrow \mu^- + \overline{\nu} . \qquad (17.75)$$

The neutrinos and antineutrinos are directed into an aluminum target, which is shielded by 13.5 m of iron. The purpose of the iron is to shield the target from muons, pions that do not decay, and other particles from the beryllium target. Only neutrinos can penetrate the iron shield. The interaction of the neutrinos is monitored in a target that is instrumented with spark chambers.

Two types of events were carefully searched for, events with muons from

$$\nu + n \rightarrow \mu^- + p \qquad (17.76)$$

or

$$\overline{\nu} + p \rightarrow \mu^+ + n , \qquad (17.77)$$

and events with electrons or positrons from

$$\nu + n \rightarrow e^- + p \qquad (17.78)$$

or

$$\overline{\nu} + p \rightarrow e^+ + n. \qquad (17.79)$$

Several events containing muons were observed, whereas no events containing electrons were observed. Recall that the origin of the neutrinos is pion decay. These neutrinos are produced in association with a muon. Such neutrinos can produce muons when they interact with matter, but not electrons. The experiment indicates that there must be two

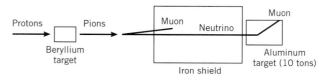

FIGURE 17-12 Experiment to demonstrate the existence of two neutrinos (muon-neutrino and electron-neutrino).

different types of neutrinos, the *muon-neutrino* (ν_μ) and the *electron-neutrino* (ν_e). The neutrinos from pion decay are muon-neutrinos. The neutrinos from beta decay are electron neutrinos. The muon and its neutrino make a second doublet of leptons. A muon lepton number (L_μ) distinct from L_e is conserved in all interactions,

$$\sum L_\mu = \text{constant}. \qquad (17.80)$$

Discovery of a Third Lepton Family

In 1975, a third lepton doublet was discovered by Martin Perl and collaborators. The third family is made up of the *tau* particle (τ) and *tau-neutrino* (ν_τ). The mass energy of the tau is

$$mc^2 = 1777 \text{ MeV}. \qquad (17.81)$$

The tau is not stable and decays by

$$\tau^- \rightarrow \mu^- + \bar{\nu}_\mu + \nu_\tau, \qquad (17.82)$$

$$\tau^- \rightarrow e^- + \bar{\nu}_e + \nu_\tau, \qquad (17.83)$$

or

$$\tau^- \rightarrow \text{hadrons} + \nu_\tau, \qquad (17.84)$$

with a mean-lifetime of about 0.3 ps. Figure 17-13 shows a tau pair event produced in electron–positron collisions, through the process

$$e^- + e^+ \rightarrow \tau^+ + \tau^-. \qquad (17.85)$$

The τ^+ has decayed into $\mu^+ + \nu_\mu + \bar{\nu}_\tau$ and the τ^- has decayed into $e^- + \bar{\nu}_e + \nu_\tau$. The four neutrinos leave no direct trace in the detector, but carry energy and momentum. The signature of the tau pair is an event with an electron and muon of opposite charge with no other particles detected. The angle between the electron and muon is *approximately* 180°, and there is a large missing energy, that is, the energy of the electron plus the energy of the muon in the final state is smaller than the energy of the electron and positron in the initial state.

The interaction of the tau-neutrino has not been directly observed.

Conservation of Lepton Numbers

Each of the three lepton families (see Table 17-4) has its own lepton quantum number. In addition to L_e conservation (17.71) and L_μ conservation (17.80), we have tau lepton number (L_τ) conservation:

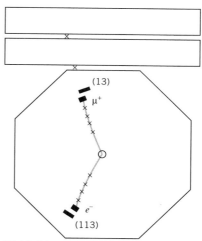

FIGURE 17-13 Discovery of the tau lepton.
A $\tau^+ + \tau^-$ pair is produced in $e^+ + e^-$ annihilations. The τ^+ decays into $\mu^+ + \nu_\mu + \bar{\nu}_\tau$ and the τ^- decays into $e^- + \bar{\nu}_e + \nu_\tau$. Only the muon and the electron are directly observable by their tracks in spark chambers. The energy deposited in an electromagnetic calorimeter is proportional to the displayed number (13 for the muon and 113 for the electron). The muon penetrates the electromagnetic calorimeter while the electron causes a shower. From the Mark-I Collaboration, reported by G. J. Feldman and M. L. Perl, *Phys. Rep.* **19C**, 233 (1975).

**TABLE 17-4
LEPTON NUMBERS.**

Lepton	L_e	L_μ	L_τ
e^-	1	0	0
ν_e	1	0	0
e^+	−1	0	0
$\bar{\nu}_e$	−1	0	0
μ^-	0	1	0
ν_μ	0	1	0
μ^+	0	−1	0
$\bar{\nu}_\mu$	0	−1	0
τ^-	0	0	1
ν_τ	0	0	1
τ^+	0	0	−1
$\bar{\nu}_\tau$	0	0	−1

$$\sum L_\tau = \text{constant}. \qquad (17.86)$$

The lepton quantum numbers are additive and are observed to be absolutely conserved in collisions. The decays of the leptons are summarized in Table 17-5. Some of the most sensitive searches for lepton number violation are a process called neutrinoless double beta decay ($\Delta L_e = 2$):

$$^A_Z N \rightarrow {^A_{Z+2}} N + e^- + e^-, \qquad (17.87)$$

neutrinoless muon decay, ($\Delta L_e = \Delta L_\mu = 1$):

$$\mu^- \rightarrow e^- + \gamma, \qquad (17.88)$$

or $\Delta L_e = -\Delta L_\mu = 1$

$$\mu^- + {^A_Z} N \rightarrow e^+ + {^A_{Z-2}} N. \qquad (17.89)$$

Another type of search for lepton number violation is

$$\nu_\mu \rightarrow \nu_e, \qquad (17.90)$$

a process called *neutrino oscillation*. None of the above processes has been observed.

> Lepton numbers are conserved in all interactions.

17-6 HEAVY QUARKS

Discovery of the Charm Quark

In November 1974, a time that is still referred to in particle physics as the "November Revolution," two groups of

TABLE 17-5
DECAY CHAINS OF THE LEPTONS.

Lepton	Mass (MeV)	Lifetime	Decay mode
Electron	0.511	Stable	
Muon	105.7	2.2 μs	$\mu^- \rightarrow e^- + \bar{\nu}_e + \nu_\mu$
Tau	1777.	3.0 ps	$\tau^- \rightarrow e^- + \bar{\nu}_e + \nu_\tau,$
			$\tau^- \rightarrow \mu^- + \bar{\nu}_\mu + \nu_\tau,$
			$\tau^- \rightarrow \text{hadrons} + \nu_\tau$
Neutrinos	0.	Stable	

experimenters led by Burton Richter at SLAC (the Mark-I collaboration) and Samuel Ting at Brookhaven announced the discovery of a new particle. The mass of the new particle was about 3.1 GeV/c^2, the heaviest known particle at the time. The SLAC group called the new particle ψ and the Brookhaven group called the new particle J, resulting in the name J/ψ. (The names J, ψ, and J/ψ are used interchangeably.) At SLAC the new particle was discovered in the process,

$$e^+ + e^- \rightarrow \psi \rightarrow \text{hadrons}, \qquad (17.91)$$

whereas at Brookhaven the particle was discovered in the process,

$$p + p \rightarrow J + X, \quad J \rightarrow e^+ + e^-. \qquad (17.92)$$

The ψ was created by electron–positron annihilation and observed by its decay into hadrons, and the J was created from hadrons and observed by its decay into electron–positron. Both groups measured the mass and determined upper limits on the width of the new particle, providing convincing evidence they were observing the same new state of matter.

The ψ particle mass spectrum measured by the Mark-I collaboration is shown in Figure 17-14. Since the reaction produces a state containing a single particle, the mass energy of the ψ particle is equal to the total energy of the annihilating electron and positron. The mass spectrum of Figure 17-14 was made by varying the total electron–positron energy and measuring the cross section for producing hadrons. The cross section becomes very large, increasing by a factor of about 100, when the electron–positron energy is equal to the mass energy of the ψ. The finite width of the mass spectrum shown in Figure 17-14, about 2 MeV, is due to the energy spread of the electron and positron beams. The natural width of the J/ψ particle is much smaller than this. Recall that the measurement of the width (Γ) of the particle is a measurement of the particle lifetime (τ) by the uncertainty relationship:

$$\Gamma\tau = \hbar. \qquad (17.93)$$

A small width corresponds to a long lifetime. The distinguishing property of the J/ψ particle is that it is extremely long-lived compared to other particles that decay by the strong interaction. The natural width of the J/ψ particle may be determined by measurement of the total $e^+ + e^-$ production cross section and fraction of the time that the J/ψ decays into $e^+ + e^-$. The result of this measurement gives $\Gamma = 63$ keV.

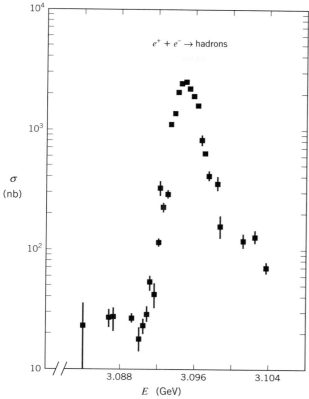

FIGURE 17-14 Discovery of the *ψ* in *e⁺*+*e⁻* collisions at SLAC.
The *ψ* particle is a spin-1 bound state of charm–anticharm quarks. The data shown here were obtained by varying the energy of the electron and positron in small steps and measuring the annihilation rate into hadrons at each step. From the Mark-I Collaboration, reported by G. J. Feldman and M. L. Perl, *Phys. Rep.* **19C**, 233 (1975).

EXAMPLE 17-8

Determine the lifetime of the *J/ψ* particle from the measurement of the natural width.

SOLUTION:

The width is

$$\Gamma = 63\,\text{keV}.$$

From the uncertainty principle, we have

$$\Gamma\tau = \hbar.$$

Therefore,

$$\tau = \frac{\hbar}{\Gamma} = \frac{6.6\times10^{-16}\ \text{eV}\cdot\text{s}}{6.3\times10^{4}\ \text{eV}} \approx 10^{-20}\ \text{s}.$$

This is about three orders of magnitude longer than the typical lifetime of a particle decaying by the strong interaction. ∎

The interpretation of the *ψ* as a bound state of a new quark and antiquark was soon proven to be correct by further detailed experimentation. The new quark has a quantum number called *charm* (*C*),

$$C = +1, \tag{17.94}$$

analogous to strangeness. (The value of *C* is zero for the up, down, and strange quarks.) The electric charge of the charm quark is

$$q = +\frac{2}{3}e. \tag{17.95}$$

The *ψ* is a *vector* meson; it has an intrinsic angular momentum quantum number of one and negative parity (*jᵖ* = 1⁻). The *ψ* is long-lived because it cannot decay into hadrons containing the charm quark. The decay into such hadrons is forbidden by energy conservation. The mass of the lightest particle that contains the charm quark is called the *D* meson. The *D* and anti-*D* mesons have equal mass (*m_D*). The mass of the *ψ* is smaller than twice the mass of the *D* meson,

$$m_\psi < 2m_D. \tag{17.96}$$

The *ψ* decays into three gluons by the process indicated in Figure 17-15.

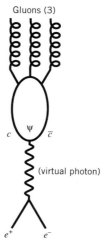

FIGURE 17-15 Feynman diagram for the process, *e⁺*+ *e⁻* → *ψ* → hadrons.

Ten days after the initial discovery, a second narrow resonance was discovered at SLAC. The second resonance has a mass of about 3.7 GeV/c^2 and is called the ψ'. The ψ' is also below threshold for decay into D and anti-D mesons. The ψ' was observed to decay

$$\psi' \rightarrow \pi^+ + \pi^- + \psi, \qquad (17.97)$$

as shown in Figure 17-16.

Charmonium and the Strong Interaction Potential

The two heavy quarks (charm and anticharm) inside the ψ particle form a bound state analogous to an atomic bound state. A bound state of charm and anticharm quarks is called *charmonium*. In charmonium the quarks are attracted by the strong interaction, analogous to the electromagnetic attraction of charges in an atom. An important difference between charmonium and an atom is the form of the potential energy of attraction. Another difference is that the quark and antiquark have identical masses. In the description of an atom, we could largely forget about the motion of the nucleus and describe the atom in terms of the quantum states of the electrons. In the case of charmonium, the quark and antiquark are equally important. A third difference is that the quark and antiquark inside the ψ particle move at relativistic speeds.

In spite of these differences, the quantum mechanics of the bound quarks is much the same as that of electrons in atoms. The energy levels are quantized and may be char-

acterized by the quantum numbers of the quarks. These quantum numbers are the principal quantum number n, the total intrinsic angular momentum quantum number s, the orbital angular momentum quantum number ℓ, and the total angular momentum quantum number j. The ψ particle is the state with $n = 1$, $\ell = 0$, and $s = 1$, making a total angular momentum quantum number $j = 1$. In the spectroscopic notation ($n^{2s+1}L_j$), the ψ particle is referred to as the 1^3s_1 state, or simply $\psi(1s)$. The ψ' is the first excited state ($n = 2$), also with $\ell = 0$, $s = 1$, and $j = 1$. The ψ' is the 2^3s_1 state, or $\psi(2s)$.

The energy diagram for charmonium is shown in Figure 17-17. Only those states below D meson pair threshold are shown. Above D meson pair threshold are many more states ($n > 2$) that have short lifetimes because they decay into charmed particles. The states shown may not decay into charmed particles because of energy conservation, resulting in long lifetimes. The lines indicate observed photon transitions between the states.

The energy levels of the charmonium states were determined by accurate measurements of the photons appearing in $\psi(2s)$ decays. These measurements are ex-

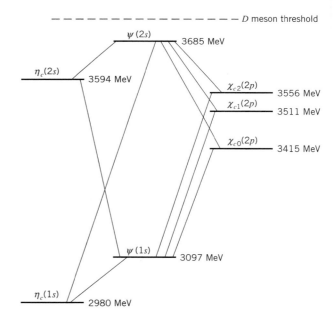

$s=0 \quad \ell=0 \quad j=0 \qquad s=1 \quad \ell=0 \quad j=1 \qquad s=1 \quad \ell=1 \quad j=0,1,2$

FIGURE 17-17 Energy levels of charmonium, bound states of charm–anticharm quarks.
The energy threshold for a pair of particles (D mesons) each with a single charm quark is indicated.

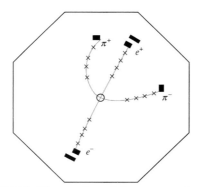

FIGURE 17-16 Observation of the process $e^+ + e^- \rightarrow \psi' \rightarrow \pi^+ + \pi^- + \psi$, with $\psi \rightarrow e^+ + e^-$.
Nature was telling the experimenters that the name of this new family of particles was "psi." From the Mark-I Collaboration, reported by G. J. Feldman and M. L. Perl, *Phys. Rep.* **19C**, 233 (1975).

perimentally challenging because of the large background of photons from π^0 decays. Data from the Crystal Ball detector at SPEAR are shown in Figure 17-18.

The measurement and analysis of charmonium was extremely important in the development of the theory of the strong interaction, quantum chromodynamics (QCD). The ψ particle is unlike any other meson discovered before it because of its large mass. Although the pion may appear simple in the quark model ($q\bar{q}$), the pion is actually a very complicated object; the light quarks are very easily created and destroyed inside the pion because their mass is so small compared to the meson mass. Light quarks are also created and destroyed inside the ψ particle, but the much heavier charm quarks are not. The speed of the heavy quarks in charmonium is also substantially smaller ($\beta \approx 1/2$) than the speed of light quarks inside the pion ($\beta \approx 1$). This means that relativistic effects are much less important. The approximate mass of the charm quark (m_c) is

$$m_c \approx \frac{m_\psi}{2}. \tag{17.98}$$

Measurement of the charmonium spectrum (Figure 17-18) allows the determination of the potential energy function of the quark–quark interaction and the accurate prediction of the masses of a complete set of bound states with various spin and orbital momentum configurations.

EXAMPLE 17-9
Estimate the speed of a charm quark in the ψ particle.

SOLUTION:
Consider a standing wave in a box that has a size of 1 fm. The minimum wavelength of the charm quark is given by

$$\frac{\lambda}{2} \approx 1 \text{ fm}.$$

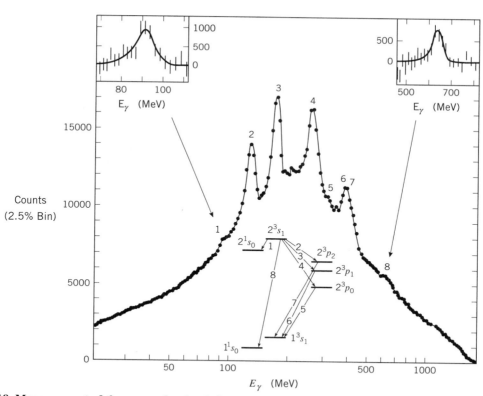

FIGURE 17-18 Measurement of the energy levels of charmonium.
The energies of photons from the $\psi(2s)$ are measured from their electromagnetic interactions in crystals of NaI(Tl). Each photon "line" corresponds to a transition between charmonium energy levels. The inserts show the transitions to the singlet states after the contributions from π^0s are subtracted. From the Crystal Ball Collaboration, reported by E. D. Bloom and C. W. Peck, "Crystal Ball Physics," *Ann. Rev. Nucl. Part. Science* **33**, 143 (1983).

The momentum of the charm quark is

$$p = \frac{h}{\lambda}$$

or

$$pc = \frac{hc}{\lambda} = \frac{1.24 \text{ GeV} \cdot \text{fm}}{2 \text{ fm}} = 0.6 \text{ GeV}.$$

The total energy is

$$E = \sqrt{\left(mc^2\right)^2 + \left(pc\right)^2}.$$

The speed of the quark is

$$v = \frac{pc}{E} c = \frac{pc}{\sqrt{\left(mc^2\right)^2 + \left(pc\right)^2}} c$$

$$\approx \frac{0.6 \text{ GeV}}{\sqrt{\left(1.5 \text{ GeV}\right)^2 + \left(0.6 \text{ GeV}\right)^2}} c$$

$$\approx 0.4 \ c. \qquad ■$$

The charmonium potential may be fit to the form

$$V = -\frac{\kappa_1}{r} + \kappa_2 r, \qquad (17.99)$$

with $\kappa_1 \approx 0.05$ GeV·fm and $\kappa_2 \approx 1$ GeV/fm. The first term corresponds to a Coulomb-like attraction, which dominates at small values of r. The linear term dominates at large values of r and is responsible for the confinement of the quarks. An infinite amount of energy would be required to separate the quarks to an arbitrarily large distance.

The strong interaction potential is shown in Figure 17-19. Examination of the energy levels of the $2p$ states (see Figure 17-17) shows that they are significantly lower in energy than the $2s$ state. Recall that for a pure $1/r$ potential, such as the hydrogen atom, the $2s$ and $2p$ states are degenerate. The wave function squared of the $2s$ states is larger than for the $2p$ state at small values of r. The presence of the linear term in charmonium makes the force stronger at larger r, giving the p states a significantly lower energy than the s state.

EXAMPLE 17-10

Use the charmonium potential (17.99) to estimate the size of the dimensionless strong interaction coupling constant (α_s) at a distance of 0.1 fm.

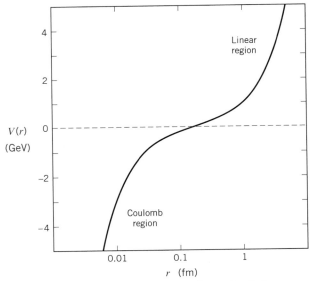

FIGURE 17-19 The strong interaction potential.

SOLUTION:

The electromagnetic coupling in terms of the electromagnetic potential V is

$$\alpha = \frac{ke^2}{\hbar c} = \frac{\dfrac{dV}{dr} r^2}{\hbar c}.$$

The analogous dimensionless strong coupling is

$$\alpha_s = \frac{\kappa_1 + \kappa_2 r^2}{\hbar c}$$

$$\approx \frac{(30 \text{ MeV} \cdot \text{fm}) + (1000 \text{ MeV/fm})(0.1 \text{ fm})^2}{200 \text{ MeV} \cdot \text{fm}}$$

$$\approx 0.2. \qquad ■$$

Discovery of the Bottom Quark

In 1977, history was repeated with the discovery of a new narrow-width resonance (Υ) with a mass of 9.4 GeV/c^2 at Fermilab by a collaboration of scientists led by Leon Lederman. The process studied was

$$p + N \rightarrow \mu^+ + \mu^- + X, \qquad (17.100)$$

where N was a copper or platinum nucleus. The spectrometer had a muon-pair mass resolution of about 2%. In analysis of muon pairs, Lederman found an excess of events at a mass of 9.4 GeV/c^2 above an exponentially falling smooth background. This particle was immediately interpreted as a bound state of a new quark–antiquark pair (bottom). The bottom quark has an electric charge of

$$q = -\frac{1}{3}e, \qquad (17.101)$$

and it has a property called *bottom* or *beauty* that no other quark has. (Compare to strangeness and charm.) A detailed study of the bottom–antibottom bound states was made at DESY and CESR directly analogous to the work done at SLAC a few years previously. The speed of a bottom quark in a upsilon particle is about 10% of c, and determination of the quark–quark interaction potential was refined from this important work.

Where Is the Top Quark?

The properties of the bottom quark indicate that it too is part of a doublet. The missing partner must have

$$q = +\frac{2}{3}e. \qquad (17.102)$$

The top quark would complete the third group (see Figure 1-14). *Why has the top quark not been observed?* The most sensitive searches for the top quark come from electron–positron annihilation. The top quark is expected to be observed through the process

$$e^+ + e^- \rightarrow t + \bar{t} \rightarrow \text{hadrons}, \qquad (17.103)$$

analogous to the manner in which the other five quarks are produced. The only credible reason that can be given for the lack of observation of the top quark is that its mass is beyond the sensitivity of the present experiments. Electron–positron collisions have been studied at a maximum total energy of about 90 GeV, so we know that the mass of the top quark must be greater than about 45 GeV/c^2. There are even higher mass limits from hadron colliders, but the sensitivity of these experiments depends on assumptions about the decay modes of the top quark. Data from a variety of experiments collectively indicate that the mass of the top quark is probably somewhere in the range from 100–200 GeV/c^2.

CHAPTER 17: PHYSICS SUMMARY

- All fermions are observed to have unique antiparticles.

- The strongly interacting free particles (hadrons) are tightly bound states of quarks that have no net color. There are two types of hadrons: (1) mesons with a net quark content of a quark–antiquark and (2) baryons with a net quark content of quark–quark–quark (antiquark–antiquark–antiquark for an antibaryon).

- Baryon number and lepton number are conserved in all interactions.

- Strong interactions conserve isospin, strangeness, charm, and bottom quantum numbers.

- Five flavors of quarks are observed; up (u), down (d), strange (s), charm (c), and bottom (b). The properties of the quarks indicate the existence of a massive sixth quark, top (t).

- The strong interaction potential is of the form

$$V(r) = -\frac{\kappa_1}{r} + \kappa_2 r,$$

where κ_1 and κ_2 are constants. The linear part is responsible for the confinement of quarks inside hadrons. At small distances or large energies the Coulomb term dominates. Quarks are asymptotically free; at short distances quarks behave as free particles.

REFERENCES AND SUGGESTIONS FOR FURTHER READING

R. B. Brode, "The Mass of the Mesotron," *Rev. Mod. Phys.* **21**, 37 (1949).

R. N. Cahn and G. Goldhaber, *The Experimental Foundations of Particle Physics,* Cambridge Univ. Press (1989).

O. Chamberlain, R. F. Mozley, J. Steinberger, and C. Wiegand, "A Measurement of the Positive π-μ-Decay Lifetime," *Phys. Rev.* **79**, 394 (1950).

O. Chamberlain, E. Segrè, C. Wiegand, and T. Ypsilantis, "Observation of Antiprotons," *Phys. Rev.* **100**, 947 (1955).

G. F. Chew, M. Gell-Mann, and A. H. Rosenfeld, "Strongly Interacting Particles," *Sci. Am.* **210**, No. 2, 74 (1964).

G. Danby et al., "Observation of High-Energy Neutrino Reactions and the Existence of Two Kinds of Neutrinos," *Phys. Rev. Lett.* **9**, 36 (1962).

S. D. Drell, "Electron-Positron Annihilation and the New Particles," *Sci. Am.* **232**, No. 6, 50 (1975).

G. Feinberg and M. Goldhaber, "The Conservation Laws of Physics," *Sci. Am.* **209**, No. 4, 36 (1963).

G. J. Feldman and M. L. Perl, "Electron-Positron Annihilation Above 2 GeV and the New Particles," *Phys. Rep.* **19C**, 233 (1975).

M. Gell-Mann, "A Schematic Model of Baryons and Mesons," *Phys. Lett.* **8**, 214 (1964).

S. W. Herb et al., "Observation of a Dimuon Resonance at 9.5 GeV in 400-GeV Proton-Nucleus Collisions," *Phys. Rev. Lett.* **39**, 252 (1977).

N. Isgur and G. Karl, "Hadron Spectroscopy and Quarks," *Phys. Today* **36**, No. 11, 36 (1983).

L. Lederman, "The Two-Neutrino Experiment," *Sci. Am.* **208**, No. 3, 60 (1963).

L. M. Lederman, "Observations in Particle Physics from Two Neutrinos to the Standard Model," *Rev. Mod. Phys.* **61**, 547 (1989).

R. Oppenheimer, "Thirty Years of Mesons," *Phys. Today* **19**, No. 11, 51 (1966).

Pions to Quarks: Particle Physics in the 1950s, ed. L. M. Brown, M. Dresden, and L. Hoddeson, Cambridge University Press (1989).

N. P. Samios and W. B. Fowler, "The Omega-Minus Experiment," *Sci. Am.* **191**, No. 4, 36 (1964).

M. Schwartz, "Feasibility of Using High-Energy Neutrinos to Study the Weak Interaction," *Phys. Rev. Lett.* **4**, 306 (1960).

M. Schwartz, "The First High-Energy Neutrino Experiment," *Rev. Mod. Phys.* **61**, 527 (1989).

R. F. Schwitters, "Fundamental Particles with Charm," *Sci. Am.* **237**, No. 4, 56 (1977).

E. Segrè and C. E. Wiegand, "The Antiproton," *Sci. Am.* **194**, No. 6, 37 (1956).

J. Steinberger, "Experiments with High Energy Neutrino Beams," *Rev. Mod. Phys.* **61**, 533 (1989).

G. Zweig, "An SU_3 Model for Strong Interaction Symmetry and Its Breaking," CERN Report 8182/Th. 401, unpublished (1964).

QUESTIONS AND PROBLEMS

Discovery of the mesons

1. Why did Occhialini take the emulsions to the top of a mountain to expose them to cosmic rays?

2. Calculate the kinetic energy of a muon produced from the decay of a pion at rest.

3. Why is the electron from muon decay not visible in Figure 17-3*a* whereas the pion and muon are clearly visible? Compare to Figure 17-3*b*.

Quantum numbers of the pion

*4. (a) A charged pion has a kinetic energy of 50 MeV when it enters a block of lead. What is its initial rate

of ionization loss? (b) Estimate the range of the pion.

5. In the experiment of Chamberlain et al. to determine the charged pion lifetime, show that the time taken for the pion to stop is much shorter than the pion lifetime.

6. What are some possible sources of error in measuring the pion lifetime by the technique of Chamberlain et al.?

7. In Chapter 12 we showed that the density of states (12.67) was proportional to the momentum squared divided by the speed. This expression may be applied to a collision ($A + B \rightarrow C + D$) in the center-of-mass frame provided that we take the speed to be the relative speed of the particles C and D. (a) Show that the cross section for $p + p \rightarrow \pi + d$ may be written as

$$\sigma_{p+p \rightarrow \pi+d} = \frac{A(2s_\pi + 1)(2s_d + 1)p_\pi^2}{v_{pp} v_{\pi d}},$$

where p_π is the pion momentum in the center-of-mass frame, $v_{\pi p}$ is the relative speed of the p and d, v_{pp} is the relative speed of the colliding protons, s_π and s_d are the spin quantum numbers of the pion and deuteron, and A is a constant. (b) Show that the cross section for $\pi + d \rightarrow p + p$ may be written as

$$\sigma_{\pi+d \rightarrow p+p} = \frac{A(2s_p + 1)2 p_p^2}{2 v_{pp} v_{\pi d}},$$

where p_p is the proton momentum in the center-of-mass frame and s_p is the spin quantum number of the proton. (c) Show that the pion spin is given by

$$s_\pi = \frac{p_\pi^2 \sigma_{p+p \rightarrow \pi+d}}{3 p_p^2 \sigma_{\pi+d \rightarrow p+p}} - \frac{1}{2}.$$

(d) The cross section for $\pi + d \rightarrow p + p$ was first measured at pion center-of-mass momentum of 90 MeV/c to be 3 mb. When stationary protons are bombarded with protons with momentum equal to 340 MeV/c, what is the expected cross section for $p + p \rightarrow \pi + d$?

*8. Show that the existence of the decay $\pi^0 \rightarrow \gamma + \gamma$ proves that the spin of the neutral pion cannot be 1.

The antiproton

9. What are some of the possible backgrounds to the antiproton experiment (besides the contribution of the π^-)?

*10. The motion of the nucleons is an important factor for the antiproton production rate when the energy is very close to threshold. (a) Estimate the energy and momentum of a nucleon in a copper nucleus. (b) If the incident proton and the target nucleon collide head-on, by how much is the center-of-mass energy increased?

Classification of the hadrons: the quark model

11. Which bosons are strongly interacting? Are they "fundamental" particles?

12. Does the existence of the Ω^- baryon constitute evidence for color?

13. A baryon ($B = 1$) has strangeness of -2 and zero electric charge. (a) What are the possible values of the isospin (I) and m_I of the particle? (b) If the particle has spin $= 1/2$, what is the common name of the particle?

14. What are the values of isospin (I) and m_I of the antiproton?

15. Consider the reaction

$$\bar{p} + n \rightarrow \pi^- + \pi^0.$$

Calculate the possible values of the isospin (I), orbital angular momentum (ℓ), and total spin (s) quantum numbers of the initial state in order that the reaction may occur, assuming that $\ell \leq 1$. (*Hint:* Parity is conserved.)

16. Which of the following strong interactions are allowed? If a process is forbidden, state the reason.

(a) $$\bar{p} + p \rightarrow \pi^0 + n$$

(b) $$\pi^- + p \rightarrow K^0 + n$$

(c) $$p + p \rightarrow \pi^+ + n + n$$

(d) $$\bar{p} + p \rightarrow \pi^0 + \pi^+ + \pi^-$$

(e) $$K^- + p \rightarrow \pi^0 + \Lambda^0$$

17. Which of the following decays are allowed? If a decay is allowed, indicate which force is at work. If a decay is forbidden, state the reason.

(a) $$\Sigma^0 \rightarrow \Lambda^0 + \gamma$$

(b) $$n \rightarrow p + \pi^-$$

(c) $$\Delta^- \rightarrow n + \pi^-$$

(d) $$K^0 \rightarrow \pi^0 + \pi^0$$

(e) $$\Xi^0 \rightarrow \Sigma^0 + \pi^0$$

Leptons

18. (a) What is the minimum momentum of an electron produced by the decay of a muon at rest? (b) What is the maximum electron momentum?

19. Muons have a weak interaction similar to the neutrinos. Why can the muons not pass through a thick iron shield, whereas the neutrinos can?

20. In the neutrino experiment proposed by Schwartz, why does the interaction rate not depend on the composition or shape of the target, but only on the target mass? What assumption is made about the size of the target relative to the size of the neutrino beam?

21. Estimate the mean free path of a one GeV neutrino in the earth.

Heavy quarks

22. For the decay $\psi \rightarrow$ hadrons, make a list of some of the allowed hadronic states. What states are not allowed? Why?

23. Why is the transition $\psi' \rightarrow \gamma + \psi$ forbidden?

24. (a) Show that the presence of the linear term in the charmonium potential (17.99) means that there is a component of force between two quarks that does not depend on their separation. (b) Evaluate the constant component of force in Newtons (J/m) and in "tons."

*25. Consider the decays, $\psi' \rightarrow \gamma + \chi$, where χ represents one of three possible charm–anticharm states with $\ell = 1$ (see Figure 17-17). Estimate the relative transition rates. (Hint: The transition rate depends on the cube of the photon energy.)

26. Use the charmonium strong interaction potential (17.99) to estimate the distance at which the dimensionless strong coupling (α_s) is unity.

*27. (a) Show that there are six states of charmonium with $n = 2$. (b) Figure 17-17 shows five observed states with $n = 2$. What are the quantum numbers (s, ℓ, j) of the unobserved state? (c) Make an estimate of the mass of the unobserved state. (d) Why has this state not easily been observed?

28. In Leon Lederman's experiment on the discovery of the upsilon particle, his spectrometer had a dimuon mass resolution of 2%. The upsilon particle was observed as a "bump" in the dimuon mass plot with a full width of 2%. What is known about the lifetime of the upsilon from these data?

29. Estimate the mass of the B meson, the lightest meson containing a b quark.

30. Make an estimate of the speed of a heavy quark in the upsilon particle.

*31. Consider bottomonium, the bound states of $b\bar{b}$. Three resonances, $\Upsilon(1s)$, $\Upsilon(2s)$, and $\Upsilon(3s)$, are observed below the B meson pair threshold. Make an energy level diagram analogous to Figure 17-17, indicating the quantum numbers of all the allowed states and the allowed photon transitions between states.

32. Consider a bound state of quark–antiquark. (a) Use the Bohr model to show that if the Coulomb term in the strong interaction potential (17.99) is neglected, the "Bohr radius" is given by

$$r_1 \approx \left(\frac{\hbar^2 c^2}{mc^2 \kappa_2} \right)^{1/3},$$

where m is the reduced mass. Evaluate r_1 for charmonium. (b) What is the corresponding quark speed? (c) When the Coulomb term is included, does the radius increase or decrease? Does the quark speed increase or decrease?

33. Consider a bound state of quark–antiquark. (a) Use the Bohr model to show that if the linear term in the strong interaction potential (17.99) is neglected, the "Bohr radius" is given by

$$r_1 \approx \frac{\hbar^2 c^2}{mc^2 \kappa_1},$$

where m is the reduced mass. Evaluate r_1 for charmonium. (b) What is the corresponding quark speed? (c) When the Coulomb term is included, does the radius increase or decrease? (d) When the Cou-lomb term is included, Does the quark speed increase or decrease?

*34. When the Schrödinger equation with the charmonium potential (17.99) is solved by numerical integration, the resulting root-mean-square quark speed for the $1s$ state is $v \approx 0.4c$, in agreement with the estimate of Example 17-9. (a) Make an estimate of the average separation between the quarks in the $1s$ state. (b) In the $2s$ state, is the average quark separation larger or smaller than in the $1s$ state? Why? (c) In the $2s$ state is the root-mean-square quark speed larger or smaller than the initial $1s$ state? Why?

Additional problems

35. Show that the process $e^+ + e^- \rightarrow \gamma(\text{virtual}) \rightarrow \pi^0 + \pi^0$ is forbidden by angular momentum conservation. Is the process $e^+ + e^- \rightarrow \pi^0 + \pi^0$, *absolutely* forbidden?

36. (a) Why does the neutral sigma baryon (Σ^0) decay electromagnetically into $\Lambda^0 + \gamma$? b) Why do the charged sigma baryons (Σ^+ and Σ^-) not have electro-magnetic decays?

*37. The Δ^{++} baryon is produced by bombarding protons at rest with π^+ mesons. Make a qualitative sketch of the Δ^{++} production cross section versus the kinetic energy of the pion. Label the energy scale. (Hint: The Δ^{++} decays into $\pi^+ + p$ with a lifetime characteristic of the strong interaction.)

38. Consider the strong interaction potential between two quarks to be quadratic, that is,

$$V = \frac{1}{2}\kappa r^2.$$

(a) Use the observed separation between the $1s$ and $2s$ states of charmonium to estimate the constant κ. (b) Estimate the size of charmonium.

HIGH-ENERGY PHYSICS: UNIFICATION OF THE FORCES

Leptons interact only with photons, and with the intermediate bosons that presumably mediate weak interactions. What could be more natural than to unite these spin-one bosons into a multiplet of gauge fields? Standing in the way of this synthesis are the obvious differences in the masses of the photon and the intermediate (boson), and in their couplings. We might hope to understand these differences by imagining that the symmetries relating the weak and electromagnetic interactions are exact symmetries of the Lagrangian but are broken by the vacuum.

Steven Weinberg

18-1 FROM QUARKS TO QUANTUM CHROMODYNAMICS

The beauty of the quark model lies in its simplicity. Many basic properties of *all* the hadrons are given by the quantum numbers of three (later to be expanded to five) constituents. The quark model correctly predicts the number of hadrons that exist with a given spin and parity. The inventors of the quark model realized, however, that the proton was *not* really composed of just three quarks and the pion was *not* composed of just one quark and one antiquark. The statement of the quark model is rather that the proton has the *quantum numbers* of three quarks and the pion has the quantum numbers of one quark and one antiquark.

In one of the most famous unpublished scientific papers, Zwieg called the hadron constituents *aces*, the mesons *deuces,* and the baryons *treys*. In the concluding remarks of this 1964 paper, Zweig ponders: "There are, however, so many unanswered questions. Are aces particles? Do aces bind to form only deuces and treys? What is the particle (or particles) that is responsible for binding the aces? ..." After pointing out that his classification of particles might well be wrong, Zwieg states: "there is also the outside chance that the model is a closer approximation to nature than we think and that fractionally charged aces abound within us."

The answers to Zweig's questions (yes, yes, and gluons) were supplied in the 1970s, with the development of the theory *quantum chromodynamics* (QCD). The path from quarks to QCD took about two decades of rigorous work by both experimentalists and theoreticians!

Properties of QCD

The quarks inside hadrons are continuously exchanging gluons. As part of this exchange process, the gluons can create new quark–antiquark pairs, especially the light quark pairs $u\bar{u}$ and $d\bar{d}$, and the quark–antiquark pairs can annihilate into gluons. Inside the hadrons, there is a condition of equilibrium,

$$g \leftrightarrow \bar{q} + q, \qquad (18.1)$$

within the limits allowed by the uncertainty principle. Thus, a hadron may be thought of as a "bag" of quarks and gluons. The reaction (18.1) does not create any *net* quark flavor because a quark is always created or annihilated with its antiquark. The quantum numbers of the hadron are not changed by the complicated strong interaction binding process. All hadrons have zero *net* strong charge (color), and their net quark content has zero strong charge even though the quark content of the hadron is continuously changing. The hadrons are *color singlets*, analogous to a spin singlet (a combination of $s = 1/2$ particles with zero net spin). In the theory of QCD, quarks come in *three* colors. There are three simple zero-color combinations of a hadron: (1) quark–antiquark, (2) three quark, or (3) three antiquark.

Asymptotic Freedom

The theory of the strong interaction, quantum chromodynamics, is analogous to the quantum theory of the electromagnetic interaction, quantum electrodynamics (QED). There is, however, a significant qualitative difference between QED and QCD. In QCD the bosons that transmit the force (the gluons) carry the quantity that they couple to, strong charge. *Gluons have strong charge.* In QED the boson that transmits the force (the photon) does not carry the quantity that it couples to, electric charge. *Photons have zero electric charge.* A consequence of this is that photons do not couple directly to photons, while gluons do couple directly to gluons. There is also a quantitative difference in the strengths of the electromagnetic and strong interactions. The value of the dimensionless strong coupling (α_s) in the present range of experiments is much larger than the electromagnetic coupling α. Perturbation theory, the technique of calculating interaction probabilities in powers of the dimensionless coupling, is more difficult in QCD and very often not possible.

One central feature of QCD that has been verified by experiment is the property of *asymptotic freedom.* Asymptotic freedom refers to the fact that while quarks are permanently bound into hadrons, at shorter and shorter distances they do behave more and more as free particles. The concept of asymptotic freedom was realized theoretically in 1973 by H. David Politzer and independently by David Gross and Frank Wilczek. In QCD the strength of the interaction (α_s) is not specified from first principles, just like α in QED cannot be determined theoretically. To determine α we must make a measurement of some quantity that is sensitive to the magnitude of the electric charge of the electron. The situation is more complicated in QCD because the strong interaction coupling is changing quite rapidly with the wavelength of the quark at distances near 1 fm. In terms of the equivalent quark energy ($E = hc/\lambda$), we have $\alpha_s \approx 1$ at $E = 1$ GeV, and $\alpha_s \approx 0.1$ at $E = 100$ GeV. Thus, α_s is often referred to as *the running coupling constant.* This is the property of asymptotic freedom: the fundamental parameter that specifies the quark-quark coupling strength decreases at shorter distances. At short distances the quarks are free of the linear portion of the potential (17.99). The qualitative reason for this is that at shorter distances, the gluon cloud surrounding the quark is penetrated. The gluons carry strong charge so that the effective strong charge of a quark is reduced at shorter distances. The variation of the strong coupling with energy is quantitatively determined in QCD. The value of α_s at a given energy depends on the number of quark flavors (n_f) that can participate in the binding process through their coupling to gluons. At energies much greater than the mass energies of the quarks, all flavors of quarks participate. In specifying the theoretical energy dependence of α_s, there remains one fundamental constant that can be determined only from experiment. The definition of this constant is arbitrary, but conventionally a constant (Λ) that has dimensions of energy is used. The constant Λ represents the energy at which the coupling becomes *infinite*, the mechanism by which quarks are confined in hadrons. The constant Λ then defines a boundary in energy, below which we have hadrons and above which we have quarks. The order-of-magnitude of Λ is the mass energy of the lightest hadron, the pion. In terms of Λ and the number of quark flavors (n_f), the energy dependence of the strong coupling parameter is

$$\alpha_s(E) \approx \frac{12\pi}{\left(33 - 2n_f\right)\ln\left(\dfrac{E^2}{\Lambda^2}\right)}, \qquad (18.2)$$

to lowest order in the energy (E). From a variety of experiments, it is determined that

$$\Lambda \approx 0.2 \text{ GeV}. \qquad (18.3)$$

There is a relatively large error on the experimental determination of Λ ($\pm 50\%$); a large error on Λ corresponds to a small error on α_s because of the logarithmic behavior. The magnitude of α_s as a function of energy is shown in Figure 18-1.

Measurement of Three Colors

A central feature of QCD is that there are three types of strong charge. One early indication that quarks come in three colors is the existence of the Δ^{++} baryon, which is a bound state of three u quarks with their spins aligned. The exclusion principle would imply that each of these three u quarks inside the Δ^{++} baryon would have to have unique quantum numbers. They do. They each have a different

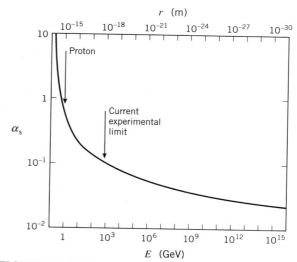

FIGURE 18-1 The running coupling parameter, α_s.

color. (Recall that we have also used the argument of color to explain the nonexistence of three identical quarks in a state of $j = 1/2$ and positive intrinsic parity, i.e., there is no nucleon with charge $+2e$ or $-e$.)

A more dramatic proof of the existence of three colors comes from electron–positron annihilation. In $e^+ + e^-$ annihilation, pairs of charged particles are produced by the electromagnetic interaction. Two Feynman diagrams for $e^+ + e^-$ annihilation is shown in Figure 18-2. The intermediate state is a virtual photon. The virtual photon couples to all fundamental charged fermion–antifermion pairs (i.e., quarks and charged leptons) with only two general restrictions: (1) energy is conserved and (2) the amplitude of the coupling is proportional to the electric charge. Conservation of energy gives a mass-energy

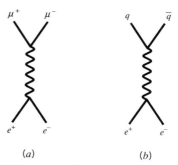

FIGURE 18-2 Feynman diagrams for electron–positron annihilation.
(*a*) Muon pair production, (*b*) quark pair production.

threshold for each pair. A particle pair is not produced if the $e^+ + e^-$ center-of-mass energy is smaller than $2mc^2$, where m is the fermion mass. Above threshold energy, there is a phase-space factor that is important only when the particles are created with small speeds. Therefore, above threshold the particle pair is produced with a probability that is directly proportional to the electric charge squared (amplitude squared).

For electron and positron with momentum p traveling in opposite directions, the annihilation probability is proportional to the amount of overlap of their wave functions. The order of magnitude of the cross section for

$$e^+ + e^- \rightarrow \mu^+ + \mu^- \qquad (18.4)$$

may be estimated from the *size* of an electron with momentum p. The electron size is given by its deBroglie wavelength ($\lambda/2\pi$). Therefore, the order of magnitude of the annihilation cross section is given by

$$\sigma_{\mu^+\mu^-} \approx \pi \left(\frac{\lambda}{2\pi} \right)^2 \alpha^2$$

$$= \pi \left(\frac{\hbar}{p} \right)^2 \alpha^2$$

$$= \frac{4\pi (\hbar c)^2 \alpha^2}{s}, \qquad (18.5)$$

where we have written the total center of mass energy (s) as

$$s = (pc + pc)^2 = 4(pc)^2. \qquad (18.6)$$

The cross section (18.5) is only an order of magnitude estimate but contains the physics of the interaction, namely the factor α^2/s. An exact QED calculation (to lowest order in α) gives

$$\sigma_{\mu^+\mu^-} = \frac{4\pi}{3}\alpha^2 \frac{\hbar^2 c^2}{s} \approx \frac{87 \text{ nb} \cdot \text{GeV}^2}{s}. \qquad (18.7)$$

Since all quarks have electric charge, they too are produced particle–antiparticle in $e^+ + e^-$ annihilations. The quarks do not appear as free particles. The force between the quarks grows without bound as they separate. In this process several hadrons, mostly pions, are created. The cross section for

$$e^+ + e^- \rightarrow q + \bar{q} \rightarrow \text{hadrons} \qquad (18.8)$$

for one color and one flavor is

$$\sigma_{q\bar{q}} = \left(\frac{q_q}{e} \right)^2 \left(\sigma_{\mu^+\mu^-} \right)$$

$$= \left(\frac{q_q}{e} \right)^2 \left(\frac{4\pi}{3}\alpha^2 \frac{\hbar^2 c^2}{s} \right), \qquad (18.9)$$

where (q_q) is the electric charge of the quark. The total annihilation cross section into hadrons (σ_{hadrons}) is obtained by summing over all available quark types,

$$\sigma_{\text{hadrons}} = \sum \sigma_{q\bar{q}}. \qquad (18.10)$$

The hadron cross section divided by the muon-pair cross section is called R. This ratio is

$$R = \frac{\sigma_{\text{hadrons}}}{\sigma_{\mu^+\mu^-}} = \frac{\sum \sigma_{q\bar{q}}}{\sigma_{\mu^+\mu^-}} = \sum \left(\frac{q_q}{e} \right)^2. \qquad (18.11)$$

Thus, for five quarks (u, d, s, c, b) each having one color, we expect

$$R \approx \frac{4}{9} + \frac{1}{9} + \frac{1}{9} + \frac{4}{9} + \frac{1}{9} = \frac{11}{9}. \qquad (18.12)$$

For five quarks each having 3 possible colors,

$$R \approx 3 \left[\frac{4}{9} + \frac{1}{9} + \frac{1}{9} + \frac{4}{9} + \frac{1}{9} \right] = \frac{33}{9}. \qquad (18.13)$$

The data on R at $e^+ + e^-$ center of mass energies from 10 to 50 GeV are shown in Figure 18-3. The data conform spectacularly to two charge 2/3 quarks and three charge 1/3 quarks each coming in three colors! The effects of gluons cause the observed value of R to be increased by a factor of $(1 + \alpha_s/\pi)$.

Observing the Quarks and Gluons

Electron–positron annihilations provided the first direct evidence for the existence of quarks. In the process

$$e^+ + e^- \rightarrow q + \bar{q} \rightarrow \text{hadrons}, \qquad (18.14)$$

the hadrons are observed to appear in two collimated *jets*. This is illustrated in Figure 18-4a. The jets become more pronounced at higher energy.

If the energy of the quark is large enough, it can radiate an energetic gluon (g) analogous to the bremsstrahlung process in QED. The process is

$$e^+ + e^- \rightarrow q + \bar{q} + g \rightarrow \text{hadrons}. \qquad (18.15)$$

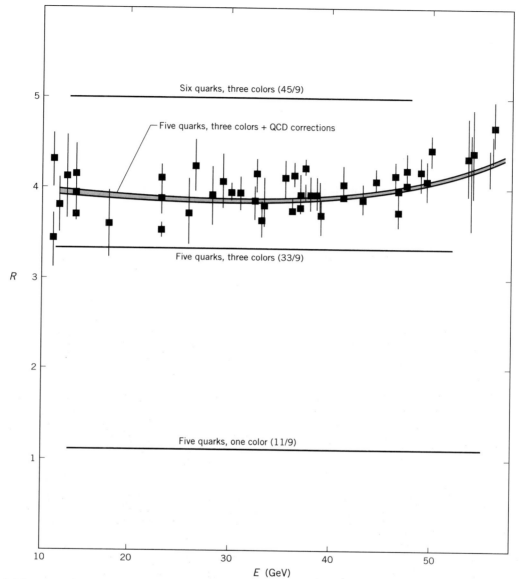

FIGURE 18-3 Measurement of R, the cross section for production of hadrons divided by the muon-pair cross section in $e^+ + e^-$ annihilations.

The value of R, apart from QCD corrections of a few percent, is equal to the sum of the squares of the charges of all quarks that are produced. These data show that there are five quarks (u, d, s, c, b), each with three colors. Adapted from the Particle Data Group, "Review of Particle Properties," *Phys. Rev.* **D45**, 1 (1992).

The energetic gluon occasionally produces a third distinct jet, as shown in Figure 18-4*b*. This process provided the first direct evidence for the strong interaction boson, the gluon.

Quarks Inside the Proton

One of the key questions that particle physicists faced was whether or not the "quarks" produced in electron–positron

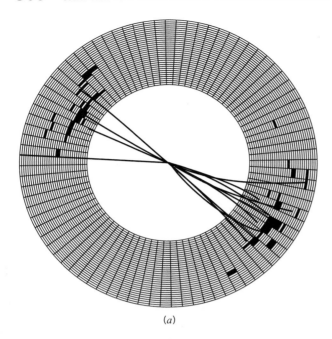

(a)

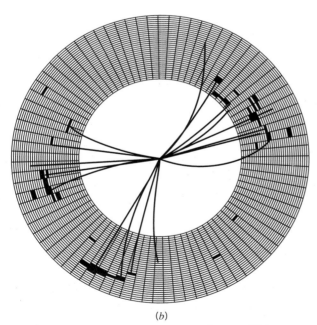

(b)

FIGURE 18-4 Jets from electron-positron collisions.
(*a*) Two-jet event from the process $e^+ + e^- \rightarrow q + \bar{q} + q$. (*b*) Three-jet event from the process $e^+ + e^- \rightarrow q + \bar{q} + g$, where the gluon is energetic. Most events are two-jet events, while three-jet events with an energetic third jet occur a few percent of the time. Detailed analysis of many such events confirmed the existence of the gluon. From B. Naroska, *Phys. Rep.* **148**, 67 (1987).

annihilations were the same as the constituents inside the proton. It is not easy to detect the presence of a particle that is permanently confined inside a hadron ($r \approx 10^{-15}$ m)! This is reflected in a quote from Feynman in his famous paper of 1969 in which he defined the variable *Feynman x*, the fraction of proton momentum carried by an individual constituent or *parton*.

I am more sure of the conclusions than of any single argument which suggested them to me for they have an internal consistency which surprises me and exceeds the consistency of my deductive arguments which hinted at their existence.

Indeed, no *single experiment* of the time could conclusively demonstrate the existence of quarks and gluons inside the proton. A long series of many varied experiments together with much theoretical work allowed the construction and the testing of QCD.

Results from the CERN ISR and Fermilab in the 1970s showed that hadrons were produced at large transverse momentum and that they were clustered into jets. Conclusive data came in 1982 from the CERN proton–antiproton collider. For the first time, quark and gluon collisions were copiously studied at momentum transfers of 50–100 GeV. The observation of jets at very high energies was observed (see Figure 18-5). The clustering of hadrons into jets is caused by the hard scattering of quarks or gluons followed by their "attempted escape" from the proton and antiproton that contained them. When the quarks or gluons scatter, they are nearly free. When the quarks are accelerated, the strong force which confines them in the proton and antiproton (see Figure 1-19) causes the quarks to fragment into jets. The energy and momentum of the jets is approximately that of the struck quarks or gluons, about 70 GeV in the collision of Figure 18-5. The unambiguous observation of jets in hadron collisions was an experimental milestone in the establishment of QCD as the theory of the strong interaction. An impressive quantitative test of quantum chromodynamics is obtained by measuring the rate of jet production in high-energy hadron–hadron scattering experiments.

Measurement of the Strong Coupling, α_s

The value of α_s has been measured in a variety of experiments at different energies. A determination of α_s requires the cross section for a strong interaction process to be both measured and calculated in QCD. The requirements for an accurate determination of α_s are twofold: (1) the experi-

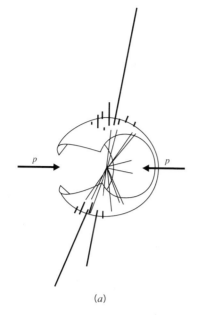

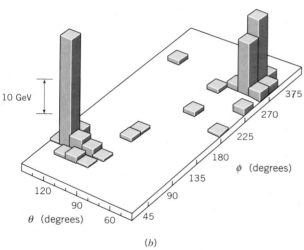

(a)

(b)

FIGURE 18-5

Jet event from the CERN UA2 experiment resulting from the collision of a proton and antiproton each with an energy of 270 GeV. Shown here are (*a*) the tracks of particles together with the energy deposited in calorimeter cells, where the height of the spike is proportional to the particle energy, and (*b*) the energy observed in each cell with the calorimeter unfolded. Theta is the angle with respect to the beam axis and phi is the azimuthal angle. From the UA2 Collaboration, reported by J. P. Repellin, Proc. of the XXIst Int. Conf. on High Energy Physics, Paris (1982).

ment must have small errors, both systematic and statistical, and (2) the QCD calculation must be accurate enough

so that higher-order uncalculated terms in α_s are small and the effects of knowledge of the quark masses are small. The second requirement automatically means that α_s can be determined only at sufficiently large energies where its value is smaller than 1. Part of the quantitative success of QCD is the fact that a variety of strong interaction processes over a wide range of energies give one consistent value of the constant Λ (18.3).

18-2 QUANTUM THEORY OF THE WEAK INTERACTION

Strength of the Weak Interaction

Consider the following two decays,

$$\Delta^+ \rightarrow p + \pi^0, \qquad (18.16)$$

which has a lifetime of 6×10^{-24} s, and

$$\Sigma^+ \rightarrow p + \pi^0, \qquad (18.17)$$

which has a lifetime of 8×10^{-11} s. The Σ^+ particle has a longer lifetime than the Δ^+ particle, by 13 orders of magnitude! This lifetime difference clearly cannot be attributed to the density of states because the Δ^+ and Σ^+ have nearly the same masses. The lifetimes are different because a different force is governing the decay in each case. The decay of the Δ^+ is governed by the strong force. The force that governs the decay of the Σ^+ is the weak force. A stronger force makes a decay happen faster.

The coupling of the quarks inside the Δ^+ particle to the quarks of its decay products (p and π^0) has a strength characterized by α_s. The coupling of the quarks inside the Σ^+ particle to the quarks of its decay products is fundamentally different because the Σ^+ contains a strange quark and there are no strange quarks in the decay products. The Σ^+ cannot decay by the strong interaction because there is no combination of hadrons with strangeness -1 and baryon number $+1$ with a total mass energy smaller than the Σ^+ mass energy (e.g., $m_\Sigma < m_K + m_p$). The Σ^+ decay (18.17) violates strangeness conservation; it is a weak interaction decay. The coupling that governs the weak decay of the Σ^+ (α_w) is much smaller than the strong coupling (α_s). The decay rate, or inverse lifetime, depends on the square of the coupling strength. Therefore, the ratio of lifetimes ($\tau_\Delta / \tau_\Sigma$) allows us to estimate the relative strengths of the strong and weak forces inside the hadrons:

$$\frac{\tau_{\Delta}}{\tau_{\Sigma}} \approx \frac{(\alpha_s)^{-2}}{(\alpha_w)^{-2}}, \qquad (18.18)$$

which gives

$$\frac{\alpha_w}{\alpha_s} \approx \sqrt{\frac{\tau_{\Delta}}{\tau_{\Sigma}}} \approx \sqrt{\frac{6 \times 10^{-24}\ \mathrm{s}}{8 \times 10^{-11}\ \mathrm{s}}} \approx 3 \times 10^{-7}. \ (18.19)$$

The weak force between quarks inside hadrons is about six orders of magnitude weaker than the strong force and about four orders of magnitude weaker than the electromagnetic force ($\alpha_w/\alpha \approx 10^{-4}$). The relative strengths of the strong, electromagnetic, and weak forces at an energy scale of about 1 GeV are

$$\alpha_s \approx 1, \qquad (18.20)$$

$$\alpha \approx \frac{1}{137}, \qquad (18.21)$$

and

$$\alpha_w \approx 10^{-6}. \qquad (18.22)$$

Weak Charge and the Fermi Constant

The weak force can be thought of as due to a *weak charge* (g), analogous to electric charge (e). The weak interaction is mediated by the exchange of massive weak bosons. The strength of the weak interaction between two weak charges is proportional to the weak charge squared divided by the weak boson mass (m_W) squared,

$$\alpha_w \propto \frac{g^2}{m_W^2}. \qquad (18.23)$$

The reason that the weak boson mass squared appears in the denominator is that the weak coupling depends on the inverse square of the momentum transfer squared. (Recall the case of Rutherford scattering where the momentum transfer squared is proportional to the factor $(1-\cos\theta)$. The cross section is proportional to $(1-\cos\theta)^{-2}$.) For the weak interaction, the mass of the weak boson dominates the momentum transfer squared unless the energies are very, very large ($E \gg m_W c^2$).

The weak interaction is "weak" at low energies not because g is small but because m_W is large. We may make an analogy with the electric and magnetic forces. Suppose we knew nothing about the relationship between electric

and magnetic fields. We perform some experiments and observe that the magnetic force on a slowly moving charge is much smaller than the electric force. We may well propose that the electric force on a particle is due to one type of charge (e_1) and the magnetic force is due to a second type of charge (e_2). It could be that the magnetic force is small because $e_2 \ll e_1$. (This *may* have been the correct description of the force!) Then from further experiments we learn how to write the force in terms of fields and that the form of the Lorentz force law is

$$F = e_1 E + e_2 \mathbf{v} \times \mathbf{B}, \qquad (18.24)$$

and that $e_1 = e_2$. The magnetic force is small only because the speed is small.

The weak charge is found to have the same order of magnitude as the electric charge,

$$g \approx e. \qquad (18.25)$$

The relationship between the weak charge and the electric charge will be discussed later in this chapter. The fundamental weak coupling is the same order of magnitude as the electromagnetic coupling. The coupling of the W and Z^0 particles to a quark is just as strong as the coupling of a photon to a quark. There are as many virtual W and Z^0 particles coupling to electrons in an atom (or to quarks inside a hadron) as there are virtual photons! The W and Z^0 particles do not produce the same momentum transfers as the photons because they have such a short range. To emphasize this important concept, we write the weak coupling constant (α_w) in terms of the electromagnetic coupling constant (α),

$$\begin{aligned}
\alpha_w &\approx \left(\frac{kg^2}{\hbar c}\right)\left(\frac{E}{m_W c^2}\right)^2 \\
&\approx \left(\frac{ke^2}{\hbar c}\right)\left(\frac{E}{m_W c^2}\right)^2 \\
&= \alpha \left(\frac{E}{m_W c^2}\right)^2, \qquad (18.26)
\end{aligned}$$

where E is the energy scale at which α_w is evaluated.

EXAMPLE 18-1

Evaluate α_w, at the energy scale of 1 GeV.

SOLUTION:

The mass of the W boson is

$$m_w = 80 \text{ GeV}/c^2.$$

Therefore,

$$\alpha_w \approx \alpha \left(\frac{E}{m_w c^2} \right)^2 = \left(\frac{1}{137} \right) \left[\frac{1 \text{ GeV}}{80 \text{ GeV}} \right]^2 \approx 10^{-6}. \quad \blacksquare$$

The Weak Interaction Constant, G_F

The strength of the weak interaction is conventionally specified in terms of the *Fermi constant* G_F. The Fermi constant is defined by

$$G_F \equiv \frac{4\pi\sqrt{2}\,\hbar^2 c^2 kg^2}{\left(m_w c^2 \right)^2}. \quad (18.27)$$

Since the weak and electromagnetic forces have essentially the same coupling strength, the approximate size of G_F is

$$G_F \approx \frac{4\pi\sqrt{2}\,\alpha(\hbar c)^3}{\left(m_w c^2 \right)^2} \approx 10^{-7} \text{ GeV} \cdot \text{fm}^3. \quad (18.28)$$

A precise value of G_F may be determined by measurement of a transition rate that can be calculated. According to Fermi's Golden Rule (11.52), a transition rate (W) is proportional to the density of states times the square of a matrix element,

$$W = \frac{2\pi}{\hbar} \rho |M|^2. \quad (18.29)$$

The muon is an example of a particle whose decay rate (matrix element) can be precisely calculated. Clearly the decay rate will depend on the square of G_F because it is through G_F that the strength of the weak interaction is specified. The calculated decay rate is

$$W_{\mu \to e} = \frac{G_F^2 m^5 c^4}{192 \pi^3 \hbar^7}, \quad (18.30)$$

where m is the mass of the muon. The muon lifetime (the inverse of the muon decay rate) is measured to be

$$\tau = \frac{1}{W_{\mu \to e}} = 2.187 \times 10^{-6} \text{ s}. \quad (18.31)$$

From this calculation and measurement, we may determine the value of G_F:

$$G_F^2 = \frac{192 \pi^3 \hbar^7}{m^5 c^4 \tau} \quad (18.32)$$

or

$$G_F = \sqrt{\frac{192 \pi^3 \hbar^7}{m^5 c^4 \tau}} = 8.96 \times 10^{-8} \text{ GeV} \cdot \text{fm}^3. \quad (18.33)$$

Decays of the Quarks

The weak interactions are mediated by the exchange of W particles, which can have positive or negative electric charge, W^+ and W^-. The W particles cause quarks to change flavor when they are exchanged. For example, a quark with an electric charge of $+2/3$ is changed to a quark with electric charge of $-1/3$ when it absorbs a W^- particle. The W particle couples to all quarks and leptons.

Beta Decay

Consider the beta-minus decay,

$$n \to p + e^- + \bar{\nu}_e. \quad (18.34)$$

We may describe this decay as a quark emitting a virtual W boson (W^*) and changing its flavor in the process,

$$d \to u + W^{*-}. \quad (18.35)$$

In this process the W^* does not exist as a real particle but is living on borrowed energy within the limits allowed by the uncertainty principle. The W^* decays

$$W^{*-} \to e^- + \bar{\nu}_e. \quad (18.36)$$

The net result of this process is

$$d \to u + e^- + \bar{\nu}_e \quad (18.37)$$

inside the neutron. The neutron has become a proton and an electron and antineutrino are emitted in the process. A down quark has changed its flavor to an up quark through the emission of a W^- particle. The exchange of a W particle is the mechanism for flavor changing in the weak interaction.

The beta-plus decay process is the inverse of beta-minus decay. The u quark can transform itself into the d quark with the emission of a W^+ particle. The simplest example is the fusion of two protons (11.81) to make deuterium. In this process a proton has been transformed into a neutron. Beta-plus decay is described at the quark level as

$$u \to d + e^+ + \nu_e. \quad (18.38)$$

Feynman diagrams for beta decay are shown in Figure 18-6.

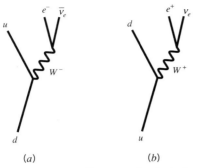

FIGURE 18-6 Feynman diagrams for beta decay.
(*a*) In beta-minus decay a down quark is transformed into an up quark. (*b*) In beta-plus decay an up quark is transformed into a down quark.

Charged Kaon Decays

We have seen that weak interaction can transform a d quark into a u quark (β^- decay) or a u quark into a d quark (β^+ decay). The W boson also couples to the other flavors of quarks. Feynman diagrams for these processes are shown in Figure 18-7. The s quark may decay spontaneously into a u quark and a virtual W. The least massive hadrons that contain a strange quark are the kaons. An example of the strange-to-up transition is the semileptonic decay of the charged kaon.

$$K^- \rightarrow \pi^0 + \mu^- + \overline{\nu}_\mu. \quad (18.39)$$

The lifetime of the charged kaon is about 10^{-8} s. An example of a charged kaon decay is shown in Figure 18-8.

* Challenging

Neutral Kaon Decays and Quark Mixing

In the late 1940s, two neutral particles were discovered, called the "τ meson" and the "θ meson". The τ meson decayed into three pions and the θ meson decayed into two pions. The τ meson and the θ meson were observed to have the same mass within experimental resolution. Both the τ meson and the θ meson were observed to have lifetimes characteristic of the weak interaction decays. The lifetime of the τ meson was measured to be about 5×10^{-8}s and the lifetime of the θ meson was measured to be about 9×10^{-11}s. The real mystery, which became known as the τ–θ puzzle, is why do the τ meson and the θ meson have the same mass yet very different decays? The τ and θ mesons are neutral kaons, the lightest neutral particles that contain the strange or antistrange quark. (The name

strangeness was quite appropriate for this mysterious quark!) The shorter-lived particle (the θ meson) is called the *K-zero-short* (K^0_S) and the longer-lived particle (the τ meson) is called the *K-zero-long* (K^0_L). The decays of these particles into pions are

$$K^0_S \rightarrow \pi^+ + \pi^- \quad \text{or} \quad \pi^0 + \pi^0, \quad (18.40)$$

and

$$K^0_L \rightarrow \pi^+ + \pi^- + \pi^0 \quad \text{or} \quad \pi^0 + \pi^0 + \pi^0. \quad (18.41)$$

Why are there two neutral kaons with the same mass and different decay properties? In the quark model, there are two light neutral particles that contain the strange or antistrange quark, the combinations $d\overline{s}$ and $s\overline{d}$. We refer to these quark model combinations as K^0 and $\overline{K}^0$. These symbols denote states of definite strangeness, $+1$ and -1, respectively. The K^0 may transform into a $\overline{K}^0$ (and vice versa) by the weak interaction as indicated in Figure 18-9,

$$K^0 \leftrightarrow \overline{K}^0. \quad (18.42)$$

The transformation between a K^0 and a $\overline{K}^0$ results from the exchange of two W bosons. Such a phenomenon is called quark mixing. A prerequisite for quark mixing to occur is that the meson live long enough for two W bosons to be exchanged. The result of quark mixing is the existence of two states, each a linear combination of K^0 and $\overline{K}^0$. Let ψ_1 represent the wave function of the $d\overline{s}$ state and let ψ_2 represent the wave function of the $s\overline{d}$ state. We may form the following two linear combinations of ψ_1 and ψ_2:

$$\psi_+ = \frac{1}{\sqrt{2}} \left(\psi_1 + \psi_2 \right) \quad (18.43)$$

and

$$\psi_- = \frac{1}{\sqrt{2}} \left(\psi_1 - \psi_2 \right). \quad (18.44)$$

Charge conjugation (**C**) is defined to be an operation that changes a particle to its antiparticle. Under the operation of parity (**P**) and charge conjugation (**C**), the state ψ_1 is turned into the state ψ_2 and the state ψ_2 is turned into the state ψ_1.

$$\mathbf{CP}\psi_1 = \psi_2 \quad (18.45)$$

and

$$\mathbf{CP}\psi_2 = \psi_1. \quad (18.46)$$

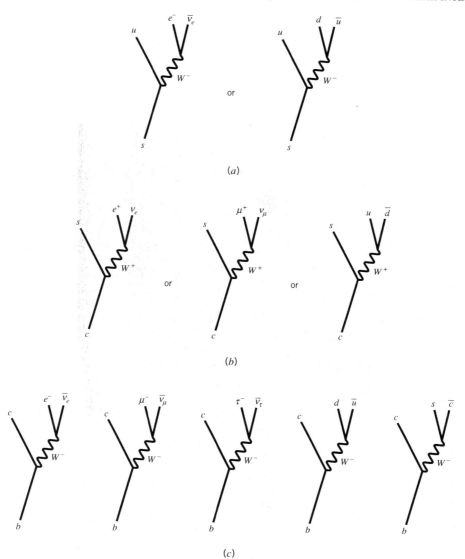

FIGURE 18-7 Feynman diagrams for decays of the s, c, and b quarks.
(a) strange-quark decay, (b) charm-quark decay, and (c) bottom-quark decay.

The operation of **CP** represents the weak transitions $K^0 \leftrightarrow \overline{K}^0$. We have

$$\mathbf{CP}\psi_+ = \mathbf{CP} \frac{1}{\sqrt{2}} \left(\psi_1 + \psi_2 \right)$$

$$= \frac{1}{\sqrt{2}} \left(\psi_2 + \psi_1 \right) = \psi_+, \quad (18.47)$$

and

$$\mathbf{CP}\psi_- = \mathbf{CP} \frac{1}{\sqrt{2}} \left(\psi_1 - \psi_2 \right)$$

$$= \frac{1}{\sqrt{2}} \left(\psi_2 - \psi_1 \right) = -\psi_-. \quad (18.48)$$

The two states ψ_+ and ψ_- have opposite values of **CP**. We say that **CP** $= +1$ for the state ψ_+ and **CP** $= -1$ for the state ψ_-.

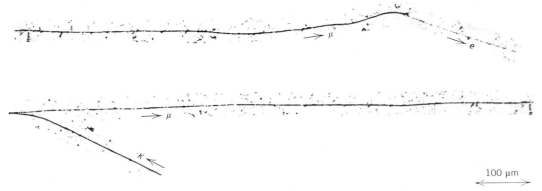

FIGURE 18-8 Decay of a charge kaon observed in emulsion.
From C. O'Ceallaigh, *Phil. Mag.* **42**, 1031 (1951).

Now consider the charge conjugation and parity operation on a state of two identical pions. The spin of the
neutral kaons is zero ($j = 0$). The two pions from K^0_S decay
must be in a state with $\ell = 0$. The operation of **CP** must not
change the wave function of the two pions because the
total wave function must be symmetric under exchange of
the two pions. Exchange of the two pions is identical to the
operation of **CP**. A state of three pions with $\ell = 0$ may be
considered as the two-pion state (**CP** $= +1$) plus a third
pion. But a single pion has (**CP** $= -1$) so that the three-pion
state has (**CP** $= -1$).

We may propose the following solution to the τ–θ
puzzle: Although parity is not conserved in the weak
interaction, the combination of charge conjugation and
parity is conserved. The state ψ_+, which has **CP** $= +1$,
corresponds to the K^0_S. The K^0_S decays into two pions
because two pions make a state of **CP** $= +1$. Similarly, the

state ψ_- corresponds to the K^0_L. The K^0_L decays into three
pions because three pions make a state of **CP** $= -1$. The
lifetimes of the two particles (K^0_S and K^0_L) are not equal
because the three-pion decay is suppressed by phase space
due to the fact that the kaon mass is only a little larger than
three times the pion.

The time-evolution (7.158) of the amplitude of the
states ψ_+ and ψ_- each contain a factor $e^{i\omega t}$. The oscillation
that occurs between the two states is due to a tiny mass
difference between the K^0_S and K^0_L. The mass difference
(Δm) is so small that it cannot be directly detected in a
spectrometer. A measurement of the oscillation frequency
(ω) determines the K^0_S and K^0_L mass difference,

$$\omega = \frac{\Delta m c^2}{\hbar}. \tag{18.49}$$

The result is

$$\Delta m = 3.52 \times 10^{-6} \text{ eV}/c^2. \tag{18.50}$$

The K^0_L mass is slightly greater than the K^0_S mass.

EXAMPLE 18-2
Compare the period of oscillation, $d\bar{s} \leftrightarrow s\bar{d}$, to the lifetime of the K^0_S and K^0_L particles.

SOLUTION:
The period (T) of the oscillation is given by

$$T = \frac{2\pi}{\omega} = \frac{2\pi\hbar}{\Delta m c^2} = \frac{h}{\Delta m c^2}$$

$$= \frac{4.14 \times 10^{-15} \text{ eV} \cdot \text{s}}{3.52 \times 10^{-6} \text{ eV}} \approx 1.2 \times 10^{-9} \text{ s}.$$

**FIGURE 18-9 Feynman diagrams for neutral kaon
mixing.**

The proper lifetimes of the particles is 5.2×10^{-8} s for the K^0_L and 8.9×10^{-11} s for the K^0_S. For energetic kaons, the particle lifetimes are longer by the Lorentz γ factor and the oscillation period is much shorter than the particle lifetime. The K^0 and $\overline{K}^0$ states are completely *mixed*. ∎

Regeneration

The neutral kaons provide a good example of the time evolution of probability amplitudes in quantum mechanics. When created by strong interactions, the neutral kaons are produced with states of definite strangeness because strangeness is absolutely conserved in the strong interaction. These states are called K^0 and $\overline{K}^0$. The K^0 and $\overline{K}^0$ are not real particles; our definition of a particle is a state with a definite lifetime. The K^0 and $\overline{K}^0$ states mix by the Feynman diagrams of Figure 18-9. The mixing occurs on the time scale of a nanosecond, resulting in a long-lived neutral kaon and a short-lived neutral kaon. The mixing is complete; there are equal numbers of K^0_L and K^0_S particles. The lifetime of the K^0_L is about 500 times the lifetime of the K^0_S. Therefore, the K^0_L travel 500 times further than the K^0_S, on the average, before they decay. After a certain decay distance (see Figure 18-10), essentially all the K^0_S have decayed and only K^0_L are left. If the K^0_L particles are allowed to pass through a piece of matter, then states of definite strangeness are regenerated! The states of definite strangeness mix to form both K^0_L and K^0_S particles. This is confirmed by observation of the K^0_S decays into two pions.

CP Violation

An elegant solution of the τ–θ puzzle has been obtained by assuming that the operation **CP** is conserved in the weak interaction. In the early 1960s, physicists were confident of their understanding of the weak interaction. Then in 1964, in an experiment led by James Cronin and Val Fitch, the K^0_L was observed to decay into two pions,

$$K^0_L \to \pi^+ + \pi^-, \tag{18.51}$$

in a small fraction (0.2%) of all decays. This decay violates **CP**, because as we have seen the K^0_L is a state of $\mathbf{CP} = -1$ and the two pions are a state of $\mathbf{CP} = +1$. Since the decay fraction or *branching ratio* is small, we see that **CP** is almost but not exactly conserved in weak interactions. The origin of **CP** violation is not understood. The neutral kaon is the only physical system where **CP** violation has been observed. ∗

Decays of Charm

The least massive hadron that contains a charm quark is called the D meson. In the quark model the D^+ meson is $c\bar{d}$, the D^- meson is $d\bar{c}$, the D^0 meson is $c\bar{u}$, and the $\overline{D}^0$ meson is $u\bar{c}$. The combination $c\bar{s}$ is called the D_s^+ and the combination $s\bar{c}$ is called the D_s^-.

The c quark may decay spontaneously into an s quark and a virtual W. An example of the $c \to s$ transition is the semileptonic decay of the D^+ meson,

$$D^+ \to K^- + \pi^+ + \mu^+ + \nu_\mu. \tag{18.52}$$

Note that a strange particle, in this example the kaon, must appear in the final state. There is also a small probability, about 5%, that the c quark will transform into a d quark rather than an s quark. In this case no strange particles appear in the final state. An example of a *Cabbibo suppressed* decay is

$$D^+ \to \pi^- + \pi^+ + \mu^+ + \nu_\mu. \tag{18.53}$$

The observation of a charged D meson decay is shown in Figure 18-11. The event was recorded in a bubble

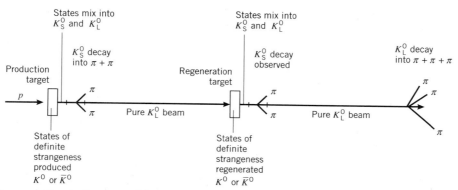

FIGURE 18-10 Regeneration in the neutral kaon system.

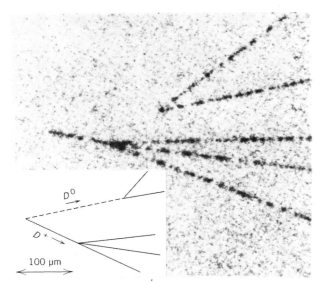

FIGURE 18-11 Example of *D* meson decays observed in a bubble chamber.
From K. Abe et al., *Phys. Rev. Lett.* **48**, 1526 (1982).

chamber at SLAC. The *D* meson was produced by the interaction of a 20-GeV photon. A beam of high-energy photons was created by Compton scattering of radiation from a YAG laser with energetic electrons from the linear accelerator. The signature of the *D* meson is that it travels 1.8 mm before decaying and that its decay products have an invariant mass that excludes the possibility of a strange particle decay. The charged *D* meson lifetime is determined to be roughly 10^{-12} s, by the measurement of the decay distance in many events.

The neutral *D* meson is observed to have a lifetime that is about a factor of two larger than the charged *D* meson. A neutral D meson is also observed in the event of Figure 18-11. The neutral *D* meson is *not* observed to mix, that is, the transitions $D^0 \leftrightarrow \overline{D}^0$ are not observed.

Decays of Bottom

The least massive hadron that contains a bottom quark is the *B* meson. In the quark model the B^- meson is $b\overline{u}$, the B^+ meson is $u\overline{b}$, the B^0 meson is $\overline{b}d$, and the $\overline{B}^0$ meson is $b\overline{d}$. The combinations $s\overline{b}$, $b\overline{s}$, $b\overline{c}$, and $c\overline{b}$, are called the B_s^0, $\overline{B}_s^0$, B_c^-, and B_c^+.

The *b* quark may decay spontaneously into a *c* quark and a virtual *W*. An example of the $b \rightarrow c$ transition is the semileptonic decay of the *B* meson

$$B^0 \rightarrow D^{*-} + e^+ + \nu_e. \qquad (18.54)$$

Note that a charm particle, in this case the *D** meson, must appear in the final state. The decay $b \rightarrow u$ is also theoretically possible, but it has not been observed. In this case no charm particles would appear in the final state. The lifetime of the *B* meson is roughly the same as that for *D* mesons, about 10^{-12} s. The *B* mesons were discovered in $e^+ + e^-$ collisions at Cornell (CESR). The decay of a pair of *B* mesons is shown in Figure 18-12.

The neutral *B* mesons have been observed to mix, that is, the transitions $B^0 \leftrightarrow \overline{B}^0$ have been observed. It is not known whether **CP** is conserved in the *B* meson system.

The Quark Transitions

The weak decays of the quarks are summarized in Table 18-1. Notice that there is a pattern with the quark transformations. A charge $+2/3$ quark (*u* or *c*) always is transformed into a charge $-1/3$ quark (*d*, *s*, or *b*) and vice versa. This is because the quark transformations are mediated by the exchange of charged *W* bosons. In addition, the quarks tend to decay into the heaviest quarks possible. The quark transition pattern is

$$b \rightarrow c \rightarrow s \rightarrow u \leftrightarrow d. \qquad (18.55)$$

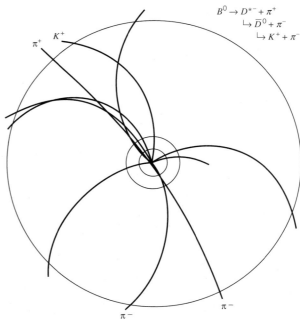

FIGURE 18-12 Example of a bottom-hadron decay.
The *B* mesons are produced in pairs as the result of the decay of the $\Upsilon(4s)$. This event was recorded in the CLEO detector at Cornell where the *B* mesons were discovered.

TABLE 18-1
TRANSFORMATION OF QUARK FLAVORS BY THE WEAK INTERACTION.

Quark	Process	Example	Lifetime (mean)
Up	$u \rightarrow d + W^{*+}$	$p + p \rightarrow d + e^{+} + v_{e}$	
Down	$d \rightarrow u + W^{*-}$	$n \rightarrow p + e^{-} + \bar{v}_{e}$	900 s
Strange	$s \rightarrow u + W^{*-}$	$K^{-} \rightarrow \pi^{0} + e^{-} + \bar{v}_{e}$	1.24×10^{-8} s
Charm	$c \rightarrow s + W^{*+}$	$D^{+} \rightarrow K^{-} + \pi^{0} + \pi^{+} + e^{+} + v_{e}$	1.1×10^{-12} s
Bottom	$b \rightarrow c + W^{*-}$	$B^{0} \rightarrow D^{*-} + e^{+} + v_{e}$	1.3×10^{-12} s

Since the b quark has a much larger mass than the c quark, the transformation from bottom to charm is "one-way." All the bottom hadrons are much more massive than the charm hadrons. (Of course, in collisions of very energetic c quarks, $E \gg m_{b}c^{2}$, the transformation $c \rightarrow b$ is possible.) The transformation from charm to strange is also one-way for the same reason. In this case there is also a small probability (about 5%) that the c quark will transform into a d quark instead of an s quark.

EXAMPLE 18-3

Estimate the fraction of time that a b quark decays into $c + e^{-} + \bar{v}_{e}$.

SOLUTION:

We see from the Feynman diagrams that there are five possibilities for the decay of the b quark:

$$b \rightarrow c + e^{-} + \bar{v}_{e},$$

$$b \rightarrow c + \mu^{-} + \bar{v}_{\mu},$$

$$b \rightarrow c + \tau^{-} + \bar{v}_{\tau},$$

$$b \rightarrow c + d + \bar{u},$$

and

$$b \rightarrow c + c + \bar{s}.$$

These first three processes occur with equal probability and the last two processes are three times more probable because there are three colors of quarks. The decay fraction of the b quark into $c + e^{-} + \bar{v}_{e}$ may be estimated to be

$$B\left(\frac{b \rightarrow c + e^{-} + \bar{v}_{e}}{b \rightarrow \text{all}}\right) \approx \frac{1}{1 + 1 + 1 + 3 + 3} \approx 0.1$$

Notice that we have ignored the effects of phase space in this simple calculation. ∎

Decays of the Leptons

The decays of the leptons are also mediated by W particles. This is the reason that the W particle mass appears in the expression for muon decay (it is contained in G_{F}). The Feynman diagrams for the decays of the leptons (μ and τ) are shown in Figure 18-13. Since the tau lepton is more

(a)

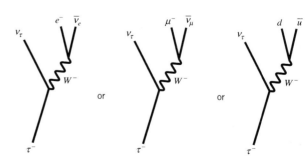

(b)

FIGURE 18-13
Feynman diagrams for lepton decays: (a) muon decay and (b) tau decay.

massive than the lightest hadrons, the tau lepton has more decays available that are forbidden by energy conservation in muon decay.

EXAMPLE 18-4

Estimate the decay fraction for $\tau^- \to \nu_\tau + e^- + \bar{\nu}_e$.

SOLUTION:

We see from the Feynman diagrams that there are three possibilities for the decay of the tau,

$$\tau^- \to \nu_\tau + e^- + \bar{\nu}_e,$$

$$\tau^- \to \nu_\tau + \mu^- + \bar{\nu}_\mu,$$

and

$$\tau^- \to \nu_\tau + d + \bar{u}.$$

These first two processes occur with equal probability and the third process is three times more probable because there are three colors of quarks. The decay fraction into $\nu_\tau + e^- + \bar{\nu}_e$ is

$$B\left(\frac{\tau^- \to \nu_\tau + e^- + \bar{\nu}_e}{\tau^- \to \text{all}}\right) \approx \frac{1}{1+1+3} = 0.2.$$

This simple calculation gives an excellent approximation to the experimental result, which is about 18%. ∎

Neutrino Interactions

The three neutrinos, and their antiparticles, have the distinction of being particles that have no strong or electromagnetic interactions. Neutrinos are therefore uniquely suited to study certain aspects of the weak interaction. Neutrino interactions are studied in fixed target experiments. Very large numbers of neutrinos are needed to get a measurable interaction rate. Neutrino beams are made by interacting intense proton beams in a metal target where pions and kaons are copiously produced by strong interactions:

$$p + N \to \pi^+ + X \tag{18.56}$$

and

$$p + N \to K^+ + X. \tag{18.57}$$

The high-momentum positively charged pions and kaons are selected with a magnetic field. The pions and kaons are allowed to decay in a long tunnel,

$$\pi^+ \to \mu^+ + \nu_\mu \tag{18.58}$$

and

$$K^+ \to \mu^+ + \nu_\mu, \tag{18.59}$$

and the muons are filtered out by their electromagnetic (dE/dx) interactions in a very thick absorber, leaving just the neutrinos. The kaons produce higher energy neutrinos than the pions because of the larger Q value of the decay. Cerenkov counters may be used to measure the relative rate of pion and kaon decays. To get a beam of muon-antineutrinos, the same technique is used with the direction of the magnetic field switched so that negatively charged pions and kaons are selected:

$$\pi^- \to \mu^- + \bar{\nu}_\mu \tag{18.60}$$

and

$$K^- \to \mu^- + \bar{\nu}_\mu. \tag{18.61}$$

Two types of neutrino beams are produced, according to the manner in which the pions and kaons are selected and focused before they are allowed to decay. In a *wide band* neutrino beam, a large range of particle momenta are allowed to decay. This produces a large number of neutrinos but they have a wide range of energies. In a *narrow band* neutrino beam, only a narrow momentum range is selected for the meson. A special magnet, called a magnetic horn because of its trumpetlike shape, is used to focus the particles. When the meson decays, a range of neutrino energies are produced but the neutrino energy is correlated with the angle that it makes with the beam axis.

The neutrino or antineutrino beams will contain a small amount of electron-neutrinos or antineutrinos from two processes:

$$K^+ \to \pi^0 + e^+ + \nu_e \tag{18.62}$$

($K^- \to \pi^0 + e^- + \bar{\nu}_e$ for antineutrinos), and

$$\mu^+ \to e^+ + \nu_e + \bar{\nu}_\mu \tag{18.63}$$

($\mu^- \to e^- + \bar{\nu}_e + \nu_\mu$ for antineutrinos). This last process also results in a small component of antineutrinos in the neutrino beam and neutrinos in the antineutrino beam.

In the interaction of a neutrino with a nucleus, or more precisely with a quark contained in a nucleus, the cross section is proportional to the weak coupling (α_w) squared and to the wavelength squared of the particles, analogous to the e^+e^- annihilation cross section (18.5). In the center-of-mass frame the quark and neutrino have the same wavelength. At neutrino energies larger than the masses of the quarks, we may write the cross section (σ) in terms of either α_w (18.26) or G_F (18.27),

$$\sigma \approx \frac{\alpha_w^2 (\hbar c)^2}{s} \approx \frac{G_F^2 s}{(\hbar c)^4}, \qquad (18.64)$$

where s is the center-of-mass energy squared. Since s is proportional to the laboratory neutrino energy (E), we have

$$\sigma \propto G_F^2 E. \qquad (18.65)$$

Neutrino interaction probabilities are directly proportional to the energy of the neutrino, directly reflecting the fact that the weak interaction is stronger at shorter distances. This energy dependence is beautifully verified by experiments in the present energy range of several hundred GeV (see Figure 18-14).

Charged Currents

Neutrino interactions that are mediated by W particles are called *charged current* interactions because the W particle has an electric charge. In a charged current interaction of a neutrino with a nucleus, the neutrino emits a W boson, which strikes a quark, and in the process the neutrino is transformed into a charged lepton. The observed reaction for the case of incident muon-neutrinos is

$$\nu_\mu + N \rightarrow \mu^- + \text{hadrons}, \qquad (18.66)$$

while for incident muon-antineutrinos,

$$\bar{\nu}_\mu + N \rightarrow \mu^+ + \text{hadrons}. \qquad (18.67)$$

In a charged current interaction a muon-neutrino always produces a negatively charged muon (see Figure 18-15). The virtual W couples to down or antiup quarks. Therefore, muon-neutrino charged current interactions are a way of probing the quark distributions in the target.

We see from the Feynman diagrams of Figure 18-15 that in charged current interactions, neutrinos strike only quarks or antiquarks with negative electric charge (e.g., d or $\bar{u}$). The physics of this is simple: lepton number conservation requires that a μ^- be produced, which means that a W^- is emitted by the quark and a W^- can be emitted only by a quark or antiquark with negative electric charge. Correspondingly, antineutrinos strike only quarks or antiquarks with positive electric charge (e.g., u or $\bar{d}$).

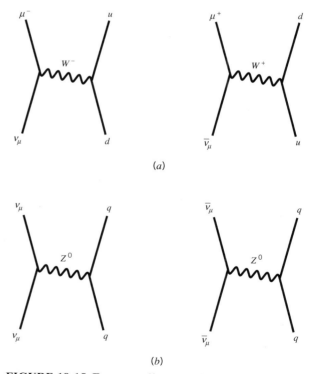

FIGURE 18-15 Feynman diagrams for neutrino interactions.
(a) charged current interactions and (b) neutral current interactions.

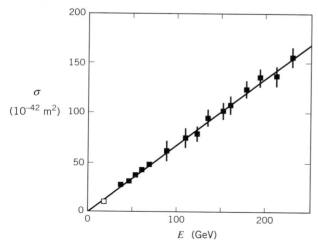

FIGURE 18-14 Neutrino interaction cross section at high energy.

The angular distribution depends on the helicity of the scattering particles. When a neutrino scatters from a quark (particle-particle scattering), the angular distribution in the center-of-mass frame is isotropic. There is no preferred direction for the muon in the final state. When a neutrino scatters from an antiquark (particle-antiparticle scattering), there is an angular dependence of the scattering due to the intrinsic angular momenta of the interacting particles. There is a vanishing probability that the quark can scatter backwards. We define y to be the fraction of energy (E) that the neutrino transfers to the quark. The muon energy (E_μ), the center-of-mass scattering angle (θ^*) and y, are related by

$$1 - y = \frac{E_\mu}{E} = \frac{1 + \cos\theta^*}{2}. \tag{18.68}$$

Thus, measurement of the muon energy is a measure of the center-of-mass scattering angle. The neutrino cross section for interaction with point-like spin 1/2 quarks and antiquarks is expected to be of the form

$$\frac{d\sigma}{dy} \approx \frac{G_F^2 s}{\pi}\left[A + B\left(1 - y\right)^2 \right], \tag{18.69}$$

where A is the contribution due to quarks and B is the contribution due to antiquarks. The antineutrino cross section is

$$\frac{d\sigma}{dy} \approx \frac{G_F^2 s}{\pi}\left[A\left(1 - y\right)^2 + B \right]. \tag{18.70}$$

The fact that there are more quarks than antiquarks in a nucleus together with the intrinsic angular momentum factor $(1-y)^2$, results in a larger neutrino interaction probability than antineutrino by a factor of two. The measured cross sections $d\sigma/dy$ agree with the predictions (18.69) and (18.70) indicating that the quarks behave as point-like particles with spin 1/2. Furthermore, comparison of neutrino cross sections with electron cross sections shows that the quarks have fractional charge and only about one-half of the nucleon momentum is carried by quarks.

Neutral Currents

A second fundamental type of weak interaction was discovered in neutrino scattering experiments in 1973. In these neutrino interactions, no charged leptons are produced. These neutrino interactions are mediated by exchange of a massive electrically neutral particle, the Z^0 boson (see Figure 18-15), and are called *neutral currents*.

In a neutral current interaction of a neutrino with a nucleus, the quark emits a Z^0 boson, which is absorbed by the neutrino. In this process the neutrino remains a neutrino and no charged leptons appear in the final state. The observed reactions are

$$\nu_\mu + N \rightarrow \nu_\mu + \text{hadrons} \tag{18.71}$$

and

$$\bar\nu_\mu + N \rightarrow \bar\nu_\mu + \text{hadrons}. \tag{18.72}$$

The virtual Z^0 boson may also come from an electron instead of a quark, in which case we observe neutrino–electron elastic scattering and no hadrons are produced:

$$\nu_\mu + e^- \rightarrow \nu_\mu + e^- \tag{18.73}$$

or

$$\bar\nu_\mu + e^- \rightarrow \bar\nu_\mu + e^-. \tag{18.74}$$

Neutrino–electron elastic scattering is independent of the type of neutrino or antineutrino because the Z^0 boson couples with equal strength to all three neutrino families.

Discovery of the Intermediate Vector Bosons

For decades, experimentalists searched for the weak interaction bosons. Until a unified theory of the electromagnetic and weak interactions was formulated, there was no prediction for the masses of the W and Z^0 bosons. The discovery of the W and Z^0 bosons provided the ultimate confirmation of our understanding of the weak interaction.

Observation of the W Bosons

The charged intermediate vector bosons, W^+ and W^-, were discovered at CERN in 1982 by the UA1 Collaboration, led by Carlo Rubbia, in energetic proton–antiproton collisions through the reaction,

$$\bar p + p \rightarrow W + X. \tag{18.75}$$

The W particle is formed by the weak interaction of a quark inside the proton with an antiquark inside the antiproton:

$$u + \bar d \rightarrow W^+ \tag{18.76}$$

and

$$\bar u + d \rightarrow W^-. \tag{18.77}$$

The W bosons were first identified by their decays:

$$W^+ \rightarrow e^+ + \nu_e \qquad (18.78)$$

and

$$W^- \rightarrow e^- + \bar{\nu}_e \ . \qquad (18.79)$$

The Feynman diagrams for W production and decay are shown in Figure 18-16.

EXAMPLE 18-5

(a) Estimate the proton–antiproton center of mass energy where a typical collision will have enough energy to produce a W particle (mass $80\,\text{GeV}/c^2$) by quark-antiquark annihilation. (b) Estimate the cross section for W boson production at such an energy. (c) Estimate the proton–antiproton luminosity needed to produce one W particle per day.

SOLUTION:

(a) The typical energy of each quark (E_q) is 15% of the proton energy (see Figure 6-18),

$$E_q \approx 0.15\, E_p\,.$$

To produce a W particle, we need

$$2\,E_q = m_W c^2 = 80\,\text{GeV}\,.$$

Therefore, the required proton energy is approximately

$$E_p \approx \frac{E_q}{0.15} = \frac{40\,\text{GeV}}{0.15} \approx 270\,\text{GeV}\,.$$

(b) The order of magnitude of the W production cross section is

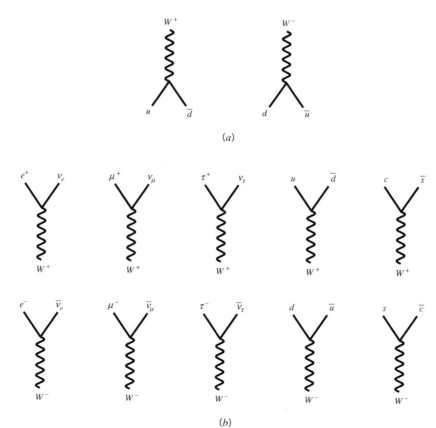

(a)

(b)

FIGURE 18-16 Feynman diagrams for (*a*) W production in proton–antiproton collisions and (*b*) W decay.

$$\sigma \approx \pi \left(\frac{\hbar c}{m_w c^2} \right) \left[\frac{G\left(m_w c^2\right)^2}{\left(\hbar c\right)^3} \right]^2$$

$$\approx \frac{\pi \left(10^{-7} \text{ GeV} \cdot \text{fm}^3 \right) \left(80 \text{ GeV} \right)^2}{\left(0.2 \text{ GeV} \cdot \text{fm} \right)^4}$$

$$\approx 10^{-7} \text{ fm}^2 = 10^{-37} \text{ m}^2 .$$

(c) The number of seconds in one day is

$$1\text{d} = \left(24 \text{ hr} \right)\left(3600 \text{ s/hr} \right) \approx 8.6 \times 10^4 \text{ s}.$$

The necessary flux (L) to produce one W per day is

$$L = \frac{(1\text{d})^{-1}}{\sigma} = \frac{\left(\dfrac{1}{8.6 \times 10^4 \text{ s}} \right)}{10^{-37} \text{ m}^2} \approx 10^{32} \text{ m}^{-2} \cdot \text{s}^{-1} . \quad \blacksquare$$

The above example illustrates the design parameters of the proton–antiproton collider at CERN. The required proton–antiproton collision rate needed for W production was achieved in 1982; however, the W detection efficiency was not known. The proton–antiproton strong interaction cross section is about $100 \text{ mb} = 10^{-29} \text{ m}^2$. The W particle cross section (see Example 18-5) is smaller by a factor of 10^8 at a center-of-mass energy of 540 GeV. (The weak interaction is "weak.") To detect the W therefore requires a large number of collisions plus the ability to detect the weak interaction among the strong interactions.

The trigger for the W was a single electron (or positron) with a large momentum in a direction orthogonal to the direction of the colliding beams. The momentum in this direction is called the *transverse* momentum. In the UA1 detector, the momentum of the electron (p) was measured by magnetic analysis and the energy of the electron (E) was measured by calorimetry. Since the electrons are relativistic, we have $pc \approx E$. This requirement provides a method for eliminating a potentially large hadronic background. (The hadronic background to an electron is a closely spaced pion pair, $\pi^+ \pi^0$, where the π^0 is more energetic than the π^- and mimics the electron signature.) In the initial search for the W particle, events were selected with an electron having a transverse momentum (p_t^e) greater than 15 GeV. The electron momentum in these events was found to be highly correlated with missing momentum. Missing momentum is defined to be the deviation from zero of the sum of the momenta of all particles that are observed to interact in the UA1 calorimeters. The missing transverse momentum is (p_t^{miss}) defined

by

$$p_t^{\text{miss}} = \sqrt{\left(0 - \sum p_x \right)^2 + \left(0 - \sum p_y \right)^2} , \quad (18.80)$$

where the summations go over all particles that are observed to interact in the detector. In the UA1 detector, the missing transverse momentum was measured with a precision of a few GeV/c. For almost all events, the missing transverse momentum is zero within experimental resolution. This is not the case for events that have a single high transverse momentum electron, where it is found that

$$p_t^{\text{miss}} \approx p_t^e . \quad (18.81)$$

This is the signature of the W particle decaying leptonically,

$$W^+ \to e^+ + v_e \text{ or } W^- \to e^- + \bar{v}_e . \quad (18.82)$$

Since the W mass energy is approximately 80 GeV and it decays into two particles, the decay products, the electron and neutrino, each have an energy of about 40 GeV in the rest frame of the W. The neutrino does not interact in the detector and its presence leaves a spectacular signature in the form of missing energy. A W event observed in the UA1 detector is shown in Figure 18-17.

Properties of the W

The momentum of the W particle depends on the momenta of the annihilating quark and antiquark. In general, the quark and antiquark do not have the same momentum. This results in the W bosons being produced with a *longitudinal* momentum, a component of momentum along the beam axis. The momentum of the W boson in a direction perpendicular to the beam axis, the *transverse momentum*, is relatively small. The transverse momentum of a W boson is small because the colliding quarks have essentially zero momentum transverse to the beam axis. The transverse momentum of the W particle (p_t^w) is not zero because in the annihilation process, the quarks emit gluons by the bremsstrahlung process. The transverse momentum of the W particle acquired from the gluon radiation process is small compared to the W mass energy,

$$p_t^w c < m_W c^2 . \quad (18.83)$$

This means that the transverse momentum of the W decay products is dominated by the mass of the W boson. The maximum transverse momentum of an electron (p_t^e) from W decay is given approximately by

$$\left(p_t^e c \right)_{\text{max}} \approx \frac{m_W c^2}{2} . \quad (18.84)$$

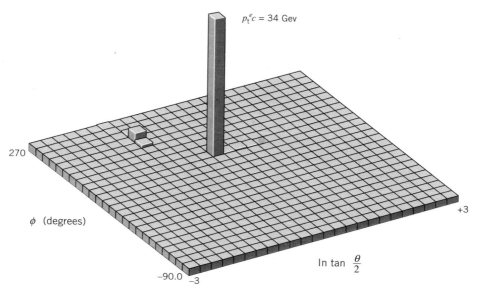

FIGURE 18-17 Discovery of the W particle.
The UA1 calorimeter measures the energy deposited by all charged particles except neutrinos. The only significant energy deposit is that due to a single electron.

Measurement of the electron transverse momentum distribution provides a determination of the W mass. The transverse momentum of electrons from W decay is shown in Figure 18-18. These data give the mass of the W particle,

$$m_{\mathrm{W}} \approx 80 \,\mathrm{GeV}/c^2. \qquad (18.85)$$

In the UA1 experiment, the W particle was also observed by its decays

$$W^+ \to \mu^+ + \nu_\mu \quad \text{or} \quad W^- \to \mu^- + \overline{\nu}_\mu, \quad (18.86)$$

and

$$W^+ \to \tau^+ + \nu_\tau \quad \text{or} \quad W^- \to \tau^- + \overline{\nu}_\tau. \quad (18.87)$$

The signature for these processes is observation of a muon or tau with large transverse momentum together with large missing transverse momentum (see color plate 18). The W also decays into quark–antiquark pairs. We know this not only because we know how the quarks decay but also more directly because the W is *produced* by quark–antiquark annihilation. The W decay into quarks is not readily observable at a hadron collider because of the very large background from quark–quark scattering. The W decay rate is proportional to the weak coupling strength (α_{w}). The energy scale of the decay process is set by the mass energy of the W ($m_W c^2$). Therefore, the order of magnitude

of the decay rate for the W boson into an electron and neutrino (18.82) is given by

$$\Gamma_{W\to e\nu} \approx \alpha_W \, m_{\mathrm{w}} c^2 \approx \alpha \, m_W c^2, \qquad (18.88)$$

where the weak coupling (18.26) must be evaluated at $E = m_w c^2$, where $\alpha_w \approx \alpha$ A precise calculation gives

$$\Gamma_{W\to e\nu} = \frac{G_F \left(m_W c^2\right)^3}{6\sqrt{2}\pi (\hbar c)^3} \approx 0.23 \,\mathrm{GeV}. \quad (18.89)$$

The decay rate into muon and neutrino (18.86) or tau and neutrino (18.87) is identical to the electron and neutrino rate. The decay rate into quarks is larger by a factor of 3 for each family of quarks because the quarks come in three colors. The W can decay into two families of quarks ($u\overline{d}$ and $c\overline{s}$ for the W^+). There is also a small QCD correction of order α_s due to the presence of gluons in the final state. The partial decay rate into hadrons is

$$\Gamma_{W\to \text{hadrons}} = (2)(3)\left(1 + \frac{\alpha_s}{\pi}\right)\Gamma_{W\to e\nu}$$
$$\approx 1.43 \,\mathrm{GeV}. \qquad (18.90)$$

The total decay width of the W is

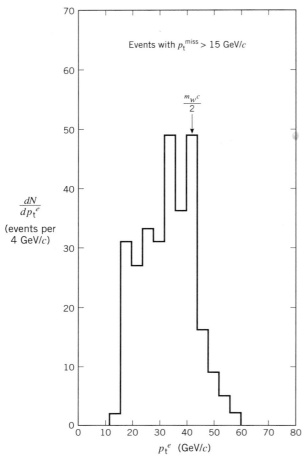

FIGURE 18-18 Measuring the mass of the W particle.
Events are selected that have an electron with a large momentum in the direction transverse to the colliding proton and antiproton ($p_t^e > 15$ GeV/c). The presence of the high transverse momentum electron is highly correlated with missing momentum. This is the signature of the W particle. For events that have a large missing transverse momentum ($p_t^{miss} > 15$ GeV/c), the electron transverse "energy" ($p_t^e c$) is plotted. The maximum value of ($p_t^e c$) is approximately equal to one-half of the W mass. Some electrons appear with slightly larger values of ($p_t^e c$) because the W is produced with a small transverse momentum.

$$\Gamma_{total} = 0.23\,\text{GeV} + 0.23\,\text{GeV} + 0.23\,\text{GeV} + 1.43\,\text{GeV}$$
$$= 2.12\,\text{GeV}. \qquad (18.91)$$

The partial decay widths of the W boson are summarized in Table 18-2.

TABLE 18-2
CONTRIBUTIONS TO THE W WIDTH.

Decay	Γ
$W \rightarrow e + \nu$	0.23 GeV
$W \rightarrow \mu + \nu$	0.23 GeV
$W \rightarrow \tau + \nu$	0.23 GeV
$W \rightarrow$ hadrons (via $u\bar{d}$)	0.71 GeV
$W \rightarrow$ hadrons (via $c\bar{s}$)	0.71 GeV
TOTAL	2.12 GeV

EXAMPLE 18-6

Estimate the decay fraction for $W^+ \rightarrow e^+ + \nu$.

SOLUTION:

The W particle couples with equal strength to each family of leptons and quarks. There are three lepton families

$$W^+ \rightarrow e^+ + \nu_e, \quad W^+ \rightarrow \mu^+ + \nu_\mu, \quad W^+ \rightarrow \tau^+ + \nu_\tau$$

and two quark families

$$W^+ \rightarrow \bar{u} + d, \quad W^+ \rightarrow \bar{c} + s,$$

but each quark family comes in three colors. The decay fraction is

$$B\left(\frac{W^+ \rightarrow e^+ + \nu_e}{W^+ \rightarrow \text{all}}\right) = \frac{1}{1+1+1+3+3} \approx 0.11. \quad \blacksquare$$

The intrinsic angular momentum quantum number of the W particle is one. This has been directly determined from the angular distribution of the leptons from W decays. Consider the production of a W particle by quark–antiquark annihilation (see Figure 18-19a). The energetic quark comes from the proton and it tends to have its intrinsic angular momentum antiparallel to the direction of the proton motion. The energetic antiquark comes from the antiproton and it tends to have its intrinsic angular momentum parallel to the direction of the antiproton motion. The intrinsic angular momentum of the W particle is therefore oriented in the antiproton direction. This is true for both W^+ and W^- particles and does not depend on the longitudinal motion of the W bosons. The W^+ particle decays into $e^+ + \nu_e$. The positron tends to have its intrinsic angular momentum parallel to the direction of motion.

Therefore, the positron tends to be emitted in the antiproton direction. The W^- particle decays into $e^- + \bar{v}_e$. The electron tends to have its intrinsic angular momentum antiparallel to the direction of motion. Therefore, the electron tends to be emitted in the proton direction. The weak interaction violates parity. The data from UA1 are shown in Figure 18-19b.

Observation of the Z^0 Boson

The intermediate vector boson, Z^0, was discovered at CERN, also by UA1, just a few months after the discovery of the W bosons. The production of the Z^0 particle in proton-antiproton collisions is also through quark–antiquark annihilation (see Figure 18-20),

$$\bar{q} + q \rightarrow Z^0. \tag{18.92}$$

The Z^0 was detected by its decays into energetic lepton pairs

$$Z^0 \rightarrow e^+ e^- \tag{18.93}$$

and

$$Z^0 \rightarrow \mu^+ \mu^-. \tag{18.94}$$

The leptons are positively identified by their electromagnetic interactions in the UA1 detector. Figure 18-21 shows the first Z^0 boson detected (see also color plates 19 and 20).

The leptonic decays of the Z^0 boson into $e^+ + e^-$ is easier to detect than the W boson decays into $e + v$, because both leptons are directly observable; however, both the production cross section and the branching ratio are smaller than for the W boson. The invariant mass distribution of high-energy electron–positron pairs is shown in Figure 18-22.

The order of magnitude of the decay width of the Z^0 boson is the same as the W boson ($\alpha\, m_z c^2$). A precise determination of the partial decay widths gives

$$\Gamma_{Z\rightarrow ee} = \Gamma_{Z\rightarrow\mu\mu} = \Gamma_{Z\rightarrow\tau\tau} = 0.083\,\text{GeV} \tag{18.95}$$

for the charged leptons,

$$\Gamma_{Z\rightarrow vv} = 0.17\,\text{GeV} \tag{18.96}$$

for each neutrino type, and

$$\Gamma_{Z\rightarrow\text{hadrons}} = 1.74\,\text{GeV} \tag{18.97}$$

for the hadrons.

The Z^0 boson has spin one just like the photon. Quantum interference between the photon and the Z^0 boson largely destroy the asymmetry that is directly observed in W decays.

Measurement of Three Neutrinos

The Z^0 boson couples to particles that carry weak charge just as the photon couples to particles that carry electric charge. There are more kinds of weakly interacting particles than electromagnetically interacting particles because neutrinos interact weakly although they have zero electric charge. The Z^0 particle couples to all quarks and leptons. Detailed study of the properties of the Z^0 boson

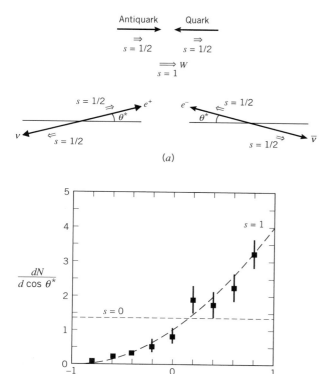

FIGURE 18-19 Measuring the intrinsic angular momentum of the W particle.
(a) Since the quarks and antiquarks are both polarized, the W spin tends to point along the direction of the antiquark. Therefore, the antiparticle from the W decay (positron or antineutrino) tends to be emitted in the direction of the antiquark. (b) Measurement of the angular distribution of electrons and positrons proves that the spin of the W particle is 1.

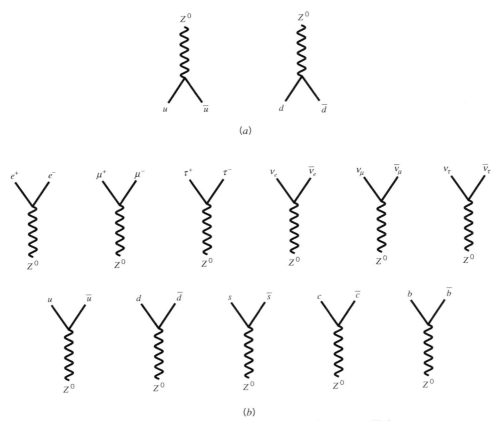

FIGURE 18-20 Feynman diagrams for (*a*) Z^0 production in proton–antiproton collisions and (*b*) Z^0 decay.

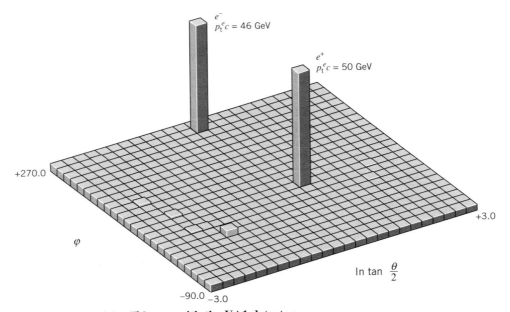

FIGURE 18-21 Observation of the Z^0 boson with the UA1 detector.
The event contains a high energy electron and positron and no other particles with large transverse momenta. Discovery of the Z^0 confirmed the unification of the electromagnetic and weak interactions.

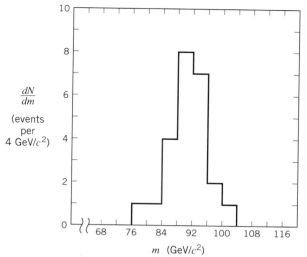

FIGURE 18-22 Mass distribution of high-energy electron–positron pairs in the UA1 detector.

tell us much about the quarks and leptons (see color plate 22). The LEP machine at CERN is dedicated to the study of the properties of the Z^0 boson through the reaction

$$e^+ + e^- \rightarrow Z^0. \qquad (18.98)$$

The decay width (Γ) of the Z^0 boson is measured by varying the electron–positron energy. The decay width of the Z^0 boson depends on the decay modes that are available. The contributions of all known decay modes to the total Z^0 width are summarized in Table 18-3. The contributions from the known particles, five quark families and three lepton families, account for the total width of the Z^0. In particular, there is no room for an additional neutrino

TABLE 18-3
CONTRIBUTIONS TO THE Z^0 WIDTH.

Decay	Γ
$Z^0 \rightarrow e^+ + e^-$	0.083 GeV
$Z^0 \rightarrow \mu^+ + \mu^-$	0.083 GeV
$Z^0 \rightarrow \tau^+ + \tau^-$	0.083 GeV
$Z^0 \rightarrow \nu_e + \bar{\nu}_e$	0.166 GeV
$Z^0 \rightarrow \nu_\mu + \bar{\nu}_\mu$	0.166 GeV
$Z^0 \rightarrow \nu_\tau + \bar{\nu}_\tau$	0.166 GeV
$Z^0 \rightarrow$ hadrons	1.74 GeV
(via $q\bar{q}$, where $q=u, d, s, c$ or b)	
Total	2.49 GeV

family. The Z^0 particle "line shape" is shown in Figure 18-23. These data show that there are three and only three families of neutrinos.

18-3 UNIFICATION OF THE FORCES

The first unification of the two apparently distinct forces of objects falling to the earth and the motion of the planets was accomplished by Sir Isaac Newton in the seventeenth century when he deduced the inverse square law of gravity from experimental data. Newton unified terrestrial and celestial mechanics. Newton also had the scientific imagination to anticipate that a whole world lay yet undiscovered before him:

I do not know what I may appear to the world, but to myself I seem to have been only like a boy playing on the seashore, and diverting myself now and then finding a smoother pebble or a prettier shell than ordinary, whilst the great ocean of truth lay all undiscovered before me.

Electromagnetism and the Weak Force

In the nineteenthth century, electrical and magnetic forces were seemingly distinct from each other. Maxwell discov-

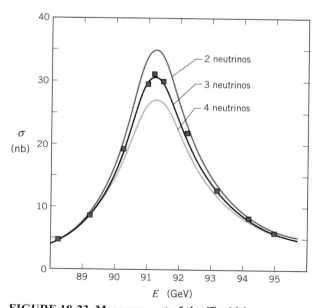

FIGURE 18-23 Measurement of the Z^0 width.
From D. Decamp et al., "Measurement of the Electroweak Parameters from Z Decays into Fermion Pairs," *Z. Phys.* **C48**, 365 (1990).

ered that the "two" forces were merely two manifestations of the same electromagnetic force. Maxwell unified electricity and magnetism.

In the period 1960–1970, various aspects of the theory of the weak interaction were deduced from experiment. The main theoretical contributions were made by Steven Weinberg, Abdus Salam, and Sheldon Glashow all working independently. A central ingredient of the "successful theory," the theory that explained the data, was that the electromagnetic and weak interactions were unified. The unified theory is called the *electroweak* theory. The motivation for the unification of the electromagnetic and weak forces is clearly stated in the opening sentences of the 1967 paper of Steven Weinberg, which is quoted at the beginning of this chapter.

In the electroweak theory the electric charge (e) and the weak charge (g) are related by a *weak mixing angle* (θ_w), which is often appropriately called the "Weinberg angle." The electroweak relationship is

$$e = g \sin \theta_w. \qquad (18.99)$$

The Weinberg angle is a parameter that is not predicted by the theory. The relationship between the Weinberg angle and the intermediate vector boson masses is

$$\cos \theta_w = \frac{m_W}{m_Z}. \qquad (18.100)$$

Since neutrino interactions involve the exchange of W and Z^0 particles, calculation of various scattering rates depends on the masses of these particles or in the electroweak theory on θ_w. A decade of detailed measurements of neutrino cross sections and electroweak calculations gives a value of

$$\sin^2 \theta_w \approx 0.22. \qquad (18.101)$$

The electroweak prediction of the W mass is

$$m_W \approx \frac{37 \text{ GeV}/c^2}{\sin \theta_w} \approx 80 \text{ GeV}/c^2. \qquad (18.102)$$

The electroweak theory contains three independent parameters, which must be determined from experiment. The parameters of choice are naturally those that are most accurately known. Currently, these are α, G_F, and m_Z. Presumably, an accurate measurement of the W mass may some day replace G_F as one of the best known fundamental parameters. Then we could use α, m_W, and m_Z to predict G_F. Currently, m_W is predicted from α, G_F, and m_Z more accurately than it is known experimentally. Any electroweak quantity may be calculated in terms of the three parameters. Since many more than three electroweak quantities have been measured, the electroweak theory is greatly overconstrained by experiment. This is the quantitative measure of its great success.

One implication of the unification of the electromagnetic and weak interactions is that the "two" forces have equal strength at very high energies, large enough so that the mass of the Z^0 boson is negligible and is on "equal footing" with the photon. We should point out that although we know that the electromagnetic and weak forces are unified, we do not know the details of *why* the photon is massless and *why* the W and Z^0 boson are very massive. Some mechanism must be responsible for breaking the symmetry between the four fundamental electroweak bosons. The mechanism for electroweak symmetry breaking is not known. Since the fundamental bosons of the weak interaction are massive, a mass scale has been introduced in the Fermi constant (G_F). The mass scale led to an energy dependence of the dimensionless weak coupling (α_w). This leads to a weak interaction cross section that increases with increasing energy, even though the wavelengths of the interacting particles are decreasing with increasing energy! The weak interaction rate cannot increase forever with increasing energy. At some very large energy, this would violate conservation of probability or *unitarity*. Unitarity is violated at an energy where the weak coupling becomes unity. Recall that

$$\alpha_w \approx \alpha \left(\frac{E}{m_W c^2} \right)^2. \qquad (18.103)$$

The energy where $\alpha_w \approx 1$ is

$$E \approx \frac{m_W c^2}{\sqrt{\alpha}} \approx 1 \text{ TeV}. \qquad (18.104)$$

A more precise calculation gives the unitarity limit to be larger by a factor of $\sqrt{3}$, or

$$E \approx \frac{\sqrt{3} m_W c^2}{\sqrt{\alpha}} \approx 1.7 \text{ TeV}. \qquad (18.105)$$

Some fundamentally new physics must appear at an energy below 1.7 TeV. The new physics should explain why the W boson has a mass of 80 GeV/c^2.

This is not the first time that we have witnessed such a phenomenon. After the advent of special relativity, it was recognized the classical electrostatic potential energy of an electron would exceed the mass energy of the electron at some very short distance (r_0). This happens when

$$mc^2 = \frac{ke^2}{r_0}. \tag{18.106}$$

The distance r_0 is called the *classical radius* of the electron,

$$r_0 = \frac{ke^2}{mc^2} = \frac{1.44\,\text{MeV}\cdot\text{fm}}{0.511\,\text{MeV}} \approx 3\,\text{fm}. \tag{18.107}$$

The classical radius was the "unitary" limit in the understanding of the electron. There is nothing special about the classical electron radius; we have performed experiments at distance scales 1000 times smaller than r_0. What *is* important is that even before we reached the unitary limit we encountered new physics: quantum mechanics. The distance scale for quantum mechanics is the Bohr radius,

$$a_0 = \frac{\hbar c}{\alpha mc^2}. \tag{18.108}$$

The Bohr radius is larger than the classical electron radius by a factor of

$$\frac{a_0}{r_0} = \frac{\left(\dfrac{\hbar c}{\alpha mc^2}\right)}{\left(\dfrac{\alpha \hbar c}{mc^2}\right)} = \frac{1}{\alpha^2}. \tag{18.109}$$

Furthermore, additional new physics (quantum electrodynamics) enters at the distance scale of the electron Compton wavelength,

$$\lambda_C = \frac{hc}{mc^2} = \frac{2\pi r_0}{\alpha}. \tag{18.110}$$

Our current understanding of the electroweak force is analogous to our understanding of the electromagnetic force in 1920. Something magical is not expected to happen at an energy of 1.7 TeV. But somewhere below this energy, possibly even significantly below this energy, there is new physics to be discovered.

Prospects for Grand Unification

Much effort was made in the 1980s in an attempt to unify the strong and electroweak interactions. We have seen that the strong interaction coupling (α_s) is getting weaker with increasing energy and that the weak coupling (α_w) is getting stronger with increasing energy. At large energies ($E \gg m_W c^2$) the electromagnetic and weak interactions are of equal strength. At large energies the electroweak coupling (α) is also changing! The magnitude of α is getting larger with energy. The magnitude of α changes *very* slowly with energy. Indeed we have been able to describe the physics of the atom by assuming that α is constant. At large distances or low energies we have

$$\alpha = \frac{1}{137}. \tag{18.111}$$

At very high energies alpha becomes slightly larger. The reason is that every charge has photons coupled to it. Consider a negative charge. The photons that couple to the charge also couple to $e^+ + e^-$ pairs, and these $e^+ + e^-$ pairs are polarized. The positron tends to be closer to the negative charge than the electron is. Thus, if we probe the charge at a large distance, then the charge that we see includes this *vacuum polarization* effect. The vacuum polarization has reduced the charge. If we probe the charge at a shorter distance, we reduce the effect of the vacuum polarization and see a larger charge. Thus, the electromagnetic force increases slightly with decreasing distance or increasing energy. At the mass energy of the Z^0, we have

$$\alpha = \frac{1}{128}, \tag{18.112}$$

and the electromagnetic force has increased in strength by about 7%.

The gravitational force between two particles depends on their masses. This actually is the low energy limit. The gravitation coupling between two particles depends on the total energy of the particles. The energy could be mass energy or kinetic energy. We may write the gravitational coupling (α_g) as

$$\alpha_g = \frac{G\left(\dfrac{E}{c^2}\right)^2}{\hbar c} = \frac{GE^2}{\hbar c^5}. \tag{18.113}$$

The energy dependence of the dimensionless couplings of the four forces are shown in Figure 18-24. When extrapolated by many orders of magnitude to roughly 10^{16} GeV (10^{-32} m), the strong and electroweak forces are expected to be of comparable strength ($\alpha \approx \alpha_s$). The simplest and most elegant grand unified theories (GUT) had as a central prediction the decay of the proton with a

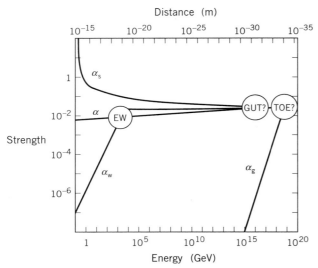

FIGURE 18-24 Energy dependence of the dimensionless couplings.

lifetime of about 10^{30} years. But the proton was not observed to decay in several experiments sensitive to this lifetime. The simplest GUT is not correct. This does not mean, of course, that the electroweak and the strong forces are not unified. We simply do not know.

Searching for Proton Decay

Because of the long lifetime of the proton, very massive detectors are needed to search for its decay. When searching for proton decay, detectors may also be sensitive to the corresponding decay of the neutron. One figure of merit is the mass of the detector. Each kiloton of detector contains about 6×10^{31} nucleons. Two general types of proton detectors have been built: (1) water Cerenkov detectors and (2) tracking calorimeters.

In a water Cerenkov proton decay detector, many phototubes are placed in a large tank of clear water. If a nucleon in either the hydrogen or the oxygen spontaneously decays, one or more of its decay products would be detected by their Cerenkov radiation. A large water Cerenkov detector, built in the Kamiokande lead mine in Japan, has been in operation since 1983. This detector originally had 3 kilotons of water and has now been expanded to 4.5 kilotons. The Cerenkov radiation is detected by 1000 specially made phototubes with very large photocathodes (see Figure 16-15). Another large water Cerenkov detector has been built in a Morton salt mine in Ohio by a group of scientists from the University of California at Irvine, the University of Michigan, and

Brookhaven National Laboratory. The IMB detector has a mass of 8 kilotons and has been operating since 1982. Because a proton decay cannot be distinguished from background if the decay occurs near the edge of the detector, the effective or "fiducial" mass of the detector sensitive to proton decay is 3.3 kilotons. If a proton decays in the middle 3300 tons of water, it should in principle be detected!

A different type of proton decay experiment uses fine-grain tracking between iron plates. This experiment was located 1780 meters under Mont Blanc in the Fréjus tunnel connecting Modane, France, to Bardonecchia, Italy. The mass of the Fréjus detector is 0.9 kilotons. The Fréjus experiment operated for four years from 1985 to 1988, and was especially sensitive to possible three-body decay modes of the proton.

All proton decay experiments are limited by cosmic ray backgrounds. Proton decay experiments are, therefore, located deep under the earth in order to be shielded from cosmic ray muons. Neutrinos, however, can penetrate all shielding and a proton decay detector must be able to differentiate between a neutrino interaction and a proton decay. These neutrinos interact at a rate of approximately 100–200 events per year per kiloton of detector. A second important background consideration in the design of a proton decay experiment is natural radioactivity.

One of the proton decay modes expected, and the easiest to detect experimentally, is

$$p \rightarrow e^+ + \pi^0. \qquad (18.114)$$

Besides the positron, which is directly visible, the two photons from the decay $\pi^0 \rightarrow \gamma + \gamma$ produce showers which are readily detectable. If we combine the results from the two large water Cerenkov experiments we arrive at a limit of

$$\tau_p > 9 \times 10^{32} \text{ y}. \qquad (18.115)$$

for the decay channel $e^+ + \pi^0$. Of course, it is possible that the proton is unstable but does not decay into $e^+ + \pi^0$. Dozens of other proton and neutron decay modes have been searched for. These decays are expected to conserve electric charge and have at least one lepton in the final state. We may summarize these data by saying that the proton lifetime is greater than about 10^{32} years.

Theory of Everything?

The lack of a successful GUT has not stopped many theorists from trying to incorporate all the forces, including gravity, into a "super-grand-unification." Such a theory is sometimes referred to as the "theory of everything" (TOE). Such theories, which are mathematically elegant, are so far unable to predict anything that may be tested by experiment. The energy scale of the TOE is given by the energy where the gravitational coupling (18.113) becomes unity. The energy (E_P) where the gravitational coupling is unity is

$$E_P = \sqrt{\frac{\hbar c^5}{G}} \approx 10^{19} \text{ GeV}. \qquad (18.116)$$

This energy is called the *Planck energy*. When the energy of two particles exceeds the Planck energy, then their gravitational potential energy of attraction at a distance defined by the particle wavelength (hc/E_P) exceeds the energy of the particles! This is a major difficulty in the formulation of a quantum theory of gravity. This difficulty does not exist for the electromagnetic force. In an atom charges are attracted to a separation of about 0.1 nm, because the electromagnetic force depends only on distance and the distance scale is defined by the electron wavelength. The gravitational force depends on both distance and energy. The unification of the forces is summarized in Figure 18-25.

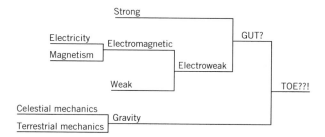

FIGURE 18-25 Unification of the forces.
The force on an apple falling to the earth and the motion of the planets about the sun were discovered to be governed by the same force of gravity by Newton in the seventeenth century. Electricity and magnetism were shown to be two aspects of the same force of electromagnetism by Maxwell in the nineteenth century. In the twentieth century, the electroweak theory of Glashow, Weinberg, and Salam, the ultimate confirmation of which came with the discovery of the W and Z bosons, unified the electromagnetic and weak forces. This is the limit of current experiments. Attempts to unify the strong and electroweak forces (GUT) have so far failed.

CHAPTER 18: PHYSICS SUMMARY

- The theory of the strong interaction is called quantum chromodynamics (QCD). The strong interaction between quarks is mediated by the exchange of massless bosons called gluons. The gluons couple to strong charge (color). The gluons also carry color. There are three types of color. The strong force between two quarks does not depend on the flavor of the quarks.

- Weak interactions provide transitions between the three families of quarks and leptons. The W and Z^0 bosons couple to all quarks and leptons. The weak interactions are mediated by the exchange of W and Z^0 bosons.

- The electromagnetic and weak interactions are unified at energies that are much larger than the masses of the W and Z^0 bosons. The mechanism for the symmetry breaking that has left the photon massless and gives a large mass to the W and Z^0 bosons is not known.

- At very, very large energies (about 10^{16} GeV), the strong interaction and the electroweak interaction are expected to have the same strength. Attempts to unify the strong and electroweak interactions are called grand unification theories (GUTs). A central prediction of the GUTs is the decay of the proton. At even higher energies, about 10^{19} GeV, gravity is expected to have a strength equal to the strong force.

REFERENCES AND SUGGESTIONS FOR FURTHER READING

C. Albajar et al., "Studies of Intermediate Vector Boson Production and Decay in UA1 at the CERN Proton-Antiproton Collider," *Zeit. Phys.* **C44**, 15 (1989).

G. Arnison et al., "Experimental Observation of Isolated Large Transverse Energy Electrons with Associated Missing Energy at $\sqrt{s}$ = 540 GeV," *Phys. Lett.* **122B**, 103 (1983).

G. Arnison et al., "Experimental Observation of Lepton Pairs of Invariant Mass Around 95 GeV/c^2 at the CERN SPS Collider," *Phys. Lett.* **126B**, 398 (1983).

J. D. Bjorken, "Feynman and Partons," *Phys. Today* **42**, No. 2, 56 (1989).

D. Decamp et al., "Measurement of the Electroweak Parameters from Z Decays into Fermion Pairs," *Z. Phys.* **C48**, 365 (1990).

F. J. Dyson, "Field Theory," *Sci. Am.* **188**, No. 4, 57 (1953).

G. J. Feldman and J. Steinberger, "The Number of Families of Matter," *Sci. Am.* **264**, No. 2, 70 (1991).

R. P. Feynman, "Very High-Energy Collisions of Hadrons," *Phys. Rev. Lett.* **24**, 1415 (1969).

H. Georgi and S. L. Glashow, "Unified Theory of Elementary-Particle Forces," *Phys. Today* **33**, No. 9, 30 (1980).

S. Glashow, "Quarks with Color and Flavor," *Sci. Am.* **233**, 38 (October 1975).

S. Glashow, *The Charm of Physics*, Simon and Schuster (1991).

M. B. Green, "Superstrings," *Sci. Am.* **255**, No. 3, 48 (1986).

D. J. Gross, "Asymptotic Freedom," *Phys. Today* **40**, No. 1, 39 (1987).

F. Halzen and A. D. Martin, *Quarks and Leptons*, Wiley (1984).

G. 't Hooft, "Gauge Theories of the Forces Between Elementary Particles," *Sci. Am.* **242**, No. 6, 104 (1980).

P. Langacker and A. K. Mann, "The Unification of Electromagnetism with the Weak Force," *Phys. Today* **42**, No. 12, 22 (1989).

T. D. Lee, "Space Inversion, Time Reversal and Particle-Antiparticle Conjugation," *Phys. Today* **19**, No. 3, 23 (1966).

J. M. LoSecco, F. Reines, and D. Sinclair, "The Search for Proton Decay," *Sci. Am.* **252**, No. 6, 54 (1985).

Y. Nambu, "The Confinement of Quarks," *Sci. Am.* **235**, 48 (November 1975).

C. Quigg, "Elementary Particles and Forces," *Sci. Am.* **252**, No. 4, 84 (1985).

C. Rubbia, P. McIntyre, and D. Cline, "Producing Massive Neutral Vector Bosons with Existing Accelerators," *Proceedings of the International Neutrino Conference*, Aachen (1976).

J. H. Schwartz, "Superstrings," *Phys. Today* **40**, No. 11, 33 (1987).

The LEP Collaborations, "Electroweak Parameters of the Z^0 Resonance and the Standard Model," *Phys. Lett.* **B276**, 247 (1992).

M. J. G. Veltman, "The Higgs Boson," *Sci. Am.* **255**, No. 5, 76 (1986).

S. Weinberg, "A Model of Leptons," *Phys. Rev. Lett.* **19**, 1264 (1967).

S. Weinberg, *Dreams of a Final Theory*, Pantheon (1992).

S. Weinberg, "Unified Theories of Elementary-Particle Interaction," *Sci. Am.* **231**, No. 1, 50 (1974).

QUESTIONS AND PROBLEMS

From quarks to quantum chromodynamics

1. Consider the hydrogen atom. The mass of the atom is *smaller* than the mass of a free proton plus a free electron. The pion has a mass of 140 MeV/c^2, but the quarks inside it have masses of only a few MeV/c^2. How can this be true?

2. Calculate the value of the strong coupling (α_s) at an energy of 1 TeV.

3. Why is α_s easier to measure than Λ?

4. (a) For e^+e^- annihilation at a center-of-mass energy of 10 GeV, what is the cross section for production of a pair of tau leptons? Give your answer in nanobarns. (b) What is the tau pair cross section at 100 GeV?

5. (a) How accurately does one need to measure R (the hadron to muon pair cross section ratio) in order to exclude the existence of a fourth quark with charge $-e/3$ at a mass below 15 GeV/c^2? Explain. (b) How accurately does one need to measure R in order to exclude the existence of a fourth lepton with charge $-e$ at a mass below 15 GeV/c^2?

6. Why are large center-of-mass energies needed (compared to a hadron mass energy) to observe jets in e^+e^- collisions?

Quantum theory of the weak interaction

7. Does the measurement of the Z^0 width exclude the existence of massive neutrinos? Explain.

8. Show that the Z^0 particle cannot decay into two identical spin-zero particles.

9. (a) Make a set of Feynman diagrams indicating the expected decay modes of the top quark. (*Hint:* How does the c quark decay?) (b) Which decays produce muons? Do not forget to include the "cascade" decays.

10. Make a set of Feynman diagrams for the decay of the antibottom quark.

11. (a) Show on dimensional grounds that the muon lifetime is inversely proportional to m^5. (b) Use the measured muon lifetime to make a prediction of the tau lifetime. (*Hint:* How many decay channels are available to the tau?)

12. (a) Why is the decay of the W into quark–antiquark so difficult to detect in an experiment at a hadron collider? (b) How do we know that the decay occurs?

13. Consider the event shown in Figure 18-11. The exact reaction is not known because all particles are not identified. Use conservation laws to write down several possibilities for the complete reaction ($\gamma + p \rightarrow ? \rightarrow ?$).

14. The D^+ meson is observed to have many possible decay modes. Why? Write down 10 possible Cabbibo-favored decays of the D^+ meson.

15. Write down 10 possible Cabbibo-favored decays of the B^0 meson.

Unification of the forces

16. (a) At an energy of 1 GeV the value of the "weak charge" (g) is *larger* than the electron charge (e) by a factor of $1/\sin\theta_w \approx 2$. Why is the electromagnetic force between an electron and a proton so much *stronger* than the weak force at a distance of 1 fm?

Additional problems

17. Use Figure 18-23 to estimate the lifetime of the Z^0 boson.

18. The B mesons are observed to mix, analogous to the K mesons. What $B^0 - \overline{B}^0$ mass difference would correspond to a oscillation period equal to the B^0 lifetime?

19. Estimate the fraction of time that a charm quark decay results in the production of a muon.

20. Neutrinos with an energy of 100 GeV are incident on a block of iron. Estimate the interaction cross section per target nucleon.

21. At Fermilab, Z^0 bosons are produced in proton-antiproton collisions at center-of-mass energy of 800 GeV. (a) If a Z^0 boson is produced at rest, what are the values of Feynman x (momentum fraction) of the annihilating quark and antiquark? (b) If a Z^0 boson is produced with a longitudinal momentum of 100 GeV, what are the approximate values of Feynman x of the annihilating quark and antiquark?

22. Do the neutral K^* mesons, the spin -1 combinations of down and antistrange or strange and antidown, mix? Explain.

23. Use the measured π^+ and π^0 lifetimes to make an order-of-magnitude estimate of α_w at an energy scale of the pion mass energy.

24. Why is the width of the W boson (18.89) proportional to the cube of its mass?

25. A Z^0 boson is produced at CERN in a proton-antiproton collision at a center of mass energy of 540 GeV and decays into leptons. In the rest frame of the Z^0, the leptons emerge at right angles to the proton–antiproton axis. If the Z^0 particle is produced with a longitudinal momentum of 140 GeV, calculate the energy and angle of the leptons?

26. An upgrade of the LEP accelerator at CERN is designed to study the reaction,

$$e^+ + e^- \rightarrow W^+ + W^-,$$

at a center-of-mass energy of 200 GeV. If the luminosity of the machine is 10^{36} m$^{-2}\cdot$s^{-1}, estimate the event rate per day.

27. Consider the prediction of the W and Z^0 masses before their experimental measurements. (a) If $\sin^2\theta_w \approx 0.22$ is known to an accuracy of 30%, what are the errors on the knowledge of the W and Z^0 masses? (b) Repeat the calculation if $\sin^2\theta_w$ is known to an accuracy of 10%. (c) If the W and Z^0 masses are measured to an accuracy of 1%, what is the resulting error on knowledge of $\sin^2\theta_w$?

The first two elements and their stable isotopes, hydrogen and helium, emerged from the first few minutes of the early high-temperature, high-density stage of the expanding universe....Where did the heavier elements originate? The generally accepted answer is that all of the heavier elements from carbon, element six, up to long-lived radioactive uranium, element ninety-two, were produced by nuclear processes in the interior of stars in our own Galaxy....Let me remind you that your bodies consist for the most part of these heavy elements. Apart from hydrogen you are 65% oxygen and 18% carbon, with smaller percentages of nitrogen, sodium, magnesium, phosphorus, sulfur, chlorine, potassium, and traces of still heavier elements. Thus it is possible to say that you and your neighbor and I, each one of us and all of us, are truly and literally a little bit of stardust.

William A. Fowler

We are all a little bit of stardust. Our "dust" comes from neighboring stars. When we look into the sky at night, it is possible to see thousands of these stars. The radiation from the stars propagates to us at the speed of light. Since the stars are far away, we observe them not as they are now, but as they were at the time the radiation was emitted. The star closest to us (apart from the sun) is Alpha Centauri at a distance of about 4×10^{16} meters. Light travels from Alpha Centauri to the earth in a time (t),

$$t \approx \frac{4 \times 10^{16} \text{ m}}{3 \times 10^8 \text{ m/s}} \approx 10^8 \text{ s}, \qquad (19.1)$$

or about four years. When we look at Alpha Centauri we are seeing it as it was four years ago. Observing a distant object necessarily involves looking back in time. Other stars are much farther away from us than Alpha Centauri and their observation allows us to look back much further in time. *How far can we look back?* We can look back about 10 billion years (see Figure 19-1)! In order to observe the universe at such early times, we must understand how to determine the distances to the stars. Before we take up the important problem of determining distances, we address an interesting paradox discovered in the eighteenth century.

19-1 THE DARK NIGHT SKY

Suppose the universe were infinite in extent and contained stars at random locations. If we look into the night sky in any random direction, our line of sight would eventually

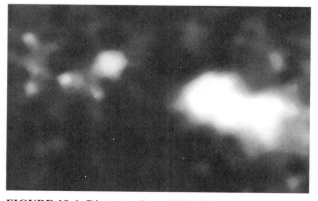

FIGURE 19-1 Distant galaxy 4C41.17.
The distance to the galaxy is about 10 billion light years. This picture was taken with the Hubble Space Telescope. Courtesy NASA.

intersect a star. *Why then is the sky dark at night?* This interesting problem is called the *bright-sky paradox*. The analysis of the bright-sky paradox had a profound influence on the development of our understanding of the early universe. In 1823, H. Wilhelm N. Olbers, following work by J. P. Loys de Chéseaux in 1744, attempted to explain the bright-sky paradox by assuming that starlight is absorbed by matter in interstellar space. (The bright-sky paradox is often called *Olbers paradox*.) Olbers didn't know about the physics of atoms and radiation. We know that such an explanation cannot be correct because the proposed matter in interstellar space would heat up from the absorbed radiation and reradiate the energy.

Suppose stars of the same luminosity and size as the sun were distributed randomly in an infinite universe. The mean free path (d) of starlight would be

$$d = \frac{1}{n \pi r_s^2}, \qquad (19.2)$$

where n is the density of stars and r_s is the radius of a star. The night sky would be bright, if we could observe light from stars as far away as d. The paradox in our model universe is resolved by understanding the size of d. A measurement n and r_s determines d. The universe contains an average of one solar mass per 10^{57} m^3,

$$n \approx 10^{-57} \text{ m}^{-3}, \qquad (19.3)$$

and the radius of the sun is

$$r_s \approx 7 \times 10^8 \text{ m}, \qquad (19.4)$$

which gives

$$d = \frac{1}{n \pi r^2}$$
$$\approx \frac{1}{\left(10^{-57} \text{ m}^{-3}\right)(\pi)\left(7 \times 10^8 \text{ m}\right)^2} \approx 10^{39} \text{ m}. \quad (19.5)$$

The time (t_d) that it takes light to travel this distance is

$$t_d = \frac{d}{c} \approx \frac{10^{39} \text{ m}}{3 \times 10^8 \text{ m/s}} \approx 10^{31} \text{ s}. \qquad (19.6)$$

The solution to the bright-sky paradox is that the universe is *evolving* on a time scale that is much shorter than t_d. Our universe is not static, but is expanding. All stars are moving away from each other, on the average. The time scale of the expansion, the time it takes for the distance between two stars to double, is about 10^{18} s, much smaller than t_d.

The solution to the bright-sky paradox requires an evolving universe. The universe is not static! This chapter is concerned with the physics of the evolving universe. There are models in which the universe has finite size and a finite age. We should stress that we simply *do not know* if our universe is finite in size or has a finite age. We *do know* that the universe is in a state of expansion and cooling. The *standard model* of the universe describes the cooling of the universe from a very hot initial stage to its present temperature of 3 K. The main observational evidence in support of the standard model is twofold: (1) distant objects are moving away from us at a speed that is directly proportional to their distance and (2) the cosmic background radiation left over from an expanding and cooling universe has been observed.

19-2 OVERVIEW OF THE UNIVERSE

Distance by Triangulation

The distances to nearby bodies, such as the planets, may be determined by the technique of *triangulation* (see Figure 19-2). We may choose a reference direction by the pattern of distant stars. As viewed from one observation point, the planet is aligned with the reference direction. As viewed from a second observation point separated by a distance x_0 from the first, the planet makes an angle θ with the reference direction. Measurement of x_0 and θ determines the distance r through the relationship,

$$\frac{x_0}{r} = \tan \theta. \qquad (19.7)$$

The distance to the object, for very small angles, is

$$r \approx \frac{x_0}{\theta}. \qquad (19.8)$$

The error in knowledge of the distance r is limited by the accuracy of the determination of the angle θ. In practice θ can be measured to an accuracy of a fraction of one arc second (1 arc second = 1/3600 of one degree, or about 4.85 microradians). If the two observation points are observatories on the earth, then the distance x_0 can be a few thousand kilometers, and the distances to planets can be measured quite accurately.

EXAMPLE 19-1

The distance to the sun ($r = 1.5 \times 10^{11}$ m) is measured by triangulation from two observatories separated by a dis-

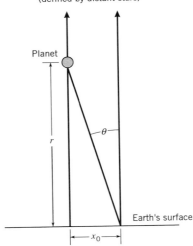

FIGURE 19-2 Distance determination by triangulation. A reference direction is chosen, for example the direction of a star that is much farther away than the distance to the planet (r). As viewed from one observation point, the planet is aligned with the reference direction. As viewed from a second observation point separated by a distance x_0 from the first, the planet makes an angle θ with the reference direction. Measurement of x_0 and θ determines the distance r through the relationship, $x_0/r = \tan\theta$.

tance of 3×10^6 meters. How accurately must the angle θ be measured in order to determine r to an accuracy of 1%?

SOLUTION:

The distance r and the angle θ are related by

$$r = \frac{x_0}{\theta}.$$

A measurement error ($\Delta\theta$) on θ makes an error (Δr) on r of

$$\Delta r = \left| \frac{\partial r}{\partial \theta} \right| \Delta\theta = \frac{x_0 \Delta\theta}{\theta^2}.$$

Therefore,

$$\Delta\theta = \frac{\theta^2 \Delta r}{x_0} = \frac{x_0 \Delta r}{r^2}$$

$$= \left(\frac{x_0}{r} \right)\left(\frac{\Delta r}{r} \right)$$

$$= \left(\frac{3 \times 10^6 \text{ m}}{1.5 \times 10^{11} \text{ m}} \right)(0.01)$$

$$= 2 \times 10^{-7}. \qquad \blacksquare$$

Suppose we use the triangulation technique to determine the distance to Alpha Centauri. If we observe the star at two laboratories separated by a distance equal to the diameter of the earth, then the angle θ is

$$\theta = \frac{1.3 \times 10^7 \text{ m}}{4 \times 10^{16} \text{ m}} \approx 3 \times 10^{-10}. \qquad (19.9)$$

This angle is too small to measure. A much larger distance between observation points can be obtained because of the motion of the earth around the sun. In the case where observations are made six months apart, the distance x_0 increases to the size of the diameter of the earth's orbit! The method of triangulation is also called the method of parallax, or apparent shift in position of an object when viewed from two positions. The astrophysical unit of distance is the parsec (parallax + second). One parsec (pc) is defined to be the distance required to cause a parallax of one second of an arc when using the radius of the earth's orbit as a baseline. For an observer at a distance of one parsec, the maximum angle between the earth and the sun would be equal to one arc second (see Figure 19-3).

The distance from the earth to the sun (D) is called one astrophysical unit. The parsec is defined by

$$\frac{D}{1\,pc} = \frac{2\pi}{(360)(60)(60)}, \qquad (19.10)$$

or

$$1\,pc = \frac{(360)(60)(60)D}{2\pi}$$

$$\approx \frac{(360)(60)(60)(1.50 \times 10^{11} \text{ m})}{2\pi}$$

$$= 3.09 \times 10^{16} \text{ m}. \qquad (19.11)$$

The distance to Alpha Centauri is determined from triangulation to be 1.3 parsecs. The method of parallax for distance determination can be used up to roughly 50 parsecs. Most stars are much farther away from us than the resolution limit of the parallax method. To determine the distances to these stars, other techniques must be used.

The relationship between the parsec and the distance that light travels in one year, the light year (ly) is

$$1\,pc = 3.26 \text{ ly}. \qquad (19.12)$$

The Cepheid Variables

When we look into the sky, some stars appear brighter than others. There are two obvious reasons for this. One is that

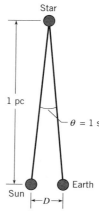

FIGURE 19-3 Definition of the parsec.
The parsec (parallax + second) is defined to be the distance at which the earth and sun have a maximum angular separation of one second, or (1/3600) of one degree.

the luminosity is not the same for every star. The second reason is that the *apparent* luminosity (radiation per area reaching our detector) depends inversely on the square of the distance to the star. The apparent luminosity is a quantity that is directly measurable. If we also measure the distance to the star, then we may calculate the actual or *absolute* luminosity.

One powerful technique for distance determination was discovered by Henrietta Leavitt and Harlow Shapley in 1912. The luminosities of a certain class of stars, called *Cepheid variables*, are observed to vary periodically. The Cepheid variables are younger and more massive than the sun. The luminosity variation is generated by an expanding and contracting stellar atmosphere. The atmospheric expansion and contraction is caused by radiation pressure, which is regulated by helium atoms. The helium atoms act as a radiation valve that is open when they are ionized and closed when they recombine. When the atmosphere expands, the escaping radiation ionizes the atoms and opens the valve, allowing radiation to escape. The decrease in pressure causes the atmosphere to contract, the ions recombine, and the radiation is contained, resulting in an increase in the radiation pressure. The increase in pressure causes an expansion, and so on. Leavitt and Shapley discovered a relationship between the period of a Cepheid variable and its absolute luminosity.

The Cepheid variable stars studied by Leavitt and Shapley are referred to as type-I. The empirically established period–luminosity relationship for Cepheid variable stars of known distances is illustrated in Figure 19-4. This relationship can then be used to infer the distance to a Cepheid variable star by measuring the apparent luminosity and calculating the absolute luminosity from the empirical formula. The problem with this technique is that all Cepheid variable stars do not fit the same empirical formula. A second class of stars (type-II) has a different period–luminosity relationship. In spite of this confusion, the Cepheid variable technique is very powerful for determining the distances to many stars.

EXAMPLE 19-2

The period of a type-I Cepheid variable is measured to be 10^6 seconds, and the radiation flux from the star is 3×10^{-13} W/m². What is the distance to the star?

SOLUTION:

From the empirical relationship shown in Figure 19-4, we see that the star has an absolute luminosity (L) of about 10 times that of the sun,

$$L \approx 4 \times 10^{27} \text{ W}.$$

If d is the distance to the star, then the flux (F) is

$$F = \frac{L}{4\pi d^2},$$

and the distance to the star is

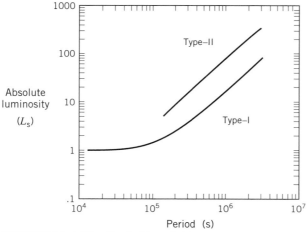

$$d = \sqrt{\frac{L}{4\pi F}}$$

$$\approx \sqrt{\frac{4 \times 10^{27} \text{ W}}{(4\pi)(3 \times 10^{-13} \text{ W/m}^2)}}$$

$$\approx 3 \times 10^{19} \text{ m} \approx 1 \text{ kpc}. \qquad \blacksquare$$

The Milky Way

The sun is just one of a huge cluster, or *galaxy*, of similar stars. The galaxy is held together by the gravitational attraction of the stars. Our galaxy is called the *Milky Way* (from the Greek *glaktik* meaning "milky"). The distances to stars within the Milky Way can be determined by the Cepheid variable technique. The distribution of the stars in the Milky Way has been mapped out in considerable detail. Most of the stars in the Milky Way are clustered in the shape of a flat disc. The radius of the disc is about 10 kpc and the thickness is about 3 kpc. Surrounding the disc is a less dense spherical distribution of stars extending to a radius of 20 kpc. From our viewpoint the Milky Way appears as a disc looked at edgewise and is readily visible at night to the naked eye as a band of hazy light crossing the entire sky. By measuring the motion of the stars and determining their distances, it is determined that the Milky Way is rotating about an axis that is perpendicular to the disc and that the sun is located in the plane of the disc about 10 kpc from the center of the galaxy. Detailed measurements show that the Milky Way has gigantic spiral arms. Our sun is located near the tip of one of the spiral arms. The rotation speed of the sun about the galactic center is about 2.5×10^5 m/s.

EXAMPLE 19-3

The sun was created approximately 10 billion years ago. Make an estimate of the number of revolutions that the sun has made about the galactic center.

SOLUTION:

Let t be the age of the sun, v be the speed of the sun, and R be the distance from the sun to the center of the Milky Way. The number of revolutions is

$$N = \frac{vt}{2\pi R}$$

$$\approx \frac{(2.5 \times 10^5 \text{ m/s})(3.16 \times 10^{17} \text{ s})}{(6.28)(10 \text{ kpc})}$$

$$= 1.26 \times 10^{21} \text{ m/kpc}.$$

FIGURE 19-4 The Cepheid variables.
Leavitt and Shapley discovered that the fluctuation in luminosity of certain stars has a period that depends on the absolute luminosity.

Using

$$1 \text{ kpc} = 3.09 \times 10^{19} \text{ m}.$$

We have

$$N = \frac{1.26 \times 10^{21}}{3.09 \times 10^{19}} \approx 41.$$

In Milky Way "years," the sun is the ripe old age of 41.

∎

EXAMPLE 19-4

The sun is gravitationally attracted by other masses in the galaxy. Use the measured orbital angular momentum of the sun to estimate the mass of the galaxy in units of the solar mass (m_s).

SOLUTION:

We assume that the mass distribution is uniform. By Gauss's law, only those objects at distances less than 10 kpc will produce a net attraction. The gravitational force equals the mass of the sun times its acceleration. Let M_g be the galactic mass:

$$\frac{m_s v^2}{r} = \frac{G m_s M_g}{r^2}.$$

Solving for M_g, we have

$$M_g = \frac{r v^2}{G}.$$

The ratio of the galactic mass to the solar mass is

$$\frac{M_g}{m_s} = \frac{r v^2}{G m_s}$$

$$= \frac{\left(3.1 \times 10^{20} \text{ m}\right)\left(2.5 \times 10^5 \text{ m/s}\right)^2}{\left(6.67 \times 10^{-11} \text{ m}^3/\text{kg} \cdot \text{s}^2\right)\left(2.0 \times 10^{30} \text{ kg}\right)}$$

$$\approx 1.5 \times 10^{11}.$$

There are about 10^{11} solar masses in the Milky Way at distances less than 10 kpc from the center. ∎

The total *mass* of the Milky Way is about 10^{11} solar masses as deduced from the motion of the sun and other stars. The total measured *luminosity* of the galaxy is smaller than expected for stars totaling 10^{11} solar masses by an order of magnitude. Therefore, approximately 90% of the matter in our galaxy is not in the form of luminous stars! This matter, the form of which is unknown, is called *dark matter*. This is found to be true of the universe as a whole; about 90% of the matter is not in the form of luminous stars. We shall return to the very important question of dark matter later in the chapter.

Discovery of the Galaxies

In the eighteenth century, William Herschel discovered the first *nebulae* appearing as small faint patches of light on his telescope. When better telescopes were built in the mid-nineteenth century, some of the closest nebulae were resolved into spiral structures. In 1924, using the 100-inch telescope at the Mt. Wilson observatory, Edwin Hubble was able to resolve many of the closest nebulae into clusters of stars. Having resolved individual stars in distant nebulae, Hubble could use the Cepheid variable technique to measure the distances to many nebulae. Hubble discovered that the star clusters were far outside of our galaxy. The existence of galaxies outside our own Milky Way had been discovered. (All observed nebulae are not galaxies; some are collections of matter not in the form of stars.) The closest large galaxy to us is Messier 31 (M31) in the Andromeda Nebula, named after Charles Messier, who classified nonstellar objects in 1781. The distance to M31 is about 700 kpc.

The Cepheid variable technique is limited to distances of about 1 Mpc because the stars are no longer bright enough to be resolved at such distances. Hubble invented a new technique for estimating the distances to galaxies. Hubble measured the apparent luminosity of the brightest star, a class called *supergiants*, in each galaxy. The supergiants are visible at much larger distances than the Cepheid variables. Hubble made the assumption that the brightest supergiant in each galaxy had nearly the same absolute luminosity, and checked that this was true for supergiants whose distances were known. Hubble then measured the supergiant apparent luminosities and used these data to infer the distances to many galaxies. The technique works to distances of about 10 Mpc, the distance where Hubble could no longer resolve the supergiants.

The galaxies were classified by Hubble according to the spatial distribution of their stars. The four main types are (see Figure 19-5): (1) ordinary spiral, (2) barred spiral, (3) elliptical, and (4) irregular.

Mapping the Galaxies

The galaxies are grouped into large clusters. The Milky Way is in a galactic cluster called the *local cluster*. Hubble extended his technique of apparent luminosity to the galaxies themselves. With the assumption that the brightest galaxy in each cluster has about the same luminosity, Hubble was able to estimate distances as far away as 500 Mpc!

The brightness of a star or a galaxy is characterized by the radiation flux reaching the earth. The brightness is characterized on a scale called *apparent magnitude* (*m*), invented by the ancient Greeks. On the apparent magnitude scale, the brightest stars in the sky are approximately $m = 1$ and the faintest that are visible to the naked eye are approximately $m = 6$. More precisely, the difference in magnitudes of two stars or galaxies is defined by

$$m_2 - m_1 = 2.5 \, \log\left(\frac{F_1}{F_2}\right), \qquad (19.13)$$

where F_1 and F_2 are the measured radiation fluxes from the two objects. If a star that has the same luminosity as the sun were observed from a distance of 10 pc, the apparent magnitude would be 4.77.

In the 1960s, the angular positions of about 3×10^4 galaxies with magnitude less than 15.7 in the northern sky were cataloged by Fritz Zwicky and collaborators. A two-

(a)

(b)

(c)

(d)

FIGURE 19-5 Classification of the galaxies.
(*a*) Spiral galaxy M74, from the Mount Wilson and Palomar Observatories. (*b*) Barred-spiral galaxy M81, from the Lick Observatory. (*c*) Eliptical galaxy M32, courtesy NASA. (*d*) Irregular galaxy M82, courtesy NASA.

dimensional map of the angular position of the galaxies is shown in Figure 19-6. The angular position of a distant object does not depend on the rotation of the earth. The units are *right ascension* (17^h to 8^h) measured from east to west from 12^h, where one hour is equal to $15°$, and a latitude *declination* ($10°$ to $50°$) measured from the equator. The cluster at ($12.5^h, 12.5°$) is due to the local cluster of galaxies. The map of Figure 19-6 does not tell us how far away the galaxies are.

EXAMPLE 19-5

Estimate the distance to a galaxy that has a luminosity 10^{11} times the sun and an apparent magnitude of $m_2 = 15.7$.

SOLUTION:

The flux from the distant galaxy (F_2) is

$$F_2 = \frac{\left(10^{11}\right)L_s}{4\pi d^2},$$

where L_s is the solar luminosity and d is the distance to the galaxy. The flux (F_1) from our "standard candle," a star with the solar luminosity at a distance of 10 parsecs, is

$$F_1 = \frac{L_s}{4\pi\left(10 \text{ pc}\right)^2}.$$

The ratio of fluxes is

$$\frac{F_2}{F_1} = \frac{\left(10^{11}\right)\left(10 \text{ pc}\right)^2}{d^2}.$$

The apparent magnitude of the reference star would be $m_1 = 4.77$. The difference in magnitudes (19.13) is

$$m_2 - m_1 = 15.7 - 4.77$$

$$= 2.5 \; \log\left(\frac{F_1}{F_2}\right)$$

$$= 2.5 \; \log\left[\frac{d^2}{\left(10^{11}\right)\left(10 \text{ pc}\right)^2}\right],$$

or

$$d = \frac{\left(10^{11}\right)\left(10 \text{ pc}\right)^2}{10^{(15.7-4.77)/2.5}} \approx 500 \text{ Mpc}. \qquad \blacksquare$$

Hubble's Law

Hubble measured the atomic spectra from many distant galaxies. The spectra are shifted, compared to known laboratory spectra, by the Doppler effect, due to the motion of the galaxies relative to the earth. Almost all galaxies are observed to have an increase in wavelength ($\Delta\lambda$) or *redshift* for all spectral lines. The redshift parameter (z) is defined:

$$z \equiv \frac{\Delta\lambda}{\lambda}. \qquad (19.14)$$

The Doppler redshift formula (4.129) with $\lambda' = \lambda + \Delta\lambda$ gives the following relationship between the relative speed ($\beta = v/c$) and the redshift:

$$1 + z = \sqrt{\frac{1+\beta}{1-\beta}}. \qquad (19.15)$$

In the special case where the speed is small ($v \ll c$), we have

$$1 + z = \sqrt{\frac{1+\beta}{1-\beta}}$$

$$= \sqrt{\frac{(1+\beta)^2}{(1+\beta)(1-\beta)}}$$

$$= \frac{1+\beta}{\gamma} \approx 1 + \beta. \qquad (19.16)$$

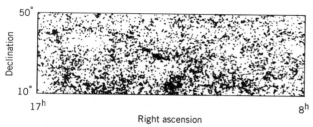

FIGURE 19-6 Galaxy map of a portion of the sky.
The map was made by F. Zwicky and collaborators. After M.J. Geller and J.P. Huchra, *Science* **246**, 897 (1989).

Hubble measured redshifts of many galaxies and deduced the speeds from the redshift formula (19.15). Hubble also measured the distance (d) to these galaxies and fit the data to a remarkably simple formula,

$$v = H_0 d, \qquad (19.17)$$

where H_0 is called the *Hubble* parameter. This result is known as *Hubble's law*: Distant galaxies are moving away from us with a speed that is proportional to their distance! The Hubble parameter has units of inverse time. The presently accepted value of the Hubble parameter is

$$H_0 \approx \frac{7 \times 10^4 \text{ m/s}}{\text{Mpc}}. \qquad (19.18)$$

By writing H_0 in this manner we immediately see that an object that is at a distance of 1 Mpc is moving away from us at a speed of 7×10^4 m/s, on the average. Since there are large uncertainties in the determination of the distances to distant galaxies, there is a rather large experimental error on Hubble's constant. We can safely say that it is in the range $(4–10) \times 10^4$ m/s per Mpc. The original measurements of Hubble did not accurately determine the value of Hubble's constant. They did, however, give a convincing demonstration of the speed versus distance relationship. Hubble's law is illustrated in Figure 19-7, where we plot the redshift of several distant galaxies as a function of their distance.

EXAMPLE 19-6

Use Hubble's law to calculate the distance to a galaxy that has a redshift of $z = 1$.

SOLUTION:

The redshift parameter (z) and $\beta = v/c$ are related by

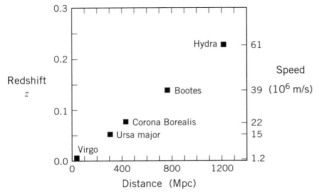

FIGURE 19-7 Experimental evidence for Hubble's law. The amount of redshift of radiation from a distant galaxy is proportional to the distance to the galaxy.

$$1 + z = \sqrt{\frac{1+\beta}{1-\beta}},$$

or

$$(1+z)^2 = \frac{1+\beta}{1-\beta}.$$

Solving for β,

$$\beta = \frac{(1+z)^2 - 1}{(1+z)^2 + 1}.$$

Since $z = 1$, we have

$$\beta = \frac{3}{5}.$$

The distance to the galaxy is

$$d = \frac{v}{H_0} = \frac{3c}{5H_0}$$

$$= \frac{(0.6)(3 \times 10^8 \text{ m/s})}{(7 \times 10^4 \text{ m} \cdot \text{s}^{-1} \cdot \text{Mpc}^{-1})}$$

$$\approx 2600 \text{ Mpc.} \qquad \blacksquare$$

The work of Hubble was continued by Allan Sandage and many others. These results show conclusively that the amount of redshift of distant galaxies is directly proportional to the distance to the galaxy. If the redshift is attributed to the Doppler effect, then all distant galaxies are observed to be moving away from us. Furthermore, the relative speed of the galaxies is directly proportional to their distance. This means that at earlier times, all the galaxies were closer together then they are at present. The universe is expanding! The expansion in no way implies that we occupy a special location in the universe. Hubble's law would be observed to hold true if measured from any galaxy in the universe. This is a statement of the *cosmological principle*.

> The cosmological principle states that on scales greater than the distance between clusters of galaxies, the universe would look the same to an observer in any galaxy. The universe has no center. The relative speed between any two galaxies is proportional to the distance between the galaxies.

The Hubble parameter has units of inverse time. We may write the Hubble parameter as

$$H_0 \approx \frac{1}{5 \times 10^{17}\,\text{s}}. \qquad (19.19)$$

Presumably, the Hubble parameter is actually changing slowly with time due to gravitational attraction. We expect that the rate of expansion of the universe is slowing down with time. We have used the subscript 0 to denote the present-day value of the Hubble parameter. The inverse of the Hubble parameter is sometimes referred to rather loosely as the "age" of the universe (t_0), the amount of time ago when all galaxies would converge at a single point if extrapolated backward

$$t_0 = \frac{1}{H_0} \approx 5 \times 10^{17}\,\text{s} \approx 15\,\text{By}. \qquad (19.20)$$

There is independent evidence that the Milky Way is between 10 and 15 billion years old and that the earth is about 4.6 billion years old, from radioactive dating (see Chapter 11).

Figure 19-8 illustrates the expansion of the universe and the Hubble parameter. A set of five galaxies is pictured when at the present time they are distances of R/4, R/2, 3R/4, and R, where R is the size of the universe. The relative speed due to expansion of the universe is proportional to the distance of the galaxy. In the past, all the galaxies were closer to us and to each other. The value of Hubble's parameter was larger. In the future, all the galaxies will be farther apart from us and from each other. The value of Hubble's parameter will be smaller. An analogy for the expanding universe may be made by considering the visible universe as the surface of a balloon. The location of galaxies may be represented as dots on the balloon. As the balloon is inflated, any two dots move away from each other at a speed directly proportional to the distance between the dots.

Figure 19-9 summarizes the large distance scales measured in the universe.

Radio Galaxies

In 1954, certain galaxies were found to be extraordinarily intense emitters of radio waves. If the position of the radio source is known accurately enough, then the radio source can be identified with a visible galaxy. If the distance to the visible galaxy is known, then the absolute luminosity of the radio source can be determined. One of the most powerful of the radio emitters was identified with the galaxy Cygnus A. The redshift of Cygnus A ($z = 0.057$) was measured in the visible spectrum. From Hubble's law the distance to Cygnus A is determined to be about 300 Mpc. The apparent radio luminosity and distance were used to calculate the

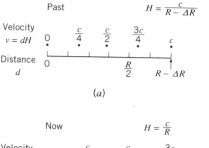

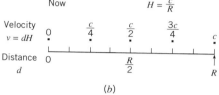

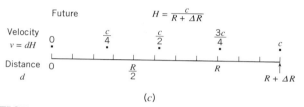

FIGURE 19-8 The expansion of the universe and the Hubble parameter.
A set of five galaxies is pictured when at the present time they are distances from us of $R/4$, $R/2$, $3R/4$, and R, where R is the size of the universe. The relative speed due to expansion is proportional to the distance of the galaxy. (a) In the past, all the galaxies were closer to us and to each other. (b) The universe as it is now. (c) In the future, all the galaxies will be farther apart from us and from each other.

absolute radio luminosity. The result was astounding. The radio luminosity of Cygnus A is about 10^7 times that of the radio luminosity of a normal galaxy!

The source of the radio waves is synchrotron radiation by electrons that are accelerated in the galactic magnetic field. The reason that we know synchrotron radiation is the cause of the radio waves is that the radiation is observed to be polarized, a characteristic of the synchrotron process. The exact source of the electrons is not known, but the electrons appear to be generated by some sort of gigantic explosion near the galactic center.

Quasars

In 1960, it was discovered that some weak radio emitters have a very compact size, closer to that of a star rather than a galaxy. These objects came to be known as quasi-stellar radio sources, or *quasars*. In 1962, the moon passed by a certain quasar (3C273) three times. Since the position of

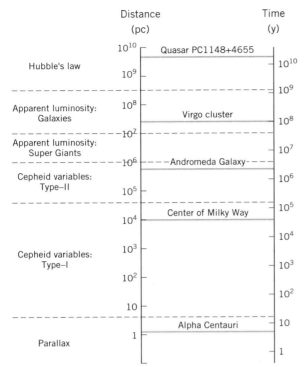

FIGURE 19-9 Large scale distances in our universe.
Distances to our neighboring star, center of our galaxy, neighboring galaxy, neighboring cluster of galaxies, and the object with the largest measured redshift are indicated together with the time taken for light to reach us from these objects. The technique for determining distance scales is indicated on the left.

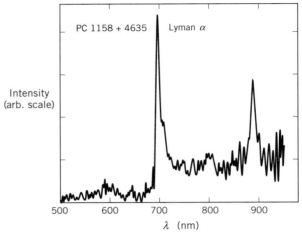

FIGURE 19-10 Spectrum of the distant quasar PC1158+4635.
The redshift is $z = 4.73$, corresponding to a shift in the Lyman-α line of hydrogen from 121.4 nm to about 696.0 nm. Courtesy of D. Schneider, M. Schmidt, and J. Gunn.

the moon is accurately known, this allowed an accurate determination of the position of the quasar. This enabled the quasar to be unambiguously identified with a bright star that was previously thought to be in the Milky Way. It was subsequently determined that the star had a large redshift ($z = 0.16$). This was an amazing result because the distance to the star according to Hubble's law would be about 1 Gpc, corresponding to an absolute magnitude 100 times brighter than the average galaxy!

With this discovery came an intensive search for other quasars with large redshifts. Soon, several were found. The quasar PC1158+4635 has a very large redshift of $z = 4.73$. The Lyman-α line in hydrogen occurs at $\lambda = 121.4$ nm for hydrogen at rest. The wavelength of the Lyman-α line from this quasar is observed at about 696.0 nm, as is shown in Figure 19-10. If this huge redshift was due to a Doppler shift, the quasar would be moving away from us at a speed of $v \approx 0.94\,c$.

Large-Scale Structure of the Universe

If the distances to galaxies are determined as well as their angular position, then a three-dimensional map of the universe may be made. We use the measured average distance between galaxies to estimate the total number of galaxies in the visible universe. There are a huge number of observable galaxies in the universe, roughly 10^{11}. There are approximately as many observable galaxies in the universe as there are stars in our galaxy.

Distances are determined by redshift measurement. It is far harder experimentally to measure a redshift than to measure a luminosity. At the time Zwicky et al. made their catalog, only about one thousand galactic redshifts were measured. By 1990, tens of thousands of redshifts had been determined. These data were then used to make a map of the universe. One way to present these data is to make polar-coordinate plots of (redshift, right ascension) for various declination angles. One such plot made by Margaret Geller and John Huchra is shown in Figure 19-11. This map corresponds to about one part in 10^5 of the visible universe. These data show the discovery of a *Great Wall* of galaxies. The shape of the Great Wall is that of a thin sheet of approximate size 100 Mpc by 250 Mpc. The thickness of the sheet is less than a few megaparsecs. The mass of the Great Wall is about equal to 2×10^{16} solar masses or about 10^5 times the mass of the Milky Way! The map of the universe also shows the presence of voids of

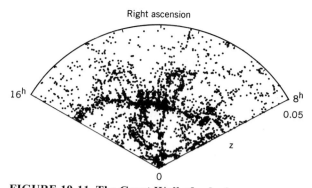

Right ascension

FIGURE 19-11 The Great Wall of galaxies.
From M. J. Geller and J. P. Huchra, "Mapping the Universe,"
Science **246**, 897 (1989).

size $\Delta z = 0.02$ corresponding roughly to 100 Mpc in size. The density of galaxies inside these voids is typically less than one-fifth of the average in the visible universe. Figure 19-12 shows the observed redshift distribution compared to that expected for a uniform distribution of galaxies.

The Cosmic Background Radiation

The universe is expanding and cooling. In the very hot, early stages of the expansion, the energy density of radiation was so large that it prevented the formation of atoms ($kT \gg 1$ eV). During this period, the radiation coupled to the free electrons and protons in the universe and remained in thermal equilibrium. When the temperature of the universe had cooled to about 3000 K, hydrogen atoms

were formed and the free charges disappeared. At that time the photons had a thermal distribution with a temperature of 3000 K. Since then, the universe has expanded by about a factor of 1000 and cooled. The radiation that is left today filling up the entire universe is a thermal spectrum with a temperature of about 3 K.

The microwave radiation was discovered in 1965 by Arnio Penzias and Robert W. Wilson (see Chapter 3). In 1990, the Cosmic Background Explorer (COBE) satellite measured the background radiation spectrum with great precision. These data are shown in Figure 19-13. The spectrum fits a thermal spectrum perfectly with a temperature of about 2.74 K. What is even *more* remarkable is that the cosmic background radiation has the same intensity and energy distribution when viewed in any direction to approximately one part in 10^6. In 1992, data from the COBE satellite showed evidence for minuscule deviations in intensity in certain directions in space. These fluctuations could provide a clue to the understanding of how galaxies formed.

The cosmic photons fill up the entire universe with a number density of

$$n_\gamma = 4 \times 10^8 \, / \mathrm{m}^3. \qquad (19.21)$$

The energy density of the cosmic background radiation is

$$\rho = 2.5 \times 10^5 \, \mathrm{eV/m}^3, \qquad (19.22)$$

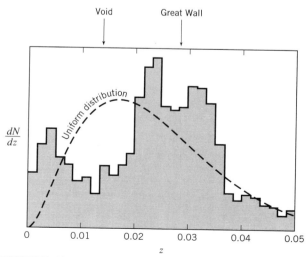

FIGURE 19-12 Redshift distribution.
From M. J. Geller and J. P. Huchra, *Science* **246**, 897 (1989).

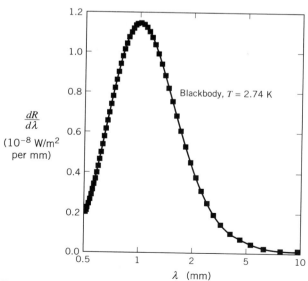

FIGURE 19-13 Results from the Cosmic Background Explorer (COBE) satellite.
After J.C. Mather et al., *Astro Jour.* **354**, L37 (1990).

which is roughly the same as the average energy density of starlight in our galaxy.

The cosmic background radiation, plus Hubble's law provide the experimental cornerstones for the understanding of the evolution of the universe from a dense radiation-dominated state to the present matter-dominated state.

19-3 EVOLUTION OF THE STARS

We live in an evolving universe as evidenced by the fact that the night sky is dark. The first stars were formed roughly 10 billion years ago. First-generation stars are fueled by proton fusion. The proton fusion creates a radiation pressure that balances the force of gravity. When enough protons are consumed, the radiation pressure drops and the star collapses. In the process of this gravitational contraction, the star heats up to a temperature that is large enough to allow α fusion. Alpha fusion requires a larger temperature and density than hydrogen burning for two reasons: (1) the α has a larger Coulomb barrier because it has an electric charge of $2e$, and (2) the intermediate state of ^{8}Be is not stable, having a half-life of only 10^{-14} seconds. The process is

$$\alpha + \alpha \rightarrow {}^8\text{Be}, \qquad (19.23)$$

followed in rapid succession by

$$\alpha + {}^8\text{Be} \rightarrow {}^{12}\text{C}. \qquad (19.24)$$

The net result is the fusion of three α particles into a carbon nucleus. This sequence of reactions gets the element-building process past the barriers at $A = 5$ and $A = 8$, where no stable nuclei are found. During the alpha fusion phase, radiation pressure balances the gravitational attraction for the core of the star but the outer portion of the star may expand by a factor of 100, creating a red giant. If the star is massive enough, then further nuclear reactions continue in the core.

White Dwarfs

When the rate of nuclear fusion reactions decreases, the radiation pressure drops and is overcome by gravity. The gravitational collapse continues until it is balanced by electron pressure generated by the Pauli exclusion principle. The star has become so dense, about 10^9 kg/m^3, that the electron repulsion due to the Pauli exclusion principle has become the main cause of resistance against further contraction. At this stage the star has become a *white*

dwarf. The typical size of a white dwarf is about the size of the earth. The typical mass of a white dwarf may range from about one-half the mass of the sun to a maximum of about 1.4 times the solar mass. (A star with a mass several times the solar mass can still end up as a white dwarf by shedding some of its mass before collapsing.) The sun is destined to become a white dwarf in a few billion years. The surface temperature of a white dwarf is about 10^4 K. Eventually, white dwarfs radiate their energy away, becoming *black dwarfs*.

Stars may be classified according to their surface temperature and luminosity as shown in Figure 19-14. This diagram is called a *Hertzprung-Russell* (HR) diagram after Henry Norris Russell and Ejnar Hertzprung. Most stars lie along a main sequence and are burning hydrogen. A typical star will leave the main sequence when its nuclear fuel is depleted, becoming first a red giant and ultimately a white dwarf. The more massive stars are on the main sequence a shorter time than the less massive stars because they consume their fuel much faster than the less massive stars. Thus, a departure of a star from the main sequence is a good indicator of the age of the star.

Neutron Stars and Black Holes

For stars with masses several times the solar mass, the gravitational collapse continues beyond the white dwarf stage. In such a case, the gravitational attraction overcomes the electron fermionic pressure, resulting in further dramatic collapse. This process is called a type-II *super-*

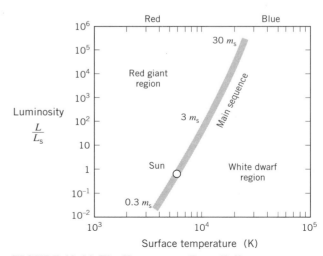

FIGURE 19-14 The Hertzprung-Russell diagram.
This diagram is used to classify stars according to their surface temperature and luminosity.

nova. (A type-I supernova is caused by the collision of two white dwarfs.) The density and temperature are large enough for the reaction,

$$e^- + p \rightarrow \nu_e + n, \qquad (19.25)$$

and a very dense neutron star is formed. The radius of a neutron star is a few kilometers. Further collapse is balanced by the neutron fermionic pressure.

The collapse of a white dwarf generates the unique conditions of high temperature and density necessary for the formation of the elements heavier than iron as well as some of the rarer isotopes of the lighter elements. In the supernova explosion, heavy elements are thrown into the interstellar gas.

EXAMPLE 19-7

A star with a mass of 10 solar masses collapses to form a neutron star with a radius of 10^4 meters. Estimate the energy released.

SOLUTION:

The energy released is the difference in the binding energies of the neutron star and the original star. Since the binding energy is inversely proportional to the radius, we neglect the binding energy of the original star. We estimate the binding energy of the neutron star by calculating how much energy is needed to "disassemble" it. The gravitational potential energy between a mass element dm and a mass m at a distance r is

$$dV = \frac{Gm\,dm}{r}.$$

This is the amount of energy that is required to separate m and dm from r to infinity. If the mass density (ρ) of the neutron star is constant (an assumption), then

$$\frac{4}{3}\pi r^3 \rho = m,$$

and we have

$$dV = \frac{Gm\,dm}{\left(\dfrac{3m}{4\pi\rho}\right)^{1/3}} = \frac{G}{\left(\dfrac{3}{4\pi\rho}\right)^{1/3}} m^{2/3}\,dm.$$

The binding energy of the neutron star is

$$E_b = \frac{G}{\left(\dfrac{3}{4\pi\rho}\right)^{1/3}} \int_0^M dm\, m^{2/3}$$

$$= \frac{G}{\left(\dfrac{3}{4\pi\rho}\right)^{1/3}} \left(\frac{3}{5}\right) M^{5/3}$$

$$= \frac{3GM^2}{5R},$$

where M is the mass of the star and R is the radius. Therefore,

$$E_b \approx \frac{3GM^2}{5R}$$

$$\approx \frac{(3)\left(6.67\times10^{-11}\ \text{m}^3/\text{kg}\cdot\text{s}^2\right)\left(2.0\times10^{31}\ \text{kg}\right)^2}{(5)\left(10^4\ \text{m}\right)}$$

$$\approx 10^{48}\ \text{J.} \qquad \blacksquare$$

An *enormous* energy (much larger than the mass energy of the earth) is released in the formation of the neutron star! About 99% percent of this energy is emitted as neutrinos in the first few seconds of the collapse. The average energy of these neutrinos is about 10 MeV. Most of the neutrinos are generated by the processes,

$$e^+ + e^- \rightarrow \overline{\nu}_e + \nu_e\,,\ \overline{\nu}_\mu + \nu_\mu\,,\ \overline{\nu}_\tau + \nu_\tau. \quad (19.26)$$

The positrons needed to fuel these reactions are supplied by:

$$p + \overline{\nu}_e \rightarrow e^+ + n. \qquad (19.27)$$

Some neutrinos are also produced directly:

$$p + e^- \rightarrow n + \nu_e. \qquad (19.28)$$

A dramatic confirmation of the theory of gravitational collapse has come from the detection of neutrinos from a supernova.

Evidence for neutron stars is detected throughout the universe, by their emissions of radio waves, x rays and γ rays. Some objects in the universe are known to emit regular periodic electromagnetic signals. These objects are known as pulsars and are believed to be spinning neutron stars. One of these objects is the Crab Pulsar, created in the famous supernova explosion of 1054. The pulsars typically have a period of roughly a second and in some cases are observed to be periodic to an accuracy of one part in 10^{-12}.

If the collapsing star is more than five times the solar mass, then even the neutron fermionic pressure will not balance gravity. In that case we may have the ultimate collapse into an object called a *black hole*. The gravitational attraction of a black hole is so strong that light cannot escape! The prediction of the existence of black holes does not depend on the details of the stellar collapse or on the unknown properties of matter at extremely large densities. There are numerous candidates for black holes throughout the universe. The various stages of stellar evolution are summarized in Table 19-1.

Supernova 1987A

On February 23, 1987, a supernova explosion was detected in the Large Magellanic Cloud (LMC). The supernova was visible to the naked eye, the first time since the supernova of 1604. The LMC, named after Ferdinand Magellan, who observed the stars in his famous voyage to the southern hemisphere in 1521, is a neighboring galaxy about 50 kpc (160,000 light years) away from us. This means of course that the star exploded 160,000 years ago! The exploding star was a blue supergiant (15–20 solar masses) named Sanduleak-69 202. The event is commonly known as *Supernova 1987A*.

The discovery was made by Ian Shelton, who had photographed the LMC from Las Campanas, Chile, in order to obtain some calibration data for his study of Halley's comet and independently by Albert Jones in New Zealand. In the photos of the LMC, a bright star was noticed that did not belong there (see Figure 19-15)! By luck, two photographs were taken on February 23 a short time apart in which the star was not visible in the earlier photo and visible in the later photo. These photos were used to pin down the time at which the visible light from the supernova reached the earth. The apparent magnitude

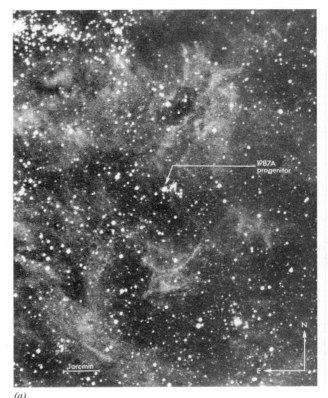

(a)

FIGURE 19-15 Supernova 1987A.
(*a*) Location of the progenitor star (Sanduleak 69 202) photographed on February 6, 1979. (*b*) Photograph of the supernova on February 26, 1987. The cross around the supernova is an optical effect of the plate holder on the telescope. Courtesy European Southern Observatory.

of Sanduleak-69 202 as a function of time is shown in Figure 19-16.

The neutrinos produced in the collapse of the supernova core to form a neutron star have a mean free path of only 0.03 meters in the center of the neutron star! The

TABLE 19-1
EVOLUTION OF A MASSIVE STAR.

Stage	Radius (approximate)	Balance against gravitational collapse
Young star	10^9 m	Radiation pressure from proton fusion
White dwarf	10^7 m	Electron pressure (Pauli exclusion principle)
Neutron star	10^4 m	Neutron pressure (Pauli exclusion principle)
Black hole	10^3 m	None

(b)

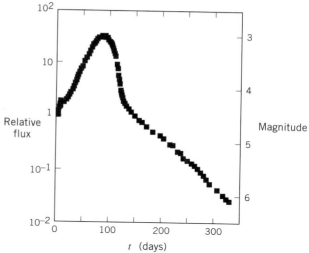

FIGURE 19-16 Brightness of SN1987A.
The apparent magnitude of Supernova 1987A as measured by the International Ultraviolet Explorer Satellite is shown as a function of time for the one-year period following the first observation of the explosion. At its brightest on May 20, 1987, it was "third magnitude," which is about as bright as the stars in the Big Dipper. After R. Kirshner, *Modern Physics in America*, AIP, (1988).

neutrinos are absorbed and reemitted over a time period of several seconds. Two large water Cerenkov radiation detectors, Kamiokande II in Japan, and IMB in Ohio, observed antineutrinos ($\bar{v}_e$) from the collapsing star through the process,

$$\bar{v}_e + p \rightarrow e^+ + n . \qquad (19.29)$$

The neutrinos were recorded (but not yet analyzed) in the detectors about 18 hours before the first optical sighting. In the Kamiokande II detector, 9 neutrino interactions were detected in a period of 2 seconds, and in IMB, 8 neutrino events were detected in 6 seconds. Positrons from the interactions of antineutrinos (19.29) are observed by their Cerenkov radiation. The Kamiokande II data are shown in Figure 19-17. The detector is sensitive to positrons above 8 MeV.

The Supernova 1987A neutrino data may be used to derive limits on the electron-neutrino mass (m) and lifetime (τ). Since the neutrinos traveled 160,000 light years to reach the earth,

$$\gamma\tau > 1.6 \times 10^5 \text{ y}, \qquad (19.30)$$

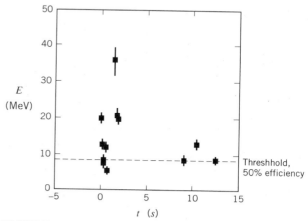

FIGURE 19-17 Observation of the neutrinos from Supernova 1987A.
The energy of the detected neutrino is plotted as a function of time. Only interactions of the highest energy neutrinos were observable in this experiment. The origin of the time scale was 07:35:35 on February 23, 1987. Nine electron events were observed in a 2-second time interval. The data were recorded in the Kamiokande detector. After K. Hirata et al., *Phys. Rev. Lett.* **58**, 1490 (1987).

CHAPTER 19 / THE EARLY UNIVERSE

where γ is the usual relativistic gamma factor (E/mc^2). The neutrinos arrive in a small time-spread, indicating that they travel from the star to the earth with a constant speed. Since the neutrinos have varying energies, this places an upper limit on the antielectron-neutrino mass of

$$m < 30 \text{ eV}/c^2. \qquad (19.31)$$

This limit is approximately as good as the best neutrino mass limits from beta decay. (To determine a limit on the neutrino mass from beta decay, the electron kinetic energy spectrum is accurately measured near its maximum. A nonzero neutrino mass makes a smaller maximum electron kinetic energy.)

EXAMPLE 19-8
Use the data of Figure 19-17 to make an estimate of the upper mass limit of the electron-antineutrino from SN1987A.

SOLUTION:
The typical energy of a detected supernova neutrino is 15 MeV. The detected neutrinos have a spread in energy of the same magnitude. Since the neutrinos are detected in a time interval of approximately 10 seconds, we need to determine the neutrino mass (m) that causes a 15-MeV neutrino to arrive 10 seconds later than a neutrino traveling at the speed of light. The distance from the earth to the supernova is

$$d \approx 50 \text{ kpc} \approx 1.6 \times 10^{21} \text{ m}.$$

The time (t_1) taken for a massless neutrino to travel to the earth is

$$t_1 = \frac{d}{c} \approx \frac{1.6 \times 10^{21} \text{ m}}{3 \times 10^8 \text{ m/s}} \approx 5 \times 10^{12} \text{ s}.$$

A massive neutrino will travel at a speed v. The time (t_2) taken to reach the earth is

$$t_2 = \frac{d}{v}.$$

The difference in times is

$$\Delta t = t_2 - t_1 = \frac{d}{v} - \frac{d}{c}$$

$$= d\left(\frac{c-v}{vc}\right) = \frac{d(1-\beta)}{v}$$

$$\approx \frac{d(1-\beta)}{c} = (1-\beta)t_1,$$

where $\beta = v/c \approx 1$. It is useful to write $1-\beta$ in terms of γ,

$$\gamma = \frac{1}{\sqrt{1-\beta^2}} = \frac{1}{\sqrt{(1-\beta)(1+\beta)}},$$

or

$$1 - \beta = \frac{1}{\gamma^2(1+\beta)} \approx \frac{1}{2\gamma^2}.$$

The relationship between γ and the neutrino mass is

$$\gamma = \frac{E}{mc^2},$$

where E is the neutrino energy. Therefore,

$$\Delta t \approx (1-\beta)t_1 \approx \frac{t_1}{2\gamma^2} = \frac{(mc^2)^2 t_1}{2E^2},$$

and

$$mc^2 \approx E\sqrt{\frac{2\Delta t}{t_1}} \approx (15 \text{ MeV})\sqrt{\frac{(2)(10 \text{ s})}{5 \times 10^{12} \text{ s}}} \approx 30 \text{ eV}.$$

The neutrino upper mass limit is about 30 eV/c^2. (A larger neutrino mass would make a time-spread longer than ±5 seconds.) ∎

19-4 THE ROLE OF GRAVITY

The Equivalence Principle

Gravity plays an important role in the evolution of the universe. The force of gravity is responsible for the observed large-scale structure in the universe. The first theory of gravity was formulated by Newton. The gravitational force between two massive objects (m_1 and m_2) is

$$F_g = \frac{Gm_1 m_2}{r^2}, \qquad (19.32)$$

where r is the distance between the two masses. The Newtonian force law is successful in describing the motion of the planets. The physics of gravity is complicated, however, at relativistic speeds. For example, photons (which have zero mass) have a measurable gravitational interaction! There is even a more fundamental problem in trying to formulate a quantum theory of gravity, that is, a theory in which the force is transmitted by the exchange of gravitational bosons (called gravitons). There is presently no accepted quantum theory of gravity, and in this sense, gravity is the least understood of the four forces.

The first extension of the theory of gravity beyond the formulation of Newton was provided by Einstein. Einstein's theory of gravity is called general relativity. The cornerstone of general relativity is Einstein's postulate of the *equivalence principle*. The equivalence principle states that uniform acceleration of an object is equivalent to a gravitational field. An object in a gravitational field is accelerated. An object freely falling toward the earth has an acceleration that just cancels the effect of the earth's gravitational field. An astronaut is weightless, not because he or she is beyond the earth's gravitational pull (he or she is not beyond the earth's gravitational field!) but rather because the spaceship is freely falling toward the earth.

The equivalence principle has profound consequences. One of these consequences is that the gravitational field causes a curvature of space-time. In special relativity, we have assumed that space-time is "flat." In a flat space-time, photons travel in a straight line. In the absence of space-time curvature, the orbit of the planet mercury is expected to be an ellipse, subject only to the corrections caused by the other planets. The gravitational effect of the sun causes space-time to be curved, and this curvature causes a precession of the perihelion (distance of closest approach) of mercury. The observation of this effect was the first triumph of general relativity.

Another consequence of general relativity is the deflection of light by a large mass. In general relativity, this may be thought of as due to space-time curvature. A large mass can bend a beam of light! The effect, as you might well imagine, is very small, but it has been observed experimentally. Light from stars have been observed to be bent by the sun. The stars, which can be viewed during a total solar eclipse, appear to have a shift in position when the starlight reaching us passes close to the sun.

Another direct observation of the curvature of space time comes from viewing a distant galaxy through a cluster of foreground galaxies. The mass of the cluster acts as a gravitational lens, and producing a double image of the distant galaxy as shown in Figure 19-18.

In Einstein's theory of gravity, gravitational force on a photon depends on the photon energy and velocity. For a photon traveling toward the mass M, the gravitational force is

$$F = \frac{GM\left(\dfrac{E}{c^2}\right)}{r^2}. \qquad (19.33)$$

(For a photon traveling perpendicular to the direction of M, the force is twice as large.)

Gravitational Redshift

The gravitational force on a photon that is propagating away from a mass M causes a decrease in the momentum of the photon. The decrease in photon momentum corresponds to an increase in the photon wavelength. This increase in photon wavelength is called *gravitational redshift*. A photon propagating toward a mass M will gain momentum and will be shifted toward the blue. Consider a photon propagating from an altitude L to the earth's surface. The change in energy of the photon is

$$\Delta E = \frac{GM\left(\dfrac{E}{c^2}\right)}{R} - \frac{GM\left(\dfrac{E}{c^2}\right)}{R+L}, \qquad (19.34)$$

where M and R are the mass and radius of the earth. The fractional change in energy of the photon is

$$\frac{\Delta E}{E} = \left(\frac{GM}{R^2 c^2}\right)\left(\frac{LR}{R+L}\right). \qquad (19.35)$$

Using

$$g \equiv \frac{GM}{R^2} \approx 9.8 \text{ m/s}^2, \qquad (19.36)$$

we have

$$\frac{\Delta E}{E} = \left(\frac{gL}{c^2}\right)\left(\frac{R}{R+L}\right) \approx 1.09 \times 10^{-16}\left(\frac{L}{R}\right), \qquad (19.37)$$

for $L \ll R$. The fractional change in energy is a very small number, even for very large values of L.

The change in energy of photons due to the effect of the earth's gravity was measured by Robert Pound and Glen Rebka, Jr. in 1960. Since the fractional change in energy ($\Delta E/E$) is so small, it can be determined only by the measurement of the average change in energy of a large number of photons that have a narrow energy distribution. For this measurement Pound and Rebka used the Mössbauer technique and a specially developed ^{57}Fe source. The source of photons used by Pound and Rebka was the 14.4-keV gamma from the decay of ^{57}Fe*, as described in Chapter 11. With special care taken in the preparation of their source to enhance the Mössbauer effect, Pound and Rebka were able to obtain a photon spectrum with a fractional width,

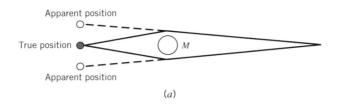

(a)

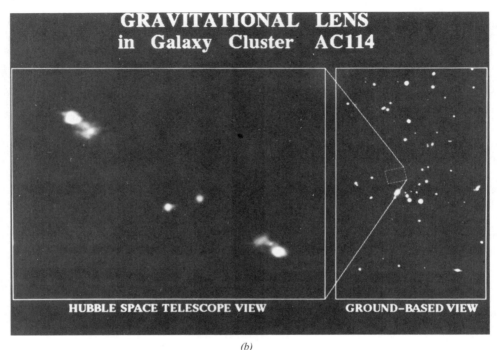

(b)

FIGURE 19-18 Detecting the curvature of space time.
Light from a distant galaxy (10 billion light years away) is viewed through a cluster of foreground galaxies AC114 (10 billion light years away). (*a*) Schematic. (*b*) Image from a 6 hour exposure with the Hubble Space Telescope. The objects near the center of the image are not related. Courtesy R. Ellis, Durham University and NASA.

$$\frac{\Delta E}{E} \approx 10^{-12}. \qquad (19.38)$$

The distance L was limited in practice to several meters by the strength of the radioactive source and the size of the absorber (a few square centimeters). This means that the expected change in energy due to the effect of gravitation was much smaller than the line width. *Statistically*, this meant that the photon spectra needed to be measured over the course of several days. To make a meaningful measurement, Pound and Rebka realized that they needed to overcome two potentially large sources of *systematic* errors: (1) the energy shifts due to hyperfine structure and (2) the energy shifts caused by small temperature varia-

tions. To be rid of the systematic error caused by differences between source and absorber, the experimenters measured the *difference in energy shifts* between the photons traveling downward and the photons traveling upward. To account for the effects of temperature variations, the experimenters monitored the temperature to 0.03 K and made the appropriate corrections to the data. (A temperature change of about 1 K caused a shift equal in magnitude to the gravitational shift.)

EXAMPLE 19-9
Calculate the speed of the source that would correspond to maximum absorption of 14.4-keV photons that travel a vertical distance of $L = 22.6$ meters.

SOLUTION:

The change in energy (ΔE) of the photons is

$$\Delta E = \frac{EgL}{c^2}.$$

This is an energy gain if the photons travel toward the earth and an energy loss if the photons travel away from the earth. The change in energy from the Doppler shift is

$$\Delta E_{\text{Doppler}} = E - E(1-\beta) = E\beta.$$

The maximum resonance absorption occurs when

$$\Delta E_{\text{Doppler}} = \Delta E_{\text{gravity}},$$

or

$$\beta = \frac{gL}{c^2} = \frac{(9.8\,\text{m/s}^2)(22.6\,\text{m})}{(3\times10^8\,\text{m/s})^2} = 2.5\times10^{-15}\,\text{m/s}.$$

This corresponds to about 1/100 of the line width in ^{57}Fe (see Figure 11-23). Note that the speed does not depend on the energy of the photon. ∎

The experimental arrangement of Pound and Rebka is indicated in Figure 19-19. The experimenters "dropped"

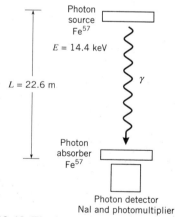

FIGURE 19-19 The experiment of Pound and Rebka to measure the gravitational attraction of a photon to the earth.
The photon source is the 14.4-keV x ray from ^{57}Fe decay produced with narrow width by the Mössbauer effect. The photon source is in motion, and the absorption condition at the target is that the change in energy by the Doppler shift is equal to the change in energy after falling a distance L in the earth's gravitational field. After R.V. Pound and G.A. Rebka, Jr., *Phys. Rev. Lett.* **4**, 337 (1960).

Doppler-shifted photons from a tower. In falling a distance $L=22.6$ meters, the photons gain an energy EgL/c^2. The photons arrive at the target with an energy E_2,

$$E_2 = E(1-\beta) + \frac{EgL}{c^2}. \tag{19.39}$$

Pound and Rebka determined the velocity at which the absorption was a maximum. Since maximum absorption occurs when $E_2 = E$ by the Mössbauer effect, we have maximum absorption when

$$\beta = \frac{gL}{c^2}. \tag{19.40}$$

The measurement of β is a direct measurement of the change in energy of the photon due to the gravitation of the earth. Pound and Rebka also "threw" photons up the tower to made sure that they were observing the effects of gravity. In this case maximum absorption was observed when the velocity of the source was in the same direction as the photon. They observed

$$\left(\frac{\Delta E}{E}\right)_{\text{down}} - \left(\frac{\Delta E}{E}\right)_{\text{up}} = (5.1\pm0.5)\times10^{-15}. \tag{19.41}$$

The prediction of relativity is

$$\left(\frac{\Delta E}{E}\right)_{\text{down}} - \left(\frac{\Delta E}{E}\right)_{\text{up}} = \frac{2gL}{c^2} = 4.9\times10^{-15}. \tag{19.42}$$

The experiment showed conclusively that the photon has a gravitational interaction in accordance with the prediction of general relativity.

19-5 THE PHYSICS OF THE ·EXPANDING UNIVERSE

The Friedmann Equation

On a large scale, we may view the universe as an expanding gas of galaxies. It is instructive to look at the expansion in terms of classical Newtonian physics. This is surely a gross oversimplification, but nevertheless, it will provide much fundamental insight into the physics of the expansion.

Consider a spherical volume of radius r, where r is much larger than the distance between two galaxies but much smaller than the size of the universe. Consider a galaxy of mass m that lies on the surface of the sphere. The galaxy has a kinetic energy because of the expansion; all galaxies are moving apart from each other. We characterize the expansion by a dimensionless time-dependent scale factor, $R(t)$, by writing

$$r = r_0 R(t), \qquad (19.43)$$

where r_0 is the size of our sphere at some arbitrary time. Differentiating with respect to time, we get

$$\frac{dr}{dt} = r_0 \frac{dR}{dt}. \qquad (19.44)$$

Dividing both sides by r,

$$\frac{\left(\dfrac{dr}{dt}\right)}{r} = \frac{\left(\dfrac{dR}{dt}\right)}{R}, \qquad (19.45)$$

or

$$\frac{dr}{dt} = \frac{\left(\dfrac{dR}{dt}\right)}{R} r. \qquad (19.46)$$

This is Hubble's law (19.17) with the identification of the Hubble parameter as

$$H(t) = \frac{\left(\dfrac{dR}{dt}\right)}{R}. \qquad (19.47)$$

If we define the expansion parameter to be unity at the present time,

$$R(t_0) \equiv 1, \qquad (19.48)$$

then

$$H_0 = \left[\frac{dR}{dt}\right]_{t=t_0}. \qquad (19.49)$$

The kinetic energy of the galaxy (assuming it is nonrelativistic) is

$$E_k = \frac{1}{2} m \left(\frac{dr}{dt}\right)^2 = \frac{1}{2} m r_0^2 \left(\frac{dR}{dt}\right)^2. \qquad (19.50)$$

Let M represent the total amount of mass that is enclosed by the sphere. This mass provides a gravitational force of attraction for the galaxy. (By Gauss's law, mass outside the sphere does not contribute to the gravitational force.) The potential energy of the galaxy at the surface of the sphere is

$$V = -\frac{mMG}{r} = -\frac{mMG}{r_0 R}. \qquad (19.51)$$

We may write the mass M in terms of the *average* mass-energy density (ρ) of matter in the universe:

$$M = \frac{4\pi\rho r^3}{3c^2} = \frac{4\pi\rho r_0^3 R^3}{3c^2}. \qquad (19.52)$$

Using this expression for M, the potential energy (19.51) becomes

$$V = -\frac{mMG}{r} = -\frac{4\pi m G r_0^2 R^2 \rho}{3c^2}. \qquad (19.53)$$

In interpreting the potential energy (19.53), it must be remembered that both the scale-factor (R) and the density (ρ) depend on time. As the universe expands, the mass of the galaxies does not change but the volume that they occupy is increasing. Obviously, $\rho(t)$ is inversely proportional to $R^3(t)$,

$$\rho(t) \propto \frac{1}{R^3}. \qquad (19.54)$$

Both the kinetic energy and the potential energy change with time as the universe expands. The total energy of the galaxy is

$$E = E_k + V = \frac{1}{2} m r_0^2 \left(\frac{dR}{dt}\right)^2 - \frac{4\pi m G r_0^2 R^2 \rho}{3c^2}. \qquad (19.55)$$

If energy is conserved, the total energy of the galaxy does not change in the expansion. This total energy is written in terms of a time-independent parameter (k),

$$E = -\frac{kmr_0^2}{2}. \qquad (19.56)$$

The k parameter has units of inverse time squared. Classically, k is a parameter that specifies the total amount of energy in the universe. In the theory of general relativity energy causes space-time curvature, so k is often referred to as the *curvature paramenter*. Conservation of energy (19.55) in terms of the k parameter gives

$$-\frac{kmr_0^2}{2} = \frac{1}{2} m r_0^2 \left(\frac{dR}{dt}\right)^2 - \frac{4\pi m G r_0^2 R^2 \rho}{3c^2} \qquad (19.57)$$

or

$$\left(\frac{dR}{dt}\right)^2 = \frac{8\pi\rho GR^2}{3c^2} - k. \qquad (19.58)$$

This expression is called the *Friedmann* equation, after Alexander Friedmann, who developed it in its general relativistic form. A useful form of the Friedmann equation (19.58) is obtained by dividing both sides by R^2 and then noting that the left-hand side is just the square of the Hubble parameter (19.47). The result is

$$H^2 = \frac{8\pi G\rho}{3c^2} - \frac{k}{R^2}. \qquad (19.59)$$

Models of the Universe

Einstein wrote down a general set of equations relating the gravitational curvature of space-time with the energy density of the universe. These mathematically complicated equations are greatly simplified with the assumption that the universe is uniform and isotropic. When general relativity is taken into account, then the Friedmann equation (19.59) holds provided that the energy density (ρ) is taken to be the total energy density in both matter and radiation. We must keep in mind, however, that the energy density has a different time dependence for the mass part and the radiation part as the universe expands. General relativity also leads to a second equation involving the radiation pressure (p):

$$\frac{\left(\frac{d^2R}{dt^2}\right)}{R} = -\frac{4\pi G}{3c^2}(\rho + 3p). \qquad (19.60)$$

The solutions (19.59) and (19.60) to the Einstein equations of gravity were discovered by Friedmann in 1922 and independently by Abbé Georges Lemaître in 1927.

The Cosmological Constant

Einstein considered the possibility of a constant term in the Friedmann equation. The constant term represents a contribution to the energy density of the universe caused by fluctuations in the vacuum. The original motivation of Einstein for adding the new term was to account for the possibility of an isotropic, homogeneous, *static* universe. With the inclusion of the Einstein term, the Friedmann equation becomes

$$H^2 = \frac{8\pi G\rho}{3c^2} - \frac{k}{R^2} + \frac{\Lambda c^2}{3}. \qquad (19.61)$$

This constant (Λ) is called the *cosmological constant*. The cosmological constant contributes to the *change* in the expansion rate. The contribution of Λ to the change in the expansion rate may be positive or negative. We do not know if the term containing the cosmological constant is needed to describe our universe, that is, it may be that Λ is exactly equal to zero. The units of Λ are inverse length squared. The interpretation of Λ is that vacuum fluctuations affect the geometry of space-time on a distance scale of $1/\sqrt{\Lambda}$. The value of Λ can be experimentally determined by measuring the volume density of distant galaxies. A nonzero value of Λ affects the volume distribution because it causes a change in the expansion rate. These measurements show that

$$|\Lambda| < 3 \times 10^{-52} \text{ m}^{-2}, \qquad (19.62)$$

or that the vacuum fluctuations do not affect space-time geometry on a distance scale of 10^{26} m. This is an amazing result because the theoretical contribution to Λ from quarks is about 10^{-6} m^{-2}. Somehow there is a most remarkable cancellation of all the contributions to the vacuum fluctuations so that their net total is zero!

With the addition of the cosmological constant term, the expression for the radiation pressure (19.60) becomes

$$\frac{\left(\frac{d^2R}{dt^2}\right)}{R} = \frac{\Lambda c^2}{3} - \frac{4\pi G}{3c^2}(\rho + 3p). \qquad (19.63)$$

The Curvature Parameter k

We now analyze the various possibilities for the constant k.

The case $k = 0$ is called the *Einstein-de Sitter universe*. In the Einstein-de Sitter universe the energy density is such that the universe expands forever with a decreasing expansion rate. A universe that expands forever is called an *open* universe. The Friedmann equation (19.58) reduces to

$$\left(\frac{dR}{dt}\right)^2 = \frac{8\pi GR^2\rho}{3c^2}. \qquad (19.64)$$

The energy density for which this equation holds is called the *critical energy density* (ρ_c). The critical energy density may be written in terms of the present value of the Hubble parameter H_0 (19.49):

$$\rho_c = \frac{3c^2 H_0^2}{8\pi G}. \qquad (19.65)$$

If the energy density of the universe is less than or equal to the critical density, then the universe expands forever. If the energy density is greater than the critical density, then gravity eventually will overcome the expansion and the universe will contract!

Since the mass-energy density (19.54) is proportional to $1/R^3$, we may write the Friedmann equation (19.64) as

$$\frac{dR}{dt} = \sqrt{\frac{8\pi G\rho_c}{3c^2}}\, R^{-1/2}, \qquad (19.66)$$

This equation may easily be integrated to give

$$R = \left(\frac{3}{2}\right)^{2/3}\left(\frac{8\pi G\rho_c}{3c^2}\right)^{1/3} t^{2/3}. \qquad (19.67)$$

EXAMPLE 19-10

Make a numerical estimate of the critical energy density.

SOLUTION:

The presently accepted value of Hubble's parameter is

$$H_0 \approx \frac{1}{5\times10^{17}\text{ s}}.$$

The critical density is given by

$$\frac{\rho_c}{c^2} = \frac{3H_0^2}{8\pi G}$$

$$= \frac{(3)\left(5\times10^{17}\text{ s}\right)^{-2}}{(8)(\pi)\left(6.67\times10^{-11}\text{ m}^3\cdot\text{kg}^{-1}\cdot\text{s}^{-2}\right)}$$

$$\approx 7\times10^{-27}\text{ kg/m}^3.$$

In units of GeV/m³ we have

$$\rho_c \approx \left(7\times10^{-27}\text{ kg/m}^3\right)\left(\frac{0.94\text{ GeV}/c^2}{1.67\times10^{-27}\text{ kg}}\right)c^2$$

$$\approx 4\text{ GeV/m}^3.$$

The knowledge of the critical density is limited by the knowledge of the Hubble parameter. The Hubble parameter could be as small as $(8\times10^{17}\text{ s})^{-1}$, corresponding to a lower limit on the critical energy density of about 2 GeV/m³. In the other extreme, the Hubble parameter could be as large as $(3\times10^{17}\text{ s})^{-1}$, corresponding to an upper limit on the critical density of about 10 GeV/m³. Thus, the critical density is somewhere in the range

$$2\text{ GeV/m}^3 < \rho_c < 10\text{ GeV/m}^3. \qquad \blacksquare$$

For $k > 0$, the universe is gravitationally bound. The universe expands until

$$\frac{dR}{dt} = 0. \qquad (19.68)$$

The Friedmann equation gives the maximum value of the expansion parameter (R_{max}):

$$R_{max} = \frac{8\pi G\rho(t_0)}{3c^2}. \qquad (19.69)$$

Gravity then causes a reversal of the expansion (dR/dt becomes negative) and the universe collapses. This situation is called a *closed* universe.

For $k < 0$, the universe expands forever. The equation for $R(t)$ cannot be integrated, but the functional form is

$$R \propto t^{2/3} \qquad (19.70)$$

for small t, and

$$R \propto t \qquad (19.71)$$

for large t.

The Acceleration Parameter

The expansion of the universe is presumably slowing down because of gravity. We do not know exactly how the rate of expansion is changing with time because we do not have precise knowledge of the total energy density of the universe and the resulting space-time geometry. The evolution of the universe is described in terms of the acceleration of the scale parameter, d^2R/dt^2. A dimensionless parameter (q) is defined by

$$q \equiv -\frac{R\dfrac{d^2R}{dt^2}}{\left(\dfrac{dR}{dt}\right)^2}. \qquad (19.72)$$

From the Friedmann equation and Hubble's law, we have

$$q = \frac{4\pi G\rho}{3c^2 H_0^2}. \qquad (19.73)$$

For the special case of $k = 0$, the parameter q does not depend on time and is equal to 1/2. For $k > 0$, we have q greater than 1/2. For $k < 0$, q must be greater than zero and less than 1/2. The expansion parameter (R) as a function of q is illustrated in Figure 19-20 for several model universes.

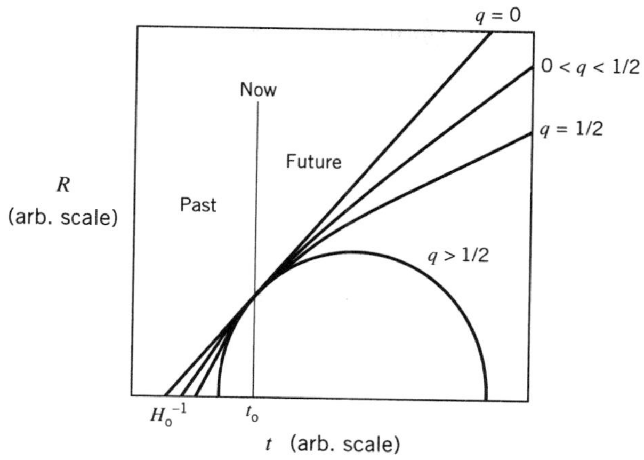

FIGURE 19-20 Time dependence of the expansion parameter, $R(t)$.
The size of the universe in the past and the future depends on the energy density of the universe which is specified by the curvature parameter k.

EXAMPLE 19-11

Show that $q = 1/2$ for the Einstein-de Sitter model of the universe.

SOLUTION:

In the Einstein-de Sitter universe, the scale parameter is

$$R(t) = Ct^{2/3},$$

where C is a constant. The derivatives are

$$\frac{dR}{dt} = \frac{2}{3}Ct^{-1/3}$$

and

$$\frac{d^2R}{dt^2} = -\frac{2}{9}Ct^{-4/3}.$$

The acceleration parameter is

$$q = -\frac{R\frac{d^2R}{dt^2}}{\left(\frac{dR}{dt}\right)^2} = \frac{\left[-Ct^{2/3}\right]\left[\left(\frac{-2}{9}\right)Ct^{-4/3}\right]}{\left[\left(\frac{2}{3}\right)Ct^{-1/3}\right]^2} = 1/2.$$

In this special case, the acceleration parameter does not change with time. ∎

Age of the Universe

The calculated age of the universe is model dependent. We do not even know if the universe had a definite beginning or if it had an infinite past. For example, in the Einstein-de Sitter universe ($k = 0$), we have

$$R(t) \propto t^{2/3}. \qquad (19.74)$$

Assuming the universe had a definite beginning, the Hubble parameter gives the age of the universe:

$$H(t) = \frac{\left(\frac{dR}{dt}\right)}{R} = \frac{\frac{2}{3}t^{-1/3}}{t^{2/3}} = \frac{2}{3}t^{-1}. \qquad (19.75)$$

Therefore, in this model the present age of the universe is

$$t_0 = \frac{2}{3}H_0^{-1} \approx 3\times10^{17}\text{ s}. \qquad (19.76)$$

Size of the Universe

The size of the universe is also model dependent. The exact space-time curvature is not known. The universe may have an infinite size. In our matter-dominated universe, the size is growing with time as

$$R \propto t^{2/3}. \qquad (19.77)$$

In the radiation-dominated early universe,

$$R \propto t^{1/2}. \qquad (19.78)$$

The "horizon" is proportional to t, so we see that for extremely small values of t, particles are beyond the horizon! At such very, very early times the universe is beyond our comprehension. We may speak of the size of the *visible* universe as the age times the speed of light. In the Einstein-de Sitter universe, the size is given by

$$R_0 \approx \left(\frac{2}{3}\right)\left(5\times10^{17}\text{ s}\right)\left(3\times10^8\text{ m/s}\right) \approx 10^{26}\text{ m}. \qquad (19.79)$$

Inflation

A model of the evolution of the universe at extremely short times was invented by Alan Guth. The *inflation* model was invented to solve two problems, commonly referred to as the *horizon* and the *flatness* problems. The horizon problem is that at very early times the universe is assumed to have been homogeneous in spite of the fact that regions of the universe had not had time to communicate with each other. The flatness problem is that the universe is at least

10 billion years old, which requires a near perfect balance (19.65) between the energy density and the Hubble parameter. The *expansion time* (t_{exp}) in the early universe is defined to be the inverse of the Hubble parameter,

$$t_{exp} \equiv \frac{1}{H(t)}. \tag{19.80}$$

The idea of inflation is that at an expansion time of somewhere in the neighborhood of 10^{-35} s, the universe underwent an extremely rapid exponential expansion for a time of about 10^{-33} s. The inflation process is due to a phase transition caused by the breaking of grand unification (unification of the strong and electroweak forces). The phase transition provides energy that maintains the energy density of the universe while the radiation propagates to distances sufficiently large to smooth out the quantum fluctuations in the early universe. Inflation creates a very uniform early universe resulting in a delicate balance between kinetic energy of expansion and the gravitational energy. This balance is needed to explain why the universe has expanded for billions of years without collapsing. Inflation provides a way to avoid the expected quantum fluctuations of the early universe and at the same time to allow enough density fluctuations to remain to allow the formation of galaxies.

Inflation may also provide a solution to the monopole problem. Magnetic monopoles are predicted to have extremely large masses (about 10^{16} GeV/c^2) and are believed to have been created at the time of grand unification of the forces. A rapid expansion of the universe would suppress the density of monopoles in the observable universe, in agreement with the fact that monopoles have not been observed.

Energy Density of the Universe

To describe the expansion of the universe we need to know the present energy density of the universe (both matter and radiation) as well as the time dependence of the energy densities. If we look around the universe, we see that the observable matter is in the form of protons and neutrons contained in nuclei that are found in stars. We can determine the mass of the stars by their gravitational interactions, determine their distances by their redshifts, sum the amount of matter in the universe, and divide by the volume that contains it. We find that the mass-energy density of observable matter (ρ_m) is given by

$$\frac{\rho_m}{c^2} \approx \frac{10^{11} \text{ solar masses}}{(\text{Mpc})^3}. \tag{19.81}$$

This mass-energy density corresponds to about one-half nucleon per cubic meter,

$$\rho_m \approx 0.5 \text{ GeV/m}^3. \tag{19.82}$$

Energy also exists in the form of radiation. The universe is filled with cosmic radiation that fits a thermal (blackbody) spectrum at a temperature of 2.7 K. The energy density in photons (u) is given by the Stefan-Boltzmann law,

$$u = \frac{4}{c}\sigma T^4$$

$$= \left(\frac{4}{3 \times 10^8 \text{ m/s}}\right)$$

$$\times \left(5.7 \times 10^{-8} \text{ W} \cdot \text{m}^{-2} \cdot \text{K}^{-4}\right)$$

$$\times (2.7 \text{ K})^4 \left(\frac{\text{eV}}{1.6 \times 10^{-19} \text{ J}}\right)$$

$$\approx 0.25 \text{ MeV/m}^3. \tag{19.83}$$

There is also expected to be radiation energy in the form of neutrinos, although these cosmic neutrinos have not been detected. The neutrinos are expected to have a slightly lower temperature (about 1.9 K) but there are more of them by a factor of 7/4. The total radiation energy density in photons plus neutrinos (ρ_r) is

$$\rho_r \approx 0.4 \text{ MeV/m}^3. \tag{19.84}$$

The radiation energy density of the universe is much smaller than the matter energy density, $\rho_r \ll \rho_m$. The energy of the universe is matter dominated. The total energy density of the universe is the sum of the matter and the radiation densities,

$$\rho = \rho_r + \rho_m. \tag{19.85}$$

The energy density of the universe has changed with time as the universe expanded. The amount of radiation energy is given by the thermal radiation (blackbody) formula because the early universe was in thermal equilibrium. The early universe was in thermal equilibrium since the interactions between photons and electrons occurred on a much shorter time scale than the time scale of expansion. Thus, the temperature dependence of the radiation energy density is given by the Stefan-Boltzmann law. The radiation density of the universe is proportional to the fourth power of the temperature,

$$\rho_r(t) \propto T^4. \tag{19.86}$$

As the universe expands, the wavelength of each photon, which is inversely proportional to the temperature, grows linearly with R,

$$\lambda \propto \frac{1}{kT} \propto R. \tag{19.87}$$

Therefore,

$$\rho_r(t) \propto T^4 \propto \frac{1}{R^4}. \tag{19.88}$$

The mass-energy density varies as the inverse of the volume, or

$$\rho_m(t) \propto \frac{1}{R^3} \propto T^3. \tag{19.89}$$

Therefore, for high temperatures, radiation energy dominates, and for low temperatures, mass energy dominates. The temperature where the radiation and matter energy densities of the universe were equal was about 3000 K.

EXAMPLE 19-12
Estimate the temperature at which the radiation and matter energy densities were equal.

SOLUTION:
Using the measured values of the energy densities, ρ_r and ρ_m, and their known temperature dependence, we may write radiation energy density as

$$\rho_r(T) \approx \left(0.4 \text{ MeV/m}^3\right)\left(\frac{T}{2.7 \text{ K}}\right)^4$$

and the matter energy density as

$$\rho_r(T) \approx \left(0.5 \text{ GeV/m}^3\right)\left(\frac{T}{2.7 \text{ K}}\right)^3.$$

The radiation and matter energy densities are equal when

$$\left(0.4 \text{ MeV/m}^3\right)\left(\frac{T}{2.7 \text{ K}}\right)^4 \approx \left(0.5 \text{ GeV/m}^3\right)\left(\frac{T}{2.7 \text{ K}}\right)^3$$

or

$$T \approx \left(\frac{0.5 \text{ GeV/m}^3}{0.4 \text{ MeV/m}^3}\right)(2.7 \text{ K}) \approx 3000 \text{ K}.$$

There is a substantial error on the temperature at which the radiation and matter energy densities are equal be-cause there is a substantial error on our knowledge of the energy density of matter. ∎

Photon-to-Baryon Ratio

The energy of the universe is dominated by the mass energy of baryons, which have an average density of about 0.5 per cubic meter. There is a much larger number of cosmic photons, about $4 \times 10^8/\text{m}^3$, but they have a very small energy. From the measured cosmic background radiation and the amount of visible matter in the universe we can calculate the photon-to-baryon ratio to be

$$\frac{N_\gamma}{N_B} \approx \frac{4 \times 10^8 \text{ m}^{-3}}{0.5 \text{ m}^{-3}} \approx 10^9. \tag{19.90}$$

This ratio has not changed as the universe has expanded because neither cosmic photons nor nucleons are created or destroyed, at least not in any significant numbers. We see that as the universe expands, the mass energy of the baryons is not changing but the energy of the photons is decreasing due to the expansion of wavelengths. From this simple geometrical argument we see that if the expansion has been occurring for a sufficiently long time, then at much earlier times the energy density of the universe was dominated by radiation. Observational evidence indicates that the expansion has been occurring for at least 10 billion years.

Critical Density
We have seen (Example 19-10) that there is a critical energy density (ρ_c) for which the force of gravitation is strong enough to stop the expansion of the universe,

$$\rho_c = \frac{3c^2 H_0^2}{8\pi G} \approx 4 \text{ GeV/m}^3. \tag{19.91}$$

The value of critical energy density is roughly an order of magnitude greater than the total observed energy density, $\rho(t_0)$. The observed energy density of the universe is described with the parameter Ω_0, defined by

$$\Omega_0 \equiv \frac{\rho(t_0)}{\rho_c}. \tag{19.92}$$

There are two contributions to error in the knowledge of Ω_0, the total observed energy density of the universe and Hubble's parameter (which affects ρ_c). The resulting bounds on Ω_0 are

$$0.1 < \Omega_0 < 0.4. \tag{19.93}$$

Dark Matter

The evidence for dark matter in our galaxy comes from measurement of the relative speeds of stars. These speeds give us a total mass of the galaxy, which is about an order of magnitude larger than the masses of the luminous stars. On a larger scale, measurement of the speeds of galaxies within clusters gives us a total mass of the cluster, which is also about an order of magnitude larger than the luminous masses of the galaxies. Thus, roughly 90% of the mass of the universe is dark. There is a large uncertainty in the amount of dark matter present in the universe. The form of the dark matter is not known, and it is not known if there is enough dark matter to close the universe. The observed amount of luminous matter corresponds to Ω_0 in the range

$$0.005 < \Omega_0 < 0.02 . \tag{19.94}$$

EXAMPLE 19-13

The number density of neutrinos in the universe is expected to be 7/4 times the number density of cosmic photons. Estimate the mass of the neutrino necessary to account for the dark matter.

SOLUTION:

The number density of photons (n_γ) is the total energy density (19.83) divided by the average energy per photon, $2.7kT$ (see Chapter 3, problem 20),

$$n_\gamma = \frac{0.25 \text{ MeV/m}^3}{(2.7)(0.026 \text{ eV})\left(\dfrac{2.7 \text{ K}}{300 \text{ K}}\right)} \approx 4 \times 10^8 \text{ m}^{-3}.$$

The number density of neutrinos (n_ν) is

$$n_\nu = \frac{7}{4} n_\gamma = \left(\frac{7}{4}\right)\left(4 \times 10^8 \text{ m}^{-3}\right) = 7 \times 10^8 \text{ m}^{-3}.$$

The energy density of dark matter (ρ_{dark}) in the universe is about 10 times the energy density of visible matter,

$$\rho_{\text{dark}} \approx 10 \, \rho_m \approx 5 \text{ GeV/m}^3.$$

The mass of the neutrino (m) would need to be

$$m = \frac{\rho_{\text{dark}}}{n_\nu} = \frac{5 \text{ GeV}/c^2/\text{m}^3}{7 \times 10^8 \text{ m}^{-3}} \approx 7 \, \text{eV}/c^2. \qquad \blacksquare$$

Time Dependence of the Temperature and Energy Density

The physics of the expanding universe is described in terms of the equilibrium temperature of the radiation. The present temperature of the radiation is measured to be

$$T_0 \approx 2.7 \text{ K}. \tag{19.95}$$

The temperature parameter provides a scale for the typical energy of a photon. The characteristic photon energy (E) in a thermal spectrum is

$$E \approx kT . \tag{19.96}$$

The present temperature corresponds to a typical energy (E_{now}) of

$$E_{\text{now}} \approx \left(\frac{1}{40}\text{eV}\right)\left(\frac{2.7 \text{ K}}{300 \text{K}}\right) \approx 2 \times 10^{-4} \text{ eV}. \tag{19.97}$$

The photon wavelengths are increasing as the universe expands. The energy of the photons is getting smaller. The universe is cooling.

The temperature of the radiation may be used to calculate the energy density. The radiation energy density as a function of temperature may be written as

$$\rho_r (T) \approx \left(0.4 \text{ MeV/m}^3\right)\left(\frac{T}{2.7 \text{ K}}\right)^4. \tag{19.98}$$

The characteristic expansion time (19.80),

$$t_{\text{exp}} = \frac{1}{H(t)}, \tag{19.99}$$

is roughly the time it takes the expanding universe to double in size. The current value (t_0) of the expansion time is

$$t_0 = H_0^{-1} = 5 \times 10^{17} \text{ s}. \tag{19.100}$$

The Friedmann equation gives the relationship between the Hubble parameter and the energy density,

$$H^2 = \frac{8\pi\rho G}{3c^2}. \tag{19.101}$$

For large temperatures such that the energy density of the universe was radiation dominated, the expansion time as a function of temperature T is

$$t_{exp}(T) = \frac{1}{H} = \sqrt{\frac{3c^2}{8\pi\rho G}}$$

$$\approx \left(\frac{2.7\ K}{T}\right)^2 \sqrt{\frac{3c^2}{8\pi G\left(0.4\ MeV/m^3\right)}}. \quad (19.102)$$

> At large temperatures ($T \gg 3000$ K), the energy density of the universe was dominated by radiation. The expansion time is inversely proportional to the square of the temperature.

EXAMPLE 19-14

Calculate the characteristic expansion time when the universe was at a temperature of 10^{12} K.

SOLUTION:

The expansion time is

$$t_{exp}(T) = \left(\frac{2.7\ K}{T}\right)^2 \sqrt{\frac{3c^2}{8\pi G\left(0.4\ MeV/m^3\right)}}.$$

The factor in the square root is

$$\sqrt{\frac{3\left(3\times10^8\ m/s\right)^2}{8\pi\left(6.7\times10^{-11}\ m^3\cdot kg^{-1}\cdot s^{-2}\right)\left(0.4\ MeV/m^3\right)}}$$
$$\times\left(1.6\times10^{-13}\ J/MeV\right)^{-1/2}$$

$$\approx 5.0\times10^{19}\ s.$$

The expansion time is

$$t_{exp} \approx \left(5.0\times10^{19}\ s\right)\left(\frac{2.7\ K}{10^{12}\ K}\right)^2 \approx 4\times10^{-4}\ s.$$

The universe was expanding extremely rapidly compared to its present rate of expansion! ∎

The relationship between expansion time and temperature (when radiation dominates the energy density) is given approximately by

$$t_{exp} \approx \frac{4\times10^{20}\ s\cdot K^2}{T^2}. \quad (19.103)$$

At low temperatures, the energy density of the universe is dominated by matter. The energy density of matter as a function of temperature is given by

$$\rho_m(T) \approx \left(0.5\ GeV/m^3\right)\left(\frac{T}{2.7\ K}\right)^3. \quad (19.104)$$

This leads to an expansion time as a function of temperature given by

$$t_{exp}(T) = \frac{1}{H} = \sqrt{\frac{3c^2}{8\pi\rho G}}$$

$$\approx \left(\frac{2.7\ K}{T}\right)^{3/2} \sqrt{\frac{3c^2}{8\pi G\rho_0}}, \quad (19.105)$$

where ρ_0 is the present mass-energy density. Taking ρ_0 to be that energy density needed to close the universe, that is, including the missing dark matter, the expansion time is equal to 5×10^{17} s.

> At low temperatures ($T \ll 3000$ K), the energy density of the universe is dominated by matter. The expansion time is inversely proportional to the temperature to the 3/2 power.

Figure 19-21 shows a plot of temperature as a function of expansion time between 10^{-43} seconds and 5×10^{17} seconds (now). Table 19-2 summarizes the main physical processes that occur at each temperature. This physics is discussed in more detail in the following sections.

The Highest Contemplatable Temperatures

The highest "particle" energy that we can contemplate is the energy at which gravity is as strong as the other forces (see Figure 18-24). Recall the definition of the dimensionless gravitational coupling parameter,

$$\alpha_g = \frac{GM^2}{\hbar c}, \quad (19.106)$$

analogous to $\alpha = ke^2/\hbar c$. The gravitation coupling becomes large for very large values of M. The definition of the Planck mass (M_P) is the mass where the gravitational coupling is unity,

$$\alpha_g = \frac{GM_P^2}{\hbar c} = 1, \quad (19.107)$$

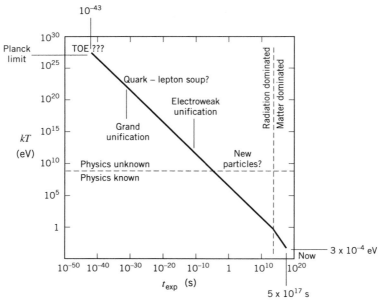

FIGURE 19-21 Temperature as a function of time.
The present time (now) is defined to be equal $H_0^{-1} = 5 \times 10^{17}$ seconds. The present temperature is measured to be 2.7 kelvin from the cosmic radiation. As we extrapolate back in time, the universe is denser and hotter. The temperature dependence is given by the expansion rate. The expansion rate depends on whether the energy of the universe is dominated by matter or radiation. The universe is presently dominated by matter, and the temperature is currently varying as $t^{-2/3}$. This was true for temperatures less than about 3000 k. At temperatures above 3000 k, the universe was radiation dominated and the temperature varied as $t^{-1/2}$.

TABLE 19-2
EVOLUTION OF THE UNIVERSE.

Time ($1/H$)	Size ($k=0$)	T	Energy/ particle	Rad. Energy Density	Physics
10^{-43} s	10^{-35} m	10^{32} K	10^{19} GeV	10^{123} GeV/m^3	Planck scale
10^{-38} s	10^{-29} m	10^{29} K	10^{16} GeV	10^{111} GeV/m^3	Grand Unification, inflation?
10^{-12} s	10^{-3} m	10^{16} K	1 TeV	10^{59} GeV/m^3	Electroweak unification
10^{-6} s	10^{3} m	10^{13} K	1 GeV	10^{47} GeV/m^3	Hadrons formed
1 s	10^{8} m	10^{10} K	2 MeV	10^{37} GeV/m^3	Neutrinos "freeze," e^+ annihilate
10^2 s	10^{11} m	10^{9} K	100 keV	10^{31} GeV/m^3	Light nuclei formed
10^{13} s	10^{22} m	3×10^3 K	0.3 eV	10^9 GeV/m^3	Atoms, galaxies, stars formed
5×10^{17} s	10^{26} m	3 K	2×10^{-4} eV	4×10^{-4} GeV/m^3	Now

or

$$M_P = \sqrt{\frac{\hbar c}{G}}. \qquad (19.108)$$

The corresponding mass energy ($M_P c^2$) is called the Planck energy (E_P) and is given by

$$E_P = M_P c^2 = \sqrt{\frac{\hbar c^5}{G}}. \qquad (19.109)$$

EXAMPLE 19-15

Calculate the numerical value of the Planck energy.

SOLUTION:

The Planck energy is given by

$$E_P = \sqrt{\frac{\hbar c^5}{G}}$$

$$= \sqrt{\frac{\left(1.05 \times 10^{-34} \text{ J} \cdot \text{s}\right)\left(3.00 \times 10^8 \text{ m/s}\right)^5}{\left(6.67 \times 10^{-11} \text{ m}^3 \cdot \text{kg}^{-1} \cdot \text{s}^{-2}\right)}}$$

$$\approx 1.95 \times 10^9 \text{ J}.$$

In electron-volts, the Planck energy is

$$E_P = \frac{1.95 \times 10^9 \text{ J}}{1.60 \times 10^{-19} \text{ J/eV}} \approx 1.22 \times 10^{28} \text{ eV}$$

$$= 1.22 \times 10^{19} \text{ GeV}.$$

The Planck mass is about 10^{19} proton masses! ∎

The force of gravity is expected to be mediated by a massless boson called the graviton. At the gigantic energies of the Planck scale, gravitons are in thermal equilibrium. The temperature (T_P) that corresponds to particle energies of E_P is

$$T_P \approx \left[\frac{1.22 \times 10^{28} \text{ eV}}{\left(\frac{1}{40} \text{ eV}\right)}\right](300 \text{ K}) \approx 10^{32} \text{ K}. \qquad (19.110)$$

The wavelength of a massless particle that has the Planck energy is called the Planck wavelength (λ_P). The value of the Planck wavelength is

$$\lambda_P = \frac{hc}{E_P} = \frac{1240 \text{ eV} \cdot \text{nm}}{1.22 \times 10^{28} \text{ eV}} \approx 10^{-34} \text{ m}. \qquad (19.111)$$

We know nothing about the universe in such a state, if indeed there even ever was such a state. If the universe is blindly extrapolated back to the gigantic temperature of 10^{32} K, it occurs at an expansion time (t_P) of

$$t_P \approx \frac{4 \times 10^{20} \text{ s} \cdot \text{K}^2}{T_P^{\,2}}$$

$$\approx \frac{4 \times 10^{20} \text{ s} \cdot \text{K}^2}{\left(10^{32} \text{ K}\right)^2} \approx 10^{-43} \text{ s}. \qquad (19.112)$$

The event horizon corresponding to this time is

$$c\,t_P \approx \left(3 \times 10^8 \text{ m/s}\right)\left(10^{-43} \text{ s}\right) \approx 10^{-35} \text{ m}. \qquad (19.113)$$

The wavelength (19.111) of a typical "particle" is larger than the size of the universe (19.113)! This has no meaning in our present understanding of the universe!

The Freezing of Gravity

It is slightly easier to contemplate the state of the universe after the gravitons have "frozen" out of equilibrium, but the strong and electroweak forces are of equal strength. The grand unification energy (E_{GUT}) is

$$E_{GUT} \approx 10^{16} \text{ GeV} \qquad (19.114)$$

(see Figure 18-24). Grand unification temperature occurs at a temperature (T_{GUT}),

$$T_{GUT} \approx 10^{29} \text{ K}. \qquad (19.115)$$

This temperature is reached at an expansion time (t_{GUT}) of

$$t_{GUT} \approx \frac{4 \times 10^{20} \text{ s} \cdot \text{K}^2}{T_{GUT}^{\,2}}$$

$$\approx \frac{4 \times 10^{20} \text{ s} \cdot \text{K}^2}{\left(10^{29} \text{ K}\right)^2} \approx 10^{-38} \text{ s}. \qquad (19.116)$$

At this stage, the world was made up of six types of quarks and antiquarks, six leptons and antileptons, all copiously created and interacting with each other by exchanging bosons (gluons, photons, Ws, and Z^0s). This hypothetical state of matter is often referred to as the *quark–lepton soup*. We really don't know if the grand unification actually takes place and if so, at what temperature. We are still in uncharted territory!

Electroweak Unification

The electromagnetic and weak forces are unified at an energy (E_{EW}),

$$E_{EW} \approx 1 \, \text{TeV} \qquad (19.117)$$

(see Figure 18-24). This corresponds to a temperature (T_{EW}),

$$T_{EW} \approx 10^{16} \, \text{K}. \qquad (19.118)$$

Such a temperature is reached at an expansion time (t_{EW}),

$$\begin{aligned} t_{EW} &\approx \frac{4 \times 10^{20} \, \text{s} \cdot \text{K}^2}{T_{EW}^2} \\ &\approx \frac{4 \times 10^{20} \, \text{s} \cdot \text{K}^2}{\left(10^{16} \, \text{K}\right)^2} \approx 10^{-12} \, \text{s}. \quad (19.119) \end{aligned}$$

The electromagnetic and weak forces are unified at a temperature of 10^{16} K. We do not know how the unification takes place. The discovery of the unification mechanism could radically affect our understanding of the universe at high temperatures (above 10^{13} K). For example, there could be one or more particles in the mass region of one TeV/c^2 that play an important role in this physics. We are not as bad off as at the grand unification scale, but the complete electroweak physics is still not known.

EXAMPLE 19-16

Make an estimate of the energy density of the universe at the electroweak unification temperature. Compare with the mass-energy density of the nucleus.

SOLUTION:

The energy density is dominated by radiation (19.98) and is given by

$$\begin{aligned} \rho_r(T) &\approx \left(0.4 \, \text{MeV/m}^3\right)\left(\frac{10^{16} \, \text{K}}{2.7 \, \text{K}}\right)^4 \\ &\approx 7 \times 10^{58} \, \text{GeV/m}^3. \end{aligned}$$

The nuclear mass-energy density is

$$\rho_{\text{nuclear}} \approx \frac{1 \, \text{GeV}}{\left(10^{-15} \, \text{m}\right)^3} \approx 10^{45} \, \text{GeV/m}^3.$$

The energy density of the universe would have been 14 orders of magnitude greater than the energy density of a nucleus! ∎

Condensation of the Quarks

Hadrons (e.g., protons and neutrons) were formed in the early universe at an energy (E_{had}),

$$E_{had} \approx 1 \, \text{GeV}. \qquad (19.120)$$

This energy marks the approximate boundary where the fundamental physics is known (see Figure 19-21). The corresponding temperature (T_{had}) is

$$T_{had} \approx 10^{13} \, \text{K}. \qquad (19.121)$$

This temperature is reached at an expansion time (t_{had}),

$$\begin{aligned} t_{had} &\approx \frac{4 \times 10^{20} \, \text{s} \cdot \text{K}^2}{T_{had}^2} \\ &\approx \frac{4 \times 10^{20} \, \text{s} \cdot \text{K}^2}{\left(10^{13} \, \text{K}\right)^2} \approx 10^{-6} \, \text{s}. \quad (19.122) \end{aligned}$$

The energy density of the universe at this time (see Table 19-2) is about equal to that of the mass-energy density inside a nucleus! The antiquarks had annihilated and the heavy quarks (s, c, b, t) had decayed. The remaining up and down quarks combined into protons and neutrons. There is at least one important question that remains unsolved from this period. *Why were there more quarks than antiquarks, leaving us in a universe made of matter?*

At temperatures smaller than about 10^{13} K, the fundamental physics is known. We can discuss much of the physics of the expanding universe with confidence. A great number of experimental observations support the standard model of the early universe at temperatures below 10^{13} K. Much interesting physics happened as the universe cooled and expanded, including the formation of light nuclei, atoms, and stars. Figure 19-22 shows the radiation and matter energy densities as a function of temperature and time during this period.

Thermal Equilibrium of Photons, Electrons, and Neutrinos

The universe is simplest to describe at a temperature corresponding to

$$kT \approx 20 \, \text{MeV}. \qquad (19.123)$$

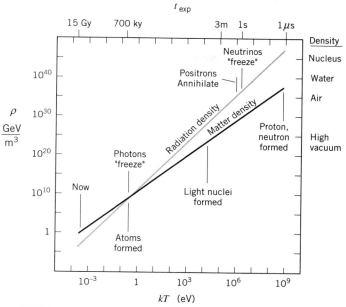

FIGURE 19-22 Energy density versus temperature in the early universe.

The radiation energy density and the matter energy density are shown versus temperature and expansion time. At some extremely high density, the universe presumably consisted of a quark–lepton soup. About 10 ms after this state, the hadrons "froze out" and the evolution of the universe is largely understood.

This energy is special because it is below the pion mass energy (140 MeV) where physics is complicated, yet well above the electron–positron pair-production threshold (1 MeV) and the neutron–proton mass difference (1.3 MeV). The energy density was dominated by radiation, and electrons and positrons were in equilibrium with radiation through

$$\gamma \leftrightarrow e^+ + e^- \qquad (19.124)$$

(with the recoil momentum balanced by any charged particle). Equilibrium is possible because the average photon energy is much larger than the electron–positron mass energy. Nuclei did not form because the photons are more energetic than the nuclear binding energy ($kT \gg 1$ MeV). The nucleons played the role of catalysts in converting electrons and positrons to neutrinos and antineutrinos. Neutrinos and antineutrinos were also in equilibrium through the following interactions:

$$e^- + p \leftrightarrow \nu_e + n \qquad (19.125)$$

and

$$e^+ + n \leftrightarrow \bar{\nu}_e + p, \qquad (19.126)$$

which readily occur because the electron and positron energies are larger than the neutron–proton mass difference. Thus, the universe was made of photons, electrons, positrons, neutrinos, antineutrinos, protons, and neutrons.

A typical radiation energy of 20 MeV corresponds to a temperature,

$$T \approx 10^{11} \text{ K}. \qquad (19.127)$$

The expansion time was

$$
\begin{aligned}
t_{\text{exp}} &\approx \frac{4 \times 10^{20} \text{ s} \cdot \text{K}^2}{T^2} \\
&\approx \frac{4 \times 10^{20} \text{ s} \cdot \text{K}^2}{\left(10^{11} \text{ K}\right)^2} \approx 10^{-2} \text{ s}. \qquad (19.128)
\end{aligned}
$$

Roughly speaking, the universe doubled in size in 10^{-2} s! This time is still very, very long compared to the interaction time scale of the photons, electrons, and neutrinos. Thus, the universe was in thermal equilibrium. The universe has never been in a simpler state. The size of the visible universe (r) was smaller than it is today (r_{now}) by the ratio of the temperatures

$$r = r_{\text{now}} \left(\frac{2.7 \text{ K}}{10^{11} \text{ K}} \right). \qquad (19.129)$$

In the words of Steven Weinberg,

The universe is simpler and easier to describe than it ever will be again. It is filled with an undifferentiated soup of matter and radiation, each particle of which collides very rapidly with the other particles. Thus, despite its rapid expansion, the universe is in a state of nearly perfect thermal equilibrium. The contents of the universe are therefore dictated by the rules of statistical mechanics, and do not depend at all on what went before. . . . All we need to know is that the temperature is 10^{11} K, and that the conserved quantities—charge, baryon number, lepton number—are all very small or zero.

Freezing of the Neutrinos

Neutrinos were no longer copiously created and destroyed at a temperature (T_ν) of

$$T_\nu \approx 10^{10} \text{ K}, \qquad (19.130)$$

corresponding to an energy (E_ν) of

$$E_\nu \approx 2\,\text{MeV}. \qquad (19.131)$$

The corresponding expansion time (t_ν) was

$$t_\nu \approx 1\,\text{s}. \qquad (19.132)$$

The energy density had dropped to

$$\rho \approx 10^{36}\ \text{GeV/m}^3, \qquad (19.133)$$

about equal to the mass-energy density of water. The neutrinos drop out of thermal equilibrium because the universe has expanded sufficiently so that the particle densities are too low to fuel the weak interactions that create and annihilate neutrinos. As the neutrinos freeze out of equilibrium, more neutrons are converted to protons than vice versa because the neutron has a greater mass than the proton. The process

$$e^+ + n \rightarrow \bar{\nu}_e + p \qquad (19.134)$$

occurs with greater probability than the process

$$e^- + p \rightarrow \nu_e + n. \qquad (19.135)$$

This results in about three times more protons than neutrons. This proton-neutron imbalance has led to more hydrogen than helium in the present universe.

The neutrinos no longer interact as the universe expands further and are left over today as a thermal distribution at a temperature of about 1.9 K. These cosmic neutrinos have not been detected, but they must be present!

Annihilation of the Positrons

At energies below 1 MeV, positrons are no longer created but they still can annihilate with electrons. The positrons disappear very quickly after the neutrinos freeze out. This energy from electron–positron annihilation serves to heat the photon distribution to a temperature slightly higher than the neutrinos.

Synthesis of the Light Nuclei

The first step in the production of nuclei is the formation of deuterium by the fusion of two protons,

$$p + p \rightarrow d + e^+ + \nu_e. \qquad (19.136)$$

The proton fusion reaction can occur only if the protons are energetic enough to overcome their electrical repulsion and if the density is large enough to make a significant

reaction rate. The deuterium may be broken apart by energetic photons through the process

$$\gamma + d \rightarrow n + p. \qquad (19.137)$$

Thus, nuclei could be made only in a short time interval when the universe was hot and dense enough for proton fusion to occur, but not so hot that the deuterium was broken apart by photo-disintegration. These conditions were met in the early universe at a temperature (T_{nuc}),

$$T_{nuc} \approx 10^9\ \text{K}. \qquad (19.138)$$

This temperature is about an order of magnitude greater than the temperature at the center of the sun. The corresponding typical photon energy (E_{nuc}) is

$$E_{nuc} = 100\ \text{keV}. \qquad (19.139)$$

The corresponding expansion time (t_{nuc}) is

$$t \approx 10^2\ \text{s}. \qquad (19.140)$$

The corresponding energy density was

$$\rho \approx 10^{31}\ \text{GeV/m}^3, \qquad (19.141)$$

about equal to the mass-energy density of a high laboratory vacuum. With the deuterium stable against photo-disintegration, α particles (helium-4 nuclei) were made through the proton cycle (via ^{3}He) just as they are in the sun:

$$p + p \rightarrow d + e^+ + \nu_e, \qquad (19.142)$$

$$p + d \rightarrow {}^3\text{He} + \gamma, \qquad (19.143)$$

and

$${}^3\text{He} + {}^3\text{He} \rightarrow p + p + \alpha. \qquad (19.144)$$

Alpha particles were also formed by the reactions

$$n + {}^3\text{He} \rightarrow \alpha + \gamma \qquad (19.145)$$

and via tritium (t) production,

$$p + d \rightarrow t + e^+ + \nu_e \qquad (19.146)$$

followed by

$$p + t \rightarrow \alpha + \gamma. \qquad (19.147)$$

The formation of heavier nuclei by proton fusion with α particles was prevented because there is no stable nucleus with atomic number five.

The neutron population was also lowered slightly by neutron decay. The neutron mean lifetime is about 10^3 s. The resulting baryon distribution was 13% neutrons and 87% protons at the time when essentially all neutrons were bound into α particles. Since an α particle is formed with two neutrons and two protons, α particles make up 26% of the nuclear abundance by weight and the protons make up 74%.

The Formation of Atoms

When the universe had expanded and cooled to a temperature (T_{atom}) of about

$$T_{atom} \approx 3000 \, \text{K}, \qquad (19.148)$$

then the photons were no longer energetic enough to break up atoms. This corresponds to a typical photon energy (E_{atom}) of

$$E_{atom} \approx 1 \, \text{eV}. \qquad (19.149)$$

The expansion time (t_{atom}) was

$$
t_{atom} \approx \frac{4 \times 10^{20} \, \text{s} \cdot \text{K}^2}{T_{atom}{}^2}
$$
$$
\approx \frac{4 \times 10^{20} \, \text{s} \cdot \text{K}^2}{(3000 \, \text{K})^2} \approx 10^{13} \, \text{s}. \qquad (19.150)
$$

At this time atoms were formed and the photons dropped out of thermal equilibrium as there were no longer free charges to which the photons coupled. The universe was made of hydrogen atoms, helium atoms, photons, neutrinos, and antineutrinos.

At this time the radiation had also cooled to a point where the energy density was no longer dominated by photons (see Figure 19-22). The photon-to-baryon ratio was fixed as neither particle species was created or destroyed. The photon wavelengths increased with the expansion of the universe while the baryon rest energies were unchanged. At this time the energy density of the universe started to be dominated by matter.

EXAMPLE 19-17

Make an estimate of the number of nucleons per cubic meter in the universe at the time when atoms formed and radiation dropped out of equilibrium.

SOLUTION:

The average visible mass-energy density of the universe currently is

$$\rho_m \approx 0.5 \, \text{GeV/m}^3.$$

At a temperature of 3000 K, the matter density (19.104) was

$$\rho_m \approx \left(0.5 \, \text{GeV/m}^3\right)\left(\frac{3000 \, \text{K}}{2.7 \, \text{K}}\right)^3 \approx 10^9 \, \text{GeV/m}^3,$$

about 10^9 nucleons per cubic meter. If we include the missing dark matter, the mass-energy density at the time atoms were formed was 10 times larger, 10^{10} GeV/m³.

■

From the First Atoms Until Now

Galaxies and clusters of galaxies were formed by gravitational attraction. The exact process of galaxy formation is not known, but there are many models. For galaxies to form, the gravitational attraction must have overcome the radiation pressure. This could have happened only after atoms had formed and the radiation pressure dramatically decreased due to the absence of free charges. For a given energy density of the universe, there is a minimum mass called the Jeans mass (M_J) that can overcome the radiation pressure.

The radiation pressure dropped by nine orders of magnitude when atoms were formed. Thus, the Jeans mass was reduced to the point where mass clusters could be formed. The value of the Jeans mass after the radiation pressure drop is nearly equal to the existing mass clusters in galaxies. The galaxies and clusters of galaxies are believed to have been formed relatively quickly, perhaps 1–2 billion years after the first atoms were formed. The nuclei heavier than ⁴He were made in the interior of collapsing stars (supernovae).

We conclude the discussion of the early universe with a quote from Steven Weinberg:

...I do not believe that scientific progress is always best advanced by keeping an altogether open mind. It is often necessary to forget one's doubts and to follow the consequences of one's assumptions wherever they may lead—the great thing is not to be free of theoretical prejudices, but to have the right theoretical prejudices. And always, the test of any theoretical preconception is in where it leads. The standard model of the early universe has scored some successes, and it provides a coherent theoretical framework for future experimental programs. This does not mean that it is correct, but it does mean that it deserves to be taken seriously.

CHAPTER 19: PHYSICS SUMMARY

- In the standard theory, the universe has evolved for the last 12 billion years by expansion and cooling. There are two major pieces of observational evidence for this: (1) the redshift of light from distant galaxies in proportion to their distance and (2) the isotropic cosmic radiation at a temperature of 2.7 kelvin.

- The fundamental physics of the early universe is known for temperatures smaller than about 10^{13} K. This is about the temperature where quarks were expected to condense into protons and neutrons.

- The expansion of the universe is parameterized with the dimensionless time-dependent scale factor, $R(t)$, that describes how the distance (r) between any two distant galaxies increases with time,

$$r = r_0 R(t).$$

Hubble's law is

$$\frac{dr}{dt} = \frac{\left(\frac{dR}{dt}\right)}{R} r = H(t) r.$$

The presently accepted value of the Hubble constant is in the range

$$\left(8 \times 10^{17} \text{ s}\right)^{-1} < H_0 < \left(3 \times 10^{17} \text{ s}\right)^{-1}$$

- The Friedmann equation describes the time dependence of the expansion parameter, $R(t)$, in terms of a time-independent space-time curvature constant,

$$H^2 = \frac{8\pi\rho G}{3c^2} - \frac{k}{R^2},$$

where H and ρ depend on time.

- The energy density of the universe exists in two forms, matter and radiation. The energy density of the universe is dominated by matter. The mass energy density is decreasing as the universe cools as the third power of the temperature. The mass-energy density is currently about 0.5 GeV/m³. The radiation energy density is currently about 0.4 MeV/m³. The radiation energy density is decreasing as the universe cools as the fourth power of the temperature. At temperatures above about 10^4 K, the energy density of the universe was dominated by radiation.

- Deuterium and helium were made by fusion during the first few minutes of the early universe and heavier elements made by fusion in collapsing stars.

REFERENCES AND SUGGESTIONS FOR FURTHER READING

L. Abbott, "The Mystery of the Cosmological Constant," *Sci. Am.* **258**, No. 5, 106 (1988).

R. A. Alpher and R. Herman, "Reflections on Early Work on 'Big Bang' Cosmology," *Phys. Today* **41**, No. 8, 24 (1988).

J. D. Barrow and F. J. Tipler, *The Anthropic Cosmological Principle*, Clarendon Press (1986).

J. Bernstein, L. S. Brown, and G. Feinberg, "Cosmological Helium Production Simplified," *Rev. Mod. Phys.* **61**, 25 (1989).

H. A. Bethe, "Supernova Mechanisms," *Rev. Mod. Phys.* **62**, 801 (1990).

H. A. Bethe and G. Brown, "How a Supernova Explodes," *Sci. Am.* **252**, No. 5, 60 (1985).

J. O. Burns, "Very Large Scale Structures in the Universe," *Sci. Am.* **255**, No. 1, 38 (1986).

S. Chandrasekhar, *Truth and Beauty: Aesthetics and Motivations in Science*, Univ. of Chicago Press (1987).

T. J.-L. Courvoisier and E. I. Robson, "The Quasar 3C 273," *Sci. Am.* **264**, No. 6, 50 (1991).

D. H. DeVorkin, "Henry Norris Russell," *Sci. Am.* **260**, No. 5, 127 (1989).

F. J. Dyson. "Energy in the Universe," *Sci. Am.* **224**, No. 3, 50 (1971).

A. Einstein, "On the Generalized Theory of Gravitation," *Sci. Am.* **182**, No. 4, 13 (1950).

W. A. Fowler, "Experimental and Theoretical Nuclear Astrophysics: The Quest for the Origin of the Elements," *Rev. Mod. Phys.* **56**, 149 (1984).

W. A. Fowler, "The Origin of the Elements," *Sci. Am.* **195**, No. 3, 82 (1956).

G. Gamow, "The Evolutionary Universe," *Sci. Am.* **195**, No. 3, 136 (1956).

M. J. Geller and J. P. Huchra, "Mapping the Universe," *Science* **246**, 897 (1989).

L. Goldberg, "Atomic Spectroscopy and Astrophysics," *Phys. Today* **41**, No. 8, 38 (1988).

S. Gulkis et al., "The Cosmic Background Explorer," *Sci. Am.* **259**, No. 1, 132 (1990).

Alan H. Guth, "Inflationary Universe: A Possible Solution to the Horizon and Flatness Problems," *Phys. Rev.* **D23**, 347 (1981).

A. H. Guth and P. J. Steinhardt, "The Inflationary Universe," *Sci. Am.* **250**, No. 5, 116 (1984).

H. J. Habbing and G. Neugebauer, "The Infrared Sky," *Sci. Am.* **251**, No. 5, 48 (1984).

J. J. Halliwell, "Quantum Cosmology and the Creation of the Universe," *Sci. Am.* **265**, No. 6, 76 (1991).

E. W. Harrison, "Why is the Sky Dark at Night," *Phys. Today* 27, No. 2., 30 (1974).

M. Hoskin, "William Herschel and the Making of Modern Astronomy," *Sci. Am.* **245**, No. 2, 106 (1986).

E.P. Hubble, "Five Historic Photographs from Palomar," *Sci. Am.* **181**, No. 5, 32 (1949).

A. D. Jeffries, P. R. Saulson, R. E. Spero, and M. E. Zucker, "Gravitational Wave Observatories," *Sci. Am.* **256**, No. 6, 50 (1987).

K. I. Kellermann and A. Richard Thompson, "The Very-Long-Baseline Array," *Sci. Am.* **258**, No. 1, 54 (1988).

R. Kirshner, "SN1987A: The Supernova of a Lifetime," *AIP Conference Proceedings on Modern Physics in America*, ed. W. Fickinger and K. Kowalski (1988).

L. M. Krauss, "Dark Matter in the Universe," *Sci. Am.* **255**, No. 6, 58 (1986).

A. Linde, "Particle Physics and Inflationary Cosmology," *Phys. Today* 40, No. 9, 61 (1987).

J. C. Mather et al., "A Preliminary Measurement of the Cosmic Microwave Background Spectrum by the *Cosmic Background Explorer (COBE)* Satellite," *Astrophys. J.* **354**, L37 (1990).

R. V. Pound and G. A. Rebka, Jr., "Apparent Weight of Photons," Phys. Rev. Lett. **4**, 337 (1960).

R. V. Pound and G. A. Rebka, Jr., "Gravitational Redshift in Nuclear Resonance," *Phys. Rev. Lett.* **9**, 439 (1959).

R. H. Price and K. S. Thorne, "The Membrane Paradigm for Black Holes," *Sci. Am.* **258**, No. 4, 69 (1988).

M. J. Rees, "Black Holes in Galactic Centers," *Sci. Am.* **263**, No. 5, 56 (1990).

A.R. Sandage, "The Red-Shift," *Sci. Am.* **195**, No. 3, 170 (1956).

D. W. Sciama, *Modern Cosmology*, Cambridge University Press (1971).

J. Shaham, "The Oldest Pulsars in the Universe," S*ci. Am.* **256**, No. 2, 50 (1987).

S. W. Stahler, "The Early Life of Stars," *Sci. Am.* **265**, No. 1, 76 (1991).

L. Stodolsky, "Neutrino and Dark-Matter Detection at Low Temperatures," *Phys. Today* 44, No. 8, 24 (1991).

C. H. Townes and R. Genzel, "What is Happening at the Center of Our Galaxy," *Sci. Am.* **262**, No. 4, 46 (1990).

V. Trimble, "1987A: The Greatest Supernova Since Kepler," *Rev. Mod. Phys.* **60**, 859 (1988).

E. L. Turner, "Gravitational Lenses," *Sci. Am.* **259**, No. 1, 54 (1988).

J. Weber, "The Detection of Gravitational Waves," *Sci. Am.* **224**, No. 5, 22 (1971).

S. Weinberg, "The Cosmological Constant Problem," *Rev. Mod. Phys.* **61**, 1 (1989).

S. Weinberg, *The First Three Minutes,* Basic Books (1977).

S. Weinberg, *Gravitation and Cosmology, Principles and Applications of the General Theory of Relativity*, Wiley (1972).

J. C. Wheeler and R. P. Harkness, "Helium-rich Supernovas," *Sci. Am.* **257**, No. 5, 50 (1987).

D. Woody and P. Richards, "Spectrum of the Cosmic Background Radiation," *Phys. Rev. Lett.* **42**, 925 (1979).

S. Woosley and T. Weaver, "The Great Supernova of 1987," *Sci. Am.* **261**, No. 2, 32 (1989).

Y. B. Zel'dovich, "Cosmology from Robertson to Today," *Phys. Today* **41**, No. 3, 27 (1988).

QUESTIONS AND PROBLEMS

The dark night sky

1. What value of stellar density would result in a bright night sky, assuming a uniform density? Explain.

2. If all the matter of the universe were suddenly converted to blackbody radiation, make an estimate of the resulting temperature. Use your result to show that conservation of energy is inconsistent with a bright night sky.

Overview of the universe

3. The sun and the moon have approximately the same apparent size in the sky. What does this tell us about their relative diameters and distances?

4. Why do several of the nearby planets look like bright stars in the sky? How can you distinguish the planets from the stars?

5. Calculate the angular resolution necessary to determine the distance to Alpha Centauri to an accuracy of 10% by triangulation from the orbit of the earth about the sun.

6. Make a rough estimate of the average distance between stars in the Milky Way.

7. What is the approximate size of a piece of matter that contains as many atoms as there are stars in the visible universe?

8. Calculate the luminosity of a galaxy that is at a distance of 100 mpc and has an apparent magnitude of 12.

9. Stars with apparent magnitudes smaller than 6 are visible with the naked eye. (a) What is the distance to

a star with $m = 6$ if it has the same luminosity as the sun? (b) What is the distance to a star with $m = 6$ if it has a luminosity 100 times that of the sun?

10. What is the apparent magnitude of the sun?

11. Two stars at equal distances from the earth have apparent magnitudes that differ by one. What is the ratio of luminosities of the stars?

12. A type-I Cepheid variable has a period of 10^6 s and an apparent magnitude of $m = 15$. How far away is the star? How far away is the star if it is a type-II Cepheid variable?

13. A type-II Cepheid variable is at a distance of 1 Mpc and has a period of 3×10^6 s. What is the apparent magnitude (m) of the star?

14. A detailed measurement of the atomic spectra from our neighboring galaxy, M31, shows that all photons from atomic spectra are observed at slightly smaller wavelengths. The fractional wavelength shift is observed to be $\Delta\lambda/\lambda = -10^{-3}$. Calculate the speed of M31 relative to the earth.

15. The redshift from a distant galaxy is $z = 1/2$. (a) What is the speed of the galaxy if the redshift is due to Doppler shift? (b) What is the distance to the galaxy from Hubble's law?

16. What is the expected redshift from a galaxy that is at a distance of 1000 mpc?

17. Show that for nonrelativistic speeds, the speed of a galaxy (v) and its redshift (z) are related by $v = cz$. At what speed is the approximate formula in error by 10%?

18. Use Hubble's law to calculate the distance to a galaxy that has a redshift of $z = 2$.

19. What speed is needed to make a redshift of $z = 10$?

Evolution of the stars

20. Estimate the radius of the sun after it has collapsed into a white dwarf.

21. Make an estimate of the mean free path of a 20-MeV neutrino inside a neutron star.

22. Nearly all the binding energy from a supernova appears in the form of neutrinos and antineutrinos of all species, which are emitted with a thermal spectrum corresponding to $kT \approx 5$ MeV. The Kamiokande detector was sensitive to electron-antineutrino interactions. The detector mass was 2×10^6 kg and the interaction detection efficiency was 90% at an energy of 14 MeV. Use the data of Figure 19-17 to make an order-of-magnitude estimate of the total energy re-

leased by SN1987A. (*Hint:* What neutrino flux is necessary to cause the observed interaction rate?) Compare your answer with the binding energy calculated in Example 19-7.

23. Make an estimate of the neutrino flux on the surface of the earth from a supernova occurring at the center of our galaxy.

The role of gravity

24. The position of a star is measured during a total solar eclipse. Light from the star passes near the sun and is deflected on its path to the earth. Calculate the apparent shift in angular position of the star. Does the shift depend on the distance to the star?

25. By how much is a visible photon gravitationally redshifted when it escapes the sun?

26. A black hole has a radius of 10^3 m. (a) *Estimate* the minimum mass necessary to cause an infinite gravitational redshift for photons. (b) In the theory of general relativity, photons cannot escape from a black hole because of space-time curvature. The gravitational or *Schwarzschild* radius (R) of a mass M is given by

$$R = \frac{2MG}{c^2}.$$

Calculate the mass corresponding to a Schwarzschild radius of 10^3 m.

The physics of the expanding universe

27. Why can blueshifted light come only from a nearby star, not a distant star?

28. At what temperature was the expansion time of the universe equal to 1 minute? What was the energy density of the universe?

29. At what temperature was the universe hot enough to produce b quarks copiously? What was the characteristic expansion time? (Compare to the b quark lifetime.)

30. Show that the time (t) taken for the universe to expand for some very high energy density ρ to a density ρ_0 $(\rho \gg \rho_0)$ is related to the expansion time (t_{exp}) by

$$t \approx \frac{t_{exp}}{2}$$

for a radiation dominated universe and

$$t \approx \frac{2t_{exp}}{3}$$

for a matter dominated universe.

Additional problems

31. (a) Make an estimate of the mean free path of starlight at a time when the stars were formed. (b) Compare your answer with the size of the universe in the Einstein-de Sitter model.

32. What is the relationship between a nonzero cosmological constant and the critical density needed for a closed universe? Explain.

33. Suppose the dark matter of the universe was in the form of Jupiter-like objects ($M \approx 2 \times 10^{27}$ kg). What density of Jupiters is needed? How many Jupiters are needed per galaxy on the average?

34. Suppose a new form of mass-energy was discovered so that the energy density of the universe was a factor of 10 larger than currently believed. How would this have affected the expansion of the universe from 10^{13} K to 3 K, that is, how would Figure 19-22 be modified? Would this have an effect on the formation of the light elements?

35. Consider the gravitational attraction of a large number of hydrogen atoms to form a star of mass and radius similar to the sun. Assuming that the gravitational binding energy appears as kinetic energy, make

an order of magnitude estimate of the temperature of the star.

*36. *Size of a neutron star.* (a) Show that the inward gravitational pressure (P_g) of a uniform sphere of N neutrons each of mass m_n is given by

$$P_g \approx \frac{3GN^2m_n^2}{20\pi R^4},$$

where R is the radius of the sphere. (b) Show that the outward Fermi pressure (P_F) is given by

$$P_F \approx \left(\frac{3}{\pi}\right)^{2/3}\left(\frac{3N}{4\pi}\right)^{5/3}\frac{h^2}{20\,m_n\,R^5}.$$

(*Hint*: Write the pressure in terms of the Fermi energy and make use of a relationship between Fermi energy and neutron number density analogous to that for electrons.) (c) Equate the gravitational and Fermi pressures to solve for the radius of the neutron star. Give a numerical answer for a neutron star with twice the solar mass.

37. It is observed that there are about seven times more protons than neutrons in the universe. (a) Use the Maxwell-Boltzmann factor to estimate the temperature (T) of the universe when the proton to neutron ratio was seven. (b) The ratio becomes frozen when deuterons form and neutrons no longer decay. Compare the average photon energy ($2.7\,kT$) from part (a) with the deuteron binding energy.

*38. *Density of a white dwarf.* Consider the collapse of the sun into a white dwarf. (a) Equate the electron Fermi pressure with the gravitational pressure to determine the radius of the white dwarf. What is the corresponding density? (b) What is the Fermi energy of the electrons?

*39. The temperature at which the radiation and mass energy densities of the universe were equal is approximately equal to the temperature at which the first atoms were formed. Do you think that this is a coincidence?

Fundamental:

Avogadro's number	N_A	6.022137×10^{23}
electric force constant	ke^2	1.439965 eV$\cdot$nm
speed of light in vacuum	c	2.99792458×10^8 m/s
electron charge	e	$1.6021773 \times 10^{-19}$ C
Boltzmann constant	k	8.61739×10^{-5} eV/K
Planck constant	h	6.626076×10^{-34} J$\cdot$s
electron mass	m	0.5109991 MeV/c^2
proton mass	m_p	938.2723 MeV/c^2
neutron mass	m_n	939.5656 MeV/c^2
gravitational constant	G	6.6726×10^{-11} m$^3\cdot$kg$^{-1}\cdot$s^2
Fermi constant	G_F	8.96197×10^{-8} GeV$\cdot$fm^3
Lamda-QCD	Λ	200 MeV
W mass	m_W	80.2 GeV/c^2
Z^0 mass	m_Z	91.17 GeV/c^2

Derived:

kT at $T = 300$ K	kT	0.0258522 eV
h times c	hc	1239.8424 eV$\cdot$nm
h-bar ($\hbar$)	$h/2\pi$	$1.0545727 \times 10^{-34}$ J$\cdot$s
$\hbar$ times c	$\hbar c$	197.32705 eV$\cdot$nm
electromagnetic coupling (α)	$ke^2/\hbar c$	$1/137.035990$
classical electron radius (r_e)	ke^2/mc^2	2.8179409 fm
Bohr radius (a_0)	$\hbar^2 c^2/mc^2 ke^2$	0.052917725 nm
Compton wavelength (λ_c)	hc/mc^2	$2.4263106 \times 10^{-12}$ m
Rydberg energy (E_0)	$mc^2\alpha^2/2$	13.605698 eV
Bohr magneton (μ_B)	$e\hbar/2m$	5.7883826×10^{-5} eV/T
nuclear magneton (μ_N)	$e\hbar/2m_p$	3.1524517×10^{-8} eV/T
Stefan-Boltzmann (σ)	$2\pi^5 k^4/15h^3 c^2$	5.6705×10^{-8} W$\cdot$m$^{-2}\cdot$K^{-4}

Wien displacement	$\lambda_m T$	2.89776×10^{-3} m·K
von Klitzing resistance (R_k)	h/e^2	25813 Ω
strong coupling (α_s), at 5 GeV	$12\pi/25 \ln(E^2/\Lambda^2)$	0.18

Astrophysical:

tropical year	y	3.15569×10^7 s
acceleration of gravity at sea level	g	9.80665 m/s^2
standard pressure	P	1.01325×10^5 N/m^2
mean radius of earth	R_e	6.378140×10^6 m
mass of earth	M_e	5.9742×10^{24} kg
mean radius of sun	R_s	6.960×10^8 m
mass of sun	M_s	1.989×10^{30} kg
mean distance to sun (astro. unit)	D	1.495979×10^{11} m
solar luminosity	L_s	3.83×10^{26} W
solar constant	S	1350 W/m^2
parsec	pc	3.085677×10^{16} m
light year	ly	9.460528×10^{15} m
Hubble parameter	H_0	40–70 km·s^{-1}·Mpc^{-1}
Cosmological constant	Λ	$|\Lambda| < 3 \times 10^{-52}$ m^{-2}

REFERENCES

E. R. Cohen and B. N. Taylor, "The 1986 Adjustment of the Fundamental Constants," *Rev. Mod. Phys.* **59**, 1121 (1987).

D. T. Goldman and R. J. Bell, eds., "The International System of Units (SI)," *Special Pub. 330*, U.S. National Bureau of Standards (1986).

B. N. Taylor, "New Measurement Standards for 1990," *Phys. Today* **42**, No. 8, 23 (1989).

Maxwell's equations are four experimentally established relationships between charges and currents and the electric and magnetic fields that they generate. One usually first encounters Maxwell's equations in *integral* form. Using two important theorems of vector calculus, the *divergence theorem* and *Stokes's theorem* (see Appendix C), Maxwell's equations may be written in *differential* form. The two forms are equivalent.

B-1 GAUSS'S LAW FOR THE ELECTRIC FIELD

The flux of electric field ($\mathbf{E}$) through any closed surface ($\mathbf{a}$) is proportional to the total electric charge (q_{tot}) contained inside the volume enclosed by that surface. Gauss's law in integral form is

$$\oiint d\mathbf{a} \cdot \mathbf{E} = 4\pi k q_{tot}. \qquad \textbf{(B.1)}$$

In applying Gauss's law to a stationary point charge q_1 and taking the surface of integration to be a sphere of radius r, we note that the electric field is everywhere normal to the surface of the sphere. For this case, we have

$$4\pi r^2 E = 4\pi k q_1 \qquad \textbf{(B.2)}$$

or

$$E = \frac{kq_1}{r^2}. \qquad \textbf{(B.3)}$$

A second charge q_2 would experience an electric force

$$\mathbf{F} = e\mathbf{E} = \frac{kq_1 q_2}{r^2}\mathbf{i}_r. \qquad \textbf{(B.4)}$$

The direction of the force ($\mathbf{i}_r$) is along the axis of the two charges.

Gauss's law may also be written as a differential equation. The *divergence theorem* (see Appendix C) relates the surface integral (B.1) to the volume integral of derivatives

$$\oiint d\mathbf{a} \cdot \mathbf{E} = \iiint_V dV \left(\frac{\partial E_x}{\partial x} + \frac{\partial E_y}{\partial y} + \frac{\partial E_z}{\partial z} \right). \qquad \textbf{(B.5)}$$

To make the notation compact, we define the divergence of $\mathbf{E}$ to be

$$\boldsymbol{\nabla} \cdot \mathbf{E} \equiv \frac{\partial E_x}{\partial x} + \frac{\partial E_y}{\partial y} + \frac{\partial E_z}{\partial z}. \qquad \textbf{(B.6)}$$

Note that the divergence is applied to a vector and the result is a scalar. From Gauss's law (B.1) with the total charge expressed as the volume integral of the charge density (ρ), we have

$$\oiint d\mathbf{a} \cdot \mathbf{E} = \iiint_V dV \, \boldsymbol{\nabla} \cdot \mathbf{E} = 4\pi k \iiint_V dV \, \rho. \qquad \textbf{(B.7)}$$

Comparing the integrands of the volume integrals, we arrive at the differential form of Gauss's law for the electric field,

$$\boldsymbol{\nabla} \cdot \mathbf{E} = 4\pi k \rho. \qquad \textbf{(B.8)}$$

The differential form of Gauss's law (B.8) is equivalent to the integral form (B.1).

B-2 GAUSS'S LAW FOR THE MAGNETIC FIELD

Nobody has ever observed a magnetic charge. Therefore, we currently assume that such magnetic *monopoles* do not exist. The flux of magnetic field (**B**) through any closed surface is zero,

$$\oiint d\mathbf{a} \cdot \mathbf{B} = 0. \qquad \textbf{(B.9)}$$

This is a statement of Gauss's law for the magnetic field. The differential from of Gauss's law for the magnetic field is

$$\nabla \cdot \mathbf{B} = 0. \qquad \textbf{(B.10)}$$

B-3 FARADAY'S LAW

The line-integral of the electric field around a closed loop (**l**) is equal to the negative of the time-rate of change of the magnetic flux through the area enclosed by the loop. This is known as Faraday's law of induction:

$$\oint d\mathbf{l} \cdot \mathbf{E} = -\frac{\partial}{\partial t} \oiint d\mathbf{a} \cdot \mathbf{B} \qquad \textbf{(B.11)}$$

By Stokes's theorem (see Appendix C) the left-hand side of Faraday's law (B.11) may be written

$$\oint d\mathbf{l} \cdot \mathbf{E} = \oiint d\mathbf{a} \cdot (\nabla \times \mathbf{E}), \qquad \textbf{(B.12)}$$

where the curl of **E** is defined to be

$$\nabla \times \mathbf{E} \equiv \left(\frac{\partial E_z}{\partial y} - \frac{\partial E_y}{\partial z} \right) \mathbf{i}_x$$

$$+ \left(\frac{\partial E_x}{\partial z} - \frac{\partial E_z}{\partial x} \right) \mathbf{i}_y$$

$$+ \left(\frac{\partial E_y}{\partial x} - \frac{\partial E_x}{\partial y} \right) \mathbf{i}_z. \qquad \textbf{(B.13)}$$

Faraday's law (B.11) becomes

$$\oiint d\mathbf{a} \cdot (\nabla \times \mathbf{E}) = -\frac{\partial}{\partial t} \oiint d\mathbf{a} \cdot \mathbf{B}. \qquad \textbf{(B.14)}$$

Comparing integrands (B.14), we arrive at Faraday's law in differential form,

$$\nabla \times \mathbf{E} + \frac{\partial \mathbf{B}}{\partial t} = 0. \qquad \textbf{(B.15)}$$

B-4 AMPÈRE'S LAW

The line-integral of the magnetic field around a closed loop is the sum of two terms, one proportional to the current (*I*) through that loop and the second proportional to the time-rate of change of the component of electric field normal to the loop. This is known as Ampère's law. In integral form, Ampère's law is written

$$\oint d\mathbf{l} \cdot \mathbf{B} = \frac{4\pi k I}{c^2} + \frac{1}{c^2} \frac{\partial}{\partial t} \oiint d\mathbf{a} \cdot \mathbf{E}. \qquad \textbf{(B.16)}$$

(Note that $4\pi k / c^2 \equiv \mu_0 = 4\pi \times 10^{-7}$ T·m/A.) We define the vector **J** to be the current per unit area, where the direction of **J** is the direction of a positive charge flow. Using Stokes's theorem, Ampère's law in differential form becomes

$$\nabla \times \mathbf{B} = \frac{4\pi k \mathbf{J}}{c^2} + \frac{1}{c^2} \frac{\partial \mathbf{E}}{\partial t}. \qquad \textbf{(B.17)}$$

B-5 CHARGE CONSERVATION

Electric charge conservation is contained in Maxwell's equations. If we take the divergence of the differential form of Ampère's law (B.17) and then use Gauss's law (B.8) and Gauss's law for the magnetic field (B.10) for the divergence of *E* and *B*, we get

$$\nabla \cdot \mathbf{J} + \frac{\partial \rho}{\partial t} = 0. \qquad \textbf{(B.18)}$$

This equation states that the current through any closed surface is equal to the time rate of change of the electric charge in the volume enclosed by the surface. Charge conservation is absolute; no experiment has been performed that has detected a nonconservation of electric charge.

B-6 THE WAVE EQUATION

Maxwell's equations (B.8), (B.10), (B.15), and (B.17) in a region free of charge reduce to

$$\nabla \cdot \mathbf{E} = 0, \qquad \textbf{(B.19)}$$

$$\nabla \cdot \mathbf{B} = 0, \qquad \textbf{(B.20)}$$

$$\nabla \times \mathbf{E} + \frac{\partial \mathbf{B}}{\partial t} = 0, \qquad \textbf{(B.21)}$$

and

$$\nabla \times \mathbf{B} - \frac{1}{c^2}\frac{\partial \mathbf{E}}{\partial t} = 0. \qquad \textbf{(B.22)}$$

If we take the curl of Faraday's law (B.21), and then use Ampère's law (B.22) for the curl of **B**, we arrive at the *wave equation* for the electric field,

$$\nabla \times (\nabla \times \mathbf{E}) + \frac{1}{c^2}\frac{\partial^2 \mathbf{E}}{\partial t^2} = 0. \qquad \textbf{(B.23)}$$

This is the wave equation in vector form, that is, it is really three equations. We now examine the wave equation for one component, the x component of E. We need to evaluate the curl of the curl. The x component of curl-curl E is the derivative with respect to y of the z component of curl E minus the derivative with respect to z of the y component of curl E,

$$\left[\nabla \times (\nabla \times \mathbf{E})\right]\cdot \mathbf{i}_x$$

$$= \frac{\partial}{\partial y}\left(\frac{\partial E_z}{\partial y} - \frac{\partial E_y}{\partial z}\right) - \frac{\partial}{\partial z}\left(\frac{\partial E_x}{\partial z} - \frac{\partial E_z}{\partial x}\right)$$

$$= -\frac{\partial^2 E_x}{\partial y^2} - \frac{\partial^2 E_x}{\partial z^2} + \frac{\partial^2 E_y}{\partial y \partial x} + \frac{\partial^2 E_z}{\partial z \partial x}$$

$$= -\frac{\partial^2 E_x}{\partial y^2} - \frac{\partial^2 E_x}{\partial z^2} + \frac{\partial}{\partial x}\left(-\frac{\partial E_x}{\partial x} - \frac{\partial E_z}{\partial z}\right) + \frac{\partial^2 E_z}{\partial z \partial x}$$

$$= -\frac{\partial^2 E_x}{\partial x^2} - \frac{\partial^2 E_x}{\partial y^2} - \frac{\partial^2 E_x}{\partial z^2}, \qquad \textbf{(B.24)}$$

where we have used the fact that the divergence of E is zero to substitute for $\partial E_y/\partial y$. The wave equation (B.23) for E_x is

$$\frac{\partial^2 E_x}{\partial x^2} + \frac{\partial^2 E_x}{\partial y^2} + \frac{\partial^2 E_x}{\partial z^2} = \frac{1}{c^2}\frac{\partial^2 E_x}{\partial t^2}. \qquad \textbf{(B.25)}$$

The wave equation (B.25) holds for each component of **E** and **B**.

B-7 RADIATION FROM AN ACCELERATED CHARGE

Consider a charge traveling with a uniform speed v that is suddenly stopped. The time taken for the charge to stop is the speed divided by the acceleration a. Figure B-1 shows the electric field of the charge at a time t after the charge has been stopped. The time t is chosen to be large compared to the time taken to stop the charge ($t \gg v/a$). At a

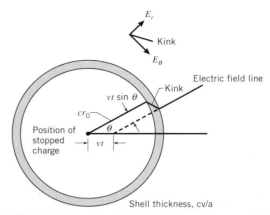

FIGURE B-1 Electric field lines from a charge that is suddenly stopped.

large distance from the charge ($r > ct$), the electric field is the same as that due to a moving charge, that is, the field lines point to the position where the charge would be if it had not been stopped. At a short distance from the charge ($r < ct$), the electric field is that due to a stationary charge. Therefore, the electric field lines have "kinks" at a distance of $r_0 = ct$ from the stopped charge. The kinks propagate outward with a speed c. In the region of the kink, the electric field in the θ direction (E_θ) divided by the electric field in the r direction (E_r) is

$$\frac{E_\theta}{E_r} = \frac{vt\sin\theta}{\left(\dfrac{cv}{a}\right)} = \frac{at}{c}\sin\theta. \qquad \textbf{(B.26)}$$

Since the radial part of the electric field is

$$E_r = \frac{ke}{r_0^2}, \qquad \textbf{(B.27)}$$

we have

$$E_\theta = \frac{at}{c}E_r\sin\theta = \left(\frac{ar_0}{c^2}\right)\left(\frac{ke}{r_0^2}\right)\sin\theta$$

$$= \frac{kea}{c^2 r_0}\sin\theta. \qquad \textbf{(B.28)}$$

The electric field component E_θ is inversely proportional to r_0 (not r_0^2). We have analyzed the electric field at a fixed time t. The electric field varies with time; this time-varying electric field generates a perpendicular magnetic field, which also varies with time. The oscillating electric and magnetic fields constitute an electromagnetic wave.

The energy per unit volume stored in a electric and magnetic fields is $(E^2 + B^2)/8\pi k$, which is equal to $E_0^{\ 2}/4\pi k$ because the fields have equal magnitudes. The energy (ΔE) stored in the spherical shell of thickness cv/a that contains the kink is

$$\Delta E = \left(\frac{cv}{a}\right)\int_0^\pi d\theta\left(2\pi r_0^{\ 2}\sin\theta\right)\frac{E_\theta^{\ 2}}{4\pi k}$$

$$= \left(\frac{v}{a}\right)\frac{ke^2a^2}{2c^3}\int_0^\pi d\theta\sin^3\theta$$

$$= \left(\frac{v}{a}\right)\left(\frac{2ke^2a^2}{3c^3}\right). \tag{B.29}$$

(The value of the integral, 4/3, may be verified by consulting a handbook.) Since the time taken for the charge to stop is v/a, the radiated power is

$$P = \frac{\Delta E}{\left(\dfrac{v}{a}\right)} = \frac{2ke^2a^2}{3c^3}. \tag{B.30}$$

SUGGESTION FOR FURTHER READING

E. M. Purcell, *Electricity and Magnetism*, McGraw-Hill (1985).

C-1 DIVERGENCE

Divergence is a mathematical operation performed on a vector. The divergence of a vector is a scalar. The notation is that of a dot product of two vectors. The *divergence* of a vector is defined to be

$$\mathbf{V} \cdot \mathbf{E} \equiv \frac{\partial E_x}{\partial x} + \frac{\partial E_y}{\partial y} + \frac{\partial E_z}{\partial z}. \qquad \text{(C.1)}$$

Consider a surface (**a**) that encloses a volume (V). The *divergence theorem* states that the integral over the surface of the component of a vector normal to the surface is equal to the volume integral of the divergence of the vector:

$$\oiint d\mathbf{a} \cdot \mathbf{E} = \iiint_V dV \, \mathbf{V} \cdot \mathbf{E}. \qquad \text{(C.2)}$$

The physical interpretation of the divergence theorem is that the flux flowing out of any volume must pass through the surface that encloses the volume. For example, all the electric field lines from a charge must pass through a surface that encloses the charge. The proof of the divergence theorem follows directly from the definition of derivative and integral. For an infinitesimal cube with a volume,

$$\Delta V = \Delta x \Delta y \Delta z, \qquad \text{(C.3)}$$

the flux of the vector **E** flowing out of the cube in the x direction (see Figure C-1) is

$$E_x (x + \Delta x, y, z) \Delta y \Delta z - E_x (x, y, z) \Delta y \Delta z$$
$$= \frac{\Delta E_x}{\Delta x} \Delta x \Delta y \Delta z. \qquad \text{(C.4)}$$

For the y and z directions, we get

$$E_y (x, y + \Delta y, z) \Delta x \Delta z - E_y (x, y, z) \Delta x \Delta z$$
$$= \frac{\Delta E_y}{\Delta y} \Delta x \Delta y \Delta z \qquad \text{(C.5)}$$

and

$$E_z (x, y, z + \Delta z) \Delta x \Delta y - E_z (x, y, z) \Delta x \Delta y$$
$$= \frac{\Delta E_z}{\Delta z} \Delta x \Delta y \Delta z. \qquad \text{(C.6)}$$

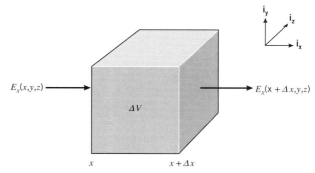

FIGURE C-1 Flux of the vector E in the x direction through a cube of volume ΔV.

The total flux flowing out of the cube is:

$$\text{Total flux} = \left(\frac{\Delta E_x}{\Delta x} + \frac{\Delta E_y}{\Delta y} + \frac{\Delta E_z}{\Delta z}\right)\Delta x \Delta y \Delta z. \quad \textbf{(C.7)}$$

In the limit where Δx, Δy, and Δz all approach zero, we have

$$\left(\frac{\Delta E_x}{\Delta x} + \frac{\Delta E_y}{\Delta y} + \frac{\Delta E_z}{\Delta z}\right) \to \nabla \cdot \mathbf{E}. \quad \textbf{(C.8)}$$

The total flux out of the cube is equal to the divergence times the volume of the cube. The physical meaning of divergence of $\mathbf{E}$ is the net outward flux of $\mathbf{E}$ per volume. Since any volume may be written as the sum of such cubes, the divergence theorem (C.2) follows. The divergence theorem is also called *Gauss's theorem*.

C-2 CURL

Curl is a mathematical operation performed on a vector. The curl of a vector is a vector. The notation is that of a cross-product of two vectors. The curl of a vector is defined to be

$$\nabla \times \mathbf{E} \equiv \left(\frac{\partial E_z}{\partial y} - \frac{\partial E_y}{\partial z}\right)\mathbf{i}_x$$
$$+ \left(\frac{\partial E_x}{\partial z} - \frac{\partial E_z}{\partial x}\right)\mathbf{i}_y$$
$$+ \left(\frac{\partial E_y}{\partial x} - \frac{\partial E_x}{\partial y}\right)\mathbf{i}_z. \quad \textbf{(C.9)}$$

Consider a loop (**l**) that encloses a surface (**a**). *Stokes's theorem* states that the integral of a vector around the loop is equal to the integral of the component of the curl of the vector normal to the surface,

$$\oint d\mathbf{l} \cdot \mathbf{E} = \oiint d\mathbf{a} \cdot (\nabla \times \mathbf{E}). \quad \textbf{(C.10)}$$

The interpretation of Stokes's theorem is that the net *circulation* of $\mathbf{E}$ around a loop is equal to the sum of the curls of $\mathbf{E}$ taken over the area of the loop. For a loop with an infinitesimal area,

$$\Delta A = \Delta x \Delta y, \quad \textbf{(C.11)}$$

the circulation of $\mathbf{E}$ around the loop (see Figure C-2) is

Circulation
$$= E_y(1)\Delta y - E_x(2)\Delta x - E_y(3)\Delta y + E_x(4)\Delta x$$
$$= \left[E_y(1) - E_y(3)\right]\Delta y + \left[E_x(4) - E_x(2)\right]\Delta x$$
$$= -\frac{\Delta E_y}{\Delta x}\Delta x \Delta y + \frac{\Delta E_x}{\Delta y}\Delta x \Delta y$$
$$= \left(\frac{\Delta E_x}{\Delta y} - \frac{\Delta E_y}{\Delta x}\right)\Delta x \Delta y. \quad \textbf{(C.12)}$$

In the limit where Δx and Δy approach zero, we have

$$\left(\frac{\Delta E_x}{\Delta y} - \frac{\Delta E_y}{\Delta x}\right) \to (\nabla \times \mathbf{E}) \cdot \mathbf{i}_z. \quad \textbf{(C.13)}$$

Stokes's theorem (C.10) follows from the fact that the area enclosed by any loop can be made up of infinitesimal squares.

C-3 GRADIENT

Gradient is a mathematical operation performed on a scalar. The gradient of a scalar is a vector. The gradient of the function f is defined to be

$$\nabla f \equiv \frac{\partial f}{\partial x}\mathbf{i}_x + \frac{\partial f}{\partial y}\mathbf{i}_y + \frac{\partial f}{\partial z}\mathbf{i}_z. \quad \textbf{(C.14)}$$

The divergence of the gradient (often called the *Laplacian*) is

$$\nabla^2 f \equiv \nabla \cdot \nabla f = \frac{\partial^2 f}{\partial x^2} + \frac{\partial^2 f}{\partial y^2} + \frac{\partial^2 f}{\partial z^2}. \quad \textbf{(C.15)}$$

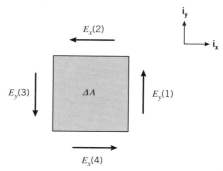

FIGURE C-2 Circulation of the vector E around a rectangular loop in the x-y plane.

D-1 THE INTEGRAL DISTRIBUTION, $n(x)$

The results of an experiment where an observable is measured may be described by defining the function $n(x)$ to be the number of occurrences or *events* with the result of the measurement less than x. If x is a discrete variable, then the function $n(x)$ is not continuous. For example, x may represent the number of pions created in a certain type of particle interaction. The results of an experiment measuring the value of x one thousand times, and counting the number of times $x < 1, 2, 3, \ldots$ was observed are shown in Figure D-1.

If x is a continuous variable, then $n(x)$ is also continuous. For example, x may represent the measurement of the position of a particle. Figure D-2 shows a graph of a distribution $n(x)$ of a continuous variable. The total number of events (N) is large.

D-2 THE DIFFERENTIAL DISTRIBUTION, dn/dx

Consider the discontinuous distribution $n(x)$ of Figure D-1. We may produce a graph of the relative occurrence of the different values of x by calculating the change in $n(x)$ at every integer boundary. If we divide each change in the distribution (Δn_i) by the step in the variable (Δx_i), we arrive at a set of numbers (g_i):

$$g_i = \frac{\Delta n_i}{\Delta x_i}, \qquad \textbf{(D.1)}$$

where i is an integer. A bar-graph of g_i versus x, as shown in Figure D-3, is called a *histogram*. The histogram of g_i

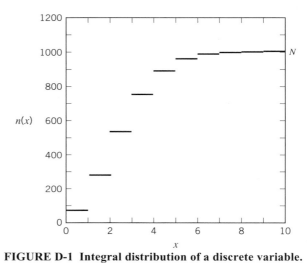

FIGURE D-1 Integral distribution of a discrete variable.
The variable x represents the number of pions created in a certain type of particle interaction. The distribution $n(x)$ gives the number of events in which the number of pions was found to be less than x. The possible values of x are nonnegative integers. The total number of events in the sample is $N = 1000$.

is a useful representation of the x distribution because the height of each bar is proportional to the probability that x has that value. (In Figure D-3 we have chosen all the histogram *bin sizes*, Δx_i, equal to one so that $g_i = \Delta n_i$. In general Δx_i does not have to be equal to unity.)

When x is continuous, we define the function $g(x)$ as the number of events where x is found to be between x and $x + \Delta x$, in the limit where $\Delta x \rightarrow 0$. The function $g(x)$ is just the derivative of $n(x)$ with respect to x,

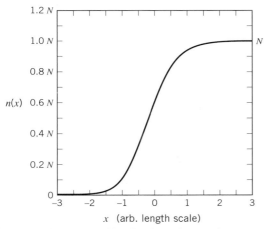

FIGURE D-2 Integral distribution of a continuous variable.
The variable x represents position of a particle. The distribution $n(x)$ gives the number of occurrences in which the value of the observable is found to be less than x. The variable x is taken to be continuous. The scales are arbitrary.

$$g(x) = \lim_{x \to 0} \frac{\Delta n}{\Delta x} = \frac{dn}{dx}. \quad \textbf{(D.2)}$$

The distribution $g(x)$ is shown in Figure D-4.

D-3 THE PROBABILITY DISTRIBUTION, *dP/dx*

The summation of the contents of all the histogram bins of Figure D-3 is equal to the total number of events,

$$\sum_{i=-\infty}^{+\infty} g_i \Delta x_i = \sum_{i=-\infty}^{+\infty} \frac{\Delta n_i}{\Delta x_i} \Delta x_i = \sum_{i=-\infty}^{+\infty} \Delta n_i = N. \quad \textbf{(D.3)}$$

This is called the *normalization* condition. If we divide the distribution g_i by the total number of events N, then the resulting distribution $(\Delta P_i/\Delta x_i)$ gives the probability that a random event has the value x,

$$\frac{\Delta P_i}{\Delta x_i} = \frac{g_i}{N} = \frac{1}{N}\frac{\Delta n_i}{\Delta x_i}. \quad \textbf{(D.4)}$$

The sum of all the probabilities is unity:

$$\sum_{i=-\infty}^{+\infty} \frac{\Delta P_i}{\Delta x_i}\Delta x_i = \frac{1}{N}\sum_{i=-\infty}^{+\infty}\frac{\Delta n_i}{\Delta x_i}\Delta x_i = \frac{1}{N}\sum_{i=-\infty}^{+\infty}\Delta n_i = 1. \quad \textbf{(D.5)}$$

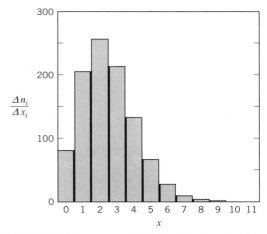

FIGURE D-3 Histogram of the distribution $\Delta n_i/\Delta x_i$ corresponding to the integral distribution of Figure D-1.
For a given value of the integer x, the height of each bin is proportional to the probability that the observable takes on that value of x. The most probable value of x is 2.

For the continuous distribution of Figure D-4, the normalization condition is

$$\int_{-\infty}^{+\infty} dx\, g(x) = N. \quad \textbf{(D.6)}$$

The probability distribution is obtained by dividing $g(x)$ by the total number of events N:

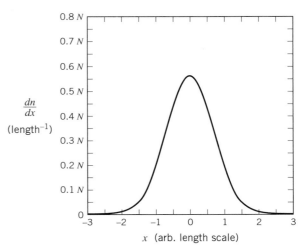

FIGURE D-4 Graph of the distribution *dn/dx* corresponding to the integral distribution of Figure D-2.
The height of the function *dn/dx* is proportional to the number of events with x between x and $x + dx$.

$$\frac{dP}{dx} = \frac{g(x)}{N} = \frac{1}{N}\frac{dn}{dx}. \tag{D.7}$$

The normalization condition is

$$\int_{-\infty}^{+\infty} dx\, \frac{dP}{dx} = 1. \tag{D.8}$$

D-4 CONSTRUCTING A DISTRIBUTION FUNCTION

Consider the distribution of masses (M) of professional baseball players. Clearly, the mass must be a positive number so that the range of M in units of kilograms is restricted to be between zero and infinity. Because M is a continuous variable, the probability of having an entry with M *exactly* equal to any specific value is zero. For example, the probability of finding a baseball player with mass exactly equal to 76 kg is zero even though we may find many baseball players having masses near 76 kg, for example, 76.0034... or 75.9672.... A meaningful question to ask is: How many baseball players have a mass in a small interval (ΔM) near M=76 kg? In practice our knowledge of the masses will be limited by the measurement accuracy. (The masses are also changing with time, but we are speaking of the masses at some fixed time.) In order to make a histogram of the mass distribution, we must choose a histogram bin size (ΔM) and then count the number of events, Δn_i, between M_i and $M_i + \Delta M$. This histogram is shown in Figure D-5 for a sample of 50 professional baseball players. The data were taken from the backs of baseball cards, selected at random.

We obtain the probability distribution, dP/dM, by dividing each histogram bin by the total number of events (N),

$$\frac{dP}{dM} = \frac{1}{N}\frac{\Delta n_i}{\Delta M}. \tag{D.9}$$

This distribution is shown in Figure D-6. The units of dP/dM are inverse kilograms. The distribution dP/dM gives us the probability that a player selected at random has a given mass. For example, we see that the dP/dM at $M = 80$ kg is about 0.04 kg^{-1}. This means that the probability for M to be in a 1-kilogram interval near 80 kg (e.g., 80–81 kg) is 4%. This is an extremely informative representation of the data.

The distributions of Figures D-5 and D-6 are not the exact mass distributions. To get the exact mass distributions, we would need to make a histogram of the masses of

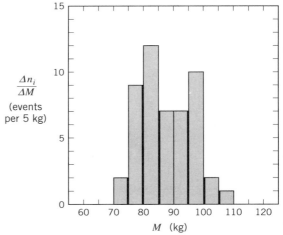

Figure D-5 Making a histogram from the data.
The variable M represents the mass of a professional baseball player. A sample of 50 players is selected at random. The bin size is chosen to be $\Delta M = 5$ kg. The height of the histogram is proportional to the number of players with mass in the interval ΔM.

every player. The distributions of Figures D-5 and D-6 represent a random sampling of players. The larger our random sample is, the closer we expect its distribution will be to the true distribution. This assumption forms the basis for statistics. In plotting the probability distribution, we place vertical bars on each datum that indicate the *statistical error*. The statistical error is defined such that each datum is expected to deviate from the true distribution by less than the statistical error 68% of the time. Except for

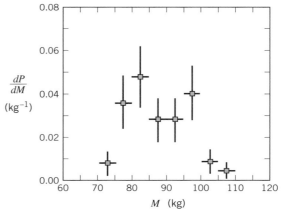

FIGURE D-6 Probability distribution function, *dP/dM*, of the mass of a professional baseball player.
The integral of the function dP/dM over all M is normalized to unity.

values of Δn_i that are very small (less than about 4), the statistical error is given by the square root of the number of events.

D-5 THE BINOMIAL DISTRIBUTION

The *binomial* distribution, $f_b(x)$, provides the foundation for the interpretation of measurements. The binomial distribution specifies the distribution of the total number of times (x) that an event occurs in n independent trials, where p is the probability of the event occurring in a single trial. The binomial distribution is determined by counting the number of ways that x events can occur.

Consider an experiment that has only two possible outcomes, for example, the flip of a coin where the outcome is equally likely to be either "heads" (H) or "tails" (T). The number of *permutations* is two. The distribution of the number of heads (x) for one flip of the coin is $f_b(0) = 1/2$ and $f_b(1) = 1/2$. If we flip the coin twice, then there are four permutations of the possible outcomes: HT, TH, TT, and HH. The distribution of the number of heads (x) for two flips is $f_b(0) = 1/4$, $f_b(1) = 1/2$, and $f_b(2) = 1/4$. If we flip the coin three times, then there are eight (2^3) permutations: HHH, HHT, HTH, THH, HTT, THT, TTH, and TTT. The distribution of the number of heads (x) for three flips is $f_b(0) = 1/8$, $f_b(1) = 3/8$, $f_b(2) = 3/8$, and $f_b(3) = 1/8$. If we flip the coin n times, then there are 2^n permutations. The number of *combinations* with x heads is obtained by counting the positions that the heads can occupy. There are n places for the first head, $n - 1$ for the second because the first head has occupied one, $n - 2$ for the third head, ...and $n - x + 1$ positions for the xth head. Then we must divide by a factor that takes into account that the heads are not distinguishable. There were x ways of choosing the first head, $x - 1$ ways to choose the second head because one head has already been chosen, ... and only one way to choose the xth head. The number of combinations (N_c) for x heads is

$$N_c = \frac{(n)(n-1)(n-2)...(n-x+1)}{(x)(x-1)(x-2)...(1)}, \quad \textbf{(D.10)}$$

which we may rewrite in the notation of factorials as

$$N_c = \frac{n!}{x!(n-x)!}, \quad \textbf{(D.11)}$$

where

$$n! \equiv (n)(n-1)(n-2)...(1). \quad \textbf{(D.12)}$$

The number of permutations (N_p) is

$$N_p = 2^n. \quad \textbf{(D.13)}$$

The binomial probability distribution is equal to the number of combinations divided by the number of permutations:

$$f_b(x) = \frac{N_c(x)}{N_p} = \frac{n!}{x!(n-x)!2^n}. \quad \textbf{(D.14)}$$

By conservation of probability, the normalization condition on $f_b(x)$ is

$$\sum_{x=0}^{n} f_b(x) = 1. \quad \textbf{(D.15)}$$

The binomial distribution (D.14) corresponds to the special case where the event probability for a single trial is equal to one-half.

Now consider the more general situation where the probability that an event occurs on a single trial is p. Then the probability that that event will not occur on a single trial is equal to $1 - p$. For n trials, the probability distribution is the number of combinations for a given value of x, times a probability factor, $p^x(1 - p)^{n-x}$. The binomial distribution becomes

$$f_b(x) = N_c(x)p^x(1-p)^{n-x}$$
$$= \frac{n!p^x(1-p)^{n-x}}{x!(n-x)!}. \quad \textbf{(D.16)}$$

EXAMPLE D-1
Calculate the average value of x for a binomial distribution.

SOLUTION:
For the binomial distribution, the allowed range of x is from zero to n. The average value of x is

$$\langle x \rangle = \sum_{x=0}^{n} x f_b(x) = \sum_{x=0}^{n} \frac{x n! p^x (1-p)^{n-x}}{x!(n-x)!}.$$

This sum will reduce to a simple expression with some careful manipulation. Since the first term in the sum is zero for $x = 0$, we may replace the limits of the sum to go from 1 to n,

$$\langle x \rangle = \sum_{x=1}^{n} \frac{x n! p^x (1-p)^{n-x}}{x!(n-x)!}.$$

Now we cancel the factor of x that appears in both the numerator and denominator,

$$\langle x \rangle = \sum_{x=1}^{n} \frac{n! p^x (1-p)^{n-x}}{(x-1)!(n-x)!}.$$

Since x is the dummy summation variable, we make the change of variables $x' = x - 1$,

$$\langle x \rangle = \sum_{x'=1}^{n-1} \frac{n! p^{x'+1} (1-p)^{n-x'-1}}{x'!(n-x'-1)!}.$$

Factoring out np, we have

$$\langle x \rangle = np \sum_{x'=1}^{n-1} \frac{(n-1)! p^{x'} (1-p)^{n-x'-1}}{x'!(n-x'-1)!}.$$

The summation is now unity since this is just the binomial distribution with $n - 1$ trials. Therefore,

$$\langle x \rangle = np.$$

We could have anticipated this result. For n trials that each have a probability p for a certain outcome, the average number of those outcomes is n times p. ■

EXAMPLE D-2
Calculate the standard deviation of x from the average for a binomial distribution.

SOLUTION:
In the previous example, we have calculated the average value of x to be

$$\langle x \rangle = np.$$

The standard deviation is

$$\sqrt{\langle (x-np)^2 \rangle} = \sqrt{\sum_{x=0}^{n} (x-np)^2 f_b(x)}.$$

We will calculate the average of $(x - np)^2$ and then take the square root. Since

$$(x-np)^2 = x^2 - 2npx + n^2 p^2,$$

we have

$$\langle (x-np)^2 \rangle = \langle x^2 \rangle - 2np\langle x \rangle + n^2 p^2 = \langle x^2 \rangle - n^2 p^2.$$

The average value of x^2 is

$$\langle x^2 \rangle = \sum_{x=0}^{n} \frac{x^2 n! p^x (1-p)^{n-x}}{x!(n-x)!}.$$

We use the same technique as in the previous example. First we change the sum to start at one,

$$\langle x^2 \rangle = \sum_{x=1}^{n} \frac{x^2 n! p^x (1-p)^{n-x}}{x!(n-x)!},$$

cancel a factor of x,

$$\langle x^2 \rangle = \sum_{x=1}^{n} \frac{x n! p^x (1-p)^{n-x}}{(x-1)!(n-x)!},$$

and then factor out np and start the sum at zero,

$$\langle x^2 \rangle = np \sum_{x=0}^{n-1} \frac{(x+1)(n-1)! p^x (1-p)^{n-x-1}}{x!(n-x-1)!}$$

$$= np \sum_{x=0}^{n-1} \frac{x(n-1)! p^x (1-p)^{n-x-1}}{x!(n-x-1)!} + np.$$

We may write

$$\langle x^2 \rangle = npS + np,$$

where

$$S \equiv \sum_{x=0}^{n-1} \frac{x(n-1)! p^x (1-p)^{n-x-1}}{x!(n-x-1)!}$$

$$= \sum_{x=1}^{n-1} \frac{x(n-1))! p^x (1-p)^{n-x-1}}{x!(n-x-1)!}.$$

We now evaluate S, by starting the sum at zero, factoring out $(n-1)p$, and noticing that the remaining sum is unity because it is the binomial distribution with $n-2$ trials:

$$S \equiv \sum_{x=0}^{n-2} \frac{(n-1)! p^{x+1} (1-p)^{n-x-2}}{x!(n-x-2)!}$$

$$= (n-1)p \sum_{x=0}^{n-2} \frac{(n-2))! p^x (1-p)^{n-x-2}}{x!(n-x-2)!}$$

$$= (n-1)p.$$

Thus, we have

$$\langle x^2 \rangle = npS + np = np(n-1)p + np$$
$$= n^2 p^2 - np^2 + np.$$

The mean-square deviation of x from the average is

$$\langle (x-np)^2 \rangle = \langle x^2 \rangle - n^2 p^2$$
$$= n^2 p^2 - np^2 + np - n^2 p^2$$
$$= np(1-p).$$

Finally, the standard deviation is

$$\sqrt{\langle (x-np)^2 \rangle} = \sqrt{np(1-p)}. \qquad \blacksquare$$

Examples D-1 and D-2 show that the average value of a binomial distribution is

$$\langle x \rangle = np, \qquad \textbf{(D.17)}$$

and the standard deviation is

$$\sigma_x = \sqrt{\langle (x-\langle x \rangle)^2 \rangle} = \sqrt{np(1-p)}. \quad \textbf{(D.18)}$$

D-6 THE GAUSSIAN DISTRIBUTION

The *Gaussian* distribution is the binomial distribution in the limit where the number of trials (n) is very large. In this case x may be regarded as a continuous variable. The distribution has a maximum at the average value of x,

$$\langle x \rangle = np. \qquad \textbf{(D.19)}$$

We are interested in the behavior of $f_b(x)$ near the maximum. We could make a Taylor expansion in $f_b(x)$, but since $f_b(x)$ is varying rapidly we get faster convergence if we expand the function,

$$g(x) \equiv \ln f_b(x) \qquad \textbf{(D.20)}$$

instead. The maximum of $g(x)$ is also the maximum of $f_b(x)$. The expansion of $g(x)$ is:

$$g(x) = [g(x)]_{x=np} + (x-np)\left[\frac{dg}{dx}\right]_{x=np}$$
$$+ \frac{(x-np)^2}{2}\left[\frac{d^2 g}{dx^2}\right]_{x=np}$$
$$+ \frac{(x-np)^3}{6}\left[\frac{d^3 g}{dx^3}\right]_{x=np} + \ldots \qquad \textbf{(D.21)}$$

Since we are expanding about the maximum, $dg/dx = 0$ and the expansion becomes

$$g(x) \approx [g(x)]_{x=np} + \frac{(x-np)^2}{2}\left[\frac{d^2 g}{dx^2}\right]_{x=np}. \quad \textbf{(D.22)}$$

Taking the exponential of both sides, we have

$$e^{g(x)} \approx \exp\left\{ [g(x)]_{x=np} + \frac{(x-np)^2}{2}\left[\frac{d^2 g}{dx^2}\right]_{x=np} \right\},$$

$$\textbf{(D.23)}$$

or

$$f_b(x) \approx f_b(np)\exp\left\{ \frac{(x-np)^2}{2}\left[\frac{d^2 \ln f_b}{dx^2}\right]_{x=np} \right\}.$$

$$\textbf{(D.24)}$$

We now make a determination of the second derivative that appears inside the exponential. From the definition of $g(x)$ as the logarithm of the binomial distribution we have

$$g(x) = \ln f_b = \ln(n!) - \ln(x!) - \ln[(n-x)!]$$
$$+ x \ln p + (n-x)\ln(1-p). \quad \textbf{(D.25)}$$

We need a determination of the second derivative of $g(x)$. The function $g(x)$ contains the term $\ln(x!)$. We note that for large values of x,

$$\frac{d(\ln x!)}{dx} \approx \frac{\ln(x+1)! - \ln x!}{(x+1)-x} \approx \ln x. \quad \textbf{(D.26)}$$

Therefore, the derivative of $g(x)$ with respect to x is

$$\frac{dg}{dx} = -\ln x + \ln(n-x) + \ln p - \ln(1-p). \quad \textbf{(D.27)}$$

The second derivative is

$$\frac{d^2 g}{dx^2} = -\frac{1}{x} - \frac{1}{(n-x)}. \quad \textbf{(D.28)}$$

The second derivative evaluated at $x = np$ is

$$\left[\frac{d^2 g}{dx^2} \right]_{x=np} = -\frac{1}{np} - \frac{1}{n-np} = -\frac{1}{np(1-p)}, \quad \textbf{(D.29)}$$

and the binomial distribution becomes

$$f_b(x) \approx f_b(np) \exp\left\{ -\frac{(x-np)^2}{2np(1-p)} \right\}. \quad \textbf{(D.30)}$$

The standard deviation (σ) of a binomial distribution is

$$\sigma = \sqrt{np(1-p)}. \quad \textbf{(D.31)}$$

Defining a to be the average of x, our large-n approximation to the binomial distribution becomes

$$f_b(x) \approx f_b(a) \, e^{-(x-a)^2/2\sigma^2}. \quad \textbf{(D.32)}$$

This is the Gaussian distribution.

The Gaussian probability distribution is usually written

$$f_g(x) = Ce^{-(x-a)^2/2\sigma^2}, \quad \textbf{(D.33)}$$

where C is the normalization constant. Normalizing to unit area, we have

$$\int_{-\infty}^{+\infty} dx f_g(x) = C \int_{-\infty}^{+\infty} dx \, e^{-(x-a)^2/2\sigma^2} = 1. \quad \textbf{(D.34)}$$

To evaluate the definite integral, we make the change of variables,

$$y \equiv \frac{x-a}{\sqrt{2}\,\sigma}. \quad \textbf{(D.35)}$$

The normalization integral becomes

$$\int_{-\infty}^{+\infty} dx f_g(x) = C\sqrt{2}\,\sigma \int_{-\infty}^{+\infty} dy \, e^{-y^2} = 1. \quad \textbf{(D.36)}$$

We now evaluate the integral

$$I_0 \equiv \int_{-\infty}^{+\infty} dy \, e^{-y^2}. \quad \textbf{(D.37)}$$

We may write the square of the integral as

$$I_0^2 = \int_{-\infty}^{+\infty} dy \int_{-\infty}^{+\infty} dz \, e^{-(y^2+z^2)}. \quad \textbf{(D.38)}$$

With the substitution

$$r = \sqrt{y^2 + z^2}, \quad \textbf{(D.39)}$$

we have

$$I_0^2 = 2\pi \int_0^{+\infty} dr \, re^{-r^2}$$
$$= 2\pi \left[\frac{-e^{-r^2}}{2} \right]_{r=0}^{r=\infty} = \pi. \quad \textbf{(D.40)}$$

Taking the square root, we get

$$I_0 = \sqrt{\pi}, \quad \textbf{(D.41)}$$

and the normalization condition becomes

$$C\sqrt{2}\,\sigma \int_{-\infty}^{+\infty} dy \, e^{-y^2} = C\sqrt{2\pi}\,\sigma = 1. \quad \textbf{(D.42)}$$

Solving for C, we get

$$C = \frac{1}{\sqrt{2\pi}\,\sigma}. \quad \textbf{(D.43)}$$

The normalized Gaussian probability distribution is

$$f_g(x) = \frac{1}{\sqrt{2\pi}\,\sigma} e^{-(x-a)^2/2\sigma^2}. \quad \textbf{(D.44)}$$

In calculations involving Gaussian distributions we often need to evaluate integrals of the type

$$I_n = \int_0^{+\infty} dx \, x^n e^{-x^2}. \quad \textbf{(D.45)}$$

The values of some of these integrals are given in Table D-1.

The area under any portion of the Gaussian distribution is obtained by integration. The fraction of events for which x is within one standard deviation of the mean is

$$f_{x<1\sigma} = \frac{1}{\sqrt{2\pi}\,\sigma} \int_{-\sigma}^{+\sigma} dx \, e^{-(x-a)^2/2\sigma^2}. \quad \textbf{(D.46)}$$

The integral may not be evaluated in closed form; however, we can easily get the answer from a computer:

TABLE D-1
DEFINITE INTEGRALS APPEARING OFTEN IN THE USE OF GAUSSIAN DISTRIBUTION FUNCTIONS.

$$\int_0^{+\infty} dx \, e^{-x^2} = \frac{\sqrt{\pi}}{2}$$

$$\int_0^{+\infty} dx \, x \, e^{-x^2} = \frac{1}{2}$$

$$\int_0^{+\infty} dx \, x^2 \, e^{-x^2} = \frac{\sqrt{\pi}}{4}$$

$$\int_0^{+\infty} dx \, x^3 \, e^{-x^2} = \frac{1}{2}$$

$$\int_0^{+\infty} dx \, x^4 \, e^{-x^2} = \frac{3\sqrt{\pi}}{8}$$

$$\int_0^{+\infty} dx \, x^5 \, e^{-x^2} = 1$$

$$f_{x<1\sigma} \approx 0.683. \tag{D.47}$$

If we add up n random numbers, where n is large, and do it many times, then the distribution of the sums is a Gaussian. This is illustrated in Figure D-7. In Figure D-7a we show a histogram of 10^4 random numbers between 0 and 1. The bin size in the plot is 0.02, and thus the average number of events per bin is 200. From bin to bin there are fluctuations about the average, which are given by $\sqrt{n_i}$, where n_i is the number of events in a bin. The distribution of the number of events per bin, for the 50 bins, is a Gaussian. By the definition of a random number, the resulting distribution is flat. The mean is 1/2 and the standard deviation is $1/\sqrt{12}$.

Now let us examine what happens if we add two random numbers. The result of 10^4 such additions of two random numbers divided by 2 is shown in Figure D-7b. The average is still 1/2, but now it is less probable to have a result that is close either to 0 or 1. The standard deviation is smaller than that of Figure D-7a by a factor of $1/\sqrt{2}$. The reason is clear: To get a result near 1 requires both random numbers to be near 1 whereas to get a result near 0.5 can occur in many combinations (0.5 + 0.5, 0.4 + 0.6, 0.3 + 0.7, ...).

Now let us add three random numbers. The result of 10^4 such additions of three random numbers divided by 3 is shown in Figure D-7c. The average is still 1/2, but now it is even less probable to have a result that is close either to 0 or 1. The distribution is more sharply peaked and the standard deviation is smaller than in Figure D-7a by a factor of $1/\sqrt{3}$.

Let us continue this process and add 4, 5, and 6 random numbers, divide by the number of random numbers added and then histogram the results for 10^4 trials. The results are shown in Figure D-7d, D-7e, and D-7f. For large values of n (the number of random numbers added), the distribution has become Gaussian. It is already a very good approximation of a Gaussian for $n = 4$ (Figure D-7d)!

We have illustrated how it is that the Gaussian distribution appears so frequently in nature. Nature is full of random processes: radioactive decays of particles, collision of gas molecules, and so on. The addition of random numbers makes a Gaussian distribution.

D-7 THE POISSON DISTRIBUTION

If the event probability p is so small that for very large values of n, the product np is small, then the binomial distribution differs appreciably from zero only when the value of x is very small. The binomial distribution may be put in a more convenient form by noting that

$$\ln\left[(1-p)^n\right] = n\ln(1-p) \approx -np, \tag{D.48}$$

for $p \ll 1$. Taking the exponential of both sides, we have

$$(1-p)^n \approx e^{-np}. \tag{D.49}$$

For small values of x ($x \ll n$), we have

$$\frac{n!}{(n-x)!} = (n)(n-1)(n-2)\ldots(n-x) \approx n^x. \tag{D.50}$$

For large values of x, $f_b(x)$ is approximately zero because p is small. Using the small-p and small-x approximations, the binomial probability distribution reduces to:

$$f_b(x) \approx \frac{(np)^x e^{-np}}{x!}. \tag{D.51}$$

This distribution is called the *Poisson* distribution, $f_p(x)$. The Poisson distribution is usually written

$$f_p(x) = \frac{e^{-a} a^x}{x!}, \tag{D.52}$$

where $a = np$ is the average value of x. The Poisson distribution is normalized to unity,

$$\sum_{x=0}^{\infty} f_p(x) = \sum_{x=0}^{\infty} \frac{e^{-a} a^x}{x!} = 1. \tag{D.53}$$

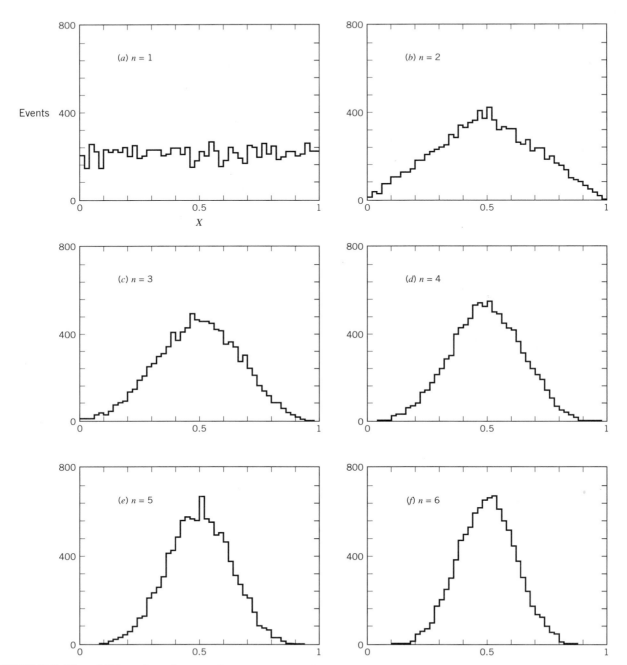

FIGURE D-7 The addition of random numbers.
For different values of n, we have made 10^4 determinations of the average of the sum of n random numbers: *(a)* $n = 1$, *(b)* $n = 2$, *(c)* $n = 3$, *(d)* $n = 4$, *(e)* $n = 5$, *(f)* $n = 6$. The distribution for $n = 1$ is flat, but it very quickly approaches a Gaussian as n increases.

E-1 DEFINITION OF THE COORDINATES

In physical systems that display spherical symmetry, it is often more convenient to replace the rectangular coordinates (x,y,z) with spherical coordinates (r,θ,ϕ). The *radial coordinate* (r) is defined to be the distance to the origin (see Figure E-1),

$$r \equiv \sqrt{x^2 + y^2 + z^2}\,. \tag{E.1}$$

The range of r is from 0 to ∞. The *colatitude angle* (θ) is defined to be the angle measured from the z axis,

$$\cos\theta \equiv \frac{z}{r} = \frac{z}{\sqrt{x^2 + y^2 + z^2}}\,. \tag{E.2}$$

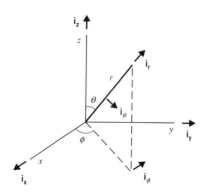

FIGURE E-1 Definition of the spherical coordinates r, θ, and ϕ.

The range of θ is from 0 to π. The *azimuthal angle* (ϕ) is defined to be the angle measured from the x axis when r is projected on to the x-y plane,

$$\cos\phi \equiv \frac{x}{r\sin\theta} = \frac{x}{\sqrt{x^2 + y^2}}\,. \tag{E.3}$$

The range of ϕ is from 0 to 2π.

E-2 VECTOR CALCULUS IN SPHERICAL COORDINATES

Volume Element

In rectangular coordinates, the differential unit lengths are dx, dy, and dz. The corresponding differential unit lengths in spherical coordinates $(dl_r, dl_\theta, \text{and } dl_\phi)$ are

$$dl_r = dr\,, \tag{E.4}$$

$$dl_\theta = r\,d\theta\,, \tag{E.5}$$

and

$$dl_\phi = r\sin\theta\,d\phi\,. \tag{E.6}$$

The differential volume element (dV) is

$$\begin{aligned} dV &= dx\,dy\,dz = dl_r\,dl_\theta\,dl_\phi \\ &= (dr)(r d\theta)(r\sin\theta\,d\phi) \\ &= r^2\sin\theta\,dr\,d\theta\,d\phi. \end{aligned} \tag{E.7}$$

Gradient

In rectangular coordinates the unit vectors $(\mathbf{i}_x, \mathbf{i}_y, \mathbf{i}_z)$ are fixed. In spherical coordinates the unit vectors $(\mathbf{i}_r, \mathbf{i}_\theta, \mathbf{i}_\phi)$ are not constants, because their directions depend on the value of the vector (r, θ, ϕ). In rectangular coordinates, the gradient of a function ψ is defined to be

$$\nabla \psi \equiv \frac{\partial \psi}{\partial x} \mathbf{i}_x + \frac{\partial \psi}{\partial y} \mathbf{i}_y + \frac{\partial \psi}{\partial z} \mathbf{i}_z. \qquad \textbf{(E.8)}$$

In spherical coordinates the gradient is

$$\nabla \psi \equiv \frac{\partial \psi}{\partial l_r} \mathbf{i}_r + \frac{\partial \psi}{\partial l_\theta} \mathbf{i}_\theta + \frac{\partial \psi}{\partial l_\phi} \mathbf{i}_\phi$$

$$= \frac{\partial \psi}{\partial r} \mathbf{i}_r + \frac{1}{r} \frac{\partial \psi}{\partial \theta} \mathbf{i}_\theta + \frac{1}{r \sin \theta} \frac{\partial \psi}{\partial \phi} \mathbf{i}_\phi. \qquad \textbf{(E.9)}$$

Divergence

In rectangular coordinates, the divergence of a vector $\mathbf{E}$ (see Appendix C) is defined to be

$$\nabla \cdot \mathbf{E} \equiv \frac{\partial E_x}{\partial x} + \frac{\partial E_y}{\partial y} + \frac{\partial E_z}{\partial z}. \qquad \textbf{(E.10)}$$

The divergence theorem may be used to arrive at the expression for $\nabla \cdot \mathbf{E}$ in spherical coordinates. The divergence theorem is

$$\oiint d\mathbf{a} \cdot \mathbf{E} = \iiint_V dV \, \nabla \cdot \mathbf{E}. \qquad \textbf{(E.11)}$$

Using the expressions for the differential volume element (E.7), the volume integral (E.11) becomes

$$\iiint_V dV \, \nabla \cdot \mathbf{E} = \int dx \int dy \int dz \nabla \cdot \mathbf{E}$$

$$= \int dr \int d\theta \, r \int d\phi \, r \sin \theta \, \nabla \cdot \mathbf{E} \ , \textbf{(E.12)}$$

and the area integral (E.11) becomes

$$\oiint d\mathbf{a} \cdot \mathbf{E} = \int dy \int dz \, E_x + \int dz \int dx \, E_y + \int dx \int dy \, E_z$$

$$= \int d\theta \, r \int d\phi \, r \sin \theta \, E_r + \int dr \int d\phi \, r \sin \theta \, E_\theta$$

$$+ \int dr \int d\theta \, r \, E_\phi. \qquad \textbf{(E.13)}$$

Comparing the integrands (E.12) and (E.13), we get

$$\nabla \cdot \mathbf{E} = \frac{1}{r^2 \sin \theta} \left[\frac{\partial}{\partial r} \left(E_r r^2 \sin \theta \right) \right.$$

$$\left. + \frac{\partial}{\partial \theta} \left(E_\theta r \sin \theta \right) + \frac{\partial}{\partial \phi} \left(E_\phi r \right) \right]. \qquad \textbf{(E.14)}$$

Divergence of the Gradient

In rectangular coordinates, the divergence of a gradient (∇^2) is

$$\nabla^2 \psi \equiv \nabla \cdot (\nabla \psi) = \frac{\partial^2 \psi}{\partial x^2} + \frac{\partial^2 \psi}{\partial y^2} + \frac{\partial^2 \psi}{\partial z^2}. \qquad \textbf{(E.15)}$$

In spherical coordinates, the expression for $\nabla^2 \psi$ follows from the expressions for gradient (E.9) and divergence (E.14):

$$\nabla^2 \psi = \nabla \cdot (\nabla \psi)$$

$$= \frac{1}{r^2 \sin \theta} \left[\frac{\partial}{\partial r} \left(r^2 \sin \theta \frac{\partial \psi}{\partial r} \right) \right.$$

$$\left. + \frac{\partial}{\partial \theta} \left(\sin \theta \frac{\partial \psi}{\partial \theta} \right) + \frac{\partial}{\partial \phi} \left(\frac{1}{\sin \theta} \frac{\partial \psi}{\partial \phi} \right) \right]. \textbf{(E.16)}$$

The Taylor expansion of a function $f(x)$ about $x = a$ is

$$f(x) = f(a) + (x-a)\left[\frac{df}{dx}\right]_{x=a}$$
$$+ \frac{(x-a)^2}{2!}\left[\frac{d^2 f}{dx^2}\right]_{x=a}$$
$$+ \ldots + \frac{(x-a)^n}{n!}\left[\frac{d^n f}{dx^n}\right]_{x=a} + \ldots \quad \text{(F.1)}$$

The Taylor expansion is particularly useful when one can neglect all but the first two terms. An example is

$$f(x) = \frac{1}{1+x}, \quad \text{(F.2)}$$

where $x \ll 1$. The Taylor expansion (F.1) about $x = 0$ gives

$$\frac{1}{1+x} \approx 1 + (x-0)\left[\frac{(-1)}{(1+x)^2}\right]_{x=0} = 1 - x. \quad \text{(F.3)}$$

We may verify the expansion (F.3) by multiplying both sides by $1 + x$, neglecting the x^2 term. Similarly, for

$$f(x) = \frac{1}{1-x}, \quad \text{(F.4)}$$

we have

$$\frac{1}{1-x} \approx 1 + x. \quad \text{(F.5)}$$

Another example is

$$f(x) = \sqrt{1+x}, \quad \text{(F.6)}$$

with $x \ll 1$. The Taylor expansion (F.1) about $x = 0$ gives

$$\sqrt{1+x} \approx 1 + (x-0)\left[\frac{\left(\frac{1}{2}\right)}{\sqrt{1+x}}\right]_{x=0} = 1 + \frac{x}{2}. \quad \text{(F.7)}$$

We may verify this result by squaring both sides. Similarly, for

$$f(x) = \sqrt{1-x}, \quad \text{(F.8)}$$

we have

$$\sqrt{1-x} \approx 1 - \frac{x}{2}. \quad \text{(F.9)}$$

Another example is

$$f(x) = \frac{1}{\sqrt{1+x}}, \quad \text{(F.10)}$$

with $x \ll 1$. The Taylor expansion (F.1) about $x = 0$ gives

$$\frac{1}{\sqrt{1+x}} \approx 1+(x-0)\left[\frac{\left(\dfrac{-1}{2}\right)}{(1+x)^{3/2}}\right]_{x=0} = 1-\frac{x}{2}. \quad \textbf{(F.11)}$$

Similarly, for

$$f(x) = \frac{1}{\sqrt{1-x}}, \quad \textbf{(F.12)}$$

we have

$$\frac{1}{\sqrt{1-x}} \approx 1+\frac{x}{2}. \quad \textbf{(F.13)}$$

TRANSFORMATION OF ELECTRIC AND MAGNETIC FIELDS

Consider electric (**E**) and magnetic fields (**B**) observed in the frame S. In the frame S', which is moving with a speed v in the x direction, the fields are given by

$$E_x' = E_x, \tag{G.1}$$

$$E_y' = \gamma E_y - \beta\gamma B_z, \tag{G.2}$$

$$E_z' = \gamma E_z + \beta\gamma B_y, \tag{G.3}$$

$$B_x' = B_x, \tag{G.4}$$

$$B_y' = \gamma B_y + \beta\gamma E_z, \tag{G.5}$$

and

$$B_z' = \gamma B_z - \beta\gamma E_y, \tag{G.6}$$

where $\beta = v/c$ and $\gamma = (1 - v^2/c^2)^{-1/2}$.

The solution to the Schrödinger equation for the hydrogen atom in spherical coordinates by separation of variables leads to three ordinary differential equations for $R(r)$, $P(\theta)$, and $F(\phi)$. The mathematics of the solution and the physical interpretation of $|\psi|^2$ as the electron probability density places constraints on the separation constants C_r and C_ϕ, and leads to the three quantum numbers n, ℓ, and m_ℓ.

H-1 THE SOLUTION FOR $F(\phi)$

The differential equation for $F(\phi)$ is

$$\frac{1}{F}\frac{d^2F}{d\phi^2} = C_\phi. \qquad \textbf{(H.1)}$$

The solution may be written as

$$F(\phi) = Ce^{im_\ell\phi}, \qquad \textbf{(H.2)}$$

where $C_\phi = -m_\ell^2$ and C is a normalization constant. It is often incorrectly stated that $F(\phi)$ must be single valued so that $F(\phi) = F(\phi + 2\pi)$ would imply that m_ℓ must be an integer. There is in fact no reason to assume that $F(\phi)$ must be single valued; only the square of the wave function must be single valued. It is true, however, that the physically allowed values of m_ℓ are integer, for a reason that we will consider later.

H-2 THE SOLUTION FOR $P(Q)$

The separated equation for $P(\theta)$ with $C_\phi = -m_\ell^2$ is

$$-m_\ell^2 P = -C_r \sin^2\theta P - \sin\theta \frac{d}{d\theta}\left(\sin\theta \frac{dP}{d\theta}\right). \qquad \textbf{(H.3)}$$

This equation becomes easier to solve with the change of variables,

$$x \equiv \cos\theta. \qquad \textbf{(H.4)}$$

Then we have

$$\frac{dx}{d\theta} = -\sin\theta = -\sqrt{1-x^2}, \qquad \textbf{(H.5)}$$

$$\frac{dP}{d\theta} = \frac{dx}{d\theta}\frac{dP}{dx} = -\sqrt{1-x^2}\,\frac{dP}{dx}, \qquad \textbf{(H.6)}$$

and

$$\sin\theta \frac{dP}{d\theta} = -\left(1-x^2\right)\frac{dP}{dx}. \qquad \textbf{(H.7)}$$

The differential equation for $P(x)$ becomes

$$-m_\ell^2 P = -C_r\left(1-x^2\right)P - \left(1-x^2\right)\frac{d}{dx}\left[\left(1-x^2\right)\frac{dP}{dx}\right],$$

(H.8)

or

$$-m_\ell^2 P = -C_r\left(1-x^2\right)P$$
$$-\left(1-x^2\right)\left[-2x\frac{dP}{dx}+\left(1-x^2\right)\frac{d^2P}{dx^2}\right]. \text{ (H.9)}$$

Rearranging terms, we have

$$\left(1-x^2\right)\frac{d^2P}{dx^2}-2x\frac{dP}{dx}+\left[C_r-\frac{m_\ell^2}{\left(1-x^2\right)}\right]P = 0.\text{(H.10)}$$

The constants C_r and $-m_\ell^2$ that appear in the differential equation (H.10) are the separation constants for the radial and phi portions of the wave function. We first analyze the case $m_\ell = 0$. The differential equation for $P(x)$ becomes

$$\left(1-x^2\right)\frac{d^2P}{dx^2}-2x\frac{dP}{dx}+C_rP = 0. \quad \text{(H.11)}$$

This equation is called *Legendre's equation*. The solutions are polynomial functions (*Legendre polynomials*). Legendre's equation may be solved by power series. Let

$$P(x) = \sum_{k=0}^{\infty}A_k x^k. \quad \text{(H.12)}$$

Substituting the power series into Legendre's equation (H.11) we get

$$\sum_{k=0}^{\infty}A_k\left[\left(1-x^2\right)k(k-1)x^{k-2}-2xkx^{k-1}+C_rx^k\right] = 0.$$

(H.13)

Collecting terms of the same power of x, we have

$$\sum_{k=0}^{\infty}A_k\left[k(k-1)x^{k-2}+\left(C_r-k-k^2\right)x^k\right] = 0. \quad \text{(H.14)}$$

In the first term inside the bracket, we make a change in the dummy summation variable, $k \to (k+2)$. The power series (H.14) becomes

$$\sum_{k=0}^{\infty}\left[A_{k+2}(k+2)(k+1)+A_k\left(C_r-k-k^2\right)\right]x^k = 0.$$

(H.15)

Since all terms must be equal to zero, this defines a recurrence relationship between the kth and the $(k+2)$th constant:

$$A_{k+2}(k+2)(k+1)+A_k\left(C_r-k-k^2\right) = 0, \quad \text{(H.16)}$$

or

$$A_{k+2} = \left[\frac{(k)(k+1)-C_r}{(k+2)(k+1)}\right]A_k. \quad \text{(H.17)}$$

For the power series to converge, there is a restriction on the allowed values of C_r. As can be seen from the numerator, the separation constant C_r must be equal to the product of two consecutive nonnegative integers. We write the power series convergence condition as

$$C_r = \ell(\ell+1), \quad \text{(H.18)}$$

where $\ell = 0, 1, 2, \ldots$ All the power series coefficients (H.17) are zero, for $k > \ell$. Therefore, $P(x)$ is a polynomial of order ℓ and we assign $P(x)$ the serial number ℓ.

For the case $\ell = 0$, the solution to Legendre's equation (H.11) is

$$P_0 = A_0 \quad \text{(H.19)}$$

and the other constants ($A_1, A_2, A_3, \ldots$) are all zero.

For the case $\ell = 1$, the solution to Legendre's equation may be written

$$P_1 = A_0 + A_1x. \quad \text{(H.20)}$$

Legendre's equation gives

$$-2xA_1 + 2A_0 + 2A_1x = 0. \quad \text{(H.21)}$$

This implies that $A_0 = 0$ and $P_1 = A_1x$. The other constants are zero.

For the case $\ell = 2$, the solution to Legendre's equation may be written

$$P_2 = A_0 + A_1 x + A_2 x^2. \qquad \textbf{(H.22)}$$

The recurrence relationship gives $A_2 = -3A_0$ and Legendre's equation gives

$$-6 A_0 \left(1 - x^2\right) - 2 x \left(A_1 - 6 A_0 x\right)$$
$$+ 6\left(A_0 + A_1 x - 3 A_0 x^2\right) = 0. \quad \textbf{(H.23)}$$

The solution implies $A_1 = 0$ and $P_2 = A_0(3x^2 - 1)$. Continuing in this fashion we may construct the entire family of solutions $P_\ell(\cos\theta)$ valid for the case $m_\ell = 0$.

If m_ℓ is not equal to zero, then the differential equation for P is called the *associated Legendre equation*:

$$\left(1 - x^2\right)\frac{d^2 P}{dx^2} - 2 x \frac{dP}{dx} + \left[C_r - \frac{m_\ell^2}{\left(1 - x^2\right)}\right] P = 0. \textbf{(H.24)}$$

The family of solutions are called *associated Legendre polynomials*.

For the case $m_\ell = 1$, one solution is

$$P = C\left(1 - x^2\right)^{1/2}. \qquad \textbf{(H.25)}$$

where C is a constant and

$$C_r = 2. \qquad \textbf{(H.26)}$$

In general the solution for any integer m_ℓ requires $C_r = \ell(\ell + 1)$ and the associated Legendre equation may be written

$$\left(1 - x^2\right)\frac{d^2 P}{dx^2} - 2 x \frac{dP}{dx} + \left[\ell(\ell+1) - \frac{m_\ell^2}{\left(1 - x^2\right)}\right] P = 0.$$
$$\textbf{(H.27)}$$

The solutions are denoted $P_{\ell, m_\ell}(x)$, where the possible values of ℓ are $|m_\ell|, |m_\ell| + 1, |m_\ell| + 2, \ldots$.

We return to the issue of the allowed values of m_ℓ. We said that the wave function does not necessarily need to be single valued but that its square, which represents the electron probability density must be single valued. The square of the wave function does not depend on ϕ, that is, $|\psi|^2$ does not depend explicitly on m_ℓ. The square of the wave function does depend on ℓ, however, and the allowed

values of ℓ depend on m_ℓ. The mathematics of the solution with a single-valued $|\psi|^2$ allows half-integer values of ℓ and m_ℓ. For example, if $m_\ell = 1/2$, then the allowed values of ℓ are $1/2, 3/2, 5/2\ldots$. The angular part of the wave function for $\ell = 1/2$ and $m_\ell = 1/2$ is

$$P_{1/2, 1/2}(\theta)F(\phi) = \sqrt{\sin\theta}\, e^{i\phi/2}, \qquad \textbf{(H.28)}$$

where $F(\phi)$ and $P_{\ell, m_\ell}(\theta)$ satisfy the differential equations (H.1) and (H.27). The half-integer values of m_ℓ that lead to half-integer values of ℓ are excluded from the description of the hydrogen atom because they are inconsistent with our interpretation of $\ell(\ell + 1)\hbar^2$ as the orbital angular momentum squared. (Mathematically, the half-integer wave functions are not well-behaved under rotations, i.e., a rotated wave function cannot be expressed as a linear combination of unrotated wave functions.) The integer solutions correspond to the three degrees of freedom. Half-integer solutions for angular momentum do appear in nature, but they are due to an internal degree of freedom that is not contained in the Schrödinger equation.

The associated Legendre polynomials may be obtained from the Legendre polynomials by the following relationship:

$$P_{\ell, m_\ell}(x) = \sqrt{\left(1 - x^2\right)^{m_\ell}}\,\frac{d^{m_\ell} P_\ell}{dx^{m_\ell}}, \qquad \textbf{(H.29)}$$

where m_ℓ is a positive integer less than or equal to ℓ. This expression follows directly from the differential equations (H.11) and (H.27). The associated Legendre polynomials for the negative values of m_ℓ are obtained by simply changing m_ℓ to $-m_\ell$ in the exponential part that contains ϕ. The product of $F(\phi)$ and $P(\theta)$ is commonly written as the *spherical harmonic*,

$$Y_{\ell, m_\ell}(\theta, \phi) = C_{\ell, m_\ell} P_{\ell, m_\ell}(\theta)F(\phi)$$
$$= C_{\ell, m_\ell} P_{\ell, m_\ell}(\theta)e^{im_\ell\phi}, \qquad \textbf{(H.30)}$$

where the normalization constants C_{ℓ, m_ℓ} are chosen so that the integral of $|\psi|^2$ over all angles is unity. These solutions are summarized in Table H-1.

H-3 THE SOLUTION FOR $R(r)$

The solution for $P(\theta)$ has restricted the radial separation constant C_r to be the product of two consecutive integers, $\ell(\ell + 1)$. We now must solve the radial equation,

TABLE H-1
SOLUTIONS TO THE ANGULAR PART OF THE
SCHRÖDINGER EQUATION.

ℓ	m_ℓ	$Y_{\ell,m_\ell}(\theta,\phi)$
0	0	$\sqrt{\dfrac{1}{4\pi}}$
1	0	$\sqrt{\dfrac{3}{4\pi}}\cos\theta$
1	1	$\sqrt{\dfrac{3}{8\pi}}\sin\theta\,e^{i\phi}$
2	0	$\sqrt{\dfrac{5}{16\pi}}\left(3\cos^2\theta-1\right)$
2	1	$\sqrt{\dfrac{15}{8\pi}}\sin\theta\cos\theta\,e^{i\phi}$
2	2	$\sqrt{\dfrac{15}{32\pi}}\sin^2\theta\,e^{2i\phi}$
3	0	$\sqrt{\dfrac{7}{16\pi}}\left(5\cos^3\theta-3\cos\theta\right)$
3	1	$\sqrt{\dfrac{21}{64\pi}}\sin\theta\left(5\cos^2\theta-1\right)e^{i\phi}$
3	2	$\sqrt{\dfrac{105}{32\pi}}\sin^2\theta\cos\theta\,e^{2i\phi}$
3	3	$\sqrt{\dfrac{35}{64\pi}}\sin^3\theta\,e^{3i\phi}$

$$-\frac{\hbar^2}{2mr^2}\frac{d}{dr}\left(r^2\frac{dR}{dr}\right)+\left[\frac{\ell(\ell+1)\hbar^2}{2mr^2}-\frac{ke^2}{r}\right]R=ER.$$
(H.31)

We make a simplification in the radial equation if we write the radial coordinate in terms of the Bohr radius,

$$x\equiv\frac{r}{a_0}=\frac{rke^2m}{\hbar^2},\qquad\textbf{(H.32)}$$

and the energy in terms of the ground state energy,

$$\varepsilon\equiv\frac{E}{E_1}=\frac{2E\hbar^2}{m\left(ke^2\right)^2}.\qquad\textbf{(H.33)}$$

With this change of variables, the radial equation becomes

$$-\frac{1}{x^2}\frac{d}{dx}\left(x^2\frac{dR}{dx}\right)+\left[\varepsilon+\frac{2}{x}-\frac{\ell(\ell+1)}{x^2}\right]R=0.\quad\textbf{(H.34)}$$

We may rewrite the radial equation as

$$\frac{d^2(xR)}{dx^2}=-\left[\varepsilon+\frac{2}{x}-\frac{\ell(\ell+1)}{x^2}\right]xR,\quad\textbf{(H.35)}$$

and make the change of variables,

$$f(x)\equiv xR(x),\qquad\textbf{(H.36)}$$

which gives

$$\frac{d^2f}{dx^2}=-\left[\varepsilon+\frac{2}{x}-\frac{\ell(\ell+1)}{x^2}\right]f.\qquad\textbf{(H.37)}$$

We may anticipate that the wave function will fall off exponentially at large values of x, and write the solution for $f(x)$ as

$$f(x)=e^{-\zeta x}g(x).\qquad\textbf{(H.38)}$$

Substituting this expression for $f(x)$ into the differential equation (H.37) gives

$$\frac{d^2g}{dx^2}-2\zeta\frac{dg}{dx}+\left[\varepsilon+\frac{2}{x}-\frac{\ell(\ell+1)}{x^2}+\zeta^2\right]g=0.\textbf{(H.39)}$$

Now we choose

$$\zeta^2\equiv-\varepsilon,\qquad\textbf{(H.40)}$$

and multiply by x^2 to arrive at

$$x^2\frac{d^2g}{dx^2}-2\zeta x^2\frac{dg}{dx}+\left[2x-\ell(\ell+1)\right]g=0.\quad\textbf{(H.41)}$$

This equation can be solved by power series,

$$g(x)=\sum_{k=0}^{\infty}A_kx^k.\qquad\textbf{(H.42)}$$

Substituting the power series (H.42) into the differential equation (H.41) gives

$$\sum_{k=0}^{\infty} A_k \left[k(k+1)x^k + 2(1-\zeta k)x^{k+1} - \ell(\ell+1)x^k \right] = 0 .$$ (H.43)

In the first term we make a change in the dummy summation variable, $k \rightarrow (k+1)$. The last term we write as

$$\sum_{k=0}^{\infty} A_k \left[-\ell(\ell+1)x^k \right] = -\ell(\ell+1)\left[A_0 + \sum_{k=0}^{\infty} A_{k+1}x^{k+1} \right],$$ (H.44)

so the sum (H.43) becomes

$$-\ell(\ell+1)A_0$$
$$+ \sum_{k=0}^{\infty} A_{k+1} \left[k(k+1) + A_k 2(1-\zeta k) - A_{k+1}\ell(\ell+1) \right]x^{k+1}$$
$$= 0.$$ (H.45)

The recurrence relation is,

$$A_{k+1} = \left[\frac{2(\zeta k-1)}{\ell(\ell+1)-k(k+1)} \right] A_k .$$ (H.46)

The series converges when the $(k+1)$th constant is zero and this happens when

$$\zeta = \frac{1}{n},$$ (H.47)

where n is a positive integer (1, 2, 3, ...). The electron energy is given by

$$\varepsilon = -\frac{1}{n^2},$$ (H.48)

or

$$E = -\frac{m(ke^2)^2}{2\hbar^2 n^2} .$$ (H.49)

The recurrence relation (H.46) says that all the coefficients up to and including A_ℓ are zero and the $(\ell+1)$th coefficient is the first that is nonzero. The series ends because $\zeta = 1/n$. Thus, the summation variable k runs from ℓ to $n-1$. The radial wave function is

$$R = xf(x) = e^{-\zeta x} \sum_{k=\ell}^{n-1} A_{k+1}x^{k+1} .$$ (H.50)

The determination of the coefficients A_{k+1} comes from substituting R into the radial equation. According to the definitions of x (H.32) and ζ (H.40) the general solution of the radial equation may be written,

$$R_{n,\ell} = r^\ell L_{n,\ell} e^{-r/na_0} ,$$ (H.51)

where $L_{n,\ell}$ is an *associated Laguerre polynomial* defined to be a solution of the differential equation

$$x\frac{d^2 L_{n,\ell}}{dx^2} + (2\ell+2-x)\frac{dL_{n,\ell}}{dx} + (n-l-1)L_{n,\ell} = 0 .$$ (H.52)

The radial wave functions are normalized such that

$$\int_0^\infty dr \, r^2 \left| R_{n,\ell} \right|^2 = 1 .$$ (H.53)

The first few solutions are given in Table H-2.

TABLE H-2
RADIAL WAVE FUNCTIONS.

n	ℓ	$R_{n,\ell}$
1	0	$\dfrac{2}{\sqrt{a_0^3}}e^{-r/a_0}$
2	0	$\dfrac{1}{\sqrt{2a_0^3}}\left(1-\dfrac{r}{2a_0}\right)e^{-r/2a_0}$
2	1	$\dfrac{1}{\sqrt{24a_0^3}}\dfrac{r}{a_0}e^{-r/2a_0}$
3	0	$\dfrac{2}{\sqrt{27a_0^3}}\left(1-\dfrac{2r}{3a_0}+\dfrac{2r^2}{27a_0^2}\right)e^{-r/3a_0}$
3	1	$\dfrac{8}{27\sqrt{6a_0^3}}\dfrac{r}{a_0}\left(1-\dfrac{r}{6a_0}\right)e^{-r/3a_0}$
3	2	$\dfrac{4}{81\sqrt{30a_0^3}}\left(\dfrac{r}{a_0}\right)^2 e^{-r/3a_0}$

FAMOUS
EXPERIMENTS
IN MODERN PHYSICS

1887	Albert Michelson, Edward Morley	Measurement of the speed of light from a moving source
1895	Wilhelm Röntgen	Discovery of x rays
1896	Antoine-Henri Becquerel	Discovery of natural radioactivity
1896	Pieter Zeeman	Discovery of the splitting of spectral lines in magnetic fields
1897	J. J. Thomson	Measurement of the charge-to-mass ratio of the electron
1901	Walter Kaufmann	Measurement of relativistic electrons in a magnetic field
1906	Charles T. R. Wilson	Observation of electron tracks in a cloud chamber
1908	Jean-Baptiste Perrin	Measurement of the size of atoms
1911	Ernest Rutherford, H. Geiger, E. Marsden	Discovery of the nucleus
1911	Heike Kamerlingh Onnes	Discovery of superconductivity
1911	Owen W. Richardson	Discovery of thermionic emission
1911	Robert Millikan	Discovery of charge quantization, measurement of the electron charge

1912	Max von Laue	Demonstration that x rays are electromagnetic waves
1912	Victor Hess	Discovery of cosmis rays
1913	W. L. Bragg, W. H. Bragg	Diffraction of x rays from crystals
1914	Henry Moseley	Measurement of x rays from the elements
1914	James Franck, Gustav Hertz	Observation of energy quantization in mercury atoms
1916	Robert Millikan	Measurement of Planck's constant in the photoelectric effect
1922	Otto Stern, Walter Gerlach	Discovery of the intrinsic angular momentum of the electron
1927	Clinton Davisson, Lester Germer, George Thomson, A. Reid	Observation of electron diffraction
1927	Arthur H. Compton	Measurement of x ray scattering
1928	C. V. Raman	Scattering of light by atoms
1929	Edwin Hubble	Discovery of the expansion of the universe
1931	Harold Urey	Discovery of deuterium
1932	James Chadwick	Discovery of the neutron

1932	Carl Anderson	Discovery of the positron	
1932	John Cockroft, Ernest Walton	Artificial disintegration of the nucleus	
1933	Otto Stern	Measurement of the magnetic moment of the proton	
1935	Pavel Cerenkov	Discovery of Cerenkov radiation	
1937	J. Street, E. Stevenson, C. Anderson, S. Neddermeyer	Discovery of the muon	
1939	L. Meitner, O. Frisch, O. Hahn, F. Strassmann	Discovery of uranium fission	
1942	E. Fermi et al.	Observation of a nuclear chain reaction and construction of a nuclear reactor	
1946	Felix Bloch, Edward Purcell	Nuclear magnetic resonance	
1947	Willis Lamb	Precision test of quantum electrodynamics by measurement of the energy difference of the $2s$ and $2p$ levels in the hydrogen atom	
1947	C. Lattes, H. Muirhead, G. Occhialini, C. Powell	Discovery of the pion	
1947	G. Rochester, C. Butler	Discovery of the kaon (particles containing the strange quark)	
1947	Polykarp Kusch	Measurement of the magnetic moment of the electron	
1947	Willard Libby	Radiocarbon dating	
1951	Charles Townes et al.	Invention of the MASER	
1952	Donald Glaser	Invention of the bubble chamber	
1955	Owen Chamberlain, Emilio Segrè et al.	Discovery of the antiproton	
1956	Clyde L. Cowan, Frederick Reines et al.	Observation of the electron neutrino	
1956	John Bardeen, Walter Brattain, William Shockley	Invention of the transistor	
1956	Robert Hofstadter	Measurement of the structure of the nucleons	
1957	C.-S. Wu	Discovery of parity violation in the weak interaction	
1957	Rudolf Mössbauer	Discovery of recoiless gamma emission from nuclei	
1958	J. Eugene Kunzler	Discovery of superconductivity in the presence of large magnetic fields	
1959	Robert Pound, Glen Rebka, Jr.	Observation of the gravitational attraction of photons to the earth	
1961	Melvin Schwartz, Leon Lederman, Jack Steinberger et al.	Discovery of the muon neutrino	
1961	B.S. Deaver, W.M. Fairbank, R. Doll, M. Näbauer	Discovery of magnetic flux quantization	
1963	P.W. Anderson, J.M. Rowell	Discovery of electron pair tunneling	
1964	Val Fitch, James Cronin	Discovery of CP violation in the weak interaction	
1964	Allan R. Sandage	Discovery of quasars	
1965	Arno Penzias, Robert W. Wilson	Discovery of the cosmic microwave radiation from the early universe	
1967	Jerome Friedman, Henry Kendall, Richard Taylor et al.	Discovery of the quark structure of the proton	
1967	Jocelyn Bell, Anthony Hewish	Discovery of pulsars	
1970	Raymond Davis, et al.	Observation of neutrinos from the sun	
1973	F.J. Hasert et al.	Discovery of weak neutral currents	
1974	Burton Richter, Samuel Ting et al.	Discovery of the charm quark	

1974	Joseph H. Taylor Russell H. Hulse	Measurements of binary neautron stars
1975	Martin Perl et al.	Discovery of the tau lepton
1976	Leon Lederman et al.	Discovery of the bottom- quark
1979	DESY experiments	Observation of gluons
1980	Klaus von Klitzing	Discovery of the quantized Hall effect
1981	Gerd Binnig, Heinrich Rohrer	Invention of the scanning-tunneling electron microscope

1983	Carlo Rubbia et al.	Discovery of the W and Z^0 particles
1986	J. Georg Bednorz, K. Alex Müller	Discovery of high-T_c superconductivity
1987	Kamiokande and IMB experiments	Observation of antineutrinos from a supernova
1989	CERN LEP experiments	Measurement of 3 neutrino families
1990	COBE experiment	Measurement of the cosmic photon spectrum

OUTSTANDING
PROBLEMS

There is something fascinating about science. One gets such wholesale returns of conjecture out of such a trifling investment of fact.

Mark Twain

The discovery of the laws of electricity and magnetism was a gigantic step in the understanding of the universe. At the end of the nineteenth century, physicists were faced with two major problems: (1) the radiation spectrum from an object in thermal equilibrium, and (2) heat capacities. The solution to both of these problems requires quantum mechanics.

The amount of knowledge of the universe that we have amassed in the last century is almost mind-boggling. When we examine the history of physics, we see that each exciting discovery has opened new questions that have stimulated the research that has led to additional discoveries!

What problems are left to be solved? We may divide these into three general classes: condensed matter, astrophysical, and particles. Our list of problems in condensed matter physics would include a deeper understanding of:

1. Critical phenomena (phase transitions),
2. Nonlinear phenomena,
3. Magnetic effects in materials, and
4. High-T_c superconductivity.

A glance at the great discoveries that have been made in the field of condensed matter physics (e.g., superconductivity, semiconductors, lasers, quantum Hall effect, etc.) leaves no doubt that more surprises are in store!

In the field of astrophysics we have:

1. Why is the cosmological constant (Λ) so small, and is it identically zero?
2. Do black holes exist?
3. What is the mechanism for galaxy formation?
4. Is the universe open or closed?
5. What is the form of the "dark" matter in the universe?
6. Why is there more matter than antimatter in the universe?
7. What mechanism provides the fantastic energy source of the quasars?

In the field of particle physics:

1. Are quarks and leptons truly fundamental or are they made up of more elementary constituents?
2. Why are there three families of quarks and leptons?
3. Why do the quarks have the masses and mixing angles that they do?
4. Are there any additional fundamental forces?
5. Do all the forces have equal strength at some extremely short distance?
6. Is the proton absolutely stable or does it merely have a very long lifetime?
7. Do magnetic monopoles exist?
8. What is the origin of the electroweak symmetry breaking?
9. What is the origin of *CP* violation?
10. Is lepton number absolutely conserved?
11. Why are neutrinos left handed?
12. Are the neutrino masses identically equal to zero?

The reader will no doubt be able to add to this list.

REFERENCES AND SUGGESTIONS FOR FURTHER READING

V. L. Ginzburg, "What Problems of Physics and Astrophysics Seem Now to be Especially Important and Interesting," *Phys. Today* **43**, No. 5, 9 (1990).

V. L. Ginzburg, *Physics and Astrophysics: A Selection of Key Problems*, Pergamon (1985).

D. J. Gross, "On the Calculation of the Fine-Structure Constant," *Phys. Today* **42**, No. 12, 9 (1989).

Element[a]	Z	A	Atomic mass (u)[b]	Nuclear mass (GeV)	E_b^c (MeV)	J^d	Natural Abundance	Half-life[e]	Decay[f]	Q^g (MeV)	A	
Hydrogen	H 1	1	1.00783	0.9383	0.00	1/2	0.99985	stable			1	H
Gr., hydro (water)		2	2.01400	1.8756	2.22	1	1.5×10^{-4}	stable			2	
+ genes (forming)		3	3.01605	2.8089	8.48	1/2		12.32 y	β^-	0.019	3	
Helium	He 2	3	3.01603	2.8084	7.72	1/2	1.38×10^{-6}	stable			3	He
Gr., helios (sun)		4	4.00260	3.7274	28.30	0	0.999999	stable			4	
		6	6.018886	5.6056	29.27	0		0.807 s	β^-	3.51	6	
		8	8.03392	7.4826	31.41	0		0.119 s	β^-	14	8	
Lithium	Li 3	6	6.015121	5.6016	32.00	1	0.075	stable			6	Li
Gr., lithos (stone)		7	7.016003	6.5339	39.25	3/2	0.925	stable			7	
		8	8.022485	7.4714	41.28	2		0.84 s	β^-	16.0	8	
		9	9.026789	8.4069	45.34	3/2		0.177 s	β^-	13.6	9	
		11	11.043908	10.2859	45.54	1/2		8.7 ms	β^-	20.7	11	
Beryllium	Be 4	7	7.016928	6.5342	37.60	3/2		53.28 d	EC	0.86	7	Be
Gr., berryllos (beryl)		9	9.012182	8.3928	58.17	3/2	1.00	stable			9	
		10	10.013534	9.3256	64.98	0		1.52 My	β^-	0.56	10	
		11	11.021658	10.2646	65.48	1/2		13.8 s	β^-	11.48	11	
		12	12.026921	11.2011	68.65	0		24 ms	β^-	11.71	12	
Boron	B 5	8	8.024605	7.4724	37.74	2		0.770 s	$\beta^+, 2\alpha$	11.15, 17.56	8	B
		10	10.012937	9.3245	64.75	3	0.199	stable			10	
		11	11.009305	10.2526	76.21	3/2	0.801	stable			11	
		12	12.014352	11.1888	79.58	1		0.0202 s	β^-	13.37	12	
		13	13.01778					0.0174	β^-	13.44	13	
Carbon	C 6	11	11.011433	10.2541	73.44	3/2		20.3 m	β^+	1.98	11	C
L., carbo (charcoal)		12	12.000000	11.1750	92.16	0	0.989	stable			12	
		13	13.003355	12.1096	97.11	1/2	0.011	stable			13	
		14	14.003241	13.0410	105.29	0		5715 y	β^-	0.016	14	

Element[a]	Z	A	Atomic mass (u)[b]	Nuclear mass (GeV/c^2)	E_b[c] (MeV)	J[d]	Natural Abundance	Half-life[e]	Decay[f]	Q^g (MeV)	A	N
Nitrogen	N 7	12	12.018613	11.1918	74.04	1		11 ms	β^+	17.34	12	N
Gr., nitron (soda) + genes		13	13.005738	12.1113	94.11	1/2		9.97 m	β^+	2.22	13	
(forming)		14	14.003074	13.0403	104.66	1	0.99634	stable			14	
		15	15.000108	13.9690	115.49	1/2	0.00366	stable			15	
		16	16.006099	14.9061	117.98	2		7.13 s	β^-	8.68	16	
Oxygen	O 8	13	13.02810					8.9 ms	β^+	17.77	13	O
Gr., oxys (acid) + genes		14	14.008595	13.0449	98.74	0		70.6 s	β^+	5.14	14	
(forming)		15	15.003065	13.9713	111.96	1/2		122 s	β^+	2.75	15	
		16	15.994915	14.8952	127.62	0	0.99762	stable			16	
		17	16.999131	15.8306	131.77	5/2	0.00038	stable			17	
		18	17.999160	16.7622	139.81	0	0.002	stable			18	
Fluorine	F 9	17	17.002095	15.8329	128.22	5/2		64.7 s	β^+	2.76	17	F
L., fluere (flow)		18	18.000937	16.7633	137.37	1		1.83 h	β^+	1.66	18	
		19	18.998403	17.6924	147.80	1/2	1.00	stable			19	
Neon	Ne 10	20	19.992436	18.6179	160.65	0	0.9051	stable			20	Ne
Gr., neos (new)		21	20.993843	19.5507	167.41	3/2	0.0027	stable			21	
		22	21.991383	20.4799	177.78	0	0.0922	stable			22	
		23	22.994465	21.4143	182.98	5/2		37.2 s	β^-	4.38	23	
		24	23.993613	22.3450	191.84	0		3.38 m	β^-	2.47	24	
Sodium	Na 11	22	21.994434	20.4822	174.15	3		2.605 y	β^+	2.84	22	Na
Medieval L., sodanum		23	22.989768	21.4094	186.57	3/2	1.00	stable			23	
(headache remedy)		24	23.990961	22.3420	193.53	4		14.97 h	β^-	5.51	24	
Magnesium	Mg 12	23	22.994124	21.4129	181.73	3/2		11.32 s	β^+	4.06	23	Mg
Gr., Magnesia (district		24	23.985042	22.3360	198.26	0	0.7899	stable			24	
in Thessaly)		25	24.985837	23.2682	205.59	5/2	0.1	stable			25	
		26	25.982594	24.1967	216.68	0	0.1101	stable			26	

Element	A	Mass (u)			I	Abundance	Half-life	Decay	Energy (MeV)	
	27	26.984341	25.1298	223.13	1/2		9.45 m	β^-	2.61	
	28	27.983877	26.0609	231.63	0		21.0 h	β^-	1.83	
Aluminum — L., alumen (Al 13)	26	25.986892	24.2002	211.90	5		0.71 My	β^+	4.01	Al
	27	26.981539	25.1267	224.95	5/2	1.00	stable			
	29	28.980446	26.9887	242.12	5/2		6.5 m	β^-	3.68	
Silicon — L., silicis (flint) (Si 14)	28	27.976927	26.0534	236.54	0	0.9223	stable			Si
	29	28.976495	26.9845	245.01	1/2	0.0467	stable			
	30	29.973770	27.9135	255.62	0	0.031	stable			
	31	30.975362	28.8464	262.21	3/2		2.62 h	β^-	1.49	
	32	31.974148	29.7768	271.41	0		100. y	β^-	0.23	
Phosphorus — Gr., phosphoros (light bearing) (P 15)	31	30.973762	28.8444	262.92	1/2	1.00	stable			P
	32	31.973907	29.7761	270.87	1		14.28 d	β^-	1.71	
	33	32.971725	30.7055	280.96	1/2		25.3 d	β^-	0.25	
Sulfer — L., sulphurium (S 16)	32	31.972071	29.7739	271.78	0	0.9502	stable			S
	33	32.971458	30.7048	280.43	3/2	0.0075	stable			
	34	33.967867	31.6329	291.84	0	0.0421	stable			
	35	34.969032	32.5655	298.83	3/2		87.2 d	β^-	0.17	
	36	35.967081	33.4952	308.71	0	0.0002	stable			
	38	37.971163	35.3620	321.06	0		2.84 m	β^-	2.94	
Chlorine — Gr., chloros (greenish-yellow) (Cl 17)	35	34.968853	32.5649	298.21	3/2	0.7577	stable			Cl
	36	35.968307	33.4958	306.79	2		0.3 My	β^+, β^-	0.071, 1.14	
	37	36.965903	34.4251	317.10	3/2	0.2423	stable			
	39	38.968004	36.2901	331.29	3/2		55.7 m	β^-	3.44	
Argon — Gr., argon (inactive) (Ar 18)	36	35.967546	33.4946	306.72	0	0.00337	stable			Ar
	38	37.962733	35.3531	327.35	0	6.3×10^{-4}	stable			
	39	38.964314	36.2861	333.95	7/2		268 y	β^-	0.57	
	40	39.962384	37.2158	343.81	0	0.996	stable			
	42	41.963049	39.0794	359.34	0		33 y	β^-	0.60	

Element[a]	Z	A	Atomic mass (u)[b]	Nuclear mass (GeV/c^2)	E_b[c] (MeV)	J[d]	Natural Abundance	Half-life[e]	Decay[f]	Q[g] (MeV)	A	
Potassium K 19		39	38.963707	36.2850	333.72	3/2	0.93258	stable			39	K
English, potash (pot ashes)		40	39.963999	37.2168	341.53	4	1.17×10^{-4}	1.26 Gy	β^-, β^+	1.31, 0.48	40	
		41	40.961825	38.1463	351.62	3/2	0.06702	stable			41	
		43	42.960717	40.0083	368.80	3/2		22.3 h	β^-	1.82	43	
Calcium Ca 20		40	39.962591	37.2150	342.06	0	0.96941	stable			40	Ca
L., calx (lime)		41	40.962278	38.1462	350.42	7/2		0.1 My EC		0.42	41	
		42	41.958618	39.0743	361.90	0	0.00647	stable			42	
		43	42.958766	40.0059	369.83	7/2	0.00135	stable			43	
		44	43.955481	40.9344	380.97	0	0.02086	stable			44	
		45	44.956185	41.8665	388.38	7/2		163.8 d	β^-	0.26	45	
		46	45.953690	42.7957	398.78	0	4.0×10^{-5}	stable			46	
		47	46.954543	43.7280	406.05	7/2		4.536 d	β^-	1.99	47	
		48	47.952533	44.6576	416.00	0	1.9×10^{-3}	stable			48	
Scandium Sc 21		41	40.969250	38.1522	343.14	7/2		0.596 s	β^+	6.50	41	Sc
L., Scandia (Scandinavia)		45	44.955910	41.8658	387.86	7/2	1	stable			45	
		46	45.955170	42.7966	396.62	4		83.8 d	β^-	2.37	46	
		47	46.952409	43.7255	407.26	7/2		3.42 d	β^-	0.60	47	
Titanium Ti 22		44	43.959690	40.9373	375.48	0		47 y	EC	0.27	44	Ti
Titans (first sons of the earth)		45	44.958124	41.8673	385.01	7/2		3.078 h	β^+	2.06	45	
		46	45.952629	42.7937	398.20	0	0.08	stable			46	
		47	46.951764	43.7244	407.08	5/2	0.073	stable			47	
		48	47.947947	44.6523	418.70	0	0.738	stable			48	
		49	48.947871	45.5838	426.85	7/2	0.055	stable			49	
		50	49.944792	46.5124	437.79	0	0.054	stable			50	
Vanadium V 23		49	48.948517	45.5839	425.46	7/2		331 d	EC	0.60	49	V
Vanadis (Scandinavian godess)		50	49.947161	46.5141	434.80	6	0.0025	$> 3.9 \times 10^{17}$ y	EC, β^-	2.21, 1.04	50	

Element	Z	A	Atomic mass	Mass (GeV)	B.E. (MeV)	Spin	Abundance	Half‑life	Decay	Energy
		51	50.943962	47.4426	445.85	7/2	0.9975	stable		
Chromium Cr 24 — Gr., chroma (color)	24	48	47.954033	44.6570	411.47	3/2		21.6 h	EC	1.65
		50	49.946046	46.5126	435.05	0	0.04345	stable		
		51	50.944768	47.4429	444.315	7/2		27.70 d	EC	0.75
		52	51.940510	48.3704	456.35	0	0.83789	stable		
		53	52.940651	49.3020	464.29	3/2	0.09501	stable		
		54	53.938883	50.2319	474.01	0	0.02365	stable		
Manganese Mn 25 — L., magnes (magnet)	25	53	52.941291	49.3021	462.9	7/2		3.7 My	EC	0.60
		54	53.940361	50.2328	471.85	3		312 d	β+	1.38
		55	54.938047	51.1621	482.08	5/2	1.00	stable		
		56	55.938907	52.0944	489.35	3		2.579 h	β-	3.70
Iron Fe 26 — L., ferrum	26	54	53.939613	50.2315	471.77	0	0.059	stable		
		55	54.938296	51.1618	481.07	3/2		2.7 y	EC	0.23
		56	55.934939	52.0902	492.26	0	0.9172	stable		
		57	56.935396	53.0221	499.91	1/2	0.021	stable		
		58	57.933277	53.9517	509.96	0	0.0028	stable		
		60	59.934077	55.8154	525.35	0		1.5 My	β-	0.24
Cobalt Co 27 — German, Kobold (goblin)	27	56	55.939841	52.0943	486.92	4		77.7 d	β+	4.57
		57	56.936294	53.0225	498.29	7/2		271 d	EC	0.84
		59	58.933198	54.8826	517.32	7/2	1.00	stable		
		60	59.933820	55.8147	524.81	5		5.272 y	β-	2.82
Nickel Ni 28 — German, Nickel (Satan)	28	56	55.942134	52.0959	484.00	0		6.10 d	EC	2.14
		58	57.935346	53.9526	506.46	0	0.6827	stable		
		59	58.934349	54.8831	515.46	3/2		76 ky	EC	1.07
		60	59.930788	55.8113	526.85	0	0.261	stable		
		61	60.931058	56.7431	534.67	3/2	0.0113	stable		
		62	61.928346	57.6720	545.27	0	0.0359	stable		
		63	62.929670	58.6048	552.11	1/2		100. y	β-	0.065

Element[a]	Z	A	Atomic mass (u)[b]	Nuclear mass (GeV/c²)	E_b[c] (MeV)	J[d]	Natural Abundance	Half-life[e]	Decay[f]	Q[g] (MeV)	A	
Copper Cu 29		64	63.927968	59.5347	561.76	0	0.0091	stable			64	Cu
	63		62.929599	58.6042	551.39	3/2	0.6917	stable			63	
L., cuprum (from island		64	63.929766	59.5359	559.31	1		12.7 h	β^+, β^-	0.58, 1.68	64	
		65	64.927793	60.4655	569.22	3/2	0.3083	stable			65	
of Cyprus)		67	66.927748	62.3285	585.40	3/2		2.58 d	β^-	0.58	67	
Zinc Zn 30		64	63.929145	59.5348	559.10	0	0.486	stable			64	Zn
German, Zink		65	64.929243	60.4664	567.08	5/2		243.8 d β^+		1.35	65	
		66	65.926035	61.3949	578.14	0	0.279	stable			66	
		67	66.927129	62.3274	585.20	5/2	0.041	stable			67	
		68	67.924846	63.2568	595.39	0	0.188	stable			68	
		70	69.925325	65.1202	611.09	0	0.006	stable			70	
		72	71.926856	66.9847	625.81	0		46.5 h	β^-	0.46	72	
Gallium Ga 31		67	66.928204	62.3279	583.41	3/2		3.26 d	EC	1.00	67	Ga
L., Gallia (France)		69	68.925580	64.1884	602.00	3/2	0.601	stable			69	
		71	70.924701	66.0506	618.96	3/2	0.399	stable			71	
		72	71.926365	66.9837	625.48	3		13.95 h	β^-	3.99	72	
Germanium Ge 32		68	67.928097	63.2588	590.80	0		288 d	EC	0.11	68	Ge
L., Germania (Germany)		70	69.924250	65.1182	610.53	0	0.205	stable			70	
		71	70.924954	66.0504	617.94	1/2		11.2 d	EC	0.24	71	
		72	71.922079	66.9792	628.69	0	0.274	stable			72	
		73	72.923463	67.9120	635.47	9/2	0.078	stable			73	
		74	73.921177	68.8413	645.68	0	0.365	stable			74	
		76	75.921402	70.7046	661.61	0	0.078	stable			76	
Arsenic As 33		73	72.923827	67.9118	634.35	3/2		80.3 d	EC	0.35	73	As
Gr., arsenikon (yellow pigment)		74	73.923928	68.8434	642.33	2		17.8 d	β^+, β^-	2.56, 1.35	74	
		75	74.921594	69.7727	652.58	3/2	1.00	stable			75	

Element	A	Atomic mass			I	%	Half-life	Decay		A	
Selenium Se 34 Gr., Selene (moon)	74	73.922475	68.8415	642.90	0	0.009	stable			74	Se
	76	75.919212	70.7015	662.08	0	0.09	stable			76	
	77	76.919913	71.6336	669.50	1/2	0.076	stable			77	
	78	77.917308	72.5627	680.00	0	0.236	stable			78	
	79	78.918498	73.4953	686.96	7/2		60 ky	β^-	0.15	79	
	80	79.916520	74.4250	696.88	0	0.499	stable			80	
	82	81.916698	76.2882	712.85	0	0.089	1.4×10^{20} y	double β^-	2.99	82	
Bromine Br 35 Gr., bromos (stench)	77	76.921377	71.6345	667.36	3/2		57.0 h	β^+	1.37	77	Br
	79	78.918336	73.4947	686.33	3/2	0.5069	stable			79	
	81	80.916289	75.3558	704.38	3/2	0.4931	stable			81	
	82	81.916802	76.2877	711.98	5		35.30 h	β^-	3.09	82	
Krypton Kr 36 Gr., kryptos (hidden)	78	77.920396	72.5646	675.56	0	0.0035	stable			78	Kr
	80	79.916380	74.4238	695.44	0	0.0225	stable			80	
	81	80.916590	75.3555	703.32	7/2		0.21 My	EC	0.28	81	
	82	81.913483	76.2841	714.28	0	0.116	stable			82	
	83	82.914135	77.2163	721.75	9/2	0.115	stable			83	
	84	83.911507	78.1453	732.27	0	0.57	stable			84	
	85	84.912532	79.0778	739.38	9/2		10.72 y	β^-	0.69	85	
	86	85.910615	80.0075	749.24	0	0.173	stable			86	
	90	89.919528	83.7418	773.23	0		32.3 s	β^-	4.39	90	
	92	91.926270	85.6111	783.09	0		1.84 s	β^-	6.16	92	
Rubidium Rb 37 L., rubidius (deepest red)	83	82.915143	77.2167	720.03	5/2		86.2 d	EC	0.96	83	Rb
	85	84.911794	79.0766	739.29	5/2	0.72165	stable			85	
	87	86.909187	80.9371	757.86	3/2	0.27835	stable			87	
Strontium Sr 38 Strontian (town in Scotland)	84	83.913430	78.1461	728.91	0	0.0056	stable			84	Sr
	85	84.912937	79.0771	737.44	9/2		64.8 d	β^+	1.08	85	
	86	85.909267	80.0052	748.93	0	0.0986	stable			86	
	87	86.908884	80.9363	757.36	9/2	0.07	stable			87	

Element[a]	Z	A	Atomic mass (u)[b]	Nuclear mass (GeV/c²)	E_b^c (MeV)	J^d	Natural Abundance	Half-life[e]	Decay[f]	Q^g (MeV)	A
		88	87.905619	81.8648	768.47	0	0.8258	stable			88
		90	89.907738	83.7298	782.64	0		29 y	β^-	0.55	90
Yttrium Y 39		88	87.909507	81.8679	764.07	4		106.61 d	EC	3.62	88
Swedish, Ytterby (a village)		89	88.905849	82.7960	775.55	1/2	1.00	stable			89
		91	90.907303	84.6604	790.34	1/2		58.5 d	β^-	1.55	91
Zirconium Zr 40		90	89.904703	83.7259	783.90	0	0.5145	stable			90
Arabic, zargun (gold color)		91	90.905644	84.6583	791.10	5/2	0.1122	stable			91
		92	91.905039	85.5893	799.73	0	0.1715	stable			92
		93	92.906474	86.5221	806.47	5/2		1.5 My	β^-	0.08	93
		94	93.906315	87.4534	814.69	0	0.1738	stable			94
		96	95.908275	89.3183	829.01	0	0.028	stable			96
Niobium Nb 41		92	91.907192	85.5907	796.95	7		37 My	EC	2.01	92
Niobe (daughter of Tantalus)		93	92.906377	86.5215	805.78	9/2	1.00	stable			93
		94	93.907281	87.4538	813.01	6		24 ky	β^-	2.04	94
Molybdenum Mo 42		92	91.906809	85.5899	796.52	0	0.1484	stable			92
Gr., molybdos (lead)		93	92.906813	86.5214	804.59	5/2		3.5 ky	EC	0.41	93
		94	93.905085	87.4513	814.27	0	0.0925	stable			94
		95	94.905841	88.3835	821.64	5/2	0.1592	stable			95
		96	95.904679	89.3139	830.79	0	0.1668	stable			96
		97	96.906020	90.2467	837.61	5/2	0.0955	stable			97
		98	97.905407	91.1776	846.26	0	0.2413	stable			98
		99	98.907711	92.1112	852.18	1/2		2.75 d	β^-	1.36	99
		100	99.907476	93.0425	860.47	0	0.0963	stable			100
Technetium Tc 43		97	96.906364	90.2465	836.51	9/2		2.6 My	EC	0.32	97
Gr., technetos (artificial)		98	97.907215	91.1788	843.79	6		4.2 My	β^-	1.79	98
Ruthenium Ru 44		96	95.907600	89.3156	826.51	0	0.0552	stable			96
L., Ruthenia (Russia)		98	97.905287	91.1764	844.80	0	0.0188	stable			98

Element	A	Decay	E (MeV)	$t_{1/2}$	Abundance	I	B (MeV)	M	Atomic mass (u)	A
	99			stable	0.127	5/2	852.27	92.1086	98.905939	99
	100			stable	0.126	0	861.94	93.0385	99.904219	100
	101			stable	0.17	5/2	868.74	93.9712	100.905582	101
	102			stable	0.316	0	877.96	94.9016	101.904349	102
	103	β^-	0.77	39.24 d		3/2	884.20	95.8349	102.906323	103
	104			stable	0.187	0	893.10	96.7656	103.905424	104
	106	β^-	0.039	1.02 y		0	907.48	98.6304	105.907322	106
Rhodium Rh 45 Gr., rhodon (rose)	101	EC	0.54	4.34 d		1/2	867.42	93.9713	100.906159	101
	102	EC	2.33	2.9 y		6	874.88	94.9034	101.906814	102
	103			stable	1.00	1/2	884.18	95.8336	102.905500	103
Palladium Pd 46 Gr., Pallas (godess of wisdom)	102			stable	0.0102	0	875.20	94.9018	101.905634	102
	103	EC	0.57	16.97 d		5/2	882.82	95.8337	102.906114	103
	104			stable	0.1114	0	892.84	96.7633	103.904029	104
	105			stable	0.2233	5/2	899.93	97.6957	104.905079	105
	106			stable	0.2733	0	909.49	98.6258	105.903478	106
	107	β^-	0.033	6.5 My		5/2	916.03	99.5588	106.905127	107
	108			stable	0.2546	0	925.25	100.4891	107.903895	108
	110			stable	0.1172	0	940.21	102.3533	109.905167	110
Silver Ag 47 L., argentum	105	EC	1.34	41.3 d		1/2	897.81	97.6966	104.906521	105
	107			stable	0.51839	1/2	915.28	99.5582	106.905092	107
	109			stable	0.48161	1/2	931.74	101.4209	108.904756	109
	111	β^-	1.04	7.47 d		1/2	947.38	103.2844	110.905295	111
Cadmium Cd 48 L., calx (lime)	106			stable	0.0125	0	905.15	98.6275	105.906461	106
	108			stable	0.0089	0	923.42	100.4884	107.904176	108
	109	EC	0.18	462.3 d		5/2	930.77	101.4206	108.904953	109
	110			stable	0.1249	0	940.67	102.3503	109.903005	110
	111			stable	0.128	1/2	947.63	103.2829	110.904182	111
	112			stable	0.2413	0	957.03	104.2131	111.902757	112

Element[a]	Z	A	Atomic mass (u)[b]	Nuclear mass (GeV/c²)	E_b^c (MeV)	J^d	Natural Abundance	Half-life[e]	Decay[f]	Q^g (MeV)	A	
		113	112.904400	105.1461	963.57	1/2	0.1222	stable			113	
		114	113.903357	106.0766	972.61	0	0.2873	stable			114	
		116	115.904755	107.9409	987.46	0	0.0749	stable			116	In
Indium	In 49	111	110.905109	103.2832	945.99	9/2		2.806 d	EC	0.86	111	
indigo		113	112.904061	105.1453	963.10	9/2	0.043	stable			113	
		115	114.903882	107.0081	979.41	9/2	0.957	4.4×10^{14} y	β^-	0.50	115	
Tin	Sn 50	112	111.904827	104.2140	953.54	0	0.0097	stable			112	Sn
L., stannum		114	113.902784	106.0751	971.58	0	0.0065	stable			114	
		115	114.903347	107.0071	979.13	1/2	0.0036	stable			115	
		116	115.901747	107.9371	988.69	0	0.1453	stable			116	
		117	116.902956	108.8697	995.64	1/2	0.0768	stable			117	
		118	117.901609	109.8000	1005.0	0	0.2422	stable			118	
		119	118.903311	110.7331	1011.4	1/2	0.0858	stable			119	
		120	119.902199	111.6635	1020.6	0	0.3259	stable			120	
		122	121.903440	113.5277	1035.5	0	0.0463	stable			122	
		123	122.905722	114.4613	1041.5	11/2		129.2 d	β^-	1.40	123	
		124	123.905274	115.3924	1050.0	0	0.0579	stable			124	
		126	125.907653	117.2576	1063.9	0		0.1 My	β^-	0.38	126	
Antimony	Sb 51	121	120.903821	112.5960	1026.3	5/2	0.573	stable			121	Sb
L., stibium (mark)		123	122.904216	114.4594	1042.1	7/2	0.427	stable			123	
		124	123.905938	115.3925	1048.6	3		60.2 d	β^-	2.91	124	
		125	124.905252	116.3234	1057.3	7/2		2.76 y	β^-	0.77	125	
Tellurium	Te 52	120	119.904047	111.6642	1017.3	0	0.00096	stable			120	Te
L., tellus (earth)		122	121.903050	113.5263	1034.3	0	0.026	stable			122	
		123	122.904271	114.4590	1041.3	1/2	0.00908	1.3×10^{13} y	EC	0.052	123	
		124	123.902823	115.3891	1050.7	0	0.04816	stable			124	

Element	A	Atomic mass			I	Abundance	Half-life	Decay		A
	125	124.904433	116.3221	1057.3	1/2	0.0714	stable			125
	126	125.903314	117.2526	1066.4	0	0.1895	stable			126
	128	127.904463	119.1166	1081.5	0	0.3169	stable			128
	130	129.906229	120.9813	1096.0	0	0.338	2.5×10^{21} y	double β^-	2.53	130
Iodine I 53	125	124.904620	116.3218	1056.3	5/2		59.9 d	β^-	0.18	125
Gr., iodes (violet)	127	126.904473	118.1846	1072.6	5/2	1.00	stable			127	I
	129	128.904986	120.0481	1088.3	7/2		17 My	β^-	0.19	129
Xenon Xe 54	124	123.905894	115.3909	1046.3	0	0.001	stable			124	Xe
Gr., xenon (stranger)	126	125.904281	117.2524	1063.9	0	0.0009	stable			126
	127	126.905182	118.1848	1071.1	1/2		36.2 d	EC	0.66	127
	128	127.903531	119.1147	1080.8	0	0.0191	stable			128
	129	128.904780	120.0474	1087.7	1/2	0.264	stable			129
	130	129.903509	120.9777	1096.9	0	0.041	stable			130
	131	130.905072	121.9107	1103.5	3/2	0.212	stable			131
	132	131.904144	122.8413	1112.5	0	0.269	stable			132
	133	132.905889	123.7745	1118.9	3/2	0.104	5.25 d	β^-	0.43	133
	134	133.905395	124.7055	1127.5	0	0.104	stable			134
	136	135.907213	126.5702	1141.9	0	0.089	stable			136
Cesium Cs 55	133	132.905429	123.7735	1118.6	7/2	1.00	stable			133	Cs
L., caesius (sky blue)	135	134.905885	125.6369	1134.3	7/2		2.3 My	β^-	0.21	135
	137	136.907074	127.5011	1149.3	7/2		30.17 y	β^-	1.17	137
Barium Ba 56	130	129.906281	120.9793	1092.8	0	0.00106	stable			130	Ba
Gr., barys (heavy)	132	131.905043	122.8411	1110.1	0	0.00101	stable			132
	133	132.905988	123.7735	1117.3	1/2		10.53 y	EC	0.52	133
	134	133.904485	124.7036	1126.7	0	0.02417	stable			134
	135	134.905665	125.6362	1133.7	3/2	0.06592	stable			135
	136	135.904553	126.5667	1142.8	0	0.07854	stable			136
	137	136.905812	127.4994	1149.7	3/2	0.1123	stable			137

Element[a]	Z	A	Atomic mass (u)[b]	Nuclear mass (GeV/c²)[b]	E_b[c] (MeV)	J[d]	Natural Abundance	Half-life[e]	Decay[f]	Q[g] (MeV)	A	
		138	137.905233	128.4303	1158.3	0	0.717	stable			138	
		140	139.910581	130.2983	1169.5	0		12.76 d	β^-	1.03	140	
		142	141.916361	132.1667	1180.2	0		10.7 m	β^-	2.13	142	
		143	142.920483	133.1020	1184.5	5/2		15 s	β^-	4.2	143	
		144	143.922845	134.0357	1190.3	0		11.5 s	β^-	3.0	144	
Lanthanum La 57		137	136.906463	127.4995	1148.3	7/2		60 ky	EC	0.61	137	La
Gr., lanthanein (hidden)		138	137.907106	128.4316	1155.8	5	0.0009	1.06×10^{11} y	β^+, β^-	1.04, 0.72	138	
		139	138.906347	129.3624	1164.6	7/2	0.9991	stable			139	
Cerium Ce 58		136	135.907139	126.5681	1138.8	0	0.0019	stable			136	Ce
Ceres (the asteroid)		138	137.905985	128.4300	1156.1	0	0.0025	stable			138	
		139	138.906631	129.3621	1163.5	3/2		137.2 d	EC		139	
		140	139.905433	130.2925	1172.7	0	0.8848	stable			140	
		142	141.909241	132.1590	1185.3	0	0.1108	stable			142	
		144	143.913643	134.0262	1197.4	0		284.4 d	β^-	0.32	144	
Praseodymium Pr 59		141	140.907647	131.2255	1177.9	5/2	1.00	stable			141	Pr
Gr., prasios (green) +		142	141.910039	132.1593	1183.8	2		19.13 h	β^-	2.16	142	
didymos (twin)		143	142.910814	133.0915	1191.1	7/2		13.58 d	β^-	0.93	143	
Neodymium Nd 60		142	141.907720	132.1566	1185.2	0	0.2713	stable			142	Nd
Gr., neos (new) +		143	142.909810	133.0901	1191.3	7/2	0.1218	stable			143	
didymos (twin)		144	143.910084	134.0218	1199.1	0	0.238	2.1×1015 y	α	1.91	144	
		145	144.912570	134.9556	1204.9	7/2	0.083	stable			145	
		146	145.913113	135.8876	1212.4	0	0.1719	stable			146	
		148	147.916889	137.7542	1225.1	0	0.0576	stable			148	
		150	149.920887	139.6209	1237.5	0	0.0564	stable			150	
Promethium Pm 61		145	144.912743	134.9553	1203.9	5/2		17.7 y	EC	0.16	145	Pm
Prometheus		146	145.914708	135.8886	1210.2	3		5.53 y	EC	1.48	146	
Samarium Sm 62		144	143.911998	134.0226	1195.8	0	0.031	stable			144	Sm

Element	Symbol Z	A	Atomic mass			Spin	Abundance	Half-life	Decay	Energy	A	
Samarskite (a mineral)		147	146.914894	136.8198	1217.3	7/2	0.15	1.06×10^{11} y	α	2.31	147	
		148	147.914819	137.7512	1225.4	0	0.113	7×10^{15} y	α	1.99	148	
		149	148.917180	138.6849	1231.3	7/2	0.138	10^{16} y	α	1.87	149	
		150	149.917272	139.6165	1239.3	0	0.074	stable			150	
		152	151.919729	141.4818	1253.1	0	0.267	stable			152	
		154	153.922205	143.3471	1267.0	0	0.227	stable			154	
Europium Eu 63		151	150.919847	140.5499	1244.2	5/2	0.478	stable			151	Eu
Europe		152	151.921742	141.4832	1250.5	3		13.4 y	EC, β^-	1.88, 1.82	152	
		153	152.921225	142.4142	1259.0	5/2	0.522	stable			153	
		154	153.922975	143.3473	1265.5	3		8.5 y	β^-	1.98	154	
Gadolinium Gd 64		150	149.918663	139.6168	1236.4	0		1.8 My	α	2.80	150	Gd
Gadolin (Finnish chemist)		152	151.919787	141.4808	1251.5	0	0.002	stable			152	
		154	153.920861	143.3448	1266.6	0	0.0218	stable			154	
		155	154.922617	144.2780	1273.1	3/2	0.148	stable			155	
		156	155.922118	145.2090	1281.6	0	0.2047	stable			156	
		157	156.923956	146.1422	1288.0	3/2	0.1565	stable			157	
		158	157.924099	147.0738	1295.9	0	0.2484	stable			158	
		160	159.927049	148.9396	1309.3	0	0.2186	stable			160	
Terbium Tb 65		157	156.924023	146.1418	1287.1	3/2		110 y	EC	0.058	157	Tb
Swedish, Ytterby (a village)		158	157.925411	147.0746	1293.9	3		180 y	EC, β^-	1.22, 0.94	158	
		159	158.925342	148.0060	1302.1	3/2	1.00	stable			159	
Dysprosium Dy 66		154	153.924428	143.3471	1261.8	0		3 My	α	2.95	154	Dy
Gr., dysprositos (hard to get at)		156	155.924277	145.2100	1278.0	0	0.0006	stable			156	
		158	157.924403	147.0731	1294.1	0	0.001	stable			158	
		159	158.925735	148.0059	1300.9	3/2		144 d	EC	0.37	159	
		160	159.925193	148.9368	1309.5	0	0.0234	stable			160	
		161	160.926930	149.8700	1315.9	5/2	0.189	stable			161	
		162	161.926796	150.8013	1324.1	0	0.255	stable			162	

Element[a]	Z	A	Atomic mass (u)[b]	Nuclear mass (GeV/c^2)	E_b[c] (MeV)	J[d]	Natural Abundance	Half-life[e]	Decay[f]	Q[g] (MeV)	A	
		163	162.928728	151.7346	1330.4	5/2	0.249	stable			163	
		164	163.929171	152.6666	1338.1	0	0.282	stable			164	
Holmium Ho 67		163	162.928731	151.7341	1329.6	7/2		4.57 ky	EC	0.004	163	Ho
L., Holmia (Stockholm)		165	164.930319	153.5986	1344.3	7/2	1.00	stable			165	
		166	165.932280	154.5319	1350.5	0		1.117 d	β^-	1.85	166	
Erbium Er 68		162	161.928774	150.8022	1320.7	0	0.0014	stable			162	Er
Swedish, Ytterby (a village)		164	163.929198	152.6656	1336.5	0	0.0161	stable			164	
		166	165.930290	154.5296	1351.6	0	0.336	stable			166	
		167	166.932046	155.4627	1358.0	7/2	0.2295	stable			167	
		168	167.932368	156.3945	1365.8	0	0.268	stable			168	
		169	168.934588	157.3281	1371.8	1/2		9.4 d	β^-	0.35	169	
		170	169.935461	158.2604	1379.1	0	0.149	stable			170	
		172	171.939353	160.1270	1391.6	0		2.05 d	β^-	0.89	172	
Thulium Tm 69		169	168.934212	157.3272	1371.4	1/2	1.00	stable			169	Tm
Thule (Scandinavia)		170	169.935798	158.2602	1378.0	1		128.6 d	β^-	0.97	170	
		171	170.936426	159.1923	1385.5	1/2		1.92 y	β^-	0.096	171	
Ytterbium Yb 70		168	167.933894	156.3949	1362.8	0	0.0013	stable			168	Yb
Swedish, Ytterby (a village)		169	168.935186	157.3276	1369.7	7/2		32.02 d	EC	0.91	169	
		170	169.934759	158.2587	1378.2	0	0.0305	stable			170	
		171	170.936323	159.1917	1384.8	1/2	0.143	stable			171	
		172	171.936378	160.1232	1392.8	0	0.219	stable			172	
		173	172.938207	161.0564	1399.2	5/2	0.1612	stable			173	
		174	173.938859	161.9886	1406.6	0	0.318	stable			174	
		175	174.941273	162.9223	1412.4	7/2		4.19 d	β^-	0.47	175	
		176	175.942564	163.8550	1419.3	0	0.127	stable			176	
Lutetium Lu 71		174	173.940336	161.9894	1404.5	1		3.31 y	EC	1.38	174	Lu
Lutetia (ancient name for Paris)		175	174.940770	162.9213	1412.1	7/2	0.9741	stable			175	

Element	Z	A	Atomic mass			Spin	Abundance	Half-life	Decay		A	
Hafnium Hf, L., Hafnia (Copenhagen)	72	176	175.942679	163.8546	1418.4	7	0.0259	3.8×10^{10} y	β^-	1.19	176	Hf
		174	173.940044	161.9886	1404.0	0	0.00162	2×10^{15} y	α	2.50	174	
		176	175.941405	163.8529	1418.8	0	0.05206	stable			176	
		177	176.943217	164.7861	1425.2	7/2	0.18606	stable			177	
		178	177.943696	165.7180	1432.8	0	0.27297	stable			178	
		179	178.945812	166.6515	1438.9	9/2	0.13629	stable			179	
		180	179.946546	167.5837	1446.3	0	0.351	stable			180	
		182	181.950550	169.4504	1458.7	0		9 My	β^-	0.37	182	
Tantalum Ta, Gr., Tantalos (father of Niobe)	73	179	178.945930	166.6511	1438.0	7/2		1.82 y	EC	0.11	179	Ta
		180	179.947462	167.5840	1444.7	1		8.15 h	EC, β^-	0.87, 0.71	180	
		181	180.947993	168.5160	1452.3	7/2	0.99998	stable			181	
		182	181.950148	169.4496	1458.3	3		115 d	β^-	1.81	182	
Tungsten W, Swedish, tung sten (heavy stone)	74	180	179.946701	167.5828	1444.6	0	0.0012	stable			180	W
		182	181.948202	169.4472	1459.4	0	0.263	stable			182	
		183	182.950220	170.3806	1465.6	1/2	0.143	stable			183	
		184	183.950929	171.3128	1473.0	0	0.3067	stable			184	
		186	185.954356	173.1790	1485.9	0	0.286	stable			186	
Rhenium Re, L., Rhenus (Rhine)	75	183	182.950817	170.3807	1464.2	5/2		70 d	EC	0.56	183	Re
		185	184.952951	172.2456	1478.4	5/2	0.374	stable			185	
		187	186.955745	174.1113	1491.9	5/2	0.626	4.2×10^{10} y	β^-	0.0025	187	
Osmium Os, Gr., osme (smell)	76	184	183.952489	171.3132	1469.9	0	0.0002	stable			184	Os
		186	185.953830	173.1775	1484.8	0	0.0158	2×10^{15} y	α	2.82	186	
		187	186.955741	174.1107	1491.1	1/2	0.016	stable			187	
		188	187.955829	175.0423	1499.1	0	0.133	stable			188	
		189	188.958137	175.9760	1505.0	3/2	0.161	stable			189	
		190	189.958436	176.9078	1512.8	0	0.264	stable			190	
		192	191.961468	178.7736	1526.2	0	0.41	stable			192	
Iridium Ir	77	189	188.958712	175.9760	1503.7	3/2		13.2 d	EC	0.54	189	Ir

Element^a	Z	A	Atomic mass (u)^b	Nuclear mass (GeV/c²)	E_b^c (MeV)	J^d	Natural Abundance	Half-life^e	Decay^f	Q^g (MeV)	A
L., iris (rainbow)		191	190.960584	177.8407	1518.1	3/2	0.373	stable			191
		192	191.962580	178.7741	1524.3	4		73.83 d	β^-	1.45	192
		193	192.962917	179.7059	1532.1	3/2	0.627	stable			193
Platinum Pt 78		190	189.959916	176.9081	1509.9	0	0.0001	6.5×10^{11} y	α	3.24	190
Sp., platina (silver)		192	191.961019	178.7721	1525.0	0	0.0079	stable			192
		193	192.962977	179.7055	1531.3	1/2		60 y	EC	0.057	193
		194	193.962655	180.6367	1539.6	0	0.329	stable			194
		195	194.964765	181.5701	1545.7	1/2	0.338	stable			195
		196	195.964927	182.5018	1553.7	0	0.253	stable			196
		198	197.967869	184.3675	1567.1	0	0.072	stable			198
Gold Au 79		195	194.965012	181.5699	1544.7	3/2		186.1 d	EC	0.23	195
L., aurum (shining dawn)		196	195.996543	182.5307	1523.4	2		6.18 d	EC, β^-	0.51, 0.69	196
		197	196.966543	183.4343	1559.4	3/2	1.00	stable			197
Mercury Hg 80		194	193.965391	180.6382	1535.5	0		520 y	EC	0.04	194
Mercury (the planet)		196	195.965807	182.5016	1551.3	0	0.0015	stable			196
		198	197.966743	184.3655	1566.5	0	0.100	stable			198
		199	198.968254	185.2984	1573.2	1/2	0.1689	stable			199
		200	199.968300	186.2299	1581.2	0	0.231	stable			200
		201	200.970277	187.1633	1587.5	3/2	0.132	stable			201
		202	201.970617	188.0951	1595.2	0	0.298	stable			202
		203	202.972848	189.0287	1601.2	5/2		46.6 d	β^-	0.49	203
		204	203.973466	189.9607	1608.7	0	0.069	stable			204
Thallium Tl 81		202	201.972085	188.0959	1593.1	2		12.23 d	β^+	1.37	202
Gr., thallos (green twig)		203	202.972320	189.0276	1600.9	1/2	0.30	stable			203
		204	203.973839	189.9606	1607.6	2		3.78 y	β^-, EC	0.76, 0.35	204
		205	204.974400	190.8926	1615.1	1/2	0.70	stable			205
		207	206.977404	192.7584	1628.5	1/2		4.77 m	β^-	1.43	207

Element	Z	A	mass		spin		half-life	decay	energy	A	symbol
		208	207.981988	193.6942	5		3.052 m	β^-	4.99	208	Pb
		209	208.985333	194.6288	1/2		2.20 m	β^-	3.98	209	
Lead	Pb 82	204	203.973020	189.9593	0	0.0014	stable			204	
L., plumbum		205	204.974458	190.8921	5/2		15.1 My	EC	0.053	205	
		206	205.974440	191.8236	0	0.241	stable			206	
		207	206.975871	192.7565	1/2	0.221	stable			207	
		208	207.976627	193.6887	0	0.524	stable			208	
		209	208.981065	194.6243	9/2		3.25 h	β^-	0.64	209	
		210	209.984163	195.5587	0		22.6 y	β^-	0.063	210	
		211	210.988735	196.4944	9/2		36.1 m	β^-	1.38	211	
		212	211.991871	197.4289	0		10.64 h	β^-	0.57	212	
		214	213.999798	199.2993	0		26.8 m	β^-	1.03	214	
Bismuth	Bi 83	207	206.978446	192.7583	9/2		32.2 y	EC	2.40	207	Bi
German, Weisse Masse		208	207.979717	193.6910	5		0.368 My	EC	2.88	208	
(white mass)		209	208.980374	194.6231	9/2	1.00	stable			209	
		210	209.984096	195.5581	1		5.01 d	β^-	1.16	210	
		211	210.987255	196.4926	9/2		2.14 m	α	8.69	211	
		212	211.991255	197.4278	1		1.009 h	β^-, α	2.25	212	
		214	213.998691	199.2977	1		19.9 m	β^-	3.27	214	
Polandium	Po 84	192	191.991443	178.7974	0		34 ms	α	7.33	192	Po
Poland		194	193.988180	180.6574	0		0.7 s	α	6.99	194	
		196	195.985539	182.5179	0		6 s	α	6.66	196	
		207	206.981570	192.7607	5/2		5.83 h	β^+	2.91	207	
		208	207.981222	193.6919	0		2.898 y	α	5.21	208	
		209	208.982404	194.6245	1/2		105 y	α	4.98	209	
		210	209.982848	195.5564	0		138.4 d	α	5.41	210	
		211	210.986627	196.4915	9/2		0.52 s	α	7.59	211	
		212	211.988842	197.4250	0		0.3 ms	α	8.95	212	

Element[a]	Z	A	Atomic mass (u)[b]	Nuclear mass (GeV/c^2)	E_b[c] (MeV)	J[d]	Natural Abundance	Half-life[e]	Decay[f]	Q[g] (MeV)	A	
		213	212.992833	198.3602	1660.2	9/2		4 ms	α	8.54	213	
		214	213.995176	199.2939	1666.1	0		163 ms	α	7.83	214	
		215	214.999418	200.2294	1670.2	9/2		1.78 ms	α	7.53	215	
		216	216.001888	201.1632	1676.0	0		0.15 s	α	6.91	216	
		218	218.008966	203.0328	1685.5	0		3.11 m	α	6.11	218	
Astatine At 85		207	206.985733	192.7641	1617.6	9/2		1.81 h	β^+, α	3.88, 5.87	207	At
Gr., astatos (unstable)		210	209.987126	195.5599	1640.5	5		8.1 h	EC, α	3.98, 5.63	210	
		211	210.987470	196.4917	1648.2	9/2		7.21 h	EC, α	0.78, 5.98	211	
		213	212.992911	198.3598	1659.3	9/2		0.11 ms	α	9.25	213	
		215	214.998638	200.2281	1670.1	9/2		0.1 ms	α	8.18	215	
		217	217.004695	202.0968	1680.6	9/2		32.3 ms	α	7.20	217	
Radon Rn 86		210	209.989669	195.5618	1637.3	0		2.4 h	α, EC	6.16, 2.37	210	Rn
L., nitens (shining)		211	210.990575	196.4941	1644.6	1/2		14.6 h	β^+, α	2.89, 5.96	211	
		212	211.990697	197.4257	1652.5	0		24 m	α	6.39	212	
		213	212.996347	198.3625	1655.3	9/2		25 ms	α	8.24	213	
		214	213.995339	199.2931	1664.3	0		0.27 ms	α	9.21	214	
		219	219.009478	203.9637	1691.5	5/2		3.96 s	α	6.95	219	
		220	220.011368	204.8970	1697.8	0		55.6 s	α	6.40	220	
		222	222.017571	206.7658	1708.2	0		3.823 d	α	5.59	222	
Francium Fr 87		220	220.012293	204.8973	1696.2	1		27.4 s	α	6.80	220	Fr
France		221	221.014230	205.8307	1702.5	5/2		4.9 m	α	6.46	221	
		222	222.017563	206.7653	1707.4	2		14.4 m	β^-	2.06	222	
		223	223.019733	207.6988	1713.5	3/2		21.8 m	β^-	1.15	223	
Radium Ra 88		206	206.003800	191.8479	1590.3	0		0.4 s	α	7.42	206	Ra
L., radius (ray)		216	216.003509	201.1626	1671.3	0		0.18 ms	α	9.53	216	
		218	218.007118	203.0290	1684.1	0		14 ms	α	8.55	218	
		220	220.011004	204.8956	1696.6	0		23 ms	α	7.59	220	

222	5.59	α	38 s		0	1708.7	206.7627	222	222.015353
223	5.98	α	11.43 d		3/2	1713.9	207.6971	223	223.018501
224	5.79	α	3.66 d		0	1720.4	208.6302	224	224.020186
225	0.37	β^-	14.8 d		3/2	1725.2	209.5649	225	225.023604
226	4.87	α	1.599 ky		0	1731.6	210.4981	226	226.025403
228	0.045	β^-	5.75 y		0	1742.5	212.3663	228	228.031064

Actinium Ac 89 Gr., aktis (ray) — Ac

225	5.94	α	10.0 d		3/2	1724.8	209.5640	225	225.023204
227	0.042, 5.04	β^-, α	21.77 y		3/2	1736.7	211.4312	227	227.027750
228	2.14	β^-	6.13 h		3	1741.8	212.3658	228	228.031014
231	2.10	β^-	7.5 m		1/2	1759.0	215.1673	231	231.038551

Thorium Th 90 Thor (Scandinavian war god) — Th

227	6.15	α	18.72 d		3/2	1736.0	211.4307	227	227.027703
228	5.52	α	1.912 y		0	1743.1	212.3631	228	228.028715
229	5.17	α	7.9 ky		5/2	1748.4	213.2975	229	229.031755
230	4.77	α	75.4 ky		0	1755.2	214.2302	230	230.033128
231	0.39	β^-	35.2 h		5/2	1760.3	215.1647	231	231.036298
232	4.08	α	1.4×10^{10} y	1.00	0	1766.7	216.0978	232	232.038051
234	0.27	β^-	24.10 d		0	1777.7	217.9660	234	234.043593

Protactinium Pa 91 Gr., protos (first) — Pa

228	2.10, 6.96	EC, α	22 h		3	1740.4	212.3645	228	228.030773
231	5.15	α	32.5 ky		3/2	1759.9	215.1638	231	231.035880
233	0.57	β^-	27.0 d		3/2	1772.0	217.0309	233	233.040242
234	2.20	β^-	6.70 h		4	1777.2	217.9652	234	234.043303

Uranium U 92 Uranus (the planet) — U

226	7.56	α	0.5 s		0	1725.0	210.4995	226	226.029170
231	0.36	EC	4.2 d		5/2	1758.7	215.1636	231	231.036270
232	5.41	α	68.9 d		0	1766.0	216.0959	232	232.037130
233	4.91	α	0.159 My		5/2	1771.8	217.0298	233	233.039628
234	4.86	α	0.245 My	5.5×10^{-5}	0	1778.6	217.9625	234	234.040947
235	4.68	α	0.704 Gy	0.0072	7/2	1783.9	218.8968	235	235.043924
236	4.57	α	23.4 My		0	1790.4	219.8298	236	236.045563

Element[a]	Z	A	Atomic mass (u)[b]	Nuclear mass (GeV/c^2)	$E_b{}^c$ (MeV)	J^d	Natural Abundance	Half-life[e]	Decay[f]	Q^g (MeV)	A	
		238	238.050785	221.6977	1801.7	0	0.99275	4.46 Gy	α	4.27	238	
		239	239.054290	222.6324	1806.5	5/2		23.54 m	β^-	1.26	239	
Neptunium	Np 93	231	231.038239	215.1650	1756.1	5/2		48.8 m	EC, α	1.84, 6.37	231	Np
Neptune (the planet)		236	236.046550	219.8302	1788.7	6		0.12 My	EC, β^-	0.99, 0.54	236	
		237	237.048168	220.7632	1795.3	5/2		2.14 My	β^-	4.96	237	
		239	239.052933	222.6307	1807.0	5/2		2.35 d	β^-	0.72	239	
Plutonium	Pu 94	239	239.052158	222.6294	1806.9	1/2		24.11 ky	β^-	5.24	239	Pu
Pluto (the planet)		242	242.058737	225.4301	1825.0	0		0.376 My	α	4.98	242	
		244	244.064198	227.2982	1836.1	0		82 My	α	4.67	244	
Americium	Am 95	241	241.056824	224.4963	1818.0	5/2		432.2 y	α	5.64	241	Am
America		243	243.061375	226.3635	1829.9	5/2		7.37 ky	α	5.44	243	
Curium	Cm 96	240	240.055503	223.5630	1810.3	0		27 d	α	6.40	240	Cm
Marie and Pierre Curie		247	247.070347	230.0974	1853.0	9/2		15.6 My	α	5.35	247	
		248	248.072343	231.0307	1859.2	0		0.34 My	α, SF	5.16	248	
		250	250.078352	232.8993	1869.8	0		9.7 ky	SF, α, β^-	5.27, 1.42	250	
Berkelium	Bk 97	247	247.070300	230.0968	1852.3	3/2		1.4 ky	α	5.89	247	Bk
Berkeley, Ca.		248	248.073107	231.0309	1857.7	1		23.7 h	β^-, EC	0.71, 0.86	248	
Californium	Cf 98	249	249.074845	231.9635	1863.4	9/2		351 y	α	6.30	249	Cf
California		251	251.079579	233.8310	1875.1	1/2		890 y	α	6.17	251	
Einsteinium	Es 99	252	252.082945	234.7651	1879.2	5		1.29 y	α, EC	6.74, 1.12	252	Es
Albert Einstein		254	254.088019	236.6328	1890.7	7		275 d	α	6.62	254	
Fermium	Fm 100	253	253.075173	235.6888	18893.8	1/2		3.0 d	EC, α	0.33, 7.20	253	Fm
Enrico Fermi		257	257.095099	239.4334	1907.5	9/2		100.5 d	α	6.86	257	
Mendelevium	Md 101	257	257.095580	239.4333	1906.3	7/2		5.5 h	EC, α	0.45, 7.60	257	Md
Dmitri Mendeleev		258	258.098570	240.3676	1911.6	8		51.5 d	α	7.40	258	
Nobelium	No 102	255	255.093260	237.5677	1891.5	1/2		3.1 m	α, EC	8.45	255	No
Alfred Nobel		258	258.098150	240.3667	1911.2	0		1.2 ms	SF		258	

Name	Symbol	Z	A	Atomic mass (u)	(b)	Binding energy (c)	Spin (d)	Half-life (e)	Decay mode (f)	Energy (g)
			259	259.100931	241.3008	1916.7	9/2	58 m	α, EC	7.79
Lawrencium	Lr	103	256	256.098490	238.5035	1894.0		28 s	α, SF	8.55
Ernest O. Lawrence			260	260.105320	242.2359	1919.9		3 m	α	8.3
Unnilquadium "one-zero-four"	Unq	104	257	257.102950	239.4387	1897.1		4.8 s	α	9.2
			261	261.108690	243.1700	1924.0		65 s	α	8.6
Unnilpentium "one-zero-five"	Unp	105	258	258.109020	240.3753	1898.7		4 s	α	9.4
			262	262.113760	244.1057	1926.6		34 s	SF, α	8.8
Unnilhexium "one-zero-six"	Unh	106	259	259.1144	241.3113	1901.0		0.5 s	α, SF	5.4
			260	260.11442	242.2397	1912.2		4 ms	α, SF	6.8
			261	261.1161	243.1719	1919.6		0.3 s	α, SF	5.9
			263	263.1182	245.0409	1929.7		0.8 s	SF, α	9.4
Unnilseptium "one-zero-seven"	Uns	107	261	261.1217	243.1806	1909.6		12 ms	α, SF	10.6
			262	262.1231	244.1133	1916.5		0.1 s	α	10.5
Unniloctium "one-zero-eight"	Uno	108	264	264.129	245.9814	1926.2		80 ms	α, SF	14.3
			265	265.1300	246.9140	1933.2		2 ms	α	14.7
Unnilennium "one-zero-nine"	Une	109	266	266.1378	247.8519	1933.5		3.4 ms	α	11.3

Notes:

(a) Gr. = Greek, L. = Latin

(b) 1 u ≡ 1/12 of ^{12}C atomic mass = 931.4943 MeV/c^2

(c) Nuclear binding energy

(d) Nuclear intrinsic angular momentum quantum number

(e) Half-life, mean lifetime multiplied by ln2

(f) EC = electron capture, SF = spontaneous fission

(g) Energy released in nuclear decay

TABLE OF
PARTICLE
PROPERTIES

FUNDAMENTAL BOSONS

		j	Mass energy	Width	Main Decay mode
Photon	γ	1	$< 3 \times 10^{-27}$ eV	stable	
Intermediate Vector Besons	W^+	1	80.2 GeV	2.1 GeV	$e^+ + \nu$ $\mu^+ + \nu$ $\tau^+ + \nu$ hadrons
	W^-	1	80.2 GeV	2.1 GeV	$e^- + \bar{\nu}$ $\mu^- + \bar{\nu}$ $\tau^- + \bar{\nu}$ hadrons
	Z^0	1	91.17 GeV	2.49 GeV	$e^+ + e^-$ $\mu^+ + \mu^-$ $\tau^+ + \tau^-$ hadrons

LEPTONS

		j	Mass energy	Lifetime	Main Decay mode
Neutrinos	ν_e $\bar{\nu}_e$	1/2	< 7 eV	stable	
	ν_μ $\bar{\nu}_\mu$	1/2	< 0.27 MeV	stable	
	ν_t $\bar{\nu}_\tau$	1/2	< 35 MeV	stable	
Electron	e^- e^+	1/2	0.511 MeV	stable	

LEPTONS

		j	Mass energy	Lifetime	Main Decay mode
Muon	μ^-	1/2	106 MeV	2.20×10^{-6} s	$e^- + \bar{\nu}_e + \nu_\mu$
	μ^+				$e^+ + \nu_e + \bar{\nu}_\mu$
Tau	τ^-	1/2	1777 MeV	3.1×10^{-13} s	$e^- + \bar{\nu}_e + \nu_\tau$
					$\mu^- + \bar{\nu}_\mu + \nu_\tau$
					$\nu_\tau +$ hadrons
	τ^+				$e^+ + \nu_e + \bar{\nu}_\tau$
					$\bar{\nu}_\tau +$ hadrons

BARYONS (STABLE AGAINST STRONG DECAYS)

		j^P	Mass energy	Lifetime	Main Decay mode
Proton	p	$1/2^+$	938.3 MeV	stable	
	$\bar{p}$	$1/2^-$			
Neutron	n	$1/2^+$	939.6 MeV	889 s	$p + e^- + \bar{\nu}_e$
	$\bar{n}$	$1/2^-$			$\bar{p} + e^+ + \nu_e$
Lambda	Λ^0	$1/2^+$	1116 MeV	2.63×10^{-10} s	$p + \pi$
	$\bar{\Lambda}^0$	$1/2^-$			$\bar{p} + e^+ + \nu_e$
Sigmas	Σ^+	$1/2^+$	1189 MeV	8.0×10^{-11} s	$p + \pi^0$
					$n + \pi^+$
	$\bar{\Sigma}^-$	$1/2^-$			$\bar{p} + \pi^0$
					$\bar{n} + \pi^-$
	Σ^0	$1/2^+$	1193 MeV	7.4×10^{-20} s	$\Lambda^0 + \gamma$
	$\bar{\Sigma}^0$	$1/2^-$			$\bar{\Lambda}^0 + \gamma$
	Σ^-	$1/2^+$	1197 MeV	1.5×10^{-10} s	$n + \pi^-$
	$\bar{\Sigma}^+$	$1/2^-$			$\bar{n} + \pi^+$
Cascades	Ξ^0	$1/2^+$	1315 MeV	2.9×10^{-10} s	$\Lambda^0 + \pi^0$
	$\bar{\Xi}^0$	$1/2^-$			$\bar{\Lambda}^0 + \pi^0$
	Ξ^-	$1/2^+$	1321 MeV	1.6×10^{-10} s	$\Lambda^0 + \pi^-$
	$\bar{\Xi}^+$	$1/2^-$			$\bar{\Lambda}^0 + \pi^+$
Omega	Ω^-	$1/2^+$	1672 MeV	8.2×10^{-11} s	$\Lambda^0 + K^-$
					$\Xi^0 + \pi^-$
					$\Xi^- + \pi^0$
	$\bar{\Omega}^+$	$1/2^-$			$\bar{\Lambda}^0 + K^+$
					$\bar{\Xi}^0 + \pi^+$
					$\bar{\Xi}^+ + \pi^0$
Lambda-c	Λ_c^+	$1/2^+$	2285 MeV	1.9×10^{-13} s	$p + K^- + \pi^+$
					$p + \bar{K}^0$
	$\bar{\Lambda}_c^-$	$1/2^-$			$\bar{p} + K^+ + \pi^-$
					$\bar{p} + K^0$
Sigma-c	Σ_c^{++}	$1/2^+$	2453 MeV		$\Lambda_c^+ + \pi^+$
	$\bar{\Sigma}_c^{--}$	$1/2^-$			$\bar{\Lambda}_c^- + \pi^-$

BARYONS (STABLE AGAINST STRONG DECAYS)

		j	Mass energy	Lifetime	Main Decay mode
	Σ_c^+	$1/2^+$	2453 MeV		$\Lambda_c^+ + \pi^0$
	$\overline{\Sigma}_c^-$	$1/2^-$			$\overline{\Lambda}_c^- + \pi^0$
	Σ_c^0	$1/2^+$	2453 MeV		$\Lambda_c^+ + \pi^-$
	$\overline{\Sigma}_c^{\,0}$	$1/2^-$			$\overline{\Lambda}_c^- + \pi^0$
Cascade-c	Ξ_c^+	$1/2^+$	2466 MeV	3×10^{-13} s	$\Lambda^0 + K^- + \pi^+ + \pi^+$
	$\overline{\Xi}_c^-$	$1/2^-$			$\overline{\Lambda}^0 + K^+ + \pi^- + \pi^-$
Lambda-b	$\Lambda_b^{\,0}$	$1/2^+$	5640 MeV		$\Lambda^0 + \psi$
	$\overline{\Lambda}_b^{\,0}$	$1/2^-$			$\overline{\Lambda}^0 + \psi$

BARYONS (STRONG DECAYS)

		j^P	Mass energy	Width (Γ)	Main Decay mode
N-star	N^{*+}	$1/2^+$	1140 MeV	200 MeV	$p + \pi^0$
					$n + \pi^+$
	$\overline{N}^{*-}$	$1/2^-$			$\overline{p} + \pi^0$
					$\overline{n} + \pi^-$
	N^{*0}	$1/2^+$			$n + \pi^0$
					$p + \pi^-$
	$\overline{N}^{*0}$	$1/2^-$			$\overline{p} + \pi^+$
					$\overline{n} + \pi^0$
Delta	Δ^{++}	$3/2^+$	1232 MeV	115 MeV	$p + \pi^+$
	$\overline{\Delta}^{--}$	$3/2^-$			$\overline{p} + \pi^-$
	Δ^+	$3/2^+$			$p + \pi^0$
					$\nu + \pi^+$
	$\overline{\Delta}^-$	$3/2^-$			$\overline{p} + \pi^0$
					$\overline{n} + \pi^-$
	Δ^0	$3/2^+$			$p + \pi^-$
					$n + \pi^0$
	$\overline{\Delta}^0$	$3/2^-$			$\overline{p} + \pi^+$
					$\overline{n} + \pi^0$
	Δ^-	$3/2^+$			$n + \pi^-$
	$\overline{\Delta}^+$	$3/2^-$			$\overline{n} + \pi^+$
Lambda-star	Λ^{*0}	$1/2^-$	1405 MeV	16 MeV	$\Sigma^0 + \pi^0$
					$\Sigma^+ + \pi^-$
					$\Sigma^- + \pi^+$
	$\overline{\Lambda}^{*0}$	$1/2^+$			$\overline{\Sigma}^0 + \pi^0$
					$\overline{\Sigma}^- + \pi^+$
					$\overline{\Sigma}^+ + \pi^-$
Sigma-star	Σ^{*+}	$3/2^+$	1383 MeV	36 MeV	$\Lambda^0 + \pi^+$
	$\overline{\Sigma}^{*-}$	$3/2^-$			$\overline{\Lambda}^0 + \pi^-$
	Σ^{*0}	$3/2^+$	1384 MeV	36 MeV	$\Lambda^0 + \pi^0$

BARYONS (STRONG DECAYS)

		j^P	Mass energy	Width (Γ)	Main Decay mode
	$\overline{\Sigma}^{*0}$	$3/2^-$			$\overline{\Lambda}^0 + \pi^0$
	Σ^{*-}	$3/2^+$	1387 MeV	39 MeV	$\Lambda^0 + \pi^-$
	$\overline{\Sigma}^{*+}$	$3/2^-$			$\overline{\Lambda}^0 + \pi^+$
Cascade-star	Ξ^{*0}	$3/2^+$	1532 MeV	9 MeV	$\Xi^0 + \pi^0$
					$\Xi^- + \pi^+$
	$\overline{\Xi}^{*0}$	$3/2^-$			$\overline{\Xi}^0 + \pi^0$
					$\overline{\Xi}^+ + \pi^-$
	Ξ^{*-}	$3/2^+$	1535 MeV	10 MeV	$\Xi^0 + \pi^-$
					$\Xi^- + \pi^0$
	$\overline{\Xi}^{*+}$	$3/2^-$			$\overline{\Xi}^0 + \pi^+$
					$\overline{\Xi}^+ + \pi^0$
Omega-star	Ω^{*-}		2250 MeV	60 MeV	$\Xi^- + K^- + \pi^+$
					$\Xi^{*0} + K^-$
	$\overline{\Omega}^{*+}$				$\overline{\Xi}^+ + K^+ + \pi^-$
					$\overline{\Xi}^{*0} + K^+$
Sigma-c	Σ_c^{++}	$1/2^+$	2452 MeV		$\Lambda_c^+ + \pi^+$
	$\overline{\Sigma}_c^{--}$	$1/2^-$			$\overline{\Lambda}_c^- + \pi^-$
	Σ_{c+}	$1/2^+$	2453 MeV		$\Lambda_c^+ + \pi^0$
	$\overline{\Sigma}_c^-$	$1/2^-$			$\overline{\Lambda}_c^- + \pi^0$
	Σ_c^0	$1/2^+$	2460 MeV		$\Lambda_c^+ + \pi^-$
	$\overline{\Sigma}_c^0$	$1/2^-$			$\overline{\Lambda}_c^- + \pi^+$

MESONS (STABLE AGAINST STRONG DECAY)

		j^P	Mass energy	Lifetime	Main Decay mode
Pions	π^+	0^-	140 MeV	2.60×10^{-8} s	$\mu^+ + \nu_\mu$
	π^-	0^-			$\mu^- + \overline{\nu}_\mu$
	π^0	0^-	135 MeV	8.4×10^{-17} s	$\gamma + \gamma$
Eta	η^0	0^-	549 MeV	6.0×10^{-19} s	$\gamma + \gamma$
					$\pi^0 + \pi^0 + \pi^0$
					$\pi^+ + \pi^- + \pi^0$
Kaons	K^+	0^-	494 MeV	1.24×10^{-8} s	$\mu^+ + \nu_\mu$
					$\pi^+ + \pi^0$
					$\pi^+ + \pi^+ + \pi^-$
					$\pi^0 + e^+ + \nu_e$
					$\pi^0 + \mu^+ + \nu_\mu$
					$\pi^+ + \pi^0 + \pi^0$
	K^-	0^-			$\mu^- + \overline{\nu}_\mu$
					$\pi^- + \pi^0$
					$\pi^- + \pi^- + \pi^+$
					$\pi^0 + e^- + \overline{\nu}_e$
					$\pi^0 + \mu^- + \overline{\nu}_\mu$

MESONS (STABLE AGAINST STRONG DECAY)

		j^P	Mass energy	Lifetime	Main Decay mode
	K_S^0	0^-	498 MeV	8.9×10^{-11} s	$\pi^- + \pi^0 + \pi^0$
					$\pi^+ + \pi^-$
					$\pi^0 + \pi^0$
	K_L^0	0^-	498 MeV	5.2×10^{-8} s	$\pi^+ + e^- + \bar{\nu}_e$
					$\pi^- + e^+ + \nu_e$
					$\pi^+ + \mu^- + \bar{\nu}_\mu$
					$\pi^- + \mu^+ + \nu_\mu$
					$\pi^0 + \pi^0 + \pi^0$
					$\pi^+ + \pi^- + \pi^0$
c mesons	D^+	0^-	1869 MeV	1.1×10^{-12} s	$\overline{K}^0 + \pi^+ + \pi^0$
					$K^- + \pi^+ + \pi^+$
	D^-	0^-			$K^0 + \pi^- + \pi^0$
					$K^+ + \pi^- + \pi^-$
	D^0	0^-	1865 MeV	4.2×10^{-13} s	$K^- + \pi^+ + \pi^0$
					$K^- + \pi^+ + \pi^0 + \pi^0$
					$\overline{K}^0 + \pi^+ + \pi^-$
					$K^- + \pi^+$
	$\overline{D}^0$	0^-			$K^+ + \pi^- + \pi^0$
					$K^+ + \pi^- + \pi^0 + \pi^0$
					$K^0 + \pi^+ + \pi^-$
					$K^+ + \pi^-$
	D_s^+	0^-	1969 MeV	4.5×10^{-13} s	$\phi + \pi^+$
	D_s^-	0^-			$\phi + \pi^-$
b mesons	B^-	0^-	5279 MeV	1.3×10^{-12} s	$D^{*+} + \pi^- + \pi^- + \pi^0$
	B^+	0^-			$D^{*-} + \pi^+ + \pi^+ + \pi^0$
	B^0	0^-	5279 MeV	1.3×10^{-12} s	$D^{*-} + \mu^+ + \nu_\mu$
					$D^{*-} + e^+ + \nu_e$
					$D^{*-} + \rho^+$
					$D^{*-} + \pi^+ + \pi^0$
	$\overline{B}^0$	0^-			$D^{*+} + \mu^- + \bar{\nu}_\mu$
					$D^{*+} + e^- + \bar{\nu}_e$
					$D^{*+} + \rho^-$
					$D^{*+} + \pi^- + \pi^0$

MESONS (STRONG DECAYS)

		j^P	Mass energy	Width (Γ)	Main Decay mode
Rho	ρ^+	1^-	770 MeV	152 MeV	$\pi^+ + \pi^0$
	ρ^-	1^-			$\pi^- + \pi^0$
	ρ^0	1^-			$\pi^+ + \pi^-$
Omega	ω	1^-	782 MeV	8.4 MeV	$\pi^+ + \pi^- + \pi^0$
					$\pi^0 + \gamma$

MESONS (STRONG DECAYS)

		j^P	Mass energy	Width (Γ)	Main Decay mode
K-star	K^{*+}	1^-	892 MeV	50 MeV	$\pi^+ + \pi^-$
					$K^+ + \pi^0$
					$K^0 + \pi^+$
	K^{*-}	1^-			$K^- + \pi^0$
					$\overline{K}^0 + \pi^-$
	K^{*0}	1^-	896 MeV	51 MeV	$K^+ + \pi^-$
					$K^0 + \pi^0$
	$\overline{K}^{*0}$	1^-			$K^- + \pi^+$
					$\overline{K}^0 + \pi^0$
Eta-prime	η'	0^-	958 MeV	200 keV	$\eta + \pi^+ + \pi^-$
					$\rho^0 + \gamma$
					$\eta + \pi^0 + \pi^0$
					$\omega + \gamma$
					$\gamma + \gamma$
Phi	ϕ	1^-	1019 MeV	4.4 MeV	$K^+ + K^-$
					$K_L^0 + K_S^0$
					$\rho^0 + \pi^0$
					$\rho^+ + \pi^-$
					$\rho^- + \pi^+$
					$\pi^+ + \pi^- + \pi^0$
$c\overline{c}$ mesons	η_c	0^-	2979 MeV	10 MeV	$\eta' + \pi^+ + \pi^-$
					$\eta' + \pi^0 + \pi^0$
	$J/\psi(1s)$	1^-	3097 MeV	86 keV	$e^+ + e^-$
					$\mu^+ + \mu^-$
					hadrons
	χ_{c0}	0^+	3415 MeV	14 MeV	$\pi^+ + \pi^- + \pi^+ + \pi^-$
					$\psi + \gamma$
	χ_{c1}	1^+	3511 MeV	0.9 MeV	$\psi + \gamma$
	χ_{c2}	2^+	3556 MeV	2.0 MeV	$\psi + \gamma$
	$\psi(2s)$	1^-	3686 MeV	278 keV	$e^+ + e^-$
					$\mu^+ + \mu^-$
					hadrons
	$\psi(3770)$	1^-	3770 MeV	24 MeV	$D^+ + D^-$
					$D^0 + \overline{D}^0$
$b\overline{b}$ mesons	$\Upsilon(1s)$	1^-	9.460 GeV	52 keV	$e^+ + e^-$
					$\mu^+ + \mu^-$
					$\tau^+ + \tau^-$
					hadrons
	χ_{b0}	$0+$	9.860 GeV		$\Upsilon + \gamma$
	χ_{b1}	$1+$	9.892 GeV		$\Upsilon + \gamma$
	χ_{b2}	$2+$	9.913 GeV		$\Upsilon + \gamma$
	$\Upsilon(2s)$	1^-	10.023 GeV	43 keV	$\Upsilon + \pi^+ + \pi^-$

MESONS (STRONG DECAYS)

	j^P	Mass energy	Width(Γ)	Decay mode
				$\Upsilon + \pi^0 + \pi^0$
χ_{b0}	0^+	10.232 GeV		$\Upsilon + \gamma$
χ_{b1}	1^+	10.255 GeV		$\Upsilon + \gamma$
χ_{b2}	2^+	10.268 GeV		$\Upsilon + \gamma$
$\Upsilon'(3s)$	1^-	10.355 GeV	24 keV	$\Upsilon + \pi^+ + \pi^-$
$\Upsilon(4s)$	1^-	10.580 GeV	24 MeV	$B^+ + B^-$
				$B^0 + \overline{B}^0$

Chapter 1
3. H_8C_3
5. (a) 6.64×10^{-27} kg
8. 3×10^{25}
10. 10^{28}
12. 10^{13}
14. (b) 0.053 nm, (c) 2.2×10^6 m/s
18. (b) 10^{-15} m
20. (a) 7×10^6 m/s
22. 0.34 m
27. (a) 6.1×10^5 V/m
29. 1.2×10^6 V/m
32. 52 kg
34. (a) 2.2 MeV, (b) 1.8 GeV
36. 4×10^{38} s^{-1}
40. 5×10^{-18}
43. 17 events
45. (b) 4×10^{29} g's

Chapter 2
5. 3.3×10^3
6. 0.3 mm
9. 0.96
12. 0.69
13. 2.3×10^6
18. 1.2 kg/m^3
20. 1030 m/s
22. 8.31 J/K
24. (a) 0.8 eV, (b) 10^4 m/s
26. O_2
28. 3.17

30. 190 K
32. 6×10^{20} m^{-2}·s^{-1}
34. 2×10^4 K
36. (b) $2kT$
37. 1.2×10^4 m/s

Chapter 3
6. 10^{20} s^{-1}
8. (a) 6000 K, (b) 10^{-6} m^2, (c) 3×10^{20} s^{-1}
9. (a) 10^8 W·m^{-2} per μm
12. 1.8×10^{-5} m^2
13. 0.5
15. 2000 s
18. (a) 1.8×10^{-5} J/m^3, (b) 0.84 J/m^3
20. (d) 4.1×10^8 m^{-2}, 0.64 meV
22. (a) 4×10^{-7} eV
23. 3.2×10^{15} s^{-1}
25. 410 nm
28. 3.3 eV
29. 4 eV
31. 308 nm
34. 7 W/m^2
36. 91.2 nm
39. $f_{orb} = 2.4 \times 10^{14}$ s^{-1}, $f_{rad} = 4.6 \times 10^{14}$ s^{-1}, 2.9×10^{15} s^{-1}
41. (a) 1.2×10^{29} m
44. 67 keV
49. 10^{36} s^{-1}
50. 25.8 kΩ

52. 2×10^{-11}
55. (a) 2.6×10^{16} W
57. 0.8 K

Chapter 4
3. (b) 240 m/s
6. 0.47 ps
8. (a) 4.6×10^3 m
11. $c\sqrt{7}/4$
13. 1.70 MeV/c, 0.88 c
15. 542 MeV/c
18. 0.16 MeV
20. 69 GeV
23. 0.43 MeV/c
25. 6.6×10^{11} GeV
27. 2.94×10^8 m/s
31. 19.4 GeV
35. (b) 0.60 MeV
36. (a) 10 GeV, 0.26 MeV
37. 0.3 keV
39. 20 GeV
43. 1.02 MeV
45. 18 GeV
48. 3 m/s
50. 5.9 eV
52. 6.6 GeV

Chapter 5
1. (a) 124 nm
2. (a) 150 eV
4. 0.15 nm
6. 1.7×10^{-18} m/s

8. (a) 3×10^{-11} m
9. (a) 0.39 nm
11. 0.86 MeV
15. 5.5 μm
17. 32 m/s
19. 3 nm
23. 1.2×10^{-12}
24. (a) 4 pm
28. 7.4×10^{-8} eV
30. 5.7×10^{-24} s
32. 10^{-1} eV

Chapter 6
7. 4.7×10^7 s
8. (a) 6 μb, (b) 16 μb
13. (a) 8.2 mb, (b) 4.9×10^{-8}
14. 3.2 mm
16. 27 MeV
18. 6.2×10^3 b
19. (a) 170 b
21. 17 MeV
24. 4.1 GeV
25. 58°
28. 3×10^{-9}
29. (a) 0.47 MeV
31. 0.5 s^{-1}

Chapter 7
3. (b) $\Delta x/L$
4. (b) 0.82
6. 50 MeV, 200 MeV
8. 0.34 nm

13. 0.5 nm
15. 10^{18} s
16. 0.1 nm
18. $5\hbar\omega/2$
24. 0.09 nm, 2.3 keV/c
25. (b) 0.1
26. (b) 25 eV
28. 0.08 nm
31 (b) 1
33. 250 MeV
35. 7 MeV

Chapter 8
1. 0
3. 0.323
4. 0.663
5. 0.56
6. (a) 0.14 nm
18. (b) 895 T/m
27. (c) 0.45 T
29. 1.45×10^{14} s^{-1}
31. (b) 0.16
32. (b) $5a_0$

Chapter 9
1. (c) 120 keV
4. (b) 15
5. $n=7$, $\ell=1$
11. 1/2, 3/2, 5/2
13. 24.6 eV
18. 200 eV
20. 6
21. 5
25. (b) 2.9 K
27. (b) $s=\ell=2$
28. (b) $j=2$

Chapter 10
5. 10 eV
10. 7 eV
11. 0.32 nm
13. ^{35}Cl, 6×10^{-6} eV
15. 7 pm
16. 0.25
17. 2.4×10^{-5}
19. 1.3 cm
21. 1.3 mm

24. 0.12 nm
25. $\ell=3$ to $\ell=4$
30. (c) 350 K
31. 0.4 eV
33. 100 kcal/mole

Chapter 11
3. (a) 140 MeV, (b) 9.7 MeV
6. (a) 218897 MeV/c^2,
    (b) 218944 MeV/c^2
7. ^{59}Co
9. ^{64}Ni
15. 1 s
16. yes, e.g. ^{64}Cu
18. 3.3 ky
19. (a) 1.2×10^4 s^{-1},
    (b) 4.1×10^{-3} s^{-1}
22. (a) ^{23}Al, (b) ^{190}Pt, (c) ^{11}B,
    (d) e^+, (e) p
27. (a) 0
29. (a) 2.3 cm/s
33. yes
34. 40 mCi
36. 2×10^{-12} s
38. 3×10^{17} m^{-2}
42. (b) 51 MeV

Chapter 12
1. 1 nm, 1 nm, no
2. (a) 3.9 keV, (b) yes
4. 2.4 meV
11. fermion
13. electrons, $\lambda_e \ll \lambda_\gamma$
20. (b) 3 K
24. 4×10^9 m^{-3}, yes

Chapter 13
4. (a) 72 μm, (b) 280 μm,
    (c) 0.96 μm
7. (a) 530 m/s, (b) 3×10^{-12} eV
8. 2.4×10^{-15} W
12. $n=3$ to $n=2$
14. 1 pm
17. (a) 0.4 K
18. 3.0 cm
23. (b) 0.67
24. (b) 2×10^{-5}

28. (b) 190 W
29. 0.1 J

Chapter 14
3. 1.4×10^{21}
4. 4.4 eV
5. 1.3×10^{29} m^{-3}, 2
7. 0.004
9. (a) 2.0×10^6 m/s
10. 3.2 K
12. 0.01 eV, 4 eV
14. 2.2×10^{-13} Ω·m
17. 3 ps
18. (b) 190 m/s
20. 130 K
26. 40 Ω^{-1}·m^{-1}
29. (b) 0.94 T
33. 3.3 eV
34. (a) n-type
35. 0.67 eV

Chapter 15
5. 0.055 T
6. (b) 1.1×10^{-6} J
7. 13 μm
8. Nb, 0.16 T
10. 6.7 K
12. 3.1×10^{28} m^{-3}
16. 2 μm
17. 1.3 meV
18. 0.49 H, 1.6×10^{-13} Ω
21. 6.5×10^{11} s^{-1}
23. 3×10^{-12} m^2
27. 4.1×10^{-7} T
28. (b) 12 T

Chapter 16
2. (a) 1.3 T
3. 140 MeV
4. (b) 20 keV
5. (b) 2×10^7 MeV/m
7. (b) 860 per day
11. 90 m
14. 2×10^4
16. 39.0 GeV
18. 3×10^{35} s^{-1}·m^{-2}
19. 0.49

22. 40 MeV

Chapter 17
2. 4.5 MeV
4. (b) 1.7 mm
7. (d) 28 mb
13. (a) 1/2, 1/2, (b) Ξ^0
16. (a) No, bayyon number
17. (a) allowed,
    electromagnetic
18. (b) 53 MeV/c
21. 3×10^{11} m
24. (b) 20 tons
28. $\tau > 3 \times 10^{-24}$ s
30. 0.1 c
38. 6.8 GeV/fm^2

Chapter 18
2. 0.11, assuming 6 quark
    flavors
4. (a) 0.87 nb, (b) 8.7 pb
5. (a) 1%
11. 0.3 ps
17. 3×10^{-25} s
18. 3.4 meV
20. 7×10^{-41} m^2
21. (a) 0.057
25. 82 GeV, 33°
26. 19 per day

Chapter 19
2. 20 K
5. 7.5×10^{-7}
8. 2 pc
9. (a) 18 pc, (b) 180 pc
11. 2.5
13. 23.6
15. 0.385 c, 1700 Mpc
18. 3400 Mpc
19. 0.984 c
21. 0.3 m
22. 10^{46} J
25. 4 μeV
29. 10^{14} K, 4×10^{-8} s
33. 1 per 2×10^{53} m^3, 10^{14}
35. 10^7 K
37. 8×10^9 K

EXAMPLE INDEX